Modern Power Systems Analysis

Xi-Fan Wang • Yonghua Song • Malcolm Irving

Modern Power Systems Analysis

Springer

Xi-Fan Wang
Xi'an Jiaotong University
Xi'an
People's Republic of China

Yonghua Song
The University of Liverpool
Liverpool
United Kingdom

Malcolm Irving
Brunel University
Middlesex
United Kingdom

ISBN 978-0-387-72852-0 e-ISBN 978-0-387-72853-7

Library of Congress Control Number: 2008924670

Printed on acid-free paper.

9 8 7 6 5 4 3 2 1

springer.com

Preface

The power industry, a capital and technology intensive industry, is a basic national infrastructure. Its security, reliability, and economy have enormous and far-reaching effects on a national economy. An electrical power system is a typical large-scale system. Questions such as how to reflect accurately the characteristics of modern electrical power systems, how to analyze effectively their operating features, and how to improve further the operating performance are always at the forefront of electrical power systems research.

Electrical power system analysis is used as the basic and fundamental measure to study planning and operating problems. In the last century, electrical power researchers have undertaken a great deal of investigation and development in this area, have made great progress in theoretical analysis and numerical calculation, and have written excellent monographs and textbooks.

Over the last 20 years, the changes in electrical power systems and other relevant technologies have had a profound influence on the techniques and methodologies of electrical power system analysis.

First, the development of digital computer technology has significantly improved the performance of hardware and software. Now, we can easily deal with load flow issues with over ten thousand nodes. Optimal load flow and static security analysis, which were once considered hard problems, have attained online practical applications.

Second, the applications of HVDC and AC flexible transmission technologies (FACTS) have added new control measures to electrical power systems, and have increased power transmission capacity, enhanced control capability, and improved operating characteristics. However, these technologies bring new challenges into the area of electrical power system analysis. We must build corresponding mathematical models for these new devices and develop algorithms for static and dynamic analysis of electrical power systems including these devices.

In addition, the rapid development of communication technology has enabled online monitoring of electrical power systems. Therefore, the demand for online software for electrical power system analysis becomes more and more pressing.

Furthermore, worldwide power industry restructuring and deregulation has separated the former vertically integrated system into various parts, and the once unified problem of power system dispatching is now conducted via complicated bilateral contracts and spot markets. New issues such as transmission ancillary service and transmission congestion have emerged.

In recent years, several power blackouts have taken place worldwide, especially the "8.13" blackout on the eastern grid of USA and Canada and the blackouts that occurred successively in other countries have attracted a great deal of attention.

All of these aspects require new theories, models, and algorithms for electrical power system analysis. It is within such an environment that this book has been developed. The book is written as a textbook for senior students and postgraduates as well as a reference book for power system researchers.

We acknowledge the support from various research funding organizations, their colleagues, and students, especially, the special funds for Major State Basic Research Projects of China "Research on Power System Reliability under Deregulated Environment of Power Market" (2004CB217905). We express our special gratitude to Professor Wan-Liang Fang and Professor Zheng-Chun Du for providing the original materials of Chaps. 5 and 6, and 7 and 8, respectively. We also express our sincere gratitude to the following colleagues for their contributions to various chapters of the book: Professor Zhao-Hong Bie for Chaps. 1 and 3; Professor Xiu-Li Wang for Chaps. 2 and 4; Dr. Ze-Chun Hu for Chap. 3; Dr. Xiao-Ying Ding for Chap. 4; Dr. Lin Duan for Chaps. 5 and 6; Professor De-Chiang Gang for Chap. 7; and Professor Hai-Feng Wang for Chaps. 6 and 8.

Xi'an, China *Xi-Fan Wang*
Liverpool, UK *Yonghuna Song*
London, UK *Malcolm Irving*

Contents

Chapter 1
Mathematical Model and Solution of Electric Network

1.1 Introduction

The mathematical model of an electric network is the basis of modern power system analysis, which is to be used in studies of power flow, optimal power flow, fault analysis, and contingency analysis. The electric network is constituted by transmission lines, transformers, parallel/series capacitors, and other static elements. From the viewpoint of electrical theory, no matter how complicated the network is, we can always establish its equivalent circuit and then analyze it according to the AC circuit laws. In this chapter, the electric network is represented by the linear lumped parameter model that is suitable for studies at synchronous frequency. For electromagnetic transient analysis, the high frequency phenomena and wave processes should be considered. In that situation, it is necessary to apply equivalent circuits described by distributed parameters.

Generally speaking, an electric network can be always represented by a nodal admittance matrix or a nodal impedance matrix. A modern power system usually involves thousands of nodes; therefore methods of describing and analyzing the electric network have a great influence on modern power system analysis. The nodal admittance matrix of a typical power system is large and sparse. To enhance the computational efficiency, sparsity techniques are extensively employed. The nodal admittance matrix and associated sparsity techniques will be thoroughly discussed in this chapter.

The nodal impedance matrix is widely applied in the fault analysis of power systems and will be introduced in Sect. 1.5.

The equivalent circuits of the transformer and phase-shifting transformer are also presented in Sect. 1.1 because they require special representation methods.

X.-F. Wang et al., *Modern Power Systems Analysis.*
doi: 10.1007/978-0-387-72853-7,

1.2 Basic Concepts

1.2.1 Node Equation and Loop Equation

There are two methods usually employed in analyzing AC circuits, i.e., the node voltage method and loop current method. Both methods require the solution of simultaneous equations. The difference between them is that the former applies node equations while the latter applies loop equations. At present, node equations are more widespread in analyzing power systems, and loop equations are used sometimes as an auxiliary tool.

In the following, we use a simple electric network as an example to illustrate the principle and characteristics of the node equation method.

As shown in Fig. 1.1, the sample system has two generators and an equivalent load, with five nodes and six branches whose admittances are $y_1 \sim y_6$.

Assigning the ground as the reference node, we can write the nodal equations according to the Kirchoff's current law,

$$\left.\begin{array}{l}
y_4(\dot{V}_2 - \dot{V}_1) + y_5(\dot{V}_3 - \dot{V}_1) - y_6\dot{V}_1 = 0 \\
y_1(\dot{V}_4 - \dot{V}_2) + y_3(\dot{V}_3 - \dot{V}_2) + y_4(\dot{V}_1 - \dot{V}_2) = 0 \\
y_2(\dot{V}_5 - \dot{V}_3) + y_3(\dot{V}_2 - \dot{V}_3) + y_5(\dot{V}_1 - \dot{V}_3) = 0 \\
y_1(\dot{V}_4 - \dot{V}_2) = \dot{I}_1 \\
y_2(\dot{V}_5 - \dot{V}_3) = \dot{I}_2
\end{array}\right\}, \tag{1.1}$$

where $\dot{V}_1 \sim \dot{V}_5$ denote the node voltages.

Combining the coefficients of node voltages, we obtain the following equations:

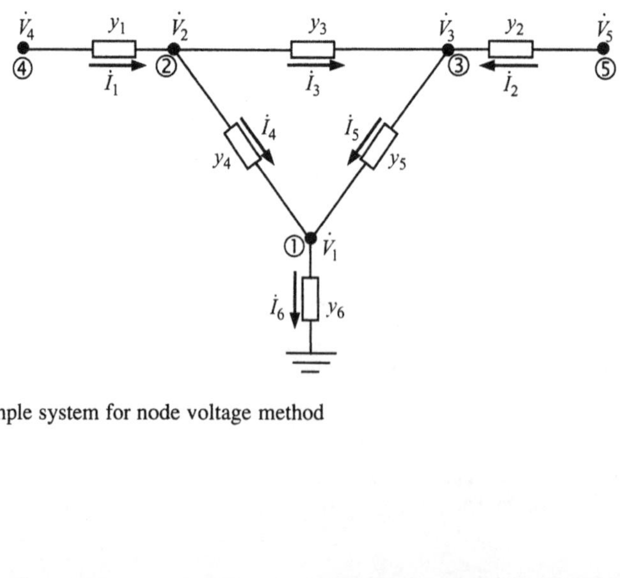

Fig. 1.1 Sample system for node voltage method

$$\left.\begin{aligned}
&(y_4 + y_5 + y_6)\dot{V}_1 - y_4\dot{V}_2 - y_5\dot{V}_3 = 0\\
&-y_4\dot{V}_1 + (y_1 + y_3 + y_4)\dot{V}_2 - y_3\dot{V}_3 - y_1V_4 = 0\\
&-y_5\dot{V}_1 - y_3\dot{V}_2 + (y_2 + y_3 + y_5)\dot{V}_3 - y_2\dot{V}_5 = 0\\
&-y_1\dot{V}_2 + y_1\dot{V}_4 = \dot{I}_1\\
&-y_2\dot{V}_3 + y_2\dot{V}_5 = \dot{I}_2
\end{aligned}\right\}. \tag{1.2}$$

In (1.2), the left-hand term is the current flowing from the node and the right-hand term is the current flowing into the node. The above equations can be rewritten in more general form as follows:

$$\left.\begin{aligned}
Y_{11}\dot{V}_1 + Y_{12}\dot{V}_2 + Y_{13}\dot{V}_3 + Y_{14}\dot{V}_4 + Y_{15}\dot{V}_5 = \dot{I}_1\\
Y_{21}\dot{V}_1 + Y_{22}\dot{V}_2 + Y_{23}\dot{V}_3 + Y_{24}\dot{V}_4 + Y_{25}\dot{V}_5 = \dot{I}_2\\
Y_{31}\dot{V}_1 + Y_{32}\dot{V}_2 + Y_{33}\dot{V}_3 + Y_{34}\dot{V}_4 + Y_{35}\dot{V}_5 = \dot{I}_3\\
Y_{41}\dot{V}_1 + Y_{42}\dot{V}_2 + Y_{43}\dot{V}_3 + Y_{44}\dot{V}_4 + Y_{45}\dot{V}_5 = \dot{I}_4\\
Y_{51}\dot{V}_1 + Y_{52}\dot{V}_2 + Y_{53}\dot{V}_3 + Y_{54}\dot{V}_4 + Y_{55}\dot{V}_5 = \dot{I}_5
\end{aligned}\right\}. \tag{1.3}$$

Comparing (1.3) with (1.2), we can see

$$\begin{aligned}
Y_{11} &= y_4 + y_5 + y_6,\\
Y_{22} &= y_1 + y_3 + y_4,\\
Y_{33} &= y_2 + y_3 + y_5,\\
Y_{44} &= y_1,\\
Y_{55} &= y_2.
\end{aligned}$$

These elements are known as nodal self-admittances.

$$\begin{aligned}
Y_{12} = Y_{21} &= -y_4,\\
Y_{13} = Y_{31} &= -y_5,\\
Y_{23} = Y_{32} &= -y_3,\\
Y_{24} = Y_{42} &= -y_1,\\
Y_{35} = Y_{53} &= -y_2.
\end{aligned}$$

Similarly, the above elements are known as mutual admittances between the connected nodes. The mutual admittances of the pair of disconnected nodes are zero.

Equation (1.3) is the node equation of the electric network. It reflects the relationship between node voltages and injection currents. Here $\dot{I}_1 \sim \dot{I}_5$ are the nodal injection currents. In this example, except $\dot{I}_4$ and $\dot{I}_5$, all other nodal injection currents are zero.

Equation (1.3) can be solved to get node voltages $\dot{V}_1 \sim \dot{V}_5$, then the branch currents can be obtained. Thus, we have obtained all the variables of the network.

Generally, for a n node network, we can establish n linear node equations in (1.3) format. In matrix notation, we have

$$\boldsymbol{I} = \boldsymbol{Y}\boldsymbol{V}, \tag{1.4}$$

where

$$\boldsymbol{I} = \begin{bmatrix} \dot{I}_1 \\ \dot{I}_2 \\ \vdots \\ \dot{I}_n \end{bmatrix}, \quad \boldsymbol{V} = \begin{bmatrix} \dot{V}_1 \\ \dot{V}_2 \\ \vdots \\ \dot{V}_n \end{bmatrix}.$$

Here $\boldsymbol{I}$ is the vector of nodal injection currents and $\boldsymbol{V}$ is the vector of nodal voltages; $\boldsymbol{Y}$ is called the nodal admittance matrix

$$\boldsymbol{Y} = \begin{bmatrix} Y_{11} & Y_{12} & \cdots & Y_{1n} \\ Y_{21} & Y_{22} & \cdots & Y_{2n} \\ \cdots & \cdots & \cdots & \cdots \\ Y_{n1} & Y_{n2} & \cdots & Y_{nn} \end{bmatrix}.$$

As we have seen, its diagonal element Y_{ii} is the nodal self-admittance and the off diagonal element Y_{ij} is the mutual admittance between node i and node j.

Now we introduce the incidence matrix that is very important in network representations.

The incidence matrix represents the topology of an electric network. Different incidence matrices correspond to different networks configurations. The elements of the incidence matrix are only 0, +1, or –1. They do not include the parameters of network branches.

For example, there are five nodes and six branches in Fig. 1.1. Its incidence matrix is a matrix with five rows and six columns.

$$\boldsymbol{A} = \begin{bmatrix} 0 & 0 & 0 & -1 & -1 & 1 \\ -1 & 0 & 1 & 1 & 0 & 0 \\ 0 & -1 & -1 & 0 & 1 & 0 \\ 1 & 0 & 0 & 0 & 0 & 0 \\ 0 & 1 & 0 & 0 & 0 & 0 \end{bmatrix}.$$

In the incidence matrix, the serial numbers of rows correspond to the node numbers and the serial numbers of columns correspond to the branch numbers. For example, the first row has three nonzero elements, which denotes node 1 is connected with three branches. These three nonzero elements are in the fourth, fifth, and sixth columns, which means the branches connected with node 1 are branches 4, 5, and 6.

If the branch current flows into the node, the nonzero element equals -1; if the branch current flows out of the node, the nonzero element equals 1. The positions of the nonzero elements in each column denote the two node numbers of the relevant branch. For example, in the fifth column the nonzero elements are in the first and third row, which means the fifth branch connects node 1 and 3. In the sixth column, there is only one nonzero element in the first row, which means the sixth branch is a grounded branch.

From the above discussion we see that an incidence matrix can uniquely determine the topology of a network configuration.

The incidence matrix has a close relationship with the network node equation. If there are n nodes and b branches in an electric network, the state equation for every branch is

$$\dot{I}_{\mathrm{B}k} = y_{\mathrm{B}k}\dot{V}_{\mathrm{B}k}, \tag{1.5}$$

where $y_{\mathrm{B}k}$ is the admittance of branch k; $I_{\mathrm{B}k}$ the current flowing in branch k; and $\dot{V}_{\mathrm{B}k}$ is the voltage difference of branch k, whose direction is determined by $I_{\mathrm{B}k}$.

If branch k includes a voltage source, as shown in Fig. 1.2a, it should be transformed to the equivalent current source as shown in Fig. 1.2b.

$$\left.\begin{aligned} y_{\mathrm{B}k} &= 1/z_{\mathrm{B}k} \\ \dot{a}_{\mathrm{B}k} &= \dot{e}_{\mathrm{B}k}/z_{\mathrm{B}k} = y_{\mathrm{B}k}\dot{e}_{\mathrm{B}k} \end{aligned}\right\}.$$

The current source can be treated as current injecting into the electric network, thus the branch can also be represented by (1.5). In matrix notation, the equation of a b branch network is

$$\boldsymbol{I}_{\mathrm{B}} = \boldsymbol{Y}_{\mathrm{B}}\boldsymbol{V}_{\mathrm{B}}, \tag{1.6}$$

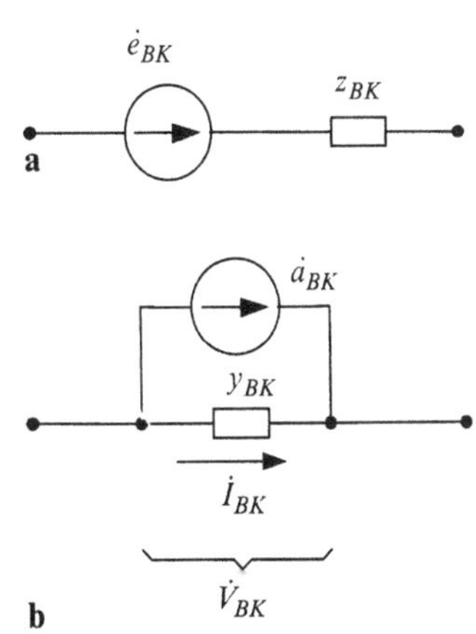

Fig. 1.2 Transformation from voltage source to current source

where $\boldsymbol{I}_{\mathrm{B}}$ is the vector of the currents in branches, $\boldsymbol{V}_{\mathrm{B}}$ the vector of the branch voltage differences, and $\boldsymbol{Y}_{\mathrm{B}}$ is a diagonal matrix constituted by the branch admittances.

According to Kirchoff's current law, the injection current $\dot{I}_i$ of node i in an electric network can be expressed as follows

$$\dot{I}_i = \sum_{k=1}^{b} a_{ik} \dot{I}_{\mathrm{B}k} \quad (i = 1, 2, \ldots, n), \tag{1.7}$$

where a_{ik} is a coefficient. If branch current $\dot{I}_{\mathrm{B}k}$ directs toward node i, $a_{ik} = -1$; if branch current $\dot{I}_{\mathrm{B}k}$ directs away from the node i, $a_{ik} = 1$; and if branch k does not connect to node i, $a_{ik} = 0$. It is easy to get the relationship between nodal current vector $\dot{\boldsymbol{I}}$ and branch current vector $\dot{\boldsymbol{I}}_{\mathrm{B}}$ as follows,

$$\boldsymbol{I} = \boldsymbol{A}\boldsymbol{I}_{\mathrm{B}}, \tag{1.8}$$

where $\boldsymbol{A}$ is the incidence matrix of the network.

Assuming the power consumed in the whole network is S, we can obtain the following equation,

$$S = \sum_{i=1}^{b} \hat{I}_{\mathrm{B}k} \dot{V}_{\mathrm{B}k} = \hat{\boldsymbol{I}}_{\mathrm{B}} * \dot{\boldsymbol{V}}_{\mathrm{B}},$$

where $\hat{I}_{\mathrm{B}k}$ and $\hat{\boldsymbol{I}}_{\mathrm{B}}$ are the conjugate of the corresponding vector and * is the scalar product of the two vectors.

From the viewpoint of the nodal input power, we have

$$S = \sum_{i=1}^{n} \hat{I}_i \dot{V}_i = \hat{\boldsymbol{I}} * \dot{\boldsymbol{V}}.$$

Obviously,

$$\hat{\boldsymbol{I}} * \dot{\boldsymbol{V}} = \hat{\boldsymbol{I}}_{\mathrm{B}} * \dot{\boldsymbol{V}}_{\mathrm{B}}. \tag{1.9}$$

From (1.8), we see

$$\hat{\boldsymbol{I}} = \hat{\boldsymbol{I}}_{\mathrm{B}} \boldsymbol{A}^{\mathrm{T}}.$$

Substituting it into (1.9), we obtain,

$$\hat{\boldsymbol{I}}_{\mathrm{B}} \boldsymbol{A}^{\mathrm{T}} \dot{\boldsymbol{V}} = \hat{\boldsymbol{I}}_{\mathrm{B}} \dot{\boldsymbol{V}}_{\mathrm{B}}.$$

Therefore,

$$A^{\mathrm{T}}\dot{V} = \dot{V}_{\mathrm{B}}. \tag{1.10}$$

Substituting (1.6) and (1.10) into (1.8) sequentially, we can get

$$\dot{I} = AY_{\mathrm{B}}A^{\mathrm{T}}\dot{V} = Y\dot{V}, \tag{1.11}$$

where Y is the nodal admittance matrix of the electric network

$$Y = AY_{\mathrm{B}}A^{\mathrm{T}}. \tag{1.12}$$

Thus the nodal equations of an electric network can be obtained from its incidence matrix.

In the following, the network shown in Fig. 1.1 is used again to illustrate the basic principle of analyzing the electric network by the loop current equations. In the loop equation method, the network elements are often represented in impedance form. The equivalent circuit is shown in Fig. 1.3. There are three independent loops in the network and the loop currents are $\dot{I}_1, \dot{I}_2$, and $\dot{I}_3$, respectively. According to Kirchoff's voltage law, the voltage equations of the loops are

$$\left.\begin{aligned} \dot{V}_4 &= (z_1 + z_4 + z_6)\dot{I}_1 + z_6\dot{I}_2 - z_4\dot{I}_3 \\ \dot{V}_5 &= z_6\dot{I}_1 + (z_2 + z_5 + z_6)\dot{I}_2 + z_5\dot{I}_3 \\ 0 &= -z_4\dot{I}_1 + z_5\dot{I}_2 + (z_3 + z_4 + z_5)\dot{I}_3 \end{aligned}\right\}. \tag{1.13}$$

Rewrite the above equation into the normative form,

$$\left.\begin{aligned} \dot{E}_1 &= Z_{11}\dot{I}_1 + Z_{12}\dot{I}_2 + Z_{13}\dot{I}_3 \\ \dot{E}_2 &= Z_{21}\dot{I}_1 + Z_{22}\dot{I}_2 + Z_{23}\dot{I}_3 \\ \dot{E}_3 &= Z_{31}\dot{I}_1 + Z_{32}\dot{I}_2 + Z_{33}\dot{I}_3 \end{aligned}\right\}, \tag{1.14}$$

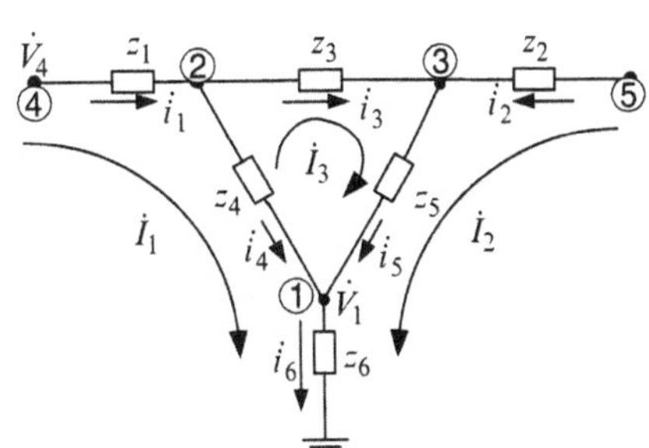

Fig. 1.3 Sample system with loop currents

where

$\dot{E}_1 = \dot{V}_4, \dot{E}_2 = \dot{V}_5, \dot{E}_1 = 0$ are voltage potentials of three loops, respectively,
$Z_{11} = z_1 + z_4 + z_6, Z_{22} = z_2 + z_5 + z_6, Z_{33} = z_3 + z_4 + z_5$ are loop self-impedances,
$Z_{12} = Z_{21} = z_6, Z_{13} = Z_{31} = -z_4, Z_{23} = Z_{32} = z_5$ are the loop mutual impedances.

If we know loop voltage $\dot{E}_1, \dot{E}_2$, and $\dot{E}_3$, we can solve the loop current $\dot{I}_1, \dot{I}_2$, and $\dot{I}_3$ from (1.14), and then obtain the branch current,

$$\dot{i}_1 = \dot{I}_1, \quad \dot{i}_2 = \dot{I}_2, \quad \dot{i}_3 = \dot{I}_3,$$
$$\dot{i}_4 = \dot{I}_1 - \dot{I}_3, \quad \dot{i}_5 = \dot{I}_2 + \dot{I}_3, \quad \dot{i}_6 = \dot{I}_1 + \dot{I}_2.$$

And the node voltages are

$$\dot{V}_1 = z_6 \dot{i}_6, \quad \dot{V}_2 = \dot{V}_4 - z_1 \dot{i}_1, \quad \dot{V}_3 = \dot{V}_5 - z_2 \dot{i}_2.$$

Thus all the variables of the electric network are solved.

Generally, an electric network with m independent loops can be formulated by m loop equations. In matrix notation, we have

$$\boldsymbol{E}_l = \boldsymbol{Z}_l \boldsymbol{I}_l, \tag{1.15}$$

where

$$\boldsymbol{I}_l = \begin{bmatrix} \dot{I}_1 \\ \dot{I}_2 \\ \vdots \\ \dot{I}_m \end{bmatrix}, \quad \boldsymbol{E}_l = \begin{bmatrix} \dot{E}_1 \\ \dot{E}_2 \\ \vdots \\ \dot{E}_m \end{bmatrix}$$

are vectors of the loop currents and voltage phasors, respectively;

$$\boldsymbol{Z}_l = \begin{bmatrix} Z_{11} & Z_{12} & \cdots & Z_{1m} \\ Z_{21} & Z_{22} & \cdots & Z_{2m} \\ \cdots & \cdots & \cdots & \cdots \\ Z_{m1} & Z_{m2} & \cdots & Z_{mm} \end{bmatrix} \tag{1.16}$$

is the loop impedance matrix, where Z_{ii} is the self-impedance of the loop i and equals the sum of the branch impedances in the loop; Z_{ij} is the mutual impedance between loop i and loop j, and equals the sum of the impedances of their common branches. The sign of Z_{ij} depends on the directions of loop currents of loop i and loop j. If their directions are identical, Z_{ij} is positive, and if their directions are different, Z_{ij} is negative.

For the example shown in Fig. 1.3 we can write the basic loop incidence matrix according to the three independent loops,

$$B = \begin{bmatrix} 1 & 0 & 0 & 1 & 0 & 1 \\ 0 & 1 & 0 & 0 & 1 & 1 \\ 0 & 0 & 1 & -1 & 1 & 0 \end{bmatrix}.$$

The serial numbers of rows correspond to the loop numbers and the serial numbers of columns correspond to the branch numbers. For example, in the third row, there are three nonzero elements in the third, fourth, and fifth columns which means loop 3 includes branches 3, 4, and 5. If the branch current has the same direction as the basic loop current, the corresponding nonzero element equals +1; if the directions of branch current and loop current are different the corresponding nonzero element equals –1.

It should be noted that a basic loop incidence matrix cannot uniquely determine a network configuration. In other words, there may be different configurations corresponding to the same basic loop incidence matrix.

Similarly to the discussion on the node incidence matrix above, we can get the basic loop equations of an electric network from its basic loop incidence matrix $\boldsymbol{B}$,

$$\boldsymbol{Z}_{\mathrm{L}} = \boldsymbol{B}\boldsymbol{Z}_{\mathrm{B}}\boldsymbol{B}^{\mathrm{T}}, \tag{1.17}$$

where $\boldsymbol{Z}_{\mathrm{B}}$ is a diagonal matrix composed of the branch impedances.

The application of incidence matrices is quite extensive. If we have the above basic concepts, network analysis problems can be dealt with more flexibly. The details will be discussed in the relevant later sections.

1.2.2 Equivalent Circuit of Transformer and Phase-Shift Transformer

The equivalent circuit of an electric network is established by the equivalent circuits of its elements such as transmission lines and transformers. The AC transmission line is often described by the nominal Π equivalent circuit which can be found in other textbooks. In this section, only the equivalent circuits of the transformer and the phase-shift transformer are discussed, especially the transformer with off-nominal turns ratios. Flexible AC Transmission Systems (FACTS) are increasingly involved in power systems, and we will discuss the equivalent circuit of FACTS elements in Chap. 5.

When the exciting circuit is neglected or treated as a load (or an impedance), a transformer can be represented by its leakage impedance connected in series with an ideal transformer as shown in Fig. 1.4a. The relation between currents and voltages can be formulated as follows:

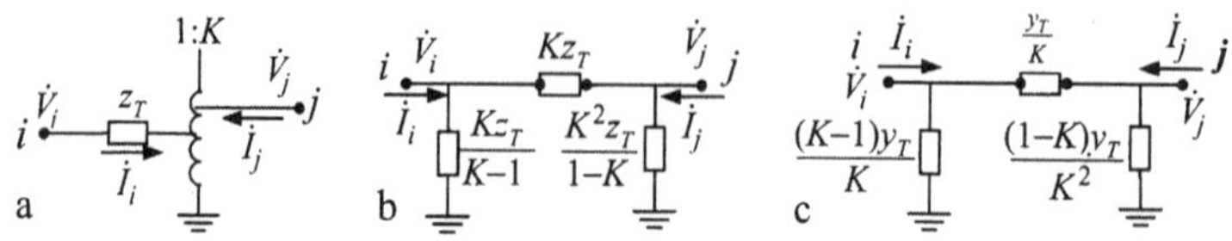

Fig. 1.4 Transformer equivalent circuit

$$\left.\begin{aligned}\dot{I}_i + K\dot{I}_j &= 0\\ \dot{V}_i - z_{\mathrm{T}}\dot{I}_i &= \frac{\dot{V}_j}{K}\end{aligned}\right\}.$$

Solving the above equation, we can obtain

$$\begin{aligned}\dot{I}_i &= \frac{1}{z_{\mathrm{T}}}\dot{V}_i - \frac{1}{Kz_{\mathrm{T}}}\dot{V}_j,\\ \dot{I}_j &= -\frac{1}{Kz_{\mathrm{T}}}\dot{V}_i + \frac{1}{K^2 z_{\mathrm{T}}}\dot{V}_j.\end{aligned} \tag{1.18}$$

Rewrite (1.18) as follows

$$\left.\begin{aligned}I_i &= \frac{K-1}{Kz_{\mathrm{T}}}\dot{V}_i + \frac{1}{Kz_{\mathrm{T}}}(\dot{V}_i - \dot{V}_j)\\ I_j &= \frac{1-K}{K^2 z_{\mathrm{T}}}\dot{V}_j + \frac{1}{Kz_{\mathrm{T}}}(\dot{V}_j - \dot{V}_i)\end{aligned}\right\}. \tag{1.19}$$

According to (1.19), we can get the equivalent circuit as shown in Fig. 1.4b. If the parameters are expressed in terms of admittance, the equivalent circuit is shown in Fig. 1.4c, where

$$y_{\mathrm{T}} = \frac{1}{z_{\mathrm{T}}}.$$

It should be especially noted in Fig. 1.4a the leakage impedance z_{T} is at the terminal where the ratio is 1. When the leakage impedance z_{T} is at the terminal where ratio is K, we should transform it to z'_{T} by using the following equation, so that the equivalent circuit shown in Fig. 1.4 also can be applied in this situation

$$z'_{\mathrm{T}} = z_{\mathrm{T}}/K^2. \tag{1.20}$$

The equivalent circuit of a two-winding transformer has been discussed above. A similar circuit can be used to represent a three-winding transformer. For example, Fig. 1.5 shows the equivalent circuit of a three-winding transformer that can be transformed into two two-winding transformers' equivalent circuits.

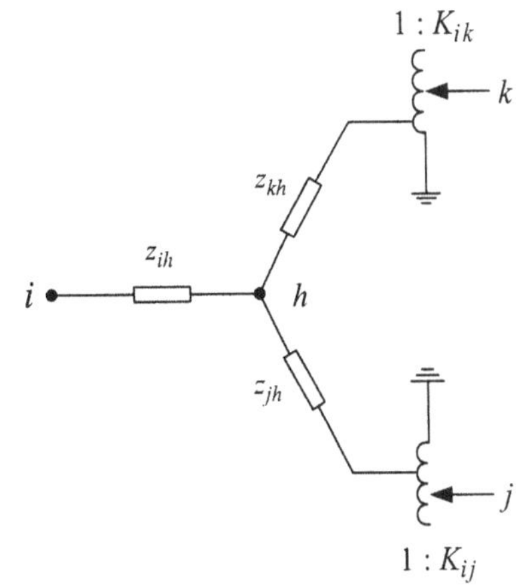

Fig. 1.5 Three-winding transformer equivalent circuit

After obtaining the transformer equivalent circuit, we can establish the equivalent circuit for a multivoltage network. For example, an electric network shown in Fig. 1.6 can be represented by the equivalent circuit shown in Fig. 1.6b or c when the leakage impedances of transformer T_1 and T_2 have been normalized to side ① and side ④. It can be proved that the two representations have an identical ultimate equivalent circuit as shown in the Fig. 1.6d.

When we analysis the operation of a power system, the per-unit system is extensively used. In this situation, all the parameters of an electric network are denoted in the per-unit system. For example, in the Fig. 1.6, if the voltage base at side ① is V_{j1}, at sides ② and ③ is V_{j2} and at side ④ is V_{j4}, then the base ratio (nominal turns ratio) of transformer T_1 and T_2 are

$$K_{j1} = \frac{V_{j2}}{V_{j1}}, K_{j2} = \frac{V_{j2}}{V_{j4}}. \tag{1.21}$$

The ratios of transformer T_1 and T_2 on a per-unit base (off-nominal turns ratio) are

$$K_{*1} = \frac{K_1}{K_{j1}}, \quad K_{*2} = \frac{K_2}{K_{j2}}. \tag{1.22}$$

Therefore, the ratio of the transformer should be K_{*1} or K_{*2} when its equivalent circuit is expressed in a per-unit system.

In modern power systems, especially in the circumstances of deregulation, the power flow often needs to be controlled. Therefore the application of the phase-shifting transformer is increasing. As we know, a transformer just transforms the voltages of its two terminals and its turn ratio is a real number. The phase-shifting transformer can also change the phase angle between voltages of its two terminals. Thus its turn ratio is a complex number. When the exciting current is neglected or treated as a load (or an impedance), a phase-shifting transformer can be represented

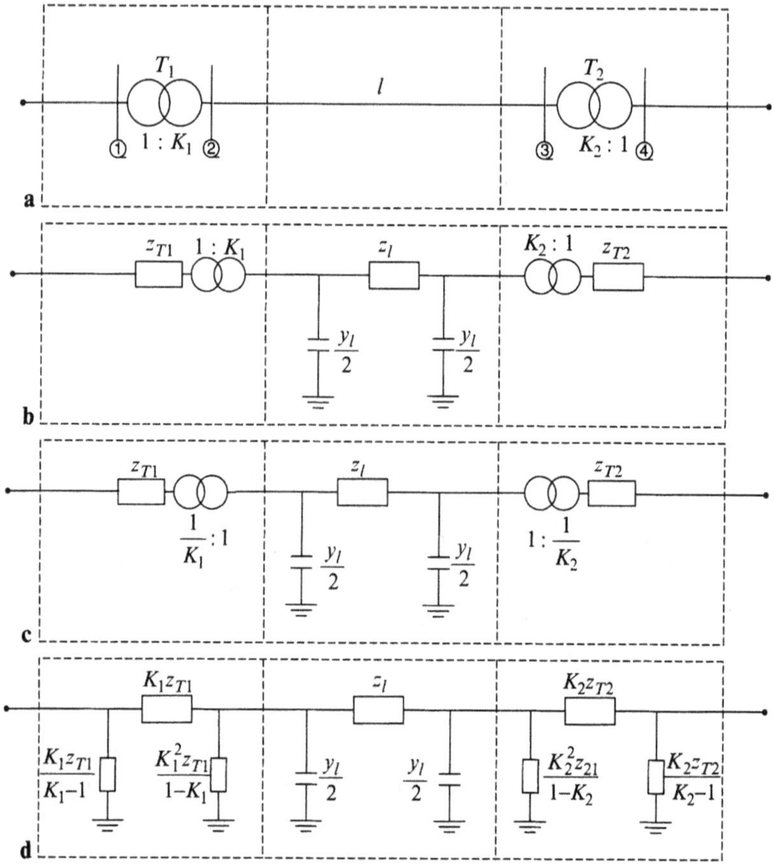

Fig. 1.6 Equivalent circuit of a multivoltage electric network

by its leakage impedance, which is connected in series with an ideal transformer having a complex turns ratio as shown in Fig. 1.7. From this figure, we can obtain the equations as follows,

$$\begin{aligned} \dot{V}_i - \dot{I}_i z_{\mathrm{T}} &= \dot{V}_j' \\ \dot{I}_i + \dot{I}_j' &= 0. \end{aligned} \tag{1.23}$$

Apparently, the two terminal voltages are related by

$$\dot{V}_j' = \dot{V}_j / \dot{K}. \tag{1.24}$$

Since there is no power loss in an ideal autotransformer,

$$\dot{V}_j' \hat{I}_j' = \dot{V}_j \hat{I}_j,$$

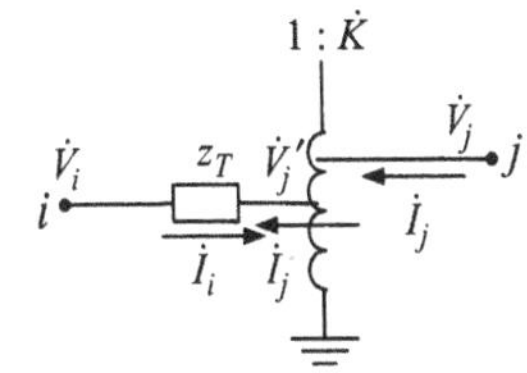

Fig. 1.7 Phase-shifting transformer representation

where $\hat{I}_j'$ and $\hat{I}_j$ are the conjugates of $\dot{I}_j'$ and $\dot{I}_j$, respectively. It follows from the above equations that

$$\dot{I}_j' = \hat{K}\dot{I}_j. \tag{1.25}$$

Substituting (1.24) and (1.25) into (1.23)

$$\begin{aligned} \dot{I}_i &= \frac{\dot{V}_i}{z_\mathrm{T}} - \frac{\dot{V}_j}{\dot{K}z_\mathrm{T}} = Y_{ii}\dot{V}_i + Y_{ij}\dot{V}_j \\ \dot{I}_j &= -\frac{\dot{V}_i}{\hat{K}z_\mathrm{T}} + \frac{\dot{V}_j}{K^2 z_\mathrm{T}} = Y_{ji}\dot{V}_i + Y_{jj}\dot{V}_j, \end{aligned} \tag{1.26}$$

where

$$Y_{ii} = \frac{1}{z_\mathrm{T}}, \quad Y_{ij} = -\frac{1}{\dot{K}z_\mathrm{T}}, \quad Y_{ji} = -\frac{1}{\hat{K}z_\mathrm{T}}, \quad Y_{jj} = \frac{1}{K^2 z_\mathrm{T}}.$$

Equation (1.26) is the mathematical model of the phase-shifting transformer. It is easy to be proved that (1.26) is the same as (1.18) when the turn ratio is a real number. This illustrates that the transformer is a particular case of the phase-shifting transformer. Because the ratio of a phase-shifting transformer is complex number, and $Y_{ij} \neq Y_{ji}$, it has no equivalent circuit and the admittance matrix of the electric network with the phase-shifting transformer is not symmetric.

1.3 Nodal Admittance Matrix

1.3.1 Basic Concept of Nodal Admittance Matrix

As mentioned above, the node equation (1.3) is usually adopted in modern power system analysis. If the number of nodes in a network is n, we have the following general simultaneous equations:

$$\left.\begin{aligned}
\dot{I}_1 &= Y_{11}\dot{V}_1 + Y_{12}\dot{V}_1 + \cdots + Y_{1i}\dot{V}_i + \cdots + Y_{1n}\dot{V}_n \\
\dot{I}_2 &= Y_{21}\dot{V}_1 + Y_{22}\dot{V}_2 + \cdots + Y_{2i}\dot{V}_i + \cdots + Y_{2n}\dot{V}_n \\
&\vdots \\
\dot{I}_i &= Y_{i1}\dot{V}_1 + Y_{i2}\dot{V}_2 + \cdots + Y_{ii}\dot{V}_i + \cdots + Y_{in}\dot{V}_n \\
&\vdots \\
\dot{I}_n &= Y_{n1}\dot{V}_1 + Y_{n2}\dot{V}_2 + \cdots + Y_{ni}\dot{V}_i + \cdots + Y_{nn}\dot{V}_n
\end{aligned}\right\}. \tag{1.27}$$

The matrix constituted by the coefficients of (1.27) is the nodal admittance matrix

$$Y = \begin{bmatrix}
Y_{11} & Y_{12} & \cdots & Y_{1i} & \cdots & Y_{an} \\
Y_{21} & Y_{22} & \cdots & Y_{2i} & \cdots & Y_{2n} \\
\vdots & \vdots & \cdots & \vdots & \cdots & \vdots \\
Y_{i1} & Y_{i2} & \cdots & Y_{ii} & \cdots & Y_{in} \\
\vdots & \vdots & \cdots & \vdots & \cdots & \vdots \\
Y_{n1} & Y_{n2} & \cdots & Y_{ni} & \cdots & Y_{nn}
\end{bmatrix}. \tag{1.28}$$

A nodal admittance matrix reflects the topology and parameters of an electric network, so it can be regarded as a mathematical abstraction of the electric network. The node equation based on the admittance matrix is a widely used mathematical model of electric networks. Next we will introduce some physical meaning of the matrix elements.

If we set a unit voltage at node i and ground other nodes, i.e.,

$$\begin{aligned}
&\dot{V}_i = 1 \\
&\dot{V}_j = 0 \quad (j = 1, 2, \ldots, n, j \neq i),
\end{aligned}$$

then the following relationships hold according to (1.27),

$$I_j = Y_{ji} \quad j = 1, 2, \ldots, n. \tag{1.29}$$

From (1.29) we can see the physical meaning of the ith column elements in the admittance matrix: the diagonal element Y_{ii} in the ith column, the self-admittance of node i, is equal to the injection current of the node i; the off-diagonal elements Y_{ij} in the ith column, the mutual-admittance of node i and node j, is equal to the injection current of node j in this situation.

We will further illustrate these concepts by a simple network shown in Fig. 1.8. The network has three nodes (plus ground), thus the dimension of its admittance matrix is 3×3,

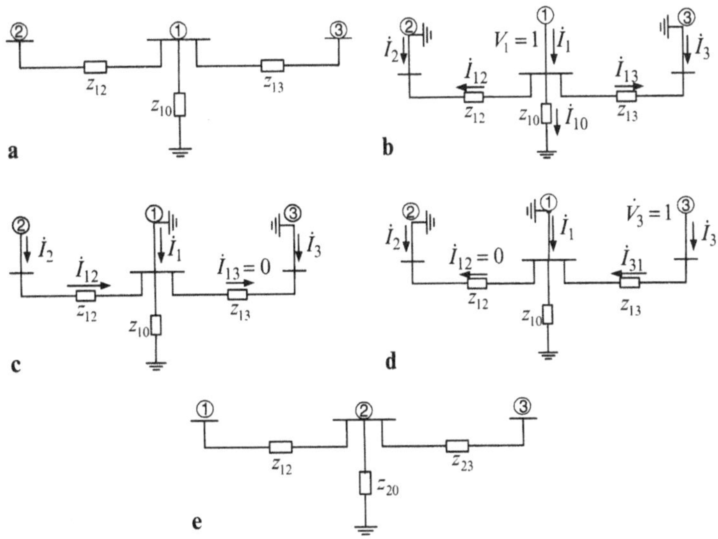

Fig. 1.8 Construction process of admittance matrix in simple electric network

$$Y = \begin{bmatrix} Y_{11} & Y_{12} & Y_{13} \\ Y_{21} & Y_{22} & Y_{23} \\ Y_{31} & Y_{32} & Y_{33} \end{bmatrix}.$$

According to the above discussion, we can get the elements of the first column: Y_{11}, Y_{21}, and Y_{31}, by setting a unit voltage on node 1 and grounding node 2 and node 3 as shown in Fig. 1.8b. Evidently,

$$\begin{aligned} \dot{I}_1 &= \dot{I}_{12} + \dot{I}_{13} + \dot{I}_{10} = \frac{1}{z_{12}} + \frac{1}{z_{10}} + \frac{1}{z_{13}} = Y_{11}, \\ \dot{I}_2 &= -\dot{I}_{12} = -\frac{1}{z_{12}} = Y_{21}, \\ \dot{I}_3 &= -\dot{I}_{13} = -\frac{1}{z_{13}} = Y_{31}. \end{aligned}$$

Similarly, setting a unit voltage at node 2 and grounding node 1 and node 3 as shown in Fig. 1.8c, we can get the elements of the second column:

$$\begin{aligned} \dot{I}_1 &= -\dot{I}_{21} = -\frac{1}{z_{12}} = Y_{12}, \\ \dot{I}_2 &= \dot{I}_{21} = \frac{1}{z_{12}} = Y_{22}, \\ \dot{I}_3 &= 0 = Y_{32}. \end{aligned}$$

For the elements of the third column we have (see Fig. 1.8d),

$$\dot{I}_1 = -\dot{I}_{31} = -\frac{1}{z_{31}} = Y_{13},$$
$$\dot{I}_2 = 0 = Y_{23},$$
$$\dot{I}_3 = \dot{I}_{31} = \frac{1}{z_{13}} = Y_{33}.$$

Finally, the admittance matrix of the above simple network becomes

$$\boldsymbol{Y} = \begin{bmatrix} \frac{1}{z_{12}} + \frac{1}{z_{10}} + \frac{1}{z_{13}} & -\frac{1}{z_{12}} & -\frac{1}{z_{13}} \\ -\frac{1}{z_{12}} & \frac{1}{z_{12}} & 0 \\ -\frac{1}{z_{13}} & 0 & \frac{1}{z_{13}} \end{bmatrix}. \tag{1.30}$$

If we change the node numbers in Fig. 1.8a, e.g., exchange the number ordering of node 1 with node 2, as shown in Fig. 1.8e, then the admittance matrix becomes,

$$\boldsymbol{Y}' = \begin{bmatrix} \frac{1}{z_{12}} & -\frac{1}{z_{12}} & 0 \\ -\frac{1}{z_{12}} & \frac{1}{z_{12}} + \frac{1}{z_{20}} + \frac{1}{z_{23}} & -\frac{1}{z_{23}} \\ 0 & -\frac{1}{z_{23}} & \frac{1}{z_{23}} \end{bmatrix}.$$

The above matrix can be obtained through exchanging the first row with the second row, and at the same time exchanging the first column with the second column of the matrix shown in (1.30). The exchange of the rows and columns of the admittance matrix corresponds to the exchange of the sequence of node equations and their variables.

The properties of the admittance matrix can be summarized as follows:

1. The admittance matrix is symmetric if there is no phase-shifting transformer in the network. From (1.30) we have

$$Y_{12} = Y_{21} = -\frac{1}{z_{12}}, Y_{13} = Y_{31} = -\frac{1}{z_{13}}, Y_{23} = Y_{32} = 0.$$

Generally, according to the reciprocity of the network,

$$Y_{ij} = Y_{ji}.$$

Therefore, the admittance matrix is symmetric. We will discuss the networks with phase-shifting transformers later.

2. The admittance matrix is sparse. From the discussion above, we know that Y_{ij} and Y_{ji} will be zero if node i does not directly connect with node j. For example, in Fig. 1.8a, node 2 does not directly connect with node 3, so both of Y_{23} and Y_{32} are zero. In general, the number of nonzero off-diagonal elements of each row is equal to the number of branches that are incident to the corresponding node. Usually, the number of branches connected to one node is 2–4, thus there are only 2–4 nonzero off-diagonal elements in each row. The property that only a few nonzero elements exist in a matrix is called sparsity. This phenomenon will be more remarkable with increase of the power system scale. For instance, for a network with 1,000 nodes, if each node directly connects three branches on average, the total number of nonzero elements for the network is 4,000, which is only 0.4% of the total elements in the admittance matrix.

The symmetry and sparsity of an admittance matrix are very important features for large-scale power systems. If we make full use of these two properties, the computation speed will be accelerated and the computer memory will be saved dramatically.

1.3.2 Formulation and Modification of Nodal Admittance Matrix

Now we discuss formulation of an admittance matrix by inspection first. When an electric network is composed of only transmission lines, the principles of constructing its admittance matrix can be summarized as follows:

1. The order of the admittance matrix is equal to the number of the nodes of the electric network.
2. The number of the nonzero off-diagonal elements in each row is equal to the number of the ungrounded branches connected to the corresponding node.
3. The diagonal elements of the admittance matrix, i.e., the self-admittance of the node, is equal to the sum of all the admittances of the incident branches of the corresponding node. Thus

$$Y_{ii} = \sum_{j \in i} y_{ij}, \tag{1.31}$$

where y_{ij} is the reciprocal of z_{ij}, which is the branch impedance between node i and node j, "j I" denotes that only the incident branches of node i (including the grounding branch) are included to the summation. For example, in Fig. 1.8, the self-admittance of node 1, i.e., Y_{11}, should be

$$Y_{11} = \frac{1}{z_{12}} + \frac{1}{z_{10}} + \frac{1}{z_{13}} = y_{12} + y_{10} + y_{13}.$$

The self-admittance of node 2, i.e., Y_{22}, should be

$$Y_{22} = \frac{1}{z_{12}} = y_{12}.$$

4. The off-diagonal element of the admittance matrix, Y_{ij}, is equal to the negative of the admittance between node i and node j

$$Y_{ij} = -\frac{1}{z_{ij}} = -y_{ij}. \tag{1.32}$$

For example, in Fig. 1.8a,

$$Y_{12} = -\frac{1}{z_{12}} = -y_{12},$$
$$Y_{13} = -\frac{1}{z_{13}} = -y_{13}.$$

Therefore, no matter how complicated the configuration of an electric network is, its admittance matrix can be established directly by inspection according to the parameters and the topology of the network.

When the electric network involves transformers or phase-shifting transformers, they need special treatment.

When branch ij is a transformer, the admittance matrix certainly can be formed following the above steps if the transformer is substituted beforehand by the Π equivalent circuit as shown in Fig. 1.4a. However, in practical application the transformer is often treated directly in forming the admittance matrix. If branch ij is a transformer, as shown in Fig. 1.4a, the elements of the admittance matrix related to the branch can be obtained as follows:

1. Add two nonzero off-diagonal elements into the admittance matrix

$$Y_{ij} = Y_{ji} = -\frac{y_T}{K}. \tag{1.33}$$

2. Add to the self-admittance of node i by,

$$\Delta Y_{ii} = \frac{K-1}{K} y_T + \frac{1}{K} y_T = y_T. \tag{1.34}$$

3. Add to the self-admittance of node j by

$$\Delta Y_{jj} = \frac{1}{K} y_T + \frac{1-K}{K^2} y_T = \frac{y_T}{K^2}. \tag{1.35}$$

When branch ij is a phase-shifting transformer, its equivalent circuit is Fig. 1.7. Then the corresponding matrix elements are obtained as follows:

1. Add two nonzero off-diagonal elements into the admittance matrix

$$Y_{ij} = -\frac{1}{\dot{K} z_T}, \tag{1.36}$$

$$Y_{ji} = -\frac{1}{\hat{K} z_{\mathrm{T}}}. \tag{1.37}$$

2. Add to the self-admittance of node i by

$$\Delta Y_{ii} = \frac{1}{z_{\mathrm{T}}}. \tag{1.38}$$

3. Add to the self-admittance of node j by

$$\Delta Y_{jj} = \frac{1}{K^2 z_{\mathrm{T}}}. \tag{1.39}$$

It can be seen from (1.36) and (1.37) that $Y_{ij} \neq Y_{ji}$, thus the admittance matrix is not symmetric any more although its structure is still symmetric.

Studies of different system operation states, such as transformer or transmission line outages, play an important part in modern power system analysis. Because the outage of branch ij only affects the self and mutual admittance of node i and node j, we can obtain the new admittance matrix for the contingency state by modifying the original admittance matrix. The modification methods for different situations are introduced as follows:

1. To add a new node with a new branch for the original network as shown in Fig. 1.9a.

   Assume that i is a node of the original network and j is the new node; z_{ij} is the impedance of the new branch. The dimension of the admittance matrix becomes $N+1$ because of the new node. There is only one branch connected to node j, therefore, its self-admittance is,

$$Y_{jj} = \frac{1}{z_{ij}},$$

   The self-admittance of node i should be modified (added) by,

$$\Delta Y_{ii} = \frac{1}{z_{ij}}.$$

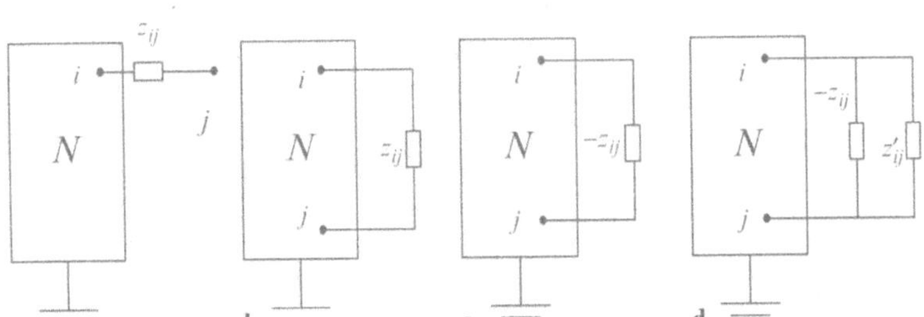

Fig. 1.9 Four cases of modifying the electric network

Two off-diagonal elements should also be created

$$Y_{ij} = Y_{ji} = -\frac{1}{z_{ij}}.$$

2. To add a new branch between node i and node j as shown in Fig. 1.9b.

   In this case, no new node is introduced and the dimension of the new admittance matrix is the same as the original one, while the following modifications should be made.

$$\left.\begin{aligned} \Delta Y_{ii} &= \frac{1}{z_{ij}} \\ \Delta Y_{jj} &= \frac{1}{z_{ij}} \\ \Delta Y_{ij} &= \Delta Y_{ji} = -\frac{1}{z_{ij}} \end{aligned}\right\}. \tag{1.40}$$

3. To remove a branch with impedance z_{ij} between node i and node j.

   In this case, it is equivalent to adding a new branch of impedance $-z_{ij}$ between node i and node j as shown in Fig. 1.9c. Therefore, the modifications of the admittance matrix are as follows:

$$\left.\begin{aligned} \Delta Y_{ii} &= -\frac{1}{z_{ij}} \\ \Delta Y_{jj} &= -\frac{1}{z_{ij}} \\ \Delta Y_{ij} &= \Delta Y_{ji} = \frac{1}{z_{ij}} \end{aligned}\right\}. \tag{1.41}$$

4. To change branch impedance z_{ij} for z'_{ij}.

This case is equivalent to removing branch impedance z_{ij} first and then adding a branch of impedance z'_{ij} between node i and node j as shown in Fig. 1.9d. Thus the modifications can be carried out according to (1.40) and (1.41).

It should be noted that the above discussion is based on the assumption that the added or removed branch is a pure impedance branch. If the branch is a transformer or a phase-shifting transformer, the modifications should be carried out according to (1.33)–(1.35) or (1.36)–(1.39).

[Example 1.1] Figure 1.10 shows an equivalent circuit of a simple electric network with two transformers. The branch impedance and grounding admittance in per unit are shown in the figure. Determine the nodal admittance matrix for the electric network.

[Solution] According to the method introduced in Sect. 1.2.2, we can assemble the elements of the admittance matrix node by node.

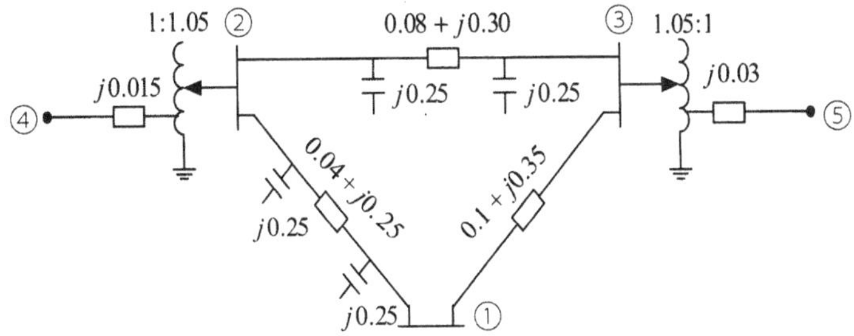

Fig. 1.10 Equivalent circuit for Example 1.1

In Fig. 1.10, parameters are in admittance for grounding branches and in impedance for other branches (branches in series connection). Using (1.31), we obtain the self-admittance of node 1 as follows:

$$Y_{11} = y_{10} + y_{12} + y_{13} = j0.25 + \frac{1}{0.04 + j0.25} + \frac{1}{0.1 + j0.35}$$
$$= 1.378742 - j6.291665.$$

The mutual admittances related to node 1 can be obtained according to (1.32),

$$Y_{21} = Y_{12} = -y_{12} = -\frac{1}{0.04 + j0.25} = -0.624025 + j3.900156$$
$$Y_{31} = Y_{13} = -y_{13} = -\frac{1}{0.1 + j0.35} = -0.754717 + j2.641509.$$

Because branch 2–4 is a transformer, the self-admittance of node 2 should be calculated according to (1.31) and (1.35) based on the equivalent circuit as shown in Fig. 1.4a

$$Y_{22} = y_{20} + y_{12} + y_{23} + \frac{y_{42}}{K_{42}^2}$$
$$= (j0.25 + j0.25) + \frac{1}{0.04 + j0.25} + \frac{1}{0.08 + j0.30} + \frac{1}{j0.015} \times \frac{1}{1.05^2}$$
$$= 1.453909 - j66.98082.$$

The mutual admittances related to node 2 are

$$Y_{23} = Y_{32} = -\frac{1}{0.08 + j0.30} = -0.829876 + j3.112033.$$

Using (1.33) we have

$$Y_{24} = Y_{42} = -\frac{y_{42}}{K_{42}} = -\frac{1}{j0.015} \times \frac{1}{1.05} = j63.49206.$$

The other elements of the admittance matrix can be calculated in a similar way. The ultimate result is

$$
Y = \begin{bmatrix}
\begin{matrix}1.378742\\-j6.291665\end{matrix} & \begin{matrix}-0.924024\\+j3.900156\end{matrix} & \begin{matrix}-0.754717\\+j2.641509\end{matrix} & & \\
\begin{matrix}-0.24024\\+j3.900156\end{matrix} & \begin{matrix}1.453909\\-j66.98082\end{matrix} & \begin{matrix}-0.829876\\+j3.112033\end{matrix} & \begin{matrix}0.000000\\+j63.19206\end{matrix} & \\
\begin{matrix}-0.754717\\+j2.641509\end{matrix} & \begin{matrix}-0.929876\\+j3.112033\end{matrix} & \begin{matrix}1.584596\\-j35.73786\end{matrix} & & \begin{matrix}0.000000\\+j31.74603\end{matrix} \\
 & \begin{matrix}0.000000\\+j63.49206\end{matrix} & & \begin{matrix}0.000000\\-j66.66667\end{matrix} & \\
 & \begin{matrix}0.000000\\+j31.74603\end{matrix} & & & \begin{matrix}0.000000\\-j33.33333\end{matrix}
\end{bmatrix},
$$

where the vacancies are zero elements.

1.4 Solution to Electric Network Equations

1.4.1 Gauss Elimination Method

At present, Gauss elimination is the most popular method to solve the electric network equations. In the initial stage of computer application in power systems, iterative methods were also been used because of the limitation of computer memory. The fatal disadvantage of the iterative methods is the convergence problem. Therefore, the Gauss elimination method almost has substituted for iterative methods after successful application of the sparse techniques [1, 2]. The Gauss elimination method is introduced in this section, and the sparse technique and sparse vector method will be described successively.

The Gauss elimination method in solving simultaneous linear equations consists of two steps, i.e., forward elimination and back substitution. Both forward elimination and back substitution can be carried out by either row or column orientation. Generally, the column-oriented forward elimination and row-oriented back substitution scheme are widely used. The related algorithm is introduced next, and other algorithms can be easily deduced similarly.

A system of n simultaneous linear equations may be written in the matrix form as $\boldsymbol{AX} = \boldsymbol{B}$ in which elements in matrix $\boldsymbol{A}$ and vector $\boldsymbol{B}$ can be either real or complex numbers. For example, the coefficient matrix of (1.3) is complex, while that of the correction equation in the Newton–Raphson method (see (2.40) in Chap. 2) is real.

Because the forward eliminations involve manipulations with matrix $\boldsymbol{A}$ and $\boldsymbol{B}$, a $n \times (n+1)$ augmented matrix is formed by appending $\boldsymbol{B}$ as the $(n+1)$th column of $\boldsymbol{A}$,

$$\bar{\boldsymbol{A}} = [\boldsymbol{A} \quad \boldsymbol{B}] = \begin{bmatrix} a_{11} & a_{12} & \cdots & a_{1n} & b_1 \\ a_{21} & a_{22} & \cdots & a_{2n} & b_2 \\ \cdots & \cdots & \cdots & \cdots & \cdots \\ a_{n1} & a_{n2} & \cdots & a_{nn} & b_n \end{bmatrix} = \begin{bmatrix} a_{11} & a_{12} & \cdots & a_{1n} & a_{1,n+1} \\ a_{21} & a_{22} & \cdots & a_{2n} & a_{2,n+1} \\ \cdots & \cdots & \cdots & \cdots & \cdots \\ a_{n1} & a_{n2} & \cdots & a_{nn} & a_{n,n+1} \end{bmatrix}.$$

In the above equation, b_j is substituted by $a_{j,n+1}$ $(j = 1, 2, \ldots, n)$ to simplify the following representation.

The process of the column-oriented forward eliminations is introduced first.

Step 1. Eliminate the first column

First, normalize the first row of the augmented matrix $\bar{\boldsymbol{A}}$,

$$1 \quad a_{12}^{(1)} \quad a_{13}^{(1)} \quad \ldots \quad a_{1,n+1}^{(1)}, \tag{1.42}$$

where

$$a_{1j}^{(1)} = \frac{a_{1j}}{a_{11}} \quad (j = 2, 3, \ldots, n+1).$$

Then the derived row as shown in (1.42) is used to eliminate the elements $a_{21}, a_{31}, \ldots, a_{n1}$ of $\bar{\boldsymbol{A}}$, and the remaining elements of the second to the nth row can be calculated by

$$a_{ij}^{(1)} = a_{ij} - a_{i1}a_{1j}^{(1)} \quad (j = 2, 3, \ldots, n+1), (i = 2, 3, \ldots, n),$$

where the superscript (1) denotes that the relative element is the result of the first manipulation. At this stage, matrix $\bar{\boldsymbol{A}}$ is changed into $\bar{\boldsymbol{A}}_1$,

$$\bar{\boldsymbol{A}}_1 = [\boldsymbol{A}_1 \quad \boldsymbol{B}_1] = \begin{bmatrix} 1 & a_{12}^{(1)} & \cdots & a_{1n}^{(1)} & a_{1,n+1}^{(1)} \\ & a_{22}^{(1)} & \cdots & a_{2n}^{(1)} & a_{2,n+1}^{(1)} \\ & \vdots & \vdots & \vdots & \vdots \\ & a_{n2}^{(1)} & \cdots & a_{nn}^{(1)} & a_{n,n+1}^{(1)} \end{bmatrix}.$$

The corresponding equation is $A_1X = B_1$ which has the same solution as the original equation. In the above matrix, the vacancies are zero elements.

Step 2. Eliminate the second column

Normalize the second row of the augmented matrix $\bar{A}$ as the following

$$0 \quad 1 \quad a_{23}^{(2)} \quad \dots \quad a_{2,n+1}^{(2)}, \tag{1.43}$$

where

$$a_{2j}^{(2)} = a_{2j}^{(1)} / a_{22}^{(1)} \quad (j = 3, 4, \dots, n+1).$$

Then the derived row shown in (1.43) is used to eliminate the elements $a_{32}^{(1)}, a_{42}^{(1)}, \dots, a_{4n}^{(1)}$ of $\bar{A}_1$ and the remaining elements of the third to the nth row can be calculated by,

$$a_{ij}^{(2)} = a_{ij}^{(1)} - a_{i2}^{(1)} a_{2j}^{(2)} \quad (j = 3, 4, \dots, n+1), (i = 3, 4, \dots, n),$$

where the superscript (2) denotes that the relative element is the result of the second manipulation. Now, matrix $\bar{A}_1$ has been transformed into $\bar{A}_2$,

$$\bar{A}_2 = [A_2 \quad B_2] = \begin{bmatrix} 1 & a_{12}^{(1)} & a_{13}^{(1)} & \cdots & a_{1n}^{(1)} & a_{1,n+1}^{(1)} \\ & 1 & a_{23}^{(2)} & \cdots & a_{2n}^{(2)} & a_{2,n+1}^{(2)} \\ & & a_{33}^{(2)} & \cdots & a_{3n}^{(2)} & a_{3,n+1}^{pg(2)} \\ & & \cdots & \cdots & & \\ & & a_{n3}^{(2)} & \cdots & a_{nn}^{(2)} & a_{n,n+1}^{(2)} \end{bmatrix}.$$

Generally, the following computation should be executed when eliminating the kth column

$$a_{kj}^{(k)} = a_{kj}^{(k-1)} / a_{kk}^{(k-1)} \quad (j = k+1, \dots, n+1), \tag{1.44}$$

$$a_{ij}^{(k)} = a_{ij}^{(k-1)} - a_{ik}^{(k-1)} a_{kj}^{(k)} \quad (j = k+1, \dots, n+1), (i = k+1, \dots, n). \tag{1.45}$$

After proceeding with the elimination n times in this manner, the elements below the diagonal of the matrix become zero, and the nth derived augmented matrix is obtained.

$$\bar{A}_n = [A_n \quad B_n] = \begin{bmatrix} 1 & a_{12}^{(1)} & a_{13}^{(1)} & \dots & a_{1n}^{(1)} & a_{1,n+1}^{(1)} \\ & 1 & a_{23}^{(2)} & \dots & a_{2n}^{(2)} & a_{2,n+1}^{(2)} \\ & & 1 & \dots & a_{3n}^{(3)} & a_{3,n+1}^{(3)} \\ & & & \ddots & \vdots & \vdots \\ & & & & 1 & a_{n,n+1}^{(n)} \end{bmatrix}. \tag{1.46}$$

The corresponding equation becomes $\boldsymbol{A}_n\boldsymbol{X} = \boldsymbol{B}_n$, that is

$$\begin{array}{llllll} x_1+ & a_{12}^{(1)}x_2+ & a_{13}^{(1)}x_3+ & \dots & +a_{1n}^{(1)}x_n = & a_{1,n+1}^{(1)} \\ & x_2 \quad + & a_{23}^{(2)}x_3+ & \dots & +a_{2n}^{(2)}x_n = & a_{2,n+1}^{(2)} \\ & & x_3 \quad + & \dots & +a_{3n}^{(3)}x_n = & a_{3,n+1}^{(3)} \\ & & & \ddots & \vdots & \vdots \\ & & & & x_n = & a_{n,n+1}^{(n)} \end{array} \tag{1.47}$$

Its solution is the same as the original equation $\boldsymbol{AX} = \boldsymbol{B}$.

For (1.47), back substitution is carried out in a bottom-up sequence. The value of x_n is obtained directly from the nth equation,

$$x_n = a_{n,n+1}^{(n)}.$$

Then substituting x_n into the $(n-1)$th equation we get the solution of x_{n-1},

$$x_{n-1} = a_{n-1,n+1}^{(n-1)} - a_{n-1,n}^{(n-1)}x_n.$$

Substituting x_{n-1} and x_n into the $(n-2)$th equation, we obtain x_{n-2}. Generally, x_i can be obtained by substituting the solved variables $x_{i+1}, x_{i+2}, \dots, x_n$ into the ith equation,

$$x_i = a_{i,n+1}^{(i)} - \sum_{j=i+1}^{n} a_{ij}^{(i)}x_j \quad (i = n, \dots, 2, 1). \tag{1.48}$$

This is the general equation of the row-oriented back substitution.

[Example 1.2] Solve the following simultaneous linear equations by using the Gauss elimination method.

$$\begin{aligned} &x_1 + 2x_2 + x_3 + x_4 = 5 \\ &2x_1 + x_2 = 3 \\ &x_1 + x_3 = 2 \\ &x_1 + x_4 = 2 \end{aligned}.$$

[Solution] Write the augmented matrix according to the original equations as below.

$$\left[\begin{array}{cccc:c} (1) & 2 & 1 & 1 & 5 \\ 2 & 1 & 0 & 0 & 3 \\ 1 & 0 & 1 & 0 & 2 \\ 1 & 0 & 0 & 1 & 2 \end{array}\right].$$

As an initial step, normalize the first row of the augmented matrix according to (1.44), i.e., divide the first row by its diagonal element.

$$\begin{bmatrix} 1 & 2 & 1 & 1 & \vdots\, 5 \\ (2) & 1 & 0 & 0 & \vdots\, 3 \\ (1) & 0 & 1 & 0 & \vdots\, 2 \\ (1) & 0 & 0 & 1 & \vdots\, 2 \end{bmatrix}.$$

Then eliminate the first column according to (1.45)

$$\begin{bmatrix} 1 & 2 & 1 & 1 & \vdots\, 5 \\ & (-3) & -2 & -2 & \vdots\, -7 \\ & -2 & 0 & -1 & \vdots\, -3 \\ & -2 & -1 & 0 & \vdots\, -3 \end{bmatrix}.$$

The next step is the elimination of the second column. When normalizing the second row, we divide the elements in the second row by the diagonal element –3

$$\begin{bmatrix} 1 & 2 & 1 & 1 & \vdots\, 5 \\ & 1 & \frac{2}{3} & \frac{2}{3} & \vdots\, \frac{7}{3} \\ & (-2) & 0 & -1 & \vdots\, -3 \\ & (-2) & -1 & 0 & \vdots\, -3 \end{bmatrix}.$$

Then eliminate the second column in terms of (1.45) to obtain

$$\begin{bmatrix} 1 & 2 & 1 & 1 & \vdots\, 5 \\ & 1 & \frac{2}{3} & \frac{2}{3} & \vdots\, \frac{7}{3} \\ & & \left(\frac{4}{3}\right) & \frac{1}{3} & \vdots\, \frac{5}{3} \\ & & \frac{1}{3} & \frac{4}{3} & \vdots\, \frac{5}{3} \end{bmatrix}.$$

Repeat the procedure for the third column. Normalize the third row through dividing the third row by the diagonal element 4/3.

$$\begin{bmatrix} 1 & 2 & 1 & 1 & \vdots\, 5 \\ & 1 & \frac{2}{3} & \frac{2}{3} & \vdots\, \frac{7}{3} \\ & & 1 & \frac{1}{4} & \vdots\, \frac{5}{4} \\ & & \left(\frac{1}{3}\right) & \frac{4}{3} & \vdots\, \frac{5}{3} \end{bmatrix}.$$

Then eliminate the third column in terms of (1.45) to obtain

$$\begin{bmatrix} 1 & 2 & 1 & 1 & \vdots\, 5 \\ & 1 & \frac{2}{3} & \frac{2}{3} & \vdots\, \frac{7}{3} \\ & & 1 & \frac{1}{4} & \vdots\, \frac{5}{4} \\ & & & \left(\frac{5}{4}\right) & \vdots\, \frac{5}{4} \end{bmatrix}.$$

The last step is normalizing the fourth row according to (1.44), that is, dividing the fourth row by the diagonal element 5/4.

$$\begin{bmatrix} 1 & 2 & 1 & 1 & \vdots\, 5 \\ & 1 & \frac{2}{3} & \frac{2}{3} & \vdots\, \frac{7}{3} \\ & & 1 & \frac{1}{4} & \vdots\, \frac{5}{4} \\ & & & 1 & \vdots\, 1 \end{bmatrix}.$$

The transformed equations after elimination become

$$\begin{aligned} x_1 + 2x_2 + x_3 + x_4 &= 5 \\ x_2 + \tfrac{2}{3}x_3 + \tfrac{2}{3}x_4 &= \tfrac{7}{3} \\ x_3 + \tfrac{1}{4}x_4 &= \tfrac{5}{4} \\ x_4 &= 1 \end{aligned}.$$

x_4, x_3, x_2, x_1 can be obtained through the back substitution according to (1.48).

$$\begin{aligned} x_4 &= 1 \\ x_3 &= \tfrac{5}{4} - \tfrac{1}{4}x_4 = 1 \\ x_2 &= \tfrac{7}{3} - \tfrac{2}{3}x_3 - \tfrac{2}{3}x_4 = 1 \\ x_1 &= 5 - 2x_2 - x_3 - x_4 = 1 \end{aligned}$$

1.4.2 Triangular Decomposition and Factor Table

In practical applications, the simultaneous equations often need to be solved repeatedly when only right-hand vector $\boldsymbol{B}$ changes while coefficient matrix $\boldsymbol{A}$ is a constant matrix. In such cases, the factor table method is often used to improve computation efficiency.

The factor table records all the operations on right-hand vector $\boldsymbol{B}$ in the Gauss elimination process. As the discussion above, The Gauss elimination method involves forward elimination and back substitution. Back substitution is determined by the upper triangular elements of the coefficient matrix after elimination operation as shown (1.46). In order to execute the elimination operation (forward elimination), the relative operation factors also need to be recorded in the elimination process. The forward elimination includes normalization and elimination operation. Take column-oriented elimination as an example, operations on the i th element of $\boldsymbol{B}$ (i.e., $b_{i,n+1}$) in the forward elimination are as follows (see (1.44) and (1.45)),

$$b_i^{(i)} = b_i^{(i-1)}/a_{ii}^{(i-1)} \quad (i = 1, 2, \dots, n), \tag{1.49}$$

$$b_i^{(k)} = b_i^{(k-1)} - a_{ik}^{(k-1)} b_k^{(k)} \quad (k = 1, 2, \dots, i-1). \tag{1.50}$$

The above operation factors $a_{i1}, a_{i2}^{(1)}, a_{i2}^{(1)}, \dots, a_{i,i-1}^{(i-2)}$ and $a_{ii}^{(i-1)}$ are to be stored in the lower triangular matrix row by row and appended to the upper triangular elements of the (1.46). Thus, we obtain the factor table as the following

$$\begin{matrix} a_{11} & a_{12}^{(1)} & a_{13}^{(1)} & a_{14}^{(1)} & \cdots & a_{1n}^{(1)} \\ a_{21} & a_{22}^{(1)} & a_{23}^{(2)} & a_{24}^{(2)} & \cdots & a_{2n}^{(2)} \\ a_{31} & a_{32}^{(1)} & a_{33}^{(2)} & a_{34}^{(3)} & \cdots & a_{3n}^{(3)} \\ a_{41} & a_{42}^{(1)} & a_{43}^{(2)} & a_{44}^{(3)} & \cdots & a_{4n}^{(4)} \\ \vdots & \vdots & \vdots & \vdots & \vdots & \vdots \\ a_{n1} & a_{n2}^{(1)} & a_{n3}^{(2)} & a_{n4}^{(3)} & \cdots & a_{nn}^{(n-1)} \end{matrix} .$$

Where the lower triangular elements are used in elimination operations on $\boldsymbol{B}$ and the upper triangular elements are used in back substitution operations. The factor table also can be denoted in the following format

$$\begin{matrix} d_{11} & u_{12} & u_{13} & u_{14} & \cdots & u_{1n} \\ l_{21} & d_{22} & u_{23} & u_{24} & \cdots & u_{2n} \\ l_{31} & l_{32} & d_{33} & u_{34} & \cdots & u_{3n} \\ l_{41} & l_{42} & l_{43} & d_{44} & \cdots & u_{4n} \\ \vdots & \vdots & \vdots & \vdots & \vdots & \vdots \\ l_{n1} & l_{n2} & l_{n3} & l_{n4} & \cdots & d_{nn} \end{matrix} , \tag{1.51}$$

where

$$d_{ii} = a_{ii}^{(i-1)},$$
$$u_{ij} = a_{ij}^{(i)} \quad (i < j),$$
$$l_{ij} = a_{ij}^{(j-1)} \quad (j < i).$$

We can see that the lower triangular elements of the factor table are exactly the operation elements used in the elimination process. Therefore, if we retain them in the original position and take the reciprocals of the diagonal elements, the lower triangular elements of the factor table can be readily obtained. The upper triangular elements of the factor table are just the upper triangular part of the coefficient matrix after the elimination operations.

If the simultaneous equations need to be solved repeatedly for different right-hand vector $\boldsymbol{B}$, we should first carry out the elimination operation on coefficient matrix $\boldsymbol{A}$ to obtain its factor table. Then the factor table can be used directly and repeatedly to solve the equations with different $\boldsymbol{B}$. In this situation, we will carry out the elimination operation on the following equations instead of (1.49) and (1.50),

$$b_i^{(i)} = b_i^{(i-1)}/d_{ii}, \tag{1.52}$$

$$b_i^{(k)} = b_i^{(k-1)} - l_{ik} \times b_k^{(k)} \quad (i = k+1, \ldots, n). \tag{1.53}$$

The back substitution will be carried out on the following equations instead of (1.48)

$$x_n = b_n^{(n)},$$

$$x_i = b_i^{(i)} - \sum_{j=i+1}^{n} u_{ij} \times x_j. \tag{1.54}$$

[Example 1.3] For the simultaneous linear equations of Example 1.2, find the factor table of its coefficient matrix $\boldsymbol{A}$ and solve the equation when $B = [-1 \quad 1 \quad 2 \quad 0]^{\mathrm{T}}$.

[Solution] Inspecting the solution process of Example 1.2, we can directly obtain the factor table of coefficient matrix $\boldsymbol{A}$,

$$\begin{matrix} 1 & 2 & 1 & 1 \\ 2 & -3 & \frac{2}{3} & \frac{2}{3} \\ 1 & -2 & \frac{4}{3} & \frac{1}{4} \\ 1 & -2 & \frac{1}{3} & \frac{5}{4} \end{matrix} \Leftrightarrow \begin{matrix} d_{11} & u_{12} & u_{13} & u_{14} \\ l_{21} & d_{22} & u_{23} & u_{24} \\ l_{31} & l_{32} & d_{33} & u_{34} \\ l_{41} & l_{42} & l_{43} & d_{44} \end{matrix}.$$

The lower triangular elements of the above factor table are just the operation factors in brackets which appeared in the elimination process, and the upper triangular elements are the upper triangular part of the coefficient matrix after elimination operation.

Now we first use the lower triangular elements of the factor table to operate column-oriented elimination on $\boldsymbol{B}$. Normalize b_1 according to (1.52),

$$b_1^{(1)} = b_1/d_{11} = (-1)/1 = -1.$$

Then operations on b_2, b_3, b_4 are carried out by using the elements of the factor table's first column in the lower triangular part according to (1.53)

$$b_2^{(1)} = b_2 - l_{21} \times b_1^{(1)} = 1 - 2 \times (-1) = 3,$$
$$b_3^{(1)} = b_3 - l_{31} \times b_1^{(1)} = 2 - 1 \times (-1) = 3,$$
$$b_4^{(1)} = b_4 - l_{41} \times b_1^{(1)} = 0 - 1 \times (-1) = 1.$$

Thus the elimination operation of the first column is completed, and we have,

$$\boldsymbol{B}^{(1)} = [-1 \quad 3 \quad 3 \quad 1]^{\mathrm{T}}.$$

Next, normalize $b_2^{(1)}$ according to (1.52),

$$b_2^{(2)} = b_2^{(1)}/d_{22} = 3/(-3) = -1.$$

The elimination operation on $b_3^{(1)}, b_4^{(1)}$ is followed by using the elements of the second column in the lower triangular part according to (1.53),

$$b_3^{(2)} = b_3^{(1)} - l_{32} \times b_2^{(2)} = 3 - (-2) \times (-1) = 1,$$
$$b_4^{(2)} = b_4^{(1)} - l_{42} \times b_2^{(2)} = 1 - (-2) \times (-1) = -1.$$

Thus the elimination operation of the second column is finished, and we have

$$\boldsymbol{B}^{(2)} = [-1 \quad -1 \quad 1 \quad -1]^{\mathrm{T}}.$$

Normalize $b_3^{(2)}$ according to (1.52) and operate $b_4^{(3)}$ according to (1.53)

$$b_3^{(3)} = b_3^{(2)}/d_{33} = 1/\tfrac{4}{3} = \tfrac{3}{4}.$$

Again, the elimination operation on $b_4^{(2)}$ is followed by using the elements of the third column in the lower triangular part according to (1.53)

$$b_4^{(3)} = b_4^{(2)} - l_{43} \times b_3^{(3)} = -1 - \tfrac{1}{3} \times \tfrac{3}{4} = -\tfrac{5}{4}.$$

Thus the elimination operation on the third column is finished, and we have

$$\boldsymbol{B}^{(3)} = \left[-1 \quad -1 \quad \tfrac{3}{4} \quad -\tfrac{5}{4}\right]^{\mathrm{T}}.$$

The last step of the elimination operation is to normalize $b_4^{(3)}$ according to (1.52)

$$b_4^{(4)} = b_4^{(3)}/d_{44} = -\tfrac{4}{5}/\left(\tfrac{4}{5}\right) = -1.$$

Now, all the elimination operations are fulfilled.

$$\boldsymbol{B}^{(4)} = \begin{bmatrix} -1 & -1 & \tfrac{3}{4} & -1 \end{bmatrix}^{\mathrm{T}}.$$

Comparing with the factor table, we obtain the following identical solution equations

$$\begin{array}{llll} x_1 + & 2x_2 + & x_3 + & x_4 \quad = -1 \\ & x_2 + & \tfrac{2}{3}x_3 + & \tfrac{2}{3}x_4 \quad = -1 \\ & & x_3 + & \tfrac{1}{4}x_4 = \tfrac{3}{4} \\ & & & x_4 \quad = -1 \end{array}.$$

Now, the unknowns could be solved using the upper triangular part of the factor table according to (1.54).

$$\begin{aligned}
&x_4 = b_4^{(4)} = -1 \\
&x_3 = b_3^{(3)} - u_{34} \times x_4 = \tfrac{3}{4} - \tfrac{1}{4} \times (-1) = 1 \\
&x_2 = b_2^{(2)} - u_{23} \times x_3 - u_{24} \times x_4 = -1 - \tfrac{2}{3} \times 1 - \tfrac{2}{3} \times (-1) = -1 \\
&x_1 = b_1^{(1)} - u_{12} \times x_2 - u_{13} \times x_3 - u_{14} \times x_4 = -1 - 2 \times (-) - 1 \times 1 - 1 \times (-1) = 1.
\end{aligned}$$

It should be pointed out that the factor table as shown in (1.50) can be established not only by the Gauss elimination method but also by the triangular decomposition method. From the above example, we can verify that the following relationship between the factor table and its coefficient matrix holds,

$$\boldsymbol{A} = \boldsymbol{L}'\boldsymbol{U}, \tag{1.55}$$

where

$$\boldsymbol{L}' = \begin{bmatrix} 1 & 0 & 0 & 0 \\ 2 & -3 & 0 & 0 \\ 1 & -2 & \tfrac{4}{3} & 0 \\ 1 & -2 & \tfrac{1}{3} & \tfrac{5}{4} \end{bmatrix} \quad \boldsymbol{U} = \begin{bmatrix} 1 & 2 & 1 & 1 \\ 0 & 1 & \tfrac{2}{3} & \tfrac{2}{3} \\ 0 & 0 & 1 & \tfrac{1}{4} \\ 0 & 0 & 0 & 1 \end{bmatrix}.$$

$\boldsymbol{L}'$ can be decomposed further,

$$\boldsymbol{L}' = \boldsymbol{L}\boldsymbol{D}. \tag{1.56}$$

In the above example, $\boldsymbol{L}$ can be obtained through dividing off-diagonal elements in each column of $\boldsymbol{L}'$ by the corresponding diagonal element,

$$\boldsymbol{L}=\begin{bmatrix}1&0&0&0\\2&1&0&0\\1&\frac{2}{3}&1&0\\1&\frac{2}{3}&\frac{1}{4}&1\end{bmatrix}\quad \boldsymbol{D}=\begin{bmatrix}1&0&0&0\\0&-3&0&0\\0&0&\frac{4}{3}&0\\0&0&0&\frac{5}{4}\end{bmatrix}.$$

Therefore the original coefficient matrix can be generally represented as follows

$$\boldsymbol{A}=\boldsymbol{LDU}. \tag{1.57}$$

From the example, we can also see the following relationship

$$\boldsymbol{L}^{\mathrm{T}}=\boldsymbol{U} \text{ or } \boldsymbol{U}=\boldsymbol{L}^{\mathrm{T}}. \tag{1.58}$$

This phenomenon is not specific to this example. The relationship in (1.58) can be proved when the coefficient matrix is symmetric.

In the following, we deduce the recursion formulae of the triangular decomposition.

Expand (1.55)

$$\begin{bmatrix}a_{11}&a_{12}&a_{13}&\cdots&a_{1n}\\a_{21}&a_{22}&a_{23}&\cdots&a_{2n}\\a_{31}&a_{32}&a_{33}&\cdots&a_{3n}\\\vdots&\vdots&\vdots&\vdots&\vdots\\a_{n1}&a_{n2}&a_{n3}&\cdots&a_{nn}\end{bmatrix}=\begin{bmatrix}l'_{11}&&&&\\l'_{21}&l'_{22}&&&\\l'_{31}&l'_{32}&l'_{33}&&\\\vdots&\vdots&\vdots&\ddots&\\l'_{n1}&l'_{n2}&l'_{n3}&\cdots&l'_{nn}\end{bmatrix}$$
$$\times\begin{bmatrix}1&u_{12}&u_{13}&\cdots&u_{1n}\\&1&u_{23}&\cdots&u_{2n}\\&&1&\cdots&u_{3n}\\&&&\ddots&\vdots\\&&&&1\end{bmatrix}. \tag{1.59}$$

Comparing two sides of the above equation, the diagonal element of the first row can be found

$$l'_{11}=a_{11}.$$

Comparing the first element of the second row and the first two elements of the second column in both sides, we can obtain

$$l'_{21}=a_{21}, l'_{11}u_{12}=a_{12}, l'_{21}u_{12}+l'_{22}=a_{22}.$$

Hence the recursion formulae are

$$l'_{21} = a_{21}, u_{12} = a_{12}/l'_{11}, l'_{22} = a_{12} - l'_{21}u_{12}.$$

The following decomposition equation can be obtained

$$\begin{bmatrix} a_{11} & a_{12} \\ a_{21} & a_{22} \end{bmatrix} = \begin{bmatrix} l'_{11} & \\ l'_{21} & l'_{22} \end{bmatrix} \times \begin{bmatrix} 1 & u_{12} \\ & 1 \end{bmatrix}.$$

Similarly, if the first $k-1$ rows of $\boldsymbol{L}'$ and the first $k-1$ columns of $\boldsymbol{U}$ have been obtained, the equation becomes

$$\begin{bmatrix} a_{11} & a_{12} & a_{13} & \cdots & a_{1,k-1} \\ a_{21} & a_{22} & a_{23} & \cdots & a_{2,k-1} \\ a_{31} & a_{32} & a_{33} & \cdots & a_{3,k-1} \\ \cdots & \vdots & \vdots & \vdots & \vdots \\ a_{k-1,1} & a_{k-1,2} & a_{k-1,3} & \cdots & a_{k-1,k-1} \end{bmatrix}$$

$$= \begin{bmatrix} l'_{11} & & & & \\ l'_{21} & l'_{22} & & & \\ l'_{31} & l'_{32} & l'_{33} & & \\ \vdots & \vdots & \vdots & \ddots & \\ l'_{k-1,1} & l'_{k-1,2} & l'_{k-1,3} & \cdots & l'_{k-1,k-1} \end{bmatrix} \times \begin{bmatrix} 1 & u_{12} & u_{13} & \cdots & u_{1,k-1} \\ & 1 & u_{23} & \cdots & u_{2,k-1} \\ & & 1 & \cdots & u_{3,k-1} \\ & & & \ddots & \vdots \\ & & & & 1 \end{bmatrix}.$$

All the elements of the two matrices in the right hand of the above equation have been solved. Comparing the first $k-1$ elements in the kth row and the first k elements in the kth column of the two sides element by element, we can get the corresponding elements by the following formulae

$$\begin{aligned} u_{ik} &= \frac{1}{l'_{ii}}\left(a_{ik} - \sum_{p=1}^{i-1} l'_{ip}u_{pk}\right) \quad (i = 1, 2, \ldots, k-1) \\ l'_{kj} &= a_{kj} - \sum_{p=1}^{j-1} l'_{kp}u_{pj} \quad (j = 1, 2, \ldots, k). \end{aligned} \tag{1.60}$$

The above are recursion formulae. Taking k from 1 to n in sequence, the triangular decomposition, $\boldsymbol{A} = \boldsymbol{L}'\boldsymbol{U}$, will be achieved by using these formulae. Furthermore, dividing the off-diagonal elements by the corresponding diagonal element, $\boldsymbol{L}$ can be obtained:

$$l_{kj} = \frac{1}{l_{jj}}\left(a_{kj} - \sum_{p=1}^{j-1} l'_{kp}u_{pj}\right) \quad (k = j+1, \ldots, n). \tag{1.61}$$

The diagonal elements of $\boldsymbol{L}'$ constitute $\boldsymbol{D}$, i.e., $d_{ii} = l'_{ii}$ $(i = 1, 2, \ldots, n)$. Now, the coefficient matrix is decomposed into the format $\boldsymbol{A} = \boldsymbol{LDU}$. It should be particularly noted that (1.58) will always be true if the coefficient matrix is symmetric.

1.4.3 Sparse Techniques

From the discussion of the above section, we know that the solution process of the electric network equation is the process of operating the right-hand constant vector successively using the elements of its factor table. In Example 1.3, there are 16 elements in its factor table: four diagonal elements, six lower triangular elements, and six upper triangular elements. Therefore the solution involves 16 multiplication operations. According to (1.53) and (1.54), if elements in the factor table are zero, the corresponding multiplication operations can be avoided (since the product will be zero) and significant computational effort can be saved. Based on this idea, so-called sparse technique is widely used in power system analysis to improve solution efficiency. The concept of the sparse technique is illustrated by an example in the following.

[Example 1.4] Solve the simultaneous linear equations in Example 1.2 by using the sparse method.

[Solution] In Example 1.2, the simultaneous linear equations are

$$\begin{array}{lllll} x_1 & 2x_2 & x_3 & +x_4 & = 5 \\ 2x_1 & +x_2 & & & = 3 \\ x_1 & & +x_3 & & = 2 \\ x_1 & & & +x_4 & = 2 \end{array}. \tag{1.62}$$

In order to make full use of the sparsity advantages of the equations, the following transformation should be made first,

$$x_1 = y_4, \; x_2 = y_2, \; x_3 = y_3, \; x_4 = y_1. \tag{1.63}$$

Then, the original equations are transformed into

$$\begin{array}{lllll} y_1 & & & +y_4 & = 2 \\ & y_2 & & +2y_4 & = 3 \\ & & y_3 & +y_4 & = 2 \\ y_1 & +2y_2 & +y_3 & +y_4 & = 5 \end{array}. \tag{1.64}$$

We will solve the equations by using its factor table. The coefficient matrix is

$$\begin{bmatrix} (1) & 0 & 0 & 1 \\ 0 & 1 & 0 & 2 \\ 0 & 0 & 1 & 1 \\ (1) & 2 & 1 & 1 \end{bmatrix}.$$

First, we normalize the first row and eliminate the first column. There are only two operations: one normalization operation and one elimination operation in this step. The elements in brackets are the computing factors. For a 4 × 4 coefficient matrix, the elimination of the first column should include one normalization operation and three elimination operations. However, because both a_{21} and a_{31} are zero, two corresponding operations are avoided. After the above operations, we obtain

$$\begin{bmatrix} 1 & 0 & 0 & 1 \\ 0 & (1) & 0 & 2 \\ 0 & 0 & 1 & 1 \\ 0 & (2) & 1 & 0 \end{bmatrix}.$$

The next step is the normalization of the second row and elimination of the second column. There are also only two operations, one normalization operation and one elimination operation in this step. The figures in the brackets of the above matrix are the computing factors. For a 4 × 4 coefficient matrix, the elimination of the second column should include one normalization operation and two elimination operations. Because $a_{32}^{(1)}$ is zero, the corresponding operation is avoided. After these operations, we obtain

$$\begin{bmatrix} 1 & 0 & 0 & 1 \\ 0 & 1 & 0 & 2 \\ 0 & 0 & (1) & 1 \\ 0 & 0 & (1) & -4 \end{bmatrix}.$$

To normalize the third row and eliminate the third column, we also need two operations, one normalization operation and one elimination operation. The computing factors are the elements in the brackets of the above matrix. After these operations, we obtain

$$\begin{bmatrix} 1 & 0 & 0 & 1 \\ 0 & 1 & 0 & 2 \\ 0 & 0 & 1 & 1 \\ 0 & 0 & 0 & (-5) \end{bmatrix}.$$

Here, the factor table of the coefficient matrix can be readily written,

$$\begin{matrix} 1 & 0 & 0 & 1 \\ 0 & 1 & 0 & 2 \\ 0 & 0 & 1 & 1 \\ 1 & 2 & 1 & -5 \end{matrix}.$$

The above factor table can also be found using (1.60) and (1.61). Because there are only six zero off-diagonal elements in the above factor table, six multiply–add operations are avoided. In the following, we will use this factor table to obtain the solution to the constant vector:

$$\boldsymbol{B} = [2 \quad 3 \quad 2 \quad 5]^{\mathrm{T}}.$$

First, eliminating $\boldsymbol{B}$ column by column is executed by using the lower triangular part of the factor table. According to (1.52), b_1 is normalized,

$$b_1^{(1)} = b_1/d_{11} = 2/1 = 2.$$

Then the operations on b_2, b_3, b_4 are continued by using the elements of the first column in the lower triangular part according to (1.53). Because l_{21} and l_{31} are zero, we have

$$b_2^{(1)} = b_2 - l_{21} \times b_1^{(1)} = b_2 = 3,$$
$$b_3^{(1)} = b_3 - l_{31} \times b_1^{(1)} = b_3 = 2.$$

The above two steps should be avoided and only the following operation is needed

$$b_4^{(1)} = b_4 - l_{41} \times b_1^{(1)} = 5 - 1 \times 2 = 3.$$

After the elimination operation of the first column, we obtain

$$\boldsymbol{B}^{(1)} = [2 \quad 3 \quad 2 \quad 3]^{\mathrm{T}}.$$

Then normalize $b_2^{(1)}$ according to (1.52)

$$b_2^{(2)} = b_2^{(1)}/d_{22} = 3/1 = 3.$$

Now, the operation on $b_3^{(1)}, b_4^{(1)}$ should use the elements of the second column in the lower triangular part according to (1.53). Because l_{32} is zero, only the operation related to l_{42} will be performed. Thus,

$$b_4^{(2)} = b_4^{(1)} - l_{42} \times b_2^{(2)} = 3 - 2 \times 3 = -3.$$

After finishing elimination operation of the second column, we have

$$\boldsymbol{B}^{(2)} = [2 \quad 3 \quad 2 \quad -3]^{\mathrm{T}}.$$

Next, we normalize $b_3^{(2)}$ according to (1.52)

$$b_3^{(3)} = b_3^{(2)}/d_{33} = 2/1 = 2.$$

And then compute $b_4^{(3)}$ according to (1.53)

$$b_4^{(3)} = b_4^{(2)} - l_{43} \times b_3^{(3)} = -3 - 1 \times 2 = -5.$$

After finishing the elimination operation of the third column, we obtain

$$\boldsymbol{B}^{(3)} = [2 \quad 3 \quad 2 \quad -5]^{\mathrm{T}}.$$

The last step of the elimination operation is to normalize $b_4^{(3)}$ according to (1.52)

$$b_4^{(4)} = b_4^{(3)}/d_{44} = -5/(-5) = 1.$$

At this stage, all of the elimination operation have been completed, the right-hand vector becomes

$$\boldsymbol{B}^{(4)} = [2 \quad 3 \quad 2 \quad 1]^{\mathrm{T}}.$$

Comparing with the factor table, we obtain the following identical solution equations of (1.64)

$$\begin{array}{rrrrl} y_1 & & & +y_4 = & 2 \\ & y_2 & & +2y_4 = & 3 \\ & & y_3 & +y_4 = & 2 \\ & & & y_4 = & 1 \end{array}.$$

Now, the unknowns can be solved using the upper triangular part of the factor table according to (1.54). Because u_{12}, u_{13}, and u_{23} are zero, corresponding operations are avoided in back substitution.

$$\begin{aligned} y_4 &= b_4^{(4)} = 1 \\ y_3 &= b_3^{(3)} - u_{34} \times y_4 = 2 - 1 \times 1 = 1 \\ y_2 &= b_2^{(2)} - u_{24} \times y_4 = 3 - 2 \times 1 = 1 \\ y_1 &= b_1^{(1)} - u_{14} \times y_4 = 2 - 1 \times 1 = 1 \end{aligned}.$$

Substituting the above results into (1.63), the solutions to original equation (1.62) can be obtained.

From the above example, we can see that the computation effort can be saved not only in the formation of the factor table but also in the forward and back substitution. The amount of computation saved by the sparse technique depends on the number of zero elements in the factor table. Therefore, the key point of improving computation efficiency is to keep the number of zero elements in the factor table as high as possible.

1.4.4 Sparse Vector Method

Nowadays, the sparse matrix techniques are adopted to solve almost all large-scale power network problems. In this section, the sparse vector method, which can further improve the computation efficiency, will be introduced [3].

Sparse vector methods are useful for solving a system of simultaneous linear equations when the independent (right-hand) vector is sparse, or only few elements in the unknown vector are wanted. To take advantage of vector sparsity is relatively simple, but the results of improving computational efficiency and saving memory can be quite dramatic. Therefore sparse vector methods are often used in the compensation method, fault analysis, optimal power flow problem and contingency analysis.

In principle, the sparse vector method can be applied to both full- and sparse-matrix equations. This section focuses only on the implementation of sparse vector methods in the sparse-matrix situation. According to the above discussion, the admittance matrix $\boldsymbol{Y}$ of an electric network without phase-shifting transformers is symmetric. If there are phase-shifting transformers in the network the sparse admittance matrix is only symmetric in its structure. Nodal voltage equations can be written as

$$\boldsymbol{YV} = \boldsymbol{I}. \tag{1.65}$$

For generality, we assume $\boldsymbol{Y}$ is an incidence-symmetric square matrix of order n and can be factorized as

$$\boldsymbol{Y} = \boldsymbol{LDU}, \tag{1.66}$$

where $\boldsymbol{L}$ and $\boldsymbol{U}$ are lower and upper triangular matrices with unity diagonals, respectively, and $\boldsymbol{D}$ is a diagonal matrix. It is easy to solve the nodal equations using the above expressions. For example, the simultaneous equations can be written in the following form

$$\boldsymbol{LDUV} = \boldsymbol{I}. \tag{1.67}$$

The above formulae can be decomposed as

$$\boldsymbol{LX} = \boldsymbol{I}, \tag{1.68}$$

$$\boldsymbol{DW} = \boldsymbol{X}, \tag{1.69}$$

$$\boldsymbol{UV} = \boldsymbol{W}. \tag{1.70}$$

$\boldsymbol{V}$ can be obtained when (1.68)–(1.70) are solved in sequence. If $\boldsymbol{Y}$ is symmetric, matrix $\boldsymbol{U}$ is the transpose of $\boldsymbol{L}$. If $\boldsymbol{Y}$ is incidence symmetric, matrix $\boldsymbol{U}$ is not the transpose of $\boldsymbol{L}$, but they are identical in the sparsity structure.

The forward substitution operations can be expressed as

$$W = D^{-1}L^{-1}I. \tag{1.71}$$

The back substitution operations can be expressed as

$$V = U^{-1}W. \tag{1.72}$$

Generally, these operations can be performed either by rows or by columns. However, for the sparse vector method, the forward elimination (1.71) must be performed by columns, while the back substitution (1.72) by rows.

Many different schemes can be used for storing and accessing $\boldsymbol{L}$ and $\boldsymbol{U}$. For the sparse vector method, the lowest-numbered, nonzero, off-diagonal element in each column of $\boldsymbol{L}$ or in each row of $\boldsymbol{U}$ must be directly accessed without search. This requirement is satisfied by most storage schemes for $\boldsymbol{L}$ and $\boldsymbol{U}$.

The independent vector $\boldsymbol{I}$ is sparse in many applications. However, the solution vector $\boldsymbol{V}$ is not sparse in general. The term ''sparse vector'' in the following refers to either a sparse vector $\boldsymbol{I}$ or a subset of vector $\boldsymbol{V}$ containing the elements of interest. The exact meaning is always clear from the context.

If the vector $\boldsymbol{I}$ is sparse, only a subset of the columns of $\boldsymbol{L}$ is needed for the forward elimination. This is called the fast forward (FF) process. If only certain elements of vector $\boldsymbol{V}$ are actually wanted, only a subset of the rows of $\boldsymbol{U}$ is needed for the backward substitution. This is called the fast backward (FB) process.

[Example 1.5] Solve the following simultaneous linear equations

$$\begin{array}{lllll} V_1 & & & +V_4 & = 0 \\ & V_2 & & +2V_4 & = 1 \\ & & V_3 & +V_4 & = 0 \\ V_1 & +2V_2 & +V_3 & +V_4 & = 0 \end{array}.$$

[Solution] The coefficient matrix of above simultaneous linear equations is the same as in (1.64) of Example 1.4. The only difference is that the right-hand vector is sparse.

$$\boldsymbol{I} = \boldsymbol{B} = [0 \quad 1 \quad 0 \quad 0]^{\mathrm{T}}.$$

Therefore, the factor table of these simultaneous linear equations is the same as that of (1.64).

$$\begin{array}{cccc} 1 & 0 & 0 & 1 \\ 0 & 1 & 0 & 2 \\ 0 & 0 & 1 & 1 \\ 1 & 2 & 1 & -5 \end{array}.$$

Decomposing the factor table, we obtain

$$L = \begin{bmatrix} 1 & & & \\ 0 & 1 & & \\ 0 & 0 & 1 & \\ 1 & 2 & 1 & 1 \end{bmatrix}; \quad D = \begin{bmatrix} 1 & & & \\ & 1 & & \\ & & 1 & \\ & & & -5 \end{bmatrix}; \quad U = \begin{bmatrix} 1 & 0 & 0 & 1 \\ & 1 & 0 & 2 \\ & & 1 & 1 \\ & & & 1 \end{bmatrix}.$$

From (1.53), we can see that all the operations related with l_{ik} $(i = k+1, \ldots, n)$ can be avoided if $b_k^{(k)}$ is equal to zero:

$$b_i^{(k)} = b_i^{(k-1)} - l_{ik} \times b_k^{(k)} \quad (i = k+1, \ldots, n).$$

In other words, the kth column in the lower triangular matrix can be ignored. In this example, b_1 is equal to zero, so we can skip the first column of $\boldsymbol{L}$. For this sparse vector, the elimination should begin from the second column. The elimination also includes the normalization and elimination operations. After this, the right-hand sparse vector is transformed into

$$\boldsymbol{B}' = [0 \quad 1 \quad 0 \quad -1]^{\mathrm{T}}.$$

The next step is elimination of the third column. Because b_3' is zero, the operations related to the third column of $\boldsymbol{L}$ are skipped, thus the elimination of the fourth column is performed directly. Here, we use d_{44} to normalize b_4', and the ultimate result vector after the elimination operation is

$$\boldsymbol{B}'' = \left[0 \quad 1 \quad 0 \quad \tfrac{1}{5}\right]^{\mathrm{T}}.$$

As we know, the backward substitution operations must be performed by rows. If only V_3 is wanted, the operations with the first and second rows of $\boldsymbol{U}$ can be neglected. If only V_2 is wanted, the operations with the first row of $\boldsymbol{U}$ can be avoided. Furthermore, the operations with the third row of $\boldsymbol{U}$ also can be omitted because $b_3' = 0$. Therefore, the back substitution is only needed to perform on the second row of $\boldsymbol{U}$. Therefore, we have

$$V_2 = b_2'' - u_{24} \times b_4'' = 1 - 2 \times \tfrac{1}{5} = \tfrac{3}{5}.$$

From the above example, we can see that the key task of sparse vector methods is to identify the active subsets of $\boldsymbol{L}$ and $\boldsymbol{U}$ for FF and FB operations. The active subset of columns for FF depends on the sparsity structure of $\boldsymbol{L}$ and $\boldsymbol{I}$ while the active subset of rows for FB depends on the sparsity structures of $\boldsymbol{U}$ and $\boldsymbol{V}$.

In order to find the active subset of FF and improve the computation efficiency, the following simple algorithm can be summarized according to the above example

1. Zero all locations in $\boldsymbol{I}$, and enter the given nonzero elements in $\boldsymbol{I}$.
2. Search the nonzero elements in $\boldsymbol{I}$ and let k be the location number of the lowest-numbered nonzero element.
3. Perform the forward eliminations defined by column k of $\boldsymbol{L}$ on $\boldsymbol{I}$.
4. If $k = n$, exit. Else, return to Step 2.

This algorithm ensures that only the necessary nonzero operations of FF are performed, but it is wasteful because of zeroing and searching. A similar algorithm can be used to FB, but it is even more wasteful.

In the following we introduce a more efficient algorithm based on the concept of the factorization path. A factorization path for a sparse vector is represented by an ordered column list of **L** for FF operations. A path is executed in forward order for FF and in reverse order for FB. The same or different paths may be used for FF and FB depending on the application.

The path for a singleton is basic to the path concept. A singleton is a vector with only one nonzero element. Assume that the nonzero element is in location k. The following algorithm determines the path of the singleton:

1. Let k be the first number in the path.
2. Get the number of the lowest-numbered nonzero element in column k of $\boldsymbol{L}$ (or row k of $\boldsymbol{U}$). Replace k with this number, and list it in the path.
3. If $k = n$, exit. Else, return to Step 2.

The path for a singleton can be determined directly from the indexing arrays without searching or testing. A general sparse vector is the sum of singleton vectors, and its path is the union of the paths of its composite singleton vectors. For any sparse system, a path can be always associated with a given sparse vector.

[Example 1.6] Find the factorization path of the electric network shown in Fig. 1.11.

[Solution] Figure 1.12 shows the sparsity structure of the incidence symmetric admittance matrix of the network as shown in Fig. 1.11 (only the lower triangular part of the matrix is labeled). Because there are 21 branches in the network, 21 •

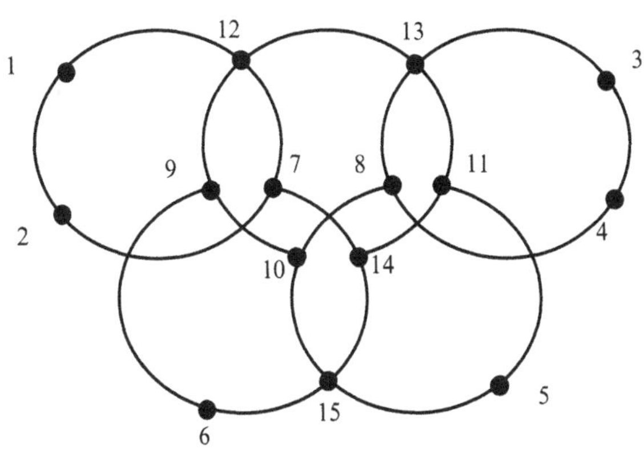

Fig. 1.11 Example electric network

represent the off-diagonal elements of the matrix. After triangular factorization, 10 fill-in elements (labeled as ◯) are added. Therefore there are altogether 31 nonzero elements in the factor table.

The factorization path of any singleton can be directly obtained from the structure of the factor table. For example

When $k = 1$, the singleton path is $1 \to 2 \to 7 \to 12 \to 13 \to 14 \to 15$
When $k = 5$, the singleton path is $5 \to 11 \to 13 \to 14 \to 15$
When $k = 6$, the singleton path is $6 \to 9 \to 10 \to 12 \to 13 \to 14 \to 15$

When a sparse vector is not a singleton, its path is the union of the paths of its composite singletons. For a sparse vector as follows

$$\boldsymbol{I} = [1 \quad 0 \quad 0 \quad 0 \quad 1 \quad 0 \quad 0 \quad 0 \quad 0 \quad 0 \quad 0 \quad 0 \quad 0 \quad 0 \quad 0]^{\mathrm{T}}$$

we have its path as the union of the paths of its composite singletons when $k = 1$ and $k = 5$,

$$1 \to 2 \to 7 \to 12 \to 5 \to 11 \to 13 \to 14 \to 15.$$

In Table 1.1 we list the entire factorization paths for the network shown in Fig. 1.12.

A pictorial view of the path table is provided by the path graph shown in Fig. 1.13. Utilizing this path graph, highly efficient algorithms for the sparse vector can be obtained. For example, assume the injected current at node 5 is I_5 while the injected currents of other nodes are zero, and the voltage at node 1 is wanted. To do so, we carry out FF operations according to the following active column sequence:

$$5 \to 11 \to 13 \to 14 \to 15.$$

And then carry out FB operations according to the following active row sequence:

$$15 \to 14 \to 13 \to 12 \to 7 \to 2 \to 1$$

In the above solution process, only the elements of five columns in lower triangular and seven rows in upper triangular elements are employed, the computation efficiency is improved dramatically. For sparse vector methods, the above path graph

Table 1.1 Path table

Node	Next node	Node	Next node
1	2	8	10
2	7	9	10
3	4	10	12
4	8	11	13
5	11	12	13
6	9	13	14
7	12	14	15

$$I = [1 \;\; 0 \;\; 0 \;\; 0 \;\; 1 \;\; 0 \;\; 0 \;\; 0 \;\; 0 \;\; 0 \;\; 0 \;\; 0 \;\; 0 \;\; 0 \;\; 0]^T$$

	1	2	3	4	5	6	7	8	9	10	11	12	13	14	15
1	•														
2	•	•													
3			•												
4			•	•											
5					•										
6						•									
7		•					•								
8				•				•							
9						•			•						
10								•	•	•					
11					•						•				
12	•	○					•		•	○		•			
13			•	○				•		○	•	•	•		
14							•				•	○	○	•	
15					•	•			○	•	○	○	○	•	•

Fig. 1.12 Sparse structure of a network's factor table

Fig. 1.13 The path graph

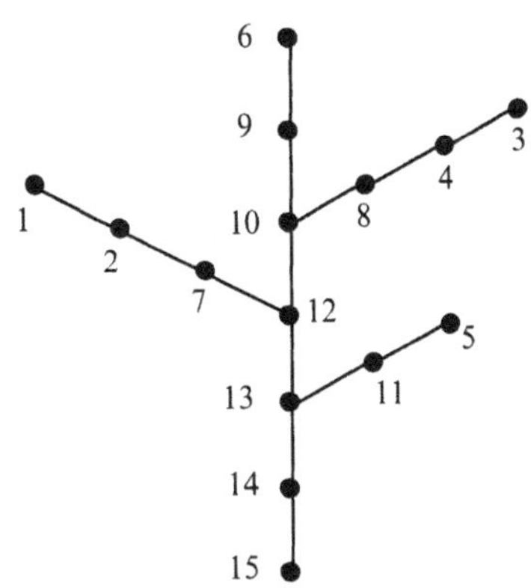

should be determined in advance and then be utilized directly, thus unnecessary zeroing and searching can be skipped.

1.4.5 Optimal Ordering Schemes of Electric Network Nodes

At present, the Gauss elimination method introduced in Sect. 1.3.1 is applied to solve the node equations $\boldsymbol{I} = \boldsymbol{YV}$ in most power system analysis programs. In order to solve the network equation repeatedly, the admittance matrix is usually

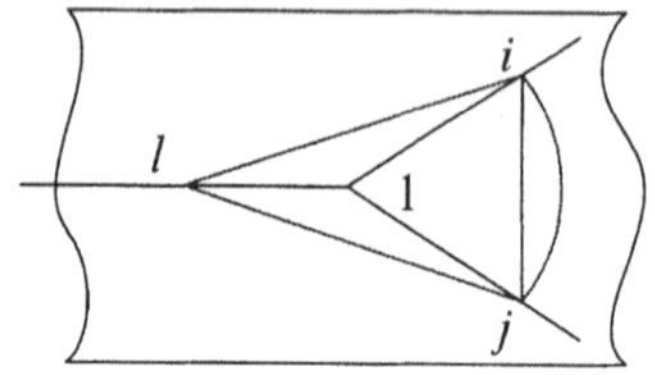

Fig. 1.14 Relationship between Gauss elimination and Y-Δ transformation

factorized first, and then the factor table can be directly used to solve the equations with different right-hand vectors.

As we know, the admittance matrix is sparse and the triangular matrices after factorization are also sparse. Generally, the distributions of nonzero elements in the admittance matrix are different from those in the factorized triangular matrix, because some new nonzero elements, i.e., the fill-in elements, may occur in the elimination or LU factorization process.

The addition of fill-ins in the elimination process can be explained intuitively by Y-Δ transformation. As shown in Fig. 1.14, node l does not directly connect with nodes i and j in the initial network, thus corresponding elements Y_{il} and Y_{lj} in its admittance matrix are zero while Y_{ij} is nonzero.

It can be proved that eliminating the first column of the admittance matrix in Gauss elimination is equivalent to eliminating node 1 by Y-Δ transformation as shown in Fig. 1.14. New branches connecting node pairs ij, il, and lj are created. Therefore, in the new admittance matrix, Y_{il}, Y_{lj}, and Y_{ij} are all nonzero elements, thus two fill-ins occur in eliminating the first column.

Generally, eliminating node k which is the central point of a star network will create a mesh network whose vertexes are nodes connecting directly with node k. If the number of nodes connecting directly with k is J_k, the branches in the mesh network should be combinations of any two nodes of J_k nodes, which is equal to $(1/2)J_k(J_k - 1)$. Assuming that there already exist D_k branches connecting these J_k nodes, the number of new branches (the number of fill-ins) after the elimination of node k is

$$\Delta b_k = \frac{1}{2} J_K(J_K - 1) - D_k. \tag{1.73}$$

The number of fill-ins highly depends on the elimination sequence or the ordering number of the nodes. In Fig. 1.15, four number ordering schemes and the corresponding fill-ins in the triangular matrix are denoted. Apparently, different number ordering schemes will result in different fill-ins.

An optimal ordering minimizes the fill-ins in the factor table during the LU factorization process. Different number ordering schemes should be compared according to the number of fill-ins. At present, several effective schemes have been developed. Among them the following three ordering schemes are widely employed:

1. *Static ordering scheme*: This scheme numbers the nodes according to the number of branches connected to them. It means that the nodes are ordered

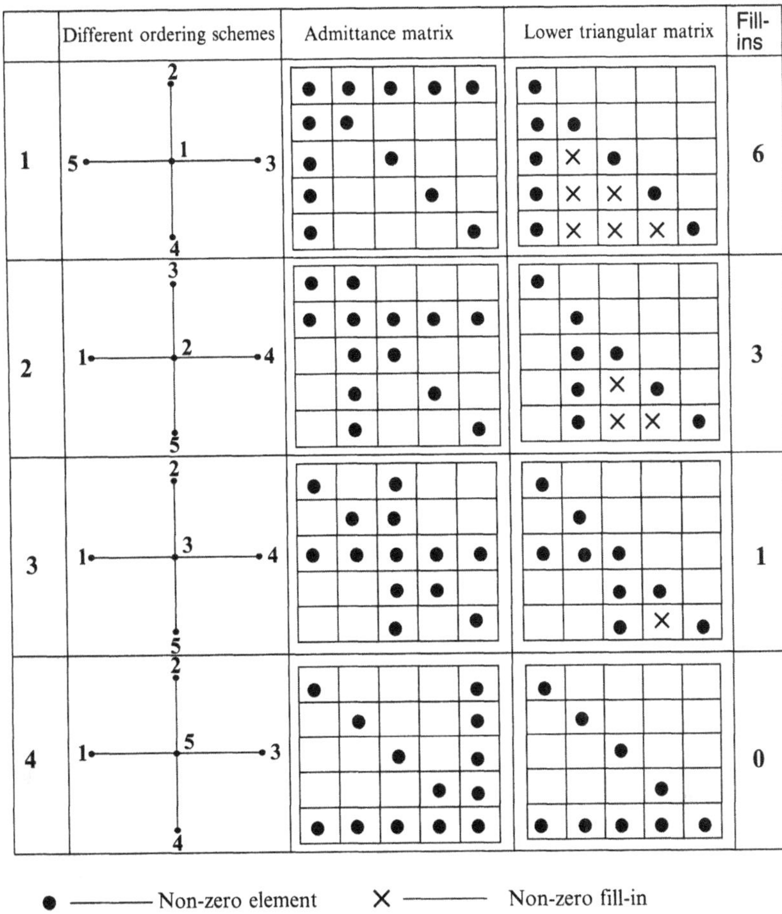

Fig. 1.15 Illustration of number ordering

from the node with fewest branches to the node with most branches. If the numbers of connected branches for more than one node are the same, any one of them can be ordered first. Before ordering, the number of the branches connected to each node needs be counted.

The scheme can be explained intuitively as follows: in the admittance matrix, the node with the fewest connected branches corresponds to the row which has the fewest nonzero elements, so the fill-ins will be generated with less possibility in the elimination operation. This scheme is very simple and suitable to be applied to small networks with fewer loops.

2. *Semidynamic ordering scheme*: In the above scheme, the number of branches connected to each node is counted based on the initial network and is constant in the ordering process. In fact, in the process of node elimination, the number of branches connected to each node will change according to $\Delta - Y$ transformation. Therefore, the number of branches of the remaining nodes should be updated

after each elimination and then they should be ordered according to new data. This ordering scheme might be expected to result in better fill-in reduction, because it considers the changing number of incident branches during the elimination process.

3. *Dynamic ordering scheme*: The above two schemes are only suboptimal, which cannot guarantee minimizing fill-in number. The more rigorous scheme numbers the node according to the principle that introduces the fewest new branches. The ordering process is as follows

   - According to $\Delta - Y$ transformation, count the number of new branches (the number of fill-ins) added after the elimination on each node, and the node with the fewest branches (including fill-ins) is numbered first.
   - Update the new number of incident branches connected to the remaining nodes.
   - It is clear that the computation complexity of this scheme is much more than the other two.

[Example 1.7] Ordering the nodes of the network as shown in Fig. 1.16.

[Solution] The above three ordering schemes are performed and compared as follows for the network as shown in Fig. 1.16.

1. *Static optimal ordering scheme*: There are eight nodes and 14 branches in this network. The number of incident branches on each node is listed in Table 1.2.

   The ordering results according to the static ordering scheme are shown in Fig. 1.17a. There are four new branches added in the process of node eliminations. When node 1 is eliminated, branch 2–7 and branch 2–8 are generated and when node 2 is eliminated, branch 3–7 and branch 4–7 are added. Factorizing the corresponding admittance matrix, we get the structure of the lower triangular matrix as shown in Fig. 1.17b. Four fill-ins, l_{72}, l_{73}, l_{74}, and l_{82} correspond with the four new added branches.
2. *Semidynamic ordering scheme*: The process of numbering is shown in Table 1.3 and the result in Fig. 1.18a.

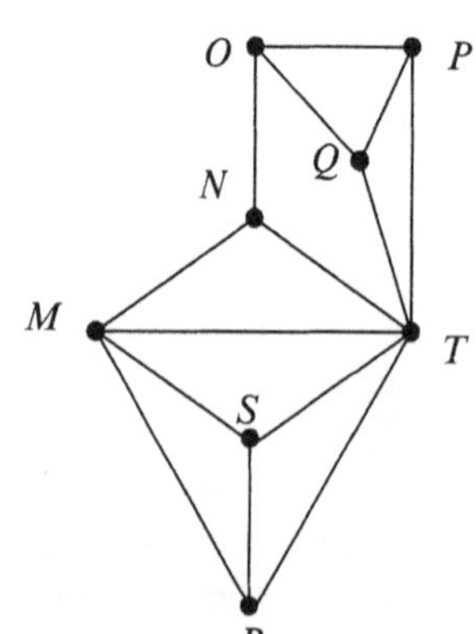

Fig. 1.16 Example of the node ordering

Table 1.2 Number of branches at each node for network shown in Fig. 1.16

Node	M	N	O	P	Q	R	S	T
Number of incident branches	4	3	3	3	3	3	3	6

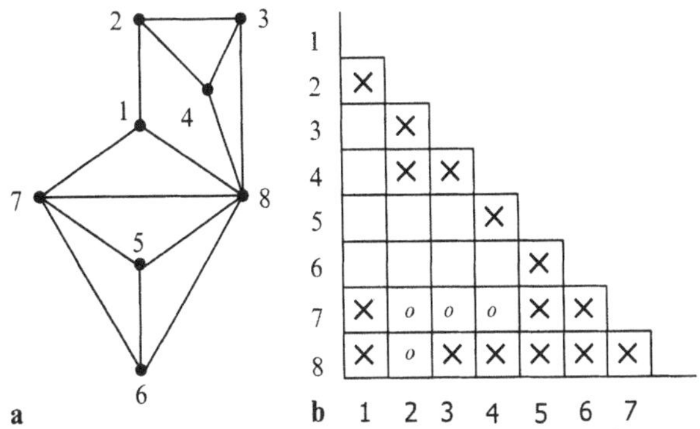

Fig. 1.17 Results of static optimal ordering

Table 1.3 Process of semidynamic ordering scheme

Node	M	N	O	P	Q	R	S	T	Node ordered	Node number
Process of	4	(3)	3	3	3	3	3	6	N	1
numbering	4		4	(3)	3	3	3	6	P	2
	4		3		(2)	3	3	5	Q	3
	4		(2)			3	3	4	O	4
	(3)					3	3	3	M	5
						(2)	2	2	R	6
							(1)	1	S	7
								(0)	T	8

In this scheme, two new branches are introduced in the elimination process, that is, when node 1 is eliminated, branch 4–5 and branch 4–8 are added.

3. *Dynamic ordering scheme*: In order to number the nodes, we need to count the number of new branches (the number of fill-ins) added after eliminating each node. The result is listed in Table 1.4. From this table, we can see that node R or S should be numbered first. Suppose that node R is selected as node 1. After this node is eliminated we count the new branch numbers when eliminating other nodes. The results are shown in Table 1.5.

   From Table 1.5, node S should be numbered as node 2. The computation is repeated until the last node has been numbered. The results are shown in Fig. 1.18b. Only one new branch is added by this scheme.

   Therefore, for complex networks, the dynamic ordering scheme can obtain more satisfactory results.

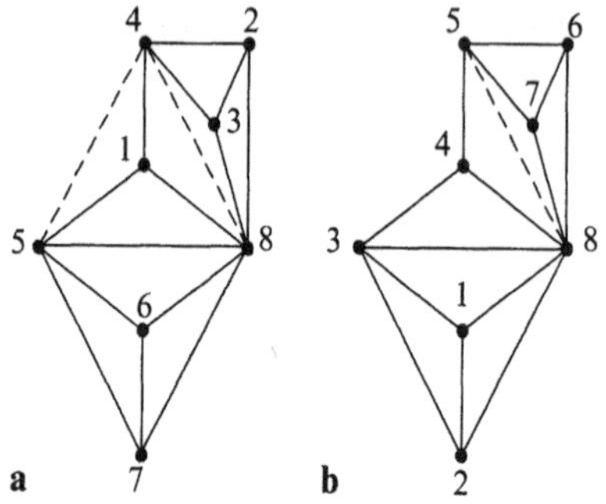

Fig. 1.18 Result of semidynamic and dynamic optimal ordering

Table 1.4 First step of dynamic ordering scheme

Node eliminated	M	N	O	P	Q	R	S	T
Number of new branches	2	2	2	1	1	0	0	10

Table 1.5 Second step of dynamic ordering scheme

Node eliminated	M	N	O	P	Q	S	T
Number of new branches	1	2	2	1	1	0	7

1.5 Nodal Impedance Matrix

1.5.1 Basic Concept of Nodal Impedance Matrix

As described above, the nodal equation of electric network can be generally represented as

$$\boldsymbol{I} = \boldsymbol{Y}\boldsymbol{V},$$

where $\boldsymbol{I}$ is the column vector of the nodal injection currents. Generally, it is the known variable in power system analysis; $\boldsymbol{V}$ is the column vector of the nodal voltages. Generally, it is unknown variable in power system analysis; and $\boldsymbol{Y}$ is the nodal admittance matrix.

The above linear simultaneous equations can be solved by various methods, such as the direct method by inverting the admittance matrix. Suppose

$$\boldsymbol{Z} = \boldsymbol{Y}^{-1}. \tag{1.74}$$

Then, the above nodal equation can be written as

$$\boldsymbol{V} = \boldsymbol{Z}\boldsymbol{I} \tag{1.75}$$

or in the expansion

$$\left.\begin{aligned}
\dot{V}_1 &= Z_{11}\dot{I}_1 + Z_{12}\dot{I}_2 + \cdots + Z_{1i}\dot{I}_i + \cdots + Z_{1n}\dot{I}_n \\
\dot{V}_2 &= Z_{21}\dot{I}_1 + Z_{22}\dot{I}_2 + \cdots + Z_{2i}\dot{I}_i + \cdots + Z_{2n}\dot{I}_n \\
&\cdots\cdots \\
\dot{V}_i &= Z_{i1}\dot{I}_1 + Z_{i2}\dot{I}_2 + \cdots + Z_{ii}\dot{I}_i + \cdots + Z_{in}\dot{I}_n \\
&\cdots\cdots \\
\dot{V}_n &= Z_{n1}\dot{I}_1 + Z_{n2}\dot{I}_2 + \cdots + Z_{ni}\dot{I}_i + \cdots + Z_{nn}\dot{I}_n
\end{aligned}\right\}. \tag{1.76}$$

Comparing (1.75) with (1.76), we can see that

$$\mathbf{Z} = \begin{bmatrix}
Z_{11} & Z_{12} & \cdots & Z_{1i} & \cdots & Z_{1n} \\
Z_{21} & Z_{22} & \cdots & Z_{2i} & \cdots & Z_{2n} \\
 & & \vdots & & & \\
Z_{i1} & Z_{i2} & \cdots & Z_{ii} & \cdots & Z_{in} \\
 & & \vdots & & & \\
Z_{n1} & Z_{n2} & \cdots & Z_{ni} & \cdots & Z_{nn}
\end{bmatrix}. \tag{1.77}$$

This is the nodal impedance matrix corresponding to the nodal admittance matrix $\boldsymbol{Y}$, and they have the same order. The diagonal element Z_{ii} is called the self-impedance or the input impedance, and the off-diagonal element Z_{ij} is called the mutual impedance or the transfer impedance between the node i and node j. When the injection currents are known, the nodal voltages of the network can be solved directly through (1.75) or (1.76).

The physical meaning of the elements in the nodal impedance matrix can be explained as follows:

If a unit current is injected into node i, and all other nodes are open, i.e.,

$$\begin{aligned}
&\dot{I}_i = 1 \\
&\dot{I}_j = 0 \quad (j = 1, 2, \ldots,\ n, j \neq i).
\end{aligned}$$

Then from (1.76), we can get

$$\begin{aligned}
\dot{V}_1 &= Z_{i1} \\
\dot{V}_2 &= Z_{i2} \\
&\cdots\cdots \\
\dot{V}_i &= Z_{ii} \\
&\cdots\cdots \\
\dot{V}_n &= Z_{in}.
\end{aligned}$$

Thus, we know that the elements in the ith column of the impedance matrix have the following physical meaning:

1. The diagonal element Z_{ii} of the impedance matrix is equal in value to the voltage of node i, when a unit current is injected into node i and all the other nodes are open. Therefore, Z_{ii} can be also regarded as the equivalent impedance between node i and the ground when all other nodes are open. If the network has some grounding branches and node i is connected to the network, Z_{ii} must be a nonzero element.
2. The off-diagonal element Z_{ij} is the mutual impedance between node i and j. When a unit current is injected into node i and all the other nodes are open, Z_{ij} is equal in value to the voltage of node j. Because there are always some electromagnetic connections (including indirect connections) among the nodes of a power network, the voltage of every node should be nonzero when node i is injected with a unit current and the other nodes are open. That is to say, all the mutual impedance elements Z_{ij} are nonzero elements. Therefore, the impedance matrix is a full matrix without zero elements.

The impedance matrix method for directly solving network voltage used to be very popular in the early stages of power system analysis by computer. But the impedance matrix is a full matrix, more memory and operations are required, which limits its applications especially for large-scale networks. Nevertheless, it is conceptually very useful in many aspects of power system analysis. This will be introduced in later chapters.

1.5.2 Forming Nodal Impedance Matrix by Using Nodal Admittance Matrix

Comparing with the admittance matrix, it is more difficult to formulae the nodal impedance matrix of an electric network. Two general methods of constructing the impedance matrix will be introduced in the next sections.

According to the discussion in Sect. 1.2.2, the admittance matrix of an electric network can be obtained directly from its configuration and parameters. So we can get the impedance matrix by inverting the admittance matrix. Several methods can be used to invert a matrix. In the following, we will illustrate one of them – inversion of a matrix through solving linear equations.

Consider an admittance matrix $\boldsymbol{Y}$ and its corresponding impedance matrix $\boldsymbol{Z}$. Solving the linear equation

$$\boldsymbol{Y}\boldsymbol{Z}_j = \boldsymbol{B}_j \tag{1.78}$$

we can get the element $\boldsymbol{Z}_j$ of the column j in the impedance matrix, where $\boldsymbol{B}_j$ is a column vector:

$$\boldsymbol{B}_j = [0 \quad \cdots \quad 0 \quad \underset{j}{1} \quad 0 \quad \cdots \quad 0]^{\mathrm{t}}.$$

Solving (1.78) successively for $j = 1, 2, \ldots, n$, we can obtain all elements of the impedance matrix. When the elements are solved column by column, only the right-hand vector $\boldsymbol{B}_j$ is changed in (1.78). Therefore, the triangular factorization algorithm is very efficient to solve (1.78) (refer to Sect. 1.3.2 for details).

Since the admittance matrix is symmetric, it can be factorized as:

$$\boldsymbol{Y} = \boldsymbol{L}\boldsymbol{D}\boldsymbol{L}^{\mathrm{T}}.$$

The elements of the unit lower triangular matrix $\boldsymbol{L}$ and the diagonal matrix $\boldsymbol{D}$ can be obtained from (1.61). Therefore, (1.78) can be rewritten as:

$$\boldsymbol{L}\boldsymbol{D}\boldsymbol{L}^{\mathrm{T}}\boldsymbol{Z}_j = \boldsymbol{B}_j. \tag{1.79}$$

Let

$$\boldsymbol{L}^{\mathrm{T}}\boldsymbol{Z}_j = \boldsymbol{W}_j, \tag{1.80}$$

$$\boldsymbol{D}\boldsymbol{W}_j = \boldsymbol{X}_j. \tag{1.81}$$

Then according to (1.79), we have

$$\boldsymbol{L}\boldsymbol{X}_j = \boldsymbol{B}_j. \tag{1.82}$$

Thus the whole process of solving (1.79) can be decomposed into three steps:

1. Solve $\mathbf{X}_j$ from (1.82)

   Expand (1.82) as

$$\begin{bmatrix} 1 & & & & & & & \\ l_{21} & 1 & & & & & & \\ l_{31} & l_{32} & 1 & & & & & \\ \vdots & \vdots & \vdots & \ddots & & & & \\ l_{j1} & l_{j2} & \cdots & l_{j,j-1} & 1 & & & \\ \vdots & \vdots & & \vdots & \cdots & \ddots & & \\ l_{n1} & l_{n2} & \cdots & l_{nj} & \cdots & l_{n,n-1} & 1 \end{bmatrix} \begin{bmatrix} x_1 \\ x_2 \\ x_3 \\ \vdots \\ x_j \\ \vdots \\ x_n \end{bmatrix} = \begin{bmatrix} 0 \\ 0 \\ \vdots \\ \vdots \\ 1 \\ 0 \\ \vdots \end{bmatrix}. \tag{1.83}$$

   Then we can get $x_1, x_2, \ldots, x_n$ sequentially from the above equation. This is the forward substitution.

2. Obtain $\boldsymbol{W}_j$ from (1.81)

   Expand (1.81) as

$$\begin{bmatrix} d_1 & & & & & & \\ & d_2 & & & & & \\ & & d_3 & & & & \\ & & & \ddots & & & \\ & & & & d_j & & \\ & & & & & \ddots & \\ & & & & & & d_n \end{bmatrix} \begin{bmatrix} w_1 \\ w_2 \\ w_3 \\ \vdots \\ w_j \\ \vdots \\ w_n \end{bmatrix} = \begin{bmatrix} x_1 \\ x_2 \\ x_3 \\ \vdots \\ x_j \\ \vdots \\ x_n \end{bmatrix}. \tag{1.84}$$

Then we can get $w_1, w_2, \ldots, w_n$ sequentially from the above equation. This is the normalization.

$$w_i = x_i/d_i \quad i = 1, 2, \ldots, n. \tag{1.85}$$

3. Obtain Z_j from (1.80)
   Expand (1.80) as

$$\begin{bmatrix} 1 & l_{21} & l_{31} & \cdots & l_{j1} & \cdots & l_{n1} \\ & 1 & l_{32} & \cdots & l_{j2} & \cdots & l_{n2} \\ & & \ddots & & & & \vdots \\ & & & 1 & \cdots & \cdots & l_{nj} \\ & & & & \ddots & & \vdots \\ & & & & & 1 & l_{n,n-1} \\ & & & & & & 1 \end{bmatrix} \begin{bmatrix} Z_{1j} \\ Z_{2j} \\ \vdots \\ Z_{jj} \\ \vdots \\ Z_{n-1,j} \\ Z_{nj} \end{bmatrix} = \begin{bmatrix} W_1 \\ W_2 \\ \vdots \\ W_j \\ \vdots \\ W_{n-1} \\ W_n \end{bmatrix}. \tag{1.86}$$

Then we can solve $Z_{nj}, Z_{n-1,j}, \ldots, Z_{jj}, \ldots, Z_{2j}, Z_{1j}$ one by one from bottom to top sequence. This is the backward substitution.

[Example 1.8] Form the impedance matrix of the electric network shown in the Example 1.1 from its admittance matrix by applying the factorization algorithm.

[Solution] The admittance matrix can be factorized by using (1.61),

$$d_1 = Y_{11} = 1.378742 - j6.291665$$

$$l_{21} = \frac{Y_{21}}{d_1} = \frac{-0.6242024 + j3.900156}{1.378742 - j64.57121} = 0.612227 + j0.034979$$

$$\begin{aligned} d_2 &= Y_{22} - l_{21}^2 d_1 \\ &= (1.453909 - j66.98082) - (0.61227 + j0.031979)^2 \times (1.378742 - j6.291665) \\ &= 1.208288 - j64.57121 \end{aligned}$$

Similarly, other elements can be found through using the recursion formulae. Then the admittance matrix is factorized as

$$L = \begin{bmatrix} 1 & & & & \\ -0.612227 + j0.034979 & 1 & & & \\ -0.425687 - j0.026671 & -0.073971 - j0.017193 & 1 & & \\ & -0.982943 - j0.018393 & -0.137743 - j0.027718 & 1 & \\ & & -0.924654 + j0.027559 & -1.189287 + j0.048151 & 1 \end{bmatrix},$$

$$D = \begin{bmatrix} 1.378742 - j6.291665 & & & & \\ & 1.208288 - j64.57121 & & & \\ & & 1.022377 - j34.30237 & & \\ & & & 0.887283 - j3.640902 & \\ & & & & 0.038964 + j1.263678 \end{bmatrix}.$$

The first step is to get the first column elements Z_1 of the impedance matrix. In this situation, (1.83) should be written as

$$\begin{bmatrix} 1 & & & & \\ -0.612227 + j0.034979 & 1 & & & \\ -0.425687 - j0.026671 & -0.073971 - j0.017193 & 1 & & \\ & -0.982943 - j0.018393 & -0.137743 - j0.027718 & 1 & \\ & & -0.924654 + j0.027559 & -1.189287 + j0.048151 & 1 \end{bmatrix} \begin{bmatrix} x_1 \\ x_2 \\ x_3 \\ x_4 \\ x_5 \end{bmatrix} = \begin{bmatrix} 1 \\ 0 \\ 0 \\ 0 \\ 0 \end{bmatrix}.$$

Therefore,

$$\begin{aligned} x_1 &= 1 \\ x_2 &= 0 - l_{21}x_1 = 0.612227 - j0.034979 \\ x_3 &= 0 - l_{31}x_1 - l_{32}x_2 = 0.471576 + j0.034609 \\ x_4 &= 0 - l_{42}x_2 - l_{43}x_3 = 0.665138 - j0.027805 \\ x_5 &= 0 - l_{53}x_3 - l_{54}x_4 = 1.226700 - j0.046890. \end{aligned}$$

From (1.85), we obtain

$$w_1 = \frac{x_1}{d_1} = \frac{1}{1.378742 - j63291665} = 0.033234 + j0.151658$$
$$w_2 = \frac{x_2}{d_2} = \frac{0.612227 - j0.034979}{1.208288 - j64.57121} = -0.000719 + j0.009468,$$
$$w_4 = \frac{x_4}{d_4} = \frac{0.665138 - j0.027805}{0.887283 - j3.640902} = 0.049233 + j0.170687$$
$$w_3 = \frac{x_3}{d_3} = \frac{0.471576 + j0.034609}{1.022377 - j34.30237} = 0.000599 + j0.013765,$$
$$w_5 = \frac{x_5}{d_5} = \frac{1.226700 - j0.046890}{0.03894 + j1.263678} = -0.006535 - j0.970940.$$

Back substitution is executed using the following equation

$$\begin{bmatrix} 1 & \begin{matrix}-0.612227\\+j0.034979\end{matrix} & \begin{matrix}-0.425687\\-j0.026671\end{matrix} & & \\ & 1 & \begin{matrix}-0.073971\\-j0.017193\end{matrix} & \begin{matrix}-0.982943\\-j0.018393\end{matrix} & \\ & & 1 & \begin{matrix}-0.137743\\-j0.027718\end{matrix} & \begin{matrix}-0.924654\\+j0.027559\end{matrix} \\ & & & 1 & \begin{matrix}-1.189287\\+j0.048151\end{matrix} \\ & & & & 1 \end{bmatrix} \begin{bmatrix} Z_{11} \\ Z_{21} \\ Z_{31} \\ Z_{41} \\ Z_{51} \end{bmatrix} = \begin{bmatrix} \begin{matrix}0.033234\\+j0.151658\end{matrix} \\ \begin{matrix}-0.000719\\+j0.009468\end{matrix} \\ \begin{matrix}0.000599\\+j0.013765\end{matrix} \\ \begin{matrix}0.049233\\+j0.170687\end{matrix} \\ \begin{matrix}-0.006535\\-j0.970940\end{matrix} \end{bmatrix}.$$

Then the first column elements of the impedance matrix are

$$\begin{aligned} Z_{51} &= -0.006535 - j0.970940 \\ Z_{41} &= -0.005290 - j0.983725 \\ Z_{31} &= -0.006862 - j1.019487 \\ Z_{21} &= -0.005555 - j1.032911 \\ Z_{11} &= 0.017972 - j0.914690. \end{aligned}$$

The computation can be performed in a similar way and the whole impedance matrix can be obtained column by column.

Thus we finally have

$$
\mathbf{Z} = \begin{bmatrix}
\begin{matrix}0.017972\\ -j0.914690\end{matrix} & \begin{matrix}-0.005555\\ -j1.032911\end{matrix} & \begin{matrix}-0.006862\\ -j1.019487\end{matrix} & \begin{matrix}-0.005290\\ -j0.983725\end{matrix} & \begin{matrix}-0.006535\\ -j0.970940\end{matrix} \\
\begin{matrix}-0.0055555\\ -j1.032911\end{matrix} & \begin{matrix}0.007781\\ -j0.961291\end{matrix} & \begin{matrix}-0.010007\\ -j1.037907\end{matrix} & \begin{matrix}0.007410\\ -j0.918658\end{matrix} & \begin{matrix}-0.009530\\ -j0.988482\end{matrix} \\
\begin{matrix}-0.006862\\ -j1.019487\end{matrix} & \begin{matrix}-0.010007\\ -j1.037907\end{matrix} & \begin{matrix}0.026875\\ -j0.90470\end{matrix} & \begin{matrix}-0.009530\\ -j0.988482\end{matrix} & \begin{matrix}0.025596\\ -j0.861619\end{matrix} \\
\begin{matrix}-0.005290\\ -j0.983725\end{matrix} & \begin{matrix}0.007410\\ -j0.918658\end{matrix} & \begin{matrix}0.007410\\ -j0.918658\end{matrix} & \begin{matrix}0.007057\\ -j0.859912\end{matrix} & \begin{matrix}-0.009076\\ -j0.941412\end{matrix} \\
\begin{matrix}-0.006535\\ -j0.970940\end{matrix} & \begin{matrix}-0.009530\\ -j0.988482\end{matrix} & \begin{matrix}-0.009530\\ -j0.988482\end{matrix} & \begin{matrix}-0.009076\\ -j0.941412\end{matrix} & \begin{matrix}0.024377\\ -j0.790589\end{matrix}
\end{bmatrix}.
$$

As described above, the elements of the jth column in the impedance matrix are equal to the nodal voltages in value when a per-unit current is injected into node j and other nodes are open. Therefore, finding the elements of the jth column from (1.78) is equivalent to solving the following nodal equation

$$\boldsymbol{YV} = \boldsymbol{I}_j, \tag{1.87}$$

where all the elements of the current column vector $\boldsymbol{I}_j$ are zero except the jth element equals 1. Obviously, $\boldsymbol{V}$ obtained from this equation is equal to $\mathbf{Z}_j$ in value.

It is worth noting that the computation burden of this method is a little too heavy in some situations; for example, if we want to derive the impedance matrix of a network with n nodes, n linear equations must be solved n times. Hence this method is only suitable for the case in which only a few elements are of interest. In power flow and short circuit analysis, the input impedance of one pair of nodes and the transfer impedance between two node pairs are often calculated using the above method. In Fig. 1.19, in order to get the input impedance of node i and j and the

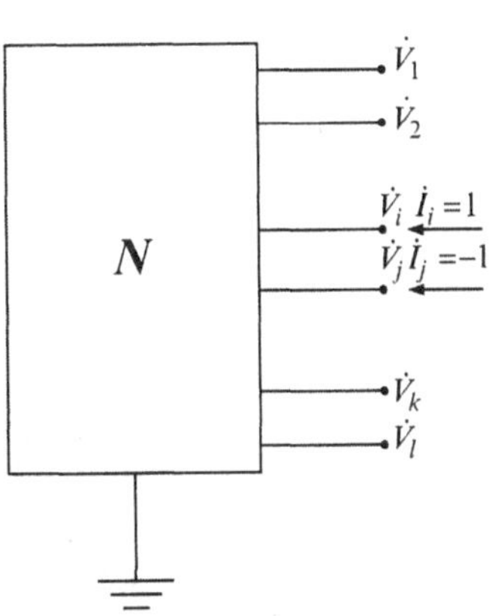

Fig. 1.19 Solving node pair's input and transfer impedance

transfer impedance between ij and kl, a unit current is injected between node i and node j, while other nodes are open. That is

$$\dot{I}_i = 1, \quad \dot{I}_j = -1.$$

In this case, solve the network equation

$$\boldsymbol{YV} = \boldsymbol{F}_{ij}, \tag{1.88}$$

where

$$\boldsymbol{F}_{ij} = \begin{bmatrix} 0 \\ \vdots \\ 1 \\ 0 \\ \vdots \\ -1 \\ 0 \\ \vdots \\ 0 \end{bmatrix} \begin{matrix} \\ \\ \xleftarrow{i} \\ \\ \\ \xleftarrow{j} \\ \\ \\ \\ \end{matrix}.$$

The nodal voltage can be obtained and the input impedance of node pair ij is

$$Z_{ij-ij} = \dot{V}_i - \dot{V}_j. \tag{1.89}$$

The transfer impedance between ij and kl is

$$Z_{kl-ij} = \dot{V}_k - \dot{V}_l. \tag{1.90}$$

1.5.3 Forming Nodal Impedance Matrix by Branch Addition Method

In the above section, we have described a method of forming the impedance matrix by using the admittance matrix. An alternative method is to form the impedance matrix directly by the branch addition method. The method is straightforward in computation and allows easy impedance matrix modification for changes in the network. Therefore it is applied widely.

The forming process is illustrated by Fig. 1.20.

We start to form the impedance matrix from a grounded branch and a matrix of order 1 is formed. In Fig. 1.20, z_{10} first is used to form this matrix. Then branch z_{12} is added and the new branch creates a new node ②. We call it adding a tree branch if a new node is generated when adding a branch. At the same time, the order of the

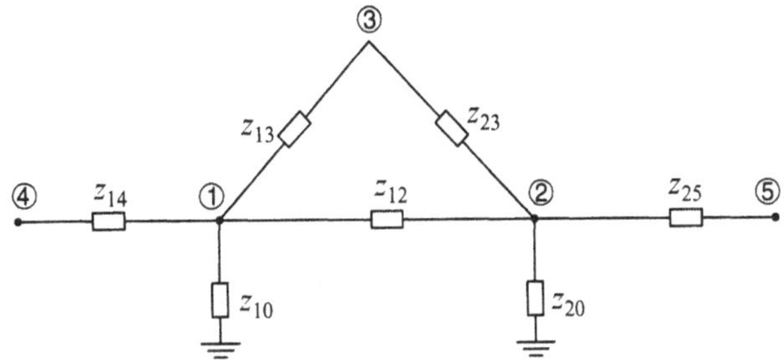

Fig. 1.20 Process of branch addition method in forming impedance matrix

corresponding matrix increases by 1. Thus after adding tree branch z_{12}, we obtain a 2×2 impedance matrix.

We next add branch z_{20}. In this situation, there is no new node generated. The order of the impedance matrix does not change. This is called adding a link branch. All the elements of the impedance matrix must be updated when a link branch is added. Repeat the operations in a similar way: after adding tree branch z_{13}, node 3 is created. Then the order of the impedance matrix becomes three.

After adding tree branch z_{14}, node 4 is created and the order of the impedance matrix becomes four.

After adding tree branch z_{25}, node 5 is generated. The order of the impedance matrix becomes five.

When adding link z_{25}, no new node is generated and the order of the impedance matrix is still five.

The impedance matrix is formed after all the branches have been added to the electric network.

It should be noted that the sequence of adding the branches is not unique. An alternative sequence is as follows:

Tree branch z_{10} $\rightarrow$ tree branch z_{20} $\rightarrow$ link z_{12} $\rightarrow$ tree branch z_{13} $\rightarrow$ link z_{23} $\rightarrow$ tree branch z_{14} $\rightarrow$ tree branch z_{25}.

Of course, there are some other schemes besides these two schemes. And it can be proved that whatever the branch adding sequence is, the impedance matrix is the same when the node number ordering is the same. However, the computation efforts under the different adding sequences are quite different. The effects of adding a tree branch or a link branch on the impedance matrix will be discussed in the following:

1. Adding a tree branch

Assume that the $m \times m$ impedance matrix of an electric network has been formed for the first m nodes.

$$
\mathbf{Z}_{\mathrm{N}} = \begin{bmatrix} Z_{11} & Z_{12} & \cdots & Z_{1i} & \cdots & Z_{1m} \\ Z_{21} & Z_{22} & \cdots & Z_{2i} & \cdots & Z_{2m} \\ & & \cdots & \cdots & & \\ Z_{i1} & Z_{i2} & \cdots & Z_{ii} & \cdots & Z_{im} \\ & & \cdots & \cdots & & \\ Z_{m1} & Z_{m2} & \cdots & Z_{mi} & \cdots & Z_{mm} \end{bmatrix}. \tag{1.91}
$$

When a tree branch Z_{ij} is added at node i, a new node j is created and the order of the impedance matrix becomes $m+1$ (see Fig. 1.21). Suppose the new impedance matrix is

$$\mathbf{Z}'_{\mathrm{N}} = \begin{bmatrix} Z'_{11} & Z'_{12} & \cdots & Z'_{1i} & \cdots & Z'_{1m} & \vdots & Z'_{1j} \\ Z'_{21} & Z'_{22} & \cdots & Z'_{2i} & \cdots & Z'_{2m} & \vdots & Z'_{2j} \\ & & \cdots & \cdots & & & \vdots & \\ Z'_{i1} & Z'_{i2} & \cdots & Z'_{ii} & \cdots & Z'_{im} & \vdots & Z'_{mj} \\ & & \cdots & \cdots & & & \vdots & \\ Z'_{m1} & Z'_{m2} & \cdots & Z'_{mi} & \cdots & Z'_{mm} & \vdots & Z'_{mj} \\ \cdots & \cdots & \cdots & \cdots & \cdots & \cdots & \vdots & \cdots \\ Z'_{j1} & Z'_{j2} & & Z'_{ji} & & Z'_{jm} & \vdots & Z'_{jj} \end{bmatrix}. \tag{1.92}$$

We first solve the $m \times m$ matrix inside the dashed lines of (1.92). In order to obtain the values of the first column $Z'_{11}Z'_{21}\cdots Z'_{i1}\cdots Z'_{m1}$, a unit current is injected in node 1 and the other nodes are open as shown in Fig. 1.21a. In this case, voltages of the node 1, 2, ..., m have nothing to do with branch z_{ij}, therefore,

$$Z'_{11} = Z_{11}, Z'_{21} = Z_{21}, \ldots, Z'_{i1} = Z_{i1}, \ldots, Z'_{m1} = Z_{m1}.$$

It means that the first column of $\mathbf{Z}'_{\mathrm{N}}$ is the same as the first column of $\mathbf{Z}_{\mathrm{N}}$. Similarly, the second column of $\mathbf{Z}'_{\mathrm{N}}$ is the same as the second column of $\mathbf{Z}_{\mathrm{N}}$. Therefore we can deduce that the $m \times m$ matrix inside the dashed lines of (1.92) is the original impedance matrix before adding the branch z_{ij}.

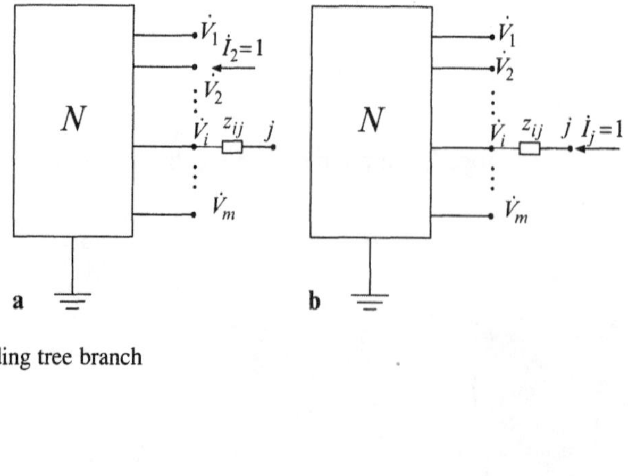

Fig. 1.21 Adding tree branch

We now solve the jth column of $\mathbf{Z}'_{\mathrm{N}}$. Similarly, j and other nodes are open as shown in Fig. 1.21b. In this situation, voltages of the node $1, 2, \ldots, i, \ldots, m$ are the same as those when a unit current is injected in node i, so we have,

$$Z'_{1j} = Z_{1i}, Z'_{2j} = Z_{2i}, \ldots, Z'_{ij} = Z_{ii}, \ldots, Z'_{mj} = Z_{mi}. \tag{1.93}$$

The voltage of node j is

$$\dot{V}_j = \dot{V}_i + z_{ij} \times 1.$$

According to the physical meaning of the impedance matrix, we obtain

$$Z_{jj} = Z_{ii} + z_{ij}. \tag{1.94}$$

Due to the symmetry of the impedance matrix, the off-diagonal elements of the jth row in $\mathbf{Z}'_N$ can be obtained as follows,

$$Z'_{j1} = Z_{1j}, Z'_{j2} = Z_{2j}, \ldots, Z'_{ji} = Z_{ij}, \ldots, Z'_{jm} = Z_{mj}. \tag{1.95}$$

Hence all the elements in the impedance matrix after adding tree branch z_{ij} are found. Additionally, although the order of the new impedance matrix increases by 1, the computation to form it is relatively simple.

1. Adding a link

The impedance matrix of the initial network is denoted as $\mathbf{Z}_{\mathrm{N}}$. When link z_{ij} is added between nodes i and j, the impedance matrix becomes $\mathbf{Z}'_{\mathrm{N}}$. The orders of these two matrices are the same because no new node is generated in the network. We now consider how to calculate the elements of new impedance matrix $\mathbf{Z}'_{\mathrm{N}}$.

As shown in Fig. 1.22, suppose the injection current vector of the new network is $\boldsymbol{I}$,

$$\boldsymbol{I} = \begin{bmatrix} \dot{I}_1 & \dot{I}_2 & \cdots & \dot{I}_i & \cdots & \dot{I}_j & \cdots & \dot{I}_m \end{bmatrix}^{\mathrm{t}}$$

and the nodal voltage vector is $\boldsymbol{V}$

$$\boldsymbol{V} = \begin{bmatrix} \dot{V}_1 & \dot{V}_2 & \cdots & \dot{V}_i & \cdots & \dot{V}_j & \cdots & \dot{V}_m \end{bmatrix}^{\mathrm{t}}.$$

Thus the following relationship holds

$$\boldsymbol{V} = \mathbf{Z}'_{\mathrm{N}}\boldsymbol{I}. \tag{1.96}$$

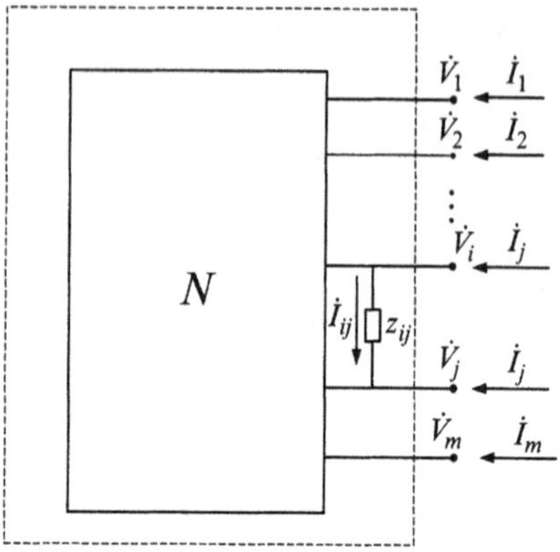

Fig. 1.22 Adding a link

From Fig. 1.22, the nodal current injected into the initial network is

$$\boldsymbol{I}' = \begin{bmatrix} \dot{I}_1 \\ \dot{I}_2 \\ \vdots \\ \dot{I}_i - \dot{I}_{ij} \\ \vdots \\ \dot{I}_j + \dot{I}_{ij} \\ \vdots \\ \dot{I}_m \end{bmatrix} = \boldsymbol{I} - \boldsymbol{A}_M \boldsymbol{I}_{ij}, \tag{1.97}$$

where $\boldsymbol{A}_M$ is a column vector related to the added link branch,

$$\boldsymbol{A}_M = \begin{bmatrix} 0 \\ \vdots \\ 1 \\ 0 \\ \vdots \\ -1 \\ 0 \\ \vdots \\ 0 \end{bmatrix} \begin{matrix} \\ \\ \leftarrow i \\ \\ \\ \leftarrow j \\ \\ \\ \\ \end{matrix} \tag{1.98}$$

According to the nodal equation of the original network,

$$\boldsymbol{V} = \mathbf{Z}_{\mathrm{N}} \boldsymbol{I}' = \mathbf{Z}_{\mathrm{N}} \boldsymbol{I} - \mathbf{Z}_{\mathrm{N}} \boldsymbol{A}_M \dot{\boldsymbol{I}}_{ij}. \tag{1.99}$$

Assume

$$\mathbf{Z}_{\mathrm{N}} \boldsymbol{A}_M = \mathbf{Z}_L. \tag{1.100}$$

We know that $\boldsymbol{Z}_L$ is a column vector

$$\boldsymbol{Z}_L = \begin{bmatrix} Z_{1i} - Z_{1j} \\ Z_{2i} - Z_{2j} \\ \vdots \\ Z_{ii} - Z_{ij} \\ \vdots \\ Z_{ji} - Z_{jj} \\ \vdots \\ Z_{mi} - Z_{mj} \end{bmatrix}. \tag{1.101}$$

Rewrite (1.99) as,

$$\boldsymbol{V} = \boldsymbol{Z}_{\mathrm{N}}\boldsymbol{I} - \boldsymbol{Z}_L\dot{\boldsymbol{I}}_{ij}. \tag{1.102}$$

The voltage difference between nodes i and j is equal to

$$\dot{V}_i - \dot{V}_j = z_{ij}\dot{I}_{ij} = \boldsymbol{A}_M^{\mathrm{T}}\boldsymbol{V}, \tag{1.103}$$

where $\boldsymbol{A}_M^{\mathrm{T}}$ is the transpose of $\boldsymbol{A}_M$. Substituting (1.102) into (1.103), we obtain

$$z_{ij}\dot{I}_{ij} = \boldsymbol{A}_M^{\mathrm{T}}\boldsymbol{Z}_{\mathrm{N}}\boldsymbol{I} - \boldsymbol{A}_M^{\mathrm{T}}\boldsymbol{Z}_L\dot{I}_{ij}.$$

$\dot{I}_{ij}$ can be solved as follows

$$\dot{I}_{ij} = \frac{1}{Z_{LL}}\boldsymbol{Z}_L^{\mathrm{T}}\boldsymbol{I}, \tag{1.104}$$

where

$$Z_{\mathrm{LL}} = \boldsymbol{A}_M^{\mathrm{T}}\boldsymbol{Z}_L + z_{ij} = Z_{ii} + Z_{jj} - 2Z_{ij} + z_{ij}, \tag{1.105}$$

$$\boldsymbol{Z}_L^{\mathrm{T}} = \boldsymbol{A}_M^{\mathrm{T}}\boldsymbol{Z}_{\mathrm{N}} = (\boldsymbol{Z}_{\mathrm{N}}\boldsymbol{A}_M)^{\mathrm{T}}.$$

Substituting (1.104) into (1.102), we have

$$\boldsymbol{V} = \left(\boldsymbol{Z}_{\mathrm{N}} - \frac{1}{Z_{LL}}\boldsymbol{Z}_L\boldsymbol{Z}_L^{\mathrm{T}}\right)\boldsymbol{I}. \tag{1.106}$$

Comparing (1.96) with (1.106), we obtain the new impedance matrix $\boldsymbol{Z}_{\mathrm{N}}'$,

$$\boldsymbol{Z}_{\mathrm{N}}' = \boldsymbol{Z}_{\mathrm{N}} - \frac{1}{Z_{LL}}\boldsymbol{Z}_L\boldsymbol{Z}_L^{\mathrm{T}}. \tag{1.107}$$

Expanding the above equation, we have the following formulae of the elements in $\mathbf{Z}'_N$

$$Z'_{kl} = Z_{kl} - \frac{Z_{Lk}Z_{Ll}}{Z_{LL}} \quad \begin{pmatrix} k = 1,2,\ldots,m \\ l = 1,2,\ldots,m \end{pmatrix}. \tag{1.108}$$

In contrast with adding a tree branch, the computation of adding a link is quite heavy and complicated in which each element of the impedance matrix must be recalculated according to (1.108). The speed of forming the impedance matrix mainly depends on computations for adding links. Therefore the sequence of adding branches affects the computation speed dramatically. For example of the network in Fig. 1.20, the computations of adding link z_{23} according to the first sequence are performed on a 5×5 matrix, but the recalculations are just executed on a 3×3 matrix according to the second sequence. Hence the more reasonable sequence of adding branches is to add links as early as possible.

If the transformer branch is involved, the Π equivalent circuit as shown in Fig. 1.4 can be used in forming the impedance matrix. Comparing with a transmission line, two more branches must be added for each transformer and in most circumstances both of them are links. Therefore the computation burden increases notably.

Now a direct method of adding a transformer branch is introduced in the following, which need not use the Π equivalent circuit.

First, we discuss the situation that the transformer added is a tree branch. In Fig. 1.23a, the leakage impedance is put at the nominal turn ratio side of the transformer. If the leakage impedance is put at the off nominal side, the formulae can be derived in a similar way.

The impedance matrix of the original network is denoted as $\mathbf{Z}_N$ (see (1.91)). When the transformer is added as a tree branch, the order of the new impedance matrix $\mathbf{Z}'_N$ increases by 1 (see (1.92)). It can be proved that the $m \times m$ block matrix in the top-left of $\mathbf{Z}'_N$ is just $\mathbf{Z}_N$.

As shown in Fig. 1.23b, the transformer is substituted for its equivalent circuit. When node j is open, the transformer's Π equivalent circuit is also opened as viewed from node i. This can be explained as follows.

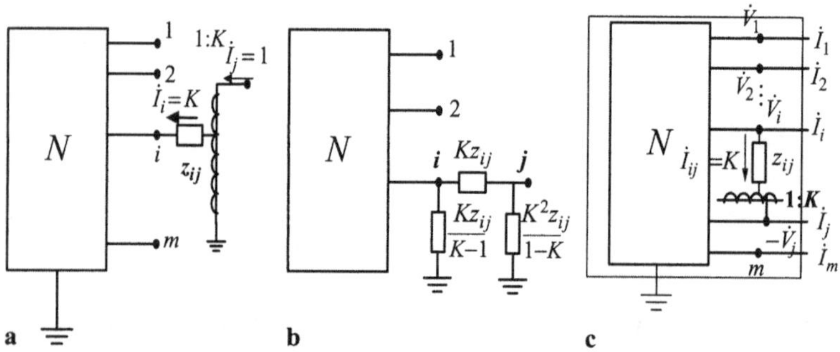

Fig. 1.23 Adding a transformer

The impedance of the loop constituted by nodes i, j and the ground is

$$z_{ij0} = Kz_{ij} + \frac{K^2}{1-K} z_{ij} = \frac{K}{1-K} z_{ij}.$$

And the impedance between node i and the ground is $z_{i0} = (K/(1-K))z_{ij0}$. The value of the parallel impedance of z_{i0} and z_{ij0} becomes infinity. When a unit current is injected at each node of the original network, the current distribution of the original network is unchanged after adding a transformer as a tree branch. Hence the nodal voltages are also unchanged.

The issue now is how to solve the new elements of $\mathbf{Z}'_{\mathrm{N}}$. Focus on this question, a unit current is injected at node j and the other nodes are open as shown in Fig. 1.23b. This is just like the injecting current K into the original network at node i. Thus the nodal voltages are

$$\dot{V}_1 = KZ_{1i}, \dot{V}_2 = KZ_{2i}, \ldots, \dot{V}_i = KZ_{ii}, \dot{V}_m = KZ_{mi}.$$

The voltage of node j is

$$\dot{V}_j = K(\dot{V}_i + Kz_{ij}) = K^2(Z_{ii} + z_{ij}).$$

Thus, we obtain

$$Z'_{1j} = Kz_{1i}, Z'_{2j} = Kz_{2i}, \ldots, Z'_{ij} = KZ_{ii}, \ldots, Z'_{mj} = KZ_{mi}, \tag{1.109}$$

$$Z'_{jj} = K^2(Z_{ii} + z_{ij}). \tag{1.110}$$

Obviously, (1.109) and (1.110) will be changed into (1.93) and (1.94) when the turn ratio $K = 1$.

The situation when the transformer added is a link branch is shown in Fig. 1.23c. Assume that the current injected into the network after adding the transformer branch is a column vector $\boldsymbol{I}$, thus the current injected into the original network

$$\boldsymbol{I}' = \begin{bmatrix} \dot{I}_1 \\ \dot{I}_2 \\ \vdots \\ \dot{I}_i - K\dot{I}_{ij} \\ \vdots \\ \dot{I}_j + \dot{I}_{ij} \\ \vdots \\ \dot{I}_m \end{bmatrix} = \boldsymbol{I} - \boldsymbol{A}'_M \boldsymbol{I}_{ij}, \tag{1.111}$$

where A'_M is a column vector.

$$A'_M = \begin{bmatrix} 0 \\ \vdots \\ K \\ 0 \\ \vdots \\ -1 \\ 0 \\ \vdots \\ 0 \end{bmatrix} \begin{matrix} \\ \\ \leftarrow i \\ \\ \\ \leftarrow j \\ \\ \\ \\ \end{matrix}.$$

The following steps are similar to that of a simple impedance link branches (see (1.99)–(1.108)). The only difference is to substitute the original A_M for A'_M. Therefore (1.101) should be changed as follows:

$$Z_L = \begin{bmatrix} KZ_{1i} - Z_{1j} \\ KZ_{2i} - Z_{2j} \\ \vdots \\ KZ_{ii} - Z_{ij} \\ \vdots \\ KZ_{ji} - Z_{jj} \\ \vdots \\ KZ_{mi} - Z_{mj} \end{bmatrix}. \tag{1.112}$$

Equation (1.103) should be rewritten as

$$K\dot{V}_i - \dot{V}_j = K^2 z_{ij} \dot{I}_{ij} = A'^{T}_{M} V. \tag{1.113}$$

Accordingly, (1.105) is changed as

$$Z_{LL} = KZ_{Li} - Z_{Lj} + K^2 z_{ij}. \tag{1.114}$$

After calculating Z_L and Z_{LL}, the elements of Z'_N can be calculated according to (1.108).

Briefly, the process of forming an impedance matrix by using the branch addition method is a process of adding branches one by one. If the configuration of a network is changed or a branch needs to be added, the impedance matrix can be modified directly according to the above formulae. For instance, if a branch z_{ij} needs to be removed, the equivalent operation is to add a branch $-z_{ij}$ into the network.

[Example 1.9] Form the impedance matrix of the electric network shown in Fig. 1.10 by using the branch addition method.

[Solution] For convenience of the computation, line-to-ground capacitances at both ends of the transmission lines are lumped to the corresponding node and denoted in the format of line-to-ground reactance. The equivalent circuit is shown in Fig. 1.24.

According to the node ordering, we can make the sequence table of branch adding as follows.

Sequence of branches added	Terminal nodes of branch $i \cdots j$	Impedance of branch
(1)	$0 \cdots 1$	$-j4$
(2)	$0 \cdots 2$	$-j2$
(3)	$1 \cdots 2$	$0.04 + j0.25$
(4)	$0 \cdots 3$	$-j4$
(5)	$1 \cdots 3$	$0.1 + j0.35$
(6)	$2 \cdots 3$	$0.08 + j0.30$
(7)	$2 \cdots 4$	$j0.015$
(8)	$3 \cdots 5$	$j0.03$

Then label the branch adding sequence on the figure as shown in Fig. 1.24.

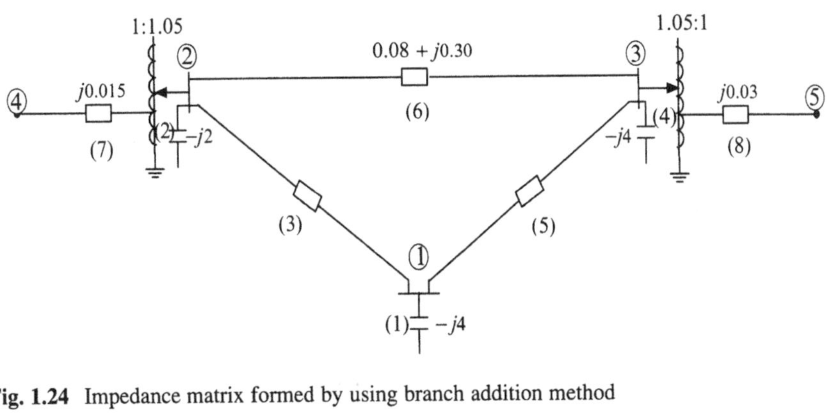

Fig. 1.24 Impedance matrix formed by using branch addition method

The procedure for forming the impedance matrix is shown as follows:

1. Start from the grounded branch z_{01} to form a 1×1 matrix. Its element is

$-j4$

2. Add branch (2): z_{02} is a tree branch, $i = 0,\ j = 2$. According to (1.93) and (1.94), the new elements are

$$Z_{12} = Z_{21} = Z_{10}, Z_{22} = Z_{00} + Z_{02}.$$

According to the physical meaning of the impedance matrix element, we have

$$Z_{10} = Z_{00} = 0.$$

Then

$$Z_{12} = Z_{21} = 0, Z_{22} = Z_{02} = -j2$$

and the 2×2 matrix is

	1	2
1	$-j4$	
2		$-j2$

3. Add branch (3): z_{12} is a link branch. The elements of $\mathbf{Z}_L$ can be obtained according to (1.101) and (1.105),

$$Z_{L1} = Z_{11} - Z_{12} = -j4$$
$$Z_{L2} = Z_{12} - Z_{22} = -j2.$$

From (1.105) we know,

$$Z_{LL} = Z_{L1} - Z_{L2} + z_{12} = -j4 - j2 + 0.04 + j0.25 = 0.04 - j5.75.$$

Modify the elements of the 2×2 matrix according to (1.108)

$$Z'_{11} = Z_{11} - \frac{Z_{L1}Z_{L1}}{Z_{LL}} = -j4 - \frac{(-j4)^2}{0.04 - j5.75} = 0.019356 - j1.217526$$
$$Z'_{12} = Z'_{21} = Z_{12} - \frac{Z_{L2}Z_{L1}}{Z_{LL}} = 0 - \frac{j2 \times (-j4)}{0.04 - j5.75} = -0.096782 - j1.301237$$
$$Z'_{22} = Z_{22} - \frac{Z_{L2}Z_{L2}}{Z_{LL}} = -j2 - \frac{(j2)^2}{0.04 - j5.75} = 0.004839 - j1.304381.$$

Thus we obtain the impedance matrix constituted by branches 1, 2, and 3

	1	2
1	0.019356 $-j1.1217526$	-0.096282 $-j1.391237$
2	-0.096282 $-j\,1.391237$	0.004839 $-j1.304381$

4. Add branch (4): z_{03} is a grounded tree branch. The computation process is the same as that in Step 2. The augmented matrix 3×3 is

	1	2	3
1	0.019356 $-j1.1217526$	-0.096282 $-j\,1.391237$	
2	-0.096282 $-j\,1.391237$	0.004839 $-j\,1.304381$	
3			$-j\,4$

5. Add branch (5) z_{13} and branch (6) z_{23}. Because both of these are links, the matrix order is unchanged. The computation process is the same as that in Step 3. The augmented matrix 3×3 is

	1	2	3
1	0.017972 $-j0.914690$	-0.005555 $-j\,1.032911$	-0.006862 $-j1.019487$
2	-0.005555 $-j\,1.0329111$	0.007781 $-j\,0.964591$	-0.010007 $-j1.037907$
3	-0.006862 $-j1.019487$	-0.010007 $-j1.037907$	0.026875 $-j0.904700$

6. Add branch (7): z_{24} is a transformer tree branch. In this network, the off normal turns ratio of the transformer is at node i. The computation cannot be performed

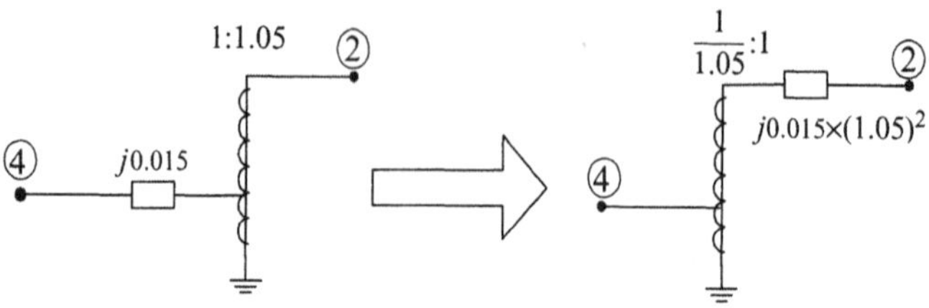

Fig. 1.25 Equivalent circuit of the transformer

directly using (1.109) and (1.110). We should transfer the off normal turns ratio to the other terminal node of the transformer. It is illustrated in the Fig. 1.25.

Then the elements of the fourth row and column can be calculated according to (1.109) and (1.110).

$$Z_{41} = Z_{14} = K'Z_{21} = \frac{1}{1.05}(-0.005555 - j1.032911) = -0.005290 - j0.983725$$

$$Z_{42} = Z_{24} = K'Z_{22} = \frac{1}{1.05}(-0.007781 - j0.964591) = 0.007410 - j0.918658$$

$$Z_{43} = Z_{34} = K'Z_{23} = \frac{1}{1.05}(-0.010007 - j1.037907) = -0.009530 - j0.988482$$

$$Z_{44} = K'^2(Z_{22} + z'_{24}) = \frac{1}{1.05^2}(-0.007781 - j0.964591) + j0.015$$
$$= 0.007057 - j0.859912.$$

We now have a 4 × 4 matrix

	1	2	3	4
1	0.017972 −j0.914690	−0.005555 −j 1.0329111	−0.006862 −j1.019487	−0.005290 −j0.983725
2	−0.005555 −j 1.0329111	0.007781 −j 0.964591	−0.010007 −j1.037907	0.007410 −j0.918658
3	−0.006862 −j1.019487	−0.010007 −j1.037907	0.026875 −j0.904700	−0.009530 −j0.988482
4	−0.005290 −j0.983725	0.007410 −j0.918658	−0.009530 −j0.988482	0.007057 −j0.859912

7. Add branch (8): z_{35} is also a transformer tree branch. Its off normal turns ratio is also at node i. The computation process is the same as that in Step 6.

   The final impedance matrix is

$$\mathbf{Z} = \begin{bmatrix} \begin{matrix}0.017972\\-j0.914690\end{matrix} & \begin{matrix}-0.0055555\\-j1.032911\end{matrix} & \begin{matrix}-0.006862\\-j1.019487\end{matrix} & \begin{matrix}-0.005290\\-j0.983725\end{matrix} & \begin{matrix}-0.006535\\-j0.970940\end{matrix} \\ \begin{matrix}-0.0055555\\-j1.032911\end{matrix} & \begin{matrix}0.007781\\-j0.964591\end{matrix} & \begin{matrix}-0.010007\\-j1.037907\end{matrix} & \begin{matrix}0.007410\\-j0.918658\end{matrix} & \begin{matrix}-0.009530\\-j0.988482\end{matrix} \\ \begin{matrix}-0.006862\\-j1.019487\end{matrix} & \begin{matrix}-0.010007\\-j1.037907\end{matrix} & \begin{matrix}0.026875\\-j0.904700\end{matrix} & \begin{matrix}-0.009530\\-j988482\end{matrix} & \begin{matrix}-0.025596\\-j0.861619\end{matrix} \\ \begin{matrix}-0.005290\\-j0.983725\end{matrix} & \begin{matrix}0.007410\\-j0.918658\end{matrix} & \begin{matrix}-0.009530\\-j988482\end{matrix} & \begin{matrix}0.007057\\-j0.859912\end{matrix} & \begin{matrix}-0.009076\\-j0.941412\end{matrix} \\ \begin{matrix}-0.006535\\-j0.970940\end{matrix} & \begin{matrix}-0.009530\\-j0.988482\end{matrix} & \begin{matrix}-0.025596\\-j0.861619\end{matrix} & \begin{matrix}-0.009076\\-j0.941412\end{matrix} & \begin{matrix}0.024377\\-j0.790589\end{matrix} \end{bmatrix}.$$

Thinking and Problem Solving

1. Prove that the incidence matrix of an electrical power network is a singular matrix
2. Is the admittance matrix generally a singular matrix? In what condition can the admittance matrix be a singular matrix?
3. What simplifications can be made to the equivalent circuit of the transformer in Fig. 1.4?
4. Why is the admittance matrix including phase shifter(s) not a symmetric matrix?
5. How many elements are there in the admittance matrix of an electrical power network with N nodes and B branches?
6. What changes will occur in the admittance matrix when the turn ratio of a transformer varies?
7. What changes will occur in the admittance matrix when a line is out of service?
8. What characteristics does the electrical power network equation have? And what requirements are there for its solution algorithm?
9. Why is the method of Gauss elimination often adopted to solve network equations?
10. How is the factor table formed? Compare the features between two methods of forming the factor tables.
11. What is the key idea behind sparse technique?
12. What fields can the sparse vector method be applied to?
13. Compare the features and application areas of three kinds of node optimal ordering methods.

14. State the significance of self-impedance, input impedance, mutual impedance, and transfer impedance.
15. How can an admittance matrix be used to find self-impedance Z_{ii} and mutual impedance Z_{ij}? Give a detailed program flowchart.
16. Describe the storage scheme of a sparse admittance matrix.

Chapter 2
Load Flow Analysis

2.1 Introduction

Load flow analysis is the most important and essential approach to investigating problems in power system operating and planning. Based on a specified generating state and transmission network structure, load flow analysis solves the steady operation state with node voltages and branch power flow in the power system. Load flow analysis can provide a balanced steady operation state of the power system, without considering system transient processes. Hence, the mathematic model of load flow problem is a nonlinear algebraic equation system without differential equations. Power system dynamic analysis (see Chaps. 5 and 6) investigates system stability under some given disturbances. Its mathematic model includes differential equations. It should be pointed out that dynamic analysis is based on load flow analysis and the algorithm of load flow analysis is also the base for dynamic analysis methods. Therefore, familiarity with the theory and algorithms of load flow analysis is essential to understanding the methodology of modern power system analysis.

Using digital computers to calculate load flow started from the middle of the 1950s. Since then, a variety of methods has been used in load flow calculation. The development of these methods is mainly led by the basic requirements of load flow calculation, which can be summed up as:

1. The convergence properties
2. The computing efficiency and memory requirements
3. The convenience and flexibility of the implementation

Mathematically, the load flow problem is a problem of solving a system of nonlinear algebraic equations. Its solution usually cannot avoid some iteration process. Thus reliable convergence becomes the prime criterion for a load flow calculation method. With the scale of power system continually expanding, the dimension of load flow equations now becomes very high (several thousands to tens of thousands). For the equations with such high dimensions, we cannot ensure that any mathematical method can converge to a correct solution. This situation requires the researchers and scholars in the power system analysis field to seek more reliable methods.

X.-F. Wang et al., *Modern Power Systems Analysis*.
doi: 10.1007/978-0-387-72853-7,

In the early stages of using digital computers to solve power system load flow problems, the widely used method was the Gauss–Seidel iterative method based on a nodal admittance matrix (it will be simply called the admittance method below) [4]. The principle of this method is rather simple and its memory requirement is relatively small. These properties made it suit the level of computer and power system theory at that time. However, its convergence is not satisfactory. When the system scale becomes larger, the number of iteration increases sharply, and sometimes the iteration process cannot converge. This problem led to the use of the sequential substitution method based on the nodal impedance matrix (also called the impedance method).

At the beginning of the 1960s, the digital computer had developed to the second generation. The memory and computing speed of computers were improved significantly, providing suitable conditions for the application of the impedance method. As mentioned in Chap. 1, the impedance matrix is a full matrix. The impedance method requires the computer to store the impedance matrix that represents the topology and parameters of the power network. Thus it needs a great amount of computer memory. Furthermore, in each iteration, every element in the impedance matrix must be operated with, so the computing burden is very heavy.

The impedance method improved convergence and solved some load flow problems that the admittance method could not solve. Therefore, the impedance method was widely applied from then on and made a great contribution to power system design, operation, and research.

The main disadvantage of the impedance method is its high memory requirement and computing burden. The larger the system is, the more serious these defects are. To overcome the disadvantage, the piecewise solution method based on impedance matrix was developed [5]. This method divides a large system up into several small local systems and only the impedance matrixes of local systems and the impedances of tie lines between these local systems are to be stored in the computer. In this way, the memory requirement and computing burden are greatly alleviated.

The other approach to overcoming the disadvantages of the impedance method is to apply the Newton–Raphson method (also called the Newton method) [6]. The Newton method is a typical method used to solve nonlinear equations in mathematics with very favorable convergence. As long as the sparsity of the Jacobean matrix is utilized in the iterative process, the computing efficiency of the Newton method can be greatly improved. Since the optimal order eliminating method [7] began to be employed in the middle of the 1960s, the Newton method has surpassed the impedance method in the aspects of convergence, memory demand, and computing speed. It is still the favored method, and is widely used in load flow calculation today.

Since the 1970s, the load flow calculating method continues to develop in various ways. Among them the most successful is the fast decoupled method, also called the $P - Q$ decoupled method [8]. Comparing with the Newton method, this method is much simpler and more efficient algorithmically, and therefore more popular in many applications.

In the recent 20 years, research on load flow calculation is still very active. Many contributions seek to improve the convergence characteristics of the Newton method and the $P-Q$ decoupled method [9–15]. Along with the development of artificial intelligent theory, the genetic algorithm, artificial neural network algorithm, and fuzzy algorithm have also been introduced to load flow analysis [16–19]. However, until now these new models and new algorithms still cannot replace the Newton method and $P-Q$ decoupled method. Because the scales of power systems continue to expand and the requirements for online calculation become more and more urgent, the parallel computing algorithms are also studied intensively now and may become an important research field [20].

This chapter mainly discusses the currently widely used Newton method and $P-Q$ decoupled method.

The degree of flexibility and convenience of load flow calculation are also very important to computer application. In practice, load flow analysis is usually part of an interactive environment, rather than a pure calculation problem. Therefore, the human–computer interface should be friendly, allowing users to monitor and control the calculation process. To obtain an ideal operation scheme, it is usually necessary to modify the original data according to the computing results. Thus, the computing method should be flexible, permitting users to readily modify and adjust their operation scheme. Input and output processes should also receive careful attention.

Power system steady state analysis includes load flow analysis and static security analysis. Load flow analysis is mainly used in analyzing the normal operation state, while static security analysis is used when some elements are out of service. Its purpose is to check whether the system can operate safely, i.e., if there are equipment overloads, or some node voltages are too low or too high. In principle, static security analysis can be replaced by a series of load flow analyses. However, usually there are very many contingency states to be checked and the computation burden is quite large if a rigorous load flow calculation method is used. Hence special methods have to be developed to meet the requirement of efficient calculation. In the first part of this chapter, the models and algorithms of load flow calculation are introduced. In the second part, the problems related to static security analysis are discussed.

2.2 Formulation of Load Flow Problem

2.2.1 Classification of Node Types

An electric power system is composed of generators, transformers, transmission lines and loads, etc. A simple power system is illustrated in Fig. 2.1. In the process of power system analysis, the static components, such as transformers, transmission lines, shunt capacitors and reactors, are represented by their equivalent circuits

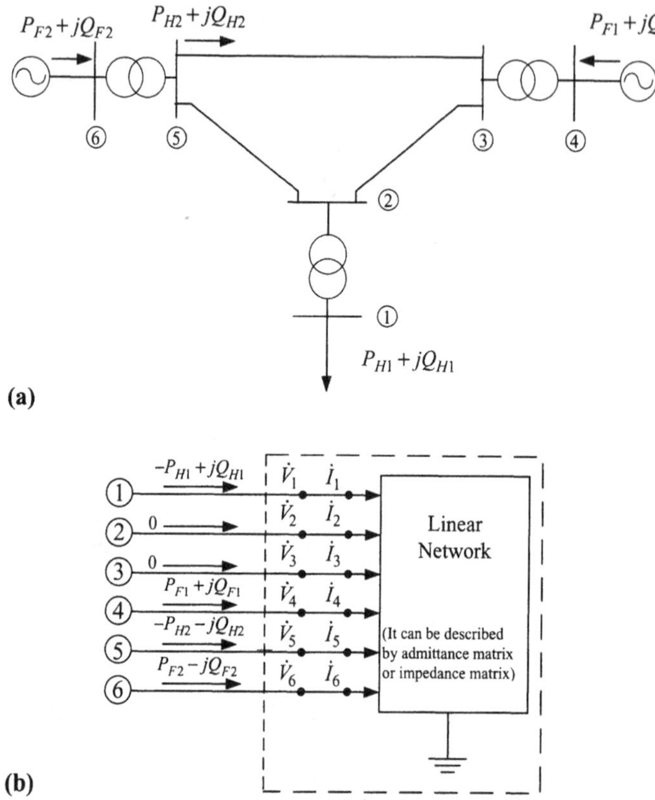

Fig. 2.1 Simple power system

consisting of R, L, C elements. Therefore, the network formed by these static components can be considered as a linear network and represented by the corresponding admittance matrix or impedance matrix. In load flow calculation, the generators and loads are treated as nonlinear components. They cannot be embodied in the linear network, see Fig. 2.1b. The connecting nodes with zero injected power also represent boundary conditions on the network.

In Fig. 2.1b, the relationship between node current and voltage in the linear network can be described by the following node equation:

$$\mathbf{I} = \mathbf{YV} \tag{2.1}$$

or

$$\dot{I}_i = \sum_{j=1}^{n} Y_{ij}\dot{V}_j \quad (i = 1, 2, \ldots, n) \tag{2.2}$$

where $\dot{I}_i$ and $\dot{V}_j$ are the injected current at bus i and voltage at bus j, respectively, Y_{ij} is an element of the admittance matrix, n is the total number of nodes in the system.

To solve the load flow equation, the relation of node power with current should be used

$$\dot{I}_i = \frac{P_i - jQ_i}{\hat{V}_i} \quad (i = 1, 2, \ldots, n) \tag{2.3}$$

where P_i, Q_i are the injected active and reactive power at node i, respectively. If node i is a load node, then P_i and Q_i should take negative values. In (2.3), $\hat{V}_i$ is the conjugate of the voltage vector at node i. Substituting (2.3) to (2.2), we have,

$$\frac{P_i - jQ_i}{\hat{V}_i} = \sum_{j=1}^{n} Y_{ij}\dot{V}_j \quad (i = 1, 2, \ldots, n)$$

or

$$\frac{P_i + jQ_i}{\dot{V}_i} = \sum_{j=1}^{n} \hat{Y}_{ij}\hat{V}_j \quad (i = 1, 2, \ldots, n) \tag{2.4}$$

There are n nonlinear complex equations in (2.4). They are the principal equations in load flow calculation. Based on different methods to solve (2.4), various load flow algorithms can be formed.

In the power system load flow problem, the variables are nodal complex voltages and complex powers: V, θ, P, Q. If there are n nodes in a power system, then the total number of variables is $4n$.

As mentioned above, there are n complex equations or $2n$ real equations defined in principal by (2.4), thus only $2n$ variables can be solved from these equations, while the other $2n$ variables should be specified as original data.

Usually, two variables at each node are assumed known, while the other two variables are treated as state variables to be resolved. According to the original data, the nodes in power systems can be classified into three types:

1. ***PQ* Nodes:** For PQ nodes, the active and reactive power (P, Q) are specified as known parameters, and the complex voltage (V, θ) is to be resolved. Usually, substation nodes are taken as PQ nodes where the load powers are given constants. When output P and Q are fixed in some power plants, these nodes can also be taken as PQ node. Most nodes in power systems belong to the PQ type in load flow calculation.
2. ***PV* Nodes:** For PV nodes, active power P and voltage magnitude V are specified as known variables, while reactive power Q and voltage angle θ are to be resolved. Usually, PV nodes should have some controllable reactive power resources and can thus maintain node voltage magnitude at a desirable value. Generally speaking, the buses of power plants can be taken as PV nodes, because voltages at these buses can be controlled with reactive power capacity of their generators. Some substations can also be considered as PV nodes when they have enough reactive power compensation devices to control the voltage.

3. **Slack Node:** In load flow studies, there should be one and only one slack node specified in the power system, which is specified by a voltage, constant in magnitude and phase angle. Therefore, V and θ are given as known variables at the slack node, while the active power P and reactive power Q are the variables to be solved. The effective generator at this node supplies the losses to the network. This is necessary because the magnitude of losses will not be known until the calculation of currents is complete, and this cannot be achieved unless one node has no power constraint and can feed the required losses into the system. The location of the slack node can influence the complexity of the calculations; the node most closely approaching a large AGC power station should be used.

We will employ different methods to treat the above three kinds of nodes in power flow calculations.

2.2.2 Node Power Equations

As described above, power system load flow calculations can be roughly considered as the problem of solving the node voltage phasor for each node when the injecting complex power is specified. If the complex power can be represented by equations of complex voltages, then a nonlinear equation solving method, such as the Newton–Raphson method, can be used to solve the node voltage phasors. In this section, node power equations are deduced first.

The complex node voltage has two representation forms – the polar form and the rectangular form. Accordingly, the node power equations also have two forms.

From (2.4), the node power equations can be expressed as

$$P_i + jQ_i = \dot{V}_i \sum_{j \in i} \hat{Y}_{ij} \hat{V}_j \quad (i = 1, 2, \ldots, n) \tag{2.5}$$

where $j \in i$ means the node j should be directly connected with node i, including $j = i$. As discussed in Chap.1, the admittance matrix is a sparse matrix, and the terms in Σ are correspondingly few. If the voltage vector of (2.5) adopts polar form,

$$\dot{V}_i = V_i e^{j\theta_i} \tag{2.6}$$

where V_i,θ_i are the magnitude and phase angle of voltage at node i. The elements of admittance matrix can be expressed as

$$Y_{ij} = G_{ij} + jB_{ij}$$

Hence (2.5) can be rewritten as

$$P_i + jQ_i = V_i e^{j\theta_i} \sum_{j \in i} (G_{ij} - jB_{ij}) V_j e^{-j\theta_j} \quad (i = 1, 2, \ldots, n) \tag{2.7}$$

Combining the exponential items of above equation and using the relationship

$$e^{j\theta} = \cos\theta + j\sin\theta$$

we have,

$$P_i + jQ_i = V_i \sum_{j \in i} V_j (G_{ij} - jB_{ij})(\cos\theta_{ij} + j\sin\theta_{ij}) \quad (i = 1, 2, \ldots, n) \tag{2.8}$$

where $\theta_{ij} = \theta_i - \theta_j$, is the voltage phase angle difference between node i and j. Dividing above equations into real and imaginary parts,

$$\left.\begin{aligned} P_i &= V_i \sum_{j \in i} V_j (G_{ij}\cos\theta_{ij} + B_{ij}\sin\theta_{ij}) \\ Q_i &= V_i \sum_{j \in i} V_j (G_{ij}\sin\theta_{ij} - B_{ij}\cos\theta_{ij}) \end{aligned}\right\} \quad (i = 1, 2, \cdots, n) \tag{2.9}$$

This is the polar form of the nodal power equations. It is not only very important in the Newton–Raphson calculation process, but also essential to establish the fast decoupled method.

When the voltage vector is expressed in rectangular form,

$$\dot{V}_i = e_i + jf_i$$

where

$$e_i = V_i \cos\theta_i \quad f_i = V_i \sin\theta_i$$

We can obtain from (2.5),

$$\left.\begin{aligned} P_i &= e_i \sum_{j \in i} (G_{ij}e_j - B_{ij}f_j) + f_i \sum_{j \in i} (G_{ij}f_j + B_{ij}e_j) \\ Q_i &= f_i \sum_{j \in i} (G_{ij}e_j - B_{ij}f_j) - e_i \sum_{j \in i} (G_{ij}f_j + B_{ij}e_j) \end{aligned}\right\} \quad (i = 1, 2, \ldots, n) \tag{2.10}$$

Let

$$\left.\begin{aligned}\sum_{j\in i}(G_{ij}e_j - B_{ij}f_j) = a_i \\ \sum_{j\in i}(G_{ij}f_j + B_{ij}e_j) = b_i\end{aligned}\right\} \tag{2.11}$$

Obviously, a_i and b_i are the real and imaginary parts of injected current at node i and (2.10) can be simplified as,

$$\left.\begin{aligned}P_i = e_i a_i + f_i b_i \\ Q_i = f_i a_i - e_i b_i\end{aligned}\right\} \quad (i = 1, 2, \ldots, n) \tag{2.12}$$

This is the rectangular form of the nodal power equations.

Both (2.9) and (2.10) are the simultaneous nonlinear equations of node voltage phasors. They are usually expressed as the following forms as mathematical models of the load flow problem:

$$\left.\begin{aligned}\Delta P_i = P_{is} - V_i \sum_{j\in i} V_j(G_{ij}\cos\theta_{ij} + B_{ij}\sin\theta_{ij}) = 0 \\ \Delta Q_i = Q_{is} - V_i \sum_{j\in i} V_j(G_{ij}\sin\theta_{ij} - B_{ij}\cos\theta_{ij}) = 0\end{aligned}\right\} \quad (i = 1, 2, \ldots, n) \tag{2.13}$$

and

$$\left.\begin{aligned}\Delta P_i = P_{is} - e_i \sum_{j\in i}(G_{ij}e_j - B_{ij}f_j) - f_i \sum_{j\in i}(G_{ij}f_j + B_{ij}e_j) = 0 \\ \Delta Q_i = Q_{is} - f_i \sum_{j\in i}(G_{ij}e_j - B_{ij}f_j) + e_i \sum_{j\in i}(G_{ij}f_j + B_{ij}e_j) = 0\end{aligned}\right\} \tag{2.14}$$
$$(i = 1, 2, \ldots, n)$$

where P_{is}, Q_{is} are the specified active and reactive powers at node i. Based on the above two simultaneous equations, the load flow problem can be roughly summarized as: for specified P_{is},Q_{is} $(i = 1, 2, \ldots, n)$, find voltage vector V_i, θ_i or e_i,f_i $(i = 1, 2, \ldots, n)$, such that the magnitudes of the power errors $\Delta P_i, \Delta Q_i$, $(i = 1, 2, \ldots, n)$ of (2.13) or (2.14) are less then an acceptable tolerance.

2.3 Load Flow Solution by Newton Method

2.3.1 Basic Concept of Newton Method

The Newton–Raphson method is an efficient algorithm to solve nonlinear equations. It transforms the procedure of solving nonlinear equations into the procedure of repeatedly solving linear equations. This sequential linearization process is the core of the Newton–Raphson method. We now introduce the Newton–Raphson method by the following nonlinear equation example,

$$f(x) = 0 \tag{2.15}$$

Let $x^{(0)}$ be the initial guess value of the above equation solution. Assume the real solution x is close to $x^{(0)}$,

$$x = x^{(0)} - \Delta x^{(0)} \tag{2.16}$$

where $\Delta x^{(0)}$ is a modification value of $x^{(0)}$. The following equation holds,

$$f(x^{(0)} - \Delta x^{(0)}) = 0 \tag{2.17}$$

When $\Delta x^{(0)}$ is known, the solution x can be calculated by (2.16). Expanding this function in a Taylor series expansion about point $x^{(0)}$yields:

$$\begin{aligned} f(x^{(0)} - \Delta x^{(0)}) = f(x^{(0)}) - f'(x^{(0)})\Delta x^{(0)} + f''(x^{(0)})\tfrac{(\Delta x^{(0)})^2}{2!} - \\ \cdots + (-1)^n f^{(n)}(x^{(0)})\tfrac{(\Delta x^{(0)})^n}{n!} + \cdots = 0 \end{aligned} \tag{2.18}$$

where $f'(x^{(0)})$,..., $f^{(n)}(x^{(0)})$ are the different order partial derivatives of $f(x)$ at $x^{(0)}$. If the initial guess is sufficiently close to the actual solution, the higher order terms of the Taylor series expansion could be neglected. Equation (2.18) becomes,

$$f(x^{(0)}) - f'(x^{(0)})\Delta x^{(0)} = 0 \tag{2.19}$$

This is a linear equation in $\Delta x^{(0)}$ and can be easily solved.

Using $\Delta x^{(0)}$ to modify $x^{(0)}$, we can get $x^{(1)}$:

$$x^{(1)} = x^{(0)} - \Delta x^{(0)} \tag{2.20}$$

$x^{(1)}$ may be more close to the actual solution. Then using $x^{(1)}$ as the new guess value, we solve the following equation similar to (2.19),

$$f(x^{(1)}) - f'(x^{(1)})\Delta x^{(1)} = 0$$

Thus $x^{(2)}$is obtained:

$$x^{(2)} = x^{(1)} - \Delta x^{(1)} \tag{2.21}$$

Repeating this procedure, we establish the correction equation in the tth iteration:

$$f(x^{(t)}) - f'(x^{(t)})\Delta x^{(t)} = 0 \tag{2.22}$$

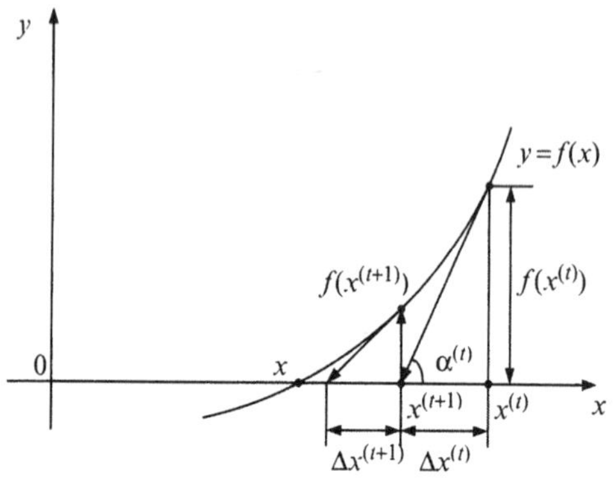

Fig. 2.2 Geometric interpretation of Newton method

or

$$f(x^{(t)}) = f'(x^{(t)})\Delta x^{(t)} \tag{2.23}$$

The left hand of the above equation can be considered as the error produced by approximate solution $x^{(t)}$. When $f(x^{(t)}) \to 0$, (2.15) is satisfied, so $x^{(t)}$ is the solution of the equation. In (2.22), $f'(x^{(t)})$ is the first-order partial derivative of function $f(x)$ at point $x^{(t)}$. It is also the slope of the curve at point $x^{(t)}$, as shown in Fig. 2.2,

$$\tan \alpha^{(t)} = f'(x^{(t)}) \tag{2.24}$$

The correction value $\Delta x^{(t)}$ is determined by the intersection of the tangent line at $x^{(t)}$ with the abscissa. We can comprehend the iterative process more intuitively from Fig. 2.2.

Now we will extend the Newton method to simultaneous nonlinear equations. Assume the nonlinear equations with variables $x_1, x_2, \ldots, x_n$;

$$\left.\begin{array}{l} f_1(x_1, x_2, \ldots, x_n) = 0 \\ f_2(x_1, x_2, \ldots, x_n) = 0 \\ \vdots \\ f_n(x_1, x_2, \ldots, x_n) = 0 \end{array}\right\} \tag{2.25}$$

Specify the initial guess values of all variables $x_1^{(0)}, x_2^{(0)}, \ldots, x_n^{(0)}$, let $\Delta x_1^{(0)}$, $\Delta x_2^{(0)}, \ldots, \Delta x_n^{(0)}$ be the correction values to satisfy the following equations,

$$\left.\begin{array}{l} f_1(x_1^{(0)} - \Delta x_1^{(0)}, x_2^{(0)} - \Delta x_2^{(0)}, \ldots, x_n^{(0)} - \Delta x_n^{(0)}) = 0 \\ f_2(x_1^{(0)} - \Delta x_1^{(0)}, x_2^{(0)} - \Delta x_2^{(0)}, \ldots, x_n^{(0)} - \Delta x_n^{(0)}) = 0 \\ \vdots \\ f_n(x_1^{(0)} - \Delta x_1^{(0)}, x_2^{(0)} - \Delta x_2^{(0)}, \ldots, x_n^{(0)} - \Delta x_n^{(0)}) = 0 \end{array}\right\} \tag{2.26}$$

Expanding the above n equations via the multivariate Taylor series and neglecting the higher order terms, we have the following equations,

$$\left.\begin{aligned}
&f_1(x_1^{(0)},x_2^{(0)},\ldots,x_n^{(0)}) - \left[\frac{\partial f_1}{\partial x_1}\Big|_0 \Delta x_1^{(0)} + \frac{\partial f_1}{\partial x_2}\Big|_0 \Delta x_2^{(0)} +,\ldots,+ \frac{\partial f_1}{\partial x_n}\Big|_0 \Delta x_n^{(0)}\right] = 0\\
&f_2(x_1^{(0)},x_2^{(0)},\ldots,x_n^{(0)}) - \left[\frac{\partial f_2}{\partial x_1}\Big|_0 \Delta x_1^{(0)} + \frac{\partial f_2}{\partial x_2}\Big|_0 \Delta x_2^{(0)} +,\ldots,+ \frac{\partial f_2}{\partial x_n}\Big|_0 \Delta x_n^{(0)}\right] = 0\\
&\vdots\\
&f_n(x_1^{(0)},x_2^{(0)},\ldots,x_n^{(0)}) - \left[\frac{\partial f_n}{\partial x_1}\Big|_0 \Delta x_1^{(0)} + \frac{\partial f_n}{\partial x_2}\Big|_0 \Delta x_2^{(0)} +,\ldots,+ \frac{\partial f_n}{\partial x_n}\Big|_0 \Delta x_n^{(0)}\right] = 0
\end{aligned}\right\} \quad (2.27)$$

here $\frac{\partial f_i}{\partial x_j}|_0$ is the partial derivative of function $f_i(x_1,x_2,\ldots,x_n)$ over independent variable x_j at point $(x_1^{(0)},\ x_2^{(0)},\ \ldots,\ x_n^{(0)})$. Rewrite the above equation in matrix form,

$$\begin{bmatrix} f_1(x_1^{(0)},x_2^{(0)},\ldots,x_n^{(0)})\\ f_2(x_1^{(0)},x_2^{(0)},\ldots,x_n^{(0)})\\ \vdots\\ f_n(x_1^{(0)},x_2^{(0)},\ldots,x_n^{(0)}) \end{bmatrix} = \begin{bmatrix} \frac{\partial f_1}{\partial x_1}\big|_0 \frac{\partial f_1}{\partial x_2}\big|_0 \cdots \frac{\partial f_1}{\partial x_n}\big|_0\\ \frac{\partial f_2}{\partial x_1}\big|_0 \frac{\partial f_2}{\partial x_2}\big|_0 \cdots \frac{\partial f_2}{\partial x_n}\big|_0\\ \vdots\\ \frac{\partial f_n}{\partial x_1}\big|_0 \frac{\partial f_n}{\partial x_2}\big|_0 \cdots \frac{\partial f_n}{\partial x_n}\big|_0 \end{bmatrix} \begin{bmatrix} \Delta x_1^{(0)}\\ \Delta x_2^{(0)}\\ \vdots\\ \Delta x_n^{(0)} \end{bmatrix} \quad (2.28)$$

This is a set of simultaneous linear equations in the variables $\Delta x_1^{(0)}, \Delta x_2^{(0)}, \ldots,$ $\Delta x_n^{(0)}$, usually called the correction equations of the Newton–Raphson method. After solving $\Delta x_1^{(0)}, \Delta x_2^{(0)}, \ldots, \Delta x_n^{(0)}$, we can get,

$$\left.\begin{aligned}
x_1^{(1)} &= x_1^{(0)} - \Delta x_1^{(0)}\\
x_2^{(1)} &= x_2^{(0)} - \Delta x_2^{(0)}\\
\vdots \quad & \quad \vdots \qquad \vdots\\
x_n^{(1)} &= x_n^{(0)} - \Delta x_n^{(0)}
\end{aligned}\right\} \quad (2.29)$$

$x_1^{(1)}, x_2^{(1)}, \ldots, x_n^{(1)}$ will approach the actual solution more closely. The updated values are used as the new guess to solve the correction equation (2.28) and to further correct the variables. In this way the iterative process of the Newton–Raphson method is formed.

Generally, the correction equation in the tth iteration can be written as,

$$\begin{bmatrix} f_1(x_1^{(t)},x_2^{(t)},\ldots,x_n^{(t)})\\ f_2(x_1^{(t)},x_2^{(t)},\ldots,x_n^{(t)})\\ \vdots \quad \vdots \quad \vdots\\ f_n(x_1^{(t)},x_2^{(t)},\ldots,x_n^{(t)}) \end{bmatrix} = \begin{bmatrix} \frac{\partial f_1}{\partial x_1}\big|_t & \frac{\partial f_1}{\partial x_2}\big|_t & \cdots & \frac{\partial f_1}{\partial x_n}\big|_t\\ \frac{\partial f_2}{\partial x_1}\big|_t & \frac{\partial f_2}{\partial x_2}\big|_t & \cdots & \frac{\partial f_2}{\partial x_n}\big|_t\\ \vdots & \vdots & & \vdots\\ \frac{\partial f_n}{\partial x_1}\big|_t & \frac{\partial f_n}{\partial x_2}\big|_t & \cdots & \frac{\partial f_n}{\partial x_n}\big|_t \end{bmatrix} \begin{bmatrix} \Delta x_1^{(t)}\\ \Delta x_2^{(t)}\\ \vdots\\ \Delta x_n^{(t)} \end{bmatrix} \quad (2.30)$$

or expressed in matrix form,

$$\mathbf{F}(\mathbf{X}^{(t)}) = \mathbf{J}^{(t)}\Delta\mathbf{X}^{(t)} \tag{2.31}$$

where

$$\mathbf{F}(\mathbf{X}^{(t)}) = \begin{bmatrix} f_1(x_1^{(t)}, x_2^{(t)}, \ldots, x_n^{(t)}) \\ f_2(x_1^{(t)}, x_2^{(t)}, \ldots, x_n^{(t)}) \\ \vdots \\ f_n(x_1^{(t)}, x_2^{(t)}, \ldots, x_n^{(t)}) \end{bmatrix} \tag{2.32}$$

is the error vector in the tth iteration;

$$\mathbf{J}^{(t)} = \begin{bmatrix} \frac{\partial f_1}{\partial x_1}\big|_t \frac{\partial f_1}{\partial x_2}\big|_t \cdots \frac{\partial f_1}{\partial x_n}\big|_t \\ \frac{\partial f_2}{\partial x_1}\big|_t \frac{\partial f_2}{\partial x_2}\big|_t \cdots \frac{\partial f_2}{\partial x_n}\big|_t \\ \vdots \\ \frac{\partial f_n}{\partial x_1}\big|_t \frac{\partial f_n}{\partial x_2}\big|_t \cdots \frac{\partial f_n}{\partial x_n}\big|_t \end{bmatrix} \tag{2.33}$$

is the Jacobian matrix of tth iteration;

$$\Delta\mathbf{X}^{(t)} = \begin{bmatrix} \Delta x_1^{(t)} \\ \Delta x_2^{(t)} \\ \vdots \\ \Delta x_n^{(t)} \end{bmatrix} \tag{2.34}$$

is the correction value vector in the tth iteration.

We also have the equation similar to (2.29),

$$\mathbf{X}^{(t+1)} = \mathbf{X}^{(t)} - \Delta\mathbf{X}^{(t)} \tag{2.35}$$

With (2.31) and (2.35) solved alternately in each iteration, $X^{(t+1)}$ gradually approaches the actual solution. Convergence can be evaluated by the norm of the correction value,

$$\left\|\Delta\mathbf{X}^{(t)}\right\| < \varepsilon_1 \tag{2.36}$$

or by the norm of the function,

$$\left\|\mathbf{F}(\mathbf{X}^{(t)})\right\| < \varepsilon_2 \tag{2.37}$$

Here ε_1 and ε_2 are very small positive numbers specified beforehand.

2.3.2 Correction Equations

In Section 2.3.1, we derived two forms of the nodal power equations. Either can be applied in the load flow calculation model.

When the polar form (2.13) is used, the node voltage magnitudes and angles V_i,θ_i $(i = 1, 2, \ldots, n)$ are the variables to be solved. For a *PV* node, the magnitude of the voltage is specified. At the same time, its reactive power Q_{is} cannot be fixed beforehand as a constraint. Therefore, the reactive equations relative to *PV* nodes should not be considered in the iterative process. These equations will be used only to calculate the reactive power of each *PV* node after the iterative process is over and all node voltages have been calculated. Similarly, the voltage magnitude and angle of the slack node are specified, hence the related power equations do not appear in the iterative process. When the iteration has converged, the active and reactive power of the slack node can be calculated by using these power equations.

Assume that total number of system nodes is n, the number of *PV* nodes is r. For convenience, let the slack bus be the last node, i.e., node n.Therefore, we have $n-1$ active power equations,

$$\left.\begin{aligned}
\Delta P_1 &= P_{1s} - V_1 \sum_{j\in 1} V_j(G_{1j}\cos\theta_{1j} + B_{1j}\sin\theta_{1j}) = 0\\
\Delta P_2 &= P_{2s} - V_2 \sum_{j\in 2} V_j(G_{2j}\cos\theta_{2j} + B_{2j}\sin\theta_{2j}) = 0\\
&\vdots\\
\Delta P_{n-1} &= P_{n-1,s} - V_{n-1} \sum_{j\in(n-1)} V_j(G_{n-1,j}\cos\theta_{n-1,j} + B_{n-1,j}\sin\theta_{n-1,j}) = 0
\end{aligned}\right\} \tag{2.38}$$

and $n-r-1$ reactive power equations.

$$\left.\begin{aligned}
\Delta Q_1 &= Q_{1s} - V_1 \sum_{j\in 1} V_j(G_{1j}\sin\theta_{1j} - B_{1j}\cos\theta_{1j}) = 0\\
\Delta Q_2 &= Q_{2s} - V_2 \sum_{j\in 2} V_j(G_{2j}\sin\theta_{2j} - B_{2j}\cos\theta_{2j}) = 0\\
&\vdots\\
\Delta Q_{n-1} &= Q_{n-1,s} - V_{n-1} \sum_{j\in(n-1)} V_j(G_{n-1,j}\sin\theta_{n-1,j} - B_{n-1,j}\cos\theta_{n-1,j}) = 0
\end{aligned}\right\} \tag{2.39}$$

In the above equations, node voltage angle θ_i and magnitude V_i are the variables to be resolved. Here the number of θ_i is $n-1$ and the number of V_i is $n-r-1$. There are $2n - r - 2$ unknown variables in total and they can be solved by the above $2n - r - 2$ equations.

Expanding (2.38) and (2.39) in a Taylor series, neglecting the high-order terms, the correction equation can be written as,

$$\begin{bmatrix} \Delta P_1 \\ \Delta P_2 \\ \vdots \\ \Delta P_{n-1} \\ \cdots\cdots \\ \Delta Q_1 \\ \Delta Q_2 \\ \vdots \\ \Delta Q_{n-1} \end{bmatrix} = \left[\begin{array}{cccc:cccc} H_{11} & H_{12} & \ldots & H_{1,n-1} & N_{11} & N_{12} & \ldots & N_{1,n-1} \\ H_{21} & H_{22} & \ldots & H_{2,n-1} & N_{21} & N_{22} & \ldots & N_{2,n-1} \\ & & \cdots & & & & \cdots & \\ H_{n-1,1} & H_{n-1,2} & \ldots & H_{n-1,n-1} & N_{n-1,1} & N_{n-1,2} & \ldots & N_{n-1,n-1} \\ \cdots\cdots & \cdots\cdots & \cdots & \cdots\cdots & \cdots\cdots & \cdots\cdots & \cdots & \cdots\cdots \\ J_{11} & J_{12} & \ldots & J_{1,n-1} & L_{11} & L_{12} & \ldots & L_{1,n-1} \\ J_{21} & J_{22} & \ldots & J_{2,n-1} & L_{21} & L_{22} & \ldots & L_{2,n-1} \\ & & \cdots & & & & \cdots & \\ J_{n-1,1} & J_{n-1,2} & \ldots & J_{n-1,n-1} & J_{n-1,1} & J_{n-1,2} & \ldots & J_{n-1,n-1} \end{array}\right] \times \begin{bmatrix} \Delta\theta_1 \\ \Delta\theta_2 \\ \vdots \\ \Delta\theta_{n-1} \\ \cdots\cdots \\ \Delta V_1/V_1 \\ \Delta V_2/V_2 \\ \vdots \\ \Delta V_{n-1}/V_{n-1} \end{bmatrix} \tag{2.40}$$

The form of the voltage magnitude correction values represented here, $\Delta V_1/V_1$, $\Delta V_2/V_2, \ldots, \Delta V_{n-1}/V_{n-1}$, allow the elements in the Jacobian matrix to have similar expressions.

Taking partial derivations of (2.38), or (2.39), and noting that both P_{is}, Q_{is} are constants, we can obtain the elements of the Jacobian matrix as,

$$H_{ij} = \frac{\partial \Delta P_i}{\partial \theta_j} = -V_i V_j (G_{ij} \sin\theta_{ij} - B_{ij}\cos\theta_{ij}) \quad j \neq i \tag{2.41}$$

$$H_{ii} = \frac{\partial \Delta P_i}{\partial \theta_i} = V_i \sum_{\substack{j \in i \\ j \neq i}} V_j (G_{ij} \sin\theta_{ij} - B_{ij}\cos\theta_{ij}) \tag{2.42}$$

or

$$H_{ii} = V_i^2 B_{ii} + Q_i \tag{2.43}$$

$$N_{ij} = \frac{\partial \Delta P_i}{\partial V_j} V_j = -V_i V_j (G_{ij}\cos\theta_{ij} + B_{ij}\sin\theta_{ij}) \quad j \neq i \tag{2.44}$$

$$N_{ii} = \frac{\partial \Delta P_i}{\partial V_i} V_i = -V_i \sum_{\substack{j \in i \\ j \neq i}} V_j (G_{ij}\cos\theta_{ij} + B_{ij}\sin\theta_{ij}) - 2V_i^2 G_{ii} = -V_i^2 G_{ii} - P_i \tag{2.45}$$

$$J_{ij} = \frac{\partial \Delta P_i}{\partial \theta_j} = V_i V_j (G_{ij}\cos\theta_{ij} + B_{ij}\sin\theta_{ij}) \quad j \neq i \tag{2.46}$$

$$J_{ii} = \frac{\partial \Delta P_i}{\partial \theta_j} = -V_i \sum_{\substack{j \in i \\ j \neq i}} V_j (G_{ij}\cos\theta_{ij} + B_{ij}\sin\theta_{ij}) = V_i^2 G_{ii} - P_i \tag{2.47}$$

$$L_{ij} = \frac{\partial \Delta Q_i}{\partial V_j} V_j = -V_i V_j (G_{ij} \sin\theta_{ij} - B_{ij} \cos\theta_{ij}) \quad j \neq i \tag{2.48}$$

$$L_{ii} = \frac{\partial \Delta Q_i}{\partial V_i} V_i = -V_i \sum_{\substack{j \in i \\ j \neq i}} V_j (G_{ij} \sin\theta_{ij} - B_{ij} \cos\theta_{ij}) + 2V_i^2 B_{ii} = V_i^2 B_{ii} - Q_i \tag{2.49}$$

The concise form of (2.40) is

$$\begin{bmatrix} \mathbf{\Delta P} \\ \mathbf{\Delta Q} \end{bmatrix} = \begin{bmatrix} \mathbf{H} & \mathbf{N} \\ \mathbf{J} & \mathbf{L} \end{bmatrix} \begin{bmatrix} \Delta\theta \\ \mathbf{\Delta V/V} \end{bmatrix} \tag{2.50}$$

Comparing (2.50) with (2.40), the meaning of elements is obvious. The correction equation can be rearranged into the following form for convenience,

$$\begin{bmatrix} \Delta P_1 \\ \Delta Q_1 \\ \Delta P_2 \\ \Delta Q_2 \\ \vdots \\ \Delta P_{n-1} \\ \Delta Q_{n-1} \end{bmatrix} = \begin{bmatrix} H_{11} & N_{11} & H_{12} & N_{12} & \cdots & H_{1,n-1} & N_{1,n-1} \\ J_{11} & L_{11} & J_{12} & L_{12} & \cdots & J_{1,n-1} & L_{1,n-1} \\ H_{21} & N_{21} & H_{22} & N_{22} & \cdots & H_{2,n-1} & N_{2,n-1} \\ J_{21} & L_{21} & J_{22} & L_{22} & \cdots & J_{2,n-1} & L_{2,n-1} \\ \vdots & \vdots & \vdots & \vdots & \vdots & \vdots & \vdots \\ H_{n-1,1} & N_{n-1,1} & H_{n-1,2} & N_{n-1,2} & \cdots & H_{n-1,n-1} & N_{n-1,n-1} \\ J_{n-1,1} & L_{n-1,1} & J_{n-1,2} & L_{n-1,2} & \cdots & J_{n-1,n-1} & L_{n-1,n-1} \end{bmatrix} \begin{bmatrix} \Delta\theta_1 \\ \Delta V_1/V_1 \\ \Delta\theta_2 \\ \Delta V_2/V_2 \\ \vdots \\ \Delta\theta_{n-1} \\ \Delta V_{n-1}/V_{n-1} \end{bmatrix} \tag{2.51}$$

When the rectangular form is adopted in the load flow model, the state variables to be solved are the real and imaginary parts of voltages, i.e., $e_1, f_1, e_2, f_2, \ldots, e_n, f_n$. Since the voltage phasor of the slack node is specified, the number of state variables is $2(n-1)$. We need $2(n-1)$ equations to solve these variables. In fact, every node has two equations except the slack bus. For PQ nodes, P_{is}, Q_{is} are given, so the equations are

$$\left.\begin{aligned} \Delta P_i &= P_{is} - e_i \sum_{j \in i} (G_{ij} e_j - B_{ij} f_j) - f_i \sum_{j \in i} (G_{ij} f_j + B_{ij} e_j) = 0 \\ \Delta Q_i &= Q_{is} - f_i \sum_{j \in i} (G_{ij} e_j - B_{ij} f_j) + e_i \sum_{j \in i} (G_{ij} f_j + B_{ij} e_j) = 0 \end{aligned}\right\} \tag{2.52}$$

For PV nodes, P_{is}, V_{is} are given, so the equations are

$$\left.\begin{aligned} \Delta P_i &= P_{is} - e_i \sum_{j \in i} (G_{ij} e_j - B_{ij} f_j) - f_i \sum_{j \in i} (G_{ij} f_j + B_{ij} e_j) = 0 \\ \Delta V_i^2 &= V_{is}^2 - (e_i^2 + f_i^2) = 0 \end{aligned}\right\} \tag{2.53}$$

There are $2(n-1)$ equations included in (2.52) and (2.53). Expanding them in a Taylor series expansion, neglecting the higher order terms, we can obtain the correction equation as follows,

$$\begin{bmatrix} \Delta P_1 \\ \Delta Q_1 \\ \Delta P_2 \\ \Delta Q_2 \\ \vdots \\ \Delta P_i \\ \Delta V_i^2 \\ \vdots \end{bmatrix} = \begin{bmatrix} \frac{\partial \Delta P_1}{\partial e_1} & \frac{\partial \Delta P_1}{\partial f_1} & \frac{\partial \Delta P_1}{\partial e_2} & \frac{\partial \Delta P_1}{\partial f_2} & \cdots & \frac{\partial \Delta P_1}{\partial e_i} & \frac{\partial \Delta P_1}{\partial f_i} & \cdots \\ \frac{\partial \Delta Q_1}{\partial e_1} & \frac{\partial \Delta Q_1}{\partial f_1} & \frac{\partial \Delta Q_1}{\partial e_2} & \frac{\partial \Delta Q_1}{\partial f_2} & \cdots & \frac{\partial \Delta Q_1}{\partial e_i} & \frac{\partial \Delta Q_1}{\partial f_i} & \cdots \\ \frac{\partial \Delta P_2}{\partial e_1} & \frac{\partial \Delta P_2}{\partial f_1} & \frac{\partial \Delta P_2}{\partial e_2} & \frac{\partial \Delta P_2}{\partial f_2} & \cdots & \frac{\partial \Delta P_2}{\partial e_i} & \frac{\partial \Delta P_2}{\partial f_i} & \cdots \\ \frac{\partial \Delta Q_2}{\partial e_1} & \frac{\partial \Delta Q_2}{\partial f_1} & \frac{\partial \Delta Q_2}{\partial e_2} & \frac{\partial \Delta Q_2}{\partial f_2} & \cdots & \frac{\partial \Delta Q_2}{\partial e_i} & \frac{\partial \Delta Q_2}{\partial f_i} & \cdots \\ \vdots & \vdots & \vdots & \vdots & \vdots & \vdots & \vdots & \vdots \\ \frac{\partial \Delta P_i}{\partial e_1} & \frac{\partial \Delta P_i}{\partial f_1} & \frac{\partial \Delta P_i}{\partial e_2} & \frac{\partial \Delta P_i}{\partial f_2} & \cdots & \frac{\partial \Delta P_i}{\partial e_i} & \frac{\partial \Delta P_i}{\partial f_i} & \cdots \\ 0 & 0 & 0 & 0 & \cdots & \frac{\partial \Delta V_i^2}{\partial e_i} & \frac{\partial \Delta V_i^2}{\partial f_i} & \cdots \\ \vdots & \vdots & \vdots & \vdots & \vdots & \vdots & \vdots & \vdots \end{bmatrix} \begin{bmatrix} \Delta e_1 \\ \Delta f_1 \\ \Delta e_2 \\ \Delta f_2 \\ \vdots \\ \Delta e_i \\ \Delta f_i \\ \vdots \end{bmatrix} \tag{2.54}$$

By differentiating (2.52) and (2.53), we can obtain elements of the Jacobian matrix. The off-diagonal elements of the Jacobian matrix for $j \neq i$ can be expressed as,

$$\left.\begin{aligned} \frac{\partial \Delta P_i}{\partial e_j} &= -\frac{\partial \Delta Q_i}{\partial f_j} = -(G_{ij}e_i + B_{ij}f_i) \\ \frac{\partial \Delta P_i}{\partial f_j} &= \frac{\partial \Delta Q_i}{\partial e_j} = B_{ij}e_i - G_{ij}f_i \\ \frac{\partial \Delta V_i^2}{\partial e_j} &= -\frac{\partial \Delta V_i^2}{\partial f_j} = 0 \end{aligned}\right\} \tag{2.55}$$

The diagonal elements of the Jacobian matrix for $j = i$,

$$\frac{\partial \Delta P_i}{\partial e_i} = -\sum_{j \in i} (G_{ij}e_j - B_{ij}f_j) - G_{ii}e_i - B_{ii}f_i$$

Using (2.11), we can rewrite the above expression as

$$\frac{\partial \Delta P_i}{\partial e_i} = -a_i - G_{ii}e_i - B_{ii}f_i$$

and can obtain the following elements similarly,

$$\left.\begin{aligned}
\frac{\partial \Delta Q_i}{\partial f_i} &= -\sum_{j\in i}(G_{ij}e_j - B_{ij}f_j) + G_{ii}e_i + B_{ii}f_i = -a_i + G_{ii}e_i + B_{ii}f_i \\
\frac{\partial \Delta P_i}{\partial f_i} &= -\sum_{j\in i}(G_{ij}f_j + B_{ij}e_j) + B_{ii}e_i - G_{ii}f_i = -b_i + B_{ii}e_i - G_{ii}f_i \\
\frac{\partial \Delta Q_i}{\partial e_i} &= \sum_{j\in i}(G_{ij}f_j + B_{ij}e_j) + B_{ii}e_i - G_{ii}f_i = b_i + B_{ii}e_i - G_{ii}f_i \\
\frac{\partial \Delta V_i^2}{\partial e_i} &= -2e_i \\
\frac{\partial \Delta V_i^2}{\partial f_i} &= -2f_i
\end{aligned}\right\} \tag{2.56}$$

The correction equations, in either polar form or rectangular form, are the basic equations that need repeatedly solving in Newton–Raphson load flow calculation. Investigating these equations, we can observe the following properties:

1. Equations (2.54) and (2.40) include $2(n-1)$ and $2(n-1)-r$ equations respectively.
2. From the expression of the off-diagonal elements of the Jacobian matrix either in polar form or in rectangular form, i.e., (2.41), (2.44), (2.46), (2.48), and (2.55), we can see that each of them is related to only one element of the admittance matrix. Therefore, if the element Y_{ij} in the admittance matrix is zero, the corresponding element in the Jacobian matrix of the correction equation is also zero. It means the Jacobian matrix is a sparse matrix, and has the same structure as the admittance matrix.
3. From the expression of the elements of the Jacobian matrix we can see that the Jacobian matrix is not symmetrical in either coordinate form. For example,

$$\frac{\partial \Delta P_i}{\partial \theta_j} \neq \frac{\partial \Delta P_j}{\partial \theta_i}, \quad \frac{\partial \Delta Q_i}{\partial V_j} \neq \frac{\partial \Delta Q_j}{\partial V_i}$$

$$\frac{\partial \Delta P_i}{\partial e_j} \neq \frac{\partial \Delta P_j}{\partial e_i}, \quad \frac{\partial \Delta Q_i}{\partial f_j} \neq \frac{\partial \Delta Q_j}{\partial f_i}, \text{etc.}$$

4. The elements in the Jacobian matrix are a function of node voltage phasors. Therefore, they will vary with node voltages during the iterative process. The Jacobian matrix must not only be updated but also be triangularized in each iteration. This has a major effect on the calculation efficiency of the Newton–Raphson method.

Many improvements of the Newton–Raphson method have focused on this problem.

For instance, when the rectangular coordinate is adopted and the injected current (see (2.4)) is used to form the load flow equations [12], the off-diagonal elements of the Jacobian matrix become constant. This property can certainly be used to improve the solution efficiency. Semlyen and de Leon [13] suggest that the Jacobian matrix elements can be updated partially to alleviate the computing burden.

Both the above two forms of coordinate system are widely used in Newton–Raphson load flow algorithms. When the polar form is used, *PV* nodes can be conveniently treated. When the rectangular form is used, the calculation of trigonometric functions is avoided. Generally speaking, the difference is not very significant. A comparison between the two coordinate systems is carried out in [14].

The fast decoupled method is derived from the Newton–Raphson method in polar form. It will be discussed in Sect. 2.4. In the next section, we mainly introduce the Newton–Raphson method based on the correction equation of (2.54) in rectangular form.

2.3.3 Solution Process of Newton Method

In the Newton–Raphson method, the electric network is described by its admittance matrix. From (2.52), (2.53), (2.55), and (2.56) we know that all operations are relative to the admittance matrix. Therefore, forming the admittance matrix is the first step in the algorithm.

The solving process of the Newton method roughly consists of the following steps.

1. Specify the initial guess values of node voltage, $e^{(0)}, f^{(0)}$;
2. Substituting $e^{(0)}, f^{(0)}$ into (2.52) and (2.53), obtain the left-hand term of the correction equation, $\Delta P^{(0)}$, $\Delta Q^{(0)}$, and $(\Delta V^2)^{(0)}$;
3. Substituting $e^{(0)}, f^{(0)}$ into (2.55) and (2.56), obtain the coefficient matrix (Jacobian matrix) of the correction equation;
4. Solving (2.54), obtain the correction variables, $\Delta e^{(0)}$ and $\Delta f^{(0)}$;
5. Modify voltages;

$$\left.\begin{aligned} e^{(1)} &= e^{(0)} - \Delta e^{(0)} \\ f^{(1)} &= f^{(0)} - \Delta f^{(0)} \end{aligned}\right\} \tag{2.57}$$

6. Substituting $e^{(1)}$ and $f^{(1)}$ into (2.52) and (2.53), obtain $\Delta P^{(1)}$, $\Delta Q^{(1)}$, and $(\Delta V^2)^{(1)}$;
7. Check whether the iteration has converged. When it has converged, calculate branch load flow and output the results; otherwise take $e^{(1)}$ and $f^{(1)}$ as the new guess value, return to step (3) and start the next iteration.

The main flowchart of the Newton–Raphson method is shown in Fig. 2.3. The above steps introduce the main principles of the solution process. There are still many details to be clarified. As mentioned above, the solution procedure of the

Newton–Raphson method is essentially the process of iteratively forming and solving the correction equations. Dealing with the correction equation has a crucial influence over the memory requirement and computing burden. This problem will be presented in the next section. First, we discuss some other important issues.

The convergence characteristic of the Newton–Raphson method is excellent. Generally, it can converge in 6–7 iterations, and the number of iteration does not depend on the scale of the power system. Theoretically speaking, the Newton–Raphson method has a quadratic convergence characteristic if the initial guess values are close to the solution. If the initial guess values are not good enough, the iterative process may not converge or may converge to a solution at which the power system cannot operate. This property stems from the Newton method itself. As described above, the substance of the Newton method is sequential linearization of nonlinear equations. It is established on the assumption that Δe and Δf are very small so that their high-order terms can be neglected. Therefore, a good initial guess value is crucial because the Newton method is very sensitive to it.

Under normal operation states of power systems, the node voltage magnitudes are usually close to their nominal voltages, and the phase angle differences between the nodes of a branch are not very large. Therefore, a "flat start" initial guess value, i.e.,

$$e_i^{(0)} = 1.0 \quad f_i^{(0)} = 0.0 \quad (i = 1, 2, \ldots, n) \tag{2.58}$$

can give satisfactory results. In Fig. 2.3, the convergence condition is

$$\left\| \Delta P^{(t)}, \Delta Q^{(t)} \right\| < \varepsilon \tag{2.59}$$

where $\left\| \Delta P^{(t)}, \Delta Q^{(t)} \right\|$ is a norm representing the maximal modulus elements in vectors $\Delta P^{(t)}, \Delta Q^{(t)}$. This convergence criterion is very intuitive, and can be used to directly control the power errors. When the calculation is based on the per unit system, we can set $\varepsilon = 10^{-4}$ or 10^{-3}. If the base value is 100 MVA, the maximum error corresponds to 0.01 MVA or 0.1 MVA.

From Fig. 2.3 we know that in the Newton–Raphson load flow calculation, the Jacobian matrix must be formed and triangularized in each iteration. Hence the computing burden in each iteration is quite heavy. From the expressions of Jacobian elements one can see that in the iteration procedure, especially when it is near convergence, the change of the elements caused by voltage variation is not significant (see Example 2.1). Therefore, to decrease the computing effort, once a Jacobian matrix is formed, it could be used in several successive iterations.

2.3.4 Solution of Correction Equations

The Newton–Raphson method, with Gauss elimination solving the correction equation, has been used in load flow calculation since the 1950s.

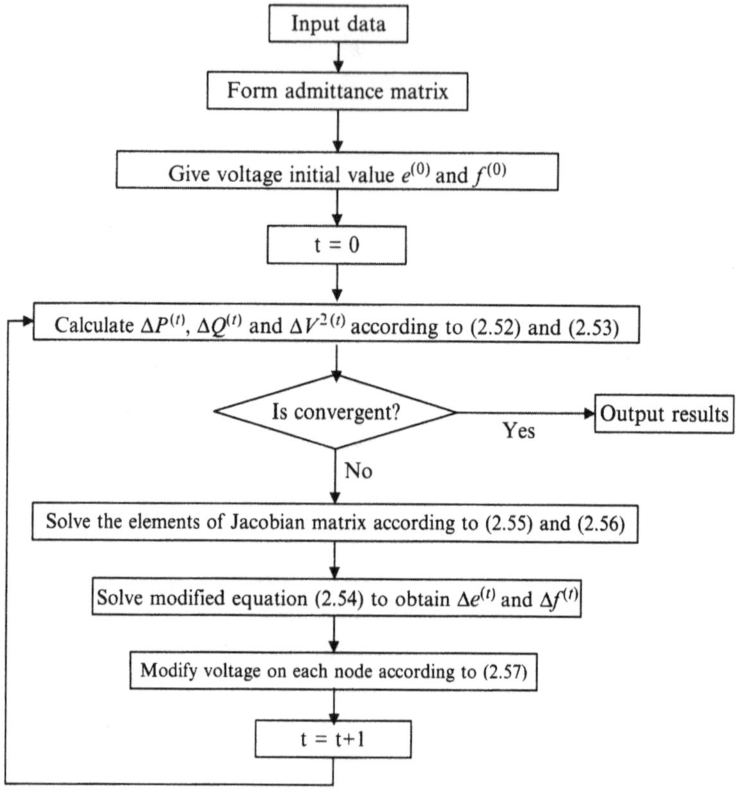

Fig. 2.3 Flowchart of Newton method

In the 1960s, the sparsity of the correction equation was fully investigated and employed in the iteration procedure. In this way, the storage and operation for zero elements in the Jacobian are avoided. When the technology of optimal node ordering is adopted, it can minimize the number of the fill-in nonzero elements in factorizing the Jacobian of the correction equation. This greatly reduces memory and computing requirements to almost proportional to the node number of the power system. Based on this sparsity technology, the Newton–Raphson method has become one of the most popular methods in power system load flow calculation [7].

With a simple system as shown in Fig. 2.4, we now illustrate some algorithmic tricks in solving the correction equation of the Newton–Raphson method. In Fig. 2.4, both node 3 and node 6 are generator nodes. We set node 3 as a *PV* node while node 6 the slack node; other nodes are all *PQ* nodes. The structure of the network admittance matrix is shown in Fig. 2.5.

The correction equation is given as (2.60). It does not include the equation related to node 6, the slack node.

Fig. 2.4 Example of simple system

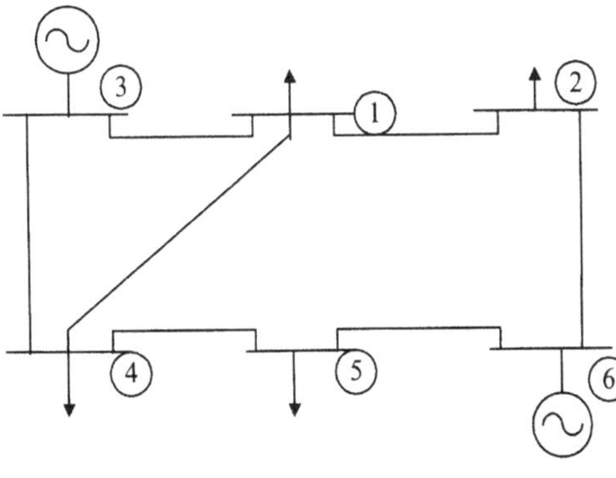

Fig. 2.5 Structure of admittance matrix

$$\begin{bmatrix} Y_{11} & Y_{12} & Y_{13} & Y_{14} & & \\ Y_{21} & Y_{22} & & & & Y_{26} \\ Y_{31} & & Y_{33} & Y_{34} & & \\ Y_{41} & & Y_{43} & Y_{44} & Y_{45} & \\ & & & Y_{54} & Y_{55} & Y_{56} \\ & Y_{62} & & & Y_{65} & Y_{66} \end{bmatrix}$$

$$\begin{bmatrix} \Delta P_1 \\ \Delta Q_1 \\ \Delta P_2 \\ \Delta Q_2 \\ \Delta P_3 \\ \Delta V_3^2 \\ \Delta P_4 \\ \Delta Q_4 \\ \Delta P_5 \\ \Delta Q_5 \end{bmatrix} = \begin{bmatrix} H_{11} & N_{11} & H_{12} & N_{12} & H_{13} & N_{13} & H_{14} & N_{14} & & \\ J_{11} & L_{11} & J_{12} & L_{12} & J_{13} & L_{13} & J_{14} & L_{14} & & \\ H_{21} & N_{21} & H_{22} & N_{22} & & & & & & \\ J_{21} & L_{21} & J_{22} & L_{22} & & & & & & \\ H_{31} & N_{31} & & & H_{33} & N_{33} & H_{34} & N_{34} & & \\ 0 & 0 & & & R_{33} & S_{33} & 0 & 0 & & \\ H_{41} & N_{41} & & & H_{43} & N_{43} & H_{44} & N_{44} & H_{45} & N_{45} \\ J_{41} & L_{41} & & & J_{43} & L_{43} & J_{44} & L_{44} & J_{45} & L_{45} \\ & & & & & & H_{54} & N_{54} & H_{55} & N_{55} \\ & & & & & & J_{54} & L_{54} & J_{55} & L_{55} \end{bmatrix} \begin{bmatrix} \Delta e_1 \\ \Delta f_1 \\ \Delta e_2 \\ \Delta f_2 \\ \Delta e_3 \\ \Delta f_3 \\ \Delta e_4 \\ \Delta f_4 \\ \Delta e_5 \\ \Delta f_5 \end{bmatrix} \tag{2.60}$$

where the constant terms ΔP_i, ΔQ_i can be obtained by (2.52),

$$\left.\begin{aligned} \Delta P_i &= P_{is} - e_i \sum_{j \in i} (G_{ij}e_j - B_{ij}f_j) - f_i \sum_{j \in i} (G_{ij}f_j + B_{ij}e_j) \\ \Delta Q_i &= Q_{is} - f_i \sum_{j \in i} (G_{ij}e_j - B_{ij}f_j) + e_i \sum_{j \in i} (G_{ij}f_j + B_{ij}e_j) \end{aligned}\right\}$$

or can be written as

$$\left.\begin{aligned} \Delta P_i &= P_{is} - (e_i a_i + f_i b_i) \\ \Delta Q_i &= Q_{is} - (f_i a_i - e_i b_i) \end{aligned}\right\} \tag{2.61}$$

From (2.56) we know the diagonal elements of the Jacobian are

$$\left.\begin{aligned} H_{ii} &= \frac{\partial \Delta P_i}{\partial e_i} = -a_i - (G_{ii}e_i + B_{ii}f_i) \\ N_{ii} &= \frac{\partial \Delta P_i}{\partial f_i} = -b_i + (B_{ii}e_i - G_{ii}f_i) \\ J_{ii} &= \frac{\partial \Delta Q_i}{\partial e_i} = b_i + (B_{ii}e_i - G_{ii}f_i) \\ L_{ii} &= \frac{\partial \Delta Q_i}{\partial f_i} = -a_i + (G_{ii}e_i + B_{ii}f_i) \end{aligned}\right\} \tag{2.62}$$

Both (2.61) and (2.62) include components of the injected current at node i, a_i and b_i. To calculate ΔP_i, ΔQ_i, and the diagonal elements of Jacobian H_{ii}, N_{ii}, J_{ii}, L_{ii}, we must first compute a_i and b_i. From (2.11) we can see, the injected current components a_i and b_i at node i only depends on the i th row elements of the admittance matrix and voltage components of corresponding nodes. Therefore, a_i and b_i can be accumulated by sequentially taking the two terms and performing multiplication plus operation.

After a_i, b_i are known, ΔP_i and ΔQ_i can be easily obtained according to (2.61).

The nondiagonal elements of the Jacobian in (2.60) can be expressed by:

$$\left.\begin{aligned} H_{ij} &= \frac{\partial \Delta P_i}{\partial e_j} = -(G_{ij}e_i + B_{ij}f_i) \\ N_{ij} &= \frac{\partial \Delta P_i}{\partial f_j} = B_{ij}e_i - G_{ij}f_i \\ J_{ij} &= \frac{\partial \Delta Q_i}{\partial e_j} = B_{ij}e_i - G_{ij}f_i = N_{ij} \\ L_{ij} &= \frac{\partial \Delta Q_i}{\partial f_j} = G_{ij}e_i + B_{ij}f_i = -H_{ij} \end{aligned}\right\} \tag{2.63}$$

Obviously, the off-diagonal elements are only related to the corresponding admittance elements and voltage components. From (2.62), the ith diagonal element consists of, besides the injecting current components at node $i(a_i$ and $b_i)$, only the arithmetic operation results of the diagonal elements of admittance matrix $G_{ii} + jB_{ii}$ and voltage components $e_i + jf_i$.

In brief, the whole correction equation can be formed by sequentially taking and arithmetically operating the elements of the admittance matrix and corresponding voltage components.

If node i is PV node, the equation of ΔQ_i should be replaced by the equation of ΔV_i^2. The constant term ΔV_i^2 on the left hand and elements R_{ii} and S_{ii} of the Jacobian can be easily obtained from (2.53) and (2.56),

$$\left.\begin{aligned} R_{ii} &= \frac{\partial \Delta V_i^2}{\partial e_i} = -2e_i \\ S_{ii} &= \frac{\partial \Delta V_i^2}{\partial f_i} = -2f_i \end{aligned}\right\} \tag{2.64}$$

Forming the correction equation is a very important step in the Newton–Raphson method which remarkably affects the efficiency of the whole algorithm. Therefore, we should investigate the above equations carefully in coding the program.

When Gauss elimination is used to solve the correction equation, we usually eliminate the correction equation row by row. The augmented matrix corresponding to (2.60) is

$$\begin{bmatrix}
H_{11} & N_{11} & H_{12} & N_{12} & H_{13} & N_{13} & H_{14} & N_{14} & & & \Delta P_1 \\
J_{11} & L_{11} & J_{12} & L_{12} & J_{13} & L_{13} & J_{14} & L_{14} & & & \Delta Q_1 \\
H_{21} & N_{21} & H_{22} & N_{22} & & & & & & & \Delta P_2 \\
J_{21} & L_{21} & J_{22} & L_{22} & & & & & & & \Delta Q_2 \\
H_{31} & N_{31} & & & H_{33} & N_{33} & H_{34} & N_{34} & & & \Delta P_3 \\
0 & 0 & & & R_{33} & S_{33} & 0 & 0 & & & \Delta V_3^2 \\
H_{41} & N_{41} & & & H_{43} & N_{43} & H_{44} & N_{44} & H_{45} & N_{45} & \Delta P_4 \\
J_{41} & L_{41} & & & J_{43} & L_{43} & J_{44} & L_{44} & J_{45} & L_{45} & \Delta Q_4 \\
 & & & & & & H_{54} & N_{54} & H_{55} & N_{55} & \Delta P_5 \\
 & & & & & & J_{54} & L_{54} & J_{55} & L_{55} & \Delta Q_5
\end{bmatrix}$$

After the equations related to node 1 and 2 are eliminated, the augmented matrix is converted as shown in Fig. 2.6. This figure tell us when the equations related to node 2 are eliminated (row 3 and row 4), all operations are independent of equations related to node 3, 4, ..., N. Therefore, in the eliminating procedure, we can eliminate the rows related to a node immediately after forming them.

In Fig. 2.6, elements such as $H''_{23}, N''_{23}, \ldots, L''_{24}$, etc. are fill-in nonzero elements created in the elimination process. To decrease the number of injected elements, we should optimize the node number ordering before load flow calculation (see Section 1.3.5). The element with superscript ($''$) represents that it has been manipulated. We need not save memory for the fill-in element in advance using this elimination procedure and thus the algorithm is simplified.

When the whole elimination procedure finished, the augmented matrix of correction equation becomes,

$$\left[\begin{array}{cccccccccc:c}
1 & N'_{11} & H'_{12} & N'_{12} & H'_{13} & N'_{13} & H'_{14} & N'_{14} & & & \Delta P'_1 \\
 & 1 & J'_{12} & L'_{12} & J'_{13} & L'_{13} & J'_{14} & L'_{14} & & & \Delta Q'_1 \\
 & & 1 & N'_{22} & H''_{23} & N''_{23} & H''_{24} & N''_{24} & & & \Delta P'_2 \\
 & & & 1 & J''_{23} & L''_{23} & J''_{24} & L''_{24} & & & \Delta Q'_2 \\
\hdashline
H_{31} & N_{31} & & & H_{33} & N_{33} & H_{34} & N_{34} & & & \Delta P_3 \\
 & & & & R_{33} & S_{33} & & & & & \Delta V_3^2 \\
H_{41} & N_{41} & & & H_{43} & N_{43} & H_{44} & N_{44} & H_{45} & N_{45} & \Delta P_4 \\
J_{41} & L_{41} & & & J_{43} & L_{43} & J_{44} & L_{44} & J_{45} & L_{45} & \Delta Q_4 \\
 & & & & & & H_{54} & N_{54} & H_{55} & N_{55} & \Delta P_5 \\
 & & & & & & J_{54} & L_{54} & J_{55} & L_{55} & \Delta Q_5
\end{array}\right]$$

Fig. 2.6 Diagram of eliminating row by row

$$
\begin{bmatrix}
1 & N'_{11} & H'_{12} & N'_{12} & H'_{13} & N'_{13} & H'_{14} & N'_{14} & & & \Delta P'_1 \\
 & 1 & J'_{12} & L'_{12} & J'_{13} & L'_{13} & J'_{14} & L'_{14} & & & \Delta Q'_1 \\
 & & 1 & N'_{22} & H''_{23} & N''_{23} & H''_{24} & N''_{24} & & & \Delta P'_2 \\
 & & & 1 & J''_{23} & L''_{23} & J''_{24} & L''_{24} & & & \Delta Q'_2 \\
 & & & & 1 & N'_{33} & H'_{34} & N'_{34} & & & \Delta P'_3 \\
 & & & & & 1 & J''_{34} & L''_{34} & & & \Delta V'^2_3 \\
 & & & & & & 1 & N'_{44} & H'_{45} & N'_{45} & \Delta P'_4 \\
 & & & & & & & 1 & J'_{45} & L'_{45} & \Delta Q'_4 \\
 & & & & & & & & 1 & N'_{55} & \Delta P'_5 \\
 & & & & & & & & & 1 & \Delta Q'_5
\end{bmatrix}
$$

Finally, using a normal backward substitution, one can get $\Delta e_1, \Delta f_1, \ldots, \Delta e_5, \Delta f_5$ from $\Delta P'_1, \Delta Q'_1, \ldots, \Delta Q'_5$.

Following to the above discussion, we can summarize the algorithm via flowchart shown in Fig. 2.7, where R represents the slack node. The correction equation can be solved by the common Gauss elimination method. The above procedure adopts the strategy of eliminating the rows related to a node immediately after forming them. At the same time, the corresponding constant terms of the correction equation are also accumulated and eliminated. Thus the operation count per iteration is significantly reduced.

[Example 2.1] Calculate the load flow of the power system shown in Fig. 2.8.

[Solution] The load flow is calculated according to the procedures of the flowchart. The first step includes forming the admittance matrix and specifying the initial voltage values.

From Example 1.1 we know the admittance matrix of this system is

$$
\mathbf{Y} = \begin{bmatrix}
1.37874 & -0.62402 & -0.75471 & & \\
-j6.29166 & +j3.90015 & +j2.64150 & & \\
-0.62402 & 1.45390 & -0.82987 & 0.00000 & \\
+j3.90015 & -j66.98082 & +j3.11203 & +j63.49206 & \\
-0.75471 & -0.82987 & 1.58459 & & 0.00000 \\
+j2.64150 & +j3.11203 & -j35.73786 & & +j31.74603 \\
 & 0.00000 & & 0.00000 & \\
 & +j63.49206 & & -j66.66667 & \\
 & & 0.00000 & & 0.00000 \\
 & & +j31.74603 & & -j33.33333
\end{bmatrix}
$$

The initial values of node voltages are given in Table 2.1.

According to (2.52) and (2.53), we can establish the expression of the constant terms (mismatch terms) of the correction equations as

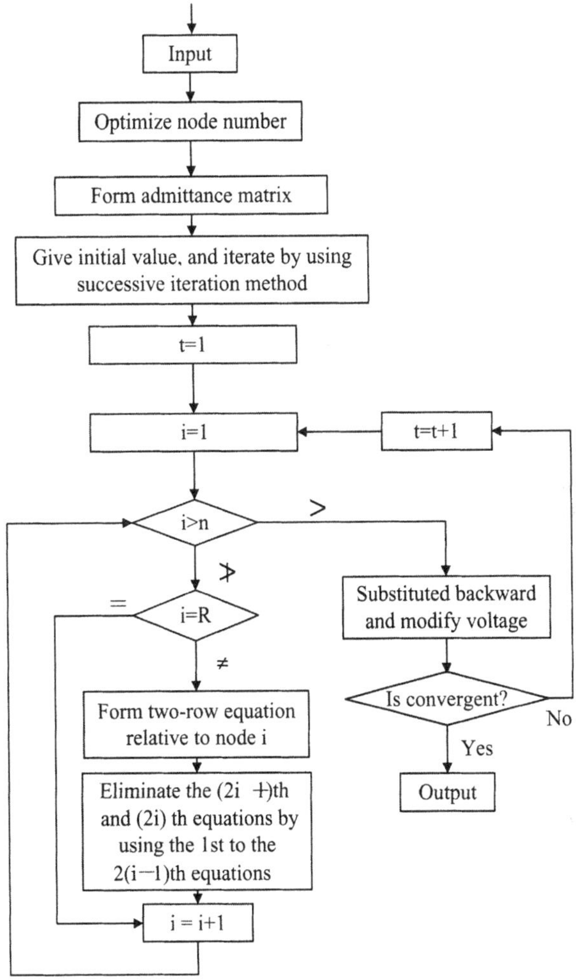

Fig. 2.7 Flowchart of Newton Method

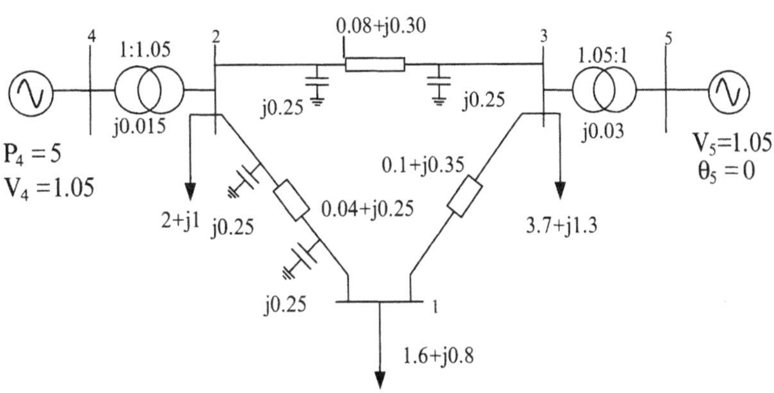

Fig. 2.8 Simple power system

Table 2.1 Voltage initial values

Node	1	2	3	4	5
$e^{(0)}$	1.00000	1.00000	1.00000	1.05000	1.05000
$f^{(0)}$	0.00000	0.00000	0.00000	0.00000	0.00000

$$\Delta P_1 = P_{1s} - e_1[(G_{11}e_1 - B_{11}f_1) + (G_{12}e_2 - B_{12}f_2) + (G_{13}e_3 - B_{13}f_3)] \\ -f_1[(G_{11}f_1 + B_{11}e_1) + (G_{12}f_2 + B_{12}e_2) + (G_{13}f_3 + B_{13}e_3)]$$

$$\Delta Q_1 = Q_{1s} - f_1[(G_{11}e_1 - B_{11}f_1) + (G_{12}e_2 - B_{12}f_2) + (G_{13}e_3 - B_{13}f_3)] + \\ e_1[(G_{11}f_1 + B_{11}e_1) + (G_{12}f_2 + B_{12}e_2) + (G_{13}f_3 + B_{13}e_3)]$$

$$\Delta P_4 = P_{4s} - e_4[(G_{42}e_2 - B_{42}f_2) + (G_{44}e_4 - B_{44}f_4)] - f_4[(G_{42}f_2 + B_{42}e_2) + \\ (G_{44}f_4 + B_{44}e_4)]$$

$$\Delta V_4^2 = V_{4s}^2 - (e_4^2 + f_4^2)$$

Using (2.55) and (2.56), we can obtain the expressions of Jacobian matrix elements:

$$\frac{\partial \Delta P_1}{\partial e_1} = -[(G_{11}e_1 - B_{11}f_1) + (G_{12}e_2 - B_{12}f_2) + (G_{13}e_3 - B_{13}f_3)] - G_{11}e_1 - B_{11}f_1$$

$$\frac{\partial \Delta P_1}{\partial f_1} = -[(G_{11}f_1 + B_{11}e_1) + (G_{12}f_2 + B_{12}e_2) + (G_{13}f_3 + B_{13}e_3)] + B_{11}e_1 - G_{11}f_1$$

$$\frac{\partial \Delta P_1}{\partial e_2} = -(G_{12}e_1 + B_{12}f_1), \quad \frac{\partial \Delta P_1}{\partial f_2} = B_{12}e_1 - G_{12}f_1$$

$$\frac{\partial \Delta P_1}{\partial e_3} = -(G_{13}e_1 + B_{13}f_1), \quad \frac{\partial \Delta P_1}{\partial f_3} = B_{13}e_1 - G_{13}f_1$$

$$\frac{\partial \Delta Q_1}{\partial e_1} = [(G_{11}f_1 + B_{11}e_1) + (G_{12}f_2 + B_{12}e_2) + (G_{13}f_3 + B_{13}e_3)] + B_{11}e_1 - G_{11}f_1$$

$$\frac{\partial \Delta Q_1}{\partial f_1} = -[(G_{11}e_1 - B_{11}f_1) + (G_{12}e_2 - B_{12}f_2) + (G_{13}e_3 - B_{13}f_3)] + G_{11}e_1 + B_{11}f_1$$

$$\frac{\partial \Delta Q_1}{\partial e_2} = \frac{\partial \Delta P_1}{\partial f_2}, \quad \frac{\partial \Delta Q_1}{\partial f_2} = -\frac{\partial \Delta P_1}{\partial e_2}$$

$$\frac{\partial \Delta Q_1}{\partial e_3} = \frac{\partial \Delta P_1}{\partial f_3}, \quad \frac{\partial \Delta Q_1}{\partial f_3} = -\frac{\partial \Delta P_1}{\partial e_3}$$

$$\frac{\partial \Delta P_4}{\partial e_4} = -[(G_{42}e_2 - B_{42}f_2) + (G_{44}e_4 - B_{44}f_4)] - G_{44}e_4 - B_{44}f_4$$

$$\frac{\partial \Delta P_4}{\partial f_4} = -[(G_{42}f_2 + B_{42}e_2) + (G_{44}f_4 + B_{44}e_4)] + B_{44}e_4 - G_{44}f_4$$

$$\frac{\partial \Delta V_4^2}{\partial e_4} = -2e_4$$

$$\frac{\partial \Delta V_4^2}{\partial f_4} = -2f_4$$

Thus according to (2.60), the correction equation of the first iteration can be written as

$$\begin{bmatrix} -1.37874 & -6.54166 & 0.62402 & 3.90015 & 0.75471 & 2.64150 & & \\ \mathit{-6.04166} & \mathit{1.37874} & 3.90015 & -0.62402 & 2.64150 & -0.75471 & & \\ 0.62402 & 3.90015 & -1.45390 & \mathit{-73.67881} & 0.82897 & 3.11203 & 0.00000 & \mathit{63.49206} \\ 3.90015 & 0.62402 & -60.28283 & 1.45390 & 3.11203 & -0.82897 & 63.49206 & 0.00000 \\ -0.75471 & 2.64150 & 0.82897 & 3.11203 & -1.58459 & \mathit{-39.98688} & & \\ 2.64150 & -0.75471 & 3.11203 & -0.82897 & \mathit{-32.38884} & 1.58459 & & \\ & & 0.00000 & 66.66666 & & & 0.00000 & -63.49206 \\ & & 0.00000 & 0.00000 & & & \mathit{-2.10000} & 0.00000 \end{bmatrix} \begin{bmatrix} \Delta e_1 \\ \Delta f_1 \\ \Delta e_2 \\ \Delta f_2 \\ \Delta e_3 \\ \Delta f_3 \\ \Delta e_4 \\ \Delta f_4 \end{bmatrix}$$

$$= \begin{bmatrix} -1.60000 \\ -0.55000 \\ -2.00000 \\ 5.69799 \\ -3.70000 \\ 2.04901 \\ 5.00000 \\ 0.00000 \end{bmatrix}$$

the above equation, the elements in italic have maximal absolute value in each row of the Jacobian matrix. Obviously, if elements are arranged this way, the maximal elements do not appear at the diagonal positions.

It should be noted that this situation is not accidental. From the above equation we can conclude that the maximal element of each row is $\frac{\partial \Delta P_i}{\partial f_i}$ or $\frac{\partial \Delta Q_i}{\partial e_i}$. This is because the active power is mainly related to the vertical component of voltage while the reactive power is mainly related to the horizontal component of voltage in high voltage power systems.

To reduce the rounding error of the calculations, the maximal elements should be located in diagonal positions. There are two methods to satisfy this requirement: the first is to exchange positions of the equations relative to ΔQ and ΔP, i.e., to exchange odd numbered rows with even numbered rows; the second method is to exchange the variables Δe and Δf, i.e., to exchange odd numbered columns with even numbered columns of the Jacobian matrix.

We now introduce the first approach. Thus the above equation will be rearranged as,

$$\begin{bmatrix} \boldsymbol{-6.04166} & 1.37874 & 3.90015 & -0.62402 & 2.64150 & -0.75471 & & \\ -1.37874 & \boldsymbol{-6.54166} & 0.62402 & 3.90015 & 0.75471 & 2.64150 & & \\ 3.90015 & 0.62402 & \boldsymbol{-60.28283} & 1.45390 & 3.11203 & -0.82897 & 63.49206 & 0.00000 \\ 0.62402 & 3.90015 & -1.45390 & \boldsymbol{-73.67881} & 0.82897 & 3.11203 & 0.00000 & 63.49206 \\ 2.64150 & -0.75471 & 3.11203 & -0.82897 & \boldsymbol{-32.38884} & 1.58459 & & \\ -0.75471 & 2.64150 & 0.82897 & 3.11203 & -1.58459 & \boldsymbol{-39.98688} & & \\ & & & & & & \boldsymbol{-2.10000} & 0.00000 \\ & & 0.00000 & 66.66666 & & & 0.00000 & -63.49206 \end{bmatrix} \begin{bmatrix} \Delta e_1 \\ \Delta f_1 \\ \Delta e_2 \\ \Delta f_2 \\ \Delta e_3 \\ \Delta f_3 \\ \Delta e_4 \\ \Delta f_4 \end{bmatrix} = \begin{bmatrix} -0.55000 \\ -1.60000 \\ 5.69799 \\ -2.00000 \\ 2.04901 \\ -3.70000 \\ 0.00000 \\ 5.00000 \end{bmatrix}$$

We can see the maximal element of each row appears in the diagonal position except for row 8.

As described in Section 2.3.4, the iteration procedure adopts the strategy of immediately eliminating the rows related to a node after forming them (see Fig. 2.7). The equations related to node 1 are formed as

$$\left[\begin{array}{cccccccc:c} -6.04166 & 1.37874 & 3.90015 & -0.62402 & 2.64150 & -0.75471 & 0 & 0 & -0.55000 \\ -1.37874 & -6.54166 & 0.62402 & 3.90015 & 0.75471 & 2.64150 & 0 & 0 & -1.60000 \end{array}\right]$$

After the elimination operation is executed, the first and second row of the upper triangular matrix can be obtained:

$$\left[\begin{array}{cccccccc:c} 1.00000 & -0.22820 & -0.64554 & 0.10328 & -0.43721 & 0.12491 & 0 & 0 & 0.09103 \\ & 1.00000 & 0.03879 & -0.58961 & -0.02215 & -0.41038 & 0 & 0 & 0.21505 \end{array}\right]$$

Then we establish the equations related to node 2, the corresponding augmented matrix is

$$\left[\begin{array}{cccccccc:c} 3.90015 & -0.62402 & -60.28283 & 1.45390 & 3.11203 & -0.82987 & 63.49206 & 0.0 & 5.69799 \\ 0.62402 & 3.90015 & -1.45390 & -73.67881 & 0.82987 & 3.11203 & 0.0 & 63.49206 & -2.0 \end{array}\right]$$

Executing the elimination operation, the third and forth rows of the upper triangular matrix become:

$$\begin{bmatrix} 1.00000 & -0.02090 & -0.08348 & 0.02090 & -1.09894 & 0.00000 & \vdots & -0.09184 \\ & 1.00000 & -0.01528 & -0.06609 & 0.01859 & -0.88943 & \vdots & 0.04253 \end{bmatrix}$$

Continuing this procedure until the eliminating operation procedure is finished, we have the upper triangular matrix:

$$\begin{bmatrix} 1.00000 & -0.22820 & -0.64554 & 0.10328 & -0.43721 & 0.12491 & & & \vdots & 0.09103 \\ & 1.00000 & 0.03879 & -0.58961 & -0.02215 & -0.41038 & & & \vdots & 0.21505 \\ & & 1.00000 & -0.02090 & -0.08348 & 0.02090 & -1.09894 & 0.00000 & \vdots & -0.09148 \\ & & & 1.00000 & -0.01528 & -0.06609 & 0.01850 & -0.88943 & \vdots & 0.04253 \\ & & & & 1.00000 & -0.03303 & -0.17246 & 0.03146 & \vdots & -0.07548 \\ & & & & & 1.00000 & -0.02816 & -0.11194 & \vdots & 0.12021 \\ & & & & & & 1.00000 & 0.00000 & \vdots & 0.00000 \\ & & & & & & & 1.00000 & \vdots & -0.45748 \end{bmatrix}$$

After the backward substitution operation, the correcting increments of node voltages can be obtained,

$$\begin{bmatrix} \Delta e_1 \\ \Delta f_1 \\ \Delta e_2 \\ \Delta f_2 \\ \Delta e_3 \\ \Delta f_3 \\ \Delta e_4 \\ \Delta f_4 \end{bmatrix} = \begin{bmatrix} 0.03356 \\ 0.03348 \\ -0.10538 \\ -0.36070 \\ -0.05881 \\ 0.06900 \\ 0.00000 \\ -0.45748 \end{bmatrix}$$

Modifying the node voltage, the voltage vector becomes:

$$\begin{bmatrix} e_1 \\ f_1 \\ e_2 \\ f_2 \\ e_3 \\ f_3 \\ e_4 \\ f_4 \end{bmatrix} = \begin{bmatrix} 0.96643 \\ -0.33481 \\ 1.10533 \\ 0.36070 \\ 1.05881 \\ -0.66900 \\ 1.05000 \\ 0.45748 \end{bmatrix}$$

Using this voltage vector as the initial voltage value, we can repeat above operations. If the tolerance is set to $\varepsilon = 10^{-6}$, the calculation converges after five iterations. The evolution process of node voltages and power mismatches is shown in Tables 2.2 and 2.3.

Table 2.2 Node voltages in iterative process

Iterating No.	e_1	f_1	e_2	f_2	e_3	f_3	e_4	f_4
1	0.96643	−0.33481	1.10538	0.36074	1.05881	−0.06900	1.05000	0.45748
2	0.87365	−0.07006	1.03350	0.32886	1.03564	−0.07694	0.97694	0.38919
3	0.85947	−0.07176	1.02608	0.33047	1.03355	−0.07737	0.97464	0.39061
4	0.85915	−0.07182	1.02600	0.33047	1.03351	−0.07738	0.97461	0.39067
5	0.85915	−0.07182	1.02600	0.33047	1.03351	−0.07738	0.97461	0.39067

Table 2.3 Node power mismatches in iterative process

Iterating No.	ΔQ_1	ΔP_1	ΔQ_2	ΔP_2	ΔQ_3	ΔP_3	ΔP_4
1	−0.55000	−1.60000	5.69799#	−2.00000	2.04901	−3.70000	5.00000
2	−0.07263	−0.03473	−6.00881#	2.10426	−0.37144	0.04904	−2.39001
3	−0.02569	−0.06011	−0.41159#	0.15764	−0.00924	0.00329	−0.16193
4	−0.00078	−0.00032	−0.0030#	−0.00054	−0.00002	0.00000	0.00069
5	0.00000	0.00000	0.00000	0.00000	0.00000	0.00000	0.00000

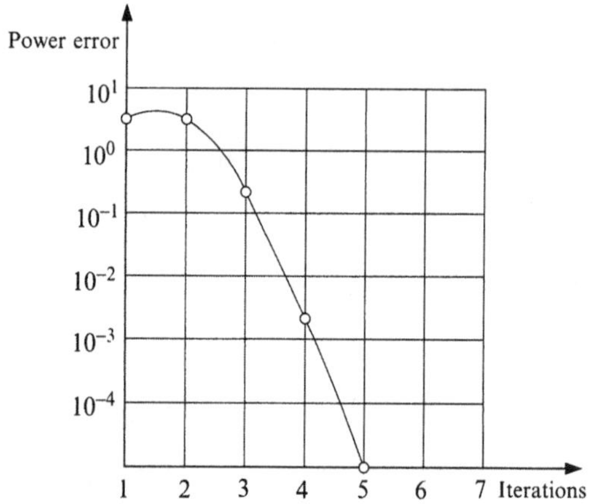

Fig. 2.9 Convergence property of Newton–Raphson method

To reveal the convergence property, the maximal power mismatches (with # in Table 2.3) in the iterative process are shown in Fig. 2.9.

In the iteration process, especially when it approaches convergence, the changes of the diagonal elements in the Jacobian are not very significant. To illustrate this point, the changes of the diagonal elements are given in Table 2.4.

The calculation results of node voltages are shown in Table 2.5.

Table 2.4 Diagonal elements of Jacobian matrix in iterative process

Iterating no.	$\frac{\partial \Delta Q_1}{\partial e_1}$	$\frac{\partial \Delta P_1}{\partial f_1}$	$\frac{\partial \Delta Q_2}{\partial e_2}$	$\frac{\partial \Delta P_2}{\partial f_2}$	$\frac{\partial \Delta Q_3}{\partial e_3}$	$\frac{\partial \Delta P_3}{\partial f_3}$	$\frac{\partial \Delta V_4^2}{\partial e_4}$	$\frac{\partial \Delta P_4}{\partial f_4}$
1	6.04166	6.54166	60.28283	73.67881	32.38884	39.08688	1.05000	63.49206
2	5.22590	6.84268	79.81886	69.30868	36.62734	38.83341	0.96259	70.18293
3	4.37415	6.42613	69.78933	69.61682	35.38612	38.39351	0.97528	65.61929
4	4.23077	6.38634	68.89682	69.52026	35.29706	38.33158	0.97463	65.14834
5	4.22720	6.38577	68.88900	69.51747	35.29572	38.33048	0.97461	65.14332

Table 2.5 Node voltage vectors

Node	Magnitude	Angle (°)
1	0.86215	−4.77851
2	1.07791	17.85353
3	1.03641	−4.28193
4	1.05000	21.84332
5	1.05000	0.00000

2.4 Fast Decoupled Method

2.4.1 Introduction to Fast Decoupled Method

The basic idea of the fast decoupled method is expressing the nodal power as a function of voltages in polar form; separately solving the active and reactive power equations [9] by using active power mismatch to modify voltage angle and using reactive power mismatch to modify voltage magnitude. In this way, the computing burden of load flow calculation is alleviated significantly. In the following, the derivation of the fast decoupled method from the Newton method is discussed.

As described previously, the core of the Newton load flow approach is to solve the correction equation. When the nodal power equation is expressed in polar form, the correction equation is (see (2.50)),

$$\begin{bmatrix} \Delta \mathbf{P} \\ \Delta \mathbf{Q} \end{bmatrix} = \begin{bmatrix} \mathbf{H} & \mathbf{N} \\ \mathbf{J} & \mathbf{L} \end{bmatrix} \begin{bmatrix} \Delta \boldsymbol{\theta} \\ \Delta \mathbf{V}/\mathbf{V} \end{bmatrix} \tag{2.65}$$

or can be written as,

$$\begin{aligned} \Delta \mathbf{P} &= \mathbf{H}\Delta\boldsymbol{\theta} + \mathbf{N}\Delta\mathbf{V}/\mathbf{V} \\ \Delta \mathbf{Q} &= \mathbf{J}\Delta\boldsymbol{\theta} + \mathbf{L}\Delta\mathbf{V}/\mathbf{V} \end{aligned} \tag{2.66}$$

This equation is derived strictly from the mathematical viewpoint. It does not take the characteristics of power systems into consideration.

We know that in high voltage power system the active power flow is mainly related to the angle of the nodal voltage phasor while reactive power flow is mainly related to its magnitude. The experiences of many load flow calculations tell us that the element values of matrix **N** and **J** in (2.66) are usually relatively small. Therefore, the first step to simplify the Newton method is to neglect **N** and **J**, and (2.66) is simplified to

$$\left.\begin{aligned}\Delta\mathbf{P} &= \mathbf{H}\Delta\boldsymbol{\theta}\\ \Delta\mathbf{Q} &= \mathbf{L}\Delta\mathbf{V}/\mathbf{V}\end{aligned}\right\} \tag{2.67}$$

Thus a simultaneous linear equation of dimension $2n$ is simplified to two simultaneous linear equations of dimension n.

The second important step to simplify the Newton method is to approximate the coefficient matrices of (2.67) as constant and symmetric matrices.

As the phase angle difference across a transmission line usually is not very large (does not exceed 10°~20°), so the following relations hold,

$$\left.\begin{aligned}\cos\theta_{ij} &\approx 1\\ G_{ij}\sin\theta_{ij} &\ll B_{ij}\end{aligned}\right\} \tag{2.68}$$

Furthermore, the admittance B_{Li} corresponding to the node reactive power is certainly far smaller than the imaginary part of the node self-admittance, i.e.,

$$B_{Li} = \frac{Q_i}{V_i^2} \ll B_{ii}$$

Accordingly,

$$Q_i \ll V_i^2 B_{ii} \tag{2.69}$$

Based on the above relationships, the element expressions of coefficient matrix in (2.67) can be represented as (see (2.41), (2.42), (2.48), and (2.49)):

$$\left.\begin{aligned}H_{ii} &= V_i^2 B_{ii}\\ H_{ij} &= V_i V_j B_{ij}\\ L_{ii} &= V_i^2 B_{ii}\\ L_{ij} &= V_i V_j B_{ij}\end{aligned}\right\} \tag{2.70}$$

Therefore, the coefficient matrix in (2.67) can be written as

$$\mathbf{H} = \mathbf{L} = \begin{bmatrix} V_1^2 B_{11} & V_1 V_2 B_{12} & \dots & V_1 V_n B_{1n}\\ V_2 V_1 B_{21} & V_2^2 B_{22} & \dots & V_2 V_n B_{2n}\\ & \vdots & & \\ V_n V_1 B_{n1} & V_n V_2 B_{n2} & \dots & V_n^2 B_{nn}\end{bmatrix} \tag{2.71}$$

It can be further represented as the product of the following matrices:

$$\mathbf{H}=\mathbf{L}=\begin{bmatrix} V_1 & & & \\ & V_2 & 0 & \\ & 0 & \ddots & \\ & & & V_n \end{bmatrix}\begin{bmatrix} B_{11} & B_{12} & \dots & B_{1n} \\ B_{21} & B_{22} & \dots & B_{2n} \\ & \vdots & \vdots & \\ B_{n1} & B_{n2} & \dots & B_{nn} \end{bmatrix}\begin{bmatrix} V_1 & & & \\ & V_2 & 0 & \\ & 0 & \ddots & \\ & & & V_n \end{bmatrix} \tag{2.72}$$

Substituting (2.72) into (2.67), we can rewrite the correction equations as follows:

$$\begin{bmatrix} \Delta P_1 \\ \Delta P_2 \\ \vdots \\ \Delta P_n \end{bmatrix}=\begin{bmatrix} V_1 & & & \\ & V_2 & 0 & \\ & 0 & \ddots & \\ & & & V_n \end{bmatrix}\begin{bmatrix} B_{11} & B_{12} & \dots & B_{1n} \\ B_{21} & B_{22} & \dots & B_{2n} \\ & \vdots & \vdots & \\ B_{n1} & B_{n2} & \dots & B_{nn} \end{bmatrix}\begin{bmatrix} V_1\Delta\theta_1 \\ V_2\Delta\theta_2 \\ \vdots \\ V_n\Delta\theta_n \end{bmatrix} \tag{2.73}$$

and

$$\begin{bmatrix} \Delta Q_1 \\ \Delta Q_2 \\ \vdots \\ \Delta Q_n \end{bmatrix}=\begin{bmatrix} V_1 & & & \\ & V_2 & 0 & \\ & 0 & \ddots & \\ & & & V_n \end{bmatrix}\begin{bmatrix} B_{11} & B_{12} & \dots & B_{1n} \\ B_{21} & B_{22} & \dots & B_{2n} \\ & \vdots & \vdots & \\ B_{n1} & B_{n2} & \dots & B_{nn} \end{bmatrix}\begin{bmatrix} \Delta V_1 \\ \Delta V_2 \\ \vdots \\ \Delta V_n \end{bmatrix} \tag{2.74}$$

Multiplying both sides of the above equation with matrix,

$$\begin{bmatrix} V_1 & & & \\ & V_2 & & \\ & & \ddots & \\ & & & V_n \end{bmatrix}^{-1}=\begin{bmatrix} \frac{1}{V_1} & & & \\ & \frac{1}{V_2} & & \\ & & \ddots & \\ & & & \frac{1}{V_n} \end{bmatrix}$$

one can obtain

$$\begin{bmatrix} \Delta P_1/V_1 \\ \Delta P_2/V_2 \\ \vdots \\ \Delta P_n/V_n \end{bmatrix}=\begin{bmatrix} B_{11} & B_{12} & \dots & B_{1n} \\ B_{21} & B_{22} & \dots & B_{2n} \\ & \vdots & \vdots & \\ B_{n1} & B_{n2} & \dots & B_{nn} \end{bmatrix}\begin{bmatrix} V_1\Delta\theta_1 \\ V_2\Delta\theta_2 \\ \vdots \\ V_n\Delta\theta_n \end{bmatrix} \tag{2.75}$$

and

$$\begin{bmatrix} \Delta Q_1/V_1 \\ \Delta Q_2/V_2 \\ \vdots \\ \Delta Q_n/V_n \end{bmatrix}=\begin{bmatrix} B_{11} & B_{12} & \dots & B_{1n} \\ B_{21} & B_{22} & \dots & B_{2n} \\ & \vdots & \vdots & \\ B_{n1} & B_{n2} & \dots & B_{nn} \end{bmatrix}\begin{bmatrix} \Delta V_1 \\ \Delta V_2 \\ \vdots \\ \Delta V_n \end{bmatrix} \tag{2.76}$$

The above two equations are the correction equations of the fast decoupled load flow method. The coefficient matrix is merely the imaginary part of the nodal admittance matrix of the system, and is thus a symmetric, constant matrix. Combining with the power mismatch equation (2.13), we obtain the basic equations of the fast decoupled load flow model

$$\Delta P_i = P_{is} - V_i \sum_{j \in i} V_j (G_{ij} \cos \theta_{ij} + B_{ij} \sin \theta_{ij}) \quad (i = 1, 2, \ldots, n) \tag{2.77}$$

$$\Delta Q_i = Q_{is} - V_i \sum_{j \in i} V_j (G_{ij} \sin \theta_{ij} - B_{ij} \cos \theta_{ij}) \quad (i = 1, 2, \ldots, n) \tag{2.78}$$

The iterative process can be briefly summarized in the following steps:

1. Specify node voltage vector initial value $\theta_i^{(0)}$, $V_i^{(0)}$
2. Calculate the node active power mismatch ΔP_i according to (2.77), and then calculate $\Delta P_i / V_i$
3. Solving correction equation (2.75), calculate the node voltage angle correction $\Delta \theta_i$
4. Modify the node voltage angle θ_i

$$\theta_i^{(t)} = \theta_i^{(t-1)} \Delta \theta_i^{(t-1)} \tag{2.79}$$

5. Calculate node reactive power mismatch ΔQ_i according to (2.78), and then calculate $\Delta Q_i / V_i$
6. Solving correction equation (2.76), calculate the node voltage magnitude correction ΔV_i,
7. Modify the node voltage magnitude V_i;

$$V_i^{(t)} = V_i^{(t-1)} - \Delta V_i^{(t-1)} \tag{2.80}$$

8. Back to step (2) to continue the iterative process, until all node power mismatches ΔP_i and ΔQ_{i} satisfy convergence conditions.

2.4.2 *Correction Equations of Fast Decoupled Method*

The main difference between the fast decoupled method and the Newton method stems from their correction equations. Comparing with correction (2.40) or (2.54) of the Newton method, the two correction equations of the fast decoupled method have the following features:

1. Equations (2.75) and (2.76) are two simultaneous linear equations of dimension n instead of a simultaneous linear equation of dimension $2n$

2. In (2.75) and (2.76), all elements of the coefficient matrix remain constant during the iterative process
3. In (2.75) and (2.76), the coefficient matrix is symmetric.

The benefit of the first feature for computing speed and storage is obvious. The second feature alleviates the computing burden in forming and eliminating the Jacobian within the iterative process. We can first form the factor table for the coefficient matrix of the correction equation (see (2.76)) by triangularization. Then we can carry out elimination and backward substitution operations for different constant terms $\Delta P/V$ and $\Delta Q/V$ through repeatedly using the factor table. In this way, the correction equation can be solved very quickly. The third feature can further improve efficiency in forming and storing the factor table.

All the simplifications adopted by the fast decoupled method only affect the structure of the correction equation. In other words, they only affect the iteration process, but do not affect the final results. The fast decoupled method and the Newton method use the same mathematical model of (2.13), if adopting the same convergence criteria we should expect the same accuracy of results.

It seems that (2.75) and (2.76) derived above have the same coefficient matrix, but in practice the coefficient matrixes of the two correction equations in the fast decoupled algorithms are different. We can simply write them as

$$\mathbf{\Delta P/V} = \mathbf{B'V\Delta\theta} \tag{2.81}$$

$$\mathbf{\Delta Q/V} = \mathbf{B''\Delta V} \tag{2.82}$$

Here $\mathbf{V}$ is a diagonal matrix with the diagonal elements being the node voltage magnitudes.

First, we should point out that the dimensions of $\mathbf{B'}$ and $\mathbf{B''}$ are different. The dimension of $\mathbf{B'}$ is $n-1$ while the dimension of $\mathbf{B''}$ is lower than $n-1$. This is because (2.82) dose not include the equations related to PV nodes. Hence if the system has r PV nodes, then the dimension of $\mathbf{B''}$ should be $n-r-1$.

To improve the convergence, we use different methods to treat $\mathbf{B'}$ and $\mathbf{B''}$, and how we treat $\mathbf{B'}$ and $\mathbf{B''}$ will result in different fast decoupled methods, are not merely the imaginary part of the admittance matrix.

As described above, (2.81) and (2.82) are the correction equations based on a series of simplifications. Equation (2.81) modifies the voltage phase angles according to the active power mismatch; (2.82) modifies the voltage magnitudes according to the reactive power mismatch. To speed up convergence, the factors that have no or less effect on the voltage angle should be removed from $\mathbf{B'}$. Therefore, we use the imaginary part of admittance to form $\mathbf{B'}$ without considering the effects of shunt capacitor and transformer's off-nominal taps. To be specific, the off-diagonal and diagonal elements of $\mathbf{B'}$ can be calculated according to following equations:

$$B'_{ij} = -\frac{x_{ij}}{r_{ij}^2 + x_{ij}^2}, \quad B'_{ii} = \sum_{j\in i} \frac{x_{ij}}{r_{ij}^2 + x_{ij}^2} = \sum_{j\in i} B'_{ij} \tag{2.83}$$

where r_{ij} and x_{ij} is the resistance and reactance of branch ij, respectively.

Theoretically, the factors that have less effect on voltage magnitude should be removed from $\mathbf{B}''$. For example, the effect of line resistance to $\mathbf{B}''$ should be removed. Therefore, the off-diagonal and diagonal elements of $\mathbf{B}''$ can be calculated according to the following equations:

$$B''_{ij} = -\frac{1}{x_{ij}},\ B''_{ii} = \sum_{j\in i} \frac{1}{x_{ij}} - b_{io}\ B''_{ii} = \sum_{j\in i} \frac{1}{x_{ij}} - b_{io} \tag{2.84}$$

where b_{io} is the shunt admittance of the grounding branch of node i.

If $\mathbf{B}'$ and $\mathbf{B}''$ are formed according to (2.83) and (2.84), the fast decoupled method is usually called the *BX* algorithm. Another algorithm opposite to *BX* method is called the *XB* algorithm in which $\mathbf{B}'$ used in the $\Delta P \sim \Delta\theta$ iteration is formed according to (2.84), while $\mathbf{B}''$ used in the $\Delta Q \sim \Delta V$ iteration is formed according to (2.83). Although these two algorithms have different correction equations, their convergence rates are almost the same. Several IEEE standard test systems have been calculated to compare the convergence of these algorithms. Table 2.6 shows the number of iterations needed to converge for these test systems.

Many load flow calculations indicate that *BX* and *XB* methods can converge for most load flow problems for which the Newton method can converge. The authors of [9, 10] explain the implications of the simplifications made in the fast decoupled method. Wong et al. [19] propose a robust fast decoupled algorithm to especially treat the possible convergence problem caused by high r/x networks. Bacher and Tinney [26] adopt the sparse vector technique to improve the efficiency of the fast decoupled method.

From the above discussion we know that the fast decoupled method uses different correction equations to the Newton method, hence the convergence properties are also different. Mathematically speaking, the iteration method based on a fixed coefficient matrix to solve a nonlinear equation belongs to "the constant slope method." Its convergence process has the characteristic of the geometric series. If the iteration procedure is plotted on a logarithmic coordinate, the convergence characteristic is nearly a straight line. In contrast, convergence of the Newton method has a quadratic property and is quite similar to a parabola. Fig. 2.10 shows the typical convergence properties of the two methods.

Figure 2.10 illustrates that the Newton method converges slower at the early stages, but once converged to some degree its convergence speed becomes very fast. The fast decoupled method converges almost at the same speed throughout the iteration procedure. If the specified convergence criterion is smaller than the errors

Table 2.6 Convergence comparison of ***BX*** method and ***XB*** method

Systems	Newton	*BX*	*XB*
IEEE-5 bus	4	10	10
IEEE-30 bus	3	5	5
IEEE-57 bus	3	6	6
IEEE-118 bus	3	6	7

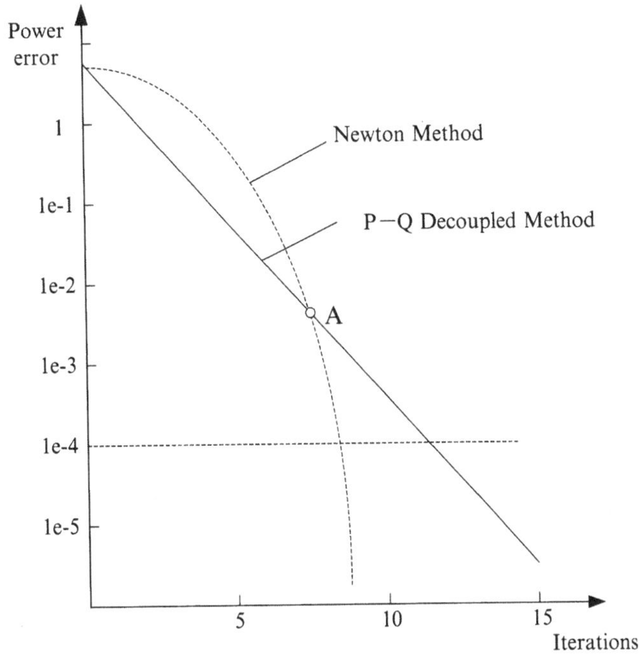

Fig. 2.10 Convergence properties of fast decoupled method and Newton method

at point A in Fig. 2.10, the iteration number of the fast decoupled method is larger than that of the Newton method. It can be roughly considered that a linear relation exists between the iteration number and the required precision when using the fast decoupled method.

Although the iteration number of the fast decoupled method is larger, its computing requirement in each iteration is far less than that of the Newton method. So the computing speed of the fast decoupled method is much higher than the Newton method.

2.4.3 *Flowchart of Fast Decoupled Method*

The principle flowchart of the fast decoupled method is shown in Fig. 2.11 which illustrates the main procedure and logical structure of the load flow calculation.

The symbols used in Fig. 2.11 are first introduced below:

t: counter for the iteration number

$K01$ a flag with “0” and “1” states, “0” indicates the active power iteration; while “1” the reactive power iteration. A whole iteration includes an active power iteration and a reactive power iteration.

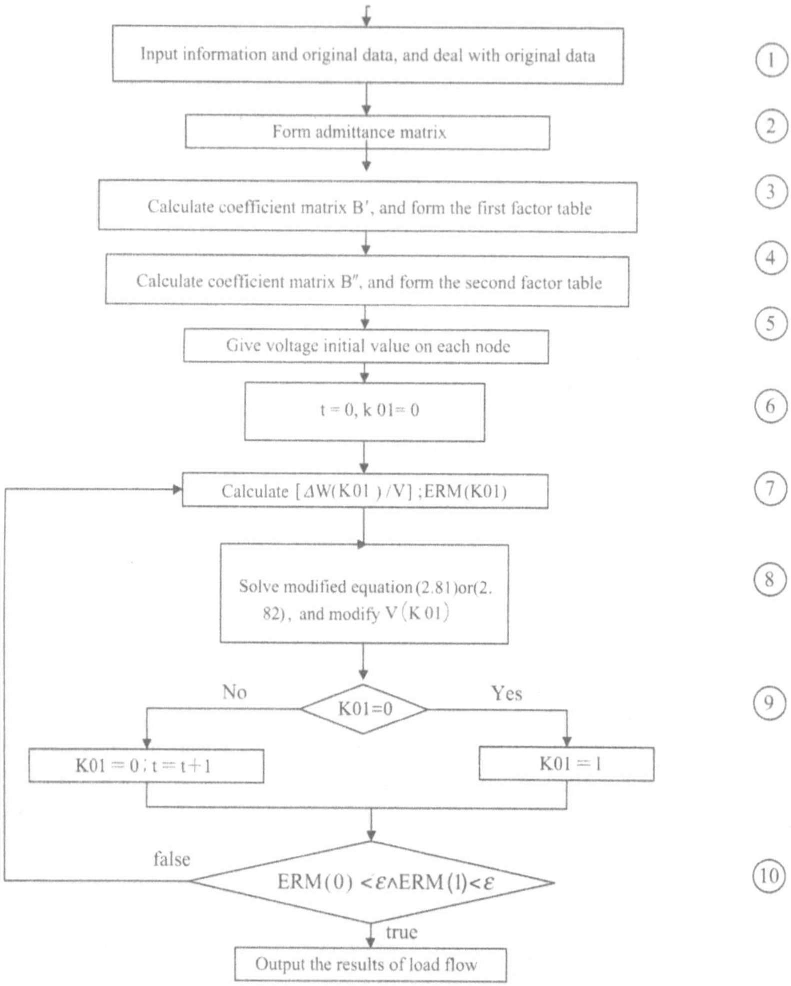

Fig. 2.11 Principle flowchart of $P - Q$ decoupled load flow program

$\Delta\mathbf{W}$: power mismatch vector: when $K01 = 0$, $\Delta\mathbf{W}(K01)$ is the mismatch of active power; when $K01 = 1$, $\Delta W(K01)$ is the mismatch of reactive power.

$\mathbf{V}$: Voltage vector: when $K01 = 0$, $\mathbf{V}(K01)$ represents voltage angle; when $K01 = 1$, $\mathbf{V}(K01)$ represents voltage magnitude.

ERM: Store the maximal power mismatch in an iteration: when $K01 = 0$, ERM ($K01$) stores the maximal active power mismatch; when $K01 = 1$, ERM($K01$) stores the maximal reactive power mismatch;

ε: Convergence criterion.

From the figure one can see, after inputting the problem data, the admittance matrix is formed. Then according to (2.83) the matrix $\mathbf{B}'$ is obtained, and triangularized to form the first factor table (block in Fig. 2.11).

After considering shunt capacitances of transmission lines and grounding branches of off-nominal taps of transformer, matrix $\mathbf{B}''$ can be formed according to (2.84), and then triangularized to form the second factor table (block in Fig. 2.11).

It should be point out that $\mathbf{B}'$ and $\mathbf{B}''$ can be formed at the same time when forming the admittance matrix. Meanwhile, the admittance matrix (block) should be stored for calculating the power mismatches according to (2.77) and (2.78).

In the flowchart, the iteration procedure is composed of blocks.

In block the initial voltage values are set accordingly for PQ nodes and PV nodes. For PQ node, the voltage magnitude can be set as the average voltage of the system; for a PV node, the voltage magnitude is set to the specified value. The voltage angle can be set to 0 as initial value for all nodes.

Block establishes the original state for iteration. The iteration procedure starts with a $P \sim \theta$ iteration, thus $K01$ is set to "0."

The iteration procedure in Fig. 2.11 follows 1θ and $1V$ mode. That is to say the iteration procedure is carries out by alternately solving $P \sim \theta$ and $Q \sim V$ correction equations.

Block calculates the node power mismatch according to (2.77) or (2.78) and records the maximal mismatch in ERM($K01$) for checking the convergence condition.

Block solves correction equations, and further modifies the voltage magnitude and angle. Block establishes the state for the next iteration and counts the iteration number.

Block checks whether the iteration procedure converges. When both the $P \sim \theta$ and $Q \sim V$ iterations converge, the iteration procedure comes to an end, otherwise the process continues to the next iteration.

[Example 2.2] Using the fast decoupled method to calculate the load flow of the system shown in Fig. 2.8.

[Solution] The calculating procedure follows the flow chart of Fig. 2.11.
The admittance matrix of the system can be found in Example 1.1. The factor table used in $P \sim \theta$ iteration is

$$\begin{bmatrix} -0.15286 & -0.59620 & -0.40379 & \\ & -0.01466 & -0.06874 & -0.93125 \\ & & -0.02769 & -0.12087 \\ & & & -0.26061 \end{bmatrix}$$

It should be pointed out that $\mathbf{B}'$ used in forming the above factor table should be calculated according to (2.83). The factor table in $Q \sim V$ iteration is

$$\begin{bmatrix} -0.15135 & -0.60541 & -0.43243 \\ & -0.01541 & -0.07804 \\ & & -0.02895 \end{bmatrix}$$

Matrix $\mathbf{B}''$ used in forming the above factor table is calculated by (2.84). Because matrix $\mathbf{B}''$ does not include the elements related to PV nodes, it is a three-dimensional matrix,

$$\mathbf{B}'' = \begin{bmatrix} -6.60714 & 4.0000 & 2.85714 \\ 4.0000 & -67.30197 & 3.33333 \\ 2.85714 & 3.33333 & -36.17480 \end{bmatrix}$$

It is easy to establish the above factor table by an elimination operation on $\mathbf{B}''$.

The initial values of node voltages are similar to example 2.1 except the polar form is used here. The average operation voltage of system is:

$$V_0 = 1.00000$$

Then the initial value of node voltage vector is:

$$V_1^{(0)} = V_2^{(0)} = V_3^{(0)} = 1.00000$$

$$V_4^{(0)} = V_5^{(0)} = 1.05000$$

$$\theta_1^{(0)} = \theta_2^{(0)} = \theta_3^{(0)} = \theta_4^{(0)} = \theta_5^{(0)} = 0$$

According to (2.77) and (2.78), the functions of node power mismatches are given as follows:

$$\begin{aligned} \Delta P_1 &= P_{1s} - V_1[V_1G_{11} + V_2(G_{12}\cos\theta_{12} + B_{12}\sin\theta_{12}) + \\ &\quad + V_3(G_{13}\cos\theta_{13} + B_{13}\sin\theta_{13})] \\ \Delta Q_1 &= Q_{1s} - V_1[-V_1B_{11} + V_2(G_{12}\sin\theta_{12} - B_{12}\cos\theta_{12}) + \\ &\quad + V_3(G_{13}\sin\theta_{13} - B_{13}\cos\theta_{13})] \\ &\cdots\cdots \\ \Delta P_4 &= P_{4s} - V_4[V_2(G_{42}\cos\theta_{42} + B_{42}\sin\theta_{42}) + V_4G_{44}] \end{aligned}$$

For the first $P \sim \theta$ iteration the node power mismatch can be calculated as

$$\Delta P^{(0)} = \begin{bmatrix} -1.60000 \\ -2.00000 \\ -3.70000 \\ 5.00000 \end{bmatrix}$$

Thus we have the right-hand term of the correction equation,

$$\left(\frac{\Delta P}{V}\right)^{(0)} = \begin{bmatrix} -1.60000 \\ -2.00000 \\ -3.70000 \\ 4.76190 \end{bmatrix}$$

Using the first factor table to execute elimination and backward substitution operations, we obtain the correcting value of node θ as

$$\Delta\theta^{(0)} = \begin{bmatrix} 0.09455 \\ -0.30580 \\ 0.07994 \\ -0.38081 \end{bmatrix}$$

Note that in the $P \sim \theta$ iteration, after solving the correction equation, we should obtain $\mathbf{V}_0\Delta\boldsymbol{\theta}$ (see (2.81)). But in this example, the calculation is based on per unit and $\mathbf{V}_0 = \mathbf{I}$, hence,

$$\mathbf{V}_0\Delta\boldsymbol{\theta}^{(0)} = \Delta\boldsymbol{\theta}^{(0)}$$

After modifying the node voltage angle, we get $\boldsymbol{\theta}^{(1)}$ as

$$\boldsymbol{\theta}^{(1)} = \boldsymbol{\theta}^{(0)} - \Delta\boldsymbol{\theta}^{(0)} = \begin{bmatrix} -0.09455 \\ 0.30580 \\ -0.07994 \\ 0.38080 \end{bmatrix}$$

The $Q \sim V$ iteration is carried out next. The node reactive power mismatches are

$$\Delta\mathbf{Q}^{(0)} = \begin{bmatrix} -1.11284 \\ 5.52890 \\ 1.41242 \end{bmatrix}$$

The right-hand term of the correction equation is

$$\left(\frac{\Delta\mathbf{Q}}{\mathbf{V}}\right)^{(0)} = \begin{bmatrix} -1.11284 \\ 5.52890 \\ 1.41242 \end{bmatrix}$$

Solving this equation, we obtain the voltage correct vector for PQ nodes:

$$\Delta\mathbf{V}^{(0)} = \begin{bmatrix} 0.10493 \\ -0.07779 \\ -0.03793 \end{bmatrix}$$

The modified node voltages can be calculated (see (2.80)):

$$V^{(1)} = V^{(0)} - \Delta V^{(0)} = \begin{bmatrix} 0.89057 \\ 1.07779 \\ 1.03793 \end{bmatrix}$$

Thus the first iteration is complete.

The iteration procedure repeats the above steps until the convergence condition is satisfied. When $\varepsilon = 10^{-5}$, the iteration procedure converges after ten iterations. The evolution of the node voltages is demonstrated in Table 2.7.

Table 2.8 shows the evolution of the maximal errors of the node powers and voltages in the iteration procedure.

The convergence property of the fast decoupled method used in this example is displayed in Fig. 2.12. From this figure we can see that the convergence characteristic of the fast decoupled method on a logarithmic coordinate is nearly a straight line. At the beginning, its convergence speed is faster than that of the Newton method.

The result of load flow calculation is shown in Fig. 2.13.

Table 2.7 Node voltage changes in the iteration process

Iterating No.	θ_1	V_1	θ_2	V_2	θ_3	V_3	θ_4
1	−0.09455	0.89507	0.30580	1.07779	−0.07995	1.03793	0.38080
2	−0.08227	0.87158	0.30728	1.07857	−0.07405	1.03743	0.37652
3	−0.08239	0.86512	0.31048	1.07813	−0.07448	1.03673	0.38010
4	−0.08316	0.86309	0.31117	1.07798	−0.07468	1.03652	0.38079
5	−0.08332	0.86244	0.31152	1.07794	−0.07471	1.03644	0.38115
6	−0.08339	0.86222	0.31162	1.07792	−0.07473	1.03642	0.38126
7	−0.08341	0.86215	0.31166	1.07791	−0.07473	1.03641	0.38129
8	−0.08342	0.86213	0.31167	1.07791	−0.07474	1.03640	0.38131
9	−0.08342	0.86212	0.31167	1.07791	−0.07474	1.03640	0.38131
10	−0.08342	0.86212	0.31168	1.07791	−0.07474	1.03641	0.38131

Note: the angles in the table are in rad

Table 2.8 Changes of maximal node power and voltage errors

Iterating No.	ΔP_M	ΔQ_M	$\Delta\theta_M$	ΔV_M
1	5.00000	5.52890	0.38080	0.10493
2	0.38391	0.15916	0.01228	0.02348
3	0.02660	0.03398	0.00358	0.00647
4	0.00898	0.01054	0.00077	0.00202
5	0.00279	0.00339	0.00036	0.00066
6	0.00095	0.00111	0.00011	0.00022
7	0.00031	0.00037	0.00004	0.00007
8	0.00010	0.00012	0.00001	0.00002
9	0.00003	0.00004	0.00000	0.00001
10	0.00001	0.00001	0.00000	0.00000

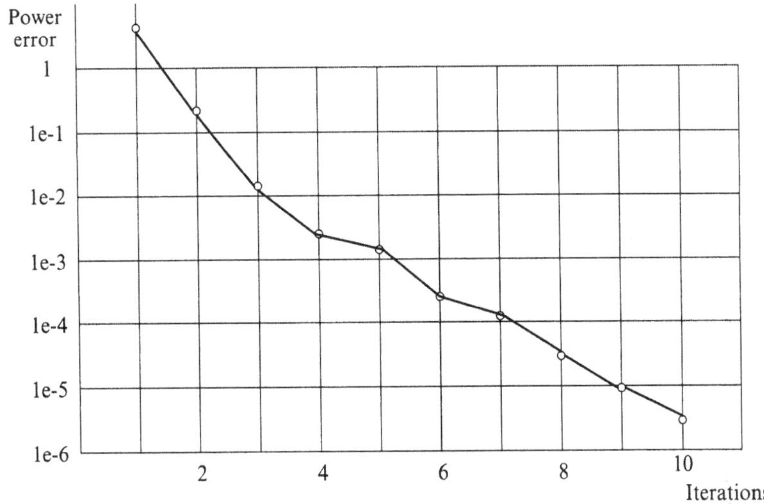

Fig. 2.12 Convergence property of $P - Q$ decoupled method

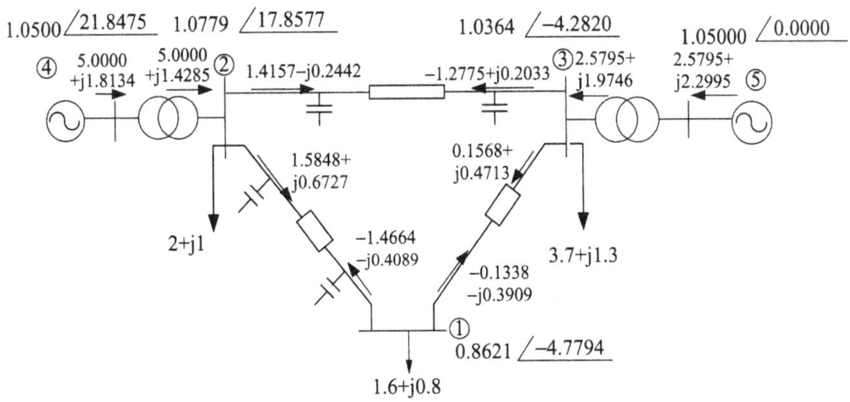

Fig. 2.13 Load flow calculation results

2.5 Static Security Analysis and Compensation Method

2.5.1 Survey of Static Security Analysis

Static security analysis is widely used in power system planning and dispatching to check the operation states when some system equipment sustains forced outages. It will answer the questions such as "what will happen if a 500 kV line is disconnected." When the results show that the power flows and voltages all are in the acceptable range, the system is static secure. When the results show that some transmission equipments are overloaded or the bus voltages of some nodes are beyond the constraints, the system is not static secure. Therefore static security analysis is a very important part in power system security analysis and is discussed in this section. The dynamic performance analysis of power systems will be presented in the last two chapters of this book.

The static security analysis can be used in evaluating the enduring capability of a planning scheme, or an operating schedule of the power system. The static security analysis usually checks the typical forced outages of generator units or transmission equipments, onefold or two-fold. Sometimes it also inspects multi-fold outages, or common mode failures, e.g., those caused by relay system failures.

In power system planning, all credible outage cases should be considered in the static security analysis. According to the result of the static security analysis the system planner usually needs to add some redundant devices or to adjust the network scheme.

In power system operation, to avoid equipment damage and large area blackouts, the static security analysis, both online and off-line, is essential [21, 22]. In particular, the power market evolution introduces many uncertain factors to system operation, and increasing demands on the security monitor and control system.

Since the dynamic performance of the power system is not involved, the static security analysis is substantially a steady analysis problem. Through load flow calculations for all possible contingencies, we can judge whether the system is secure or not. Unfortunately, since the number of possible contingencies in static security analysis is very large, it is almost impossible to complete the task by the conventional load flow analysis method in a reasonable period of time for on-line or real-time use. Therefore, many special methods for static security analysis have been developed, such as the compensation method, DC load flow model and the sensitivity method, etc. These methods will be presented below.

2.5.2 Compensation Method

When a minor change of the network topology occurs in a power system, we can still use the original admittance matrix, even the original factor table to calculate the load flow after such a change. To accomplish this we usually use the compensation method.

The compensation method is a very useful tool in power system analysis, not only used in the static security evaluation but also widely applied in the dynamic performance study and short circuit current calculation.

We first introduce the basic principles of the compensation method.

Assume the admittance matrix and the factor table of network N shown in Fig. 2.14 have been formed, and the currents injected into the nodes are known,

$$I = \left[\dot{I}_1 \quad \ldots \quad \dot{I}_i \quad \ldots \quad \dot{I}_j \quad \ldots \quad \dot{I}_n\right]^{\mathrm{T}}$$

The problem in question is: when an impedance Z_{ij} is added across nodes i and j, how to solve the voltage $\boldsymbol{V}$ under the new condition by using the factor table of the original network N:

$$\dot{V} = \left[\dot{V}_1 \quad \dot{V}_2 \quad \ldots \quad \ldots \quad \dot{V}_n\right]^{\mathrm{T}}$$

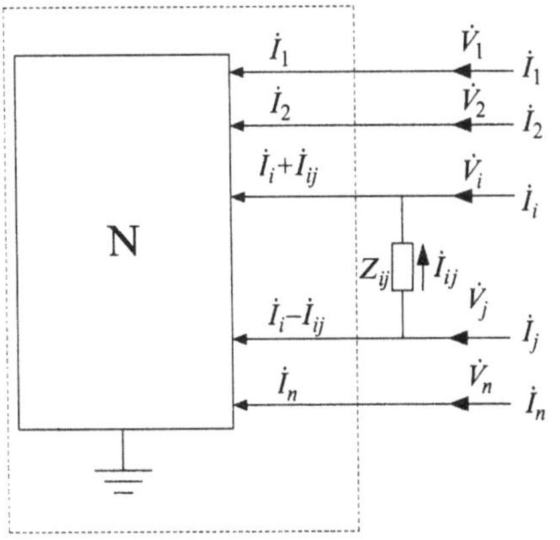

Fig. 2.14 Equivalent circuit for network branch changing

If we can get the current injecting into network N,

$$\boldsymbol{I} = \begin{bmatrix} \dot{I}_1 \\ \dot{I}_2 \\ \vdots \\ \dot{I}_i + \dot{I}_{ij} \\ \vdots \\ \dot{I}_j - \dot{I}_{ij} \\ \vdots \\ \dot{I}_n \end{bmatrix} \tag{2.85}$$

Thus the node voltage vector **V** can be calculated by an elimination and substitution manipulation on **I**′ employing the original factor table. But before the node voltage vector is obtained, the current $\dot{I}_{ij}$ flowing into branch Z_{ij} is unknown. Therefore, the node voltage cannot be calculated directly according to **I**′.

On the basis of the superposition principle, we can decompose network N shown in Fig. 2.14 into two equivalent networks, as showing in Fig. 2.15a, b. The node voltage vector **V** can be decomposed as

$$\mathbf{V} = \mathbf{V}^{(0)} + \mathbf{V}^{(1)} \tag{2.86}$$

where $\mathbf{V}^{(0)}$ is related to the original network without the added line, see Fig. 2.15a. Since the node injecting current vector **I** is known, $\mathbf{V}^{(0)}$ can be easily calculated by using the factor table of original network N:

$$V^{(0)} = \begin{bmatrix} \dot{V}_1^{(0)} & \dot{V}_2^{(0)} & \cdots & \dot{V}_i^{(0)} & \cdots & \dot{V}_j^{(0)} & \cdots & \dot{V}_n^{(0)} \end{bmatrix}^{\mathrm{T}} \tag{2.87}$$

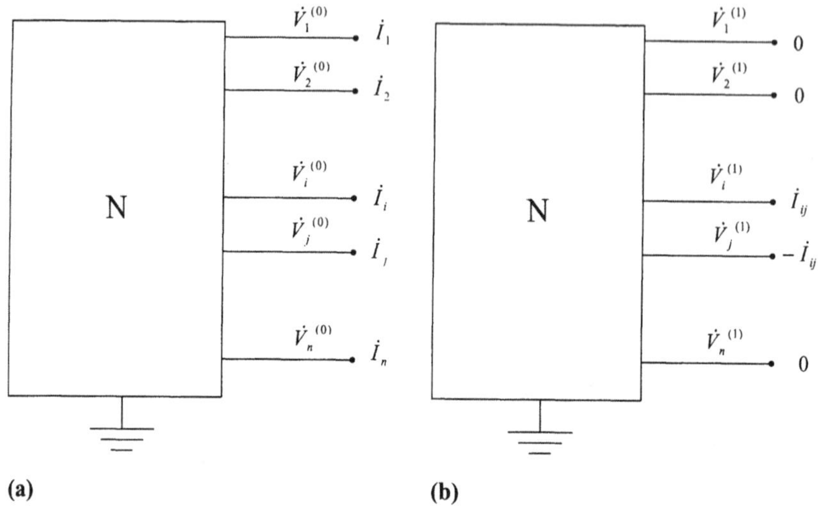

Fig. 2.15 Principle of compensation method

Now we discuss how to calculate $\mathbf{V}^{(1)}$ in Fig. 2.15b. In this figure, the current vector injected into the original network is

$$\mathbf{I}^{(1)} = \begin{bmatrix} 0 \\ \vdots \\ \vdots \\ \dot{I}_{ij} \\ 0 \\ \vdots \\ -\dot{I}_{ij} \\ 0 \\ \vdots \end{bmatrix} = \dot{I}_{ij} \begin{bmatrix} 0 \\ \vdots \\ \vdots \\ 1 \\ 0 \\ \vdots \\ -1 \\ 0 \\ \vdots \end{bmatrix} \begin{matrix} \\ \\ \\ \leftarrow i \\ \\ \\ \leftarrow j \\ \\ \\ \end{matrix} \tag{2.88}$$

where $\dot{I}_{ij}$ is an unknown variable at this stage. But let $\dot{I}_{ij} = 1$, the node voltage can be calculated by using the original factor table:

$$\mathbf{V}^{(ij)} = \begin{bmatrix} \dot{V}_1^{(ij)} & \dot{V}_2^{(ij)} & \ldots & \dot{V}_i^{(ij)} & \ldots & \dot{V}_j^{(ij)} & \ldots & \dot{V}_n^{(ij)} \end{bmatrix}^{\mathrm{T}} \tag{2.89}$$

Because the network is linear, if the $\dot{I}_{ij}$ can be obtained, then the final voltage vector can be calculated by the following equation:

$$\mathbf{V} = \begin{bmatrix} \dot{V}_1^{(0)} \\ \dot{V}_2^{(0)} \\ \vdots \\ \vdots \\ \dot{V}_n^{(0)} \end{bmatrix} + \dot{I}_{ij} \begin{bmatrix} \dot{V}_1^{(ij)} \\ \dot{V}_2^{(ij)} \\ \vdots \\ \vdots \\ \dot{V}_n^{(ij)} \end{bmatrix} \tag{2.90}$$

Therefore, the problem we face now is to get $\dot{I}_{ij}$. Here we need utilize the equivalent generator principle.

As mentioned before, $\mathbf{V}^{(0)}$ is the node voltage when branch Z_{ij} is open. If we consider the whole system as the equivalent source of branch Z_{ij}, then the no-load voltage of this source is

$$\dot{E} = \dot{V}_i^{(0)} - \dot{V}_j^{(0)} \tag{2.91}$$

The equivalent internal impedance, Z_T, is

$$Z_\mathrm{T} = \dot{V}_i^{(ij)} - \dot{V}_j^{(ij)} \tag{2.92}$$

where $(\dot{V}_i^{(ij)} - \dot{V}_j^{(ij)})$ is the voltage drop between nodes i and j due to injecting positive and negative unit current into these nodes. Thus we have the equivalent circuit shown in Fig. 2.16, and can obtain $\dot{I}_{ij}$ directly:

$$\dot{I}_{ij} = -\frac{\dot{V}_i^{(0)} - \dot{V}_j^{(0)}}{Z'_{ij}} \tag{2.93}$$

where

$$Z'_{ij} = Z_\mathrm{T} + Z_{ij} \tag{2.94}$$

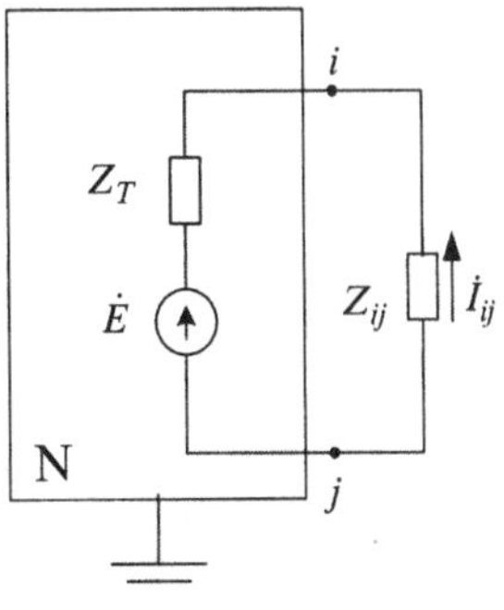

Fig. 2.16 Equivalent circuit to get $\boldsymbol{I}_{ij}$

Substituting $\dot{I}_{ij}$ into (2.90), we finally obtain node voltage vector **V**.

The basic principle of compensation method has been introduced. In the practice, the compensation method can be used according to the following steps:

1. Find $\mathbf{V}^{(ij)}$ for the injecting unit current vector by using the factor table of the original normal network.

$$\mathbf{I}_{ij} = \begin{bmatrix} 0 \\ \vdots \\ \vdots \\ 1 \\ 0 \\ \vdots \\ -1 \\ 0 \\ \vdots \end{bmatrix} \begin{matrix} \\ \\ \\ \leftarrow i \\ \\ \\ \leftarrow j \\ \\ \\ \end{matrix} \tag{2.95}$$

2. Calculate the internal impedance, Z_T, by (2.92), and then obtain Z'_{ij} by (2.94).
3. Calculate $\mathbf{V}^{(0)}$ by using the original factor table for the injected current vector **I** (see Fig. 2.15a).
4 Obtain the current $\dot{I}_{ij}$ flowing into branch Z_{ij} by (2.93).
5. Solving node voltage vector **V** according to (2.90).

In theory, the compensation method can also be used when more than one operation occurs simultaneously in the network. In this case, the above calculation steps should be used recursively.

Now, we will show how to use the compensation method to analysis the contingency state in the fast decoupled method.

The correction of (2.81) and (2.82) can be considered as the node equations of the network based on "admittance matrix" $\mathbf{B}'$ and $\mathbf{B}''$, and the injecting currents $\Delta\mathbf{P}/\mathbf{V}$ and $\Delta\mathbf{Q}/\mathbf{V}$, respectively. The node voltages $\mathbf{V}_0\Delta\boldsymbol{\theta}$ and $\Delta\mathbf{V}$ are the variables to be solved. In this way, the above calculation process can be followed directly. When branch ij trips, the branch impedances added between i and j for $\mathbf{B}'$ and $\mathbf{B}''$ should be (see Fig. 2.14):

$$Z'_{ij} = \frac{-1}{B_{ij}}, \quad Z''_{ij} = -x_{ij}, \tag{2.96}$$

If the tripped branch is a nonnominal tap transformer, the current representation in (2.95) should be changed as

$$\mathbf{I}^{(ij)} = \begin{bmatrix} 0 \\ \vdots \\ \vdots \\ n_{\mathrm{T}} \\ 0 \\ \vdots \\ -1 \\ 0 \\ \vdots \end{bmatrix} \begin{matrix} \\ \\ \\ \leftarrow i \\ \\ \\ \leftarrow j \\ \\ \\ \end{matrix} \tag{2.97}$$

where n_{T} is the nonnominal tap on the node j side of the transformer. In this situation, (2.91), (2.92), and (2.93) should be revised, respectively, as

$$\dot{E} = n_{\mathrm{T}}\dot{V}_i^{(0)} - \dot{V}_j^{(0)} \tag{2.98}$$

$$Z_{\mathrm{T}} = n_{\mathrm{T}}\dot{V}_i^{(ij)} - \dot{V}_j^{(ij)} \tag{2.99}$$

$$\dot{I}_{ij} = -\frac{n_{\mathrm{T}}\dot{V}_i^{(0)} - \dot{V}_j^{(0)}}{Z'_{ij}} \tag{2.100}$$

where $Z'_{ij} = Z_T + Z_{ij}$.

It should be noted, in above line outage operation, only the series branch of the opened line (or transformer) is considered in (2.96). Rigorously speaking, the shunt branches for line charging capacitance (and transformer ground branches) should also be tripped simultaneously. However, tripping three branches at the same time makes the calculation too complicated. Fortunately, practice indicates that the errors caused by neglecting grounding branches are not very significant. Therefore, the grounding branches can be neglected when the compensation method is used to analyze line outage states.

2.6 DC Load Flow Method

The DC load flow simplifies the AC load flow to a linear circuit problem. Consequently, it makes the steady state analysis of the power system very efficient. The main shortcoming of the DC load flow model is that it cannot be used in checking voltage limit violations. Because the DC load flow uses a linear model, it is not only suitable to efficiently treat the problem of line outages, but is also suitable to form linear optimization problems. Therefore, the DC load flow method has been widely used in power system planning and operating problems.

2.6.1 Model of DC Load Flow

The node active power equations of an AC load flow are given by (2.9),

$$P_i = V_i \sum_{j \in i} V_j (G_{ij} \cos\theta_{ij} + B_{ij} \sin\theta_{ij}) \quad (i = 1, 2, \ldots, n) \tag{2.101}$$

Branch active power is

$$P_{ij} = V_i V_j (G_{ij} \cos\theta_{ij} + B_{ij} \sin\theta_{ij}) - t_{ij} G_{ij} V_i^2 \tag{2.102}$$

where t_{ij} is the circuit transformer ratio per unit of branch ij, θ_{ij} is the phase angle difference across branch ij; G_{ij}, B_{ij} are the real and imaginary parts of corresponding elements of the node admittance matrix, respectively.

$$\theta_{ij} = \theta_i - \theta_j \tag{2.103}$$

$$G_{ij} + jB_{ij} = -\frac{1}{r_{ij} + jx_{ij}} = \frac{-r_{ij}}{r_{ij}^2 + x_{ij}^2} + j\frac{x_{ij}}{r_{ij}^2 + x_{ij}^2} \tag{2.104}$$

where, r_{ij}, x_{ij} are resistance and reactance of line ij. When $i = j$,

$$G_{ii} = -\sum_{\substack{j \in i \\ j \neq i}} G_{ij}$$

$$B_{ii} = -\sum_{\substack{j \in i \\ j \neq i}} B_{ij}$$

Under assumptions applied in the fast decoupled method, the above AC load flow equations can be simplified to the following equations.

$$P_i = \sum_{j \in i} B_{ij} \theta_{ij} \quad (i = 1, 2, \ldots, n)$$

which can be rewritten as,

$$P_i = \sum_{j \in i} B_{ij} \theta_i - \sum_{j \in i} B_{ij} \theta_j \quad (i = 1, 2, \ldots, n)$$

From (2.104), we know the first term in the right hand of the above equation is 0, thus we have,

$$P_i = -\sum_{j \in i} B_{ij} \theta_j \quad (i = 1, 2, \ldots, n) \tag{2.105}$$

The DC flow model usually has no negative sign, thus we redefine B_{ij} as,

$$B_{ij} = -\frac{1}{x_{ij}} \tag{2.106}$$

thus

$$B_{ii} = \sum_{\substack{j \in i \\ j \neq i}} \frac{1}{x_{ij}} \tag{2.107}$$

Finally, we establish the DC flow equation,

$$P_i = \sum_{j \in i} B_{ij}\theta_j \quad (i = 1, 2, \ldots, n) \tag{2.108}$$

or in matrix form,

$$\mathbf{P} = \mathbf{B}\boldsymbol{\theta} \tag{2.109}$$

where $\mathbf{P}$ is $\boldsymbol{\theta}$ the node injection power vector and its ith element is given by $P_i = P_{Gi} - P_{Di}$, here P_{Gi} and P_{Di} are the generator output and load at node i, respectively; is the phase angle vector and $\mathbf{B}$ is the matrix whose elements are defined by (2.106) and (2.107).

Equation (2.109) can also be expressed as follows

$$\boldsymbol{\theta} = \mathbf{X}\mathbf{P} \tag{2.110}$$

where $\mathbf{X}$ is the inverse of matrix $\mathbf{B}$,

$$\mathbf{X} = \mathbf{B}^{-1} \tag{2.111}$$

Similarly, substituting the simplifying conditions into (2.102), one obtains the active power flowing into branch ij,

$$P_{ij} = -B_{ij}\theta_{ij} = \frac{\theta_i - \theta_j}{x_{ij}} \tag{2.112}$$

or in matrix form,

$$\mathbf{P}_{\mathrm{l}} = \mathbf{B}_l\boldsymbol{\Phi} \tag{2.113}$$

If the number of branches is l, $\mathbf{B}_l$ is an $l \times l$ diagonal matrix whose elements are branch admittance; $\mathbf{P}_{\mathrm{l}}$ is the branch active power vector; $\boldsymbol{\Phi}$ is the end terminal phase angle difference vector.

Assuming that the network incidence matrix is **A**, then one arrives at

$$\mathbf{\Phi} = \mathbf{A}\boldsymbol{\theta} \tag{2.114}$$

Equations (2.109), (2.110), and (2.113) are basic DC load flow equations which are linear. Under given system operation conditions, the state variable θ may be obtained through triangularizition or matrix inversion from (2.110), then branch active power can be obtained from (2.113).

2.6.2 Outage Analysis by DC Load Flow Method

From the above discussion, it can be seen that it is very simple to solve system state and active power flow by DC load flow equations. It will also be shown that because these equations are linear, it is possible to carry out fast load flow computation after adding or tripping a line.

Assuming that the original network nodal impedance matrix is $\boldsymbol{X}$ and a branch k is connected between nodes i and j. When a line with reactance x_k is added in parallel with branch k, a new network is formed. We now demonstrate how to obtain the new network state vector in this situation from the original network impedance matrix and state vector.

Assuming the new network impedance matrix is $\boldsymbol{X}'$, it can be obtained according to the branch adding principle of section 1–4 (see (2.1–2.107)),

$$\mathbf{X}' = \mathbf{X} - \frac{\mathbf{X}_\mathrm{L}\mathbf{X}_\mathrm{L}^\mathrm{T}}{\mathbf{X}_\mathrm{LL}} \tag{2.115}$$

where $\mathbf{X}_\mathrm{L} = \mathbf{X}\mathbf{e}_k$,

$$\mathbf{e}_k = \begin{bmatrix} 0 \\ \vdots \\ 1 \\ \vdots \\ -1 \\ 0 \\ \vdots \end{bmatrix} \begin{matrix} \\ \\ \leftarrow i \\ \\ \leftarrow j \\ \\ \\ \end{matrix} \tag{2.116}$$

$$\mathbf{X}' = \mathbf{X} - \frac{\mathbf{X}\mathbf{e}_k\mathbf{e}_k^T\mathbf{X}}{x_k + \mathbf{e}_k^\mathrm{T}\mathbf{X}\mathbf{e}_k} \tag{2.117}$$

Equation (2.117) can be further reduced to,

$$\mathbf{X}' = \mathbf{X} + \beta_k\mathbf{X}\mathbf{e}_k\mathbf{e}_k^\mathrm{T}\mathbf{X} \tag{2.118}$$

where

$$\beta_k = \frac{-1}{x_k + \chi_k} \tag{2.119}$$

$$\chi_k = \mathbf{e}_k^{\mathrm{T}} \mathbf{X} \mathbf{e}_k = X_{ii} + X_{jj} - 2X_{ij} \tag{2.120}$$

where X_{ii}, X_{jj}, X_{ij} are elements of $\mathbf{X}$.

From (2.118), the incremental change of the nodal impedance matrix is given by:

$$\Delta\mathbf{X} = \mathbf{X}' - \mathbf{X} = \beta_k \mathbf{X} \mathbf{e}_k \mathbf{e}_k^{\mathrm{T}} \mathbf{X} \tag{2.121}$$

According to (2.121) and (2.110), under constant nodal injection power conditions, the change in original state vector after adding line k is

$$\Delta\boldsymbol{\theta} = \Delta\mathbf{X}\mathbf{P} = \beta_k \mathbf{X} \mathbf{e}_k \phi_k \tag{2.122}$$

where $\phi_k = \mathbf{e}_k^{\mathrm{T}} \boldsymbol{\theta}$, is the terminal phase angle difference of branch k before the change. The new network state vector is given by

$$\boldsymbol{\theta}' = \boldsymbol{\theta} + \Delta\boldsymbol{\theta} = \boldsymbol{\theta} + \beta_k \mathbf{X} \mathbf{e}_k \phi_k \tag{2.123}$$

Thus after adding line k, the new network nodal impedance matrix and the new state vector can be obtained by (2.118) and (2.123) using the original network parameters. When line k trips, the above equations can still be applied with x_k being replaced by $-x_k$.

If the outage of branch k causes system disconnection, the new impedance matrix $\mathbf{X}'$ does not exist and β_k of (2.119) becomes infinite, i.e., $-x_k + \chi_k = 0$. Therefore, it is very easy to check whether the outage of a branch will cause system disconnection by using the DC load flow equation. However, it is impossible to carry out line outage analysis directly.

2.6.3 *N*-1 Checking and Contingency Ranking Method

A network design has to satisfy certain operational security requirements. As discussed earlier, a common network operational security requirement is to satisfy N-1 checking, i.e., when one of N branches fails, the system operation criteria remain within given requirements. At the initial stage of forming a network configuration, the principle is to ensure that there is no overloading in the network; i.e., the network satisfies the requirements for securely transmitting power. To this end, one has to carry out the overload check for successive line outages. If the outage of a line causes overloading or system disconnection, then the network does not satisfy N-1 checking.

The strict N-1 checking on all branches needs N line outage analyses, resulting in a large amount of computing. In practice, some line outages do not cause system overloading. Therefore, a contingency ranking is carried out according to the probability of system overload being caused by a line outage, then the checking is first performed on the lines with higher probability. If the checking of a line indicates that its outage does not cause overloading, the lines with lower rank are not subjected to any further checking, which significantly reduces the amount of computing. Such a process is also called contingency selection. A number of contingency ranking methods are available in the literature [23, 24], each having a different criterion for assessing the system contingency. This section describes a contingency ranking method based on the criterion of system overloading.

To reflect the overall system overloading, a system performance index (PI) is defined as follows:

$$PI = \sum_{l=1}^{L} \alpha_l w_l \left(\frac{P_l}{\bar{P}_l} \right)^2 \tag{2.124}$$

where P_l, active power of line l
$\bar{P}_l$, transmission capacity of branch l
α_l, number of parallel lines for branch l
w_l, weighting factor of line l, which reflects the influence of a fault
L, number of branches in the network

It can be seen from (2.124) that when there is no overloading, $P_l/\bar{P}_l$ is not greater than 1, the PI is small. When there is overloading in the system, $P_l/\bar{P}_l$ for the overloaded line is greater than 1 and the positive exponential element makes the PI relatively large. Therefore, this index generally reflects the system security. It may also be possible to use a higher order exponential instead of a square element in the equation to further obviate the overloading problem.

A sensitivity analysis of the PI with respect to the change of a line admittance will reveal the impact of an outage on the system security. When line k fails, the change in the PI is given by

$$\Delta PI_k = \frac{\partial PI}{\partial B_k} \Delta B_k \tag{2.125}$$

where $\Delta B_k = B_k$, is the admittance of line k. The bigger ΔPI_k is, the larger the increase in the PI will be, which indicates that the probability of a faulted line k causing system overloading becomes higher.

ΔPI_k may be calculated from Telegen's theorem and the adjoint network method. In the following study, we will derive a formula to calculate ΔPI_k directly using nominal load flow results.

Assuming that after line k fails other line flows become $P'(l = 1, 2, \ldots, L;$ $l \neq k)$, the system performance index becomes,

$$PI' = \sum_{l=1}^{L} \alpha_l w_l \left(\frac{P'_l}{\bar{P}_l} \right)^2 \quad (2.126)$$

Hence

$$\Delta PI_k = PI' - PI \quad (2.127)$$

For the sake of simplicity, we change the system index to a function of voltage angles and express it in the matrix form. From (2.113),

$$\mathbf{P}_l = \mathbf{B}_l \phi_l \quad (2.128)$$

Substituting the above equation into (2.124) and defining

$$PI_\phi = PI = \sum_{l=1}^{L} \alpha_l w_l \frac{(B_l \phi_l)^2}{\bar{P}_l^2} = \phi^{\mathrm{T}} w_d \phi \quad (2.129)$$

where

$$\phi^{\mathrm{T}} = [\phi_1, \ldots, \phi_k, \ldots, \phi_L]$$

and

$$\mathbf{w}_d = \begin{bmatrix} \frac{\alpha_1 w_1 B_1^2}{\bar{P}_1^2} & \cdots & & 0 \\ & & \frac{\alpha_k w_k B_k^2}{\bar{P}_k^2} & \\ & & & \cdots \\ 0 & & & \frac{\alpha_L w_L B_L^2}{\bar{P}_L^2} \end{bmatrix}$$

Substituting (2.114) into (2.129), one obtains

$$PI_\phi = \boldsymbol{\theta}^{\mathrm{T}} \mathbf{A}^{\mathrm{T}} \mathbf{w}_d \mathbf{A} \boldsymbol{\theta} = \boldsymbol{\theta}^{\mathrm{T}} \omega \boldsymbol{\theta} \quad (2.130)$$

where

$$\mathbf{w} = \mathbf{A}^{\mathrm{T}} \mathbf{w}_d \mathbf{A} \quad (2.131)$$

is a symmetric matrix. Matrix **w** has the same structure as matrix **B**. Thus its formation is equivalent to directly forming the admittance matrix using element $\alpha_l w_l B_l^2 / \bar{P}_l^2$ to replace B_l. Similarly, PI'_ϕ can be expressed as

$$PI'_\phi = \boldsymbol{\theta}'^{\mathrm{T}} \mathbf{w} \boldsymbol{\theta}' \quad (2.132)$$

where $\boldsymbol{\theta}'$ is the voltage angle vector after the line k fails.

Equation (2.132) contains all elements relevant to line k which should not appear in the new system performance index PI'. Thus

$$PI' = PI'_{\phi} - w_k \frac{(B_k \phi'_k)^2}{\bar{P}_k^2} \tag{2.133}$$

Substituting (2.130) and (2.133) into (2.127), one obtains

$$\Delta PI_k = PI'_{\phi} - PI_{\phi} - \frac{w_k B_k^2}{\bar{P}_k^2}(\phi'_k)^2 = \boldsymbol{\theta}'^{\mathrm{T}} \mathbf{w} \boldsymbol{\theta}' - \boldsymbol{\theta}^{\mathrm{T}} \mathbf{w} \boldsymbol{\theta} - \frac{w_k B_k^2}{\bar{P}_k^2}(\phi'_k)^2 \tag{2.134}$$

From (2.123), we know

$$\begin{aligned} \boldsymbol{\theta}' &= \boldsymbol{\theta} + \beta_k \mathbf{X} \mathbf{e}_k \phi_k \\ \phi'_k &= \mathbf{e}_k \theta' = (1 + \beta_k \chi_k) \phi_k \end{aligned}$$

Substituting the above two equations into (2.134), we have

$$\begin{aligned} \Delta PI_k &= (\boldsymbol{\theta} + \beta_k \mathbf{X} \mathbf{e}_k \phi_k)^{\mathrm{T}} \mathbf{w} (\boldsymbol{\theta} + \beta_k \mathbf{X} \mathbf{e}_k \phi_k) - \boldsymbol{\theta}^{\mathrm{T}} \mathbf{w} \boldsymbol{\theta} - \frac{w_k B_k^2}{\bar{P}_k^2}(1 + \beta_k \chi_k)^2 \phi_k^2 \\ &= \beta_k \phi_k (\boldsymbol{\theta}^{\mathrm{T}} \mathbf{w} \mathbf{X} \mathbf{e}_k + \mathbf{e}_k^{\mathrm{T}} \mathbf{X} \mathbf{w} \boldsymbol{\theta}) + \beta_k^2 \phi_k^2 \mathbf{e}_k^{\mathrm{T}} \mathbf{X} \mathbf{w} \mathbf{X} \mathbf{e}_k - \frac{w_k B_k^2}{\bar{P}_k^2}(1 + \beta_k \chi_k)^2 \phi_k^2 \end{aligned} \tag{2.135}$$

Taking into account the symmetry of matrices $\mathbf{X}$ and $\mathbf{w}$, let

$$\begin{aligned} \gamma_k &= \boldsymbol{\theta}^{\mathrm{T}} \mathbf{w} \mathbf{X} \mathbf{e}_k = \mathbf{e}_k^{\mathrm{T}} \mathbf{X} \mathbf{w} \boldsymbol{\theta} = \mathbf{e}_k^{\mathrm{T}} \mathbf{R} \\ \tau_k &= \mathbf{e}_{\mathbf{k}}^{\mathrm{T}} \mathbf{X} \mathbf{w} \mathbf{X} \mathbf{e}_k = \mathbf{e}_{\mathbf{k}}^{\mathrm{T}} \mathbf{T} \mathbf{e}_k \end{aligned} \tag{2.136}$$

where

$$\begin{aligned} \mathbf{R} &= \mathbf{X} \mathbf{w} \boldsymbol{\theta} \\ \mathbf{T} &= \mathbf{X} \mathbf{w} \mathbf{X} \end{aligned} \tag{2.137}$$

Substituting (2.136) into (2.135), one obtains

$$\Delta PI_k = 2\beta_k \phi_k \gamma_k + \beta_k^2 \phi_k^2 \tau_k - \frac{w_k B_k^2}{\bar{P}_k^2}(1 + \beta_k \chi_k)^2 \phi_k^2 \tag{2.138}$$

When line k fails, β_k in the above equations becomes

$$\beta_k = \frac{-1}{-x_k + \chi_k} = \frac{B_k}{1 - B_k\chi_k}$$

Substituting the above equation into (2.138) gives,

$$\Delta PI_k = \frac{2B_k\phi_k\gamma_k}{1 - B_k\chi_k} + \frac{B_k^2\phi_k^2\tau_k}{(1 - B_k\chi_k)^2} - \frac{w_k B_k^2\phi_k^2}{(1 - B_k\chi_k)P_k^2} \tag{2.139}$$

Because $P_k = B_k\phi_k$,

$$\Delta PI_k = \frac{2P_k\gamma_k}{1 - B_k\chi_k} + \frac{P_k^2\tau_k}{(1 - B_k\chi_k)^2} - \frac{w_k P_k^2}{(1 - B_k\chi_k)P_k^2} \tag{2.140}$$

Equations (2.138), (2.139), and (2.140) have no essential difference except for different expressions. Variables in these equations are obtained from the normal load flow calculation. Under the condition that matrices **X**, **w**, **R**, **T** have been formed, it is very convenient to compute ΔPI after a line outage.

The process of contingency ranking is essential to compute the values of ΔPI from (2.138) [or (2.139) and (2.140)] for all lines and arrange them in descending order of magnitude of ΔPI. During the line outage analysis, load flow calculation and overload checking are first carried out on the line with the largest value of ΔPI, and then the procedure is continued until there is no overload caused by the outage of certain lines. The lines with smaller values of ΔPI are not subjected to further analysis because the probability of overload caused by other outages is very small. However, the use of this system performance index may cause a "screening" effect. For example, the value of ΔPI for the situation where there is overloading in some lines and the flow in the other lines is very small may be smaller than that for the situation where there is no overloading but line flows are large. Therefore, the contingency ranking by this index may introduce some error. In practice, one may decide that the line outage analysis is terminated only after a number of consecutive line outages do not cause system overloading.

Thinking and Problem Solving

1. What functions do the swing bus and *PV* buses in load flow calculations have? How should they be selected?
2. Compare the advantages and disadvantages of nodal power equations with polar coordinates and rectangular coordinates.
3. Give the characteristics of Newton method based modified equations in load flow calculation.
4. Give the physical meaning of Jacobian elements of the modified equation.

5. Give the flowchart of the Newton method based load flow calculation by polar coordinate equations.
6. Design the storage modes of the Jacobian matrix elements of the Newton method based load flow calculation with two kinds of coordinates.
7. How should node conversion, such as changing a *PV* node into a *PQ* node, or changing a *PQ* node into a *PV* node, be implemented in the design of a load flow program?
8. What simplified suppositions are considered in the $P - Q$ decomposition method? Why is it that the $P - Q$ decomposition method can obtain the same calculation accuracy as the Newton method after so many suppositions?
9. How can we improve the convergence of the $P - Q$ decomposition method when the ratio *R*/*X* is very big?
10. How can the compensation method be applied to the case with two branches out of service?
11. Prove that the DC load flow model has the same solution as that of the following optimizing problem:

    a. $obj \min \sum_{ij \in B} P_{ij} X_{ij}^2$ P_{ij} and X_{ij} are the active load flow and reactance of branch ij, B is the branch set.
    b. $s.t. \sum_{ij \in i} P_k = 0$ $ij \in i$ denotes all branches that connect to node i.

12. Discuss the issues raised by the *N*-1 checking method being used as static security analysis tool for electrical power systems.

Chapter 3
Stochastic Security Analysis of Electrical Power Systems

3.1 Introduction

Recently, worldwide power blackouts have attracted great attention to the reliability of electrical power systems. In the power market environment, the operating modes of electrical power systems vary rapidly so that operators must closely monitor system states and transmission configurations, consider all kinds of stochastic factors, adjust system operation away from critical margins, and avoid potential cascade failures. In this situation, the traditional deterministic security analysis methods have limitations, consequently the concept of operational reliability has been presented [28, 29].

According to deterministic security analysis methods, the system should operate safely under various prescribed contingencies. When overload and abnormal voltages occur in the system under certain contingencies, measures must be taken to make the system operate securely. The advantage of the deterministic method is that the theory is simple, but its obvious subjectivity generally makes the system security level inconsistent. Because this method does not consider the occurrence probability and consequences of various contingencies synthetically, some high-risk contingencies may be ignored.

The failure probability of electrical power systems is a function of many stochastic factors, such as the fluctuation of load, the random failure of generator units, the random failure of transmission and distribution components, etc. [30–32]. In a power market, a greater number of probabilistic factors will influence the operation, due to separation of generation and the grid and competition among generators.

When considering synthetically the occurrence probability and seriousness of consequences of various contingencies, the idea of system operation risk is introduced.

Assuming the contingency j occurs in system operating mode i, the risk of state x_{ij} can be assessed by the following equation,

$$r_{ij} = p_r(x_{ij}) \cdot \text{Sev}(x_{ij}),$$

X.-F. Wang et al., *Modern Power Systems Analysis*.
doi: 10.1007/978-0-387-72853-7, © Springer Science + Business Media, LLC 2008

where $p(x_{ij})$ is probability of this sate, $\text{Sev}(x_{ij})$ is the severity degree of the consequences of this contingency, which can be measured by the amount of load shedding. Assuming the set of stochastic events for system operating mode i is Ω_i, then the risk of this operating mode is

$$R_i = \sum_{x_{ij} \in \Omega_i} r_{ij} = \sum_{x_{ij} \in \Omega_i} p_r(x_{ij}) \cdot \text{Sev}(x_{ij}).$$

The risk-evaluation-based operational decision making is obviously more objective and reasonable. In fact, this risk is one measurement of the reliability of an electrical power system. Although there is no recognized risk standard at present, a risk evaluating standard system can be set up according to each specific application in electrical power systems.

At present, the Monte Carlo simulation method and probabilistic load flow method can be used as risk analysis algorithms. The Monte Carlo simulation method can consider comprehensively more complex stochastic factors [33], but these factors are always neglected in analytical methods because of their inherent complication. The drawback of Monte Carlo simulation is its large computational complexity. If we only study the effect of some key system factors on the reliability, the probabilistic load flow method [34–36] may be more effective. In this chapter, we will introduce these two methods in detail and present a system reliability evaluation method based on probabilistic network-flow models. These developments involve the measurement, transformation, and calculation of random variables; therefore probability theory is first briefly introduced [37].

3.2 Basic Concepts of Probability Theory

3.2.1 Probability of Stochastic Events

In certain conditions, stochastic events are things that may or may not occur, and are referred to concisely as events. The measurement of the possibility of occurrence of stochastic events is probability. Therefore, each event has one related probability value. Probability value is between 0 and 1, in which 1 denotes an inevitable event and 0 denotes an impossible event. The probability of event A is defined as $P(A)$, which must meet the requirements as follows:

$$0 \le P(A) \le 1, \tag{3.1}$$

$$P(\Omega) = 1, \; P(\varphi) = 0, \tag{3.2}$$

where Ω is the sample space and is an inevitable event; φ is the empty set and is an impossible event.

In addition, suppose event A and event B are incompatible with each other, that is, $A \cap B = \varphi$, then,

$$P(A \cup B) = P(A) + P(B). \tag{3.3}$$

Equations (3.1)–(3.3) represent the basic attributes of probability or are referred to as axiomatic definitions of probability. According to this definition, we get,

(a) If two events A and B are independent, then,

$$P(A \cap B) = P(A)P(B). \tag{3.4}$$

(b) If two events A and B are incompatible with each other (mutually exclusive), then,

$$P(A \cap B) = 0. \tag{3.5}$$

(c) If two events A and B are independent but not mutually exclusive, then,

$$P(A \cup B) = P(A) + P(B) - P(A)P(B). \tag{3.6}$$

Furthermore, the probability value of an event can be obtained by reasoning only for very few cases (prior probability); in most situations, it is described as frequency value by repeated experiments (posterior), which is referred to as a statistical definition of probability. The reliability parameters of electrical components, such as failure rate and so on, belong to the latter.

Conditional probability is an important concept in probability theory. The conditional probability of an event A is the probability that the event will occur given the knowledge that an event B has already occurred. The definition is

$$P(A|B) = \frac{P(A \cap B)}{P(B)}, \quad P(B) > 0. \tag{3.7}$$

Several important formulas can be deduced according to conditional probability.

1. *Multiplication probability theorem*. Let $A_1, A_2, \ldots, A_n$ be n arbitrary events, the probability of their intersection set is

$$\begin{aligned} P(A_1 \cap A_2 \cap \cdots \cap A_n) = {} & P(A_1)P(A_2|A_1)P[A_3|(A_1 \cap A_2)] \\ & \cdots P[A_n|(A_1 \cap A_2 \cap \cdots \cap A_{n-1})]. \end{aligned} \tag{3.8}$$

   However, when $A_1, A_2, \ldots, A_n$ are independent, we have

$$P(A_1 \cap A_2 \cap \cdots \cap A_n) = P(A_1)P(A_2) \cdots P(A_n). \tag{3.9}$$

2. *Formula of total probability*. Let event A occur according to the given condition of events $B_1, B_2, \ldots, B_n$. A can only occur at the same time as one of $B_1, B_2, \ldots,$

B_n occurs, and any two of B_i are mutually exclusive, but their union sets consist of the sample space of one event, that is, $B_iB_j = \varphi(i \neq j), \sum_{i=1}^{n} B_i = \Omega, P(B_i) > 0$, then the total probability of event A, $P(A)$, is

$$P(A) = \sum_{i=1}^{n} P(B_i)P(A/B_i). \tag{3.10}$$

3. *Bayes' Formula*. Assume the occurring condition of event B_i (i = 1,2,..., n) is same as that in (2), then the probability of occurrence of event B_i after the event A occurred, is denoted by

$$P(B_i/A) = \frac{P(B_i)P(A/B_i)}{\sum_{i=1}^{n} P(B_i)P(A/B_i)} \quad (i = 1, 2, \ldots). \tag{3.11}$$

Equation (3.11) is Bayes' Formula. It means that once event A occurred in experiment, (3.11) is used to reassess the cause B_i, so the probability $P(B_i/A)$ is called posterior probability.

3.2.2 Random Variable and its Distribution

If the outcome of a random experiment can be described by one numerical variable, and this numerical value is determined by a certain probability, then the variable is named a random variable. In mathematical terms, it can be described that the set Ω of all sample points e is one sample space in a random experiment, and X is a real-valued function defined on the sample space, that is,

$$e \in \Omega, \ X(e) \in R.$$

If there exist real values $a < b$, such that the set of sample points satisfies

$$\{e|a \leq X(e) \leq b\},$$

then this set is an event, and the function $X(e)$ is referred to as a random variable. If $\alpha = -\infty$, event $\{e| - \infty \leq X(e) \leq b\}$ can be described by $\{X \leq b\}$ for short. Its probability measurement,

$$F(x) = P(X \leq x) \tag{3.12}$$

is defined as the distribution function of random variable X. x can be any given real value.

The general random variable X can be classified into a discrete random variable and a continuous random variable according to its different possible values.

For continuous random variables, another function to express its probability is the probability density function $f(x)$, which is defined by,

$$f(x) = \lim_{\Delta x \to 0} \frac{1}{\Delta x} P(x < X < x + \Delta x), \tag{3.13}$$

which can also written in incremental format,

$$P(x < X < x + \Delta x) \approx f(x)\Delta x. \tag{3.14}$$

Formula (3.14) can be interpreted as the probability under the condition that random variable X is in the interval $(x, x + \Delta x)$ and $\Delta x \to 0$. Obviously, the probability of random variable X between a and b is,

$$P(a < X \leq b) = \int_a^b f(x)\mathrm{d}x \tag{3.15}$$

and the relationship between (3.15) and distribution function $F(x)$ in formula (3.12) can be written as,

$$F(x) = \int_{-\infty}^{x} f(x)\mathrm{d}x \tag{3.16}$$

and

$$f(x) = \frac{\mathrm{d}F(x)}{\mathrm{d}x}. \tag{3.17}$$

For a discrete random variable (as shown in Fig. 3.1), X may be x_i ($i = 1, 2, \ldots, n$), then its probability density function is defined as

$$p(x) = \begin{cases} P(X = x_i) & x = x_i \\ 0 & x \neq x_i \end{cases} \tag{3.18}$$

and the distribution function is

$$F(x) = \sum_{x_i \leq x} p(x_i). \tag{3.19}$$

3.2.3 Numeral Characteristics of Random Variable

In many practical problems, we can specify the characteristics of random variables by finding the average value of random variables and the degree of value dispersion. The two most commonly used methods are introduced as follows.

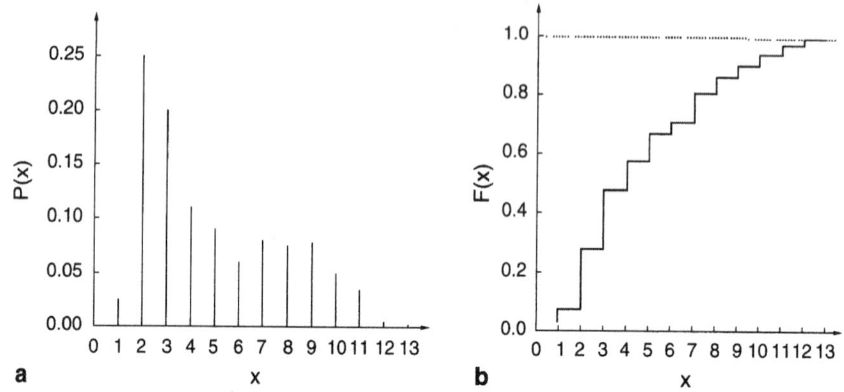

Fig. 3.1 The relative function of discrete random variable (**a**) probability density function; (**b**) distribution function

3.2.3.1 Mathematical Expectation (Mean Value)

Discrete random variable X can be $x_1, x_2, \ldots$, and its corresponding probability is

$$P(X = x_i) = p_i \quad i = 1, 2, \ldots$$

Then mathematical expectation or expectation, $E(X)$, is defined as

$$E(X) = \sum_{i=1}^{\infty} x_i p_i. \tag{3.20}$$

For a continuous random variable X, when its density function is $f(x)$, we have

$$E(X) = \int_{-\infty}^{\infty} x f(x) \mathrm{d}x. \tag{3.21}$$

For the mathematical expectation of a set of random variables X_i ($i = 1, 2, \ldots, n$), there are characteristics such as described as follows

$$E\left(\sum_{i=1}^{n} X_i\right) = \sum_{i=1}^{n} E(X_i). \tag{3.22}$$

3.2.3.2 Variance

Discrete random variable X is denoted as σ^2, which is defined by,

$$\sigma^2 = \sum_{i=1}^{n} (x_i - m)^2 p_i, \tag{3.23}$$

where $m = E(X)$, that is average value. Obviously, σ^2 represents the degree of dispersion of its value deviating from the average value m.

For a continuous random variable X, we get,

$$\sigma^2 = \int_{-\infty}^{\infty} (x - m)^2 f(x) \mathrm{d}x. \tag{3.24}$$

Some properties and applications related to other numerical characteristics of random variables will be discussed in Sect. 3.5.1.

3.2.4 Convolution of Random Variable

Suppose two random variables X and Y are independent, and they have probability density functions $f_1(x)$ and $f_2(y)$, respectively, then $Z = X + Y$ is still a random variable. The probability density function of Z is

$$\begin{aligned} \phi(z) &= \int_{-\infty}^{\infty} f(x, z - x) \mathrm{d}x \\ &= \int_{-\infty}^{\infty} f_1(x) f_2(z - x) \mathrm{d}x. \end{aligned} \tag{3.25}$$

Its distribution function is

$$F(z) = \int_{-\infty}^{z} \int_{-\infty}^{\infty} f_1(x) f_2(z - x) \mathrm{d}x \, \mathrm{d}z. \tag{3.26}$$

If X and Y are discrete random variables, then the distribution function is

$$F(z) = \sum_{x_i + y_j < z} P(X = x_i,\ Y = y_j) = \sum_{x_i + y_j < z} p(z)$$

and

$$p(z) = \sum_{i=0}^{\infty} P(X = x_i) P(Y = z - x_i). \tag{3.27}$$

Equations (3.25) and (3.27) are the convolution formula of probability density functions.

3.2.5 Several Usual Random Variable Distributions

Now, we introduce several probability distributions that are often used in probability analysis of electrical power systems.

3.2.5.1 Binomial Distribution

Let $P(X = 1) = p$ and $P(X = 0) = 1 - p$, then the random variable X is a Bernoulli distribution with parameter p. It is used for describing random phenomena with only two states. Components, such as transformer, transmission lines, etc., have random phenomena with only two states: running (denoted as 1) and outage (denoted as 0). In an experiment repeated n times, the number of occurrences of event A (success) is r, and the probability is p^r, then nonoccurrence number (failure) is $(n - r)$, and the probability is $(1 - p)^{n-r}$. The probability of random variable X is

$$P_r(n,p) = \binom{n}{r} p^r (1-p)^{n-r} \quad r = 0, 1, 2, \ldots, n, \tag{3.28}$$

where

$$\binom{n}{r} = \frac{n!}{r!(n-r)!}$$

is called the binomial distribution and n, p are known constants (parameters).

The mean and variance of the binomial distribution are, respectively,

$$E(X) = np, \tag{3.29}$$

$$\sigma^2 = np(1-p). \tag{3.30}$$

3.2.5.2 Uniform Distribution

Let continuous random variable X have probability density

$$f(x) = \begin{cases} \dfrac{1}{b-a}, & a < x < b \\ 0, & \text{otherwise} \end{cases}. \tag{3.31}$$

Then X is uniformly distributed in the interval (a, b), it is written as $X \sim U(a, b)$.

It is easily known that $f(x) \geq 0$, and $\int_{-\infty}^{\infty} f(x)\mathrm{d}x = 1$.

The random variable X, which is uniformly distributed in interval (a, b), has equal probability to be in any subinterval with equal length in the interval (a, b).

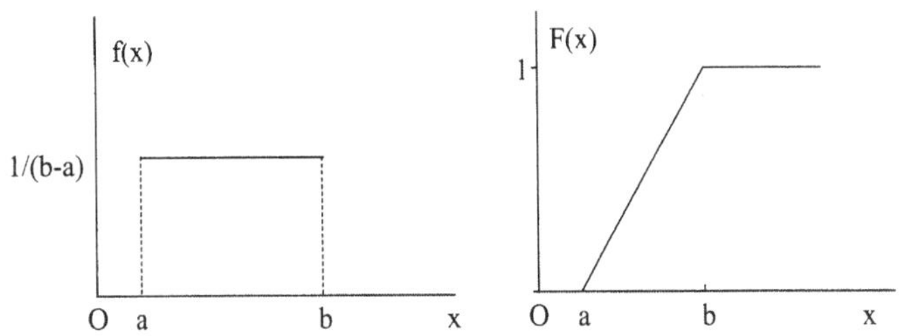

a Probability density function **b** Distribution function

Fig. 3.2 Uniform distribution

In other words, the probability that it is in a subinterval in (a, b) depends on the length of the subinterval, and is independent of the location of the subinterval. In fact, for any subinterval, $(c, c + l)$ and $a \le c < c + l \le b$, with length l, we get

$$P(c < X \le c + l) = \int_c^{c+l} f(x)\mathrm{d}x = \int_c^{c+l} \frac{1}{b-a}\mathrm{d}x = \frac{l}{b-a}. \tag{3.32}$$

The corresponding distribution function is

$$F(x) = \begin{cases} 0, & x < a \\ \dfrac{x-a}{b-a}, & a \le x < b. \\ 1, & x \ge b \end{cases} \tag{3.33}$$

The figures of $f(x)$ and $F(x)$ is shown in Fig. 3.2.

3.2.5.3 Normal Distribution

When a continuous random variable X has such a probability density function, described as follows

$$f(x) = \frac{1}{\sigma\sqrt{2\pi}}\mathrm{e}^{-(x-m)^2/2\sigma^2} \quad -\infty < x < \infty \tag{3.34}$$

then X is referred to as normally distributed, and is written as $N(m, \sigma^2)$ for short. Where, σ is positive value and m can be any constant.

Two parameters of the normal distribution, m and σ, are its mean value and variance. m is also the position parameter, and it is depended on the movement of density curve on horizontal axis. σ is also referred as a scale parameter and is depended on the shape of the curve. A normal distribution with parameter $m = 0$ and

$\sigma = 1$, $N(0, 1)$, is named a standard normal distribution. Its probability density function is

$$f(x) = \frac{1}{\sqrt{2\pi}} e^{-x^2/2}.$$

The distribution function of a normal distribution is

$$F(x) = \frac{1}{\sigma\sqrt{2\pi}} \int_{-\infty}^{x} e^{-(u-m)^2/2\sigma^2} du. \tag{3.35}$$

Formula (3.35) cannot be approximately represented by general elementary function. Let $= \frac{u-m}{\sigma}$, then (3.35) is changed into a standard normal distribution

$$F(x) = \frac{1}{2\pi} \int_{-\infty}^{\frac{u-m}{\sigma}} e^{-z^2/2} dz. \tag{3.36}$$

The value of a standard normal distribution can be obtained by the integration. Figure 3.3 is the density function curve of the normal distribution.

The following formula can be deduced from the normal distribution function $F(x)$

$$R(x) = 1 - F(x) = \int_{x}^{\infty} f(u) du.$$

3.2.6 Markov Process

Many random phenomena encountered in probability evaluation of electrical power systems can be described by time-dependent random variables, which are called random processes, and written as $X(t)$.

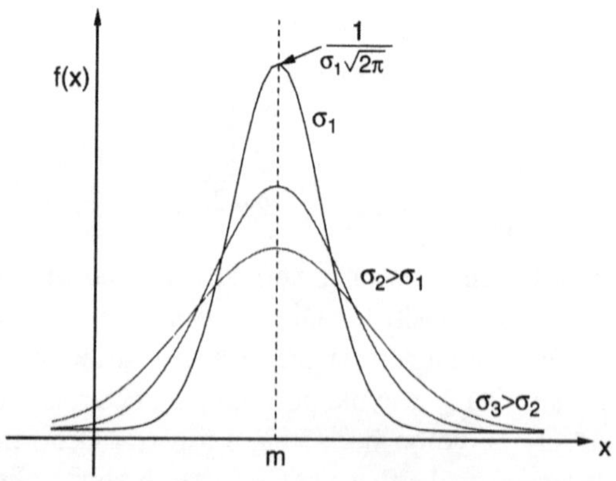

Fig. 3.3 The density function of normal distribution

The random process, $X(t)$, can be described by their conditional probability. The distribution of random variable $X(t_n)$ at any time t_n is related to all random variable $X(t_i)$ at all past time instants t_i $(1 \leq i \leq n-1)$. In discrete time, $X(t_i)$ will be written as X_i for short hereinafter, and its probability can be described in the following form

$$P[X_n = x_n | (X_1 = x_1, X_2 = x_2, \ldots, X_{n-1} = x_{n-1})]. \tag{3.37}$$

If the distribution of X_n is only related to the most recent state, and not to any previous states, i.e.,

$$P(X_n = x_n | (X_{n-1} = x_{n-1}). \tag{3.38}$$

It is called a Markov Process or memory-less process.

Generally, the Markov process with discrete time and discrete state space is referred to as a Markov chain. This process can be described by the conditional probability of (3.24). For brevity, we denote the present state as i and the next state as j, then (3.38) can be written as

$$P(X_n = j | X_{n-1} = i) = p_{ij}. \tag{3.39}$$

In which p_{ij} is the transition probability from state i to state j. If the transition probability in one transition process is independent of time t, and is also a constant, i.e.,

$$P(X_n = j | X_{n-1} = i) = P(X_k = j | X_{k-1} = i) = p_{ij}$$

then this Markov chain is time homogeneous. In the later study of probability problems of electrical power systems, we are only concerned with the time-homogeneous Markov chain.

As the transition from state i to state j is completed in one step, p_{ij} is referred to as the one-step transition probability. If there are n states, then the one-step transition probability can be written in matrix format

$$P = \begin{bmatrix} p_{11} & p_{12} & \cdots & p_{1n} \\ p_{21} & p_{22} & \cdots & p_{2n} \\ \cdots & \cdots & \cdots & \cdots \\ p_{n1} & p_{n2} & \cdots & p_{nn} \end{bmatrix}, \tag{3.40}$$

where $p_{ij} \geq 0$ $i, j = 1, 2, \ldots, n$,

$$\sum_{j=1}^{n} p_{ij} = 1 \quad i = 1, 2, \ldots, n. \tag{3.41}$$

Equation (3.41) describes the property of a stochastic matrix.

If the process of transferring from state i to state j is completed by m steps, then

$$p_{ij}^{(m)} = P(X_{n+m} = j | X_n = i).$$

$p_{ij}^{(m)}$ is m-step transition probability, which can be described as a function of the one-step transition probability matrix

$$P^{(m)} = P^m. \tag{3.42}$$

If the initial state probability vector of process is

$$P(0) = [p_1(0),\ p_2(0), \ldots,\ p_n(0)],$$

the probability in state j after m-steps can be obtained by the following formula

$$P^{(m)} = P(0)P^m. \tag{3.43}$$

In each component of $P(0)$, when a process begins from one specified state i, it is often set $P_i(0) = 1$, and other components are all 0.

3.3 Probabilistic Model of Power Systems

There are many random factors that influence the operation and planning of electrical power systems. When assessing system operation security, we mainly consider such random factors as load fluctuation, random failure of generator units, and random failure of transmission and distribution components, etc. Now, we introduce their models respectively as follows.

3.3.1 Probabilistic Model of Load

The loads in an electrical power system vary continuously, so the system operation must adapt to this kind of variation at any time. Therefore, to build an appropriate load model is very important for system security assessment. According to the requirement of security assessment, the load model can be divided into time-instance load model and time-period load model.

For the time-instance load model of node i, we generally describe it by a normal distribution $N(\mu_i, \sigma_i)$, in which parameter μ_i is the numerical expectation of this distribution and also is usually the forecast load value of node i at this time. Parameter σ_i is the variance of this distribution, describing the degree that the real load value w_i deviates from the forecast value μ_i of node i. To obtain a load sample value w_i on node i, a random number y_i of standard normal distribution $N(0, 1)$ is first generated, which is then used to modify the forecast load value for the node

$$w_i = u_i + \sigma_i \cdot y_i, \tag{3.44}$$

where w_i is the load sample value of node i, μ_i is the load forecasting value, σ_i is the variance of load distribution of the node i.

When security assessment requires a load model over a certain time period T, we should first obtain the load duration curve for this time period, then change it into a probability distribution of load. Suppose point (x, t) of the load duration curve represents the time duration t for which the load is greater than or equal to x, i.e.,

$$t = F(x). \tag{3.45}$$

Dividing both sides of the above equation by time period T we have

$$f(x) = F(x)/T, \tag{3.46}$$

where $f(x)$ can be regarded as the probability that load is larger than or equal to x, i.e., distribution of the load.

3.3.2 *Probabilistic Models of Power System Components*

When assessing the security and reliability of power system operation, we should first build the probabilistic model of operating components, which usually include such components as transmission lines, transformers, generator units, and so on.

Each transmission line has two operating states, normal operating state and failure state. The basic parameter to represent the characteristic of a transmission line operation is its forced outage rate (FOR), which is generally denoted as q. Suppose the transmission capacity of transmission line is c, we have,

$$P(X = x_i) = \left\{ \begin{matrix} 1 - q & x_i = c \\ q & x_i = 0 \end{matrix} \right\}. \tag{3.47}$$

This means that probability of the available capacity of this transmission line being c is equal to $1 - q$, and the probability of its available capacity being 0 is equal to q. The relative cumulative probability (that is the distribution function) is

$$P(X \leq x_i) = \left\{ \begin{matrix} 1 & x_i = c \\ q & x_i = 0 \end{matrix} \right\}. \tag{3.48}$$

This denotes the probability 1 of the available capacity of this transmission line being less than or equal to c, and the probability q of the available capacity of this transmission line being less than or equal to 0.

When assessing the reliability of electrical power systems, it is more convenient to use outage capacity of a component. Comparing with the expression for available capacity X, outage capacity is denoted by $\overline{X}$. Then we get,

$$p(\overline{X} = x_i) = \begin{Bmatrix} q & x_i = c \\ 1 - q & x_i = 0 \end{Bmatrix}, \tag{3.49}$$

which denotes the probability is q of the outage capacity of this transmission line being equal to c, and the probability is $1 - q$ of the outage capacity of this transmission line being equal to 0. The cumulative probability of outage capacity is

$$P(\overline{X} \geq x_i) = \begin{Bmatrix} 1 & x_i = 0 \\ q & x_i = c \end{Bmatrix}, \tag{3.50}$$

which denotes the probability 1 of the outage capacity of this transmission line being greater than or equal to 0, and the probability q of the outage capacity of this transmission line being greater than or equal to c.

The probabilistic models of the transformer and generator, which are usually considered as two-state components, are similar to that of the transmission line. Their probabilistic model can be constructed according to (3.47) and (3.48).

Sometimes the generator needs to be treated more fastidiously, because the generator may include prime mover. The boiler and prime mover of thermal power generator units are comparatively more complex, and failure of the ancillary components will influence the output power of the generator unit. Therefore, besides normal operation and failure state, a generator unit may also have degraded operating states. When building the probabilistic model of generator unit, it may be required to know the probability of degraded operation.

3.3.3 *Outage Table of Power System Components*

The states of power system components, such as the state of a generator unit, can be described in the form of an outage table. Table 3.1 gives an outage table of a two-state generator unit. This table can be set up according to (3.47) and (3.48).

It can be seen from above table that the probability is 1 of the outage capacity of the generator unit being greater than or equal to 0, and the probability is q of the outage capacity of the generator unit being greater than or equal to$c$. In other words, the probability is 1 of the available capacity of the generator unit being less than or equal to c, and the probability is q of the available capacity of the generator unit being less than or equal to 0.

In real power systems, a power generation plant always has many generator units with different types and capacities, we often combine these generator units by using

Table 3.1 The outage table of a generator unit

Available capacity (MW)	Outage capacity (MW)	Deterministic probability, p	Cumulative probability, P
c	0	$1-q$	1
0	c	q	q

the convolution formula, and hence obtain the outage table of this power generation plant. The outage table gives the probability of various available capacities of power generation plant under a definite step size Δx.

Suppose there is an outage table of $n-1$ generator units, the outage capacity $\overline{X}$ is a random variable of a discrete distribution, and its probability is $p_{n-1}(\overline{X})$. When the nth new generator unit, with available capacity c_n and forced outage rate q_n, is added, the new probability $p_n(\overline{X})$ of the outage capacity can directly deduced using the convolution formula, (3.27)

$$p_n(\overline{X}) = p_{n-1}(\overline{X})p(0) + p_{n-1}(\overline{X} - c_n)p(c_n).$$

The two-state probability of the nth generator unit can be deduced by (3.49)

$$p(c_n) = q_n, \quad p(0) = 1 - q_n.$$

Substituting them into the above equation, we can obtain the recursive formula

$$p_n(\overline{X}) = p_{n-1}(\overline{X})(1 - q_n) + p_{n-1}(\overline{X} - c_n)q_n. \tag{3.51}$$

Equation (3.51) can recursively calculate the state probability, and can also calculate the cumulative probability $P_n(\overline{X})$. To do so we only need to change $P_n(\overline{X})$ and $P_{n-1}(\overline{X})$ in the formula into the corresponding cumulative probabilities. But the initial condition of the calculation is that probability $p_{n-1}(\overline{X} - c_n) = 0$, and cumulative probability $P_{n-1}(\overline{X} - c_n) = 1$ when $\overline{X} \leq c_n$,

$$p_n(\overline{X}) = p_{n-1}(\overline{X})p(0) + p_{n-1}(\overline{X} - c_n)p(c_n).$$

[Example 3.1] A power generation system has two generator units, and their power capacities are 30 MW and 40 MW, respectively, their forced outage rates (FOR) are 0.04 and 0.06, respectively. Form the outage table of this power generation system.

[Solution] The outage table of the power generation systems refers to the probability table of various power capacity states of total system, and can be formed by using the recursive formula described above, according to the outage table of each generator unit. Therefore, first we should set up the outage table of each generator unit. Let step size be $\Delta X = 10$ MW, then the outage tables of these two generator units are shown in Tables 3.2 and 3.3.

Then the outage table of the power generation system can be formed by recursive formula, and is shown in Table 3.4.

Table 3.2 Outage table of 30-MW generator unit

i	$\overline{X}_i$ (MW)	P_i	p_i
0	0	$0.100000 \times 10^{+01}$	$0.960000 \times 10^{+00}$
1	10	0.400000×10^{-01}	$0.000000 \times 10^{+00}$
2	20	0.400000×10^{-01}	$0.000000 \times 10^{+00}$
3	30	0.400000×10^{-01}	0.400000×10^{-01}

Table 3.3 Outage table of 40-MW generator unit

i	$\overline{X}_i$ (MW)	P_i	p_i
0	0	$0.100000 \times 10^{+01}$	$0.940000 \times 10^{+00}$
1	10	0.600000×10^{-01}	$0.000000 \times 10^{+00}$
2	20	0.600000×10^{-01}	$0.000000 \times 10^{+00}$
3	30	0.600000×10^{-01}	$0.000000 \times 10^{+00}$
4	40	0.600000×10^{-01}	0.600000×10^{-01}

Table 3.4 Outage table of 30-MW and 40-MW generator units

i	$\overline{X}_i$ (MW)	P_i	p_i
0	0	$0.100000 \times 10^{+01}$	0.902400E + 00
1	10	0.976000×10^{-01}	$0.000000 \times 10^{+00}$
2	20	0.976000×10^{-01}	$0.000000 \times 10^{+00}$
3	30	0.976000×10^{-01}	0.376000×10^{-01}
4	40	0.600000×10^{-01}	0.576000×10^{-01}
5	50	0.240000×10^{-02}	$0.000000 \times 10^{+00}$
6	60	0.240000×10^{-02}	$0.000000 \times 10^{+00}$
7	70	0.240000×10^{-02}	0.240000×10^{-02}

In order to illustrate computing process, we introduce the application of the recursive formula for the case with $P_3 = 0.0976$ when $i = 3$, $\overline{X} = 30$ MW in Table 3.4. Let 30-MW power generator unit be component a and 40-MW power generator unit be component b. Then, we know $n_a = 3$. According to (3.51), the cumulative probability of combined equivalent generator unit c in condition $k = 3$ is

$$\begin{aligned} P_c(3) &= \sum_{i=0}^{3} p_a(i)P_b(k-i) \\ &= p_a(0)P_b(3) + p_a(1)P_b(2) + p_a(2)P_b(1) + p_a(3)P_b(0). \end{aligned}$$

Substituting the corresponding numerical values in outage table into above equation, we have

$$P_c(3) = 0.96 \times 0.06 + 0 \times 0.06 + 0 \times 0.06 + 0.04 \times 1.0 = 0.0976.$$

As to parallel transmission lines and transformers, we can also form their outage tables. Once outage tables of components of power systems have been obtained, we can simplify the security or reliability evaluation process of system operation.

3.4 Monte Carlo Simulation Method

3.4.1 Fundamental Theory of Monte Carlo Simulation Method

Monte Carlo simulation method [33, 38] is a simulation approach based on probability and statistics theory and methodology. At present, the Monte Carlo simulation method has been applied to many fields of engineering and scientific theory, with the advantage of simple principles and realization, insensitivity to the dimension of problems, avoidance of any constraining assumptions, and strong adaptability.

In the Monte Carlo simulation method, the state of each component in the system is obtained by sampling. The components include various system equipment, such as generators, transmission lines, transformers, etc., and different load levels. Let the state of a power system be represented by the vector $x = (x_1, x_2, x_3, \ldots, x_m)$ where x_i is the state of the ith component. The set of all possible states x, arising from combinations of component states is denoted by **X**, the state space. Suppose $F(x)$ is one experiment for a given state x. The objective of the test is to verify whether that specific configuration of generators and circuits is able to supply that specific load. The reliability indices correspond to the expected value, $E(F)$, of various Types of test functions over all possible states:

$$E(F) = \sum_{x \in X} F(x) \cdot P(x). \tag{3.52}$$

For example, the system loss of load probability (LOLP) is equal to the numerical expectation of a dual-value function as follows

$$F(x) = \begin{cases} 1, & \text{if } x \text{ is a failed state} \\ 0, & \text{if } x \text{ is a normal state} \end{cases}.$$

The expectation $\hat{E}(F)$ of experiment function $F(x)$ can be estimated by (3.53)

$$\hat{E}(F) = \frac{1}{\mathrm{NS}} \sum_{i=1}^{\mathrm{NS}} F(x_i), \tag{3.53}$$

where $\hat{E}(F)$ is the estimate of the expected value of the test function, NS the number of samplings, x_i the ith sampled value, and $F(x_i)$ is the test result for the ith sampled value.

It can be seen from (3.53) that $\hat{E}(F)$ is the estimated value rather than the real value of $E(F)$. Since x and $F(x)$ are random variables, $\hat{E}(F)$ is also a random variable. $\hat{E}(F)$ is the average value of the function $F(x)$. The variance of $\hat{E}(F)$ is

$$V(\hat{E}(F)) = \frac{V(F)}{\mathrm{NS}}, \tag{3.54}$$

where $V(F)$ is the variance of the test function F, and its estimated value $\hat{V}(F)$ is

$$\hat{V}(F) = \frac{1}{\mathrm{NS}-1}\sum_{i=1}^{\mathrm{NS}}(F(x_i) - \hat{E}(F))^2. \tag{3.55}$$

Now we will discuss the convergence, convergence rate, calculation error, and computing process of the Monte Carlo simulation method.

3.4.1.1 The Convergence of the Monte Carlo Simulation Method

From (3.53), we know that $\hat{E}(F)$ is often used as the estimate of the reliability indices when we evaluate a system's reliability using Monte Carlo simulation method.

It can be understood from the Kolmogorov Strong Law of Large Numbers that if $\{F(x_i), i = 1, 2, \ldots\}$ is a sequence of independent and identically distributed random variables and its numeral expectation exists, then,

$$P\left(\lim_{\mathrm{NS}\to\infty}\left|\frac{1}{\mathrm{NS}}\sum_{i=1}^{\mathrm{NS}}F(x_i)\right| = E(F)\right) = 1, \tag{3.56}$$

i.e.,

$$P\left(\lim_{\mathrm{NS}\to\infty}\hat{E}(F) = E(F)\right) = 1. \tag{3.57}$$

In other words, the estimated value $\hat{E}(F)$ of the reliability indices converges to $E(F)$ with probability 1 when $\mathrm{NS} \to \infty$ by using the Monte Carlo simulation method.

3.4.1.2 The Convergence Rate of the Monte Carlo Simulation Method

According to the Central Limit Theorem, if $\{F(x_i), i = 1, 2, \ldots\}$ is a sequence of independent and identically distributed random variables and its numerical expectation exists, and finite variance $V(F) \neq 0$, then, when $\mathrm{NS} \to \infty$, the random variable,

$$Y_{\mathrm{NS}} = \frac{(\hat{E}(F) - E(F))}{(\sqrt{V(F)}/\sqrt{\mathrm{NS}})} \tag{3.58}$$

tends to the standard normal distribution $N(0, 1)$, hence,

$$P(Y_{\mathrm{NS}} < x_\alpha) \to \frac{1}{\sqrt{2\pi}}\int_{-\infty}^{x_\alpha} \mathrm{e}^{-\frac{1}{2}x^2}\,\mathrm{d}x. \tag{3.59}$$

It is obvious that the random sample $\{F(x_i), i=1, 2, \ldots\}$ of test function $F(x)$ satisfies above condition according to the Monte Carlo simulation method. Thus for any $x_\alpha > 0$, we have

$$P(|Y_{NS}| < x_\alpha) = P\left(\left|\hat{E}(F) - E(F)\right| < \frac{x_\alpha\sqrt{V(F)}}{\sqrt{NS}}\right)$$
$$\doteq \frac{1}{\sqrt{2\pi}}\int_{-\infty}^{x_\alpha} e^{-\frac{1}{2}x^2}dx = 1 - \alpha. \tag{3.60}$$

This indicates that

$$\left|\hat{E}(F) - E(F)\right| < \frac{x_\alpha\sqrt{V(F)}}{\sqrt{NS}} \tag{3.61}$$

has probability $1 - \alpha$. In general, α is also referred as the confidence of estimation, $1 - \alpha$ is the confidence level. α and x_α have the relation of one to one correspondence. Equation (3.61) indicates that the convergence order with which estimated value $\hat{E}(F)$ converges to the real value $E(F)$ is $O(NS^{-1/2})$.

3.4.1.3 The Error of the Monte Carlo Simulation Method

In general, variance coefficient β is often used to represent the estimated error

$$\beta = \frac{\sqrt{V(\hat{E}(F))}}{\hat{E}(F)}. \tag{3.62}$$

Equation (3.54) is substituted into the above equation, we obtain

$$\beta = \frac{\sqrt{V(F)/NS}}{\hat{E}(F)}. \tag{3.63}$$

Hence,

$$NS = \frac{V(F)}{(\beta\hat{E}(F))^2}. \tag{3.64}$$

Equation (3.64) shows that the computational effort of Monte Carlo sampling, given by the required sample size NS, does not depend on the number of states x or on the complexity of the test function F. The required number of samplings NS depends on the variance of the random variable and on the desired accuracy β. If we wish to reduce the number of samplings while maintaining the same accuracy, we must find ways to reduce the variance.

3.4.1.4 The Flowchart of the Monte Carlo Simulation Method

The flowchart of Monte Carlo simulation method is shown in Fig. 3.4.

3.4.2 *Sampling of System Operation State*

3.4.2.1 Random Number

When the operating process of electrical power systems is simulated by using the Monte Carlo simulation method, it is required to generate random variables with various probability distributions, in which the simplest, the most basic, and the most important random variable is the random variable with uniform distribution on interval [0, 1]. Generally, the sample value of a random variable with uniform distribution on interval [0, 1] is referred to as a random number, and the samples of random variables with other distributions can be obtained by means of a random number with uniform distribution on interval [0, 1]. Thus we can say that the random number is a basic tool of random sampling.

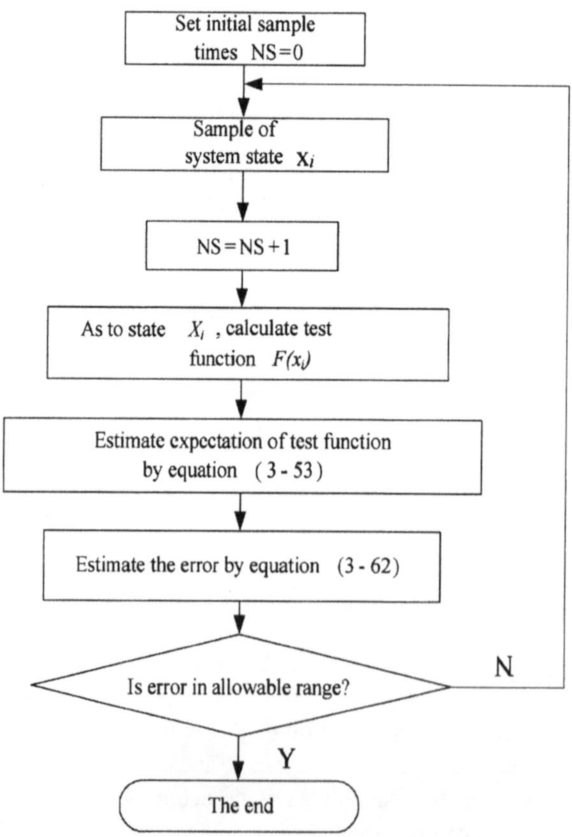

Fig. 3.4 The computing flowchart of the Monte Carlo simulation method

One approach to generate random numbers is to use physical simulation. But this is not very practical due to the high cost and the difficulty of generating random numbers repeatedly.

Now the most widely used approach to generate random numbers is by using a mathematical method with the aid of a computer. The advantages of this method include low memory requirement, fast generating speed, and convenient regenerating without restriction on computer conditions. But these random numbers are obtained by a deterministic recursive equation, and periodic phenomena exist. Also, once the initial value is determined, all random numbers will be uniquely determined, which does not meet the requirement for a true random number. Consequently, random numbers generated using the mathematical method are called pseudorandom numbers. In application, as long as these pseudorandom number sequences are tested by a series of statistical methods, they can be applied as a surrogate for "true" random numbers.

There are many methods to generate pseudorandom numbers, such as mid-square method, mid-product method, and multiplicative and additive congruence method. At present, the standard algorithms to generate uniform distribution can be directly found in many books that introduce advanced computer languages, and the existing functions also can be found directly. Therefore, we only need to call these functions when sampling electrical power systems states.

3.4.2.2 Sampling of System Operation States

In the Monte Carlo simulation method, the power system state is sampled according to element distribution functions. Though there are a large number of uncertainties that affect the system reliability indices, we only consider some primary uncertainties associated with the generator availability, the branch availability, and the system load variation.

According to the Monte Carlo simulation method, the state of the electrical power systems is determined by samples from the probability distribution functions of each component. In reliability evaluation of electrical power systems, uncertain factors such as random failure of generator, random failure of transmission line, and random fluctuation of load are always considered.

For generators and branches, two operating states are employed. Their probability distribution functions are two-point distribution according to Sect. 3.3.2. We can generate a random number with uniform distribution $U(0,1)$, compare this random number with device forced outage rate q, then determine whether device state is in failure state or running state, as shown by (3.65).

$$
\begin{aligned}
\eta_g^i &= \begin{cases} 1, & x \geq q_{\mathrm{G}i}, \ \text{generator } i \text{ is in running state} \\ 0, & x < q_{\mathrm{G}i}, \ \text{generator } i \text{ is in failure state} \end{cases} \\
\eta_b^i &= \begin{cases} 1, & x \geq q_{\mathrm{L}i}, \ \text{transmission line } i \text{ is in running state} \\ 0, & x < q_{\mathrm{L}i}, \ \text{transmission line } i \text{ is in failure state} \end{cases}
\end{aligned}
\tag{3.65}
$$

where, η_g^i and η_b^i is the state of generator i and transmission line i, respectively, q_{Gi} and q_{Li} are corresponding forced outage rates of ith generator and ith transmission line, and x is random number following uniform distribution $U(0,1)$.

For the power generator unit, because a power generation plant usually includes many generator units with different Types and capacities, we can take each power generation plant as a multistate generator unit, and directly sample by using the outage table of the power generation plant (shown in Sect. 3.3.3) in order to decrease computing burden.

For the load, the sample value of load can be obtained directly by considering random fluctuation factors according to the method introduced in Sect. 3.3.1.

Thus, we have obtained each component state of system. All these states consist of state vector x, the set of all possible state x, $\mathbf{X}$, is referred to as the state space. Having obtained the system state by sampling, we can now evaluate the system state.

3.4.3 State Evaluation Model

3.4.3.1 Load Flow Model

System state evaluation (that is the process to calculate test function $F(x)$) can use the AC load flow model or DC load flow model. An introduction to AC load flow can be found in Sects. 2.3 and 2.4, and an introduction to DC load flow can be found in Sect. 2.6.1. The reliability evaluation of electrical power systems is generally based on the linearized load flow model, i.e., the DC load flow model. In practice, the accuracy and computing load of the DC load flow mode is satisfactory.

3.4.3.2 Model of Load Curtailment

A considerable number of contingency states in which no load curtailment exists will have been excluded after the contingency analysis of all the sampled contingency states. For those contingency states which may have load curtailment, generation outputs at some buses cannot be maintained due to generating unit contingencies and/or there are some line overloads due to transmission component outages. Generation outputs should be rescheduled to maintain generation demand balance and alleviate line overloads and, at the same time, to avoid load curtailment if possible or to minimize total load curtailment if unavoidable. The following minimization model of load curtailment can be used for this purpose:

$$\text{Min} \sum_{j=1}^{N_\text{D}} P_{\text{D}i}$$

$$
\begin{aligned}
&\text{s.t.}\, \boldsymbol{B\theta} + \boldsymbol{P}_{\mathrm{L}} = \boldsymbol{P}_{\mathrm{G}} + \boldsymbol{P}_{\mathrm{D}}, \\
&\sum_{i=1}^{N_{\mathrm{G}}} P_{\mathrm{G}i} + \sum_{i=1}^{N_{\mathrm{D}}} P_{\mathrm{D}i} = \sum_{i=1}^{N_{\mathrm{D}}} P_{\mathrm{L}i}, \\
&\underline{\boldsymbol{P}}_{\mathrm{G}} \le \boldsymbol{P}_{\mathrm{G}} \le \overline{\boldsymbol{P}}_{\mathrm{G}}, \\
&\boldsymbol{0} \le \boldsymbol{P}_{\mathrm{D}} \le \boldsymbol{P}_{\mathrm{L}}, \\
&|\boldsymbol{T}| \le \overline{\boldsymbol{T}},
\end{aligned}
\tag{3.66}
$$

where, $\boldsymbol{P}_{\mathrm{D}}$ is active power vector of virtual generators representing the amount of load shedding on nodes; $\boldsymbol{P}_{\mathrm{G}}$ is injecting active power vector of generator; $\boldsymbol{P}_{\mathrm{L}}$ is load vector of nodes; $\overline{\boldsymbol{P}}_{\mathrm{G}}$ is the upper limit vector of active power of generators; $\underline{\boldsymbol{P}}_{\mathrm{G}}$ is the lower limit vector of active power of generators; $\boldsymbol{T}$ is the active load flow vector of transmission lines; $\overline{\boldsymbol{T}}$ is the upper vector of active load flow of transmission lines; N_{G} is the number of generator nodes; and N_{D} is the number of load nodes.

The above rescheduled-model is a standard form of linear program and can be solved by linear programming algorithms. The details of the solution are omitted here.

3.4.4 Indices of Reliability Evaluation

There are many possible indices which can be used to measure the adequacy of power systems. Most adequacy indices are basically expected values of a random variable. An expected value is not a deterministic parameter. It is a long-run average of the phenomenon under study. Expectation indices provide valid adequacy indicators which reflect various factors such as system component availability and capacity, load characteristics, etc. In this book, the following indices are calculated.

3.4.4.1 Loss of Load Probability (LOLP)

LOLP represents the probability of load shedding due to various reasons. Its defining equation is

$$
\mathrm{LOLP} = \sum_{C_i \neq 0} p_i, \tag{3.67}
$$

where p_i is the probability of load curtailment value C_i.

3.4.4.2 Expected Energy Not Supplied (EENS)

Expected Energy Not Supplied (EENS) is the expectation of energy-shortage in a period of time. If we want to calculate the expectation of energy-shortage in a year, the defining equation is

$$\text{EENS} = \sum_{C_i \neq 0} C_i \times p_i \times 8{,}760, \tag{3.68}$$

where C_i is the amount of load curtailment and p_i is the corresponding probability.

Since EENS is an energy index, it is significant for economical assessment of reliability, optimizing reliability, and system planning. Thus EENS is a very important index for reliability assessment of the whole system.

3.4.5 Flowchart of Composite System Adequacy Evaluation

Figure 3.5 shows the flowchart of composite system adequacy evaluation using Monte Carlo simulation.

[Example 3.2] To evaluate the reliability of the 5-node system shown in Fig. 2.6, using the Monte Carlo simulation method. The capacity and reliability parameters of system component are shown in Tables 3.5 and 3.6.

[Solution] This system consists of five nodes, seven branches, in which there are two power generation plants, and total installed capacity is 11 and load is 7.3 represented as per unit values.

According to the process shown in Fig. 3.5, we can calculate the reliability indices of system.

At first, system state $\boldsymbol{x}_i$ is sampled by the Monte Carlo simulation method. Then we generate a random number for every component with the aid of a computer, and determine the device state according to this random number according to the approach introduced in Sect. 3.4.2. The random number is generated for each component and the state vector $\boldsymbol{x}_i$ of components is determined. The states of generation and network branch are determined as shown in Tables 3.7 and 3.8.

Thus, we can get a sample state $\boldsymbol{x}_i$. First, we analyze the network topology in this state, and judge whether the system is connected or not. It can be seen from Figs. 3.2–3.6 that the available capacity of system is only the power output, 5.0 p.u., of power generation plant G1 after a contingency occurs on transformer 5 (branch 3.5) and power generation plant G2 is separated from the system. Comparing power output and load, we can find that total power generation plant output (5.0 p.u.) is less than total load (7.3 p.u.), so it cannot meet the balance condition of active power, and partial load must be shed. Here, the state contributes to the reliability indices. System blackout occurs once, and the amount of load shedding is 2.3 p.u.

Thus we have complete one sampling. By resampling, we can get a new system state $\boldsymbol{x}_j$. In this state, the output of power generation plant G1 and G2 are all 5.0 p.u.,

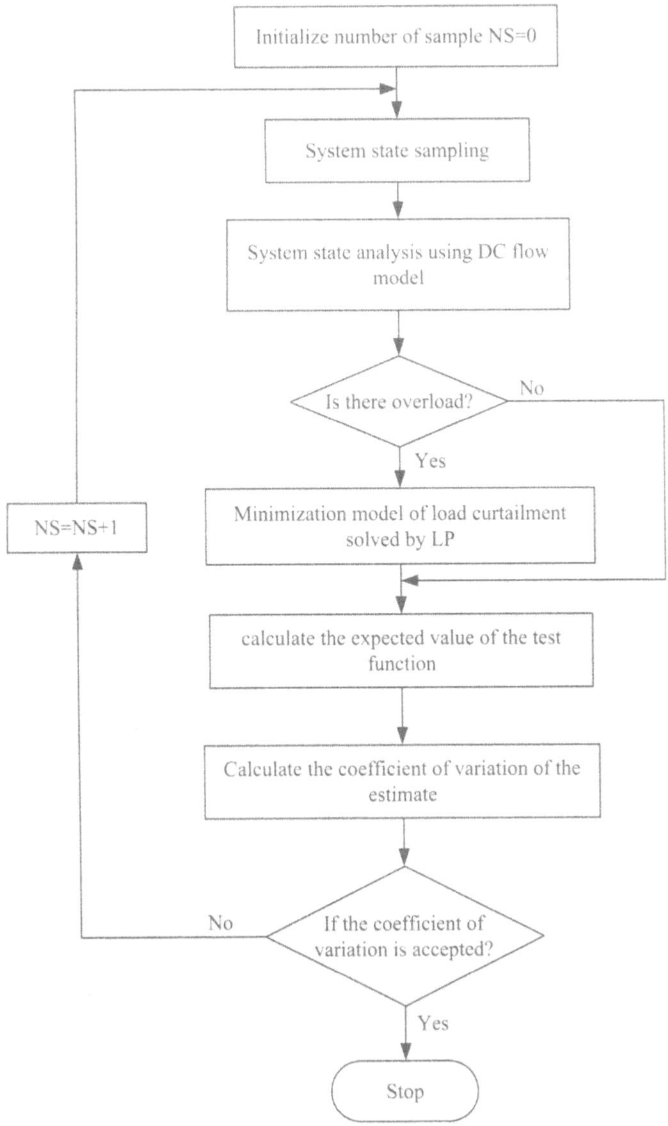

Fig. 3.5 The flowchart of Monte Carlo simulation

and all other branches are running except fault branch 1–2. For x_j, the state evaluation proceeds:

1. Analyze network topology in state x_j, and judge its connectivity. It can be seen from Figs. 3.2–3.6 that system is still connected after contingency occurs on transmission line 1–2.
2. Judge whether active power of system is balanced or not. In state x_j, active power output of system is 10.0 p.u. and load is 7.3 p.u., so the active power output is

Table 3.5 Reliability parameter of power generation component

Power generation plant G1		Power generation plant G2	
Available capacity (p.u.)	Cumulative probability	Available capacity (p.u.)	Cumulative probability
5.0	1.00	6.0	1.00
4.0	0.06	5.0	0.08
3.0	0.04	4.0	0.06
2.0	0.02	3.0	0.04
1.0	0.01	2.0	0.02
0.0	0.01	1.0	0.01
–	–	0.0	0.01

Table 3.6 Reliability parameter of transmission component

Branch node number	Capacity	FOR
1–2	2.0	0.05
1–3	2.0	0.05
2–3	2.0	0.05
2–4	5.0	0.05
3–5	5.5	0.05

Table 3.7 Power generation output determined according to random number with uniform distribution $U(0, 1)$

Power generation plant	$U(0, 1)$ Random number, x	Available capacity of power generation plant (p.u.)
G1	0.6502	5.0
G2	0.1325	6.0

Table 3.8 Branch state determined according to random number with uniform distribution $U(0, 1)$

Branch	1–2	1–3	2–3	2–4	3.5
$U(0, 1)$ Random number	0.32	0.2	0.46	0.75	0.017
Fault rate of branch	0.05	0.05	0.05	0.05	0.05
Branch state	Up	Up	Up	Up	Down

larger than load. Therefore, it can meet the balance demand of active power of system.

3. Judge whether branch load flow can satisfy the restriction of transfer capacity or not. Here, the DC load flow model is used to calculate branch load flow. The calculated power outputs of power generation plant G1 and G2 are 5.0 p.u. and 2.3 p.u., respectively, and branch load flow data are shown in Table 3.9.
   It can be seen from Table 3.9 that overload has occurred on branch 2–3, and it is required to adjust power output of power generation plant.
4. Modify system state by adjusting output of power generation plant.

System state correction can be adjusted according to the state evaluation model of (3.68) introduced in Sect. 3.4.3. After adjusting, the power outputs of power

Table 3.9 The load flow of branches

Branch	1–3	2–3	2–4	3.5
Transfer power (p.u.)	1.6	3	5	2.3
Transfer capacity (p.u.)	2	2	5	5.5
Is restriction condition satisfied	Satisfied	Not satisfied	Satisfied	Satisfied

Table 3.10 The load flow on each branch after adjusting

Branch	1–3	2–3	2–4	3.5
Transfer power (p.u.)	1.6	2	4	3.3
Transfer capacity (p.u.)	2	2	5	5.5
Is restriction condition satisfied	Satisfied	Satisfied	Satisfied	Satisfied

Table 3.11 The reliability indices of five-node system

Reliability indices of system	Calculation results
LOLP	0.13345
EENS (MW h)	300,384.78

generation plant G1 and G2 are 4.0 p.u. and 3.3 p.u. respectively, and branch load flow is shown in Table 3.10.

It can be seen from above calculation that although overload has occurred on a certain branch in this sample state, the overload on the transmission line can be eliminated after adjusting power outputs of power generation plant, thus no load is shed for this sample state.

Repeating the above processes, sampling time after time, summarizing each calculation result, and adding up the times and the amount of load curtailment, we can obtain the reliability indices of the system. From (3.67), we can get LOLP. And from (3.68), we can get EENS.

The calculation results for the 5-node system are shown in Table 3.11.

It can be seen from the above table that LOLP is 0.13345, and expectation of power cut is 300384.78 MW h each year which is about 4.7% energy for the total system. The reliability indices of the system are not satisfactory, and strengthening measures must be adopted to improve system reliability.

Next, we further carry out the statistics and analysis of the convergence rate of reliability indices.

The convergence curve of reliability index EENS is shown in Figs. 3.6 and 3.7. LOLP has similar characteristics which need not be discussed further.

It can be seen from Fig. 3.6 that EENS will converge to a stable numerical value by 20,000 samplings, which is more clearly recognized from the convergence curve of EENS relative error. When the number of samples reaches 20,000, the relative error of EENS is 0.02. If we want to improve the precision of calculation further, and reduce the error of calculation, we should increase the number of samples. For example, when the number of samples is 40,000, the relative error of EENS index is 0.014. It can be seen that the computing load of the Monte Carlo simulation method is in inverse proportion to the square of estimated error. For a definite precision, the only way to reduce the number of samples is to reduce the variance.

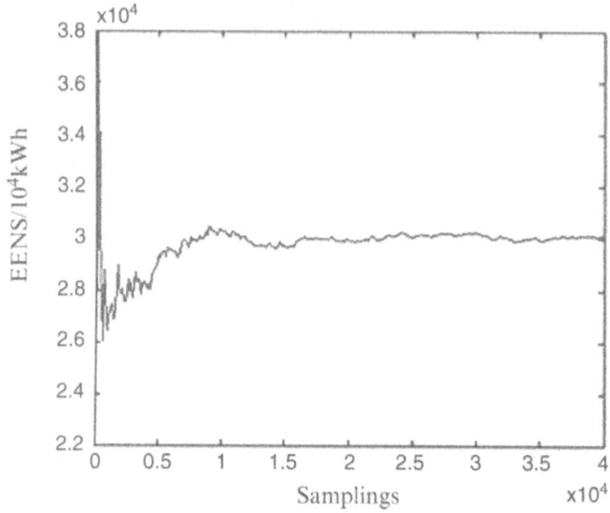

Fig. 3.6 The convergence curve of EENS

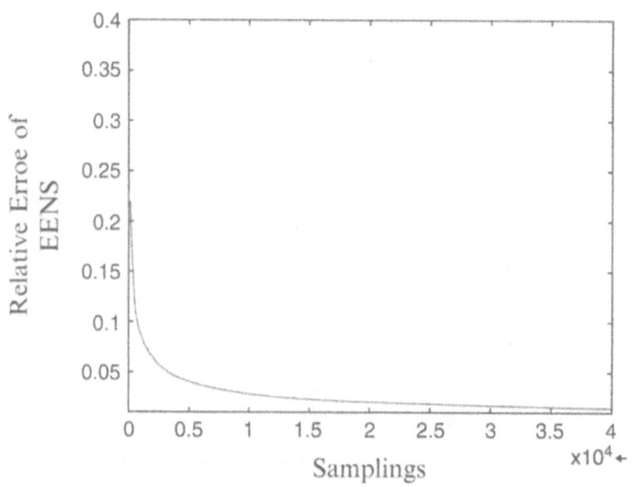

Fig. 3.7 The convergence curve of EENS relative error

3.4.6 Markov Chain Monte Carlo (MCMC) Simulation Method

Markov Chain Monte Carlo (MCMC) was presented in the 1950s. The MCMC method, as a computer-intensive tool, has enjoyed an enormous upsurge in interest over the last few years. MCMC methods are widely used to estimate expectations of functions with respect to complex, high-dimensional distributions, especially in Bayesian analysis [38], statistical physics, and estimation of Value-at-Risk. But there is no report of the MCMC method being applied to reliability evaluation of power systems so far.

As discussed above, the Monte Carlo simulation method is a kind of statistical experimental method, and is more flexible in reliability evaluation of large-scale electrical power systems. It is inefficient because of its proportional characteristic with the required index accuracy. Moreover, Monte Carlo methods are based on sampling independent sequences, which cannot reflect the relativities in the changes of the system states. Additionally, it is difficult to obtain the independent sequences from multivariate probability distributions.

For the above reasons, a new Monte Carlo simulation is proposed to evaluate reliability indices of large-scale system based on Markov chains. First, a Markov chain, whose distribution corresponds to the target probability distribution, is constructed by sampling. Then, this Markov chain is used to sample the state to compute reliability indices [39]. MCMC is a special Monte Carlo method, which applies a Markov chain in the stochastic process to implement dynamic Monte Carlo simulation.

Consider a sequence X_0, X_1, X_2, ... such that X_{k+1} is generated from the conditional distribution for $\{X_{k+1} \mid X_k$ and X_0 represents some initial condition. By the form of the conditional distribution, knowledge of X_k provides the information required to probabilistically characterize the behavior of the state X_{k+1}. That is, the distribution for X_{k+1} depends only on the most recent state, not on the earlier states $X_0, X_1, X_2, \ldots, X_{k-1}$. Hence, $X_0, X_1, X_2, \ldots$ is a Markov chain.

Under standard conditions for Markov chains, the dependence of X_k on any fixed number of early states, say $X_0, X_1, X_2, \ldots, X_m, M < \infty$, disappears as $k \to \infty$. Hence, the density of X_k will approach a stationary form, say $p^*(\cdot)$. That is, as k gets large, the random vectors in the Markov chain will become a dependent sequence with a common density $p^*(\cdot)$. Ignoring the first M iterations in the chain called the "burn-in" period whose density distribution is not $p^*(\cdot)$, we can form an ergodic average

$$E[F(x)] = \frac{1}{n-M} \sum_{k=M+1}^{n} F(X_k). \tag{3.69}$$

Equation (3.69) is a practical realization of the famous ergodic theorem of stochastic processes. The variance of $\hat{E}(f)$ (the estimate of $E(f)$) is

$$V[\hat{E}(F)] = V(F)/n. \tag{3.70}$$

The estimate of $V(f)$ is

$$\hat{V}(F) = \frac{1}{n-1} \sum_{i=1}^{n} [F(x_i) - \hat{E}(F)]^2. \tag{3.71}$$

From the above discussion, we can see that the key idea in the MCMC method is to design Markov chains that have stationary distribution $p^*(\cdot)$. That is, the limit of the ergodic mean in (3.69) will correspond to the desired value $E[f(x)]$ computed with respect to $p(x)$. It is very easy to construct such a Markov chain. One most popular algorithm of the MCMC method is the Gibbs sampler. The Gibbs sampler obtains samples from the full conditional distributions without the difficult task of adjusting the acceptance rate, so it is used in this paper to generate the desired Markov chain.

In the Gibbs sampler for the MCMC method, the system states are obtained from the devices' full conditional probability distributions. In this section, generators and lines in the system are considered with the two-state model. p_i is the failure probability of the components in the system. Loads are fixed to the maximum load of the whole year.

$\boldsymbol{X}$ represents the state variables of generators and lines and is a collection of m univariate components. Number m is the sum of the number of sampled generators and lines. The kth sample $\boldsymbol{X}$ from the Gibbs sampler is

$$\boldsymbol{X}_k = [X_{k1} \quad X_{k2} \cdots \quad X_{km}]^{\mathrm{T}},$$

where X_{ki} denotes the state of the ith component for the kth replicate of $\boldsymbol{X}$ generated via the sampler.

$$X_{ki} = \begin{cases} 1 & \text{the } i\text{th component is in the running state when the } k\text{th sampling,} \\ 0 & \text{the } i\text{th component is in the failure state when the } k\text{th sampling,} \end{cases}$$
$$i = 1, 2, \ldots, m.$$

From the initial state of the components $\boldsymbol{X} = 1$ and initial relative prior probability p_{one}, the process of obtaining $X_{k+1,i}$ is as follows:

1. Under the current system state $\boldsymbol{X}$, the relative posterior probability p_{one} or p_{zero} for the next state is obtained from the full distribution $p\{X_{k+1,i} \mid X_{k\backslash i}\}$ according to the current state of the component, where

$$X_{k\backslash i} = \{X_{k+1,1}, X_{k+1,2}, \ldots, X_{k+1,i-1}, X_{k,i+1}, \ldots, X_{k,m}\}.$$

   The first $i-1$ elements of $X_{k\backslash i}$ represent the sample points at the same $(k+1)$th iteration, whereas the remaining $m-i$ elements are points available from the kth iteration. So p_{one} or p_{zero} is

$$\ln\left[\prod_{j=1}^{i-1} p_j^{\,1-X_{k+1,j}}(1-p_j)^{X_{k+1,j}} \prod_{l=i+1}^{m} p_j^{\,1-X_{k,l}}(1-p_j)^{X_{k,l}}\right],$$

   where the purpose of logarithm is to make the value of p_{one} or p_{zero} between 0 and 1, and convenient for calculating.
2. Calculating the probability of this component for the next state which is in a running state (taking 1),

$$\eta = 1/[\exp(p_{\text{zero}} - p_{\text{one}}) + 1],$$

   where exp is corresponding to the logarithm taken in step 1.
3. Generating a random number u with the uniform distribution and determining the next state of generator or line by comparing η and u.

$$X_{k+1,i} = \begin{cases} 1 & u < \eta \quad \text{the } i\text{th component is in running state} \\ 0 & u \geq \eta \quad \text{the } i\text{th component is in failure state} \end{cases}.$$

4. If the state of the component changes, the p_{one} or p_{zero} obtained from full conditional distribution is used as the relative probability for the next component keeping its current state.

Repeat these steps to obtain a Markov chain whose stationary distribution is the system probability distribution. Then utilizing the convergent Markov chain as system state samples, the reliability indices of the electrical power system is obtained by state evaluation and adjustment.

The computing process of the state evaluation model and active power optimal adjustment model for generators is the same as that of the traditional Monte Carlo simulation method, which can be found in Sect. 3.3.2, and is not repeated here.

It can be proved that the computing load of MCMC method (sampling times) is the same as that of the traditional MC method, and is not influenced by system scale and complexity, thus this method is also suitable for dealing with various complex factors, such as relative load and various operating control strategies and so on. In addition, the techniques for reducing variance in the MC method can also be used in the MCMC method.

In essence, the MCMC method is a kind of special MC method, which applies Markov chains in the random process for MC simulation. This not only inherits the advantages of traditional Monte Carlo simulation methods, with computing load being approximately linearly proportion to increasing system scale, but also achieves a dynamical simulation of the Monte Carlo method.

[Example 3.3] Calculate the reliability evaluation of the IEEE-TRS 24-node reliability test system [40] by the MCMC method.

[Solution] The Gibbs sampler iterates for 55,000 times, in which the first 5,000 iterations are used for "annealing" to remove the effect of initial values. The results of the remaining 50,000 iterations are used as samples to evaluate reliability indices.

The reliability indices obtained by use of the MCMC method are shown in Table 3.12. The results of comparing the MCMC method with other methods are shown in Table 3.13.

The data in Table 3.13 show that the results of the proposed method are very close to the other methods. It is illuminated that the proposed method in this chapter is more effective.

Next, we will discuss the convergence rate and stability of the MCMC method, respectively, according to calculation results of the example.

Table 3.12 The reliability indices of IEEE-RTS 24-node system

Reliability indices of system	Value
LOLP	0.08464
EENS (MW h)	127,859.73

Table 3.13 The results of comparing different calculating methods on IEEE-RTS 24-node system

Reliability indices	Convolution method	State enumeration method	MC method based on random sampling	MCMC method
LOLP	0.084578	0.084575	0.084420	0.084640
EENS (MW h)	128,716.62	128,695.3	129,781.86	127,859.7

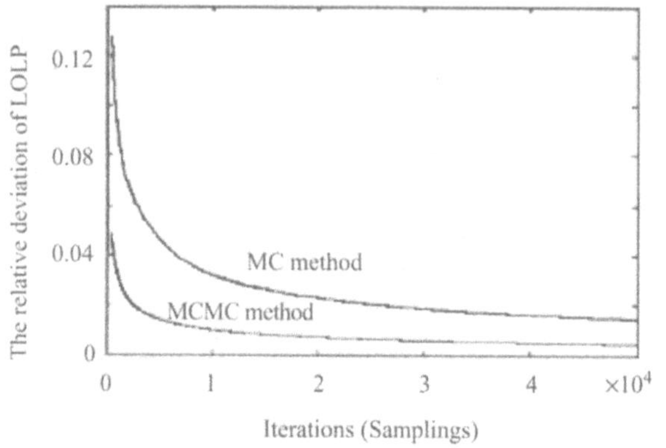

Fig. 3.8 The convergence curves of LOLP relative error to two methods

3.4.6.1 Convergence Comparison of the Algorithms

Figure 3.8 is the convergence rate diagram of index LOLP relative error. From Fig. 3.8, we can see that the variance coefficient of LOLP for the MCMC method is about 0.35 than that of MC method with same number of samples. In other words, the convergence rate of LOLP in the MCMC method has been improved seven times compared to that of the MC method with the same number of samples. At the same time, we can see from Fig. 3.8 that the variance coefficient of the index LOLP already reaches 0.01 after sampling 10,000 times when using the MCMC method, which also shows that we can obtain more accurate results by only sampling 10,000 times when utilizing the MCMC method, which decreases sampling time and speeds up the evaluation process.

3.4.6.2 Stability Comparison of the Algorithms

In order to check the stability of the algorithms, the reliability index LOLP has been calculated, using the MCMC method and the MC method, ten times each. The results are shown in Figs. 3.9 and 3.10. Comparing Fig. 3.9 with Fig. 3.10, we can see that the stability of the MCMC method is higher than that of the MC method. The variance of LOLP over ten calculations using the MCMC method is 1.89% of that obtained from ten calculation results using the MC method. That is because the MCMC method is based on the dependent sequence whose convergence is not related to the initial value, and the sampled Markov chains converge to the same objective probability distribution. Therefore, the calculated reliability indices are very stable.

From Fig. 3.9, we can see that the reliability indices converge to stable values after 10,000 iterations (samplings) for each Markov chain. However, because the MC method is based on independent sequences and the sampled sequences have greater differences between each other for each sampling, the calculated indices are very different and not stable enough. To obtain steadier results, the sampling number must be increased.

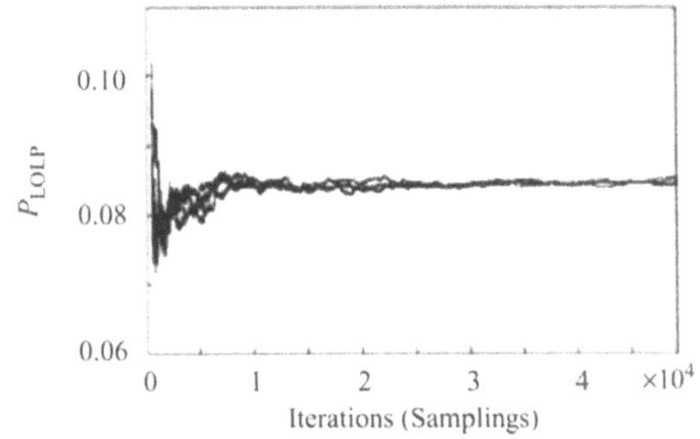

Fig. 3.9 The distribution of LOLP for ten calculations by using of MCMC method

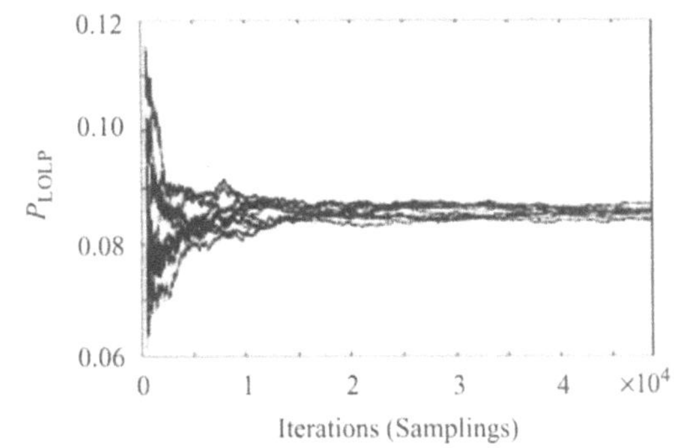

Fig. 3.10 The distribution of LOLP for ten calculations by using of MC method

3.5 Probabilistic Load Flow Analysis

The probabilistic load flow method is a macroscopic stochastic method under steady-state power system operation, which takes into account various random factors in system operation, such as load fluctuation, generator faults, and the impact of transmission component faults on steady-state operation. Therefore, the probabilistic load flow method is more capable of revealing power system operation characteristics than the conventional load flow method. It enables us to find the potential crisis and weak points of the system operation. For example, it can give the probability of line overloading, over voltage, steady-state instability, and so on.

Network planning involves a large number of uncertainties, such as inaccurate load forecasts in long-term planning, generation capacity, alteration to the dispatch plans, etc. These factors have a large impact on network planning schemes. In order to assess overall system performance, network planners usually carry out load flow calculations for many possible system operation modes, resulting in a large amount of computing and also difficulty in reflecting global situations. The use of the probabilistic load flow calculation method effectively solves these problems, providing decision-making evaluation of planning schemes with more global information.

Generally speaking, probabilistic load flow calculations consist of two parts: linearization of load flow equations and convolution calculations.

In this section, we will describe a probabilistic load flow calculation method which is based on the linearized model of the Newton-Raphson load flow calculation method. This model is characterized by its simplicity. It can provide line flow and nodal voltage probability distribution, and also estimate the probability of steady-state instability under a studied operation mode, making probabilistic load flow results more valuable.

The convolution calculation of random variables is a very important part of probabilistic load flow calculation. The probabilistic load flow calculation method described in this section uses the cumulant method to perform convolution computation of random variables and the Gram–Charlier series expansion to compute their distributions. Therefore, the overall probabilistic calculation is very efficient and its accuracy satisfies the requirements of planning and operation analysis.

The cumulant method has attracted wide application and deep research investigation because it is high in computation efficiency and flexible in solving problems. This section will put emphasis on the method. First of all the cumulant method for the probabilistic production simulation is introduced in combination with random distribution cumulant and Gram–Charlier series and Edgeworth series. Then explanations are made of how to deal with multistate generating units and sectionalized generating units. Finally, the errors of the algorithm are analyzed.

3.5.1 Cumulants of Random Distribution

When a random variable's distribution is known, its moment of every order can then be obtained. Suppose the density function of a continuous random variable x is $g(x)$, then its v-order moment α_v can be calculated by the following equation:

$$\alpha_v = \int_{-\infty}^{\infty} x^v g(x) \mathrm{d}x. \tag{3.72}$$

When $v = 1$, we have the expectation of the random variable x,

$$\mu = \alpha_1 = \int_{-\infty}^{\infty} x g(x) \mathrm{d}x. \tag{3.73}$$

From the expectation μ, the central moment of every order M_v can be calculated.

Then the central moment of every order M_v can be solved by the expectation μ,

$$M_v = \int_{-\infty}^{\infty} (x-\mu)^v g(x) \mathrm{d}x. \tag{3.74}$$

For discrete random variables, the equations for the moment of every order are different from the above equations in form. Suppose the probability of the discrete random variable x having a value x_i is p_i. Then the v-order moment is defined as

$$\alpha_v = \sum_i p_i x_i^v \tag{3.75}$$

in which $\sum$ means the summing up of all the possible value points of the random variable x.

The expectation of random variable x should be

$$\mu = \alpha_v = \sum_i p_i x_i. \tag{3.76}$$

Therefore the central moment of every order M_v can be expressed as

$$M_v = \sum_i p_i (x_i - \mu)^v. \tag{3.77}$$

A random variable's moment of every order is a numerical characteristic and to some extent represents the nature of the random distribution. The cumulant is also a kind of numerical characteristic, which can be calculated from the moments of all orders not higher than the corresponding order. The relationships between the first eight cumulants and the moments of every order are given below:

$$\left.\begin{aligned}
K_1 &= \alpha_1 \\
K_2 &= \alpha_2 - \alpha_1^2 \\
K_3 &= \alpha_3 - 3\alpha_2\alpha_1 + 2\alpha_1^3 \\
K_4 &= \alpha_4 - 4\alpha_3\alpha_1 - 3\alpha_2^2 + 12\alpha_2\alpha_1^2 - 6\alpha_1^4 \\
K_5 &= \alpha_5 - 5\alpha_4\alpha_1 - 10\alpha_3\alpha_2 + 20\alpha_3\alpha_1^2 + 30\alpha_2^2\alpha_1 - 60\alpha_2\alpha_1^3 + 24\alpha_1^5 \\
K_6 &= \alpha_6 - 6\alpha_5\alpha_1 - 15\alpha_4\alpha_2 + 30\alpha_4\alpha_1^2 - 10\alpha_3^2 + 120\alpha_3\alpha_2\alpha_1 - 120\alpha_3\alpha_1^3 \\
&\quad + 30\alpha_2^3 - 270\alpha_2^2\alpha_1^2 + 360\alpha_2\alpha_1^4 - 120\alpha_1^6 \\
K_7 &= \alpha_7 - 7\alpha_6\alpha_1 - 21\alpha_5\alpha_2 + 42\alpha_5\alpha_1^2 - 35\alpha_4\alpha_3 + 210\alpha_4\alpha_2\alpha_1 \\
&\quad - 210\alpha_4\alpha_1^3 + 140\alpha_3^2\alpha_1 + 210\alpha_3\alpha_2^2 - 1260\alpha_3\alpha_2^2 + 840\alpha_3\alpha_1^4 \\
&\quad - 630\alpha_2^3\alpha_1 + 2520\alpha_2^2\alpha_1^2 - 2520\alpha_2\alpha_1^5 + 720\alpha_1^7 \\
K_8 &= \alpha_8 - 8\alpha_7\alpha_1 - 28\alpha_6\alpha_2 + 56\alpha_6\alpha_1^2 - 56\alpha_5\alpha_3 + 336\alpha_5\alpha_2\alpha_1 - 336\alpha_5\alpha_1^3 \\
&\quad - 35\alpha_4^2 + 560\alpha_4\alpha_3\alpha_1 + 420\alpha_4\alpha_2^2 - 2520\alpha_4\alpha_2\alpha_1^2 + 1680\alpha_2\alpha_1^4 \\
&\quad + 560\alpha_3^2\alpha_2 - 1680\alpha_3^2\alpha_1^2 - 5040\alpha_3\alpha_2^2\alpha_1 + 13440\alpha_3\alpha_2\alpha^3 \\
&\quad - 6720\alpha_3\alpha_1^5 - 630\alpha_2^4 + 10080\alpha_2^3\alpha_1^4 - 25200\alpha_2^2\alpha_1^4 - 20160\alpha_2\alpha_1^6 \\
&\quad - 5040\alpha_1^8
\end{aligned}\right\}. \tag{3.78}$$

From (3.74) and (3.77), we can see that the first-order central moment of random variables is exactly equal to zero

$$M_1 = 0.$$

If the above equation is substituted into (3.78), we obtain the relationships between cumulants and central moments:

$$\left.\begin{aligned} K_2 &= M_2 \\ K_3 &= M_3 \\ K_4 &= M_4 - 3{M_2}^2 \\ K_5 &= M_5 - 10M_3M_2 \\ K_6 &= M_6 - 15M_4M_2 - 10M_3^2 + 30{M_2}^3 \\ K_7 &= M_7 - 21M_3M_2^2 - 35M_4M_3 + 210M_3{M_2}^3 \\ K_8 &= M_8 - 28M_3^2M_2 - 56M_5M_3 - 35{M_4}^2 + 420M_4{M_2}^2 \\ &\quad + 560{M_3}^2M_2 - 630{M_2}^4 \end{aligned}\right\}. \qquad (3.79)$$

A cumulant has the following important quality. If random variables $x^{(1)}$ and $x^{(2)}$ are independent of each other, and each has its own k-order cumulants $K_v^{(1)}$ and $K_v^{(2)}$ $(v = 1, 2, \ldots, k)$, then the v-order cumulant of the random variable $x^{(t)} = x^{(1)} \bigoplus x^{(2)}$ is

$$K_v^{(t)} = K_v^{(1)} + K_v^{(2)} \ (v = 1, 2, \ldots, k). \qquad (3.80)$$

The above quality can be generalized to the situation when there are n-independent random variables $x^{(i)}$ (i = 1, 2, ..., n). Now the v-order cumulant of the sum of n independent random variables can be expressed as

$$K_v^{(t)} = \sum_{i=1}^{n} K_v^{(i)} \ (v = 1, 2, \ldots, k). \qquad (3.81)$$

Equations (3.80) and (3.81) are called the sum of "cumulants," which moments or central moments do not possess, and which is the reason why it is sometime also called semi-invariate.

Now, we can calculate the every order moment and cumulants of random distributions for loads and generator units according to above equations.

In general, the load curve of a power system is usually taken to be a step-wise curve. The moments of each order can be calculated using by (3.75),

$$\alpha_{Lv} = \sum_i p_i x_i^v \ (v = 1, 2, \ldots, k), \qquad (3.82)$$

where α_{Lv} is the v-order moment of the load curve and p_i is the probability when the load has the value of x_i,

$$p_i = t_i/T,$$

where t_i is the duration of the load x_i and T is the investigated period.

For a generating unit, the moments of each order can be calculated using the following equation, when it is a multistate generating unit and the probability of outage capacity $\overline{C}_{is}$ is p_{is}:

$$\alpha_{iv} = \sum_{s=1}^{N_{is}} p_{is}\overline{C}_{is}^{v} \ (v = 1, 2, \ldots, k), \tag{3.83}$$

where N_{is} is the number of states of generating unit i. For a dual-state generating unit, N_{is} = 2, under this circumstance, the probability when the outage capacity equals the rated capacity C_i is q_i and the probability when the outage capacity equals zero is $1 - q_i$. Therefore,

$$\alpha_{iv} = q_i C_i^{v}. \tag{3.84}$$

When the moments of every order of the load curve and the generating unit's outage capacity distribution are known, all the cumulants of the respective order K_{Lv} and K_{iv} can be obtained by using (3.78). However, to simplify the computation, the central moments M_{Lv} and M_{iv} of the load curve and the generating unit's outage capacity distribution are usually first calculated and then the corresponding cumulants are obtained by using the relatively simple (3.79). The moments of each order can also be transformed into the central moment of each order by using (3.74) and (3.77),

$$M_v = \sum_{j=0}^{v} \binom{v}{j} \alpha_{v-j}(-\alpha_i)^j, \tag{3.85}$$

where $\binom{v}{j}$ is the combination of "v and j,"

$$\binom{v}{j} = v(v-1)(v-2)\cdots(v-j+1)/j!$$

From (3.85), we get,

$$M_2 = \alpha_2 - \alpha_1^2$$
$$M_3 = \alpha_3 - 3\alpha_2\alpha_1 + 2\alpha_1^3$$
$$M_4 = \alpha_4 - 4\alpha_3\alpha_1 + 6\alpha_1^2\alpha_2 - 3\alpha_1^4$$
$$\ldots \qquad \ldots \qquad \ldots$$

etc.

Equation (3.79) can be used to compute the corresponding cumulant when the moments of each order are known. If only (3.82)–(3.84) are used to obtain the moments of each order of the load curve and the generating unit's outage capacity, then (3.78) should be used to compute the cumulants of each order.

A method to calculate cumulants of every order from the random variable distribution was discussed above. It was shown that the convolution of independent random variables could be simplified to the addition of cumulants according to the summability of cumulants.

The following problem to be dealt with is how to form the distribution of a random variable from its moments or cumulants. There are many methods to solve this problem. The Gram–Charlier series expansion and Edgeworth series expansion are mainly used in probabilistic production simulation. These two series both represent the random variable's distribution function by using the derivatives of the random variable. The coefficients of the series are formed by the random variable's cumulants.

To simplify the form of series, we define

$$g_v = K_v/\sigma^v = K_v/K_2^{v/2} \ (v = 1, 2, \ldots, 8), \tag{3.86}$$

where g_v is the normalized cumulant of the order v and σ is the standard deviation.

From (3.36) and (3.38), we get,

$$\begin{aligned}
\sigma &= \sqrt{K_2}, \\
g_1 &= 0, \\
g_2 &= 1, \\
g_3 &= M_3/\sigma_3, \\
g_4 &= M_4/\sigma_4 - 1,
\end{aligned}$$

where g_3 is the deviation coefficient of random variable distribution and g_4 is its transcending coefficient.

The random variable's distribution function can be expressed as the following expansion of the Gram–Charlier series using the normalized cumulants:

$$\begin{aligned}
f(x) = &\int_x^{\infty} N(x)\mathrm{d}x + \frac{g_3}{3!}N^{(2)}(x) - \frac{g_4}{4!}N^{(3)}(x) - \frac{g_5}{5!}N^{(4)}(x) \\
&- \frac{g_6 + 10g_3^2}{6!}N^{(5)}(x) + \frac{g_7 + 35g_3g_4}{7!}N^{(6)}(x) \\
&- \frac{g_8 + 56g_3g_5 + 35g_4^2}{8!}N^{(7)}(x) + \cdots
\end{aligned} \tag{3.87}$$

or as the following expansion of the Edgeworth series

$$
\begin{aligned}
f(x) = & \int_x^{\infty} N(x)\mathrm{d}x + \frac{g_3}{3!}N^{(2)}(x) - \frac{g_4}{4!}N^{(3)}(x) + \frac{g_5}{5!}N^{(4)}(x) - \frac{10g_3^2}{6!}N^{(5)}(x) \\
& + \frac{35g_3g_4}{7!}N^{(6)}(x) + \frac{280g_3^3}{9!}N^{(8)}(x) - \frac{g_6}{6!}N^{(5)}(x) \\
& - \frac{35g_4^2}{8!}N^{(7)}(x) - \frac{56g_3g_5}{8!}N^{(7)}(x) - \frac{2{,}100g_3^2g_4}{10!}N^{(9)}(x) \\
& - \frac{15{,}400g_3^4}{12!}N^{(11)}(x) + \cdots,
\end{aligned}
\tag{3.88}
$$

where $f(x)$ is the probability when the random variable adopts a value greater than or equal to x. From (3.49) and (3.50), it can be seen that the Gram–Charlier series and the Edgeworth series are expanded according to different rules, which will be further discussed in Sect. 3.3.5. $N(x)$ is the standard normal distribution density function:

$$
N(x) = \frac{1}{\sqrt{2\pi}}\,\mathrm{e}^{-\frac{1}{2}x^2},
$$

where $N^{(\gamma)}(x)$ ($\gamma = 1, 2, \ldots$) is the γ-order derivative of $N(x)$.

$$
N^{(\gamma)}(x) = \left(\frac{\mathrm{d}}{\mathrm{d}x}\right)^{\gamma} N(x).
$$

After the differentiation

$$
N^{(\gamma)}(x) = (-1)^{\gamma} H_{\gamma}(x) N(x), \tag{3.89}
$$

where $H_{\gamma}(x)$ is Hermite polynomial, the first ten Hermite polynomials are

$$
\left.
\begin{aligned}
H_0(x) &= 1 \\
H_1(x) &= x \\
H_2(x) &= x^2 - 1 \\
H_3(x) &= x^3 - 3x \\
H_4(x) &= x^4 - 6x^2 + 3 \\
H_5(x) &= x^5 - 10x^3 + 15x \\
H_6(x) &= x^6 - 15x^3 + 45x^2 - 15 \\
H_7(x) &= x^7 - 21x^5 + 105x^3 - 105x \\
H_8(x) &= x^8 - 28x^6 + 210x^4 - 420x^2 + 105 \\
H_9(x) &= x^9 - 36x^7 + 378x^5 - 1{,}260x^3 + 945x \\
H_{10}(x) &= x^{10} - 45x^8 + 630x^6 - 3{,}150x^4 + 4{,}725x^2 - 945
\end{aligned}
\right\}. \tag{3.90}
$$

The general expression and recursive equation for the Hermite polynomial can be seen in [37, 41]. Substituting the above equations into (3.87) and (3.88), we get the Gram–Charlier expansion:

$$\begin{aligned} f(x) = \int_x^{\infty} N(x)\mathrm{d}x + N(\bar{x})\Bigg[& \frac{g_3}{3!}H_2(\bar{x}) + \frac{g_4}{4!}H_3(\bar{x}) + \frac{g_5}{5!}H_4(\bar{x}) \\ & + \frac{g_6 + 10g_3^2}{6!}H_5(\bar{x}) + \frac{g_7 + 35g_3g_4}{7!}H_6(\bar{x}) \\ & + \frac{g_8 + 56g_3g_5 + 35g_4^2}{8!} \times H_7(\bar{x}) + \cdots \Bigg] \end{aligned} \tag{3.91}$$

and the Edgeworth expansion

$$\begin{aligned} f(x) = \int_x^{\infty} N(x)\mathrm{d}x + N(\bar{x})\Bigg[& \frac{g_3}{3!}H_2(\bar{x}) + \frac{g_4}{4!}H_3(\bar{x}) + \frac{g_5}{5!}H_4(\bar{x}) \\ & + \frac{10g_3^2}{6!}H_5(\bar{x}) + \frac{35g_3g_4}{7!}H_6(\bar{x}) + \frac{280g_3^3}{9!}H_8(\bar{x}) + \frac{g_6}{6!}H_5(\bar{x}) + \frac{35g_4^2}{8!}H_7(\bar{x}) \\ & + \frac{56g_3g_5}{8!}H_7(\bar{x}) + \frac{2100g_3^2g_4}{10!}H_9(\bar{x}) + \frac{15400g_3^4}{12!}H_{11}(\bar{x}) + \cdots \Bigg] \end{aligned} \tag{3.92}$$

For the convenience of using the standard normal distribution table and the Hermite polynomials derived from the standard normal distribution density function, the normalized random variable x is used on the right-hand side of (3.91) and (3.92):

$$\bar{x} = \frac{x - \mu}{\sigma}, \tag{3.93}$$

where μ and σ are the expectation and standard variance of the random distribution.

The Gram–Charlier expansion in (3.91) uses the first eight orders of cumulants and the first seven Hermite polynomials, whereas the Edgeworth expansion in (3.92) uses only the first six orders of cumulants but the first 11 Hermite polynomials.

3.5.2 *Linearization of Load Flow Equation*

In Sect. 2.7.1, we used the Taylor series expansion in deriving the linear (2.150) from the nodal power (2.141) under the condition that a disturbance occurs to injection powers, i.e.,

$$\Delta X = S_0 \Delta W, \tag{3.94}$$

where (3.91) is linear. The distribution of the random state variable ΔX may be obtained by convolution from the distribution operation of the ΔW.

Following a similar argument, we can also obtain the linear relationship between the branch flow (2.153) and the nodal injection power ΔW. Now, rewrite (2.153) as

$$\left.\begin{aligned} P_{ij} &= V_iV_j(G_{ij}\cos\theta_{ij} + B_{ij}\sin\theta_{ij}) - t_{ij}G_{ij}V_i^2 \\ Q_{ij} &= V_iV_j(G_{ij}\sin\theta_{ij} - B_{ij}\cos\theta_{ij}) + (t_{ij}B_{ij} - b_{ij})V_i^2 \end{aligned}\right\}.$$

The above equation can be written as follows

$$Z = g(X).$$

Expanding the above equations according to the Taylor series, we obtain

$$Z = Z_0 + \Delta Z = g(X_0 + \Delta X) = g(X_0) + G_0\Delta X + \cdots. \tag{3.95}$$

Since Z_0 is obtained from normal load flow calculation and

$$Z_0 = g(X_0). \tag{3.96}$$

Ignoring high-order terms of (3.92), we get

$$\Delta Z = G_0\Delta X, \tag{3.97}$$

where

$$G_0 = \left.\frac{\partial Z}{\partial X}\right|_{X=X_0}. \tag{3.98}$$

G_0 is $2b \times 2N$ matrix (b is the number of branches, N is the number of nodes), with its elements given by,

$$\begin{aligned}
&\frac{\partial P_{ij}}{\partial\theta_i} = -H_{ij}, && \frac{\partial P_{ij}}{\partial\theta_j} = H_{ij}, && \frac{\partial P_{ij}}{\partial\theta_k} = 0\ k \notin \{i,j\} \\
&V_i\frac{\partial P_{ij}}{\partial V_i} = 2P_{ij} - N_{ij}, && V_j\frac{\partial P_{ij}}{\partial V_j} = N_{ij}, && V_k\frac{\partial P_{ij}}{\partial V_k} = 0\ k \notin \{i,j\} \\
&\frac{\partial Q_{ij}}{\partial\theta_i} = -J_{ij}, && \frac{\partial Q_{ij}}{\partial\theta_j} = J_{ij}, && \frac{\partial Q_{ij}}{\partial\theta_k} = 0\ k \notin \{i,j\} \\
&V_i\frac{\partial Q_{ij}}{\partial V_i} = 2Q_{ij} - H_{ij}, && V_j\frac{\partial Q_{ij}}{\partial V_j} = H_{ij}, && V_k\frac{\partial Q_{ij}}{\partial V_k} = 0\ k \notin \{i,j\},
\end{aligned} \tag{3.99}$$

where H_{ij}, N_{ij}, J_{ij} are elements of the Jacobian matrix (2.164), thus G_0 can easily be obtained.

ΔZ of (3.94) is the random fluctuation component of branch flows. In order to meet the independence requirements of random variable convolution operation, substituting (3.91) into (3.94) we obtain

$$\Delta Z = G_0 S_0 \Delta W = T_0 \Delta W, \tag{3.100}$$

where

$$T_0 = G_0 S_0 \tag{3.101}$$

is a transformation matrix. Equation (3.100) is linear. The use of the convolution operation gives probability distribution of branch flows by injecting independent random variable ΔW.

The random disturbance of the injection power ΔW is mainly comprised of random factors of nodal load and generator failure. For the nodal load power, its random component arises from forecast errors or random fluctuations of load. It can generally be described by a normal distributed random variable. When the load varies according to a load curve, we can simulate it using a discrete distribution, thus reflecting operation modes of several load profiles in the load flow model. For details of the probability distribution of generator power output and its cumulant solution, see Sect. 3.3.

The effect of random failure of transmission and distribution components on system can be simulated by using the corresponding injecting power ΔW_y (see (2.151) in Sect. 2.7). Suppose FOR of the transmission line is q_1, then the probability of injecting power (being ΔW_y) of the corresponding transmission line outage is q_1, and the probability of injecting power (being 0) is $1 - q_1$. Hence we can get corresponding cumulant.

The random variable of nodal injection power is given by

$$\Delta W = \Delta W_{\mathrm{g}} \oplus \Delta W_{\mathrm{l}} \oplus \Delta W_{\mathrm{t}}, \tag{3.102}$$

where ΔW_{g} and ΔW_{l} are random variables of generator and load and ΔW_{t} is the injecting power of transmission and distribution components corresponding to random failure. Symbol $\oplus$ means convolution operation. Therefore, in order to obtain the random variable of the nodal injection power ΔW it is necessary to perform a random variable convolution. From ΔW it is also necessary to perform a linear transformation on the random variable ΔW according to (3.91) and (3.97) in order to obtain the probability distributions, ΔX and ΔZ, of state variables and branch flows.

In order to use the Gram–Charlier series expansion to approximate the distributions of random variables, it is necessary to know every order of cumulant of this variable. However, the use of a semi-invariant has to satisfy the independent requirements of variables. Therefore, we assume that all nodal injection power random variables are independent.

Through the cumulants' properties of random variable (see Sect. 3.5.1), the kth order semi-invariant of nodal injection power $\Delta W^{(k)}$ may be obtained from the corresponding semi-invariant of load injection power $\Delta W_{\mathrm{l}}^{(k)}$ and generator injection power $\Delta W_{\mathrm{g}}^{(k)}$, i.e.,

$$\Delta W^{(k)} = \Delta W_{\mathrm{g}}^{(k)} + \Delta W_{\mathrm{l}}^{(k)} + \Delta W_{\mathrm{t}}^{(k)}. \tag{3.103}$$

From the cumulants' properties, we could obtain cumulants of the state variable ΔX and branch flows ΔZ from respective cumulants of ΔW according to the linear relationships (3.94) and (3.100), i.e.,

$$\Delta X^{(k)} = S_0{}^{(k)} \cdot \Delta W^{(k)}, \tag{3.104}$$

$$\Delta Z^{(k)} = T_0{}^{(k)} \cdot \Delta W^{(k)}, \tag{3.105}$$

where $S_0^{(k)}$ and $T_0^{(k)}$ are obtained as k exponents of matrix S_0 and T_0, respectively, i.e.,

$$S_0{}^{(k)}(i,j) = [S_0(i,j)]^k,$$
$$T_0{}^{(k)}(i,j) = [T_0(i,j)]^k.$$

It can be seen from the above discussion that after the random variables have been transformed to the form of cumulants, the convolution and linear transformation operations given by (3.100)–(3.102) become very simple. Therefore, after the cumulants of the distributions of nodal load and generator powers have been obtained, it is very easy to solve for cumulants, $\Delta X^{(k)}$ and $\Delta Z^{(k)}$, of the state variable ΔX and branch flows ΔZ based upon which the probability distribution of ΔX and ΔZ can be obtained by the Gram–Charlier series.

Sometimes, in order to obtain the probability density function of a state variable or line flow, it is necessary to expand the density function using the Gram–Charlier series:

$$\begin{aligned} p(x) = N(x)\Bigg(& 1 + \frac{g_3}{3!}H_3(\bar{x}) + \frac{g_4}{4!}H_4(\bar{x}) + \frac{g_5}{5!}H_5(\bar{x}) + \frac{g_6 + 10g_{3^2}}{6!}H_6(\bar{x}) \\ &+ \frac{g_7 + 35g_3g_4}{7!}H_7(\bar{x}) - \frac{g_8 + 56g_3g_5 + 35g_4^2}{8!}H_8(\bar{x}) + \ldots\Bigg). \end{aligned} \tag{3.106}$$

The above equation is obtained by differentiating (3.91) in the above section with respect to x.

3.5.3 Computing Process of Probabilistic Load Flow

According to the above linearized probabilistic load flow model and convolution operation of random variables, we have the flowchart for probabilistic load flow calculations as shown in Fig. 3.11.

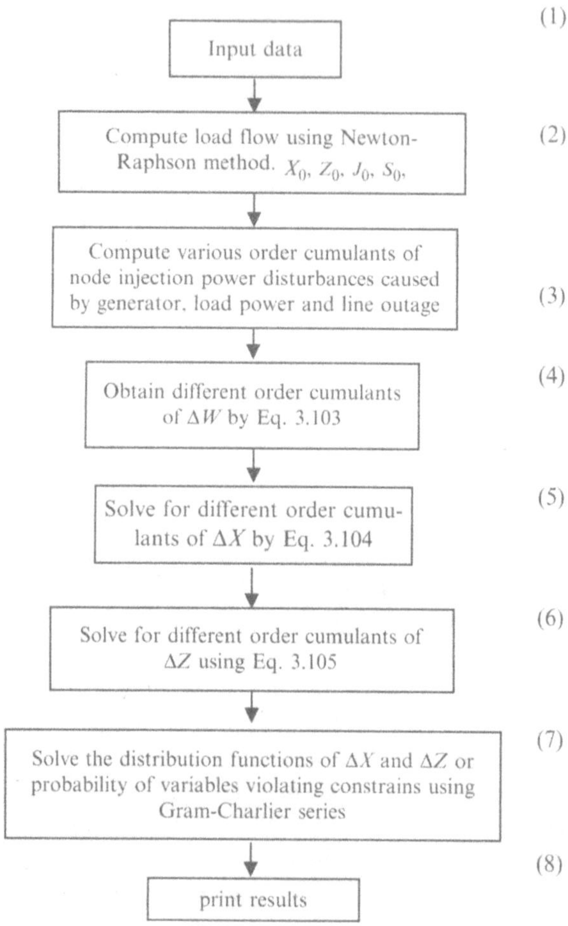

Fig. 3.11 Flowchart of probabilistic load flow calculations

The meaning of each block in Fig. 3.11 is briefly explained as follows:

1. Raw data required by the probabilistic load flow calculation include data required by the normal load flow calculation and information about the probability distribution of relevant nodal injection variables, such as the means and standard deviations of normally distributed load, load power, and the corresponding probability at each discrete point for discretely distributed load and the number, capacity and forced outage rate of the generators in order to compute probability distribution of the generator output.
2. Use the Newton-Raphson method to calculate normal load flow distributions, obtaining the state variable X_0 branch flow Z_0 and Jacobian matrix J_0. The sensitivity matrix S_0 is obtained through triangularization of J_0.
3. Compute moments of nodal injection power random variables and their cumulants $\Delta W_1^{(k)}$ and $\Delta W_g^{(k)}$.

4. Superimpose cumulants of generator outputs on those of loads to obtain cumulants of nodal injection powers.
5. Compute cumulants of state variables $\Delta X^{(k)}$ from $\Delta W^{(k)}$ according to (3.104), preparing for the subsequent calculation of statistical distribution.
6. After having solved for G_0 according to (3.98), form T_0 according to (3.101), and solve for semi-invariants of branch flows $\Delta Z^{(k)}$ from $\Delta W^{(k)}$ according to (3.105). It should be worth noting that ΔX is not an independent random variable any more; therefore, $\Delta X^{(k)}$ cannot be used to obtain $\Delta Z^{(k)}$ according to the linear (3.97).
7. Use the Gram–Charlier series expansion to obtain the distributions of ΔX and ΔZ (3.87). The mathematical expectations of X and Z are X_0 and Z_0, respectively. Their probability distributions are equivalent in shift to those of ΔX and ΔZ by the expectation values X_0 and Z_0. If it is only necessary to calculate the probability of a variable being out of its limits, only the corresponding distribution function needs to be computed. On the other hand, if further study of the distribution of a variable is needed, the expansion form of its density function, (3.106), is used.

The flowchart of probabilistic flow calculations (Fig. 3.11) takes full advantage of results obtained by the Newton-Raphson method during the formation of the linearized model. When random variable convolution operations are carried out it uses highly efficient cumulants and Gram–Charlier series. The whole model is characterized by clear concepts and rapid computation. However, due to the solving of the distribution of random variables by use of the semi-invariant, it is inevitable to generate errors in the calculation results. To reduce the errors, we can deal with continuous distribution and discrete distribution of random variables separately. For the discrete random variable, the Von Mises method is used. For the continuous random variable, a Gram–Charlier series is adopted. Thereby, the calculation error can be greatly reduced [36, 41].

[Example 3.4] Perform probabilistic load flow calculations on the IEEE-14 bus system. To stress the total computing process of probabilistic load flow, the case with branch outage is not considered in this example.

Statistical data of nodal injection powers are as follows. The relative data of generator units are shown in Table 3.15.

The load at node 9 is discretely distributed and its values are given in Table 3.16. Loads at the other nodes are normally distributed; their expectation values and standard deviations are given in Table 3.17.

[Solution] According to the flowchart shown in Fig. 3.11, we have the following procedures of computation:

1. Use the Newton-Raphson method to calculate the normal load flow. State variables X_0 and branch flows Z_0 under normal conditions have been obtained as shown in Table 3.18. X_0 and Z_0 will be regarded as expected values for probabilistic load flow calculation. Similarly, the Jacobian matrix Jo and sensitivity matrix So have also been obtained (they are not listed here because of limited space).

Table 3.14 The data of nodes and branches (on the 100 MV A base)

Branch				Node			
Node of branch	Resistance	Reactance	b_{ij0} or t_{ij}	Node	Active power	Reactive power	Voltage
1–2	0.01938	0.05917	0.01320	1[a]	2.324	0	1.06
1–3	0.05403	0.22304	0.01320	2	0.183	0	1.04
2–3	0.04699	0.19797	0.01095	3	−0.942	0	1.01
2–4	0.05811	0.17632	0.00935	4	−0.478	0.039	–[b]
2–5	0.05695	0.17388	0.00850	5	−0.076	−0.016	–
3–4	0.06701	0.17103	0.00865	6	−0.112	0	1.07
4–5	0.01335	0.04211	0.00320	7	−0.0	0	–
4–7[c]	0.00000	0.20912	0.97800	8	−0.0	0	1.09
4–9[c]	0.00000	0.55618	0.96900	9	−0.295	0.046	–
5–6[c]	0.00000	0.25202	0.93200	10	−0.090	−0.058	–
6–11	0.09498	0.19890	0.00000	11	−0.035	−0.018	–
6–12	0.12291	0.25581	0.00000	12	−0.061	−0.016	–
6–13	0.06615	0.13027	0.00000	13	−0.135	−0.058	–
7–8	0.00000	0.17615	0.00000	14	−0.149	−0.050	–
7–9	0.00000	0.11001	0.00000				
9–10	0.03181	0.08450	0.00000				
9–14	0.12711	0.27038	0.00000				
10–11	0.08205	0.19207	0.00000				
12–13	0.22092	0.19988	0.00000				
13–14	0.17093	0.34802	0.00000				

[a]Represents that node 1 is slack node
[b]"–"Represents PQ node, and its voltage is unknown
[c]Represents transformer branch, the value in the last column is the ratio

Table 3.15 The relative data of generator units

Node	Capacity (MW)	Number	FOR	Expectation of power output (MW)
1	2.5	10	0.08	23.00
2	22	2	0.09	40.04

Table 3.16 Probability distribution of load at node 9

Active load (MW)	13.4	19.6	30.2	34.8	37.3
Probability	0.10	0.15	0.30	0.25	0.20
Reactive load (MVAR)	7.5	11.0	17.0	19.6	21.0
Probability	0.10	0.15	0.30	0.25	0.20

2. Calculate the cumulants of the nodal injection power. Based upon the method of solving for statistically distributed cumulants described in Sect. 3.3.1, we could obtain eight cumulants of generator nodes 1 and 2 and discretely distributed load at node 9, as shown in Table 3.19 (where all values are in per unit).

For normally distributed injection powers, their first-order cumulant is equal to their expected value, the second cumulant is equal to their variances, while the third-order to eighth-order cumulants are equal to zero. For example, the load at

Table 3.17 Statistical data on nodal loads

Node	Active load (MW)		Reactive load (MVAR)	
	Expectation	Deviation (%)	Expectation	Deviation (%)
1	0.0	0.0	0.0	0.0
2	21.74	0.09	12.7	0.092
3	94.20	0.10	19.0	0.105
4	47.80	0.11	−3.9	0.097
5	7.60	0.05	1.6	0.05
6	11.20	0.06	7.5	0.063
7	0.0	0.0	0.0	0.0
8	0.0	0.0	0.0	0.0
10	9.0	0.10	5.8	0.10
11	3.5	0.095	1.8	0.095
12	6.1	0.076	1.6	0.086
13	13.5	0.105	5.8	0.095
14	14.9	0.086	5.0	0.086

Table 3.18 The results of load flow calculation by Newton-Raphson method

Node of branch	Load flow on branch				Nodal voltage		
	P_{ij}	Q_{ij}	P_{ij}	Q_{ji}	Node	Amplitude	Angle
1–2	1.5694	−0.1893	−1.5264	0.2914	1	1.06000	0.00000
1–3	0.7547	0.0550	−0.7271	0.0305	2	1.04500	−4.98429
2–3	0.7327	0.0475	−0.7095	0.0273	3	1.01000	−12.73054
2–4	0.5614	−0.0041	−0.5446	0.0351	4	1.01714	−10.30872
2–5	0.4152	0.0259	−0.4062	−0.0164	5	1.01873	−8.76485
3–4	−0.2325	0.0455	0.2363	−0.0537	6	1.07000	−14.21900
4–5	−0.6110	0.1608	0.6161	−0.1511	7	1.06128	−13.35621
4–7	0.2806	−0.0983	−0.2806	0.1154	8	1.09000	−13.35621
4–9	0.1607	−0.0049	−0.1607	0.0179	9	1.05571	−14.93501
5–6	0.4411	0.1210	−0.4411	−0.0769	10	1.05080	−15.09401
6–11	0.0737	0.0361	−0.0731	−0.0349	11	1.05681	−14.78788
6–12	0.0779	0.0251	−0.0772	−0.0236	12	1.05517	−15.07369
6–13	0.1776	0.0724	−0.1754	−0.0682	13	1.05035	−15.15407
7–8	0.0000	−0.1730	0.0000	0.1777	14	1.03539	−16.03092
7–9	0.2506	0.0576	−0.2806	−0.0496			
9–10	0.5212	0.0418	−0.0520	−0.0414			
9–14	0.0942	0.0358	−0.0930	−0.0334			
10–11	−0.0380	−0.0166	0.0381	0.0169			
12–13	0.0162	0.0076	−0.0161	−0.0075			
13–14	0.0565	0.0178	−0.0560	−0.0166			

Table 3.19 Cumulants of nodal injection powers at nodes 1, 2, and 9

Order	Node 1	Node 2	Node 9	
			Active power	Reactive power
1	$0.230000 \times 10^{+1}$	$0.230000 \times 10^{+0}$	$-0.295000 \times 10^{+0}$	$-0.166000 \times 10^{+0}$
2	0.460000×10^{-1}	0.792792×10^{-2}	0.599600×10^{-2}	0.191550×10^{-2}
3	$-0.9660000 \times 10^{-2}$	-0.143020×10^{-2}	0.430640×10^{-2}	0.778925×10^{-4}
4	0.160540×10^{-2}	0.195156×10^{-3}	-0.168809×10^{-4}	-0.172255×10^{-5}
5	-0.705180×10^{-4}	-0.119061×10^{-5}	-0.142591×10^{-4}	-0.824467×10^{-6}
6	-0.100259×10^{-3}	-0.121103×10^{-4}	-0.163633×10^{-5}	-0.535577×10^{-7}
7	0.553146×10^{-4}	0.502300×10^{-5}	0.714816×10^{-6}	0.132089×10^{-7}
8	-0.132800×10^{-4}	-0.736492×10^{-6}	0.335819×10^{-6}	0.351225×10^{-8}

Table 3.20 Expectation and standard deviation values of state variables

Node	Voltage (p.u)		Angle (°)	
	Expectation	Deviation	Expectation	Deviation
1	1.06000	0.00000	0.00000	0.00000
2	1.04500	0.00000	−4.98429	0.44298
3	1.01000	0.00000	−12.73054	0.99757
4	1.01714	0.00202	−10.30872	0.68979
5	1.01873	0.00164	−8.76485	0.57883
6	1.07000	0.00000	−14.21900	0.84952
7	1.06128	0.00286	−13.35621	0.97527
8	1.05571	0.00000	−13.35621	0.97527
9	1.09000	0.00519	−14.93501	1.114956
10	1.05080	0.00441	−15.09401	1.09751
11	1.05681	0.00231	−14.78788	0.97113
12	1.05517	0.00069	−15.07369	0.88307
13	1.05035	0.00120	−15.15407	0.90842
14	1.03539	0.00368	−16.03092	1.06123

node 2 has an expected value of 0.2174 and variance (in percent) of 0.009; its cumulants are

$$K_1 = 0.2174,$$
$$K_2 = (0.2174 \times 0.09)^2 = 0.000382828, \cdot$$
$$K_j = 0 \quad j = 3, 4, \ldots, 8$$

The sum of cumulants of the load and generator output at node 2 are displayed in the third column of Table 3.19. Similarly, cumulants of the nodal injection powers may be obtained.

3. Compute the cumulants of the state variables. Since the sensitivity matrix S_0 has been obtained from normal load flow calculation, the cumulants of the state variables may be obtained directly from those of the nodal injection powers according to (3.104). Table 3.20 shows the expected values and standard

deviations of nodal voltages and angles, where expected values are the first-order cumulants, and the standard deviations are the square root of the second-order cumulants.

4. Compute the cumulants of branch flows. When the Jacobian matrix and branch flows under normal conditions are given, G_0 is easily obtained from (3.99). It is a sparse matrix. Matrix T_0 is formed according to (3.101). For example, for line 5–6, its position in T_0 is the 19th and 20th rows with each row having 28 elements, of which the first 14 correspond to active power and the last 14 to reactive power, i.e.,

$$\begin{aligned} T_0(19) = {} & [0.0, -0.00523, -0.0209, -0.0370, 0.01435, -0.6871 \\ & -2.07, -2.07, -0.2967, -0.3687, -0.5273, \\ & -0.6689, -0.6441, -0.4591; 0.0, 0.0, 0.0, -0.00814, \\ & -0.0129, 0.0, -0.0393, 0.0, -0.0736, -0.0658, \\ & -0.0367, -0.0030, -0.0168, -0.0548] \\ T_0(20) = {} & [0.0, -0.0013, -0.0084, -0.0547, 0.0747, -0.0124 \\ & -0.0556, -0.0556, -0.0542, -0.0478, -0.0309, \\ & -0.0153, -0.0194, -0.0413; 0.0, 0.0, 0.0, -0.1166, \\ & 0.1935, 0.0, 0.4848, 0.0, 0.0445, 0.0368, 0.0186, 0.00384, \\ & 0.0065, 0.0286] \end{aligned}$$

Eight cumulants of the branch flow are obtained from those of the nodal injection power according to (3.105). The eight cumulants corresponding to line 5–6 are

$$\begin{aligned} & K_1 = 0.44111 \times 10^{+0} \quad K_2 = 0.70949 \times 10^{-3} \\ & K_3 = -0.11284 \times 10^{+4} \quad K_4 = -0.13095 \times 10^{-6} \\ & K_5 = 0.32814 \times 10^{-7} \quad K_6 = -0.11174 \times 10^{-8} \\ & K_7 = -0.14485 \times 10^{-0} \quad K_8 = 0.20194 \times 10^{-10} \end{aligned}$$

Table 3.21 shows the expected values and standard deviations of branch flows.

5. Compute the probability distributions of the state variables. We take node 4 as an example to solve for the probability density function of voltages.

Eight cumulants of the voltage at node 4 are given as

$$\begin{aligned} & K_1 = 0.10171 \times 10^{+1} \quad K_2 = 0.40674 \times 10^{-5} \\ & K_3 = 0.41810 \times 10^{-8} \quad K_4 = -0.32393 \times 10^{-11} \\ & K_5 = -0.55467 \times 10^{-13} \quad K_6 = -0.13057 \times 10^{-15} \\ & K_7 = 0.11764 \times 10^{-17} \quad K_8 = 0.11428 \times 10^{-19} \end{aligned}$$

Table 3.21 Expectation and standard deviation values of branch flows

Branch	Active load flow (MW)		Reactive load flow (MVAR)	
	Expectation	Mean Variance (%)	Expectation	Mean Variance (%)
1–2	156.9366	13.3943	−18.9334	3.1295
1–5	75.4682	4.7871	5.5020	0.5054
2–3	73.2721	5.7571	4.7525	0.5651
2–4	56.1419	3.3318	−0.4093	0.6566
2–5	41.5220	2.4094	2.5914	0.5004
3–4	−23.2535	4.4619	4.5501	2.0654
4–5	−61.0946	4.4898	16.0791	1.4391
4–7	28.0606	3.5716	−9.8291	0.8584
4–9	16.0705	2.0367	−0.4891	0.7895
5–6	44.1110	2.6636	12.1028	0.5586
6–11	7.3663	1.4758	3.6053	1.0306
6–12	7.7890	0.4155	2.5089	0.1953
6–13	17.7556	1.2340	7.2398	0.6805
7–8	0.0000	0.0000	−17.3021	1.6740
7–9	28.0607	3.5716	5.763*9	2.3431
9–10	5.2150	1.5654	4.1753	1.0012
9–14	9.4161	1.2503	3.5818	0.6851
10–11	−3.7978	1.4467	−1.6587	1.0088
12–13	1.6172	0.3685	0.7594	0.1822
13–14	5.6540	1.1001	1.7751	0.6725

Taking the discrete step length as 0.01, we obtain the probability density function of the voltage at node 4, as shown in Fig. 3.12, according to the Gram–Charlier series method and (3.106), as described in Sect. 3.3.2. If the upper limit of the voltage is 1.02, then the probability of this voltage being greater than its upper limit is

$$P(V_4 > 1.02) = 0.106232.$$

The dashed line in Fig. 3.12 is the result of using the sixth-order semi-invariant.

Following the same argument, we could obtain probability density distribution functions of other state variables and branch flows and their over-limit probability.

3.6 Probabilistic Network-Flow Analysis

3.6.1 *Introduction*

The adequacy of the transmission system is basic to guarantee the secure operation of power systems. The fundamental cause of the "8.13" Blackout in USA and Canada is that many transmission components of the power system operated in their limit states. Several blackouts in USA were caused by grounding fault induced by

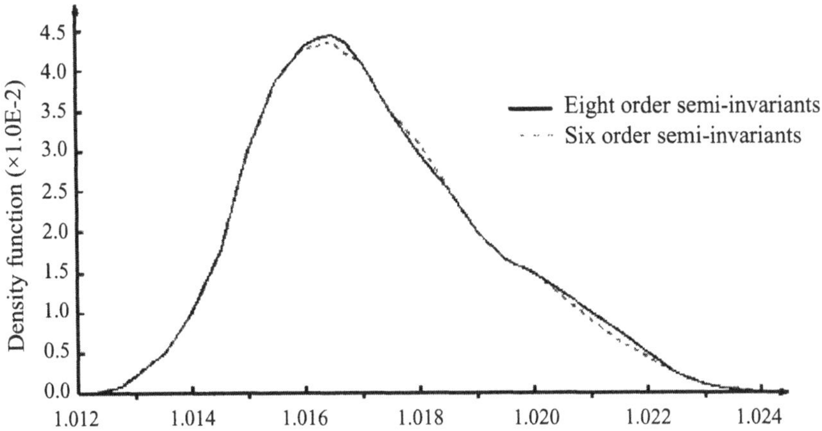

Fig. 3.12 Voltage probability density function at node 4

increasing sag of some over heated conductor. When adequacy of the transmission system is not enough, after one transmission component goes out of service, load transfer can easily cause cascading overload, which will expand the extent of outages. A series of worldwide blackout events show that they are usually caused by cascading outages.

The occurrence of cascading outages has a very close relation with power network structure. When a power network may induce cascading outages in a certain operating condition, we define the power grid structure as not stable in this operating situation. And when the power network cannot induce cascading outages in a certain operating condition, we define the power grid structure as stable in this operating situation. An example of whether a power grid structure is stable or not is given in Fig. 3.13.

There are three sources and three loads in the power network shown in Fig. 3.13, and the capacity limit of each transmission line is also given in this figure. The power network structure shown in Fig. 3.13a is stable in the corresponding load condition, and satisfies the requirement of continuous power supply for load in normal and $N - 1$ conditions. Due to the increasing load, the power network structure shown in Fig. 3.13b becomes unstable. In such a condition, any outage of an outlet line of the three sources will cause a blackout of the total system. For instance, as shown in Fig. 3.13d, after the outage of the outlet line of source C, overload of outlet line of source A and B occurs because of load transfer, which therefore induces blackout of the whole power grid. But, when the connection lines of DE and EF do not exist, the power network is stable because the outage of any component in the power grid will not cause cascading outage, which is shown in Fig. 3.13c.

In Fig. 3.13b, the blackout of whole power network caused by the outage of transmission line has no relation with angular stability and voltage stability. In fact, even though the length of all transmission lines in this power network approaches 0, the blackout of whole power grid will still occur. For an unstable power grid,

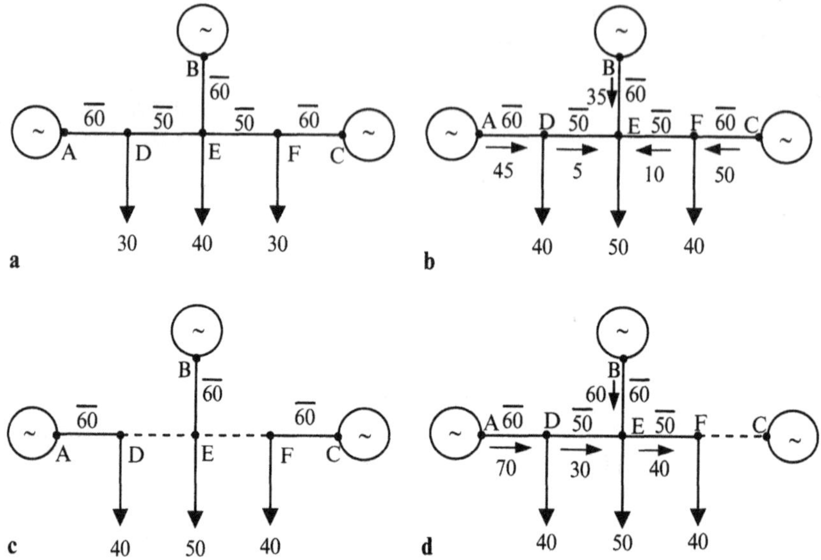

Fig. 3.13 The stability of power grid structure

measures to prevent cascading outages should be carried out. For instance, when a transmission line trips, we should adopt such measures as load shedding to restrict outage scope to a minimum area.

In this section, we will analyze power network structure by using a probabilistic network-flow model, and discriminate components that induce cascading outages. On this basis, we consider random outages of generation and transmission components, calculate the probability (i.e., reliability) that a given load is satisfied, and quantify the importance of each component in the power grid. This model can provide useful information for transmission system planning, generation and transmission device maintenance, contingency setting, and countermeasures.

3.6.2 Network-Flow Model

The "network" is a graph with capacity constrained branches [42], and nodes can also be set with capacity restrictions under certain conditions. Communication networks, transportation networks, and power networks are all typical networks [43].

The transmission system, shown in Fig. 3.14a, with N nodes, N_b branches, N_s generators, and N_l loads, can be represented by the network illustrated in Fig. 3.14b.

In Fig. 3.14b, each component ij of the transmission system is assigned a capacity C_k. The generators are represented by branches connected to the fictitious source node S, whose capacities $C_{g1}, C_{g2}, \ldots, C_{gs}$ are determined by generators'

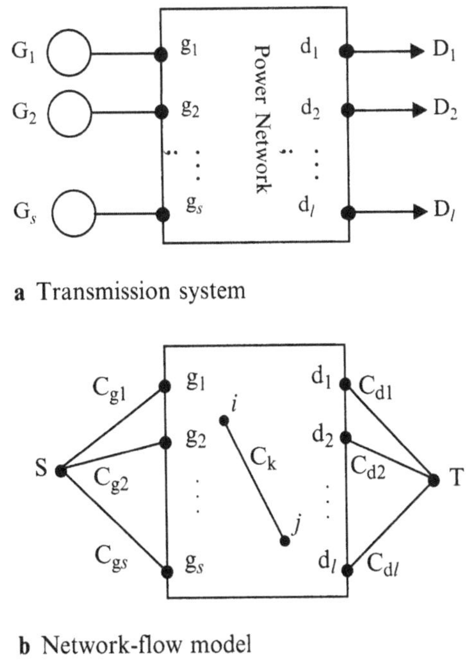

Fig. 3.14 The network-flow model of transmission system

available capacities. The loads are represented by branches connected to the fictitious sink node T, whose capacities $C_{\rm d1}, C_{\rm d2}, \ldots, C_{{\rm d}l}$ are determined by loads' demands. As a consequence, the number of branches in Fig. 3.14b is

$$N_\Sigma = N_{\rm b} + N_{\rm s} + N_{\rm l}.$$

Assume the sum of loads is D, i.e.,

$$\sum_{k=1}^{N_{\rm l}} C_{{\rm d}k} = D.$$

Then the feasible network-flow model can be described by

$$\sum_{k \in j} F_k = 0, \quad j = 1, 2, \ldots, N;\ j \neq S, T, \tag{3.107}$$

$$\sum_{k \in t} F_k = D, \tag{3.108}$$

$$F_k \leq C_k. \tag{3.109}$$

$$N_\Sigma = N_{\rm b} + N_{\rm s} + N_{\rm l}.$$

Where F_k is flow on branch k. Equation (3.107) means that the inflow and outflow power at each node should be balanced; i.e., the feasible flows must be balanced at each node, or satisfy the Kirchhoff's First Law of electric circuits, which is the basic property of network-flow. Equation (3.108) shows that flows must satisfy the load demand D. Equation (3.109) presents the branch capacity constraints. Since $F_{kt} \leq C_{dk}$, (3.108) is equivalent to

$$F_{kt} = C_{dk} = D_k, \quad k = 1, 2, \ldots, N_1. \tag{3.110}$$

Each feasible solution corresponds to a possible operating state satisfying transmission requirements. All feasible flow solutions represent all possible operating states under the given generation and transmission resources.

To thoroughly dissect the grid structure, one should solve and analyze all feasible flow solutions [42]. To do so, we need find the minimal path set from S to T in Fig. 3.14b, $\boldsymbol{P} = \{P_1, P_2, \ldots, P_m\}$, where m is number of all minimal paths. Assume $\{f_1, f_2, \ldots, f_m\}$ are flows on these paths, the feasible solutions should satisfy the following constraints

$$\sum_{P_j \in k} f_j \leq C_k, \quad k = 1, 2, \ldots, N_\Sigma, \tag{3.111}$$

$$\sum_{j=1}^{m} f_j = D. \tag{3.112}$$

Equation (3.111) is the capacity constraint for branch k, $P_j \in k$ denotes the minimal paths that pass branch k; Equation (3.112) shows that flow on all minimal paths should satisfy the load demand. Because a feasible solution is described by accumulation of flows on minimal paths, flow balance condition represented by (3.107) holds automatically. Therefore, the feasible flow model of (3.111) and (3.112) is equivalent to that of (3.107), (3.108), and (3.109).

To understand the algorithm of network-flow model conveniently, the process for finding a feasible solution is explained in detail by a numeral example.

[Example 3.5] Form the minimal path set and solve the feasible flow solution of the sample system shown in Fig. 3.15.

[Solution] The sample system shown in Fig. 3.15a contains five nodes, five branches, two generators and three loads (①, ②, ③). Its electrical parameters are illustrated in Example 1.1. The network model is demonstrated in Fig. 3.15b, and the numbers in parentheses present capacities of respective branches. To improve computational efficiency, the capacities of all branches are enlarged ten times in this example, thus floating computation can be replaced by integer computation. In Fig. 3.15b, besides the original five branches, we add two branches (6 and 7) connected with source point S, representing generators G1 and G2, respectively.

First, find the minimal path of arriving at each load node ① ② ③ of the network shown in Fig. 3.15b from its incident matrix [44, 45]. Altogether ten minimal paths are fund as illustrated in Table 3.22.

Fig. 3.15 The network-flow model of sample system

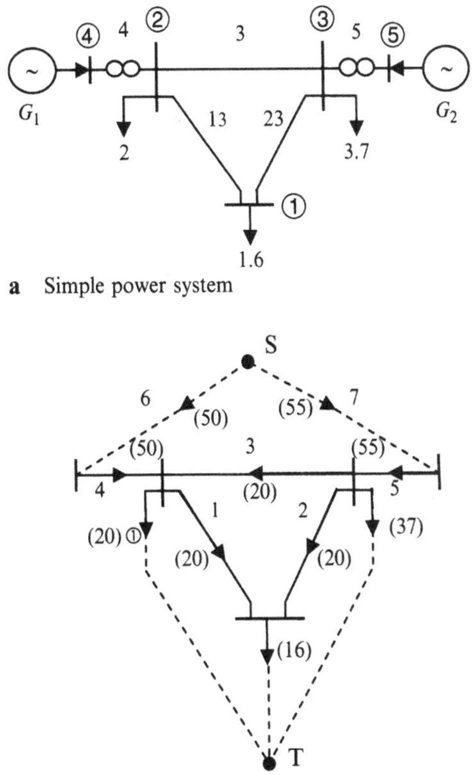

a Simple power system

b Network.flow model

Table 3.22 Set of minimal paths

No.	Including paths	Supplying load
1	1 3 5 7	Load ①
2	1 4 6	
3	2 −3 4 6	
4	2 5 7	
5	−1 2 5 7	Load ②
6	3 5 7	
7	4 6	
8	−2 1 4 6	Load ③
9	−3 4 6	
10	5 7	

*±sign is relative to given direction of the branch

Then the feasible solution set is solved by an implicit enumeration method according to (3.111) and (3.112). Forty-five feasible solutions have been found as shown in Table 3.23. Because system load is 73, so the sum of the flows on the minimal paths in each feasible set is 73. Flows on branches of each feasible solution are shown in Table 3.24. The table demonstrates all possible operating states and connecting modes that give satisfactory power supply, and provides comprehensive

Table 3.23 Set of feasible solutions

No.	$f1$	$f2$	$f3$	$f4$	$f5$	$f6$	$f7$	$f8$	$f9$	$f10$
1	16				20			24	13	
2	16				20			24		13
3	16				20				36	1
4	16					4	16	4	30	3
5	16					4	16	4		33
6	16					4	16		34	3
7	16						20	4	26	7
8	16						20	4		33
9	16						20		30	7
10	16						20			37
11		16			20			24	10	3
12		16			20			24		13
13		16			20				20	17
14		16				20		4	30	3
15		16				20		4		33
16		16				20			34	3
17		16					20	4	10	23
18		16					20	4		33
19		16					20		14	23
20		16					20			37
21			16		4	16		24	10	3
22			16		4	16		24		13
23			16		4	16			20	17
24			16		4		16	18		19
25			16		4		16		4	33
26			16		4		16			37
27			16			20		20	14	3
28			16			20		20		17
29			16			20			24	13
30			16				20	14		23
31			16				20		4	33
32			16				20			37
33				16	4	16		24	13	
34				16	4	16		24		13
35				16	4	16			36	1
36				16	4		16	24	10	3
37				16	4		16	24		13
38				16	4		16		20	17
39				16		20		20	17	
40				16		20		20		17
41				16		20			37	
42				16			20	20	10	7
43				16			20	20		17
44				16			20		20	17
45				16			20			37

Table 3.24 Branch states of feasible solutions

No.	b_1	B_2	b_3	b_4	b_5	No.	b_1	b_2	b_3	B_4	b_5
1	20	4	3	37	36	24	14	2	16	50	23
2	20	4	16	24	49	25	4	20	20	36	37
3	4	20	20	36	37	26	4	20	16	32	41
4	20	4	10	50	23	27	20	4	10	50	23
5	20	4	20	20	53	28	20	4	4	36	37
6	16	■	14	50	23	29	■	16	20	40	33
7	20	4	10	50	23	30	14	2	16	50	23
8	20	4	16	24	49	31	■	16	20	40	33
9	16	■	14	50	23	32	■	16	16	36	37
10	16	■	16	20	53	33	20	4	3	37	36
11	20	4	10	50	23	34	20	4	16	24	49
12	20	4	■	40	33	35	4	20	20	36	37
13	4	20	20	36	37	36	20	4	10	50	23
14	20	4	10	50	23	37	20	4	■	40	33
15	20	4	20	20	53	38	4	20	20	36	37
16	16	■	14	50	23	39	20	4	3	37	36
17	20	4	10	50	23	40	20	4	20	20	53
18	20	4	■	40	33	41	■	16	17	37	36
19	16	■	14	50	23	42	20	4	10	50	23
20	16	■	■	36	37	43	20	4	■	40	33
21	20	4	10	50	23	44	■	16	20	40	33
22	20	4	■	40	33	45	■	16	■	20	53
23	4	20	20	36	37						

information of operating states under given generation and transmission condition. The flow presented by shadow area denotes no flow on that relative branch, which means that there still is a feasible solution even if corresponding branch is tripped. For example, when branch b_1 is tripped by an incident, there are still feasible solutions numbered 29, 31, 32, 41, 44; when branch b_2 is tripped, there are still feasible solutions numbered 6, 9, 10, 16, 19; when branch b_3 is tripped, feasible solutions numbered 12, 18, 22, still can satisfy load requirements. Even if branches b_2 and b_3 are simultaneously tripped, solution numbered 20 still can satisfy load requirements. When branches b_1 and b_3 are tripped simultaneously solution numbered 45 still can satisfy load requirements.

Table 3.24 can be simplified to Table 3.25 according to different outputs of generators and configuration of the transmission system.

From Table 3.24 and 3.25, we can see that when branches 1, 2, and 3 tripped separately, and branches 2 and 3, as well as branches 1 and 3 tripped simultaneously, feasible solutions still exist. This implies that under such conditions the transmission system still can satisfy load requirement.

In all feasible solutions, branch 4 and 5 are indispensable, their failures certainly cause other branches to overload or induce cascade failures. In other words, these two branches are most important to maintain reliability of the transmission system. By contrast, failures of other branches (1, 2, 3) have no essential impact on

Table 3.25 Feasible solutions for different network configurations

Outage branches	Generator output		Ordinary number of feasible solution
	G1	G2	
No outage branch	37	36	1, 33, 39
	24	49	
	36	37	2, 8, 34
	50	23	3, 13, 23, 25, 28, 35, 38,
	20	53	4, 7, 11, 14, 17, 21, 24
	32	41	27, 30, 36, 42
			5, 15, 40
			20
Branch 1	40	33	29, 31, 44
	36	37	32
	37	36	41
Branch 2	50	23	6, 9, 16, 19
	20	53	10
Branch 3	40	33	12, 18. 22, 37, 43
Branch 2&3	36	37	20
Branch 1&3	20	53	45

continuity of power supply, thus they are less important than branches 4 and 5. Branch 3 has the least contribution to reliability of the transmission system. When it is tripped, the transmission system can still satisfy load demand even if branch 1 or 2 is further tripped.

Therefore, from a network structure perspective, branches 4, 5 are the most valuable branches, branches 1, 2 are less valuable branches, and branch 3 is the least valuable branch.

Valuation of each branch in a transmission system can be further quantified, see the example in Sect. 3.6.4.

It should be pointed out, 45 feasible solutions in Table 3.24 and 3.25 are valid only under the condition that available capacities of generators G_1 and G_2 are 5.0 and 5.5, respectively. In such condition, reserve of the power system is big enough to re-dispatch when contingency occurs. From the tables we can see that power regulation range of G_1 is 2.0–5.0, while that of G_2 is 2.3–5.3. When limiting the available capacity of these two generators to 4.0, the number of the feasible solution decreases to 22, which can be found in Tables 3.24 and 3.25. In this situation, no feasible solution exists when branches 1 and 3 are simultaneously tripped.

3.6.3 Lower Boundary Points of Feasible Flow Solutions

The network-flow model cannot include the Kirchhoff's Second Law in its constraints. Therefore, multiple solutions may exist under the same generation and transmission condition. For instance, feasible solutions 21 and 27, 25 and 35 in Table 3.24 are multiple solutions. In order to efficiently evaluate reliability of the transmission system, the feasible solutions need to be sifted so that there is only one feasible solution for each operating condition.

The sifting principle in this paper is to retain the feasible solution with the least "total reactive power loss" ΔQ under the same operation condition

$$\Delta Q = \sum_{ij} P_{ij}^2 X_{ij}, \tag{3.113}$$

where P_{ij}, X_{ij} are the active flow and reactance of branch ij; $\sum$ sums up "loss" of all branches in the network. This principle can obtain the solution that best approximates the DC power flow solution. The assertion is proved below.

Let us investigate the following optimization model,

$$\text{obj} : \text{Min} \frac{1}{2} \sum_{ij \in \text{Network}} P_{ij}^2 X_{ij}, \tag{3.114}$$

$$\text{s.t.} \sum_{ij \in i} P_{ij} = 0, \quad i \in N. \tag{3.115}$$

The problem is to minimize "total loss," with the constraint of Kirchhoff's First Law, i.e., the conservative balance of the flows at each node. The Lagrangian function of the problem is

$$L = \frac{1}{2} \sum_{ij \in \text{Network}} P_{ij}^2 X_{ij} - \sum_{i=1}^{N} \theta_i \sum_{ij \in i} P_{ij}, \tag{3.116}$$

where θ_i is the Lagrangian multiplier. Taking derivative of L respective to P_{ij}, and setting it to 0, we have

$$\frac{\partial L}{\partial p_{ij}} = P_{ij} X_{ij} - (\theta_i - \theta_j) = 0. \tag{3.117}$$

Taking θ_i, θ_j as the phase angles of voltages at nodes i, j, one can see that (3.117) represents Kirchhoff's Second Law in the DC power flow model. For any loop L in a transmission network, from (3.117) the next equation holds

$$\sum_{ij \in L} P_{ij} X_{ij} = 0. \tag{3.118}$$

It is clear that the solution of above optimization model of (3.114) and (3.115) is equivalent to that of the DC power flow model. Therefore, the solution retained according to the least loss principle will best approximate DC power flow solution.

By use of above sifting principle, after solving feasible solution of each network flow in each operating state, we can calculate ΔQ according to (3.113) at once, and remain flow solution relative to minimal ΔQ by comparing with original feasible solution ΔQ. Thus same feasible solutions are sifted at the same time.

After sifting solutions, we have a unique feasible solution corresponding to each feasible operating state. Define the branch flow vectors of a feasible solution as the lower boundary point $\Phi_{(i)}$,

$$\Phi_{(i)} = \{F_{(i)1}, F_{(i)2}, \ldots, F_{(i)N_\Sigma}\}, \quad i = 1, 2, \ldots, k_0, \tag{3.119}$$

where $F_{(i)j}$ is flow on branch j of the ith feasible operating state and k_0 is the total number of feasible operating states. When the capacity of each branch is larger than or equal to the corresponding component of this vector, the transmission system can satisfy the load demand.

3.6.4 Reliability of Transmission System

The transmission system can be considered as a stochastic network, and the capacities of generators, transmission lines, and transformers are treated as stochastic variables because of random outages. Generally speaking, load demands are also stochastic variables. But in reliability study, the load demands D are often given as a standard to measure reliability, so that a reliability index can be defined as the probability of satisfying load demand D by the transmission system.

When the lower boundary points of all feasible solutions $\Phi_{(i)}$ ($i = 1, 2, \ldots, k_0$) have been obtained, we can calculate the probability of satisfying load demand D according to the distribution of each branch capacity.

Let us define event B_i,

$$B_i = \{Y | Y \geq \Phi_{(i)}\}, \tag{3.120}$$

where $Y \geq \Phi_{(i)}$ means

$$Y_j > \Phi_{(i)j}, \quad j = 1, 2, \ldots, N_b. \tag{3.121}$$

Then probability that B_i occurs is

$$P(B_i) = \prod_{j=1}^{N_b} P(F_{(i)j}), \tag{3.122}$$

where $P(F_{(i)j})$ is cumulative probability which can be obtained from the outage table of component j (see Sect. 3.3.3).

The event A_D that satisfies load demand D is

$$A_D = \bigcup_{i=1}^{k_0} B_i. \tag{3.123}$$

Therefore, the probability that load demand D is satisfied or the reliability of the transmission system R_D is described by,

$$R_D = P\left(\bigcup_{i=1}^{k_0} B_i\right) \tag{3.124}$$

The above indices R_D can be calculated by the inclusion–exclusion rule [46].

In this section, the probabilistic network-flow model and algorithm of reliability evaluation are introduced. Its feasible solution set illustrates the whole scenario of possible operating states of transmission systems, which can be used to analyze power network structure, and to find components whose outages cause system cascade failures. The importance of each component in a power grid can be quantified by probabilistic network flow, and the probability (i.e., reliability) to satisfy given load is calculated. It provides abundant quantified information for transmission system operation and planning, generation and transmission component maintenance, the dispatch and purchase of operational reserve, contingency setting, countermeasures, and so on.

When solving for network-flow, it is enough to give the available capacities instead of output power. Therefore, the feasible operation solution obtained by the proposed model automatically takes redispatch into consideration. As a static security analysis tool, the efficiency is much higher than load flow analysis.

From the viewpoint of application, probabilistic network-flow model and algorithm have provided a powerful tool for the evaluation of power networks and tie-lines between areas and the reliability of interconnected systems.

There are two shortcomings related to the proposed model that need to be further improved. Firstly, the model neglects Kirchhoff's Second Law, and the load flow is therefore erroneous. Although treated by a remedial measure, this problem is still worth paying attention to in applications. Secondly, the application of the stochastic network-flow model needs the development of effective algorithms. The stochastic network-flow model is a NP complexity problem, and the key algorithm adopts the implicit enumeration technique. This problem needs further investigation to develop various effective simplified algorithms.

[Example 3.6] Evaluating the reliability of transmission system shown in Fig. 3.15, in which the capacity and reliability parameter of components are given in Table 3.5 and 3.6.

[Solution] Firstly, we analyze the effect of each branch on system reliability.

After enumerating and sifting feasible solutions we can calculate composite reliability R_D according to (3.124). Assume maximum available capacities of and are 5 and 5.5, respectively. The transmission system reliability R_D is 0.8685586. When neglecting random outages of generators, i.e., FOR is 0 for all generators, reliability R_D increases to 0.9002437.

Let us now analyze the contribution of each branch on transmission system reliability. The results are shown in Table 3.26. The basic scheme in the table corresponds to the configuration shown in Fig. 3a. Its reliability index R_D has been mentioned above. To evaluate the contribution of each branch on reliability, in the table we demonstrate R_D for the schemes of the transmission system without branches b1, b2, and b3, respectively.

We can see the impact of each branch on composite reliability from the table. The reliability index of the transmission system without branch b1 or b2 are significantly worse than the basic case, R_D decreases to 0.8121741 from 0.8685586. Branch b3 has less impact on composite reliability. When the scheme excludes it, R_D decreases to 0.8516820. When neglecting random outages of generators (FOR = 0), branch b3 has almost no contribution to transmission reliability, see the third column of the table.

Impacts on composite reliability reflect the values of branches in the transmission system.

When load demand increases, the reliability deteriorates. The reliability indices for different load levels from 1.0 to 1.4 times the original demand are demonstrated in Fig. 3.16. When all loads are up to 1.4 times their original demands, the

Table 3.26 R_D for different transmission schemes

Schemes	FOR > 0	FOR = 0
Basic scheme	0.8685586	0.9002437
Without b1	0.8121741	0.8573750
Without b2	0.8121741	0.8573750
Without b3	0.8516820	0.9002437

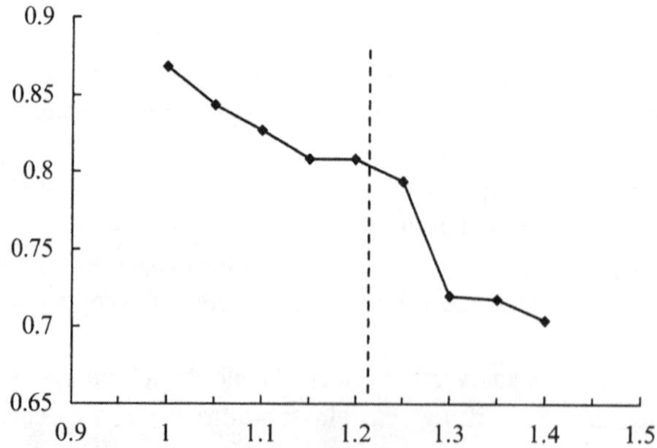

Fig. 3.16 Variations of reliability index with load level

Table 3.27 Simulation results

Cases	Available Capacity		Reliability
	G1	G2	R_D
1	4	4	0.827689
2	4	4.5	0.827689
3	4.5	4	0.843002
4	5.5	5	0.843002
5	5	5.5	0.868558

reliability index R_D decreases from 0.8685586 to 0.704385. It is worth noting that reliability drops significantly, from 0.793561 to 0.704385, between 1.25 and 1.3 times loads. The reason behind this is that when loads up to 1.3 times, the load at node 1 increases to 2.08, which exceeds the transmission capacity of branch 1 and 2. Therefore, outage of either of these two branches will cause shortage of power supply for node 1. That is to say, at such a load level the transmission system cannot satisfy the $N-1$ check for branch 1 and 2. We can conclude that below 1.25 times load level, system reliability mainly relates to generation adequacy, beyond that load level the transmission network is heavily stressed, and transmission adequacy becomes the main reason for blackouts.

Generally speaking, the more the generation reserve, the more reliable the composite system. However, because of capacity limitation of tie-lines, the same amount of generation reserve at different generators may have a different influence on composite reliability. The reliability indices with different reserves for the sample system are demonstrated in Table 3.27. When available capacities of both generators are 4.0, composite reliability R_D is 0.827689 (see row 1). Adding 0.5 reserve capacity to G2, i.e., making its available capacity 4.5, has no influence on R_D (see row 2). By contrast, Adding 0.5 reserve capacity to G1, R_D increases to 0.843002 (see row 3). Similarly, if now we increase available capacity of G1 to 5.5, R_D is unchanged (see row 4); if we increase available capacity of G2 to 5.5, R_D increases to 0.868558 (see row 5).

In conclusion, we can use the stochastic network-flow model not only to comprehensively evaluate transmission system reliability but also to efficiently purchase reserve to improve system reliability in a market environment.

Thinking and Problem Solving

1. What are the important characteristics of deterministic methods and stochastic methods, when evaluating the security level of electrical power systems?
2. What stochastic factors influence the operation of electrical power systems? How are these stochastic factors simulated?
3. What is understood by the "randomness" of the load? When can the randomness of load be simulated by discrete stochastic variables?

4. What does the component state of electrical power systems refer to? What does the available capacity of a component refer to? Why are transmission lines and transformers generally dual-state components, while generator units may be the multistate components?
5. Two transmission lines operating in parallel can be considered as a triple-state component. How can the triple-state model be obtained by using convolution formulae?
6. Deduce the formulae of accumulative probability in an outage table.
7. What are the basic theory and main characteristics of the Monte Carlo simulation method?
8. How can the sampling state of the system be obtained effectively when using the Monte Carlo simulation method to evaluate the reliability of an electrical power system?
9. Outline the simplified supposition conditions of the optimal adjustment model for power output of generator units during reliability evaluation of electrical power systems.
10. How can we evaluate the adequacies of a power generation system and power transmission system, respectively, using the Monte Carlo simulation method?
11. What is the basic principle of the Markov Chain Monte Carlo (MCMC) method? How can we obtain the sampling state of the system according to the Gibbs sampler in the MCMC simulation method?
12. Why are random numbers with the uniform distribution used to determine the component states in Monte Carlo sampling?
13. Outline the characteristics of the semi-invariant method in probabilistic load flow.
14. What are the factors that influence the errors of probabilistic load flow?
15. What are the factors that influence the calculation speed of probabilistic load flow?
16. What is the stability of an electrical power grid structure? What are cascade failures?
17. What are the features of the network-flow method used to analyze the operational security of electrical power systems?
18. Design the calculation process to obtain all feasible solutions in the network-flow method.
19. Design and realize the calculation process shown in formula (3–124) when k_0 is uncertain.

Chapter 4
Power Flow Analysis in Market Environment

4.1 Introduction

Around the world, the electric industry is undergoing sweeping restructuring. The trend is toward increased competition and reduced regulation.

According to economic theory, we can define the power market as a collection of electricity buyers and electricity sellers that interact, resulting in the possibility for exchange. It should be noted that besides electricity, different kinds of ancillary services are also included in the goods of the power market. Ancillary services are mainly employed to maintain the power system operating securely and efficiently, comprising electricity transmission, providing capacity reserve, reactive power and voltage regulation, etc.

A variety of restructuring models are being proposed, considered and experimented with in different countries. Because production and consumption of electric power occur simultaneously, transmission and relative services become a salient feature of the power market. Because of its scale effects transmission service is regarded as a natural monopoly. Therefore, the unbundling of generation from transmission as a separate business is prevalent in different countries, and in general the transmission sector remains regulated to permit a competitive environment for generation and retail services.

Figure 4.1 shows the main structural components of the power market [48]. Generating companies (G) and electricity power market (PM) form market sellers; retail service providers (R), and distribution service providers (D) form market buyers. The transmission part (T) of the power market consists of five components: transmission owners (TO), independent system operators (ISOs), ancillary service (AS) providers, power exchanges (PXs), and scheduling coordinators (SCs), as briefly explained below.

4.1.1 Transmission Owner

The basic premise of transmission open access is that the transmission owners treat all transmission users on a nondiscriminatory and comparable basis regarding

X.-F. Wang et al., *Modern Power Systems Analysis*.
doi: 10.1007/978-0-387-72853-7,

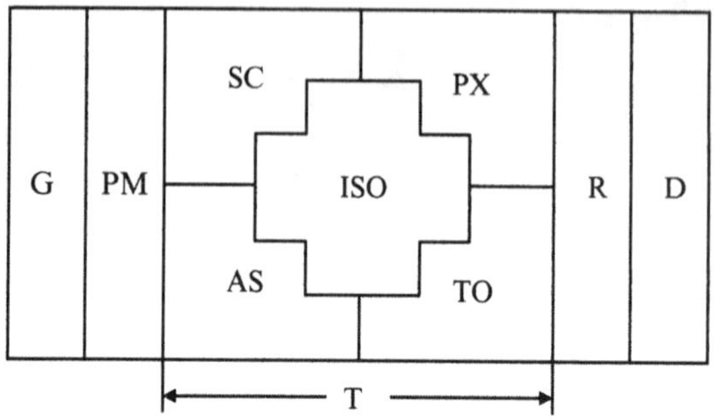

Fig. 4.1 Structure components of power market

access to and use of the transmission system and services. This requirement could be difficult to ensure if the transmission owners have any financial interests in energy generation or supply. A general trend is, therefore, to designate an ISO to operate the transmission system and facilitate provision of transmission services. Maintenance of the transmission system generally remains the responsibility of the transmission owners.

4.1.2 Independent Operator

The ISO operates the transmission grid and provides transmission services to all transmission customers. The basic requirement of an ISO is lack of financial interest in generation resources and load market. The responsibilities and scope of the different ISOs existing or emerging different countries around the world vary widely, but mainly include the following areas:

(1) Operations planning/scheduling
(2) Dispatching
(3) Control and monitoring
(4) Online network security analysis
(5) Market administration
(6) Transmission planning

4.1.3 Power Exchange

The basic function of a power exchange is to provide a forum to match electrical energy supply and demand in the forward power markets. The market horizon may range from an hour to a few months. The most usual situation is a day-ahead market to facilitate energy trading 1 day before each operating day. Depending on the market design, the day-ahead market may be preceded by a longer term market and supplemented by hour-ahead markets. An hour-ahead market provides energy trading opportunities to 1 or 2 h before the operating hour. Usually, the power exchange is to act as a pool for energy supply and demand bids, and establish a market clearing price (MCP). The MCP is then the basis for the settlement of forward market commitments.

The above-mentioned components: TO, ISO, PX provide a platform for energy transaction, and should not have any financial interests in the market.

4.1.4 Ancillary Service

Ancillary service providers supply the transmission network support services that are needed for reliable operation of the power system. The majority of ancillary services are, in fact, real or reactive power resources needed to operate the transmission system in a secure and reliable manner.

Depending on the market design and activity rules, ancillary services may be traded in the PX or ISO. The ancillary services may be provided in a bundled manner or as an unbundled menu. Some ancillary services may be self-provided by users of the transmission system, such as capacity reserve, energy imbalance, etc.

4.1.5 Scheduling Coordinator

Scheduling coordinators (SCs) are entities that put together supply and demand energy schedules, outside the power exchange. Some structures restrict forward schedule coordination to a center pool and do not permit other SCs to operate. In some other structures no center pool or regulated power exchange exists; schedule coordination is done in a decentralized manner often by the existing control areas. In many new and emerging structures SCs are integral components of the market.

The structural components mentioned above may not be present in a specific restructuring model. In some cases one or more of the segments are missing. In other cases, two or more of these structural components are merged and delegated to a single entity. In any case the relative functions cannot be eliminated.

The ongoing restructuring of the electric industry has imposed tremendous challenges on both economic and technical aspects of power systems under this new environment.

The theories of electricity pricing and transaction mechanisms are at the core of economic problems of the power market. Electricity pricing can be traced to the 1980s, when the effect of electricity spot pricing on the efficient location of resources was theoretically proved [49]. The studies on electricity price include cost analysis and market clearing mechanism. The bidding strategy of electricity sellers is another hot topic of investigation. This problem not only affects the benefits for generating companies, but also efficiency and stability of market operation.

The new technical issues in the market-oriented power system mainly relate to security and reliability problems. Regardless of the market structures that may emerge in various parts of the world, system security, reliability and quality of supply must be maintained.

Transmission congestion is an important phenomenon in power markets. It can be defined as the condition that occurs when there is insufficient transmission capability to simultaneously implement all preferred transactions. Congestion can impose a significant barrier with respect to trading electricity. Therefore, congestion management is a major function of the ISO in any type of power market. In addition, for better transmission services support and full utilization of transmission assets, an ISO should accurately evaluate the transfer capability remaining in the system for further transactions, which is termed as available transfer capability (ATC).

To price the transmission service in the market environment, steady-state analysis should provide not only power flow in each branch of the system, but also the contribution of every generating company or consumer to the branch flow. Thus, the so-called power flow tracing problem becomes a new challenge.

This chapter will focus on some new developments in the steady-state analysis field, including optimal power flow (OPF) and its application in spot pricing and congestion management, power flow tracing and ATC problems.

4.2 Optimal Power Flow

In 1962, Carpentier introduced a generalized nonlinear programming (NLP) formulation of the economic dispatch (ED) problem [50] including voltage and other constraints. The problem was later named OPF. The OPF procedure consists of determining the optimal steady-state operation of a power system, which simultaneously minimizes the value of a chosen objective function and satisfies certain physical and operating constraints. Today OPF has been playing a very important role in power system operation and planning: different classes of OPF problems, tailored towards special-purpose applications are defined by selecting different function to be minimized, different sets of controls and different sets of constraints.

4.2.1 General Formulation of OPF Problem

The OPF problem is setup on the following basis:

1. The operating generating units are predetermined
2. The power outputs of hydro units are predetermined by reservoir dispatching
3. The structure of transmission network is predetermined, which means the network reconfiguration problem is not considered in OPF

The variables of OPF problem consist of a set of dependent variables and a set of control variables. The dependent variables include node voltage magnitudes and phase angles, as well as MVAr output of generators performing node voltage control. The control variables might include real and reactive power output of generators, voltage settings of voltage control nodes, LTC transformer tap positions, phase shifter angles, operating capacities of shunt capacitors, reactors, etc.

The constraints of OPF include

1. Power flow equations
2. Upper and lower bounds on the generator active power outputs
3. Upper and lower bounds on the generator reactive power outputs
4. Capacity constraints on shut capacitors and reactors
5. Upper and lower bounds on the transformer or phase shifter tap positions
6. Branch transfer capacity limits
7. Node voltage limits

Except constraint (1), the other constraints are all inequality constraints. Among these constraints, (1) and (6) are functional type constraints. When the rectangular coordinates format is used to describe node voltages, constraint (7) is also of the functional type. The others are constraints on variables.

The objective function of an OPF problem may take many different forms according to the different applications. The general nonlinear OPF problem can be formulated as (4.1)–(4.6)

Objective function:

$$\min \sum_{i \in S_G} (a_{2i} P_{Gi}^2 + a_{1i} P_{Gi} + a_{0i}), \tag{4.1}$$

where P_{Gi} is the active generation of i th unit. a_{0i}, a_{1i}, a_{2i} are the fuel cost coefficients of unit i, S_G is the set of generating units in the system.

Constraints:

$$\left.\begin{aligned} P_{Gi} - P_{Di} - V_i \sum_{j=1}^{n} V_j (G_{ij} \cos\theta_{ij} + B_{ij} \sin\theta_{ij}) = 0 \\ Q_{Gi} - Q_{Di} + V_i \sum_{j=1}^{n} V_j (G_{ij} \sin\theta_{ij} - B_{ij} \cos\theta_{ij}) = 0 \end{aligned}\right\} \quad i \in S_B, \tag{4.2}$$

$$\underline{P}_{Gi} \le P_{Gi} \le \overline{P}_{Gi} \quad i \in S_G, \tag{4.3}$$

$$\underline{Q}_{Ri} \le Q_{Ri} \le \overline{Q}_{Ri} \quad i \in S_R, \tag{4.4}$$

$$\underline{V}_i \le V_i \le \overline{V}_i \quad i \in S_B, \tag{4.5}$$

$$|P_l| = |P_{ij}| = |V_i V_j (G_{ij} \cos\theta_{ij} + B_{ij} \sin\theta_{ij}) - V_i^2 G_{ij}| \le \overline{P}_l \quad l \in S_l, \tag{4.6}$$

where the power flow (4.2) is equality constraints, constraints (4.3)–(4.6) are inequality constraints, they are upper and lower bounds on the active sources, reactive sources, node voltages, and branch flows respectively. S_B is the set of nodes in the system. S_R is the set of reactive sources. S_l is the set of transmission lines. P_{Gi}, Q_{Gi} are the active and reactive power output, P_{Di}, Q_{Di} are the active and reactive demand at node i. V_i, θ_i are the voltage magnitude and angle of node i; $\theta_{ij} = \theta_i - \theta_j$. G_{ij}, B_{ij} are the real and imaginary parts of the transfer admittance between nodes i and j. P_l is the power flow on transmission line l that connects nodes i and j. In the above OPF formulation the node voltage is expressed in polar coordinates. The rectangular coordinate expression is also often used in formulating the OPF problem.

4.2.2 Approaches to OPF

OPF development has been closely following the progress in numerical optimization techniques and computer technology. Many different approaches have been proposed to solve the OPF problem. These techniques include NLP, quadratic programming (QP), linear programming (LP), mixed programming (MP), as well as interior point and artificial intelligence algorithms.

4.2.2.1 Nonlinear Programming

NLP deals with problems involving nonlinear objective and nonlinear constraint functions. It includes unconstrained programming and constrained programming. It is well known that many simple and effective methods have been developed and used successfully to solve unconstrained programming. For constrained programming, we usually first transform the problem into unconstrained programming by constructing an augmented Lagrangian objective function, and can then use various different optimization algorithms to solve this unconstrained programming problem.

In 1968, Dommel and Tinney [51] developed an NLP model to minimize fuel cost and active power losses. Based on the Newton–Raphson method of power flow calculation, Kuhn–Tucker equations are solved by using a combination of the gradient method for a fixed set of independent variables and penalty function for violated dependent constraints. Their work has served as the guiding pioneer work for commercial OPF tools. Dealing with inequality constraints via penalty functions may cause ill-conditioning leading to very slow convergence. Sasson [52] presented a modified method which implements step correction by the Fletcher–Powell algorithm and checks convergence at each step in the optimization process. However, because of inherent limitations of the gradient method the oscillation phenomena cannot be avoided completely. In 1970, Sasson [53] extended Dommel and Tinney's work, trying to improve convergence of the Newton-based approach. These methods are useful only for problems of limited size or using a nonsparse formulation, because they generate dense Hessians. In 1982, Divi and Kesavan [54] presented a shifted penalty-function approach which overcomes the ill-conditioning of the penalty-function method in solving constrained NLP problems. The method exploited a reduced gradient concept and adapted Fletcher's quasi-Newton technique for optimization of shifted penalty functions, which further improves convergence and accuracy. This method was validated on three synthetic systems with an 11-node system being the largest. The method saves about 30% of the computational cost over standard penalty-function methods. In the same year (1982), Talukdar et al. [55] presented a quasi-Newton (variable metric) method for solving general OPF problems. The method is attractive for the following reasons: (1) it can accommodate OPF constraints in a straightforward manner, (2) it is robust and will attain a feasible solution from infeasible initial starting points, and (3) it appears to be several times faster than its competitors.

4.2.2.2 Quadratic Programming

The quadratic programming is a special form of NLP with quadratic objective function and linear constraints. In 1973, Reid and Hasdorf [56] presented a quadratic programming method to solve the economic dispatching (ED) problem. In this work, the fuel cost function is approximated to a quadratic function by introducing slack variables, constraints are linearized by Taylor expansion, and then Wolfe's algorithm is used to solve the ED problem. Convergence is very fast and does not depend upon the selection of gradient step size or penalty factors. However, the CPU time dramatically increased as the system size increased. More papers about applications of quadratic programming for the OPF problem appeared in 1980s. In 1982, Burchett et al. [57] presented a method which can obtain a feasible solution from an infeasible starting point even if the power flow diverges. The method creates a sequence of quadratic programming problems that converge to the optimal solution of the original nonlinear problem. Comparing with the older algorithm which uses an augmented Lagrangian, the method has advantages in terms of CPU time and robustness.

4.2.2.3 Linear Programming

Linear programming is another useful technique for solving OPF problems. Usually, the original OPF problem is first decomposed into active power and reactive power sub-problems, then these two sub-problems are solved alternatively or separately. Piecewise linear techniques and sequential approximation techniques are often used to solve LP-based OPF problem. In 1968, Wells [58] developed a linear programming approach to dispatching an economic operation plan including network security requirements. The objective function and its constraints were linearized and solved using the simplex method. The limitations of this method are: (1) the final results for an infeasible situation obtained may not be optimum and (2) rounding errors caused by digital computers may cause constraints to appear overloaded. In 1970, Shen and Laughton [59] presented a dual linear programming technique. Both primal and dual problems were proposed and the solutions were obtained using the revised simplex method. This work has been well tested and has shown more promising online performance than the NLP technique.

4.2.2.4 Mixed Programming

Because active and reactive power sub-problems are provided with different characteristic, two or more optimization techniques are often combined to solve OPF problem. Nabona and Ferris [60] presented a method which involved quadratic and linear programming for optimizing the economic dispatch objective function. The minimum loss problem was solved using a linear programming approach, and the minimum cost and reactive power problems were solved using either a quadratic or a linear programming approach. In [61], the LP-based methods have been shown to be effective for problems where the objective functions are separable and convex functions, but not so effective for nonseparable objective functions, especially when the objective function is the minimization of transmission losses. Both the QP-based and Newton methods are second-order methods, and programs based on these methods appear to overcome the shortcomings that exist in the LP-based programs. But second-order methods require the calculation of second partial derivatives of the Lagrangian (i.e., the Hessian matrix), which may be unavailable if the original generation cost curves are given as piecewise linear functions or discrete segments by which generators' input–output characteristic with valve-point loadings are modeled. This hybrid feature makes the approach very flexible while preserving the efficiency of decoupling.

4.2.2.5 Interior Point Method (IPM)

Linear programming is one of the most widely applied mathematical techniques. Until very recently, the standard method for solving LP problems was the simplex method, first proposed by Dantzig. Since then, it has been routinely used to solve problems in business, logistics, economics, and engineering. All forms of simplex method reach the optimum by traversing a series of basic solutions. Since each basic solution represents an extreme point of the feasible region, the track followed by the algorithm moves around the boundary of the feasible region. In the worse case, it may be necessary to examine most if not all of the extreme points. This could be very inefficient given that the number of extreme points grows exponentially with the number of constraints and variables of the problem. Fortunately, the worst-case behavior has not been experienced for practical problems.

Ever since the simplex method was first presented, many researchers have applied themselves to create an algorithm for solving LP problems that proceeded on a path through the polytope rather than around its perimeter. One of the earliest IPMs, originally proposed by Frish [62] in 1954, was a Barrier Method leading to the solution of an unconstrained optimization problem. Early IPMs also include that of Huard (1967) [63], this Center Method is able to solve constrained NLP. In the same year, Dikin [64] presented a method which is known as the Affine Scaling Method. The advantage of this idea is that the steepest descent step in the original space can be extremely short if the current iterate is close to the boundary, whereas "long" steps are always possible in the transformed space. Though theoretically efficient, code developers were never able to realize an implementation that matched the performance of the current simplex method. While IPMs were put aside in the 1970s, significant advances were made in numerical linear algebra and, of course, in computational capacity and speed. Interest in IPMs was then revitalized by the announcement of Karmarkar [65] that he had developed an IPM which had provable polynomial complexity and was competitive with the simplex method. In fact, he claimed a factor of 100 speed-up, comparing with the contemporary state-of-the-art simplex solver MPSX. A few years later, IPMs were extended to apply to NLP problems by Gill [66].

Karmarkar's method and its relatives, including the affine scaling algorithm, logarithmic barrier function algorithm, and path following algorithm, were studied intensively. Since then, thousands of papers have been written on both the theoretical and computational aspects of IMP for LP and on the extension of these ideas to quadratic and more general NLP problems. The most successful IPMs are based on using a primal–dual formulation and applying Newton's method to the system of equations arising from the barrier method, or in other words, arising by perturbing the optimality conditions. This method has been widely used in power system optimization problems because of its favorable convergence, robustness, and insensitivity to infeasible starting points. The IPM for solving OPF problems will be introduced in detail in Sect. 4.2.3.

4.2.2.6 Artificial Intelligence Method (AIM)

Though the optimization techniques described above gradually overcame the difficulties in calculation speed, convergence, and starting points; discrete problem could not be readily dealt with. Furthermore, these optimization methods are essentially based on the idea of neighborhood search (also called local search). The methods rely on convexity to obtain the global optimum and as such are forced to simplify relationships to ensure convexity. However, the OPF problem is in general nonconvex and as a result, many local optima may exist. Artificial intelligence methods can resolve some of the above problems. The major AI methods include: Evolutionary Algorithms (EA), Fuzzy Set Theory (FST), Simulated Annealing (SA), etc.

The evolutionary algorithms, in part, emulate biological evolution and operate on a population of candidate solutions to a problem. They chiefly include Evolutionary Programming and Genetic Algorithms. An EA is often used to solve reactive power optimization problems because it is good at dealing with discrete variables. EA is a stochastic optimization method, which can obtain the global optimum with high probability (in theory). Moreover, EA has many advantages, such as global convergence, parallel processing, universality, and robustness.

Fuzzy set theory represents an attractive tool to aid research in optimization techniques when uncertain relationships or inconsistent measurements among model parameters limit the specification of model objective functions and constraints. Recently, fuzzy set theory has been successfully applied in solving power system optimization problems, because it provides a new approach to coordinating multiple conflicting objectives of the problem. In [67], constraints are partitioned into two parts: soft constraints and hard constraints. The OPF problem is formulated with fuzzy objective and fuzzy soft constraints. An efficient successive linear programming method is then modified to solve this new formulation. The numerical results show that the fuzzy OPF can be equivalent to the crisp OPF where a feasible solution exists. When there is no feasible solution for the crisp OPF, the fuzzy OPF can obtain a more realistic solution that "evenly" distributes violations of the limits, rather than violate a single normal limit excessively.

The simulated annealing algorithm is an iterative, stochastic search process, which simulates the physical annealing mechanics of melting metal. The working principle of SA is very simple, the only difference with a general search method is that it allows some nonimproving solutions to be accepted with some probability in every iteration step. In [68], SA is used to solve the reactive power optimization problem. Theoretically, the global optimum can be obtained after many trials, but this inevitably consumes a lot of calculation time.

The major drawback of artificial intelligence methods is usually their poor computation efficiency. Due to the complexity of power system optimization problems, hybridization of these with other optimization algorithms would be a way forward to develop more powerful approaches to produce some particular properties.

4.2.3 Interior Point Method for OPF Problem

The main idea of IPM is to approach the optimal solution from the strict interior of the feasible region. Two conditions must be noticed: (1) start from a feasible point and (2) construct a barrier that prevents any variables from reaching a boundary [69]. But it is very difficult to find a feasible start point for large-scale practical problems. For many years researchers endeavored to weaken the "feasible start point" condition, to improve the IPM performance. The center path following IPM discussed in this section is a successful example. The "feasible start point" condition is replaced by simple inequality constraints that request nonzero slack variables and Lagrangian multipliers.

To explain the problem explicitly, the OPF problem expressed by (4.1)–(4.6) is expressed as the following generalized nonlinear optimal model,

$$\text{obj } \min f(\boldsymbol{x}), \tag{4.7}$$

$$\text{s.t.}\ \mathbf{h}(\mathbf{x}) = \mathbf{0}, \tag{4.8}$$

$$\underline{\mathbf{g}} \le \mathbf{g}(\mathbf{x}) \le \overline{\mathbf{g}}, \tag{4.9}$$

where objective function (4.7), corresponding to (4.1) in the OPF formulation, is a nonlinear function. Equality constraint (4.8), corresponding to (4.2) in OPF formulation, is also nonlinear. Equation (4.9) is nonlinear inequality constraints, for which the upper bound is $\overline{\mathbf{g}} = [\bar{g}_1, \ldots, \bar{g}_r]^{\mathrm{T}}$ and the lower bound is $\underline{\mathbf{g}} = [\underline{g}_1, \ldots, \underline{g}_r]^{\mathrm{T}}$. It is assumed that there are n variables, m equality constraints, and r inequality constraints in this formulation.

First, inequality constraint (4.9) is translated into an equality constraint by introducing slack variables:

$$\mathbf{g}(\mathbf{x}) + \mathbf{u} = \overline{\mathbf{g}}, \tag{4.10}$$

$$\mathbf{g}(\mathbf{x}) - \mathbf{l} = \underline{\mathbf{g}}, \tag{4.11}$$

where slack variables $\mathbf{l} = [l_1, \ldots, l_r]^{\mathrm{T}}$, $\mathbf{u} = [u_1, \ldots, u_r]^{\mathrm{T}}$ must be positive:

$$\mathbf{u} > 0, \quad \mathbf{l} > 0. \tag{4.12}$$

Thus the original problem becomes optimization problem A:

$$\begin{aligned}
&\text{obj}\ \min f(\mathbf{x}),\\
&\text{s.t.}\ \mathbf{h}(\mathbf{x}) = \mathbf{0},\\
&\mathbf{g}(\mathbf{x}) + \mathbf{u} = \overline{\mathbf{g}},\\
&\mathbf{g}(\mathbf{x}) - \mathbf{l} = \underline{\mathbf{g}},\\
&\mathbf{u} > 0, \quad \mathbf{l} > 0.
\end{aligned}$$

Then " $\log(l_j)$" and " $\log(u_j)$" are added to the objective function of problem A to construct a barrier function objective which is equivalent to $f(\mathbf{x})$ when l_j and $u_j (j = 1, \ldots, r)$ are more than zero. The barrier function objective would become very large if any l_j or $u_j (j = 1, \ldots, r)$ approaches zero. Thus barrier function optimization problem B is obtained

$$\text{obj}\ \min f(\mathbf{x}) - \mu \sum_{j=1}^{r} \log(l_j) - \mu \sum_{j=1}^{r} \log(u_j),$$

$$\text{s.t.} \quad \mathbf{h}(\mathbf{x}) = \mathbf{0},$$
$$\mathbf{g}(\mathbf{x}) + \mathbf{u} = \overline{\mathbf{g}},$$
$$\mathbf{g}(\mathbf{x}) - \mathbf{l} = \underline{\mathbf{g}},$$

where factor (or barrier parameter) μ satisfies $\mu > 0$. An inequality-constrained optimization problem A is transformed into an equality-constrained problem B by incorporating the inequality constraints in a logarithmic barrier function that imposes a growing penalty as the boundary ($u_j = 0$, $l_j = 0$ for all j) is approached. Therefore, the Lagrangian multiplier method of classical calculus can be used to solve problem B.

The Lagrangian function of problem B is

$$L = f(\mathbf{x}) - \mathbf{y}^{\mathrm{T}}\mathbf{h}(\mathbf{x}) - \mathbf{z}^{\mathrm{T}}[\mathbf{g}(\mathbf{x}) - \mathbf{l} - \underline{\mathbf{g}}] - \mathbf{w}^{\mathrm{T}}[\mathbf{g}(\mathbf{x}) + \mathbf{u} - \overline{\mathbf{g}}] - \mu \sum_{j=1}^{r} \log(l_r)$$
$$- \mu \sum_{j=1}^{r} \log(u_r), \tag{4.13}$$

where $\boldsymbol{y} = [y_1, \ldots, y_m]$, $\boldsymbol{z} = [z_1, \ldots, z_r]$, $\boldsymbol{w} = [w_1, \ldots, w_r]$ are Lagrange multipliers.

The necessary conditions for a stationary point of the constrained optimization problem B are that the partial derivatives of the Lagrangian function (4.13) with respect to each variable must be zero.

$$\mathbf{L}_x = \frac{\partial L}{\partial \mathbf{x}} \equiv \nabla_x f(\mathbf{x}) - \nabla_x \mathbf{h}(\mathbf{x})\mathbf{y} - \nabla_x \mathbf{g}(\mathbf{x})(\mathbf{z} + \mathbf{w}) = 0, \tag{4.14}$$

$$\mathbf{L}_y = \frac{\partial L}{\partial \mathbf{y}} \equiv \mathbf{h}(\mathbf{x}) = 0, \tag{4.15}$$

$$\mathbf{L}_z = \frac{\partial L}{\partial \mathbf{z}} \equiv \mathbf{g}(\mathbf{x}) - \mathbf{l} - \underline{\mathbf{g}} = 0, \tag{4.16}$$

$$\mathbf{L}_w = \frac{\partial L}{\partial \mathbf{w}} \equiv \mathbf{g}(\mathbf{x}) + \mathbf{u} - \overline{\mathbf{g}} = 0, \tag{4.17}$$

$$\mathbf{L}_l = \frac{\partial L}{\partial \mathbf{l}} = \mathbf{z} - \mu \mathbf{L}^{-1}\mathbf{e} \Rightarrow \mathbf{L}_l^{\mu} = \mathbf{LZe} - \mu \mathbf{e} = 0, \tag{4.18}$$

$$\mathbf{L}_u = \frac{\partial L}{\partial \mathbf{u}} = -\mathbf{w} - \mu \mathbf{U}^{-1}\mathbf{e} \Rightarrow \mathbf{L}_u^{\mu} = \mathbf{UWe} + \mu \mathbf{e} = 0, \tag{4.19}$$

where $\mathbf{L} = \mathrm{diag}(l_1, \ldots, l_r)$, $\mathbf{U} = \mathrm{diag}(u_1, \ldots, u_r)$, $\mathbf{Z} = \mathrm{diag}(z_1, \ldots, z_r)$, $\mathbf{W} = \mathrm{diag}(w_1, \ldots, w_r)$. From (4.18) and (4.19) we obtain

$$\mu = \frac{\mathbf{l}^{\mathrm{T}}\mathbf{z} - \mathbf{u}^{\mathrm{T}}\mathbf{w}}{2r}. \tag{4.20}$$

We now define a duality gap as: Gap $= \mathbf{l}^{\mathrm{T}}\mathbf{z} - \mathbf{u}^{\mathrm{T}}\mathbf{w}$, then we have

$$\mu = \frac{\mathrm{Gap}}{2r}. \tag{4.21}$$

Fiacco and McCormick [70] proved that under certain conditions, if $\mathbf{x}^*$ is the optimal solution of problem A, $\mathbf{x}(\mu)$ is the optimal solution of problem B while μ is fixed, and sequence $\{\mathbf{x}(\mu)\}$ becomes sufficiently close to $\mathbf{x}^*$ as Gap $\to 0$, i.e., $\mu \to 0$. When solving the sequence of problems B, as the strength of the barrier function is decreased, the optimum follows a well-defined path (hence the term "path following") that ends at the optimal solution to the original problem. As shown in [71], when μ is set according to (4.21), the convergence of the algorithm is sometimes very weak, so the following modification is suggested:

$$\mu = \sigma\frac{\mathrm{Gap}}{2r}, \tag{4.22}$$

where $\sigma \in (0,1)$ is called center parameter, usually one can get satisfactory convergence by setting σ around 0.1. Because $\mu > 0$ and $\mathbf{l} > 0$, $\mathbf{u} > 0$ from (4.18) and (4.19), we know that $\mathbf{z} > 0$, $\mathbf{w} < 0$ must be satisfied.

The necessary conditions for optimality, (4.14)–(4.19), can be solved by the Newton method. The direction of the Newton update can be obtained by solving the following linearized equations:

$$\begin{aligned}-[\nabla_x^2 f(\mathbf{x}) - \nabla_x^2\mathbf{h}(\mathbf{x})\mathbf{y} - \nabla_x^2\mathbf{g}(\mathbf{x})(\mathbf{z}+\mathbf{w})]\Delta\mathbf{x} + \nabla_x\mathbf{h}(\mathbf{x})\Delta\mathbf{y}\\ + \nabla_x\mathbf{g}(\mathbf{x})(\Delta\mathbf{z}+\Delta\mathbf{w}) = \mathbf{L}_x,\end{aligned} \tag{4.23}$$

$$\nabla_x\mathbf{h}(\mathbf{x})^{\mathrm{T}}\Delta\mathbf{x} = -\mathbf{L}_y, \tag{4.24}$$

$$\nabla_x\mathbf{g}(\mathbf{x})^{\mathrm{T}}\Delta\mathbf{x} - \Delta\mathbf{l} = -\mathbf{L}_z, \tag{4.25}$$

$$\nabla_x\mathbf{g}(\mathbf{x})^{\mathrm{T}}\Delta\mathbf{x} + \Delta\mathbf{u} = -\mathbf{L}_w, \tag{4.26}$$

$$\mathbf{Z}\Delta\mathbf{l} + \mathbf{L}\Delta\mathbf{z} = -\mathbf{L}_l^{\mu}, \tag{4.27}$$

$$\mathbf{W}\Delta\mathbf{u} + \mathbf{U}\Delta\mathbf{w} = -\mathbf{L}_u^{\mu}. \tag{4.28}$$

The above equations can be rewritten in the matrix form:

$$\begin{bmatrix} \mathbf{H} & \nabla_x\mathbf{h}(\mathbf{x}) & \nabla_x\mathbf{g}(\mathbf{x}) & \nabla_x\mathbf{g}(\mathbf{x}) & \mathbf{0} & \mathbf{0} \\ \nabla_x^{\mathrm{T}}\mathbf{h}(\mathbf{x}) & \mathbf{0} & \mathbf{0} & \mathbf{0} & \mathbf{0} & \mathbf{0} \\ \nabla_x^{\mathrm{T}}\mathbf{g}(\mathbf{x}) & \mathbf{0} & \mathbf{0} & \mathbf{0} & -\mathbf{I} & \mathbf{0} \\ \nabla_x^{\mathrm{T}}\mathbf{g}(\mathbf{x}) & \mathbf{0} & \mathbf{0} & \mathbf{0} & \mathbf{0} & \mathbf{I} \\ \mathbf{0} & \mathbf{0} & \mathbf{L} & \mathbf{0} & \mathbf{Z} & \mathbf{0} \\ \mathbf{0} & \mathbf{0} & \mathbf{0} & \mathbf{U} & \mathbf{0} & \mathbf{W} \end{bmatrix} \begin{bmatrix} \Delta\mathbf{x} \\ \Delta\mathbf{y} \\ \Delta\mathbf{z} \\ \Delta\mathbf{w} \\ \Delta\mathbf{l} \\ \Delta\mathbf{u} \end{bmatrix} = \begin{bmatrix} \mathbf{L}_x \\ -\mathbf{L}_y \\ -\mathbf{L}_z \\ -\mathbf{L}_w \\ -\mathbf{L}_l^{\mu} \\ -\mathbf{L}_u^{\mu} \end{bmatrix}, \tag{4.29}$$

where $\mathbf{H} = -[\nabla_x^2 f(\mathbf{x}) - \nabla_x^2\mathbf{h}(\mathbf{x})\mathbf{y} - \nabla_x^2\mathbf{g}(\mathbf{x})(\mathbf{z}+\mathbf{w})]$.

The order of the above matrix is $(4r+m+n)\cdot(4r+m+n)$. The most computationally intensive task involves forming the left-hand coefficient matrix of (4.29) and then solving the equation. To reduce the calculation effort, (4.29) can be transformed into the following form by exchanging some rows and columns:

$$\begin{bmatrix} \mathbf{L} & \mathbf{Z} & \mathbf{0} & \mathbf{0} & \mathbf{0} & \mathbf{0} \\ \mathbf{0} & -\mathbf{I} & \mathbf{0} & \mathbf{0} & \nabla_x^{\mathrm{T}}\mathbf{g}(\mathbf{x}) & \mathbf{0} \\ \mathbf{0} & \mathbf{0} & \mathbf{U} & \mathbf{W} & \mathbf{0} & \mathbf{0} \\ \mathbf{0} & \mathbf{0} & \mathbf{0} & \mathbf{I} & \nabla_x^{\mathrm{T}}\mathbf{g}(\mathbf{x}) & \mathbf{0} \\ \nabla_x\mathbf{g}(\mathbf{x}) & \mathbf{0} & \nabla_x\mathbf{g}(\mathbf{x}) & \mathbf{0} & \mathbf{H} & \nabla_x\mathbf{h}(\mathbf{x}) \\ \mathbf{0} & \mathbf{0} & \mathbf{0} & \mathbf{0} & \nabla_x^{\mathrm{T}}\mathbf{h}(\mathbf{x}) & \mathbf{0} \end{bmatrix} \begin{bmatrix} \Delta\mathbf{z} \\ \Delta\mathbf{l} \\ \Delta\mathbf{w} \\ \Delta\mathbf{u} \\ \Delta\mathbf{x} \\ \Delta\mathbf{y} \end{bmatrix} = \begin{bmatrix} -\mathbf{L}_l^{\mu} \\ -\mathbf{L}_z \\ -\mathbf{L}_u^{\mu} \\ -\mathbf{L}_w \\ \mathbf{L}_x \\ -\mathbf{L}_y \end{bmatrix}.$$

By simple mathematical manipulation among rows and columns, the above equation can be rewritten as

$$\left[\begin{array}{cccc:cc} \mathbf{I} & \mathbf{L}^{-1}\mathbf{Z} & \mathbf{0} & \mathbf{0} & \mathbf{0} & \mathbf{0} \\ \mathbf{0} & \mathbf{I} & \mathbf{0} & \mathbf{0} & -\nabla_x^{\mathrm{T}}\mathbf{g}(\mathbf{x}) & \mathbf{0} \\ \mathbf{0} & \mathbf{0} & \mathbf{I} & \mathbf{U}^{-1}\mathbf{W} & \mathbf{0} & \mathbf{0} \\ \mathbf{0} & \mathbf{0} & \mathbf{0} & \mathbf{I} & \nabla_x^{\mathrm{T}}\mathbf{g}(\mathbf{x}) & \mathbf{0} \\ \hdashline \mathbf{0} & \mathbf{0} & \mathbf{0} & \mathbf{0} & \mathbf{H}' & \nabla_x\mathbf{h}(\mathbf{x}) \\ \mathbf{0} & \mathbf{0} & \mathbf{0} & \mathbf{0} & \nabla_x^{\mathrm{T}}\mathbf{h}(\mathbf{x}) & \mathbf{0} \end{array}\right] \begin{bmatrix} \Delta\mathbf{z} \\ \Delta\mathbf{l} \\ \Delta\mathbf{w} \\ \Delta\mathbf{u} \\ \Delta\mathbf{x} \\ \Delta\mathbf{y} \end{bmatrix} = \begin{bmatrix} -\mathbf{L}^{-1}\mathbf{L}_l^{\mu} \\ \mathbf{L}_z \\ -\mathbf{U}^{-1}\mathbf{L}_u^{\mu} \\ -\mathbf{L}_w \\ \mathbf{L}'_x \\ -\mathbf{L}_y \end{bmatrix}, \quad (4.30)$$

where

$$\begin{aligned} \mathbf{L}'_x &= \mathbf{L}_x + \nabla_x\mathbf{g}(\mathbf{x})[\mathbf{L}^{-1}(\mathbf{L}_l^{\mu} + \mathbf{Z}\mathbf{L}_z) + \mathbf{U}^{-1}(\mathbf{L}_u^{\mu} - \mathbf{W}\mathbf{L}_w)]\mathbf{H}' \\ &= \mathbf{H} - \nabla_x\mathbf{g}(\mathbf{x})[\mathbf{L}^{-1}\mathbf{Z} - \mathbf{U}^{-1}\mathbf{W}]\nabla_x^{\mathrm{T}}\mathbf{g}(\mathbf{x}), \\ \mathbf{H}' &= \mathbf{H} - \nabla_x\mathbf{g}(\mathbf{x})[\mathbf{L}^{-1}\mathbf{Z} - \mathbf{U}^{-1}\mathbf{W}]\nabla_x^{\mathrm{T}}\mathbf{g}(\mathbf{x}). \end{aligned}$$

At this stage, major calculation effort of solving (4.30) is the LDL^{T} decomposition of $\begin{bmatrix} \mathbf{H}' & \nabla_x\mathbf{h}(\mathbf{x}) \\ \nabla_x^{\mathrm{T}}\mathbf{h}(\mathbf{x}) & \mathbf{0} \end{bmatrix}$ matrix which is of order $(m+n)\cdot(m+n)$, and is much smaller than the left-hand coefficient matrix of (4.30). The other variables are easily obtained by back substitution. This approach can be implemented in a very efficient manner because $\begin{bmatrix} \mathbf{H}' & \nabla_x\mathbf{h}(\mathbf{x}) \\ \nabla_x^{\mathrm{T}}\mathbf{h}(\mathbf{x}) & \mathbf{0} \end{bmatrix}$ is a highly sparse matrix.

The Newton direction for the kth iteration is obtained by solving (4.30), and the new approximations for the optimal solution are

$$\mathbf{x}^{(k+1)} = \mathbf{x}^{(k)} + \alpha_{\mathrm{p}}\Delta\mathbf{x}, \quad (4.31)$$

$$\mathbf{l}^{(k+1)} = \mathbf{l}^{(k)} + \alpha_{\mathrm{p}}\Delta\mathbf{l}, \quad (4.32)$$

$$\mathbf{u}^{(k+1)} = \mathbf{u}^{(k)} + \alpha_{\mathrm{p}}\Delta\mathbf{u}, \quad (4.33)$$

$$\mathbf{y}^{(k+1)} = \mathbf{y}^{(k)} + \alpha_\mathrm{d}\Delta\mathbf{y}, \tag{4.34}$$

$$\mathbf{z}^{(k+1)} = \mathbf{z}^{(k)} + \alpha_\mathrm{d}\Delta\mathbf{z}, \tag{4.35}$$

$$\mathbf{w}^{(k+1)} = \mathbf{w}^{(k)} + \alpha_\mathrm{d}\Delta\mathbf{w}, \tag{4.36}$$

where α_p and α_d are primal-step length and dual-step length respectively, they can be set by the following equations to ensure $\mathbf{u} > 0$ and $\mathbf{l} > 0$:

$$\left.\begin{aligned}
\alpha_\mathrm{p} &= 0.9995 \min\left\{ \min_i \left(\frac{-l_i}{\Delta l_i}, \quad \Delta l_i < 0; \quad \frac{-u_i}{\Delta u_i}, \quad \Delta u_i < 0 \right), 1 \right\} \\
\alpha_\mathrm{d} &= 0.9995 \min\left\{ \min_i \left(\frac{-z_i}{\Delta z_i}, \quad \Delta z_i < 0; \quad \frac{-w_i}{\Delta w_i}, \quad \Delta w_i > 0 \right), 1 \right\}
\end{aligned}\right\}$$

$$i = 1, 2, \ldots, r. \tag{4.37}$$

The flow chart of the IPM for OPF is shown in Fig. 4.2, with the initialization including:

1. Set $\mathbf{l}, \mathbf{u}$ conforming to $[\mathbf{l}, \mathbf{u}]^\mathrm{T} > \mathbf{0}$
2. Set Lagrangian multipliers $\mathbf{z}, \mathbf{w}, \mathbf{y}$ conforming to $[\mathbf{z} > \mathbf{0}, \quad \mathbf{w} < \mathbf{0}, \quad \mathbf{y} \neq \mathbf{0}]^\mathrm{T}$

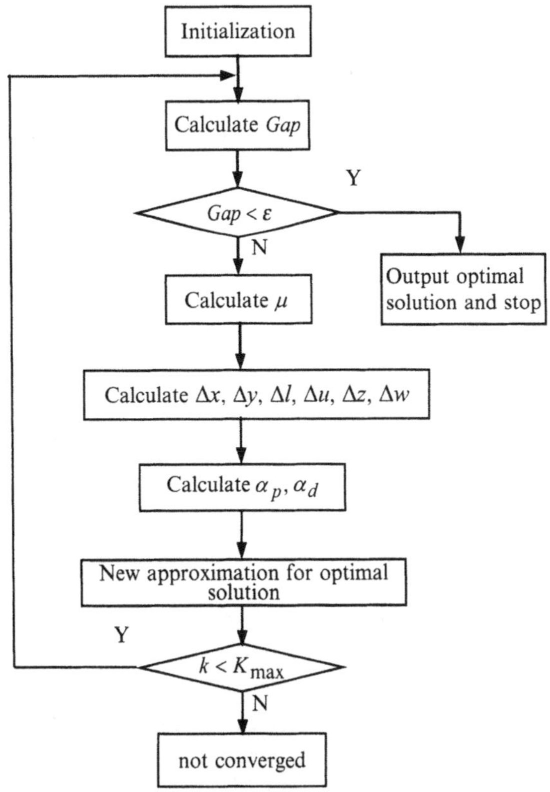

Fig. 4.2 Flow chart of OPF

3. Set initial value of variables of the original OPF problem
4. Set $\sigma \in (0, 1)$, calculation precision $\varepsilon = 10^{-6}$, number of iterations $k = 0$, and the maximum number of iterations $K_{\max} = 50$.

A five-node power system shown in Fig. 2.6 is used for demonstrating the implementation of the IPM for the OPF problem. More details and programming techniques can be found in [71].

[Example 1] Solve the OPF problem of a five-node power system shown in Fig. 2.6.

[Solution] In addition to the parameters shown in Fig. 2.6, the data for the power flow limits of transmission lines, the output limits of active and reactive sources, and the fuel cost curves of each generator are listed in Tables 4.1 and 4.2. The data listed in these tables are in per unit unless otherwise specified. The base power of the system is 100 MVA, the upper and lower bound of node voltages are 1.1 and 0.9 respectively.

The fuel costs F_i of generators are represented by

$$F_i = (a_{21}P_{Gi}^2 + a_{11}P_{Gi} + a_{0i}) \quad i = 1, 2.$$

There are five nodes, two generators, and five lines in this system. Accordingly, the state variables are

$$\tilde{\mathbf{x}} = \{\theta_1 \quad V_1 \quad \theta_2 \quad V_2 \quad \theta_3 \quad V_3 \quad \theta_4 \quad V_4 \quad \theta_5 \quad V_5\}.$$

The control variables are

$$\tilde{\mathbf{u}} = \{P_{G1} \quad P_{G2} \quad Q_{R1} \quad Q_{R2}\}.$$

It should be noted that numbering of the active and reactive sources is independent of numbering of the node that they belong to; for instance, the numbering of the generator in node 4 is G1 in this example. The total number of variables in this system is 14:

$$\mathbf{x} = \{P_{G1} \quad P_{G2} \quad Q_{R1} \quad Q_{R2} \quad \theta_1 \quad V_1 \quad \theta_2 \quad V_2 \quad \theta_3 \quad V_3 \quad \theta_4 \quad V_4 \quad \theta_5 \quad V_5\}.$$

Table 4.1 Power flow limit of transmission lines

Line number	Line terminal	Power flow limit
1	1–2	2
2	1–3	0.65
3	2–3	2
4	2–4	6
5	4.5	5

Table 4.2 Parameters of generators

Generator	Node	Upper bound		Lower bound		Coefficient of fuel cost curve		
		Active power	Reactive power	Active power	Reactive power	a_{2i}	a_{1i}	a_{0i}
1	4	8	3	1	−3	50.4395	200.4335	1,200.6485
2	5	8	5	1	−2.1	200.550	500.746	1,857.201

The OPF of this system can be formulated as the following:

1. *Objective Function:*

$$\min(a_{21}P_{\text{G1}}^2 + a_{11}P_{\text{G1}} + a_{01}) + (a_{22}P_{\text{G2}}^2 + a_{12}P_{\text{G2}} + a_{02})$$

2. *Constraints:* Each node has two power balance equations. Therefore, we have ten equality constraints for this OPF formulation.
For nongenerator node:

$$\left.\begin{aligned} -P_{\text{D}i} - V_i\sum_{j=1}^{5} V_j(G_{ij}\cos\theta_{ij} + B_{ij}\sin\theta_{ij}) = \Delta P_i \\ -Q_{\text{D}i} - V_i\sum_{j=1}^{5} V_j(G_{ij}\sin\theta_{ij} - B_{ij}\cos\theta_{ij}) = \Delta Q_i \end{aligned}\right\} \quad (i = 1, 2, 3).$$

For generator node:

$$\left.\begin{aligned} \sum_{k\in i} P_{\text{G}k} - P_{\text{D}i} - V_i\sum_{j=1}^{5} V_j(G_{ij}\cos\theta_{ij} + B_{ij}\sin\theta_{ij}) = \Delta P_i \\ \sum_{k\in i} Q_{\text{G}k} - Q_{\text{D}i} - V_i\sum_{j=1}^{5} V_j(G_{ij}\sin\theta_{ij} - B_{ij}\cos\theta_{ij}) = \Delta Q_i \end{aligned}\right\} \quad (k = 1, 2;\; i = 4, 5),$$

where $k \in i$ notes that kth generator belongs to node i, so $k = 1 \in 4$, $k = 2 \in 5$.

There are 14 inequality constraints:

$$\underline{P}_{\text{G}i} \le P_{\text{G}i} \le \overline{P}_{\text{G}i} \quad (i = 1, 2),$$

$$\underline{Q}_{\text{R}i} \le Q_{\text{R}i} \le \overline{Q}_{\text{R}i} \quad (i = 1, 2),$$

$$\underline{V}_i \le V_i \le \overline{V}_i \quad (i = 1, \ldots, 5),$$

$$-\overline{P_{ij}} \le P_{ij} \le \overline{P_{ij}} \quad (\text{for five lines}),$$

where $P_{ij} = V_iV_j(G_{ij}\cos\,\theta_{ij} + B_{ij}\sin\,\theta_{ij}) - V_i^2G_{ij}$.

Based on the above formulation the left-hand coefficient matrix and right-hand constant vector of (4.30) can be established as follows.

4.2.3.1 Left-Hand Coefficient Matrix

Left-hand coefficient matrix of (4.30) is made up of four basic parts: the Jacobian matrix of equality constraints $\nabla_x\mathbf{h}(\mathbf{x})$, Jacobian matrix of inequality constraints $\nabla_x\mathbf{g}(\mathbf{x})$, diagonal matrix $\mathbf{L}^{-1}\mathbf{Z}$, $\mathbf{U}^{-1}\mathbf{W}$, and Hessian matrix $\mathbf{H}'$. The structures of these matrices are shown as the following:

– The Jacobian matrix of equality constraints $\nabla_x \mathbf{h}(\mathbf{x})$

$$\nabla_x \mathbf{h}(\mathbf{x}) = \begin{bmatrix} \dfrac{\partial \mathbf{h}}{\partial \mathbf{P}_{\mathrm{G}}} \\ \dfrac{\partial \mathbf{h}}{\partial \mathbf{Q}_{\mathrm{R}}} \\ \dfrac{\partial \mathbf{h}}{\partial \tilde{\mathbf{x}}} \end{bmatrix}_{14\times 10}$$

There are three sub-matrices in $\nabla_x \mathbf{h}(\mathbf{x})$:

$$\frac{\partial \mathbf{h}}{\partial \mathbf{P}_{\mathrm{G}}} = \begin{bmatrix} \dfrac{\partial \Delta P_1}{\partial P_{\mathrm{G1}}} & \dfrac{\partial \Delta Q_1}{\partial P_{\mathrm{G1}}} & \cdots & \dfrac{\partial \Delta P_5}{\partial P_{\mathrm{G1}}} & \dfrac{\partial \Delta Q_5}{\partial P_{\mathrm{G1}}} \\ \dfrac{\partial \Delta P_1}{\partial P_{\mathrm{G2}}} & \dfrac{\partial \Delta Q_1}{\partial P_{\mathrm{G2}}} & \cdots & \dfrac{\partial \Delta P_5}{\partial P_{\mathrm{G2}}} & \dfrac{\partial \Delta Q_5}{\partial P_{\mathrm{G2}}} \end{bmatrix}_{2\times 10},$$

$$\text{where} \begin{cases} \dfrac{\partial \Delta Q_j}{\partial P_{Gi}} = 0 \\ \dfrac{\partial \Delta P_j}{\partial P_{Gi}} = \begin{cases} 1 & i \in j \\ 0 & i \notin j \end{cases} \end{cases}$$

$$\frac{\partial \mathbf{h}}{\partial \mathbf{Q}_{\mathrm{R}}} = \begin{bmatrix} \dfrac{\partial \Delta P_1}{\partial Q_{\mathrm{R1}}} & \dfrac{\partial \Delta Q_1}{\partial Q_{\mathrm{R1}}} & \cdots & \dfrac{\partial \Delta P_5}{\partial Q_{\mathrm{R1}}} & \dfrac{\partial \Delta Q_5}{\partial Q_{\mathrm{R1}}} \\ \dfrac{\partial \Delta P_1}{\partial Q_{\mathrm{R2}}} & \dfrac{\partial \Delta P_1}{\partial Q_{\mathrm{R2}}} & \cdots & \dfrac{\partial \Delta P_5}{\partial Q_{\mathrm{R2}}} & \dfrac{\partial \Delta P_5}{\partial Q_{\mathrm{R2}}} \end{bmatrix}_{2\times 10},$$

$$\text{where} \begin{cases} \dfrac{\partial \Delta P_j}{\partial Q_{\mathrm{R}i}} = 0 \\ \dfrac{\partial \Delta Q_j}{\partial Q_{\mathrm{R}i}} = \begin{cases} 1 & i \in j \\ 0 & i \notin j \end{cases} \end{cases}$$

In the above equations, i is the numbering of generators, j is the numbering of nodes. $i \in j$ denotes that ith generator belongs to node j.

$$\frac{\partial \mathbf{h}}{\partial \tilde{\mathbf{x}}} = \begin{bmatrix} \dfrac{\partial \Delta P_1}{\partial \theta_1} & \dfrac{\partial \Delta Q_1}{\partial \theta_1} & \cdots & \dfrac{\partial \Delta P_5}{\partial \theta_1} & \dfrac{\partial \Delta Q_5}{\partial \theta_1} \\ \dfrac{\partial \Delta P_1}{\partial V_1} & \dfrac{\partial \Delta Q_1}{\partial V_1} & \cdots & \dfrac{\partial \Delta P_5}{\partial V_1} & \dfrac{\partial \Delta Q_5}{\partial V_1} \\ \vdots & & \ddots & & \vdots \\ \dfrac{\partial \Delta P_1}{\partial \theta_5} & \dfrac{\partial \Delta Q_1}{\partial \theta_5} & \cdots & \dfrac{\partial \Delta P_5}{\partial \theta_5} & \dfrac{\partial \Delta Q_5}{\partial \theta_5} \\ \dfrac{\partial \Delta P_1}{\partial V_5} & \dfrac{\partial \Delta Q_1}{\partial V_5} & \cdots & \dfrac{\partial \Delta P_5}{\partial V_5} & \dfrac{\partial \Delta Q_5}{\partial V_5} \end{bmatrix}_{10\times 10}$$

Obviously, $\dfrac{\partial \mathbf{h}}{\partial \tilde{\mathbf{x}}}$ is the Jacobian matrix of the power flow problem.

Jacobian matrix of inequality constraints $\nabla_x \mathbf{g}(\mathbf{x})$

$$\nabla \mathbf{g}(x) = \begin{bmatrix} \frac{\partial \mathbf{g}_1}{\partial \mathbf{P}_G} & \frac{\partial \mathbf{g}_2}{\partial \mathbf{P}_G} & \frac{\partial \mathbf{g}_3}{\partial \mathbf{P}_G} & \frac{\partial \mathbf{g}_4}{\partial \mathbf{P}_G} \\ \frac{\partial \mathbf{g}_1}{\partial \mathbf{Q}_R} & \frac{\partial \mathbf{g}_2}{\partial \mathbf{Q}_R} & \frac{\partial \mathbf{g}_3}{\partial \mathbf{Q}_R} & \frac{\partial \mathbf{g}_4}{\partial \mathbf{Q}_R} \\ \frac{\partial \mathbf{g}_1}{\partial \tilde{\mathbf{x}}} & \frac{\partial \mathbf{g}_2}{\partial \tilde{\mathbf{x}}} & \frac{\partial \mathbf{g}_3}{\partial \tilde{\mathbf{x}}} & \frac{\partial \mathbf{g}_4}{\partial \tilde{\mathbf{x}}} \end{bmatrix}_{14\times 14}$$

where $\mathbf{g}_1$, $\mathbf{g}_2$, $\mathbf{g}_3$, and $\mathbf{g}_4$ denote active power output constraints, reactive power output constraints, voltage magnitude constraints, and line flow constraints, respectively.

$$\frac{\partial \mathbf{g}_1}{\partial \mathbf{P}_G} = \mathbf{I}_{2\times 2}, \quad \frac{\partial \mathbf{g}_1}{\partial \mathbf{Q}_R} = \mathbf{0}_{2\times 2}, \quad \frac{\partial \mathbf{g}_1}{\partial \tilde{\mathbf{x}}} = \mathbf{0}_{10\times 2}$$

$$\frac{\partial \mathbf{g}_2}{\partial \mathbf{P}_G} = \mathbf{0}_{2\times 2}, \quad \frac{\partial \mathbf{g}_2}{\partial \mathbf{Q}_R} = \mathbf{I}_{2\times 2}, \quad \frac{\partial \mathbf{g}_2}{\partial \tilde{\mathbf{x}}} = \mathbf{0}_{10\times 2}$$

$$\frac{\partial \mathbf{g}_3}{\partial \mathbf{P}_G} = \mathbf{0}_{2\times 5}, \quad \frac{\partial \mathbf{g}_3}{\partial \mathbf{Q}_R} = \mathbf{0}_{2\times 5}$$

$$\frac{\partial \mathbf{g}_4}{\partial \mathbf{P}_G} = \mathbf{0}_{2\times 10}, \quad \frac{\partial \mathbf{g}_4}{\partial \mathbf{Q}_R} = \mathbf{0}_{2\times 10}$$

$$\frac{\partial \mathbf{g}_3}{\partial \tilde{\mathbf{x}}} = \begin{bmatrix} 0 & 0 & \dots & 0 & 0 \\ 1 & 0 & \dots & 0 & 0 \\ \vdots & & \ddots & & \vdots \\ 0 & 0 & \dots & 0 & 0 \\ 0 & 0 & \dots & 0 & 1 \end{bmatrix}_{10\times 5}$$

In the above equation, only element of row $2 \times i$ and column i of matrix $\partial \mathbf{g}_3 / \partial \tilde{\mathbf{x}}$ are 1 ($i = 1, \dots, 5$), all other elements are 0.

$$\frac{\partial \mathbf{g}_4}{\partial \tilde{\mathbf{x}}} = \begin{bmatrix} \frac{\partial g_{4,1}}{\partial \theta_1} & \frac{\partial g_{4,2}}{\partial \theta_1} & \dots & \frac{\partial g_{4,10}}{\partial \theta_1} \\ \frac{\partial g_{4,1}}{\partial V_1} & \frac{\partial g_{4,2}}{\partial V_1} & \dots & \frac{\partial g_{4,10}}{\partial V_1} \\ \vdots & & \ddots & \vdots \\ \frac{\partial g_{4,1}}{\partial \theta_5} & \frac{\partial g_{4,2}}{\partial \theta_5} & \dots & \frac{\partial g_{4,10}}{\partial \theta_5} \\ \frac{\partial g_{4,1}}{\partial V_5} & \frac{\partial g_{4,2}}{\partial V_5} & \dots & \frac{\partial g_{4,10}}{\partial V_5} \end{bmatrix}_{10\times 5}$$

The elements in matrix $\frac{\partial \mathbf{g}_4}{\partial \tilde{\mathbf{x}}}$ can be calculated by the following equations:

$$\frac{\partial P_{ij}}{\partial \theta_i} = -V_i V_j (G_{ij} \sin \theta_{ij} - B_{ij} \cos \theta_{ij}),$$

$$\frac{\partial P_{ij}}{\partial \theta_j} = V_i V_j (G_{ij}\sin\theta_{ij} - B_{ij}\cos\theta_{ij}),$$

$$\frac{\partial P_{ij}}{\partial V_i} = V_j (G_{ij}\cos\theta_{ij} + B_{ij}\sin\theta_{ij}),$$

$$\frac{\partial P_{ij}}{\partial V_j} = V_i (G_{ij}\cos\theta_{ij} + B_{ij}\sin\theta_{ij}).$$

- Diagonal matrix $\mathbf{L}^{-1}\mathbf{Z}$, $\mathbf{U}^{-1}\mathbf{W}$

$$\mathbf{L}^{-1}\mathbf{Z} = \mathrm{diag}(z_1/l_1, \cdots, z_{14}/l_{14}),$$

$$\mathbf{U}^{-1}\mathbf{W} = \mathrm{diag}(w_1/u_1, \cdots, w_{14}/u_{14}).$$

- Hessian matrix $\mathbf{H}'$

$$\mathbf{H}' = -\nabla_x^2 f(\mathbf{x}) + \nabla_x^2 \mathbf{h}(\mathbf{x})\mathbf{y} + \nabla_x^2 \mathbf{g}(\mathbf{x})(\mathbf{z}+\mathbf{w}) - \nabla_x \mathbf{g}(\mathbf{x})[\mathbf{L}^{-1}\mathbf{Z} - \mathbf{U}^{-1}\mathbf{W}]\nabla_x^{\mathrm{T}}\mathbf{g}(\mathbf{x}).$$

Clearly, this is the most complex component of the left-hand coefficient matrix in (4.30). Matrix $\mathbf{H}'$ includes four items, the last item $\nabla_x \mathbf{g}(\mathbf{x})[\mathbf{L}^{-1}\mathbf{Z} - \mathbf{U}^{-1}\mathbf{W}]\nabla_x^{\mathrm{T}}\mathbf{g}(\mathbf{x})$ can be obtained by manipulating matrices: $\nabla_x \mathbf{g}(\mathbf{x})$, $\mathbf{L}^{-1}\mathbf{Z}$, and $\mathbf{U}^{-1}\mathbf{W}$, which are already obtained above. The first item is the Hessian matrix of the objective function $\nabla_x^2 f(\mathbf{x})$:

$$\nabla_x^2 f(\mathbf{x}) = \begin{bmatrix} 2\mathbf{A}_2 & \mathbf{0} & \mathbf{0} \\ \mathbf{0} & \mathbf{0} & \mathbf{0} \\ \mathbf{0} & \mathbf{0} & \mathbf{0} \end{bmatrix}_{14\times 14},$$

where $\mathbf{A}_2$ is a diagonal matrix with $A_{2ii} = a_{2i}$ $(i \in S_{\mathrm{G}})$, the coefficient of the fuel cost curve of the ith generator.

The second term of Hessian matrix $\mathbf{H}'$ is $\nabla_x^2 \mathbf{h}(\mathbf{x})\mathbf{y}$: the product of the Hessian matrix of equality constraints and Lagrangian multiplier vector $\mathbf{y}$. It can be described by

$$\sum_{i=1}^{n} \begin{bmatrix} \left(\frac{\partial^2 \Delta P_i}{\partial \mathbf{P}_{\mathrm{G}}^2} y_{2i-1} + \frac{\partial^2 \Delta Q_i}{\partial \mathbf{P}_{\mathrm{G}}^2} y_{2i}\right) & \left(\frac{\partial^2 \Delta P_i}{\partial \mathbf{P}_{\mathrm{G}} \partial \mathbf{Q}_{\mathrm{R}}} y_{2i-1} + \frac{\partial^2 \Delta Q_i}{\partial \mathbf{P}_{\mathrm{G}} \partial \mathbf{Q}_{\mathrm{R}}} y_{2i}\right) & \left(\frac{\partial^2 \Delta P_i}{\partial \mathbf{P}_{\mathrm{G}} \partial \tilde{\mathbf{x}}} y_{2i-1} + \frac{\partial^2 \Delta Q_i}{\partial \mathbf{P}_{\mathrm{G}} \partial \tilde{\mathbf{x}}} y_{2i}\right) \\ \left(\frac{\partial^2 \Delta P_i}{\partial \mathbf{Q}_{\mathrm{R}} \partial \mathbf{P}_{\mathrm{G}}} y_{2i-1} + \frac{\partial^2 \Delta Q_i}{\partial \mathbf{Q}_{\mathrm{R}} \partial \mathbf{P}_{\mathrm{G}}} y_{2i}\right) & \left(\frac{\partial^2 \Delta P_i}{\partial \mathbf{Q}_{\mathrm{R}}^2} y_{2i-1} + \frac{\partial^2 \Delta Q_i}{\partial \mathbf{Q}_{\mathrm{R}}^2} y_{2i}\right) & \left(\frac{\partial^2 \Delta P_i}{\partial \mathbf{Q}_{\mathrm{R}} \partial \tilde{\mathbf{x}}} y_{2i-1} + \frac{\partial^2 \Delta Q_i}{\partial \mathbf{Q}_{\mathrm{R}} \partial \tilde{\mathbf{x}}} y_{2i}\right) \\ \left(\frac{\partial^2 \Delta P_i}{\partial \tilde{\mathbf{x}} \partial \mathbf{P}_{\mathrm{G}}} y_{2i-1} + \frac{\partial^2 \Delta Q_i}{\partial \tilde{\mathbf{x}} \partial \mathbf{P}_{\mathrm{G}}} y_{2i}\right) & \left(\frac{\partial^2 \Delta P_i}{\partial \tilde{\mathbf{x}} \partial \mathbf{Q}_{\mathrm{R}}} y_{2i-1} + \frac{\partial^2 \Delta Q_i}{\partial \tilde{\mathbf{x}} \partial \mathbf{Q}_{\mathrm{R}}} y_{2i}\right) & \left(\frac{\partial^2 \Delta P_i}{\partial \tilde{\mathbf{x}}^2} y_{2i-1} + \frac{\partial^2 \Delta Q_i}{\partial \tilde{\mathbf{x}}^2} y_{2i}\right) \end{bmatrix}$$

$$= \begin{bmatrix} \mathbf{0}_{2\times 2} & \mathbf{0}_{2\times 2} & \mathbf{0}_{2\times 10} \\ \mathbf{0}_{2\times 2} & \mathbf{0}_{2\times 2} & \mathbf{0}_{2\times 10} \\ \mathbf{0}_{10\times 2} & \mathbf{0}_{10\times 2} & \mathbf{A}_{10\times 10} \end{bmatrix}_{14\times 14}$$

In the above equation, matrix $\mathbf{A} = \sum_{i=1}^{5} (y_{2i-1}\mathbf{A}_{Pi} + y_{2i}\mathbf{A}_{Qi})$, where matrix $\mathbf{A}_{Pi}$ and $\mathbf{A}_{Qi}$ are:

$$\mathbf{A}_{Pi} = \begin{bmatrix} \frac{\partial^2 \Delta P_i}{\partial \theta_1^2} & \frac{\partial^2 \Delta P_i}{\partial \theta_1 \partial V_1} & \cdots & \frac{\partial^2 \Delta P_i}{\partial \theta_1 \partial \theta_5} & \frac{\partial^2 \Delta P_i}{\partial \theta_1 \partial V_5} \\ \vdots & & \ddots & & \vdots \\ \frac{\partial^2 \Delta P_i}{\partial V_5 \partial \theta_1} & \frac{\partial^2 \Delta P_i}{\partial V_5 \partial V_1} & \cdots & \frac{\partial^2 \Delta P_i}{\partial V_5 \partial \theta_5} & \frac{\partial^2 \Delta P_i}{\partial V_5^2} \end{bmatrix}.$$

According to the equation of ΔP_i, the elements in the above matrix are

$$\frac{\partial^2 \Delta P_i}{\partial \theta_i^2} = V_i \sum_{j \neq i} V_j (G_{ij} \cos \theta_{ij} + B_{ij} \sin \theta_{ij}),$$

$$\frac{\partial^2 \Delta P_i}{\partial \theta_i \partial \theta_j} = -V_i V_j (G_{ij} \cos \theta_{ij} + B_{ij} \sin \theta_{ij}),$$

$$\frac{\partial^2 \Delta P_i}{\partial \theta_i \partial V_i} = \sum_{j \neq i} V_j (G_{ij} \sin \theta_{ij} - B_{ij} \cos \theta_{ij}),$$

$$\frac{\partial^2 \Delta P_i}{\partial \theta_i \partial V_j} = V_i (G_{ij} \sin \theta_{ij} - B_{ij} \cos \theta_{ij}), \text{ etc.}$$

Similarly, for

$$\mathbf{A}_{Qi} = \begin{bmatrix} \frac{\partial^2 \Delta Q_i}{\partial \theta_1^2} & \frac{\partial^2 \Delta Q_i}{\partial \theta_1 \partial V_1} & \cdots & \frac{\partial^2 \Delta Q_i}{\partial \theta_1 \partial \theta_5} & \frac{\partial^2 \Delta Q_i}{\partial \theta_1 \partial V_5} \\ \vdots & & \ddots & & \vdots \\ \frac{\partial^2 \Delta Q_i}{\partial V_5 \partial \theta_1} & \frac{\partial^2 \Delta Q_i}{\partial V_5 \partial V_1} & \cdots & \frac{\partial^2 \Delta Q_i}{\partial V_5 \partial \theta_5} & \frac{\partial^2 \Delta Q_i}{\partial V_5^2} \end{bmatrix}.$$

It is easy to calculate their elements, for example:

$$\frac{\partial^2 \Delta Q_i}{\partial \theta_i^2} = V_i \sum_{j \neq i} V_j (G_{ij} \sin \theta_{ij} - B_{ij} \cos \theta_{ij}),$$

$$\frac{\partial^2 \Delta Q_i}{\partial \theta_i \partial \theta_j} = -V_i V_j (G_{ij} \sin \theta_{ij} - B_{ij} \cos \theta_{ij}),$$

$$\frac{\partial^2 \Delta Q_i}{\partial \theta_i \partial V_i} = -\sum_{j \neq i} V_j (G_{ij} \cos \theta_{ij} + B_{ij} \sin \theta_{ij}),$$

$$\frac{\partial^2 \Delta Q_i}{\partial \theta_i \partial V_j} = -V_i(G_{ij}\cos\,\theta_{ij} + B_{ij}\sin\,\theta_{ij}), \text{ etc.}$$

Finally, we obtain the elements of matrix **A** as

$$\begin{aligned}\sum_{k=1}^{5}\left(\frac{\partial^2 \Delta P_k}{\partial \theta_i^2}y_{2k-1} + \frac{\partial^2 \Delta Q_k}{\partial \theta_i^2}y_{2k}\right) = V_i\sum_{j\neq i}V_j[&G_{ij}(\cos\theta_{ij}y_{2i-1} + \sin\theta_{ij}y_{2i}\\ &+ \cos\theta_{ij}y_{2j-1} - \sin\theta_{ij}y_{2j})\\ &+ B_{ij}(\sin\theta_{ij}y_{2i-1} - \cos\theta_{ij}y_{2i}\\ &- \sin\theta_{ij}y_{2j-1} - \cos\theta_{ij}y_{2j})],\end{aligned}$$

$$\begin{aligned}\sum_{k=1}^{5}\left(\frac{\partial^2 \Delta P_k}{\partial \theta_i \partial V_i}y_{2k-1} + \frac{\partial^2 \Delta Q_k}{\partial \theta_i \partial V_i}y_{2k}\right) = \sum_{j\neq i}V_j[&G_{ij}(\sin\theta_{ij}y_{2i-1} - \cos\theta_{ij}y_{2i}\\ &+ \sin\theta_{ij}y_{2j-1} + \cos\theta_{ij}y_{2j})\\ &+B_{ij}(-\cos\theta_{ij}y_{2i-1} - \sin\theta_{ij}y_{2i}\\ &+\cos\theta_{ij}y_{2j-1} - \sin\theta_{ij}y_{2j})],\end{aligned}$$

$$\begin{aligned}\sum_{k=1}^{5}\left(\frac{\partial^2 \Delta P_k}{\partial \theta_i \partial \theta_j}y_{2k-1} + \frac{\partial^2 \Delta Q_k}{\partial \theta_i \partial \theta_j}y_{2k}\right) = V_iV_j[&G_{ij}(-\cos\theta_{ij}y_{2i-1} - \sin\theta_{ij}y_{2i} - \cos\theta_{ij}y_{2j-1}\\ &+ \sin\theta_{ij}y_{2j}) + B_{ij}(-\sin\theta_{ij}y_{2i-1} + \cos\theta_{ij}y_{2i}\\ &+ \sin\theta_{ij}y_{2j-1} + \cos\theta_{ij}y_{2j})],\end{aligned}$$

$$\begin{aligned}\sum_{k=1}^{5}\left(\frac{\partial^2 \Delta P_k}{\partial \theta_i \partial V_j}y_{2k-1} + \frac{\partial^2 \Delta Q_k}{\partial \theta_i \partial V_j}y_{2k}\right) = V_i[&G_{ij}(\sin\theta_{ij}y_{2i-1} - \cos\theta_{ij}y_{2i} + \sin\theta_{ij}y_{2j-1}\\ &+\cos\theta_{ij}y_{2j}) + B_{ij}(-\cos\theta_{ij}y_{2i-1} - \sin\theta_{ij}y_{2i}\\ &+\cos\theta_{ij}y_{2j-1} - \sin\theta_{ij}y_{2j})],\end{aligned}$$

$$\sum_{k=1}^{5}\left(\frac{\partial^2 \Delta P_k}{\partial V_i^2}y_{2k-1} + \frac{\partial^2 \Delta Q_k}{\partial V_i^2}y_{2k}\right) = -2(G_{ii}y_{2i-1} - B_{ii}y_{2i}),$$

$$\begin{aligned}\sum_{k=1}^{5}\left(\frac{\partial^2 \Delta P_k}{\partial V_i \partial \theta_i}y_{2k-1} + \frac{\partial^2 \Delta Q_k}{\partial V_i \partial \theta_i}y_{2k}\right) = \sum_{j\neq i}V_j[&G_{ij}(\sin\theta_{ij}y_{2i-1} - \cos\theta_{ij}y_{2i}\\ &+ \sin\theta_{ij}y_{2j-1} + \cos\theta_{ij}y_{2j})\\ &+ B_{ij}(-\cos\theta_{ij}y_{2i-1} - \sin\theta_{ij}y_{2i}\\ &+ \cos\theta_{ik}y_{2j-1} - \sin\theta_{ij}y_{2j})],\end{aligned}$$

$$\sum_{k=1}^{5}\left(\frac{\partial^2 \Delta P_k}{\partial V_i \partial \theta_j} y_{2j-1} + \frac{\partial^2 \Delta Q_k}{\partial V_i \partial \theta_j} y_{2k}\right) = V_j[G_{ij}(-\sin\theta_{ij} y_{2i-1} + \cos\theta_{ij} y_{2i} - \sin\theta_{ij} y_{2j-1} - \cos\theta_{ij} y_{2j}) + B_{ij}(\cos\theta_{ij} y_{2i-1} + \sin\theta_{ij} y_{2i} - \cos\theta_{ij} y_{2j-1} + \sin\theta_{ij} y_{2j})],$$

$$\sum_{k=1}^{5}\left(\frac{\partial^2 \Delta P_k}{\partial V_i \partial V_j} y_{2k-1} + \frac{\partial^2 \Delta Q_k}{\partial V_i \partial V_j} y_{2k}\right) = -\ [G_{ij}(\cos\theta_{ij} y_{2i-1} + \sin\theta_{ij} y_{2i} + \cos\theta_{ij} y_{2j-1} - \sin\theta_{ij} y_{2j}) + B_{ij}(\sin\theta_{ij} y_{2i-1} - \cos\theta_{ij} y_{2i} - \sin\theta_{ij} y_{2j-1} - \cos\theta_{ij} y_{2j})],$$

The third term of Hessian matrix $\mathbf{H}'$ is $\nabla_x^2 \mathbf{g}(\mathbf{x})(\mathbf{z}+\mathbf{w})$. Assume that $\mathbf{z}+\mathbf{w} = \mathbf{c}$, then:

$$\nabla_x^2 \mathbf{g}(\mathbf{x})(\mathbf{z}+\mathbf{w}) = \nabla_x^2 g(x)c = \sum_{i=1}^{2}\begin{bmatrix} \frac{\partial^2 g_{1i}}{\partial \mathbf{P}_G^2} c_i & \frac{\partial^2 g_{1i}}{\partial \mathbf{P}_G \partial \mathbf{Q}_R} c_i & \frac{\partial^2 g_{1i}}{\partial \mathbf{P}_G \partial \tilde{\mathbf{x}}} c_i \\ \frac{\partial^2 g_{1i}}{\partial \mathbf{Q}_R \partial \mathbf{P}_G} c_i & \frac{\partial^2 g_{1i}}{\partial \mathbf{Q}_R^2} c_i & \frac{\partial^2 g_{1i}}{\partial \mathbf{Q}_R \partial \tilde{\mathbf{x}}} c_i \\ \frac{\partial^2 g_{1i}}{\partial \tilde{\mathbf{x}} \partial \mathbf{P}_G} c_i & \frac{\partial^2 g_{1i}}{\partial \tilde{\mathbf{x}} \partial \mathbf{Q}_R} c_i & \frac{\partial^2 g_{1i}}{\partial \tilde{\mathbf{x}}^2} c_i \end{bmatrix}$$

$$+ \sum_{i=1}^{2}\begin{bmatrix} \frac{\partial^2 g_{2i}}{\partial \mathbf{P}_G^2} c_{2+i} & \frac{\partial^2 g_{2i}}{\partial \mathbf{P}_G \partial \mathbf{Q}_R} c_{2+i} & \frac{\partial^2 g_{2i}}{\partial \mathbf{P}_G \partial \tilde{\mathbf{x}}} c_{2+i} \\ \frac{\partial^2 g_{2i}}{\partial \mathbf{Q}_R \partial \mathbf{P}_G} c_{2+i} & \frac{\partial^2 g_{2i}}{\partial \mathbf{Q}_R^2} c_{2+i} & \frac{\partial^2 g_{2i}}{\partial \mathbf{Q}_R \partial \tilde{\mathbf{x}}} c_{2+i} \\ \frac{\partial^2 g_{2i}}{\partial \tilde{\mathbf{x}} \partial \mathbf{P}_G} c_{2+i} & \frac{\partial^2 g_{2i}}{\partial \tilde{\mathbf{x}} \partial \mathbf{Q}_R} c_{2+i} & \frac{\partial^2 g_{2i}}{\partial \tilde{\mathbf{x}}^2} c_{2+i} \end{bmatrix}$$

$$+ \sum_{i=1}^{5}\begin{bmatrix} \frac{\partial^2 g_{3i}}{\partial \mathbf{P}_G^2} c_{2+2+i} & \frac{\partial^2 g_{3i}}{\partial \mathbf{P}_G \partial \mathbf{Q}_R} c_{2+2+i} & \frac{\partial^2 g_{3i}}{\partial \mathbf{P}_G \partial \tilde{\mathbf{x}}} c_{2+2+i} \\ \frac{\partial^2 g_{3i}}{\partial \mathbf{Q}_R \partial \mathbf{P}_G} c_{2+2+i} & \frac{\partial^2 g_{3i}}{\partial \mathbf{Q}_R^2} c_{2+2+i} & \frac{\partial^2 g_{3i}}{\partial \mathbf{Q}_R \partial \tilde{\mathbf{x}}} c_{2+2+i} \\ \frac{\partial^2 g_{3i}}{\partial \tilde{\mathbf{x}} \partial \mathbf{P}_G} c_{2+2+i} & \frac{\partial^2 g_{3i}}{\partial \tilde{\mathbf{x}} \partial \mathbf{Q}_R} c_{2+2+i} & \frac{\partial^2 g_{3i}}{\partial \tilde{\mathbf{x}}^2} c_{2+2+i} \end{bmatrix}$$

$$+ \sum_{i=1}^{5}\begin{bmatrix} \frac{\partial^2 g_{4i}}{\partial \mathbf{P}_G^2} c_{2+2+5+i} & \frac{\partial^2 g_{4i}}{\partial \mathbf{P}_G \partial \mathbf{Q}_R} c_{2+2+5+i} & \frac{\partial^2 g_{4i}}{\partial \mathbf{P}_G \partial \tilde{\mathbf{x}}} c_{2+2+5+i} \\ \frac{\partial^2 g_{4i}}{\partial \mathbf{Q}_R \partial \mathbf{P}_G} c_{2+2+5+i} & \frac{\partial^2 g_{4i}}{\partial \mathbf{Q}_R^2} c_{2+2+5+i} & \frac{\partial^2 g_{4i}}{\partial \mathbf{Q}_R \partial \tilde{\mathbf{x}}} c_{2+2+5+i} \\ \frac{\partial^2 g_{4i}}{\partial \tilde{\mathbf{x}} \partial \mathbf{P}_G} c_{2+2+5+i} & \frac{\partial^2 g_{4i}}{\partial \tilde{\mathbf{x}} \partial \mathbf{Q}_R} c_{2+2+5+i} & \frac{\partial^2 g_{4i}}{\partial \tilde{\mathbf{x}}^2} c_{2+2+5+i} \end{bmatrix}.$$

It is obvious that the first, second, and third items in the above expression are zero matrices. The elements of the fourth item can be obtained in a similar way to the above.

4.2.3.2 Right-Hand Constant Vector

Vectors $\mathbf{L}_y$, $\mathbf{L}_z$, $\mathbf{L}_w$, $\mathbf{L}_l^{\mu}$, and $\mathbf{L}_u^{\mu}$ can be calculated by using (4.15)–(4.19), while vector $\mathbf{L}'_x$ can be described as the following:

$$\begin{aligned}\mathbf{L}'_x = \frac{\partial \mathbf{L}}{\partial x} &= \nabla_x f(\mathbf{x}) - \nabla_x \mathbf{h}(\mathbf{x})\mathbf{y} - \nabla_x \mathbf{g}(\mathbf{x})(\mathbf{z}+\mathbf{w}) \\ &+ \nabla_x \mathbf{g}(\mathbf{x})[\mathbf{L}^{-1}(\mathbf{L}_l^{\mu} + \mathbf{Z}\mathbf{L}_z) + \mathbf{U}^{-1}(\mathbf{L}_u^{\mu} - \mathbf{W}\mathbf{L}_w)],\end{aligned}$$

where the gradient vector of objective function is

$$\nabla_x f(\mathbf{x}) = \begin{bmatrix} \dfrac{\partial f}{\partial \mathbf{P}_\mathrm{G}} \\ \dfrac{\partial f}{\partial \mathbf{Q}_\mathrm{R}} \\ \dfrac{\partial f}{\partial \tilde{\mathbf{x}}} \end{bmatrix}_{14\times 1} = \begin{bmatrix} 2a_{21}P_{\mathrm{G1}} + a_{11} \\ 2a_{22}P_{\mathrm{G2}} + a_{12} \\ \mathbf{0} \\ \mathbf{0} \end{bmatrix}.$$

The other terms of $\mathbf{L}'_x$ are ready to be calculated based on the Jacobian matrices of the equality and inequality constraints discussed above.

Up to now, all the formulas that may be used in this algorithm are enumerated.

The five-node system shown in Fig. 2.6 now is used to numerically demonstrate the process for searching the solution of OPF problem. It is assumed that the power output of generators at node 4 and 5 are adjustable. The initial values of the variables are set as: nonslack node voltages are $V_i = 1$, $\theta_i = 0$ $(i = 1, 2, 3, 4)$; the slack node voltage is $V_5 = 1.05$, $\theta_5 = 0$. The active and reactive power outputs of sources are set to the average value between their upper bound and lower bound. Slack variables are $l_i = 1$ and $u_i = 1$ $(i = 1, \ldots, 14)$, Lagrangian multipliers are $z_i = 1$, $w_i = -0.5$ $(i = 1, \ldots, 14)$, $y_{2i-1} = 1E-10$, and $y_{2i} = -1E-10$ $(i = 1, 2, 3, 4, 5)$. Based on the flow chart shown in Fig. 4.2, the optimal solution is reached after 17 iterations with tolerance ε being 10^{-6}. The value of $\mathbf{L}_x$, $\mathbf{L}_y$, $\mathbf{L}_z$, and $\mathbf{L}_w$ in the first iteration are listed in Tables 4.3–4.6.

The corrections of node voltages, active, and reactive output power quantities at each iteration are listed in Tables 4.7 and 4.8.

The change of dual gap in the iteration process is shown in Fig. 4.3, which demonstrates the convergence characteristic of IPM for OPF problems.

Comparisons between the OPF solution and conventional power flow are shown in Tables 4.9–4.11. From Table 4.9 we find that the power output of generator at node 4 increased, and that of generator in node 5 decreased because the fuel cost of

Table 4.3 Value of $\mathbf{L}_x$ after the first iteration

L_{x1}	L_{x2}	L_{x3}	L_{x4}	L_{x5}	L_{x6}	L_{x7}
653.889	2305.1960	−0.5000	−0.5000	0	−0.5000	0
L_{x8}	L_{x9}	L_{x10}	L_{x11}	L_{x12}	L_{x13}	L_{x14}
−0.5000	0	−0.5000	0	−0.5000	0	−0.5000

Table 4.4 Value of $\mathbf{L}_y$ after the first iteration

L_{y1}	L_{y2}	L_{y3}	L_{y4}	L_{y5}
4.5000	−3.1746	4.5000	−1.9667	−1.6000
L_{y6}	L_{y7}	L_{y8}	L_{y9}	L_{y10}
−0.5500	−3.700	2.0490	−2.000	2.5234

Table 4.5 Value of $\mathbf{L}_z$ after the first iteration

L_{z1}	L_{z2}	L_{z3}	L_{z4}	L_{z5}	L_{z6}	L_{z7}	L_{z8}	L_{z9}	L_{z10}
2.500	2.500	2.000	2.550	−0.900	−0.850	−0.900	−0.900	−0.900	1.000
L_{z11}	L_{z12}	L_{z13}	L_{z14}	L_{z15}	L_{z16}	L_{z17}	L_{z18}	L_{z19}	
1.000	−0.350	−0.350	1.000	1.000	5.000	5.000	4.000	4.000	

Table 4.6 Value of $\mathbf{L}_w$ after the first iteration

L_{w1}	L_{w2}	L_{w3}	L_{w4}	L_{w5}	L_{w6}	L_{w7}	L_{w8}	L_{w9}	L_{w10}
−2.500	−2.500	−2.000	−2.550	0.900	0.950	0.900	0.900	0.900	−1.000
L_{w11}	L_{w12}	L_{w13}	L_{w14}	L_{w15}	L_{w16}	L_{w17}	L_{w18}	L_{w19}	
−1.000	0.350	0.350	−1.000	−1.000	−5.000	−5.000	−4.000	−4.000	

the former is lower than that of the latter. To satisfy the voltage magnitude constraint of node 1, the network losses and the reactive output increased. Nevertheless, the total fuel cost in OPF solution decreases $243.76.

If the active power output of generator in node 4 is fixed at 5.0, then OPF can be used as a tool to decrease network loss through reactive power optimization. The results of such optimization are shown in Tables 4.12–4.14. From Table 4.12 we find that the network loss decreases 1.78 MW, which results in $27.27 saving of fuel cost. At the same time, the voltage magnitude of node 1 is improved to 0.9129 (see Table 4.13).

4.3 Application of OPF in Electricity Market

4.3.1 Survey

As mentioned before, the main aim of the OPF is to determine the optimal steady-state operation of a power system, which simultaneously minimizes the value of a chosen objective function and satisfies certain physical and operating constraints. Thus the economic dispatch (ED) and power flow (PF) calculation have been ideally integrated into OPF problem. OPF is a constrained NLP with both active and reactive power variables, which implements the integration of economics and

Table 4.7 Corrections of node voltages

Iteration number	$\Delta\theta_1$	ΔV_1	$\Delta\theta_2$	ΔV_2
1	1.578×10^{-1}	4.392×10^{-1}	6.724×10^{-1}	5.668×10^{-1}
2	-1.508×10^{-1}	8.101×10^{-1}	-5.343×10^{-1}	3.873×10^{-1}
3	-3.494×10^{-4}	-3.101×10^{-1}	7.388×10^{-2}	-2.867×10^{-1}
4	-2.866×10^{-2}	-3.042×10^{-1}	3.188×10^{-2}	-2.326×10^{-1}
5	-3.948×10^{-2}	-2.880×10^{-1}	1.980×10^{-2}	-1.623×10^{-1}
6	-4.262×10^{-2}	-2.322×10^{-1}	1.105×10^{-3}	-1.615×10^{-1}
7	-2.439×10^{-2}	-4.229×10^{-2}	-7.738×10^{-3}	-1.010×10^{-2}
8	-7.035×10^{-3}	-9.491×10^{-3}	-2.730×10^{-3}	-2.328×10^{-3}
9	-5.185×10^{-3}	2.251×10^{-3}	-5.593×10^{-3}	-2.221×10^{-3}
10	-6.356×10^{-3}	-3.512×10^{-5}	-9.307×10^{-3}	6.195×10^{-3}
11	-4.284×10^{-2}	2.069×10^{-3}	-6.595×10^{-2}	-2.035×10^{-3}
12	-4.046×10^{-2}	1.932×10^{-3}	-6.229×10^{-2}	-1.932×10^{-3}
13	-1.846×10^{-2}	7.766×10^{-4}	-2.852×10^{-2}	-7.766×10^{-4}
14	-5.974×10^{-4}	3.890×10^{-7}	-9.428×10^{-4}	-3.830×10^{-7}
15	-9.147×10^{-7}	-1.847×10^{-9}	-1.432×10^{-7}	6.73×10^{-10}
16	1.507×10^{-9}	-1.92×10^{-10}	2.454×10^{-7}	1.56×10^{-10}
17	1.99×10^{-10}	-1.94×10^{-11}	3.21×10^{-10}	1.58×10^{-11}

Iteration number	ΔV_3	$\Delta\theta_4$	ΔV_4	ΔV_5
1	4.444×10^{-1}	7.727×10^{-1}	4.637×10^{-1}	3.195×10^{-1}
2	1.185×10^{0}	-5.874×10^{-1}	3.009×10^{-1}	1.194×10^{0}
3	-9.723×10^{-2}	8.757×10^{-2}	-2.600×10^{-1}	-3.064×10^{-2}
4	-1.804×10^{-1}	4.053×10^{-2}	-2.020×10^{-1}	-1.361×10^{-1}
5	-2.549×10^{-1}	2.432×10^{-2}	-1.285×10^{-1}	-2.303×10^{-1}
6	-1.469×10^{-1}	5.845×10^{-3}	-1.371×10^{-1}	-1.173×10^{-1}
7	-1.812×10^{-2}	-7.739×10^{-3}	-4.859×10^{-3}	-1.298×10^{-2}
8	-5.358×10^{-4}	-2.734×10^{-3}	-1.305×10^{-3}	9.482×10^{-4}
9	1.783×10^{-2}	-5.521×10^{-3}	-2.950×10^{-3}	1.957×10^{-2}
10	2.122×10^{-5}	-1.045×10^{-2}	6.650×10^{-3}	-4.706×10^{-4}
11	2.206×10^{-3}	-6.978×10^{-2}	-3.451×10^{-3}	-2.184×10^{-3}
12	2.082×10^{-3}	-6.592×10^{-2}	-3.259×10^{-3}	-2.040×10^{-3}
13	9.166×10^{-4}	-3.023×10^{-2}	-1.306×10^{-3}	-8.197×10^{-4}
14	2.149×10^{-5}	-1.009×10^{-3}	8.609×10^{-7}	-4.108×10^{-7}
15	3.474×10^{-8}	-1.530×10^{-6}	4.715×10^{-9}	3.171×10^{-9}
16	9.749×10^{-11}	2.590×10^{-9}	2.353×10^{-10}	3.204×10^{-10}
17	8.141×10^{-12}	3.398×10^{-10}	2.357×10^{-11}	3.233×10^{-11}

security of the full power system. Transforming the reliability and quality of power supply into corresponding economic index OPF can realize optimization of source location and reduction of generation and transmission cost.

The emerging competitive electricity market brings more requirements and challenges to OPF. In this section we will briefly review some important application of extended OPF techniques in the deregulated electricity market, such as energy and ancillary market pricing, transmission pricing, congestion management, and available transmission capacity evaluation.

Table 4.8 Corrections of active and reactive power outputs of generators

Iteration number	Correction of active power output		Correction of reactive power output	
	ΔP_{G1}	ΔP_{G2}	ΔQ_{G1}	ΔQ_{G2}
1	1.868×10^{0}	-3.568×10^{0}	-4.225×10^{-1}	-6.253×10^{-1}
2	1.926×10^{-2}	-7.906×10^{-1}	-6.139×10^{0}	3.959×10^{0}
3	5.205×10^{-1}	-5.646×10^{-1}	1.411×10^{0}	2.726×10^{0}
4	1.121×10^{-1}	-1.923×10^{-1}	1.793×10^{0}	1.442×10^{0}
5	-3.941×10^{-2}	-3.246×10^{-2}	2.107×10^{0}	2.068×10^{-1}
6	-6.956×10^{-2}	-1.318×10^{-2}	1.357×10^{0}	7.985×10^{-1}
7	-7.135×10^{-3}	1.597×10^{-2}	1.997×10^{-1}	6.519×10^{-2}
8	-2.662×10^{-3}	6.659×10^{-3}	3.086×10^{-2}	4.090×10^{-2}
9	-6.003×10^{-3}	7.330×10^{-3}	-9.541×10^{-2}	1.275×10^{-1}
10	-1.554×10^{-3}	-3.253×10^{-3}	3.206×10^{-2}	-3.285×10^{-2}
11	-3.016×10^{-1}	2.273×10^{-1}	-1.653×10^{-1}	-1.651×10^{-1}
12	-2.853×10^{-1}	2.156×10^{-1}	-1.550×10^{-1}	-1.546×10^{-1}
13	-1.323×10^{-1}	1.038×10^{-1}	-6.626×10^{-2}	-6.307×10^{-2}
14	-4.717×10^{-3}	4.584×10^{-3}	-1.069×10^{-4}	-2.733×10^{-4}
15	-7.174×10^{-6}	6.991×10^{-6}	9.323×10^{-9}	-3.002×10^{-7}
16	1.217×10^{-8}	-8.857×10^{-9}	7.474×10^{-9}	8.755×10^{-9}
17	1.596×10^{-9}	-1.251×10^{-9}	7.566×10^{-10}	9.026×10^{-10}

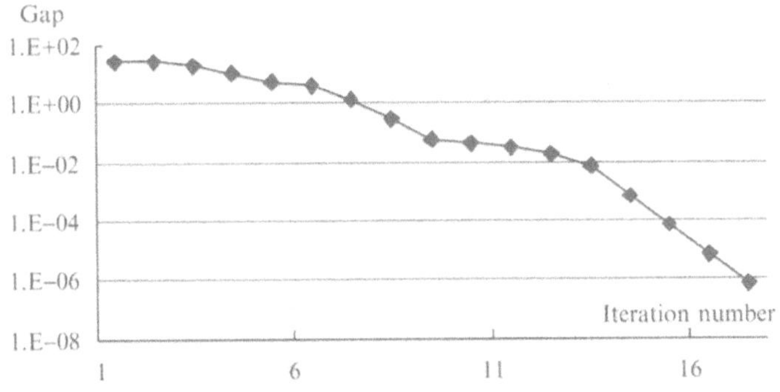

Fig. 4.3 Convergence performance of IPM for OPF problem

Table 4.9 Outputs of each active and reactive source

Generator index	Node	Active power output		Reactive power output		Fuel cost ($)	
		OPF	PF	OPF	PF	OPF	PF
1	4	5.5056	5.0000	1.7780	1.8311	3,833.06	3,463.80
2	5	2.1568	2.5794	2.6194	2.2994	3,870.13	4,483.15
Total		7.6624	7.5794	4.3974	4.1305	7,703.19	7,946.95

The spot pricing of electricity was established by Scheweppe et al. [72] in 1988. In their work, the concept of marginal price of the microeconomics has been extended to power systems. The spot pricing idea stresses on that price of electricity

Table 4.10 Voltage magnitude and angle of each node

Node	Voltage magnitude		Voltage angle (radian)	
	OPF	PF	OPF	PF
1	0.90000	0.8622	−0.00697	−0.08340
2	1.10000	1.0779	0.40491	0.31160
3	1.08175	1.0364	−0.057126	−0.07473
4	1.06970	1.05000	0.47867	0.31160
5	1.10000	1.05000	0	0

Table 4.11 Power flow in each branch

Line number	Terminal node	Line power			
		P_{ij}		P_{ji}	
		OPF	PF	OPF	PF
1	1–2	−1.6064	−1.4662	1.7347	1.5845
2	1–3	−0.0064	−0.1338	−0.0203	0.1569
3	2–3	1.7709	1.4155	−1.5635	−1.2774
4	2–4	−5.5056	−5	5.5056	5
5	4.5	−2.1568	−2.5794	2.1568	2.5794

Table 4.12 Active and reactive outputs of generators

Generator index	Node	Active power output		Reactive power output		Fuel cost ($)	
		OPF	PF	OPF	PF	OPF	PF
1	4	5.0000	5.0000	2.3585	1.8311	3,463.80	3,463.80
2	5	2.5616	2.5794	1.5381	2.2994	4,455.88	4,483.15
Total		7.5616	7.5794	3.8966	4.1305	7,919.68	7,946.95

Table 4.13 Voltage magnitude and angle of each node

Node	Voltage magnitude		Voltage angle (radian)	
	OPF	PF	OPF	PF
1	0.9129	0.8622	−0.06917	−0.0834
2	1.1000	1.0779	0.30003	0.31160
3	1.0855	1.0364	−0.06787	−0.07473
4	1.0669	1.0500	0.36718	0.38123
5	1.0960	1.0500	0	0

will vary with different time and different space. Theoretically, the spot price can improve production efficiency and yield maximum social benefit. Due to the historical limitation, this model cannot be directly applied in practical system in despite of rigorous mathematic provenance [73, 74].

The spot pricing theory based on OPF model is developed following development of OPF technique. Ray and Alvarado [75] use a modification of OPF model to analyze the effects of spot pricing policies. This is the first application of OPF for

Table 4.14 Power flow in each branch

Line number	Terminal node	Line power			
		P_{ij}		P_{ji}	
		OPF	PF	OPF	PF
1	1–2	−1.4777	−1.4662	1.5840	1.5845
2	1–3	−0.1223	−0.1338	0.1448	0.1569
3	2–3	1.4160	1.4155	−1.2832	−1.2774
4	2–4	−5	−5	5	5
5	4.5	−2.5616	−2.5794	2.5616	2.5794

spot pricing analysis. In paper [76], spot pricing model is extended by introducing reactive power pricing and reveals that λ_p and λ_q, Lagrangian multipliers corresponding to node power balance equations in OPF, represent the marginal costs of node power injections, which have the same economic meanings as active and reactive spot prices, respectively. The OPF technique becomes more promising in spot pricing calculation based on these works.

The ancillary services include services such as operating reserves, frequency control, loss compensation, energy imbalances, reactive power, black-start capability, etc. Siddiqi and Baughman [77] extend the reliability differentiated pricing (RDP) model to derive an optimal price for spinning reserve and an optimal level of spinning reserve from a social welfare point of view. Rather than incorporating a minimum spinning reserve constraint, RDP model includes customer's outage costs into the overall objective function to reflect the idea that outages created by insufficient generation or transmission actually cause a loss of welfare. Zobian and llic [78] focus on the ancillary service of energy imbalance compensation. Considering most ancillary services and incorporating constraints on power quality and environment, an advanced pricing prototype is introduced in [79, 80], which combines the dynamic equations for load–frequency control with the static equations of constrained OPF. Xie et al. [81] develop OPF pricing formulation with more system operation constraints, and the primary-dual interior point algorithm is employed to solve it. This chapter reveals that λ_p and λ_q not only have the similar economic meanings as spot price, but also can be further decomposed into four components reflecting the effects of various ancillary services: the first part is the marginal generation cost; the second part is the loss compensation cost; the third part is concerned with the coupling between active and reactive power; and the last part is associated with security cost, for active power it only represents congestion alleviation cost, for reactive power, it also includes voltage support cost.

Transmission open access plays a key role in making the competitive electricity market work. An important step of power industry restructuring is the transmission open access. However, as a natural monopoly, the transmission sector remains being more or less regulated to permit a competitive environment for generation and retail services. The operating and planning for transmission network and the pricing of transmission services are still retained as challenge on both theoretical and practical aspects in the development of electricity markets. Many models and algorithms are developed to account for transmission pricing. Yu and Ilic, and Wang et al. [82, 83] are concerned with transmission right evaluation and action. In [82], the idea of priority insurance contracts is introduced for selling long-term transmission right. A customer is obligated to obtain a transmission right before implementing the long-term bilateral transaction. The main difficulty with the physical right is its physical interpretation. The financial transmission right (FTR) path is defined by the transmission reservation from the point where the power is scheduled to be injected into the grid (source) to the point that is scheduled to be withdrawn (sink). Unlike the traditional idea for transmission pricing, independent system operators (ISO) are not required to know the flow paths of the transactions. A FTR's economic value is based on the megawatt reservation level multiplied by the difference between the location marginal price (LMP) of the source and sink points. In [83], the impacts of controlling different FACTS devices on FTR action are taken account for.

Security and reliability are the major concerns in the deregulated and unbundled electricity supply industry due to the increased number of market participants and the changing demand patterns. Congestion management has been debated much for increasing competition electricity power generation in both pool and bilateral dispatch models.

Congestion can be corrected by applying controls (corrective actions) such as phase shifter [85], tap transformers and FACTS devices [86]. A fast relief of congestion may be possible by removing congested lines to prevent severe damages to system [84]. All these strategies are based on physical principle to control network flows to increase trade possibilities. Physical curtailment without any economic considerations could and should be considered as the last resort option when it is impossible to wait for the system users to respond according to their economic criteria. In the power market environment, congestion management focus on increasing or curtailing transaction based on economic signal to alleviate overload of transmission lines. Various congestion management schemes for the different restructuring paradigms, different policy system and different technique level have appeared, such as transaction curtailment, transmission capacity reservation, and system redispatch. In the real world in order to manage transmission congestion efficiently, market participants must have freedom to engage in various mechanisms to protect their business. The best solution might always be a combination of several of the basic methods for different time scales. A framework for real-time congestion management under a market structure similar to the newly proposed UK trading arrangement is presented in [87]. Based on the uplift bid, not only resources in balancing market but also some bilateral contracts can be dispatched if necessary to relief congestion. The linearized model of a modified OPF is proposed to implement such a framework (see details in Sect. 4.3.2).

A new phenomenon in power market is a sudden increase interest in available transfer capabilities (ATC) evaluation. Investment in new transmission facilities is slow, hampered by environmental constraints and economic considerations. This has led to a more intensive use of the existing transmission corridors. These aspects have motivated the development of methodologies to evaluate the ATC and transmission margins. On technique views, ATC can instruct system operators controlling power system safely and reliably. From economical viewpoint, market participants can make decisions to obtained maximal benefit based on ATC information. ATC is a measure of the transfer capability remaining in the physical transmission network for future commercial activity over and above already committed uses. Mathematically, ATC can be obtained by solving an optimization problem. Hamoud [88] describes a method for determining the ATC between any two locations in a transmission system (single area or multiarea) under a given set of selected transmission operating conditions. Linear programming can be used to solve this problem when network is simulated by a DC flow model. A few recent attempts have been made to include AC power flow constraints because reactive power or voltage level has a great impact on transmission limit. In [89], based on the OPF formulation, a neural network approach is proposed to the problem of real power transfer capability calculations. In [90], a combined "OPF + MAT (maximum allowable transfer)" scheme is suggested for computing online power system ATC. The cycle:

(1) Identifying the dangerous contingencies and corresponding critical machines
(2) Computing active power decrease on each one of them
(3) Reallocating (almost) the same amount of power among noncritical machines, is repeated until stabilizing all dangerous contingencies, so as to ensure maximum power transfer on the tie–lie of concern.

OPF techniques enter the scene with their explicit recognition of network characteristic within the broader context of power system optimization. As mentioned above, OPF problems tailored towards special-purpose applications are defined by selecting different objective function to be minimized, different sets of controls and different sets of constraints. The potential applications of OPF in power markets are shown in Table 4.15.

4.3.2 Congestion Management Method Based on OPF

Power markets with different models have been developed in many countries all over the world in the last decade. Generally speaking, these models can be summarized as pool model or bilateral contract model. Congestion management is one of the most important tasks of the independent system operator (ISO) in the market. No matter which model is adopted, the best way to solve the problem of real-time congestion management under a deregulated environment is to establish a special real-time balancing market (RBM) and to encourage more market participants to take part in the competition in such a balancing market.

In power markets, there is a balance between the competition in power market and the real-time dispatch of power system. With the trend that more and more bilateral contracts are signed to trade electricity, the new problem might be what we can do if the resources in the balancing market are not enough to eliminate network congestion. When the congestion problem is very severe, not only the contracts in pool balancing market but also some bilateral contracts should be rescheduled.

Table 4.15 Applications of OPF in power markets

Application in electricity markets	Extended OPF problems			
	Objective function	Network model	Special constraints	Special controls
Spot market clearing and pricing	Maximize the social welfare	DC/AC	Ramp-rate constraints and reserve-related constraints	Supply offers and demand bids (mostly piece-wise linear bidding functions)
Transmission pricing	Minimize generation cost/maximize consumer net benefit	DC/AC	Contingency constraints	Generation, load and FACTS devices
Congestion management	Minimize the cost of congestion management	DC/AC	Operating constraints, contingency constraints, stability constraints	Incremental and decremental adjustment, FACTS devices, curtailment on bilateral contracts
ATC evaluation	Maximize the TTC	AC	Contingency, stability, stochastic operating constraints	FACTS devices
Ancillary services procurement	Minimize the cost of ancillary services	DC	Reserve-related constraints	Reserve capacity from generation and demand
Transmission right allocation	Maximize the revenue of transmission rights auction	DC	Bids-related, contingency constraints	Injections and withdraws for bids of transmission rights, FACTS devices

Moreover, demand side participants are encouraged to take an active role in the competition of real-time balancing market. An OPF can be used to solve this problem. The task is to minimize the absolute MW of rescheduling, in light of the cost which is determined by the submitted incremental bids in balancing market and the compensative prices submitted by participants of bilateral contracts.

The simplified curve of bidding prices of a generator at node i is given in Fig. 4.4, in which b_i^+ and b_i^- are incremental and decremental bidding prices, respectively; $b_i^{i,j}$ is the curtailment price of bilateral contract between node i and j; P_i^0 is its current output; $P_i^{i,j}$ is the total amount of bilateral contracts between i and j. $P_i^{\min}$ and $P_i^{\max}$ are the power output lower and higher limit of the resource at node i, respectively. From this curve, it can be seen that the incremental bidding price is higher than decremental bidding price while the curtailment price of bilateral contracts is much higher than other two. The reason is that increasing output needs more fuel cost, and the curtailment of a bilateral contract will affect the financial interests of both participants. The bidding price curves of customers have the similar form as that of the generators.

The congestion management model can be written as

Objective Function:

$$\min \sum_{i=1}^{N} \max[b_i^+(\hat{P}_i - \hat{P}_i^0), 0, b_i^-(\hat{P}_i^0 - \hat{P}_i)] + \sum_{i=1}^{N}\sum_{j=1}^{N} b_i^{i,j}(P_i^{i,j} - P_i^{i,j0}), \tag{4.38}$$

where N is the number of total nodes. In the objective function, the first item represents the total adjustment cost of generators and load in RBM. The second item notes the total adjustment cost of bilateral contract. It must be noted that $\widehat{P}_i$ in (4.38) is the difference between the total output power and the output power of the bilateral contract in node i, which is not equal to the P_i in Fig. 4.4, so is $\hat{P}_i^0$. Their relationship can be written as

$$\hat{P}_i = P_i - \sum_{j=1,\, j\neq i}^{N} P_i^{i,j}. \tag{4.39}$$

Constraints:

For generator nodes ($i \in S_G$)

$$\hat{P}_i + \sum_{\substack{j=1 \\ j\neq i}}^{N} P_i^{i,j} - [V_i^2 G_{ii} + V_i \sum_{\substack{j\in i \\ j\neq i}}^{N} V_j(G_{ij}\cos\theta_{ij} + B_{ij}\sin\theta_{ij})] = 0, \tag{4.40}$$

$$Q_i - [-V_i^2 B_{ii} + V_i \sum_{\substack{j\in i \\ j\neq i}}^{N} V_j(G_{ij}\sin\theta_{ij} - B_{ij}\cos\theta_{ij})] = 0, \tag{4.41}$$

$$P_i^{\min} \leq \hat{P}_i + \sum_{j=1,\, j\neq i}^{N} P_i^{i,j} \leq P_i^{\max}, \tag{4.42}$$

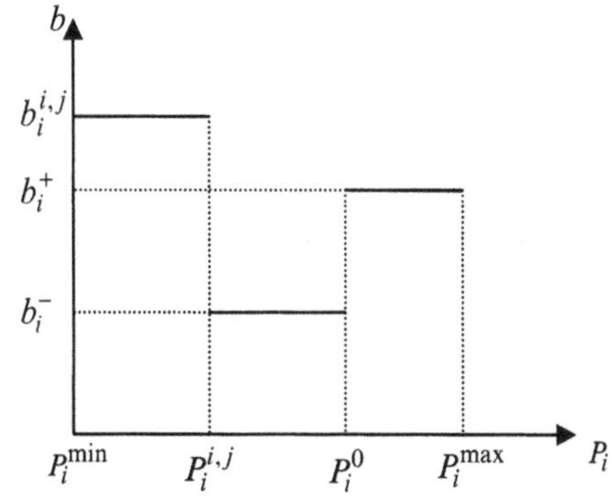

Fig. 4.4 Curve of bidding price in Balancing Market

$$0 \leq \sum_{j=1,\, j\neq i}^{N} P_i^{i,j} \leq P_i^0, \tag{4.43}$$

$$0 \leq \hat{P}_i \leq P_i^{\max}, \quad Q_i^{\min} \leq Q_i \leq Q_i^{\max}. \tag{4.44}$$

For load nodes ($i \in S_\mathrm{L}$):

$$V_i^2 G_{ii} + V_i \sum_{\substack{j\in i \\ j\neq i}}^{N} V_j (G_{ij}\cos\theta_{ij} + B_{ij}\sin\theta_{ij}) + \hat{P}_i + \sum_{\substack{j=1 \\ j\neq i}}^{N} P_i^{i,j} = 0, \tag{4.45}$$

$$-V_i^2 B_{ii} + V_i \sum_{\substack{j\in i \\ j\neq i}}^{N} V_j (G_{ij}\sin\theta_{ij} - B_{ij}\cos\theta_{ij}) + Q_i = 0, \tag{4.46}$$

$$0 \leq \hat{P}_i + \sum_{\substack{j=1 \\ j\neq i}}^{N} P_i^{i,j} \leq P_i^0. \tag{4.47}$$

For all nodes ($i \in N$):

$$V_i^{\min} \leq V_i \leq V_i^{\max}, \tag{4.48}$$

$$P_{ij} = V_i^2 G_{ii} - V_i V_j (G_{ij}\cos\theta_{ij} + B_{ij}\sin\theta_{ij}) \leq P_{ij}^{\max}, \tag{4.49}$$

where S_G is the set of generator nodes; S_L is the set of load nodes; Q_i is the total reactive output of node i; $Q_i^{\min}$ and $Q_i^{\max}$ are the lower limit and higher limit of reactive output of node i, respectively; V_i, θ_i are the magnitude and angle of voltage of node i, $\theta_{ij} = \theta_i - \theta_j$; $V_i^{\min}$ and $V_i^{\max}$ are the lower limit and higher limit of voltage magnitude of node i; G_{ij}, B_{ij} are the real and imaginary part of transfer admittance between nodes i and j.

The objective is to minimize the cost of congestion management. All the resources that are adjusted to mitigate congestion will be paid by ISO. $\hat{P}$, $P_i^{i,j}$, and Q_i are treated as independent control variables during the optimization process; V_i and θ_i are dependent variables. Equations (4.39)–(4.41) and (4.49) are power balance equations for generator nodes and load nodes, respectively. Equations (4.47), (4.48), (4.43), and (4.44) are variable inequality constraint and functional inequality constraint, (4.45) is branch power flow constraint.

To solve this optimization problem more reliably and efficiently, the OPF is implemented with an interior point approach (see Sect. 4.2.3 for the detail). The steps of this congestion management process are listed as follows.

Step 1: Original generation schedule is produced through bilateral contract market (BCM) and pool day-ahead auction market (PDAM).

Step 2: Both generators and consumers submit their incremental and decremental energy bids to the ISO. Newton power flow is run to get the initial state of the power system.

Step 3: If there is any congestion in the network? If yes, go to Step 4, otherwise, stop program and output "There is no congestion in the system, it is a feasible generation schedule!".

Step 4: Run OPF to solve the problem of congestion management.

Step 5: Obtain the optimal control strategy for congestion management, and stop.

[Example 2] A five-node test system is used to demonstrate the procedure of congestion management. The network and the power flow in the nominal condition are shown in Fig. 4.5. In this system there are two generators which are G1 at node 5 and G2 at node 4, and three loads which are L1 at node 1, L2 at node 2, and L3 at node 3. One bilateral contract of 300 MW is signed between G1 and L3 in BCM. All the other electricity supplies in this system are arranged by ISO in PAM.

[Solution] In RBM, G1, G2, and L2 submit their incremental and decremental bids to the ISO to take part in the real-time dispatching competition. The bilateral contract $P^{3,4}$ also submits a curtailment price $b^{3,4}$ to ISO:

$$
\begin{aligned}
&b_{G1}^{+} = 20\,\$\,\mathrm{MW}^{-1}, \quad b_{G1}^{-} = 8\,\$\,\mathrm{MW}^{-1},\\
&b_{G2}^{+} = 15\,\$\,\mathrm{MW}^{-1}, \quad b_{G2}^{-} = 5\,\$\,\mathrm{MW}^{-1},\\
&b_{L2}^{-} = 30\,\$\,\mathrm{MW}^{-1},\\
&P^{3,4} = 300\,\mathrm{MW}, \quad b^{3,4} = 50\,\$\,\mathrm{MW}^{-1}.
\end{aligned}
$$

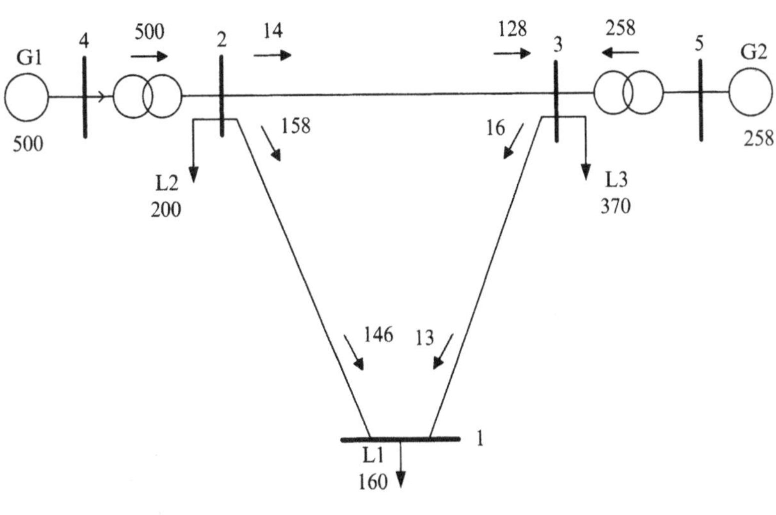

Fig. 4.5 Five-node system example of congestion management

From above bidding prices, it can be found that the price for adjustment of loads is higher than that of generators, because the latter are more flexible to adjust. The curtailment price of bilateral contract is rather high because both sides of the contract are apt not to change their agreement. Therefore, only in the case of very severe situation when the balance market cannot remove the congestion, adjustment of bilateral contracts may be considered.

Case 1: Congestion management without changing bilateral contracts. Assume that the MW limit on line 2–3 is reduced to 100 MW due to some reason. To mitigate this congestion, the best solution is decreasing the output of G1 to 441.8 MW and increasing the output of G2 to 308 MW. As a result, the active power flow of line 2–3 is reduced to 99.67 MW. The total cost of this adjustment is $1,253. The bilateral contract $P^{3,4}$ is carried out without curtailment because the available resource of G1 in RBM is enough to eliminate the congestion.

Case 2: Congestion management with changing bilateral contract. Assume that the MW limit on transformer 4–2 is reduced to 250 MW. Obviously, the output of G1 should be reduced to 250 MW. But there is 300 MW of the bilateral contract $P^{3,4}$ between G1 and L3, the curtailment to $P^{3,4}$ must be done in this case. The best strategy to solve this problem is as the following:

- Reducing 200 MW from the output of G1 in RBM
- Curtailing 50 MW from bilateral contract $P^{3,4}$, both G1 and L3 will be reduced 50 MW. Thus the power output of G1 is totally used to satisfy the bilateral contract between G1 and L3
- Increasing the output of G2 to 442.5 MW

The total cost of this adjustment is $7,192.

Case 3: Congestion management with curtailing load. In both above two cases, load need not to be curtail in congestion management. When congestion becomes more severe, load adjustment must be done. Assume that the transmission limit on line 2–3 is reduced to 100 MW and the MW limit on branch 4–2 is reduced to 250 MW due to some reason. To mitigate this congestion, the cheapest solutions are as follows:

- Decreasing the output of G2 to 392.9 MW
- Curtailing 50 MW from bilateral contract $P^{3,4}$, which means both G1 and L3 will be reduced 50 MW
- Limiting the power output of G2 at 250 MW
- Decreasing load L2 to 47.4 MW

The total cost of this strategy in this case is $7,385.

4.4 Power Flow Tracing

In the market environment, transmission becomes a special business that provides services to independent power producers (IPPs) and electricity suppliers, or provides wheeling services to other utilities. The IPPs, suppliers and utilities become the users of transmission services. Therefore, there is a growing need to identify the items and cost of transmission services. In such a situation, dispatchers should not only supervise power flow of the system, but also answer questions such as "how much use is this generator (or load) making of this transmission line?" or "what proportion of the network losses is allocated to this generator (or load)?" Solution to these problems is very important to measure the services supplied by transmission systems, and has a direct influence on wheeling cost.

Many methods have been used to determine the cost of transmission services, such as the Postage stamp method (allocates transmission cost based on an average embed cost and the magnitude of transacted power, the users are not differentiated by the extend of use of transmission resources), the Contract path method (assumes the transacted power would be confined to flow along an artificially specified path through the involved transmission systems), the MW-mile method (based on the flow distribution of a given transaction ignoring the other transactions), and so on. However, these methods cannot accurately measure the transmission usage of the users, and cannot obtain reasonable and equitable transmission cost allocation. And what is more, these methods cannot send correct signals to the transmission users, which may result overload and imperil system operation. Therefore, the power flow distribution of transmission users in different operation modes should be correctly evaluated, leading into power tracing problem.

In recent years, many papers have contributed on these topics. Kirschen et al. [91] propose an active power tracing method. By this method, the active power flow from a generator to a load and the transmission usage of a generator (or load) can be calculated, thus transmission cost and network losses can be allocated. However, this method cannot be used to solve the problem if transmission network contains loop circuits. Wang and Wang [92] introduce two current decomposition axioms, and solve the current tracing and distribution problems that cannot be solved by only using the Kirchoff's laws. Based on the two axioms, a new active power tracing method is proposed and used to solve network loss allocation and transmission usage problems. This method can treat networks with loops well and overcome the drawbacks of [91]. Based on active power tracing and marginal cost theory, [93] also provides a method to calculate transmission cost including fixed cost, losses, and congestion cost.

As the theory basis of transmission pricing, the power flow tracing problem is discussed in this section. First, the two current decomposition axioms as the fundamentals of load flow analysis are introduced. Then the mathematical models of the distribution factor problem and loss allocation problem are established. To solve theses problem, series of theorems based on the graph theory are presented and a very simple and efficient algorithm to solve above problems is suggested.

In transmission pricing, network loss allocation and usage of transmission facilities mainly depend on the active power, for simplification the affect of reactive power is ignored in the following discussion. The further development of the power tracing problem can be found in [94].

4.4.1 Current Decomposition Axioms

Up to now, the research on electric circuits is limited to analyzing the currents of circuit elements and their respective effects, without considering components of the currents and allocation of their effects. Similarly, in a conventional load flow study of power systems, concern is usually given to the powers (or currents) flowing along branches (lines or transformers), and respective losses in the branches. However, when considering the problems of wheeling cost, we need to identify the power (or current) components in each branch and allocate the effects such as losses to its components. To solve this kind of problems it is not enough to use only Kirchoff's laws of electric circuit. Therefore, we need first to introduce two axioms [92].

Assume the current of branch k, $I_{(k)}$, consists of L current components $I_{(k)l}(l = 1,2,\ldots,L)$ supplied by L generators,

$$I_{(k)} = \sum_{l=1}^{L} I_{(k)l}, \tag{4.50}$$

where $I_{(k)}$ and $I_{(k)l}$ are the effective or rms values of the currents, which can be either "active" or "reactive" component. Similarly, in the following description, the term "power" can also be replaced by either "active power" or "reactive power" as most papers in this field do.

Axiom 1 Components of current in a branch are conservative.

The axiom states that each component $I_{(k)l}$ maintains the same at the initial and terminal node of a branch.

$$I'_{(k)l} = I''_{(k)l} = I_{(k)l} \quad l = 1,2,\ldots,L, \tag{4.51}$$

where $I'_{(k)l}$ and $I''_{(k)l}$ are the component currents shared by generator l at the initial and terminal node, respectively, L is the number of generators in the system.

The usage of branch k making by generator l denoted by $f_{(k)l}$ is termed distribution factor, and defined by

$$f_{(k)l} = \frac{I_{(k)l}}{\sum_{l=1}^{L} I_{(k)l}} = \frac{I_{(k)l}}{I_{(k)}}. \tag{4.52}$$

Corollary 1 *Distribution factors are the same at the two nodes of a branch.*

Assume the voltage at the initial and terminal nodes of branch k are V_{S} and V_{R}. Thus, the respective powers are $V_{\mathrm{S}}I_{(k)}$ and $V_{\mathrm{R}}I_{(k)}$. The powers at the two nodes supplied by source l are $V_{\mathrm{S}}I'_{(k)l}$ and $V_{\mathrm{R}}I''_{(k)l}$. Hence we have

$$\frac{V_{\mathrm{S}}I'_{(k)l}}{V_{\mathrm{S}}I_{(k)}} = \frac{V_{\mathrm{R}}I''_{(k)l}}{V_{\mathrm{R}}I_{(k)}} = \frac{I_{(k)l}}{I_{(k)}} = f_{(k)l}. \tag{4.53}$$

Corollary 2 *The loss of a branch should be allocated proportionally to the current (or power) component.*

The loss of a branch k, $\Delta P_{(k)}$ can be represented by

$$\Delta P_{(k)} = V_{\mathrm{S}} I_{(k)} - V_{\mathrm{R}} I_{(k)} = (V_{\mathrm{S}} - V_{\mathrm{R}}) \sum_{l=1}^{L} I_{(k)l}. \tag{4.54}$$

According (4.50), the loss caused by current component $I_{(k)l}$ is

$$\Delta P_{(k)l} = V_{\mathrm{S}} I'_{(k)l} - V_{\mathrm{R}} I''_{(k)l} = (V_{\mathrm{S}} - V_{\mathrm{R}}) I_{(k)l}. \tag{4.55}$$

Combining (4.54) and (4.55), we obtain

$$\Delta P_{(k)l} = \Delta P_{(k)} \frac{I_{(k)l}}{\sum_{l=1}^{L} I_{(k)l}} = \Delta P_{(k)} \frac{I_{(k)l}}{I_{(k)}}. \tag{4.56}$$

In some reference, the principle of loss allocation on the basis of demand squared was also suggested, i.e., the loss allocated to component current $I_{(k)l}$ should be calculated according to

$$\Delta P_{(k)l} = \Delta P_{(k)} \frac{I^2_{(k)l}}{\sum_{l=1}^{L} I^2_{(k)l}}. \tag{4.57}$$

This principle is not economically reasonable, because it can cause inefficient resource allocation.

The next axiom is the proportional sharing assumption commonly used by researchers in this field.

Axiom 2 The current component in an outgoing line of an injected current at a node is proportional to the current of the outgoing line.

Assume the total current injected at node i is I_i, this axiom states that when the current injected by generator l at node i is I_l, its component current $I_{(k)l}$ in outgoing line k is

$$I_{(k)l} = I_l \frac{I_{(k)}}{\sum_{k=1}^{L_0} I_{(k)}}, \tag{4.58}$$

where L_0 is the number of the outgoing lines at node i.

This axiom can be interpreted as that the injected current is randomly distributed to the outgoing lines, and the distribution probability is proportional to their currents. To deduce the next corollary, the definition of "node loss" is first given as the following.

Definition *The whole loss caused by transmitting power from generators to a node is called "loss of the node." We will denoted the loss of node i by* δP_i.

Obviously, when electric power of node i is directly transmitted through its incoming lines from generators, the loss of node i, δP_i is equal to the total loss of these incoming lines. To allocate δP_i to the outgoing lines of node i we have the following corollary.

Corollary 3 *The factor of node loss allocated to an outgoing line is equal to its allocation factor.*

Proof. Assume that node i has L_i incoming lines all directly connected with the generators. Then the loss of node i is

$$\delta P_i = \sum_{m=1}^{L_i} \Delta P_{(m)}, \tag{4.59}$$

where ΔP_m is the loss of incoming line. According to Corollary 2 of Axiom 1 and Axiom 2, the loss allocated to outgoing line k of ΔP_m is

$$\Delta P_{(m)(k)} = \Delta P_{(m)} \frac{I_{(k)}}{\sum_{k=1}^{L_o} I_{(k)}}.$$

Therefore, the total loss allocated to outgoing line k can be calculated by

$$DP_{(k)} = \sum_{m=1}^{L_i} \Delta P_{(m)(k)} = \sum_{m=1}^{L_i} \Delta P_{(m)} \frac{I_{(k)}}{\sum_{k=1}^{L_o} I_{(k)}} = a_{i(k)} \delta P_i, \tag{4.60}$$

where $a_{i(k)}$ is the allocation factor

$$a_{i(k)} = \frac{I_{(k)}}{\sum_{k=1}^{L_o} I_{(k)}} = \frac{P_{(k)}}{\sum_{k=1}^{L_o} P_{(k)}}. \tag{4.61}$$

When the incoming lines of node i all are not connected to the generators, (4.60) and (4.61) can be proved by the recursive reasoning method. □

4.4.2 *Mathematical Model of Loss Allocation*

After running a traditional load flow, the total system loss and the loss of each branch ΔP_{ij} can be obtained. The question now is how to allocate the losses to each generator or each load. In the following we first introduce the mathematical model of loss allocation to each load.

Here, the key issue is to calculate the node loss $\delta P_i (i = 1, 2, \ldots, N)$. δP_i consists of the following two parts:

1. The sum of branch loss ΔP_{ji} that belongs to incoming line set $ji \in \Gamma_-(i)$
2. The total transmitted loss δP_j from node j through lines in the incoming line set of node i, $ji \in \Gamma_-(i)$

For a N node system, the network loss balance equation can be formulated as follows:

$$\delta P_i = \sum_{ji \in \Gamma_-(i)} (\Delta P_{ji} + a_{j(k)} \delta P_j) \quad i = 1, 2, \ldots, N, \tag{4.62}$$

where $a_{j(k)}$ is the factor that δP_j is transmitted from node j to node i, along the outgoing line ji; k is the index of branches ji. From (4.61),

$$a_{j(k)} = \frac{P_{ji}}{\sum\limits_{ji \in \Gamma_+(j)} P_{ji} + P_j^{(\mathrm{L})}}, \tag{4.63}$$

where $P_j^{(\mathrm{L})}$ is the load power at node j; $\Gamma_+(j)$ is the set of outgoing lines at node j; (4.62) consists of N linear equations and N variables $\delta P_i (i = 1, 2, \ldots, N)$, the common solution method can be used to solve this equations.

After obtaining all node loss δP_j, the loss allocated to each load can be calculated as the follows:

$$DP_j = \delta P_j \frac{P_j^{(\mathrm{L})}}{\sum\limits_{ji \in \Gamma_+(j)} P_{ij} + P_j^{(\mathrm{L})}}. \tag{4.64}$$

If the loss is to be allocated to each generator, the model is similar. In such a situation, the network loss balance equation becomes:

$$\delta P_i = \sum_{k = ij \in \Gamma_+(i)} (\Delta P_{ij} + a_{i(k)} \delta P_j) \quad i = 1, 2, \ldots, N, \tag{4.65}$$

where k is the index of branches ij; $a_{i(k)}$ is the factor that δP_j is transmitted from node j to node i, along the incoming line ij,

$$a_{i(k)} = \frac{P_{ij}}{\sum\limits_{ij \in \Gamma_-(j)} P_{ij} + P_j^{(\mathrm{G})}}, \tag{4.66}$$

where $P_j^{(\mathrm{G})}$ is the generator power at node j.

After all the node loss δP_j are calculated, the loss allocated to each generator can be obtained by the following equations:

$$DP_j = \delta P_j \frac{P_j^{(G)}}{\sum\limits_{ij \in \Gamma_-(j)} P_{ij} + P_j^{(G)}}. \tag{4.67}$$

Equation (4.62) or (4.65) in above models can be solved by a conventional linear equation solution method. After getting node loss $\delta P_j (j = 1, 2, \ldots, N)$, one can then allocate loss to each load or generator $DP_j (j = 1, 2, \ldots, N)$ by (4.54) or (4.67).

However, we can use a more efficient and simple algorithm to allocate loss to loads or generators avoiding solving (4.62) or (4.65). The algorithm is based on the graph theory which can be found in Sect. 4.4.4. The following is the algorithm for loss to allocate to loads.

Step 1: Set the initial condition including a load flow run, and forming the set $\Gamma_+(j)$, $\Gamma_-(j)$, $d_+(j)$, $d_-(j)$, $(j = 1, 2, \ldots, N)$

Step 2: Search node j satisfying $d_-(j) = 0$. Since no incoming lines at node j, $\delta P_j = 0$; or δP_j having been cumulated according to (4.62), and set node j as the node to be eliminated

Step 3: Calculate loss allocation to the load at node j by (4.64)

Step 4: For all the node i $(ji \in \Gamma_+(j))$, cumulate loss δP_i, and decrease $d_-(i)$ of node i by 1

Step 5: Set $d_-(j)$ as -1, flagging the node has been eliminated

Step 6: Back to Step 2, Search for the next node without incoming lines, until all nodes are eliminated.

In the following, an example is introduced based on the 24-node IEEE-RTS System [95]. In the load flow calculation, node 23 is set as the balance node, all other generator nodes are set as PV node. The total network loss is 40.731 MW, or per unit is 0.40731.

First, we allocate the loss to all load nodes. The order of node eliminating is showed in Table 4.16. From the table, we can find that the first node to be eliminated is node 1, and branches 1–3 are eliminated at the same time; the second eliminated node is 2, and branches 4 and 5 are eliminated at the same time; the third eliminated node is 7, and branch 11 is eliminated at the same time. After all nodes are eliminated according to the order, the final loss allocation results are obtained as presented in Table 4.17.

For comparison, the results for allocating loss to generators are shown in Table 4.18.

4.4.3 Usage Sharing Problem of Transmission Facilities

The theories and algorithms of usage sharing problems of transmission facilities are discussed in this section. First, based on the two current decomposition axioms in

Table 4.16 Calculation process of loss allocation to load

Eliminated order	Node	Branch	Eliminated order	Node	Branch	Eliminated order	Node	Branch
1	1	1, 2, 3	9	16	23, 28	17	24	7
2	2	4, 5	10	14	19	18	3	6
3	7	11	11	19	32	19	9	8, 12
4	22	30, 34	12	23	21, 22, 33	20	4	
5	21	25, 31	13	13	18, 20	21	8	13
6	15	24, 26	14	11	14, 16	22	10	9, 10
7	18	29	15	12	15, 17	23	5	
8	17	27	16	20		24	6	

Table 4.17 Allocated loss and loss rate to loads

Load node	Allocated loss	Allocated loss rate	Load node	Allocated loss	Allocated loss rate
3	0.057469	0.031666	10	0.039106	0.020054
4	0.015793	0.021342	14	0.069593	0.035873
5	0.011249	0.015844	15	0.016692	0.005266
6	0.049263	0.036223	19	0.049815	0.027522
8	0.047754	0.027926	20	0.019094	0.015578
9	0.031485	0.017991	Total	0.407310	–

Table 4.18 Allocated loss and loss rate to generators

Generator node	Allocated loss	Allocated loss rate	Generator node	Allocated loss	Allocated loss rate
1	0.012743	0.006673	16	0.017297	0.011159
2	0.025549	0.025773	18	0.036400	0.009100
7	0.052097	0.017366	21	0.073100	0.018275
13	0.052158	0.009135	22	0.105882	0.035294
15	0.017548	0.011321	23	0.014536	0.002202

Sect. 4.4.1, the model of sharing cost of transmission facilities is proposed. The characteristics of directed load flow graph are investigated to find out the efficient algorithm. The research shows that there exists no directed circuit in a load flow directed graph. According to this theorem, a very efficient solution to this distribution factor problem is developed.

For a specified operation mode, the power flow along branches (transmission lines and transformers) can be achieved by a load flow calculation. The main problem at present is how to determine the users' (IPPs or utilities) power distribution in each branch.

In the following, we will discuss the generator power distribution problem, i.e., the generators' distribution factors on transmission facilities. The load power distribution factor model and algorithm can be deduced in the similar way.

Assume a power system with N nodes, N_G generators and N_B branches. Thus our problem is to determine a $N_B \times N_G$ matrix $\mathbf{F}$, its elements are defined:

$$f_{(k)m} = P_{(k)m}/P_{(k)} = I_{(k)m}/I_{(k)}, \tag{4.68}$$

where $P_{(k)}$, $I_{(k)}$ are the active power and current of branch k; $P_{(k)m}$, $I_{(k)m}$ are the active power and current components supplied by generator m in branch k.

Obviously, the element $f_{(k)m}$ in matrix $\mathbf{F}$ is the distribution factor of generator m in branch k. Therefore, we have

$$\sum_{m=1}^{N_G} f_{(k)m} = 1 \quad k = 1, 2, \ldots, N_B. \tag{4.69}$$

To establish the mathematic model, the axioms in Sect. 4.4.1 need to be utilized. Based on axiom 1, the distribution factor $f_{(k)m}$ of generator m are the same at the two nodes of branch k in (4.68). Thus the powers $P_{(k)m}, P_{(k)}$ and currents $I_{(k)m}, I_{(k)}$ can be set the value of either node. The axiom 2 in Sect. 4.4.1 states that the current component of an injected current in an outgoing line at the node is proportional to the current of the outgoing line. The axiom 2 can be expressed as

$$I_{(k)m} = I_{mi}\frac{I_{(k)}}{I_i}, \tag{4.70}$$

where I_{mi} is the injection current of generator m in node i, I_i is the total injection current in node i. Multiplying the voltage of node i, (4.70) can be expressed as

$$P_{(k)m} = P_{mi}\frac{P_{(k)}}{P_i},$$

where P_{mi} is the injection power of generator m at node i. P_i is the total injection power at node i. Substituting the above equation into (4.68), we have

$$f_{(k)m} = \frac{P_{(k)m}}{P_{(k)}} = \frac{P_{mi}}{P_i}. \tag{4.71}$$

Therefore, if we can obtain injection power P_{mi}, the distribution factor of generator m at node i can be calculated through the above equation. To do so, we first establish the relationship between each generator power and total injection power at each node.

For the load flow distribution of N node system, we have the following relationship.

$$\mathbf{P}_g = \mathbf{A}\mathbf{P}_n, \tag{4.72}$$

where $\mathbf{P}_{\mathrm{g}} = [P_{1g}, P_{2g}, \ldots, P_{ng}]^{\mathrm{T}}$ is the vector of generator powers; $\mathbf{P}_{\mathrm{n}} = [P_1, P_2, \ldots, P_n]^{\mathrm{T}}$ is the vector of total node injection powers; $\mathbf{A}$ is a $N \times N$ matrix, elements of which are defined by

$$a_{ji} = \begin{cases} 1 & i = j \\ P_{ji}/P_i & ji \in \Gamma_{-}(i), \\ 0 & \text{otherwise} \end{cases} \tag{4.73}$$

where $ji \in \Gamma_{-}(i)$ means that branch ji belongs to the incoming lines of node i.

After running a load flow, the total injection power at each node and the load flow in each branch are known, hence the elements of matrix $\mathbf{A}$ are known. Equation (4.72) can be reversed:

$$\mathbf{P}_{\mathrm{n}} = \mathbf{A}^{-1}\mathbf{P}_{\mathrm{g}}, \tag{4.74}$$

where $\mathbf{A}^{-1}$ represents the contribution rates of each generator to total injection power at each node. We can further obtain the relationship between total injection power at each node and the outgoing line powers:

$$\mathbf{P}_{\mathrm{B}} = \mathbf{C}\mathbf{P}_{\mathrm{n}}, \tag{4.75}$$

where $\mathbf{P}_{\mathrm{B}} = [P_{(1)}, P_{(2)}, \ldots P_{(k)}, \ldots, P_{(N_{\mathrm{B}})}]^{\mathrm{T}}$ is the vector of initial branch powers.

$\mathbf{C}$ is $N_{\mathrm{B}} \times N$ matrix, elements of which are defined by

$$C_{(k)i} = \begin{cases} P_{(k)}/P_i & k \in \Gamma_{+}(i) \\ 0 & \text{otherwise} \end{cases}, \tag{4.76}$$

where $k \in \Gamma_{+}(i)$ means branch k belongs to the outgoing line of node i.

Substituting (4.74) into (4.75), we have

$$\mathbf{P}_{\mathrm{B}} = \mathbf{C}\mathbf{A}^{-1}\mathbf{P}_{\mathrm{g}} = \mathbf{B}\mathbf{P}_{\mathrm{g}}, \tag{4.77}$$

where

$$\mathbf{B} = \mathbf{C}\mathbf{A}^{-1}. \tag{4.78}$$

According to (4.77), the contribution of generators at node i to the load flow of branch k can be calculated by

$$P_{(k)i} = b_{(k)i}P_{ig}. \tag{4.79}$$

Therefore, according to (4.53), the distribution factor of source i to branch k is

$$f_{(k)i} = P_{(k)i}/P_{(k)} = b_{(k)i}P_{ig}/P_{(k)}. \tag{4.80}$$

In summary, to get **F**, matrix **A** should be first established and inversed, then the matrix **B** can be achieved based on (4.78). Finally the elements of **F** can be obtained according to (4.80).

In the next section the methodology of the graph theory will be introduced. The method not only can quickly calculate the distribution factors of each generator to braches, but also can conveniently solve the loss allocation problem discussed in Sect. 4.4.2.

4.4.4 Methodology of Graph Theory

A marked-load flow distribution graph is a directed graph, and its circuit configuration is the corresponding base graph. Different operation modes correspond to different directed graph. At this stage, the direction of each branch is determined by the direction of its active power flow. Each branch has its initial node and terminal node, while each node has its outgoing lines and incoming lines. The number of outgoing lines at node i is denoted by $d_{+}(i)$, the number of incoming lines by $d_{-}(i)$. The set of outgoing lines is denoted by $\Gamma_{+}(i)$ and the set of incoming lines by $\Gamma_{-}(i)$.

A directed path is formed along the direction of branches. When the initial node and terminal node of a directed path are identical, we have a directed circuit. In a load flow graph, we use $R_{(k)}, X_{(k)}, P_{(k)}, Q_{(k)}$ to denote the resistance, reactance, active, and reactive power flow of branch k, and we have the following theorem.

Theorem 1 *If the following relationship holds for each branch along the direction of its active power in a load flow graph,*

$$P_{(k)}X_{(k)} > Q_{(k)}R_{(k)}$$

the graph has no directed circuit.

Proof. We use the methodology of reduction to absurdity. If there is a directed circuit in the graph, then the following relationship holds

$$\sum_{k \in L} \Delta\theta_{(k)} = 0, \tag{4.81}$$

where $\Delta\theta_{(k)}$ is the phase angle difference between the two nodes of branch k and can be expressed by

$$\Delta\theta_{(k)} = \arctan\left\{\frac{(P_{(k)}X_{(k)} - Q_{(k)}R_{(k)})/V_{(k)}}{V_{(k)} + (P_{(k)}X_{(k)} + Q_{(k)}R_{(k)})/V_{(k)}}\right\}. \tag{4.82}$$

The variables in (4.82) take the values at the terminal node of branch k. When $P_{(k)}X_{(k)} > Q_{(k)}R_{(k)}$, then $\Delta\theta_{(k)} > 0$, thus (4.81) cannot hold. Therefore, the directed circuit cannot exist in the situation.

It should be noted that the condition is a sufficient condition, which is satisfied for most load flow distributions. In case there is a branch not satisfied the condition, it does not mean that there exists a directed circuit. In this situation the power flow along the branch is certainly negligible, thus we can consider the branch is open. □

Theorem 2 *When a directed graph has no directed circuit, there are at least two nodes, which satisfied $d_+(i) = 0$ and $d_-(j) = 0$ respectively.*

Proof. The proof is also based on the methodology of reduction to absurdity. Assume $d_+(i) > 0$ holds for all nodes, i.e., each node has at least one outgoing line. Thus setting out from any node n_1, we can travel to next node n_2 along its outgoing line. And from n_2 we can travel further to n_3 by similar reason, and so on. Thus there are only two possible outcomes: one is that we have an infinite travel. This is impossible for a finite graph; the other is that there exist directed circuits. This contradicts the condition of the theorem. Hence we can conclude that there is at least a node with $d_+(i) = 0$. Similarly, we can prove the other half of the theorem, i.e., there is at least one node that satisfies $d_-(j) = 0$. □

Combining Theorem 1 and 2, we obtain the next corollary.

Corollary 4 *On a load flow graph, there at least exist one node without outgoing line and one node without incoming line.*

Before the next theorem is presented, we give the following definition. Assume i is a node with $d_-(i) = 0$ on a load flow graph D, the process of eliminating node i and its outgoing lines $\Gamma_+(i)$ is called eliminating process for node i.

Theorem 3 *In a load flow graph, all branches can be eliminated through a recursive node eliminating process.*

Proof. Denote the node set of a load flow graph by V, and the branch set by U. Because directed graph D(V, U) has no directed circuit, there at least exists one node i_1 with $d_-(i_1) = 0$. Carrying out an eliminating process for node i_1 in $\Gamma_+(i_1)$, we get subgraph $D'(\mathrm{V}\backslash i_1, \mathrm{U}\backslash\Gamma_+(i_1))$. Because $D \supset D'$, D' also has no directed circuit. Hence, there is at least one node, say i_2 with $d_-(i_2) = 0$ in D'. Then we can carry out an eliminate process for node i_2, and so on. Thus we can eliminate all branches by a finite (less than N step) recursive eliminating process.

Now we explain the eliminating process by a simple example, as shown in Fig. 4.6a. This directed load flow graph has no directed circuit and $d_-(1) = 0$. Thus we can first eliminate node 1 and its outgoing lines (1), (2), (3). After eliminating node 1, the subgraph D' is formed as shown in Fig. 4.6b, in which $d'_-(2) = 0$. Therefore we can further eliminate node 2 and its outgoing line (4). After node 2 is eliminated, subgraph D'' is formed, as shown in Fig. 4.6c, where $d^{(}_-4) = 0$. Hence we can eliminate node 4 and its outgoing line (5), and thus we complete the eliminating process.

In the above eliminating process, the node with $d_-(i) = 0$ and its outgoing lines $\Gamma_+(i)$ are successively eliminated, which is called eliminating outgoing lines process. Eliminating process can also be carried out by successively eliminating the node with $d_+(i) = 0$ and its incoming lines $\Gamma_-(i)$, which is called eliminating incoming lines process. The corresponding definitions and theorems are similar to

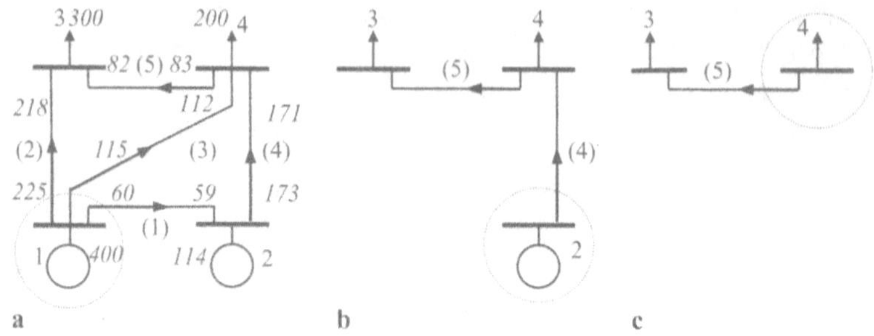

Fig. 4.6 Eliminating outgoing line process

the discussion above. We will not discuss it in detail. The two eliminating processes are the basic framework of distribution factor problem and loss allocation problem. Usage of the two algorithms can allocate the sharing cost of transmission facilities and loss to generators or loads. □

[Example 3] Solve the distribution factor problem of the simple system shown in Fig. 4.6a. This system includes four nodes and five branches, two loads and two generators. The power flow on each branch is illustrated in the figure.

[Solution] From Fig. 4.6a we can see that node 1 has no incoming line, so it can be eliminated first. Here $P_1 = 400$, $P_1^G = P_{1,1} = 400$, $\Gamma_+(1) = \{(1), (2), (3)\}$, According to (4.80), we have

$$f_{(1)1} = f_{(2)1} = f_{(3)1} = {}^{400}\!/_{400} = 1.$$

According to (4.73), the injection power $P_{1,1}$ at node 1 can transfer to other nodes 2–4 through transfer coefficients a_{ji},

$$P_{2,1} = 400 \times {}^{59}\!/_{400} = 59,$$

$$P_{3,1} = 400 \times {}^{218}\!/_{400} = 218,$$

$$P_{4,1} = 400 \times {}^{112}\!/_{400} = 112.$$

Thus we complete eliminating node 1 and its outgoing lines. The system is now simplified to the subsystem as shown in Fig. 4.6b.

In Fig. 4.6b, node 2 has no incoming line because of $d_-(2) = 0$. Hence we now eliminate node 2. Here $P_2 = 173$, $P_{2,1} = 59$, $P_{2,2} = 114$, $\Gamma_+(2) = \{(4)\}$.

According to (4.71), we have

$$f_{(4)1} = {}^{59}\!/_{173} = 0.34104,$$

$$f_{(4)2} = {}^{114}\!/_{173} = 0.65896.$$

Table 4.19 Distribution factors of simple system in Fig. 4.6

Lines	Generator 1	Generator 2
(1)	1.0	0.0
(2)	1.0	0.0
(3)	1.0	0.0
(4)	0.34104	0.65896
(5)	0.60183	0.39817

According to (4.73), the injection power $P_{2,1}$, $P_{2,2}$ can transfer to other node 4 through transfer coefficients a_{ji},

$$P_{4,1} = 59 \times {}^{171}\!/_{173} = 58.31792,$$

$$P_{4,2} = 114 \times {}^{171}\!/_{173} = 112.68208.$$

Thus we complete eliminating node 2 and its outgoing lines. The system is now simplified to the subsystem as shown in Fig. 4.6c.

Finally, node 4 has no incoming line in Fig. 3c, thus we now eliminate node 4.

Here $P_{4,1} = 58.31792 + 112 = 170.31792$, $P_{4,2} = 112.68208$, $P_4 = 283$, $\Gamma_+(4) = \{(5)\}$.

According to (4.80), we have

$$f_{(5)1} = {}^{170.31792}\!/_{283} = 0.60183,$$

$$f_{(5)2} = {}^{112.68208}\!/_{283} = 0.39817.$$

Thus we complete the whole eliminating process. The distribution factors of the simple system are shown in Table 4.19.

4.5 Available Transfer Capability of Transmission System

4.5.1 *Introduction to Available Transfer Capability*

The ATC is very important for secure and reliable operation of the power systems. In traditional vertical-regulation situation, ATC is considered by system operators as a measure of margin between constraints and the current operating state. In the power market environment, the uncertainties of power system operation increase, and electricity trading, hence operation mode, varies frequently. Accidents that cause overload of branches and voltage violation are more likely to happen. How to evaluate the remaining transfer capability in the transmission network accurately and efficiently becomes hot topic in power research.

Research on the transfer capability of transmission system can be traced back to 1970s, but it did not catch enough attention of researchers and engineers until FERC issued an order in 1996 that requires electric utilities calculate ATC and post ATC information to power market participants [96]. NERC defines ATC as the measure of the transfer capability remaining in the physical transmission network for further commercial activity above already committed uses [97]. This shows that in power market ATC is not merely the traditional energy-exchange capability between regions but the maximal transfer capability from a node in a region to a node in another region considering the existing transmission contracts and the constraints of power system's safety and reliability. ATC is the measure of actual transfer capability of transmission network over and above the existing transmission contracts, and it can be conceptually defined as

$$\mathrm{ATC} = \mathrm{TTC} - \mathrm{TRM} - \mathrm{CBM} - \mathrm{ETC},$$

where, TTC is the total transmission capability, it denotes the maximal transfer capability of interconnection line between regions subject to the safety and reliability constraints of power systems; TRM is transmission reliability margin, as amount of transmission transfer capability needed to ensure that the interconnected-transmission network is secure under a reasonable range of uncertainties in system conditions. Capacity benefit margin (CBM) is the amount of transmission transfer capability reserved by load serving entities to ensure access to generation form interconnected systems to meet generation reliability requirements; existing transmission commitments (ETC) means the transfer capability occupied by the existing transmission contract. Transmission contracts can be further described as "nonrecallable," "recallable," "reserved," and "scheduled" according to the stability of ETC contracts. If large energy-exchange exits between interconnected network and system reliability could be threatened by random disturbance, part of transmission contract should be curtailed. Thus congestion occurs.

From the above definition of ATC, it can be seen that the uncertainty of power system have significant impacts on ATC. Contingency of transmission line or generator can result in a tremendous decrease of ATC. Therefore, how to treat the impact of the uncertainty and stochastic factors on ATC is the vital problem in calculating ATC.

ATC is used to measure the transfer capability remainder in the transmission system for a period of time (1 h, 1 day, or a longer period) in the future. According to the requirement on the length of this time period, ATC can be classified as online ATC and off-line ATC calculation.

When evaluating off-line ATC, uncertainties have more influence on ATC calculation. Generally speaking, the longer is the time period, the more influence has the uncertainty. Therefore, the probabilistic model is usually adopted in off-line ATC study.

When calculating online ATC, only the several most serious contingencies are considered, thus the computation burden decreases greatly. The deterministic model is often used in online ATC calculation.

The above two models of ATC are briefly introduced as follows.

4.5.1.1 ATC Calculation Based on Deterministic Model

At present, the calculation methods based on deterministic model mainly includes

(1) *Linear programming model [98]*. Considering the constraints imposed on the transmission network due to thermal and stability limitation, the method uses linear programming to calculate ATC based on DC flow model. DC flow model ignores the impacts of voltage and reactive power; thus, the method cannot be applied in heavily loaded system where reactive power support and effective voltage control means may be not sufficient. Furthermore, the computation time of linear programming increases tremendously with expand of power systems, so this method is not suitable for calculating ATC of large scale systems.
(2) *Continuation power flow (CPF) model [99]*. CPF can trace the change of power flow. From a basic operation point, increasing the power flow between the two regions, we solve the resulting power flow problem step by step, until static voltage stability limit is reached, and thus the critical maximal power flow is obtained. The impacts of voltage, reactive power, and other nonlinear factors are considered in CPF. Compared with linear programming model, the ATC value obtained by CPF is more accurate. But the load factor in CPF is not changed when the load and generation power output increase, ignoring the optimal distribution of load and generation output. This may cause the ATC result conservative.
(3) *OPF model [100]*. ATC evaluation based on OPF is an improvement on CPF. OPF can deal with all kinds of system constraints, static contingencies and optimize resource dispatch. OPF is suitable to calculate ATC. However, CPF must solve nonlinear equations and OPF must find the optimal dispatch in ATC calculation, their computational burden is forbidding.
(4) *Distribution factor model [101]*. It is also called sensitivity analysis model. This model overcomes the computational burden at cost of computation accuracy, and can get approximate ATC value very fast.
(5) *Genetic algorithm (GA) model [102]*. This model utilizes the ability of searching global optimum of GA to find the maximal transmission capability between regions. Generally, GA is better than CPF in ATC calculation from the point view of computation time and accuracy.
(6) *Online ATC evaluation software package (TRACE) [103]*. The soft package is developed by EPRI associated with some electricity companies in late 1996. It is the first software package that can be used in real power system to calculate ATC. TRACE can calculate ATC and TTC of the specified paths with the real-time state-estimation data from energy management system (EMS). Thus, TRACE can optimize the energy trading of the system. The fast contingency-capture program embedded in TRACE can identify the emergent contingencies, so TRACE is a very efficient tool for online ATC evaluation.

The framework for online ATC calculation is shown in Fig. 4.7. The ATC program exchange information with the following modules in EMS: state estimator (SA), security analysis (SA), current operating plan (COP), and open access

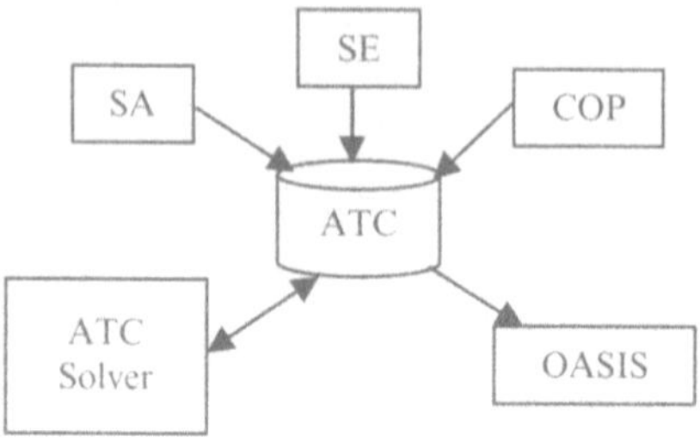

Fig. 4.7 Framework of online ATC calculation

same-time information system (OASIS). The current system state is obtained from SE; the contingency list is obtained from SA and the load forecast, generation schedules and information of outage equipment come from the COP system. The obtained ATC values are transferred for posting at the OASIS.

4.5.1.2 ATC Calculation Based on Probabilistic Model

Off-line ATC calculation needs to consider large amount of uncertain factors. If dealing with all the factors one by one, the computation burden is forbidding for practical application. Therefore, off-line ATC calculation is usually based on probabilistic approach.

There are three models based on probabilistic approach for ATC calculation:

(1) *Stochastic programming model [104]*. This algorithm considers the uncertainty of the availability of generators and transmission lines as well as load forecast error. The availability of generators and transmission lines is considered as random variables of the binomial distribution, and the load forecast error is represented by random variables of the normal distribution. The original discrete variables are first transformed into continuous random variables based on two-stage stochastic programming with recourse (SPR). Then chance constrained programming (CCP) is employed to treat the continuous variables. At last, the ATC and corresponding probability distribution can be obtained. Stochastic programming involves the stochastic power flow, discrete and continuous variables, so its computing speed is usually not satisfactory.
(2) *Enumeration method [105]*. This model combines the system state enumeration and optimization to evaluate ATC value. Because the enumeration method has exponential complexity, it cannot be applied in large scale power system.
(3) *Monte Carlo simulation method [106]*. As mentioned in Sect. 3.5, the algorithm is composed of the Monte Carlo simulation and optimization algorithm. The Monte Carlo simulation can generate and deal with huge scenarios resulting from large number of uncertain factors. With the increase of system scale and network complexity the computing time of this method does not increase

significantly. Therefore, the Monte Carlo simulation can be applied in evaluating ATC of large-scale power systems.

With the ATC calculation method based on probabilistic model, we can get the expected ATC value, its probability density and distribution function curve. The confidence interval of ATC under a certain confidence level can be estimated. The stochastic information of ATC can guide the arrangement of power system's operation and forecast the electricity price.

In the following, an off-line ATC calculation algorithm is introduced. This algorithm is the combination of Monte Carlo simulation and sensitivity analysis method, which can compute the ATC value between the generation node and the load node. The Monte Carlo simulation is used to sample system state. In this way, huge amount of uncertain factors can be treated effectively and lots of stochastic information of ATC can be obtained. Sensitivity-analysis method is a well understood method and can be implemented easily. It is used to solve the optimization problem involved in ATC calculation. The algorithm can provide the accurate ATC value within a reasonable computing time.

4.5.2 Application of Monte Carlo Simulation in ATC Calculation

Usually, three kinds of uncertainties are considered in the off-line ATC calculation, including uncertainties of generator outages, transmission line outages and nodal load forecasting errors. Generators and transmission lines have two states: operation and failure, and the availability of generators and transmission lines are considered as random variables of the binomial distribution. The load is simulated by the normal distribution, $N(\mu, \sigma^2)$, μ is the expected value of node load and can be represented by forecasted load; σ is the standard variance of load.

There is a probability $P(x)$ corresponds to a certain system state x. The ATC value for sampled system state x is denoted by $\mathrm{ATC}(x)$. According to the probability theory, for a given load level, the expected ATC value can be calculated by

$$E(\mathrm{ATC}) = \sum_{x \in X} \mathrm{ATC}(x)P(x). \tag{4.83}$$

The expected value of $\mathrm{ATC}(x)$ in the Monte Carlo simulation, $\hat{E}(\mathrm{ATC})$, is defined as

$$\hat{E}(\mathrm{ATC}) = \frac{1}{N}\sum_{i=1}^{N} \mathrm{ATC}(x_i), \tag{4.84}$$

where x_i is the ith sampled state, N is the number of total sampled states, $\mathrm{ATC}(x_i)$ is the ATC value for state x_i.

From (4.84) it can be seen that $\hat{E}(\mathrm{ATC})$ is only the expected value of ATC(x), not the real value. Because x and ATC(x) are both stochastic variables, $\hat{E}(\mathrm{ATC})$ is also a stochastic variable. The variance of $\hat{E}(\mathrm{ATC})$ is formulated as

$$V[\hat{E}(\mathrm{ATC})] = \frac{V(\mathrm{ATC})}{N}, \tag{4.85}$$

here $V(\mathrm{ATC})$ is the variance of ATC, its estimate value is $\hat{V}(\mathrm{ATC})$.

$$\hat{V}(\mathrm{ATC}) = \frac{1}{N-1}\sum_{i=1}^{N}[\mathrm{ATC}(x_i) - \hat{E}(\mathrm{ATC})]^2. \tag{4.86}$$

According to the probability theory, if the probability density curve is bell shaped, the confidential interval of $\hat{E}(\mathrm{ATC})$ with a confidence $1 - 2\alpha$ can be approximated by

$$[\hat{E}(\mathrm{ATC}) - u_\alpha\sqrt{\hat{V}(\mathrm{ATC})}, \hat{E}(\mathrm{ATC}) + u_\alpha\sqrt{\hat{V}(\mathrm{ATC})}], \tag{4.87}$$

here u_α is the α upside fractile of standard normal distribution $N(0,1)$. u_α can be obtained from the distribution table.

The variance coefficient β is usually used to evaluate error of the expected value.

$$\beta = \frac{\sqrt{V[\hat{E}(\mathrm{ATC})]}}{\hat{E}(\mathrm{ATC})} = \frac{\sqrt{V(\mathrm{ATC})/N}}{\hat{E}(\mathrm{ATC})}. \tag{4.88}$$

Equation (4.88) indicates that the computational burden of the Monte Carlo simulation is not sensitive to the system scale and complexity. Therefore Monte Carlo simulation is effective to treat the uncertainties and suitable to calculate ATC for large-scale power systems. Furthermore, it can be seen that decrease of $V(\mathrm{ATC})$ and increase of N can reduce the variance coefficient β and improve the computation accuracy.

The flow chart of ATC calculation is shown in Fig. 4.8.

4.5.3 ATC Calculation with Sensitivity Analysis Method

Sensitivity analysis method is a useful tool widely used in power system planning and control. It evaluates the impact of a particular variable on power system performance by analyzing the relation of the operation index to this variable. Furthermore, measures to improve the operation index can be suggested based on the analysis. The sensitivity method for ATC calculation is based on the relation of power flow in transmission line to nodal input power.

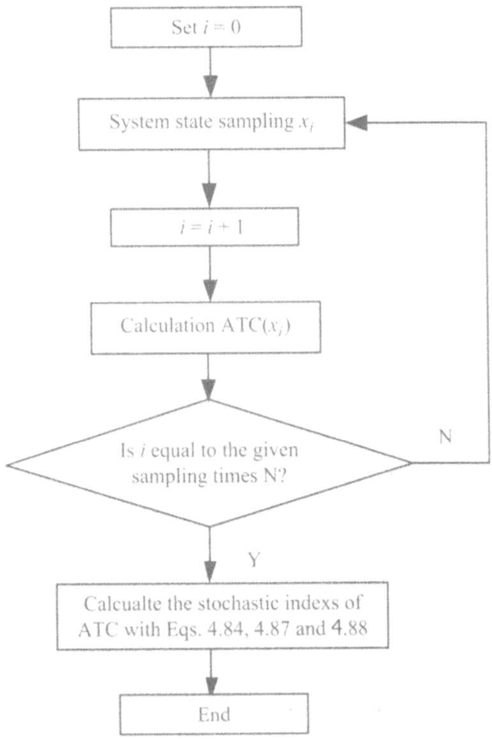

Fig. 4.8 Flow chart of ATC calculation with Monte Carlo simulation

According to the DC power flow and branch power flow equations, the sensitivity equation of the branch power flow to nodal input power can be obtained (suppose the branch number is b and node number is n),

$$\Delta \mathbf{L}_k = \mathbf{S}'\Delta \mathbf{P}, \tag{4.89}$$

$$\mathbf{S}' = \mathbf{B}_l \mathbf{A} \mathbf{B}^{-1}, \tag{4.90}$$

where $\Delta \mathbf{L}_k$ is the incremental vector of the branch flows; $\Delta \mathbf{P}$ is the incremental vector of the nodal power injection (including the active power change at the slack node); $\mathbf{B}_l$ is the diagonal matrix composing of the line admittance $b \times b$; $\mathbf{B}$ is the square matrix composing of imaginary part of the nodal admittance matrix; $\mathbf{A}$ is the branch-node incidence matrix.

$\mathbf{S}'$ is a $b \times (n - 1)$ sensitivity matrix, and it represents relation between the node power injection and the line flow, its element $S'_{k,i}$ is the incremental of power flow in the line k due to increasing a unit of power injection in the node i.

Matrix $\mathbf{S}'$ is calculated taking the slack node as reference, thus when we expend it to a $b \times n$ matrix $\mathbf{S}$, its elements corresponding to the slack node are all zeros. The element $S_{k,i}$ of $\mathbf{S}$ has the same meaning like the element $S'_{k,i}$ in $\mathbf{S}'$.

Sensitivity analysis model can solve the following problems related to ATC evaluation:

1. Adjusting system state
2. Calculating ATC between two given nodes on the condition that the input power at other nodes keeps constant (noted by ATC1 in the following discussion)
3. Calculating ATC between two given nodes when considering adjustment of generators' outputs (noted by ATC2 in the following discussion)

Now, let us discuss these problems in some detail.

4.5.3.1 Adjusting System State

In the Monte Carlo simulation, a sampled system state x_i is assessed to check that whether state x_i meets system security requirements. When investigating ATC1, if state x_i does meet security constraints, ATC between two given nodes is zero. We set ATC1 $= 0$. When investigating ATC2, necessary correcting measure should be used to adjust the system to normal state before ATC2 is calculated. The correcting principals include

(1) When the load is higher than the generators' output because of generator outage or nodal load fluctuation, curtail system load until power balance is reached. In power market, the nodal load should be curtailed according to the signed electricity contracts. To make problem simple, we curtail all nodal load in proportion.
(2) If some transmission lines overload due to network branch failure, adjust the nodal active power using sensitivity analysis method, or curtail system load until overload is eliminated.

In the following we introduce a heuristic approach based on analyzing the influence of "adjusting generators output" and "reducing system load" on eliminating overload. Adjusting output of generators should be used first to reducing load curtailment. The adjusting process of generator output is as follows:

(1) Select the most serious overloaded line k.
(2) Select a pair of generators at node i and j to be adjusted: When line k overloads in positive direction, decrease the output of generator i with the maximal positive sensitivity value, and increase the output of generator j with the maximal negative sensitivity value. When line k overloads in negative direction, the adjusting actions are converse.
(3) Determine adjustment amount Δp of generators at node i and j. To eliminate overload of branch k, Δp can be derived from (4.89),

$$\Delta p = \frac{|L_k| - \bar{L}_k}{S_{k,i}^{+\max} - S_{k,j}^{-\max}}, \tag{4.91}$$

where $S_{k,i}^{+\max}$ is the positive sensitivity value of node i, $S_{k,j}^{-\max}$ denotes the negative sensitivity value of node j, L_k is the load flow of branch k, $\bar{L}_k$ is the transmission capacity of branch k.

If the adjust amount computed by (4.91) is too large, violation of the generator's maximal and minimal output constraints may occur. In this situation, Δp should be reduced appropriately:

(4) Adjust the output of generators at node i and j
(5) Update power flow according to (4.89) and (4.90)
(6) Repeat steps (1)–(5) until overload is eliminated or adjusting output has no effect on alleviating the overload

If overload still exists after the above adjusting, curtailing system load must be carried out to eliminate overload. The curtailing process is similar to generators' output adjustment.

4.5.3.2 Calculating ATC1

For each transmission line, calculate the possible maximal power change at the given node pair (A, B) when the power flows through the transmission line reach its limit. The calculating formula is

$$\Delta p_{g,A} = \Delta p_{l,B} = \begin{cases} \dfrac{\overline{L_k} - L_k}{S_{k,A} - S_{k,B}} & S_{k,A} - S_{k,B} > 0 \\ \dfrac{-\overline{L_k} - L_k}{S_{k,A} - S_{k,B}} & S_{k,A} - S_{k,B} < 0 \end{cases} \quad (k = 1, \ldots, b), \qquad (4.92)$$

where, $\Delta p_{g,A}$ is the amount of generator output increment at node A, $\Delta p_{l,B}$ is the amount of the load increment at node B.

Branch k that makes $\Delta p_{g,A}$ or $\Delta p_{l,B}$ the minimal is the bottle-neck line influencing ATC1. Accordingly, the minimal $\Delta p_{g,A}$ or $\Delta p_{l,B}$ is ATC1.

4.5.3.3 Calculating ATC2

ATC between a given pair of nodes can be increased by adjusting the generators' output. In the following, we give the steps of calculating ATC2 between a given pair of nodes (A, B) with the sensitivity analysis method:

(1) Increase all generators' output at node A by Δp, to their maximal output. At the same time, increase the load at node B by the same amount Δp.
(2) Updating the branch power flow according to (4.89) and (4.90), find out whether there is overload. If no branch overloads, the ATC2 between the given node pair (A, B) is Δp. Otherwise, adjust other generators' output to eliminate the

overload, as discussed above. If the adjustment eliminates the overload, the ATC2 between given node pair (A, B) is also Δp. Otherwise, go to step (3).
(3) Properly reduce Δp, set other generators' outputs to their initial values. Then go to step (2).

From the above calculating process, it can be seen that sensitivity analysis method can calculate ATC1 between two given nodes very fast, while calculating ATC2 may need to adjust the generators' outputs repeatedly, thus its calculating time may be longer than calculating ATC1.

[Example 4] Calculate ATC1 between generator node 5 and load node 2 in a five-node power system as shown in Fig. 4.9.

[Solution] The system has five nodes, seven branches, and nine generators. Its total generation capability is 1,164 MW. Suppose that the variance of forecasted load is $\sigma^2 = 0.02$, i.e., load of each node obeys normal distribution $N(u,\ 0.02)$. The sampling number is 10,000, variance coefficient $\beta < 0.002$. Node 5 is the slack node. The initial data are shown in Tables 4.20 and 4.21.

The Monte Carlo simulation is used to sample system state x_i. The state of equipment is determined by a random number which is generated by computer. The random numbers of equipments and the corresponding equipment states in one system state sampling are shown in Tables 4.22–4.24. The system state x_i is composed of the states of all the equipments. Then the system state is assessed by the following three steps:

(1) Analyzing the network topology under system state x_i, check whether node 5 and node 2 are connected. It is obviously that the node 5 and node 2 are connected if no branch failure occurs.
(2) Evaluate whether the generators' output is less than the system load. Under system state x_i, the system output is 919 MW including the AGC generator's 297 MW, and the total load of the system is 703.29 MW. Thus generator output is larger than load. The system can keep balance by adjusting the output of AGC generator.
(3) Check whether the power flows along branches are within their transmission capacity limits. Here DC power flow model is used to calculate the power flow for simplicity. The sensitivity matrix S of branch power flow to nodal injected power can be obtained according to (4.90) (the last column is the extended column of slack node, so all elements are zero).

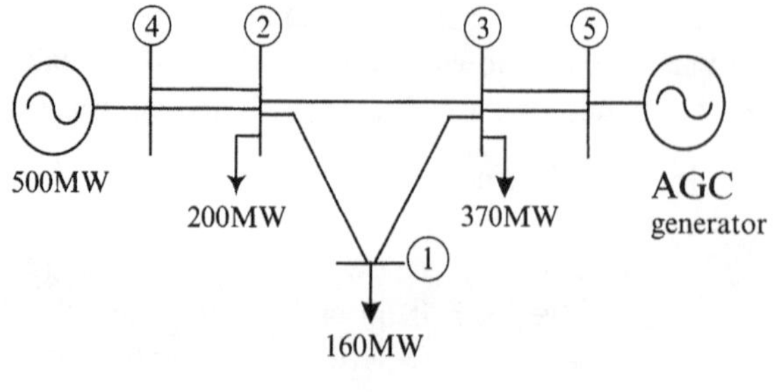

Fig. 4.9 The graph of five-node system

Table 4.20 Parameters of generator

Generator number	Generator node number	Rated power (MW)	Actual output (MW)	Generator fault probability
1	4	155	125	0.04
2	4	155	125	0.04
3	4	155	125	0.04
4	4	155	125	0.04
5	5	197	197	0.05
6	5	197	AGC generator	0.05
7	5	50	33	0.01
8	5	50	AGC generator	0.01
9	5	50	AGC generator	0.01

Table 4.21 Parameters of branch

Branch number	Nodes of branch	Branch reactance	Branch capacity (MW)	Branch fault probability
1	1–2	0.2500	305	0.000438
2	1–3	0.3500	175	0.000582
3	2–3	0.3000	305	0.000445
4	2–4	0.0300	400	0.001653
5	2–4	0.0300	400	0.001653
6	4.5	0.0300	400	0.001653
7	4.5	0.0300	400	0.001653

Table 4.22 Generator state defined by $U(0, 1)$ random number

No. of generator	1	2	3	4	5	6	7	8	9
Random number	0.038	0.531	0.435	0.371	0.332	0.286	0.774	0.509	0.977
FOR of generator	0.04	0.04	0.03	0.03	0.05	0.05	0.01	0.01	0.01
Generator state	Down	Up	Up	Up	Up	Up	Up	Up	Up

Table 4.23 Branch state defined by uniform distribution $U(0, 1)$ random number

Branch number	1	2	3	4	5	6	7
Random number	0.040282	0.531725	0.858605	0.601367	0.113788	0.122979	0.329386
Branch fault probability	0.000438	0.000582	0.000445	0.001653	0.001653	0.001653	0.001653
Branch state	Operation	Operation	Operation	Operation	Operation	Operation	Operation

Table 4.24 Correction of node load with random number obeying $N(0,1)$

Number of node	1	2	3
Random number	−2.529600	0.480062	0.161125
Node load's correction amount (p.u.)	−0.3578	0.0679	0.0228
The node load after correction (p.u.)	1.2422	2.0679	3.7228
The node load after correction (MW)	124.22	206.79	372.28

The sensitivity matrix $\mathbf{S}'$ is

$$\begin{bmatrix} 0.3889 & -0.3333 & -0.0000 & -0.3333 & 0.0000 \\ 0.6111 & 0.3333 & 0.0000 & 0.3333 & 0.0000 \\ 0.3889 & 0.6667 & 0.0000 & 0.6667 & 0.0000 \\ 0.0000 & -0.0000 & -0.0000 & -0.5000 & 0.0000 \\ 0.0000 & -0.0000 & -0.0000 & -0.5000 & 0.0000 \\ 0.5000 & 0.5000 & 0.5000 & 0.5000 & 0.0000 \\ 0.5000 & 0.5000 & 0.5000 & 0.5000 & 0.0000 \end{bmatrix}.$$

The load flow of each branch is calculated by (4.89) (the base of per unit is 100 MW):

$P_1 = -1.043804$, $P_2 = -0.198456$, $P_3 = 0.638305$, $P_4 = -1.875000$, $P_5 = -1.875000$, $P_6 = -1.641469$, $P_7 = -1.641469$.

These values are all less than the transmission capacity limits of the corresponding lines.

In the above state assessment process, if the two nodes do not connect, or the total generator output is less than system load, or a transmission line is overload, the power system cannot securely operate under state x_i. Thus $\mathrm{ATC1}(x_i) = 0$ and sensitivity analysis algorithm is not needed. In this example, the system can operate normally under state x_i.

The next step is to calculate ATC1 from node 5 to node 2 using sensitivity analysis method. According to (4.92), estimate the allowable amount of power exchange between node 5 and node 2 when each branch reaches its capability limit:

$$\Delta P_1 = \frac{3.05000 - (-1.043804)}{0 - (-0.33333)} = 12.281412,$$

$$\Delta P_2 = \frac{-1.75 - (-0.198456)}{0 - 0.33333} = 4.654631,$$

$$\Delta P_3 = \frac{-3.05000 - (0.638305)}{0 - 0.66666} = 5.532457,$$

$$\Delta P_4 = \infty,$$

$$\Delta P_5 = \infty,$$

$$\Delta P_6 = \frac{-3.50000 - (-1.641469)}{0 - 0.50000} = 6.717062,$$

$$\Delta P_7 = \frac{-3.50000 - (-1.641469)}{0 - 0.50000} = 6.717062.$$

Of the seven branches, branch 2 hits its transmission capacity limit, the maximal amount of power increment between node 5 and node 2 is 465.46 MW. Then, consider the influence capability of generator at node 5 on this power increment: the total installed capacity at node 5 is 544 MW and its output is 328.29 MW, thus the maximal amount of increment is 215.71 MW. This value is less than 465.46 MW, so the maximal amount of power increment between node 5 and node 2 is 215.71 MW when considering both branch and generator capacity constraints. According to the definition of ATC, power increment 215.71 MW is ATC1 from node 5 to node 2 under system state x_i, and the generator at node 5 is the bottle-neck equipment. The bottle-neck equipment denotes the equipment whose constraint is active in ATC calculating. The information of bottle-neck equipment is very important to system operating in power market.

ATC1 of all pair of generator nodes to load nodes is calculated and shown in Table 4.25. For example, the expected ATC1 from generator node 5 to load node 2 is 273.00 MW, its standard variance is 80.67 MW. The probability of the sampling ATC1 value belongs to interval [139.89 MW, 406.11 MW] is 97.5%. And the output constraint of generator at node 5 is active with the probability 93.52%. The data on bottle-neck equipment shown in Table 4.25 imply that the shortage of installed generator capacity is the main cause hindering increase of ATC between nodes.

Table 4.25 ATC1 from generator node to load node (unit: MW)

Generator node	Load node	ATC1 under initial state	$\hat{E}$(ATC1)	S(ATC)	Confidence interval of $\hat{E}$(ATC1) at confidence 97.5%	Bottle-neck equipment (active probability P_M)
4	1	120.00	113.61	13.55	[88.80, 120.00]	Generator at node 4 (98.37%)
4	2	120.00	113.67	18.31	[88.80, 120.00]	Generator at node 4 (93.52%)
4	3	120.00	113.61	13.55	[88.80, 120.00]	Generator at node 4 (98.37%)
4	5	120.00	113.61	13.55	[88.80, 120.00]	Generator at node 4 (98.37%)
5	1	290.00	254.88	69.90	[139.67, 354.38]	Branch 1–3 (69.68%) Generator at node 5 (28.74%)
5	2	314.00	273.00	80.67	[139.89, 406.11]	Generator at node 5 (93.52%)
5	3	314.00	273.00	80.67	[139.89, 406.11]	Generator at node 5 (93.54%)
5	4	314.00	273.00	80.67	[139.89, 406.11]	Generator at node 5 (93.52%)

Note: The $P_M = (N_1/N)100\%$, where N denotes the sampling times; N_1 is the times that the constraint of equipment is active.

Thinking and Problem Solving

1. Discuss the effect of electricity markets on the operation of electrical power systems.
2. Why can the optimal load flow not solve such issues as startup–shutdown of generator units and the economical dispatch of hydropower stations?
3. What characteristics do optimal load flow problems have from the viewpoint of mathematical optimal algorithms? What are the requirements of the optimal load flow algorithm?
4. State the basic idea and key algorithm of an interior point algorithm for the optimal load flow.
5. Give the optimal dispatching model when a system loses a power resource, according to the congestion management method introduced in this chapter.
6. Discuss the effect of congestion on real-time prices.
7. Set up the real-time price model based on the optimal load flow.
8. Discuss the reasonability of loss allocated according to load–power proportional and load–power-squared allocation methods.
9. Design an approach to determine loss coefficients for each transmission customer during a certain period (such as each month or each year) based on a loss allocation model.
10. Propose several schemes for determining transmission pricing.
11. What is the effect of ATC evaluation on system dispatching under an electricity market environment?
12. Compare the advantages and disadvantages of deterministic and probabilistic models of ATC calculation.
13. What measures are there to improve ATC?
14. Discuss the limitation of ATC calculations based on the sensitivity method.

Chapter 5
HVDC and FACTS

5.1 Introduction

The bulk power transmission from generators to load centers is one of the main study areas in electric power engineering. Electrical power engineers and researchers have been exploring new ways of power transmission for decades while making efforts to enhance the capability of existing power grids. Multiple-phase transmission was proposed by American researchers in 1972. The transmission capacity can be increased dramatically by utilizing multiple three-phase lines, for example, 6, 9, or 12 phases. The main benefits are lower phase-to-phase voltage than a single three-phase system, and the reduction of line-to-line spacing to save the land usage. Compact transmission was introduced by ex-Soviet researchers in 1980. The surge impedance loading can be improved by optimizing line and tower structures: increasing the number of sub-conductors in each phase to smooth the electric field distribution around conductors, and reducing the phase-to-phase spacing. The concept of fractional frequency transmission system (FFTS), presented by one of the authors in 1995 [107], is under investigation. It uses low frequency on transmission system, for example, one third of normal frequency, to reduce impedance – so-called electrical distance. Dutch scientists discovered superconductors in 1911. Superconductor technology could also be applied in power transmission. Small capacity generators, transformers, and cables have been manufactured using superconductors. However, there is a long way to go before industrial application. Wireless transmission has the possibility to transport electric power without transmission lines. This concept can be traced back to Tesla's test in 1899. Current study and feasible industrialized application are mainly on microwave, laser, and vacuum tube transmission. Wireless transmission has been studied for over 30 years. There are a lot of technical problems to be solved, and it is far from industrial utilization.

High voltage direct current (HVDC) and flexible AC transmission system (FACTS) are electric power transmission facilities utilizing electronic technology.

In the early stages of power industry development, Thomas Alva Edison (1847–1931) was leading the group supporting a complete DC system from generation to transmission; while George Westinghouse (1846–1914) was in favor of an AC system. AC systems finally dominated the power industry because of their economical

X.-F. Wang et al., *Modern Power Systems Analysis*.
doi: 10.1007/978-0-387-72853-7,

and technical advantages: the successful synchronized operation of multiple generators, the development of transformers, and the invention of three-phase induction motors. With respect to generation and voltage transformation, AC systems have well-known advantages over DC systems. While for transmission, DC has its merits over AC. DC systems have three advantages compared to AC: (1) The stability due to synchronization phenomena in AC increases the cost of building long, high capacity transmission lines dramatically. As the transmission length reaches a certain distance, DC transmission becomes economically attractive over AC. Converter stations are the major cost of building DC lines. As power electronics technology advances, converting valves, the key element of converter station, have much higher voltage and current ratings, thus driving down the price of DC transmission. (2) With the development of modern control technology, power transported through DC lines can be adjusted very quickly (in milliseconds) through valve control. This can be applied to enhance the stability of the AC system. (3) DC lines can be used to connect power systems with different frequencies. This is called a back-to-back DC system and is the most convenient way to connect multiple power grids while maintaining their independent operation. The above three advantages have increased the competitively of DC transmission. Today, HVDC is to be found in more and more power networks around the world. Modern power systems often include both AC and DC transmission.

To meet the ever growing demand for bulk power transmission over long distance, one solution is to build HVDC transmission lines. On the other hand, increasing the capacity of existing transmission lines can achieve the same objective. There are many more AC lines than DC lines in today's power networks. Technical renovation could greatly enhance the utilization of existing AC lines, which can be more cost effective than building new lines.

The concept of FACTS was introduced by N.G. Higorani in the late 1980s [108–109]. FACTS does not have a strict and well-recognized definition. FACTS uses power electronic equipment to control and adjust operation and network parameters to optimize system operation and transmission capacity. HVDC could be seen as an example of FACTS according to the above definition. However, since it has been developed into a specific transmission technique, HVDC is not classified as FACTS equipment in present day terminology.

The main driving force for FACTS development is as follows. The increasing system load makes the existing power system incapable of carrying sufficient power over long distances. The difficulties of obtaining new rights of way, as well as environmental protection requirements, constrain the building of new lines. One of the main approaches to meet the transmission requirements is to improve the usage and capacity of existing lines. The improving manufacturing technology for high power electronic apparatus is leading to lower prices, which make FACTS a feasible solution for renovating existing grids. Rapid development in computation and control techniques, as well as the widespread use of computers, opens the way to FACTS implementation for fast, flexible, and secure control action. Furthermore, deregulation of electric power markets complicates the system operation. Power systems demand stronger self-control capability to meet various technical and economic requirements of market participants.

For a conventional power system, without FACTS equipment, the network parameters are fixed. The possible adjustment and control during system operation relies primarily on the control of the real and reactive power of generators. There are other types of adjustment measure such as tap changing transformers, series capacitor compensators, and shunt capacitor or reactor compensators to change network parameters. Transmission lines being put into or out of service can also change network topology. All these measures are accomplished through mechanical means. Their response speed is not compatible with the fast system transients. There is no corresponding control mechanism for the conventional transmission network comparable to various fast governor and voltage control facilities for generators. Because of the above restrictions, power flows are determined by Kirchhoff's current and voltage laws, and Ohm's law, given the system generation and load. This is called natural power flow. Generators and power networks have been gradually established for a long time. It is very difficulty to achieve reasonable operation for all system conditions through planning. In fact, natural power flows are hardly the optimal power flows for the existing conventional power systems. For example, the loop flow will increase the system loss in many circumstances. For parallel flows, current distributes in inverse proportion to the line impedance. It is often the case that one line has reached its thermal limit while the other line is still below its normal loading. The adjustment of generation can hardly optimize system operation or even obtain the desired operation mode. The thermal limits of transmission lines are determined by the size of cross-section area of the conductors. It is usually the relatively short lines that reach their thermal limits in conventional power systems. Another constraint on transporting energy through the network is the stability limit of synchronization. It is much more complicated to determine the stability limits than thermal limits. They depend on network structure, operation modes, control measurements, line location, and possible fault types and locations. Conventional power systems do not have fast and flexible control over the transmission networks. Their stability limits are usually less or far less than their thermal limits. This means that the power transport capabilities of existing power networks are not fully utilized. FACTS has emerged to implement fast and flexible control over transmission networks through power electronic devices. FACTS can increase the capacity of the transmission network in coordination with various fast control measures for generators.

Conventional power systems provide some control mechanisms for transmission networks. For example, there is series capacitor compensation in appropriate locations to reduce line impedance. The installation of shunt capacitors/reactors, static VAR compensator, and on-load tap changing transformers are used to control bus voltage. Phase shifters are applied to change phase angle differences between the voltages at the two terminals of transmission lines, and hence adjust the sharing of load among parallel paths. However, the increase of transmission capability is limited by the lack of rapid and continuous adjustment of controlled parameters for these devices.

There are various types of FACTS equipment. The literature [110] has recommended the terminology and definition of FACTS. Conventional power systems have some experience in the use of series and shunt compensation, phase shifters and tap changing transformers. The above apparatus equipped with power electronic devices represents the relatively mature FACTS technology. The power electronic devices help execute the predesigned control schemes through fast, continuous adjustment of their parameters. The objective is to achieve the desired dynamic performance of power systems, raise stability limits and transmission capacity. For the past 20 years, FACTS technology has been under continual development, and has more and more industrial application. The analysis methodology and control mechanisms of HVDC and FACTS under various operation modes, as well as suitable power flow calculation methods, have become important research fields in power engineering.

5.2 HVDC Basic Principles and Mathematical Models

This section will introduce the basic principles and modeling of HVDC by the analysis of normal operation modes of converters.

5.2.1 HVDC Basic Principles

The diagram of a basic HVDC system is shown in Fig. 5.1. This is a simple system with two converters C1 and C2, and one DC line. Based on the polarity of DC lines, HVDC systems are classified as monopolar, bipolar, and homopolar systems. The system in Fig. 5.1, a monopolar system, has only one DC line that normally has negative polarity and uses the ground as the current return path. The monopolar system is mainly implemented to reduce the cost of line construction. The physical conditions of earth will affect the current flow. It may also have a negative impact on underground facilities, causing heavy erosion to various metal pipelines in the current path. Bipolar systems can eliminate these problems by using two DC lines, one positive and the other negative. In Fig. 5.1, the converter has a single bridge. Multibridge converters, comprising several bridges connected in series, can increase the line voltage and reduce harmonics. Multibridge converters can use bipolar and homopolar connections. Figure 5.2a, b shows the bipolar and homopolar connections. All lines have the same polarity in the homopolar connection. A monopolar connection is often used as the first phase of bipolar or homopolar

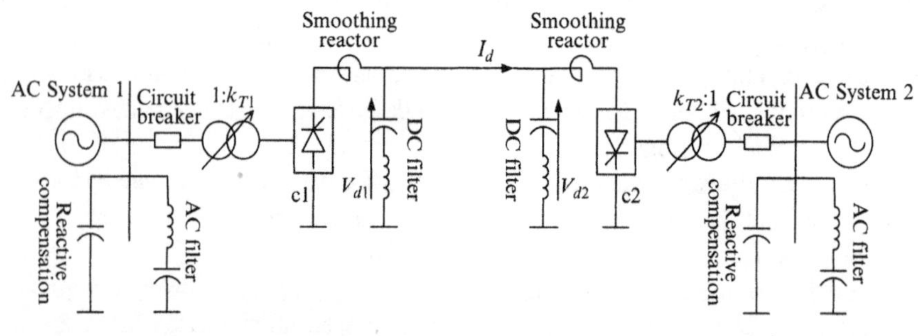

Fig. 5.1 HVDC basic connection diagram

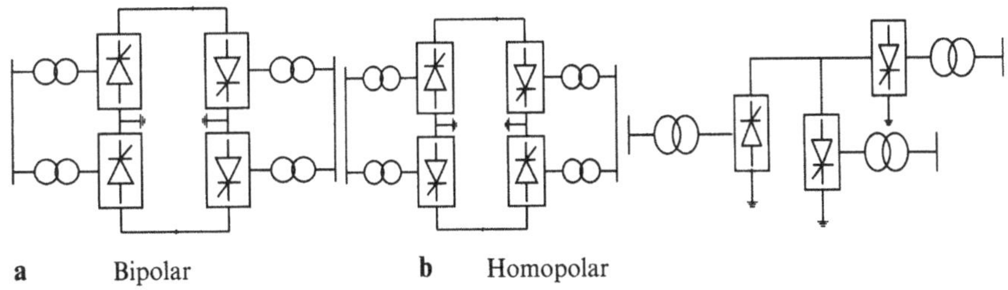

Fig. 5.2 HVDC connection

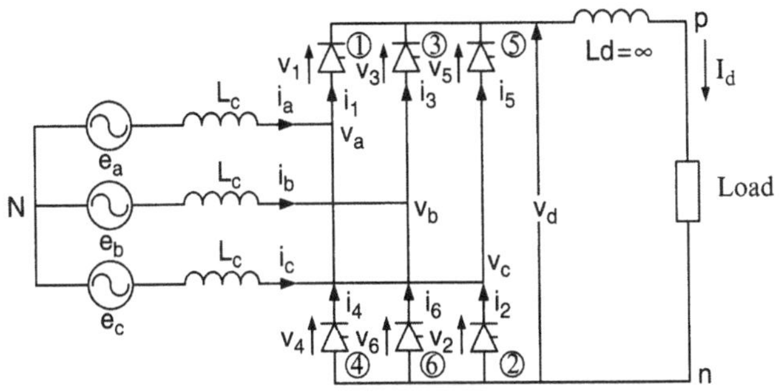

Fig. 5.3 Equivalent circuits of a three-phase whole-wave bridge

connections. A converter station is usually called a terminal of the DC transmission system. Monopolar two-terminal, bipolar two-terminal, and homopolar two-terminal systems are shown in Figs. 5.1 and 5.2a,b, respectively. HVDC can actually have multiple terminals connecting three or more AC systems. Figure 5.2c shows a monopolar three-terminal system.

The main equipment in a converter station includes converters, converter transformers, smoothing reactors, AC filters, DC filters, reactive power compensation devices, and circuit breakers. The converter is to transfer energy between AC and DC. Converting AC to DC is implemented by rectifiers while converting DC to AC is by inverters. The main element in a converter is the valve. Modern HVDC systems use thyristors as converter valves. Thyristors are usually rated 3–5 kV in voltage and 2.5–3 kA in current. Due to the limited ratings, converters usually consist of one or more converter bridges connected in series or parallel. The HVDC converter bridge is a three-phase converter bridge circuits as shown in Fig. 5.3. Each converter bridge has six branches of valves. The DC terminals of a converter connect to DC lines and the AC terminals to AC lines. The converter transformer is a conventional transformer with on-load tap changers. The turns-ratio of the converter transformer can then vary to manage the converter operation. The "DC side" of the converter transformer is usually delta or Y connected with ungrounded neutral, so that the DC line can have an independent voltage reference relative to the AC network. Harmonic voltages and currents arise during converter operation. Harmonics deteriorate the power quality, interfere with wireless communication, and should be filtered out with filters having appropriate parameters. The inductance of the smoothing reactor is very large and can reach 1 H. Its main function is to reduce the harmonic voltages and currents on the DC lines, to prevent commutation failure of inverters, to maintain continuous current under light loading, and to curtail short-circuit current in converters during faults. Converters consume a great amount of reactive power in operation. The reactive power in steady operation can be 50% of the real power transmitted on the DC lines, with much more consumption during system transients. Reactive power compensation near the converters is used to provide the reactive power source for converter operation.

HVDC is used to convert energy from AC to DC, to transport the energy as DC, and to convert it back from DC to AC. When AC system 1 transports energy to AC system 2 through DC lines, C1 runs in rectifying mode and C2 in inverting mode. Hence C1 can be seen as source and C2 as load. Given the resistance R of DC line, the line current is

$$I_{\mathrm{d}} = \frac{V_{\mathrm{d1}} - V_{\mathrm{d2}}}{R}. \tag{5.1}$$

Thus the power sent out from C1 and received at C2 is

$$\left.\begin{aligned} P_{\mathrm{d1}} &= V_{\mathrm{d1}} I_{\mathrm{d}} \\ P_{\mathrm{d2}} &= V_{\mathrm{d2}} I_{\mathrm{d}} \end{aligned}\right\}. \tag{5.2}$$

The difference between the two represents the energy loss in the DC line. The power transmitted through DC lines is purely real power. Note that the DC voltage V_{d2} of converter C2 has the opposite direction to the DC current I_{d}. If V_{d1} is greater than V_{d2}, there will be a DC current going through DC line as indicated in (5.1). The adjustment of DC voltage can control the quantity of power transported on the line. It is important to point out that if the polarity of V_{d2} remains unchanged, a higher V_{d2} than V_{d1} cannot make power flow from C2 to C1. The current in (5.1) cannot be negative since converters only allow current to flow in one direction. To change the power flow direction, the polarities of both converters at the two terminals should be reversed at the same time by the converter control, making C1 operate as an inverter and C2 as a rectifier.

As indicated in (5.1) and (5.2), the current and power on DC lines are determined by the DC voltage at the two terminals, irrespective of frequencies and voltage phase angles of the connected AC systems. The adjustment of DC voltage is achieved through the control of the firing angle of the converter bridge instead of the voltage magnitude of the AC systems. The adjustment ranges of DC voltage are much greater than AC. HVDC can provide a high power carrying capacity over long distance without any stability constraints, while an AC system would face more difficulties under the same situation. The HVDC control uses electronics to achieve rapid control action. During system transients, fast and large changes of transmission power result in system frequency variation, while generators in the AC system do not take up all the power imbalances. For example, increasing the power transmission will lower the frequency in AC system 1 and raise the frequency in AC system 2. This converts the rotating kinetic energy in AC system 1 into electric energy and passes it to AC system 2. Eventually, the frequency control devices in AC system 1 will trigger output increases of generators in the system to restore the frequency. The fast response of HVDC is therefore very important when AC system 2 requires emergency power support.

Here we are going to introduce the working principles of converters and their basic equations with the following general assumptions:

1. AC systems are three-phase symmetric without neutral shift and sinusoidal with a single frequency and no harmonics.
2. DC current is constant without fluctuation.
3. Converter transformer is an ideal transformer without magnetizing reactance, copper loss, or saturation.
4. The distributed characteristics of DC line parameters are not considered.

If readers are interested only in power flow calculation of AC/DC interconnected system and not in the operation principles and introduction of the basic equations, they should directly browse to Fig. 5.15 and the basic equations (5.37)–(5.39).

5.2.2 Converter Basic Equations Neglecting L_c

Figure 5.3 demonstrates the equivalent circuits and valve symbols of a three-phase whole-wave bridge converter. The normal operation of valves has two modes: on and off. There are two conditions for a valve to be switched from off to on: first having a higher anode voltage than its cathode or a positive voltage across the valve; secondly receiving a firing pulse. A positive valve voltage without a firing pulse keeps the valve in the off state. A normal diode is unable to be off with a positive voltage. The valve will maintain conducting once it is turned on even without any firing pulse. The valve will be changed to the off state after the valve current reduces to zero and the valve voltage has been negative for a short period (milliseconds). The equivalent resistance of an on state valve is approximately zero. In the off state, the valve can withstand very high positive or negative voltage between its anode and cathodes without being turned on (having very small leakage current). The equivalent resistance of an off state valve is approximately infinity. The valve is an ideal valve when neglecting its on state voltage drop and the off state leakage current.

Based on the general assumptions, the AC system (including converting transformer) consists of an ideal voltage source having constant frequency and voltage and a series connected reactance L_c. The instantaneous voltage of the ideal voltage source is

$$\left.\begin{aligned} e_a &= E_{\mathrm{m}} \cos(\omega t + \pi/3) \\ e_b &= E_{\mathrm{m}} \cos(\omega t - \pi/3) \\ e_c &= E_{\mathrm{m}} \cos(\omega t - \pi) \end{aligned}\right\}. \tag{5.3}$$

The phase-to-phase voltage is

$$\left.\begin{aligned} e_{ac} &= e_a - e_c = \sqrt{3}E_\text{m}\cos(\omega t + \pi/6) \\ e_{ba} &= e_b - e_a = \sqrt{3}E_\text{m}\cos(\omega t - \pi/2) \\ e_{cb} &= e_c - e_b = \sqrt{3}E_\text{m}\cos(\omega t + 5\pi/6) \end{aligned}\right\}. \tag{5.4}$$

Figure 5.4a shows the waveforms of (5.3) and (5.4).

First we will analyze zero-delayed firing angle α. A zero α means once the anode voltage of the valve is higher than its cathode, a firing pulse is sent to its control gate. The valve is turned on immediately, when neglecting L_c. In Fig. 5.3, the valves in the upper bridge are labeled as 1, 3, 5 and the lower bridge as 4, 6, 2. The number actually indicates the sequence of valve conduction from the analysis below. The cathodes of valves 1, 3, and 4 are connected together. When the phase-to-ground voltage of phase a is higher than the phase-to-ground voltages of phases b and c, valve 1 is turned on. Since the voltage drop across a conducting valve is ignored, the cathode voltages of valves 3 and 5 equal phase voltage e_a due to a conducting valve 1. It is higher than the anode voltages of valves 3 and 5, so they are turned off. Similarly, the anodes of valves 2, 4, and 6 are connected. When the phase-to-ground voltage of phase c is lower than phases a and b, valve 2 is turned on while valves 4 and 6 are turned off.

Based on the waveforms of Fig. 5.4a, when $\omega t \in [-120°, 0°]$, e_a is greater than e_b as well as e_c. In this period, valve 1 in the upper bridge is on. When $\omega t \in [-60°, 60°]$, e_c is less than e_a as well as e_b. In this period, valve 2 in the lower bridge is on. With zero L_c, there is one conducting valve in both upper and lower bridges under normal converter operation. During the period of $\omega t \in [-60°, 0°]$, valve 1 in the upper bridge and valve 2 in the lower bridge are on while all other valves are off. In the same period, the cathode voltage of the upper bridge is e_a and the anode voltage of the lower bridge is e_c. Apparently, the source voltage in the DC loop is $e_{ac} = e_a - e_c$; AC current $i_a = -i_c = I_\text{d}$; $i_b = 0$. The same analysis applies to other time periods.

Before $\omega t = 0°$, valve 1 is on. After $\omega t = 0°$, once e_b is greater than e_a, valve 3 is triggered to conduct. The cathode voltage of valve 1 is e_b after valve 3 is on. Valve 1 is then turned off due to its negative valve voltage because e_b is greater than e_a. The turn-on of valve 1 and turn-off of valve 3 occur at the same time of $\omega t = 0°$ with a zero L_c. In Fig. 5.4a, e_a is always greater than e_b when $\omega t \in [0°, 120°]$ to keep valve 3 conducting. Note the conducting status of the lower bridge. When $\omega t \in [0°, 60°]$, valve 3 is on in the upper bridge and valve 2 is on in the lower bridge. The source voltage of the DC loop is $e_{bc} = e_b - e_c$; AC current $i_b = -i_c = I_d$; $i_a = 0$. It is referred to as commutation when valve 1 is switched from on state to off state and valve 3 from off state to on state. The source voltage in the DC loop is changed from e_{ac} to e_{bc}.

Before $\omega t = 60°$, valve 2 is on. After $\omega t = 60°$, once e_a is less than e_c, valve 4 is triggered to conduct. The anode voltage of valve 2 is e_a after valve 4 is on. Valve 2 is then turned off due to its negative valve voltage. The turn-on of valve 4 and turn-off of valve 2 occur at the same time of $\omega t = 60°$ with a zero L_c. In Fig. 5.4a, e_a is

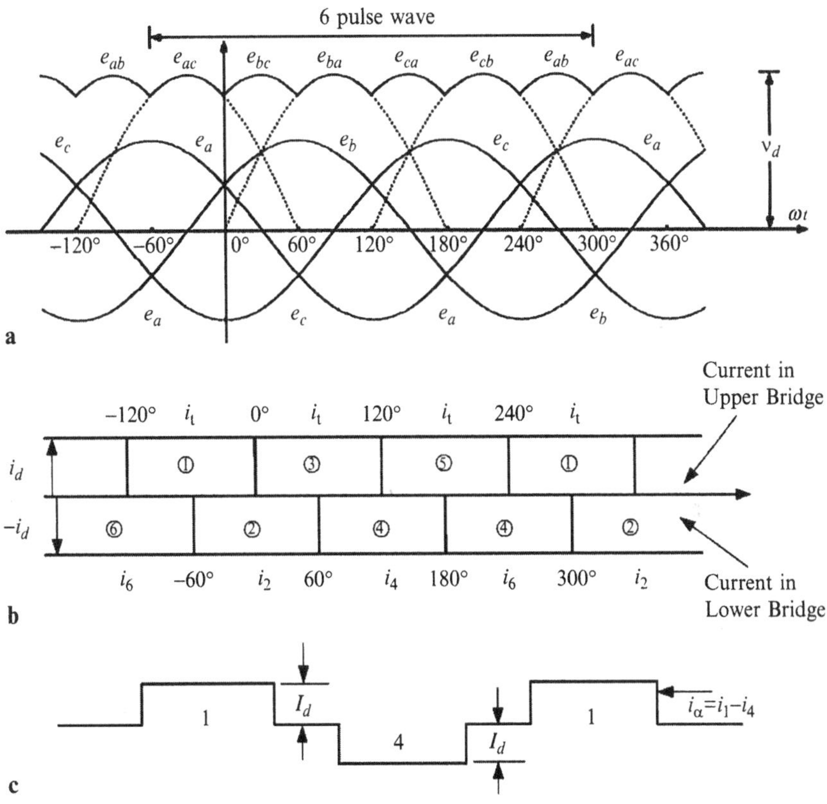

Fig. 5.4 (**a**) AC phase-to-ground, phase-to-phase voltage, DC instant voltage v_d; (**b**) conducting vales in different time periods; (**c**) phase a current

always less than e_c when $\omega t \in [60°, 180°]$ to keep valve 4 conducting. Note the conducting status of the upper bridge. When $\omega t \in [60°, 120°]$, valve 3 is on in the upper bridge and valve 4 is on in the lower bridge. The source voltage in the DC loop is $e_{ba} = e_b - e_a$; AC current $i_b = -i_a = I_d$; $i_c = 0$.

Before $\omega t = 120°$, valve 3 is on. After $\omega t = 120°$, once e_c is greater than e_b, valve 5 is triggered to conduct. Valve 3 is then turned off. When $\omega t \in [120°, 240°]$, valve 5 is always on. When $\omega t \in [120°, 180°]$, valve 5 is on in the upper bridge and valve 4 is on in the lower bridge. The source voltage in the DC loop is $e_{ca} = e_c - e_a$; AC current $i_c = -i_a = I_d$; $i_b = 0$.

Before $\omega t = 180°$, valve 4 is on. After $\omega t = 180°$, once e_b is less than e_a, valve 6 is triggered to conduct. Valve 4 is then turned off. When $\omega t \in [180°, 300°]$, valve 6 is always on. When $\omega t \in [180°, 240°]$, valve 5 is on in the upper bridge and valve 6 is on in the lower bridge. The source voltage in the DC loop is $e_{cb} = e_c - e_b$; AC current $i_c = -i_b = I_d$; $i_a = 0$.

Before $\omega t = 240°$, valve 5 is on. After $\omega t = 240°$, once e_a is greater than e_c, valve 1 is triggered to conduct. Valve 5 is then turned off. When $\omega t \in [240°, 360°]$, valve 1 is always on. When $\omega t \in [240°, 300°]$, valve 1 is on in the upper bridge and valve 6 is on in the lower bridge. The source voltage in the DC loop is $e_{ab} = e_a - e_b$; AC current $i_a = -i_b = I_d$; $i_c = 0$.

The above process will repeat periodically. The upper and lower parts in Fig. 5.4b represent the valves in the upper and lower bridges. The source in the DC loop changes phases when one valve is switched from on state to off state and another valve from off state to on state. For example, at $\omega t = 0°$ valve 1 is off and valve 3 is on. The source voltage in the DC loop is changed from $e_{ac} = e_a - e_c$ to $e_{bc} = e_b - e_c$. The commutation instantly completes with a zero L_c. There are only two valves numbered in sequence being on at any moment, one in the upper bridge; the other in the lower bridge. When L_c is not zero, current cannot undergo a sudden change due to inductance. Commutation cannot then complete. The electric angle corresponding to the time needed is called commutation angle γ. We will discuss the case with nonzero commutation angles.

The above analysis can generate the waveforms of three phase currents on the AC side. Figure 5.4c shows the current of phase *a*. The smoothing reactors, filters, and zero L_c produce the rectangular waves. In fact, I_d is the average DC current. We will study its magnitude later. Figure 5.4b displays the conducting period of each valve. Figure 5.4a shows the waveform of the DC instantaneous voltage v_d. This is the voltage across the anode of the upper bridge and the cathode of the lower bridge. From the above analysis, there are six commutations in a complete AC cycle of $\omega t \in [0°, 360°]$ and the waveform of DC instantaneous voltage v_d has six ripples of equal length. Hence the three-phase whole-wave converter is also called six-pulse converter. Rippled DC voltage v_d can be mathematically transformed by Fourier analysis to obtain its DC component voltage V_d. The DC component V_d is the average of v_d.

The average DC voltage is V_{d0} when the delayed firing angle α is zero and commutation angle γ is zero, as indicated by the waveforms of instantaneous voltage v_d in Fig. 5.4a

$$V_{d0} = \frac{1}{2\pi}\int_{0°}^{360°} v_d \mathrm{d}\theta = \frac{3\sqrt{6}}{\pi}E, \tag{5.5}$$

where E is the r.m.s. (root mean square) value of phase-to-ground voltage of the AC source.

The above is the analysis of the DC voltage and current when the firing angle delay is zero. If there is a time delay τ_a in firing after the valve voltage becomes positive, the electric angle $\omega\tau_a = \alpha$ is the delayed firing angle. Figure 5.4 shows the firing moment of each valve when $\alpha = 0°$. If α is not zero, the electric angles at the moment of conducting for valves 3, 4, 5, 6, 2, and 1 are $0° + \alpha$, $60° + \alpha$, $120° + \alpha$, $180° + \alpha$, $240° + \alpha$, and $300° + \alpha$. When $\alpha \neq 0°$ and neglecting L_c, the waveforms of DC instantaneous voltage v_d are shown in Fig. 5.5a. The conducting periods of each valve are displayed in Fig. 5.5b. As stated in the two necessary conditions of valve conduction, the range of the delayed firing angle α is $[0°, 180°]$ in order to turn the valve on. If the delayed firing angle α is beyond this range, the valve cannot be triggered to conduct, as the valve voltage is negative as seen from the waveforms of the AC source. Take valve 3 as an example. Valve 1 is on before valve 3 is turned on. The cathode voltage of valve 3 is e_a and its anode voltage is e_b. The waveforms

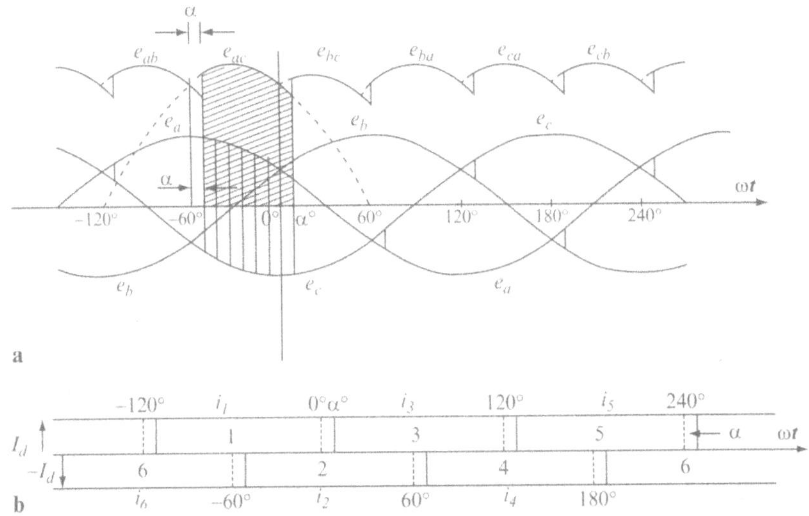

Fig. 5.5 **(a)** $\alpha \neq 0, \gamma = 0$, Waveform of DC instant voltage v_d **(b)** conducting periods of each valve

of e_a and e_b in Fig. 5.4 indicate that e_b is greater than e_a when $\omega t \in [0°, 180°]$ and valve 3 has positive valve voltage e_{ba}. Valve 3 can be triggered to conduct if α is less than 180°. When α is greater than 180°, e_b is less than e_a so valve 3 does not meet the conduction condition and cannot be turned on. The same analysis applies to other valves.

Based on Fig. 5.5a, when the delayed firing angle $\alpha \in [0°, 180°]$, the average of DC voltage is

$$V_\mathrm{d} = \frac{1}{2\pi}\int_{-60°}^{300°} v_\mathrm{d}\mathrm{d}\theta = \frac{6}{2\pi}\int_{-60°+\alpha}^{0°+\alpha} e_{ac}\mathrm{d}\theta = V_{\mathrm{d}0}\cos\alpha. \tag{5.6}$$

From the above equation, the average DC voltage V_d is less than $V_{\mathrm{d}0}$ when α is not zero. When α increases from zero to 90°, V_d decreases from $V_{\mathrm{d}0}$ to zero; when α increases from 90° to 180°, V_d decreases from zero to $-V_{\mathrm{d}0}$. When DC voltage becomes negative, the direction of DC current does not change due to the unidirectional valve characteristics. In this case, the product of DC voltage and current is negative, i.e., the power consumption from the AC system is negative. The real power actually flows from the DC system to the AC system under this operation mode. When a converter provides real power for the AC system, it converts DC energy into AC energy and passes the energy into the AC system. This type of converter operation is called inversion.

Below we will explore the phase-angle relationship between AC fundamental frequency current i_{a1} and AC voltage e_a to find out how the converter operation changes from rectification to inversion as the firing angle delay increases.

The commutation in the above analysis instantly completes when neglecting L_c. By comparing Figs. 5.4b and 5.5b, the electric angle corresponding to the conducting period for each valve is 120° regardless whether the delayed firing angle α is zero or not. The valve current is a rectangular wave lasting for 120° with a magnitude of I_d. Figure 5.4c shows the current waveform of i_a with $\alpha = 0°$. Let us consider the phase relationship between i_a and AC source e_a. When α increases from zero, the waveform of i_a is unchanged and shifted to the right. Based on Fourier analysis, it is not difficult to obtain the fundamental frequency i_{al} from i_a. The delayed phase angle of i_{al} with regard to the AC source e_a is the delayed firing angle α. The r.m.s. value of the fundamental frequency current is

$$I = \frac{2}{\sqrt{2}\pi} \int_{-30^\circ}^{30^\circ} I_d \cos x \, dx = \frac{\sqrt{6}}{\pi} I_d. \quad (5.7)$$

We assume that there are ideal filters on both AC and DC sides. The harmonic power is zero. Neglecting the power loss of the converter, AC fundamental frequency power equals DC power

$$3EI \cos\varphi = V_d I_d, \quad (5.8)$$

where φ is the leading phase angle of the fundamental frequency voltage over current and is called power factor angle of the converter. Substituting (5.6) and (5.7) into (5.8), we have

$$3E\frac{\sqrt{6}}{\pi} I_d \cos\varphi = I_d \frac{3\sqrt{6}}{\pi} E \cos\alpha \quad (5.9)$$

Hence

$$\cos\varphi = \cos\alpha. \quad (5.10)$$

The above verifies that the phase angle difference between AC fundamental frequency voltage and current is the firing angle delay α. From the above analysis, the complex power of the AC system is

$$P + jQ = \frac{3\sqrt{6}}{\pi} EI_d(\cos\alpha + j\sin\alpha). \quad (5.11)$$

Equations (5.6) and (5.7) indicate that the ratio between AC fundamental frequency current and DC current is fixed during conversion from AC to DC or from DC to AC; while the ratio between voltages is determined by the firing angle delay α. Equation (5.11) gives the complex power that the AC system sends to the DC system through the converter. In other words, it is the complex power that the DC system obtains from the AC system. This power is controlled by the firing angle delay. When $\alpha \in [0°, 90°]$, real power is positive, the DC system obtains real power from AC, converting energy from AC to DC. When $\alpha \in [90°, 180°]$, real power is negative, the converter provides real power to AC system, converting energy from DC to AC. Further, (5.11) indicates that the rectifier ($\alpha \in [0°, 90°]$) and inverter

($\alpha \in [90°, 180°]$) consume reactive power from AC system, although the DC system only transports real power.

5.2.3 Converter Basic Equations Considering L_c

In Fig. 5.3, the inductance L_c is the equivalent inductance of the converter transformer, which is nonzero in real systems. L_c prevents the current from undergoing sudden changes. The converter cannot switch its connection with AC source from one phase to another instantly and needs a period of time τ_γ to complete the commutation. τ_γ is called the commutation period. The electrical angle corresponding to commutation period, $\gamma = \omega\tau_\gamma$, is named the commutation angle. During commutation, the current of the valve turning on will increase from zero to I_d, and the current of the valve turning off will decrease from I_d to zero. Under normal operation, commutation angle is less than 60°. The typical value of the commutation angle is around 15–25° during full load. For $\gamma \in [0°, 60°]$, there are three conducting valves in the converter during commutation. Among them, one is in the noncommutation state with valve current of I_d; the second is to be conducting with valve current migrating from zero to I_d; the third one is to stop its conduction with valve current migrating from I_d to zero. There is one valve conducting in both the upper bridge and the lower bridge in between commutation. Figure 5.6a,b shows the valve conducting states under two conditions when fire angle delay a is zero:

1. The commutation angle r is zero
2. The commutation angle γ is nonzero

The electric angle difference between the start of two consecutive commutations is 60°; the noncommutation period is $60°-\gamma$. If γ is greater than 60°, the noncommutation period becomes negative. In other words, a new commutation begins before the previous one completes. This is the case where there are more than three valves conducting at the same time and is an abnormal operation. The average

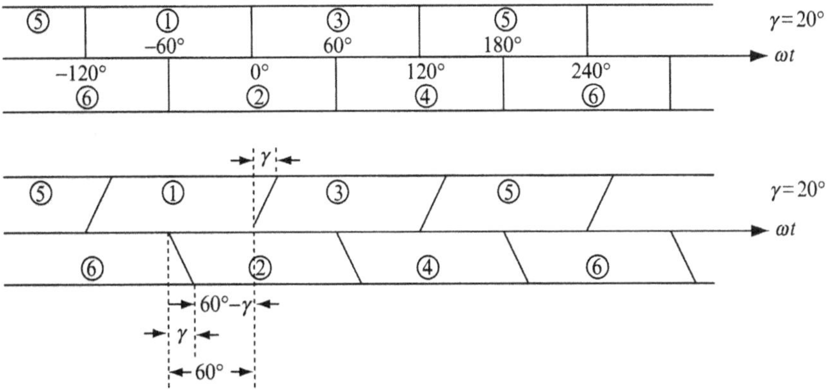

Fig. 5.6 Valve conducting state (**a**) $\alpha = 0, \gamma = 0$ (**b**) $\alpha = 0, \gamma \neq 0$

DC voltage will decrease as the DC current increases due to the existence of the commutation angle.

Now we are going to analyze the factors affecting the magnitude of the commutation angle and the impact of the commutation angle on DC voltage.

Because of commutation angles, the ranges of delayed firing angle adjustment decrease to $[0°, 180° - \gamma]$ as seen from waveforms of the AC source. We will explain the reason later.

Figure 5.7 shows the valve conducting states when the delayed firing angle $\alpha \in [0°, 180° - \gamma]$ and the commutation angle $\gamma \in [0°, 60°]$. Take valve 3 as an example. At $\omega t = 0° + \alpha$, valve 1 starts the commutation to valve 3. At $\omega t = 0° + \alpha + \gamma = 0° + \delta$, the commutation completes. δ is the sum of the delayed firing angle and the commutation angle, called extinction angle. Note that at the beginning of the commutation ($\omega t = \alpha$), the current of valve 1 i_1 is I_d and i_3 of valve 3 is zero. At the end of the commutation ($\omega t = \alpha + \gamma = \delta$), i_1 is zero and i_3 is I_d. In Fig. 5.7, we can see that valves 1, 2, and 3 are conducting during the commutation, $\omega t \in [\alpha, \delta]$. The equivalent circuit of the converter is shown in Fig. 5.8.

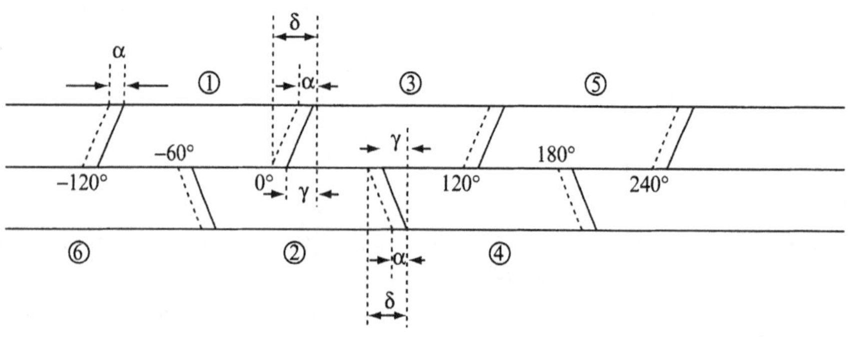

Fig. 5.7 Conducting states when α $[0°, 180° - \gamma]$ and $\gamma[0°, 60°]$

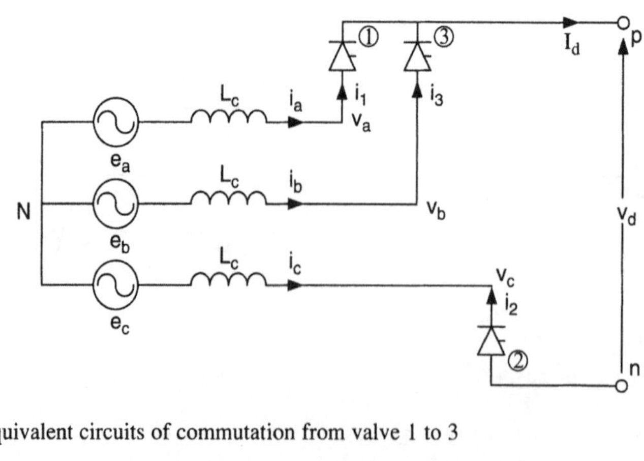

Fig. 5.8 Equivalent circuits of commutation from valve 1 to 3

The loop voltage equation in the one made of valves 1 and 3 is

$$e_b - e_a = L_c \frac{di_3}{dt} - L_c \frac{di_1}{dt}.$$

We call voltage $e_b - e_a$ the commutation voltage and current i_3 the commutation current. Since $I_d = i_1 + i_3$, the above equation can be rewritten by combining with (5.5).

$$\sqrt{3}E_m \sin \omega t = 2L_c \frac{di_3}{dt}. \tag{5.12}$$

Based on boundary conditions, the solution for current is

$$\begin{aligned} \int_0^{i_3} di_3 &= \int_{\alpha/\omega}^{t} \frac{\sqrt{3}E_m}{2L_c} \sin \omega t \, dt, \\ i_3 &= \frac{\sqrt{3}E_m}{2\omega L_c} \cos \omega t \Big|_t^{\alpha/\omega} = I_{s2}(\cos\alpha - \cos\omega t), \end{aligned} \tag{5.13}$$

where

$$I_{s2} = \frac{\sqrt{3}E_m}{2\omega L_c}. \tag{5.14}$$

The commutation current has two components based on (5.13). One of them is a constant component and the other is sinusoidal. The constant component is related to the firing angle delay α. The phase angle of the sinusoidal component lags the commutation voltage e_{ab} by 90°. It is not difficult to understand this case. As shown Fig. 5.8, valves 1 and 3 are conducting during the commutation. This corresponds to a two-phase short circuit between phase a and b through two L_c. The commutation current i_3 is actually the short circuit current of AC source e_b. The constant component is the free component of the short circuit current, resulting from the fact that current cannot suddenly change in an inductive circuit. The sinusoidal component is the forced component of the short circuit current. Since the short circuit loop is purely inductive, the phase angle of the current lags its source voltage by 90°. i_{s2} is the peak of the forced component of the short circuit current. Hence the steady-state converter operation is that two-phases are short circuits in commutation and one-phase is an open line during the noncommutation period.

The commutation completes with $i_3 = I_d$ when $\omega t = \alpha + \gamma = \delta$. The magnitude of the commutation angle reflects the time for the commutation current to rise from zero to I_d. From (5.13), we have

$$I_d = \frac{\sqrt{3}E_m}{2\omega L_c}[\cos\alpha - \cos(\alpha + \gamma)]. \tag{5.15}$$

The above equation indicates that the commutation angle γ is related to operation parameters I_d, E_m, α, and network parameter L_c. The greater the I_d, the greater the commutation angle will be; the greater E_m, the less the commutation angle. When

$\alpha = 0°$ or approaching 180°, the commutation angle reaches its maximum value with regard to α. At $\alpha = 90°$, the commutation angle obtains its minimum value. Besides, the greater L_c, the greater the commutation angle is. As L_c approaches zero, so does the commutation angle, giving rise to the case of neglecting L_c as we discuss before. It is necessary to point out that the sum of the currents of valves 1 and 3 is I_d during commutation, so the magnitude of the commutation angle has no direct impact on DC current I_d. The relationship between AC fundamental frequency current and DC current, as stated in (5.7), is still valid.

We will examine the impact of the commutation angle on DC voltage below.

During the commutation, as shown in Fig. 5.8

$$v_p = v_a = v_b = e_b - L_c \frac{di_3}{dt}.$$

As indicated in (5.12)

$$L_c \frac{di_3}{dt} = \frac{\sqrt{3}E_m \sin \omega t}{2} = \frac{e_b - e_a}{2}.$$

Thus

$$v_p = v_a = v_b = e_b - \frac{e_b - e_a}{2} = \frac{e_a + e_b}{2}.$$

Note that if neglecting the commutation angle, the voltage of converter negative pole v_p is equal to e_b once the valve is on. While considering the commutation angle, v_p equals $(e_a + e_b)/2$ during commutation. v_p will equal to e_b after the commutation completes. Figure 5.9 shows the voltage waveform during commutation from valve 1 to 3. Note the three areas below when $\omega t \in [0°, 60°]$.

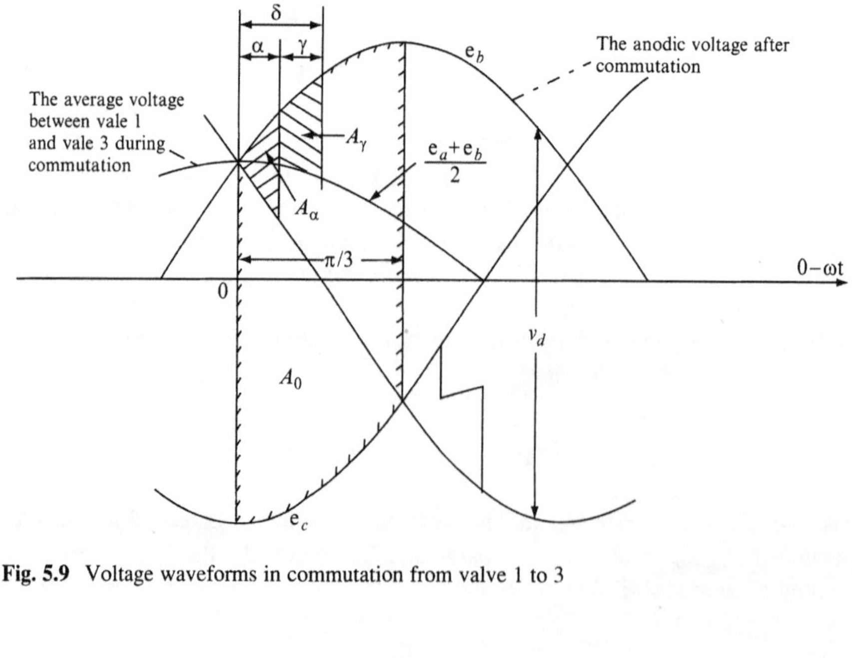

Fig. 5.9 Voltage waveforms in commutation from valve 1 to 3

$$A_0 = \int_{0^\circ}^{60^\circ} (e_b - e_c)\mathrm{d}\omega t = \int_{0^\circ}^{60^\circ} \sqrt{3}E_\mathrm{m}\cos(\omega t - 30^\circ)\mathrm{d}\omega t = \sqrt{3}E_\mathrm{m}, \tag{5.16}$$

$$A_\alpha = \int_{0^\circ}^{\alpha} (e_b - e_a)\mathrm{d}\omega t = \int_{0^\circ}^{\alpha} \sqrt{3}E_\mathrm{m}\cos(\omega t - 90^\circ)\mathrm{d}\omega t = \sqrt{3}E_\mathrm{m}(1 - \cos\alpha), \tag{5.17}$$

$$\begin{aligned} A_\gamma &= \int_{\alpha}^{\delta} \left(e_b - \frac{e_a + e_b}{2}\right)\mathrm{d}\omega t = \frac{1}{2}\int_{\alpha}^{\delta} (e_b - e_a)\mathrm{d}\omega t \\ &= \frac{\sqrt{3}}{2}E_\mathrm{m}(\cos\alpha - \cos\delta). \end{aligned} \tag{5.18}$$

Based on (5.16), the average DC voltage having no firing delay and a zero commutation angle is

$$V_\mathrm{d0} = \frac{A_0}{(\pi/3)} = \frac{3\sqrt{3}}{\pi}E_\mathrm{m}.$$

This is consistent with (5.5). Based on (5.16) and (5.17) with firing delay and zero commutation angle, the average DC voltage is

$$V_\mathrm{d} = \frac{A_0 - A_\alpha}{(\pi/3)} = \frac{\sqrt{3}E_\mathrm{m}\cos\alpha}{(\pi/3)} = V_\mathrm{d0}\cos\alpha.$$

This is consistent with (5.6). As shown in Fig. 5.9, nonzero commutation angle reduces the average DC voltage. In a full cycle, there is one commutation every 60°. From (5.18), we can find the voltage drop due to the commutation angle as

$$\Delta V_\mathrm{d} = \frac{6A_\gamma}{2\pi} = \frac{V_\mathrm{d0}}{2}(\cos\alpha - \cos\delta). \tag{5.19}$$

In the above equation, use (5.15) to substitute $\cos\alpha - \cos\delta$

$$\Delta V_\mathrm{d} = \frac{3}{\pi}\omega L_\mathrm{c} I_\mathrm{d} = R_\gamma I_\mathrm{d}, \tag{5.20}$$

where

$$R_\gamma = \frac{3}{\pi}\omega L_\mathrm{c} = \frac{3}{\pi}X_\mathrm{c}. \tag{5.21}$$

R_γ is the equivalent resistance. Please note that R_γ does not represent a real resistor. It does not consume real power. Its magnitude represents the ratio with which the average DC voltage decreases as DC current increases. R_γ is a network parameter and does not change with the operation modes.

The average DC voltage, taking into consideration both firing angle delay and commutation angle, is

$$V_{\mathrm{d}} = \frac{A_0 - A_\alpha - A_\gamma}{(\pi/3)} = V_{\mathrm{d}0}\cos\alpha - \Delta V_{\mathrm{d}} = V_{\mathrm{d}0}\cos\alpha - R_\gamma I_{\mathrm{d}}. \tag{5.22}$$

The above equation explains that the DC voltage of the converter is a function of firing angle delay α, DC current I_{d} and AC source voltage E_{m}. In HVDC operation, DC voltage can be controlled by adjusting the firing angle delay and the voltage of AC system. From (5.1) we know that the DC current I_{d} is determined by the DC voltages generated by the converters at the two terminals. The introduction of R_γ makes the commutation angle γ disappear from (5.22). The effect of commutation is reflected in the product of commutation resistor and DC current. Please note that a valid equation (5.22) requires $\alpha \in [0°, 180° - \gamma]$ and $\gamma \in [0°, 60°]$. A larger DC current can force the commutation angle beyond 60° and the converter into abnormal operation.

As discussed previously, neglecting commutation angle, the converter is a rectifier when $\alpha \in [0°, 90°]$ and an inverter when $\alpha \in [90°, 180°]$. Considering the commutation angle, combining (5.19) with (5.22) yields

$$V_{\mathrm{d}} = V_{\mathrm{d}0}\cos\alpha - \Delta V_{\mathrm{d}} = \frac{V_{\mathrm{d}0}}{2}(\cos\alpha + \cos\delta). \tag{5.23}$$

The way to identify a converter as a rectifier or inverter is whether DC voltage V_{d} is positive or negative. We use α_t to represent α when V_{d} is zero. From the above equation

$$V_{\mathrm{d}} = \frac{V_{\mathrm{d}0}}{2}[\cos\alpha_t + \cos(\alpha_t + \gamma)] = 0.$$

Obtaining

$$\alpha_t = \frac{\pi - \gamma}{2}. \tag{5.24}$$

When considering the commutation angle, the angle separating rectification from inversion is reduced by $\gamma/2$ from 90°.

As mentioned earlier, the normal operating range of the firing angle delay decreases when the commutation angle is considered. Here we identify the cause of the change by analyzing the commutation from valve 1 to 3. Note that valve 1 is conducting before valve 3 is triggered to conduct, so the cathode voltage of valve 3 is v_a. The conduction condition of valve 3 is that v_b is greater than v_a. From Fig. 5.9 we can see that $v_b > v_a$ when $\omega t \in [0°, 180°]$. Due to the existence of commutation angle, valve 1 cannot be immediately turned off when valve 3 is triggered to conduct. The turn-off of valve 1 is delayed to $\omega t = \delta = \alpha + \gamma$. To guarantee a successful commutation and turn-off of valve 1, the extinction angle δ must be less than 180°. Otherwise, v_b is less than v_a, which makes the valve voltage of valve 3 negative, turning off valve 3, and keeping valve 1 on. This is a commutation failure. Hence $0° \leq \alpha \leq 180° - \gamma$.

5.2.4 Converter Equivalent Circuits

In the above analysis, there are three angles: the delayed firing angle α, the commutation angle γ, and the extinction angle δ. In engineering analysis, α and δ are dedicated to represent rectifier operation while the leading firing angle β and the leading extinction angle μ for inverter operation. The commutation angle γ is used for both rectifier and inverter. There are the following relationships

$$\left.\begin{aligned} \beta &= \pi - \alpha \\ \mu &= \pi - \delta \\ \gamma &= \delta - \alpha = \beta - \mu \end{aligned}\right\}. \tag{5.25}$$

When the converter is an inverter, α is between 90° and 180°. So β and μ are between 0° and 90°. In this way, the leading firing angle and the leading extinction angle of inverters have similar values as the firing angle delay of rectifiers. We can apply the previous equations directly to rectifiers. For inverters, applying (5.25) in (5.22)

$$V_{\mathrm{d}} = V_{\mathrm{d0}} \cos(\pi - \beta) - R_{\gamma} I_{\mathrm{d}} = -(V_{\mathrm{d0}} \cos\beta + R_{\gamma} I_{\mathrm{d}}). \tag{5.26}$$

On the basis of Fig. 5.1, we denote V_{d1} and V_{d2} the voltages of the rectifier and the inverter. Note that the voltage reference direction of rectifier and inverter are opposite to each other. We have

$$V_{\mathrm{d1}} = V_{\mathrm{d0}} \cos\alpha - R_{\gamma} I_{\mathrm{d}}, \tag{5.27}$$

$$V_{\mathrm{d2}} = V_{\mathrm{d0}} \cos\beta + R_{\gamma} I_{\mathrm{d}}. \tag{5.28}$$

From the above, we can derive the equivalent circuits for converters in rectification and inversion modes as shown in Fig. 5.10a, b. Both the voltage and current in the diagram are average values. No matter whether the converter is in rectification mode or inversion mode, DC current always flows from valve anode to cathode so the commutation resistor carries a negative sign in Fig. 5.10b. We know from (5.5) that V_{d0} is related to AC system voltage. DC current is determined by (5.1). The control variables in DC system are AC system voltage and converter firing angle. In

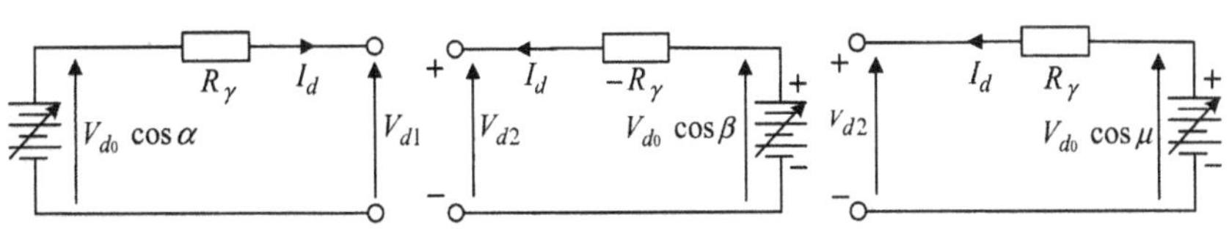

a Rectification

b Inversion Using Leading Firing Angle β

c Inversion Using Leading Extinction Angle μ

Fig. 5.10 Converter equivalent circuits

(5.28) the inverter control variable is the leading firing angle. The voltage representation of rectifier and inverter is different. When using the leading extinction angle for inverter, their representations have the same appearance. Substituting (5.25) into (5.23) and the extinction angle δ with leading extinction angle μ, we obtain

$$V_{\mathrm{d}} = \frac{V_{\mathrm{d0}}}{2}(\cos\alpha - \cos\mu).$$

Use the above equation in (5.23) to eliminate α

$$-V_{\mathrm{d}} = V_{\mathrm{d0}}\cos\mu - R_{\gamma}I_{\mathrm{d}}.$$

Since the voltage reference direction of an inverter is opposite to a converter, we have

$$V_{\mathrm{d2}} = V_{\mathrm{d0}}\cos\mu - R_{\gamma}I_{\mathrm{d}}. \tag{5.29}$$

The corresponding equivalent circuit is shown in Fig. 5.10c.

Figure 5.11 shows the voltage waveforms and valve conducting state when the converter operates in inverter mode.

We will identify the relation between AC and DC quantities when considering the commutation angle below.

After considering the commutation angle, the AC current is no longer rectangular. Figure 5.12 gives the waveform of phase b current. The waveforms of the other two phases can be drawn similarly. The representation of its positive increasing side is (5.13). Its positive decreasing side can be represented by the current of valve 3 during commutation from valve 3 to 5. From (5.13)

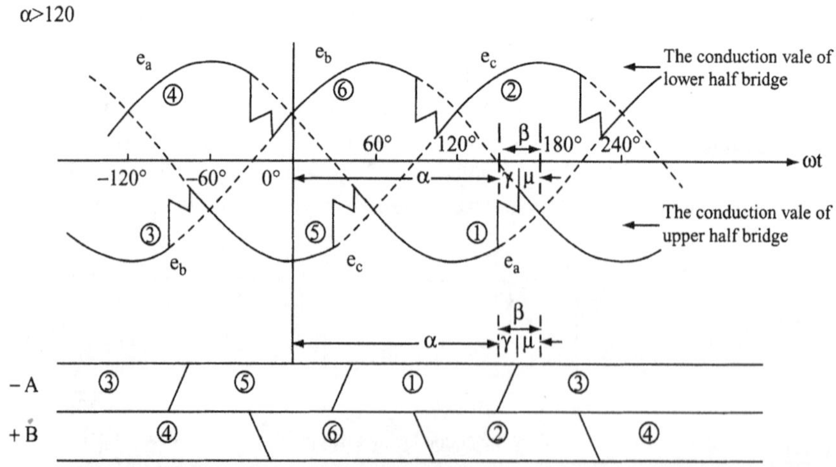

Fig. 5.11 Voltage waveforms and valve conducting state when converter operates in inverter mode

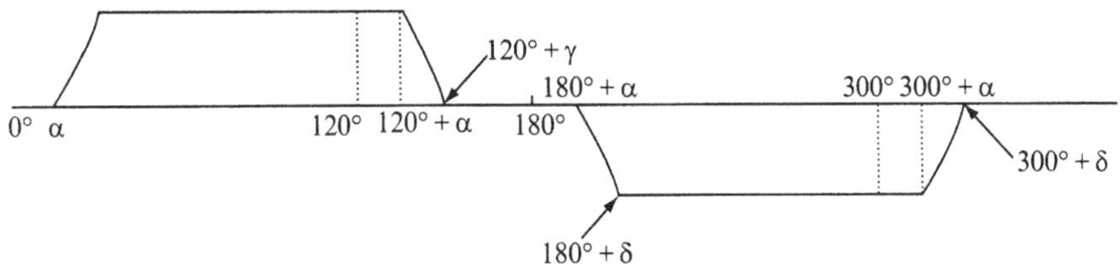

Fig. 5.12 Phase *b* current with commutation angle

$$
\begin{aligned}
i_5 &= I_{s2}[\cos\alpha - \cos(\omega t - 120°)] &\quad \omega t \in [120° + \alpha,\ 120° + \delta] \\
i_3 &= I_d - i_5 = I_d - I_{s2}[\cos\alpha - \cos(\omega t - 120°)] &\quad \omega t \in [120° + \alpha,\ 120° + \delta]
\end{aligned}.
$$

Fourier analysis gives rise to AC fundamental frequency current when considering the commutation angle

$$I = k(\alpha, \gamma)\frac{\sqrt{6}}{\pi}I_d, \tag{5.30}$$

where

$$k(\alpha,\gamma) = \frac{1}{2}[\cos\alpha + \cos(\alpha+\gamma)] \times \sqrt{1 + [\gamma \csc\gamma \csc(2\alpha+\gamma) - \mathrm{ctg}(2\alpha+\gamma)]^2}. \tag{5.31}$$

Under normal operation, the values of α and γ make $k(\alpha,\gamma)$ approach 1. For simplicity of analysis, we take $k(\alpha,\gamma)$ approximately as constant, $k_\gamma = 0.995$. The relationship between AC fundamental frequency current and DC current is

$$I = k_\gamma \frac{\sqrt{6}}{\pi}I_d. \tag{5.32}$$

From (5.23) and (5.5), we obtain the relationship between DC voltage and AC voltage

$$V_d = \frac{3\sqrt{6}}{\pi}\frac{\cos\alpha + \cos\delta}{2}E. \tag{5.33}$$

Using the same logic as (5.8), AC real power equals DC power. From (5.32) and (5.33), we have

$$3\left(k_\gamma \frac{\sqrt{6}}{\pi}I_d\right)E\cos\varphi = \left(\frac{3\sqrt{6}}{\pi}\frac{\cos\alpha + \cos\delta}{2}E\right)I_d.$$

Hence

$$k_\gamma \cos\varphi = \frac{\cos\alpha + \cos\delta}{2}. \tag{5.34}$$

Substituting the above into (5.33), we obtain the relationship between DC voltage and AC voltage when considering commutation angle

$$V_\mathrm{d} = k_\gamma \frac{3\sqrt{6}}{\pi} E \cos\varphi. \tag{5.35}$$

In the above we deduce the operation conditions of a single bridge and its basic (5.27), (5.29), (5.32), and (5.35).

5.2.5 Multiple Bridge Operation

Real HVDC systems usually apply multiple bridges to achieve higher DC voltage. Multiple bridge converters generally have an even number of bridges connected in series on the DC side while connected in parallel on the AC side. Figure 5.13 shows the connection of a double-bridge converter. The dotted lines are drawn for ease of understanding. Since the currents on the two dotted lines are equal and have opposite direction, they do not exist in a real system. With the dotted lines, the double-bridge converter can be treated as two independent single bridge converters connected in parallel. Below we identify the characteristics of a double-bridge converter. In Fig. 5.13, the connection of two converter transformers is different, one Y/Y; the other Y/Δ. This connection makes the phase-angles of AC voltages on the two bridges have a difference of 30°. We use v_du and v_dl to represent the pulsed

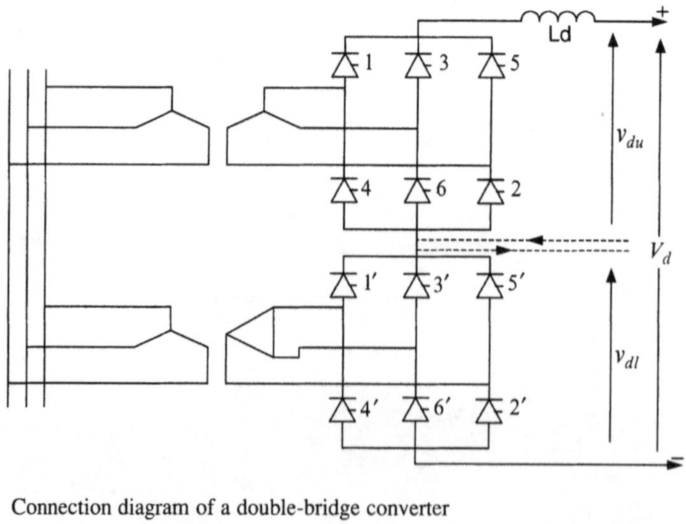

Fig. 5.13 Connection diagram of a double-bridge converter

DC voltages of the upper and lower bridge, respectively. The pulsed DC voltage v_d of the double-bridge is the sum of v_{du} and v_{dl}. Based on the previous analysis, the waveform of v_{du} is as shown in Fig. 5.7a. Note that there is a 30° phase-angle difference between the upper bridge and lower bridge. Thus, the waveform of v_{dl} is that of v_{du} shifted to the right by 30°. The waveform of v_d is the sum of v_{du} and v_{dl}. Apparently, the sum of two 6-pulse waveforms with 30° phase-angle difference produces a 12-pulse waveform and results in smaller magnitudes of pulses. It is easy to understand that the DC voltage of a double-bridge converter is the sum of its two single bridge converters. The DC voltage fluctuation is reduced in comparison with a single-bridge converter. A double-bridge converter is also called 12-pulse converter.

Now we show the AC current analysis. The AC current of the upper bridge is shown in Fig. 5.4c. The current waveform on the secondary side of the lower bridge converter transformer is the current of the upper bridge shifted to the right by 30°. Since the transformer has Y/Δ connection, its phase currents are the combination of line currents

$$\left.\begin{aligned} i_{ap} &= (2i_{bl} + i_{cl})/3 \\ i_{bp} &= (2i_{cl} + i_{al})/3 \\ i_{cp} &= (2i_{al} + i_{bl})/3 \end{aligned}\right\}, \tag{5.36}$$

where subscript p and l indicate phase and line components. The current waveforms on the primary side of the converter transformer have the same shape as the

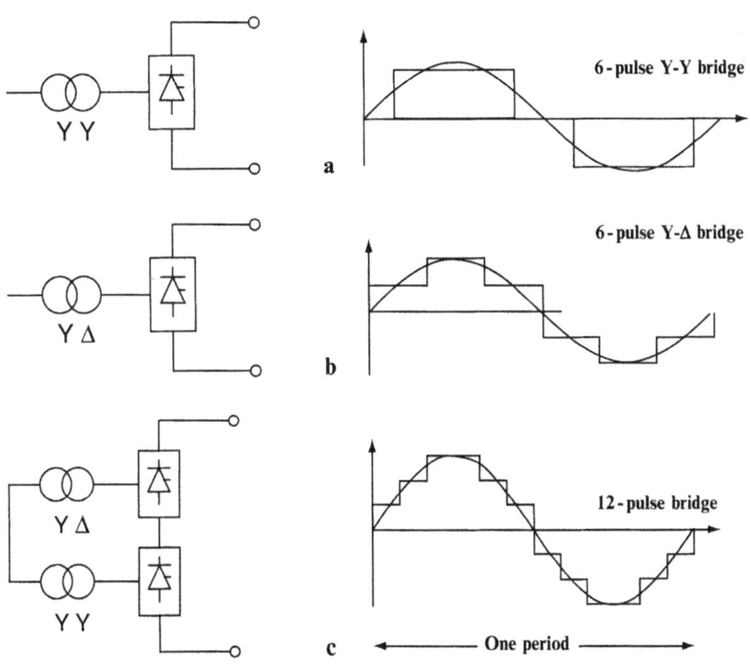

Fig. 5.14 AC current waveforms in multibridge converters

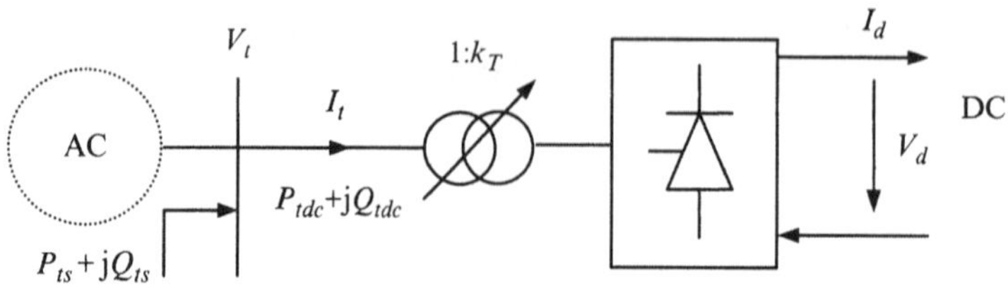

Fig. 5.15 AC/DC interconnected systems and converter

secondary except being shifted to the left by 30°. Taking phase *a* as an example, Fig. 5.14 gives the waveforms of AC currents of multibridge converters. From Fig. 5.14c we can see that the combined current of a double-bridge converter is more like a sinusoidal wave. This will greatly reduce the harmonic current on the AC side, saving on AC filter investment. Using a similar analysis as the double-bridge converter, three or four bridge converters are known as 18 or 24 pulse converters. The number of pulses in their DC voltage is 18 or 24, respectively. The more bridges there are, the less will be the harmonic components and their magnitudes, and the less DC voltage fluctuation. However, the connection of converter transformers and the DC control systems become very complicated when using more than two bridges. Hence the 12-pulse converters are most commonly used in the industry.

Below we are going to derive an equivalent single-bridge converter model from the multibridge converter analysis, as shown in Fig. 5.15. The converter transformer is now assumed to be an ideal transformer with k_{T} as the transformer ratio. The equivalent reactance X_{c} of the converter transformer is represented in the converter basic equations. V_{t} and I_{t} are the fundamental frequency voltage and current on the primary side of the converter transformer. $P_{\mathrm{tdc}} + jQ_{\mathrm{tdc}}$ is the power that the DC system takes from AC system. $P_{\mathrm{ts}} + jQ_{\mathrm{ts}}$ is the power injected into the AC bus. Suppose that the converter has n_{t} bridges. Based on the analysis of a double-bridge converter, the DC voltage output of a multibridge converter is the sum of individual bridge DC voltage outputs. AC fundamental frequency current is the sum of the fundamental frequency currents of individual bridges. It is important to note that the physical variables mentioned are those on the secondary side of the converter transformer. E in (5.5) is the phase voltage on the secondary side of the converter transformer. Thus $E = k_{\mathrm{T}}V_{\mathrm{t}}/\sqrt{3}$. Corresponding to the equations of single-bridge converter 5.27, 5.29, 5.35, and 5.32, the multibridge converter has

$$V_{\mathrm{d}} = n_{\mathrm{t}}\left(V_{\mathrm{d0}}\cos\theta_{\mathrm{d}} - R_{\gamma}I_{\mathrm{d}}\right) = \frac{3\sqrt{2}}{\pi}n_{\mathrm{t}}k_{\mathrm{T}}V_{\mathrm{t}}\cos\theta_{\mathrm{d}} - \frac{3}{\pi}n_{\mathrm{t}}X_{\mathrm{c}}I_{\mathrm{d}}, \tag{5.37}$$

$$V_{\mathrm{d}} = n_{\mathrm{t}}\left(k_{\gamma}\frac{3\sqrt{6}}{\pi}E\cos\varphi\right) = \frac{3\sqrt{2}}{\pi}k_{\gamma}n_{\mathrm{t}}k_{\mathrm{T}}V_{\mathrm{t}}\cos\varphi, \tag{5.38}$$

$$I_t = k_\gamma n_t k_t \frac{\sqrt{6}}{\pi} I_d. \tag{5.39}$$

Note that (5.37) is a common equation that can be used for both rectifier and inverter. θ_d is generalized to represent the control angle of the converter. In fact, it is the firing angle delay α of a rectifier and the leading firing angle μ of an inverter. Besides, note that the reference direction of the inverter DC voltage is opposite to the rectifier. As shown in Fig. 5.15, we will use the reference direction of the rectifier DC voltage when we do not distinguish a rectifier and an inverter. For an inverter, multiplying the value obtained from (5.37) by -1 results in the same as shown in Fig. 5.15.

The above three equations are the general basic equations of the converter and will play an important role in the analysis of AC/DC interconnected transmission systems.

5.2.6 *Converter Control*

Take the two-terminal DC transmission systems in Fig. 5.1. Substituting (5.37) into (5.1) yields

$$I_d = G_\Sigma (k_{T1} V_{t1} \cos \alpha_1 - k_{T2} V_{t2} \cos \mu_2), \tag{5.40}$$

where

$$G_\Sigma = \sqrt{2} \bigg/ \left(X_{c1} + \frac{\pi}{3 n_t} R + X_{c2} \right) \tag{5.41}$$

is a constant parameter with the unit of inductance. From the above equations and (5.2) we know that the adjustment of converter control angles (α_1, μ_2) and the ratios (k_{T1}, k_{T2}) of converter transformers can control the power transported on the DC lines. The AC system voltages V_{t1}, V_{t2} at the two terminals of the DC lines have direct impact on the transmitted power. It is far more convenient to adjust the transported power by changing the converter control angles and transformer ratios than directly controlling AC voltages V_{t1}, V_{t2}. When fast power control is required, the control of V_{t1}, V_{t2} is not normally used. The control of transformer ratios is carried out by adjusting tap changers. Note that the manufactured design requires that tap changers of transformers operate in steps. This makes the ratio a discrete variable. Also, tap changers are mechanical devices, with one step change taking 5–6 s. The adjustment of converter firing angles is carried out by modifying the electrical parameter of the control circuits. The response speed of firing angle control is very high, ranging from 1 to 10 ms. Due to this characteristic of firing angle control, HVDC can adjust the transported power accurately and rapidly, and play an important role in emergency power support for AC systems. The general control process in power system operation is as follows. First, use an automatic control system to adjust firing angles (α_1, μ_2) to make power systems operate in a proper state. Then, set the tap changers of converter transformers to put the converter firing angles into appropriate ranges. Finally, make the system run in an ideal condition through AC system optimization.

We need to pay attention to the following issues regarding the operation control of the DC system in steady state.

First, since G_Σ in (5.41) is very large, a small variation of AC voltage V_{t1}, V_{t2} in system operation can give rise to a huge change in DC current. It is necessary that fast adjustment of converter firing angles is applied, to follow the variation of AC voltages and to prevent large fluctuations of DC current, in order to achieve normal operation of the DC system.

Next, the converter steady-state operation adjustment should maintain DC voltage around its rated value. Although converter bridges can withstand high over-voltage; DC voltage in operation should not be higher than its rated value for a long period to safeguard the equipment of the whole DC system. On the other hand, DC voltage should not be much lower than its rated value. Given a certain amount of transmitted power, lower voltage requires higher current. As shown in (5.39), AC fundamental frequency current is proportional to DC current. A higher DC current increases the loss on DC lines as well as AC system. From (5.15) and (5.24), a higher DC current also enlarges the commutation angle and reduces the range of firing angle adjustment.

Also, the converter steady-state operation should maintain a power factor as high as possible. The benefit is obvious. First, it reduces the reactive power compensation requirements of the AC system. Next, it increases the capacity of real power that can be processed by converters and converter transformers as well as the transmission capacity. It also reduces power losses. To operate with a higher power factor, a rectifier should have a small firing angle delay and an inverter should have small leading extinction angle, as seen in (5.34). However, in real operation, there is a minimum constraint for rectifier firing angle delay α, $\alpha > \alpha_{\min}$. This is to make sure that valves have adequate positive voltage before firing to provide sufficient energy for the firing generation circuits. For 50 Hz systems, $\alpha_{\min}$ is around 5°. Normal system operation uses 15°–20° to leave a certain margin. Similarly, inverters need to have adequate time to complete commutation under positive valve voltage after firing. As shown in Fig. 5.11, the leading extinction angle μ of zero is a theoretical critical value. In fact, the commutation angle γ varies with AC voltage and DC current (refer to (5.15)). There must be some margin left for leading extinction angles. Thus, there is a minimum constraint for μ, normally $\mu > \mu_{\min}$. For 50 Hz systems, $\mu_{\min}$ is 15°.

Finally, AC/DC interconnected system operation requires designating control schemes for every converter based on operation requirements. The most common control scheme is: adjust the rectifier firing angle to maintain a constant DC current, called fixed current control; adjust the inverter firing angle to maintain a constant leading extinction angle, called fixed extinction angle control. Power flow calculation usually includes the following control schemes:

1. Fixed current control

$$I_d - I_{ds} = 0 \tag{5.42}$$

2. Fixed voltage control

$$V_d - V_{ds} = 0 \tag{5.43}$$

3. Fixed power control

$$V_d I_d - P_{ds} = 0 \tag{5.44}$$

4. Fixed angle control

$$\cos\theta_d - \cos\theta_{ds} = 0 \tag{5.45}$$

5. Fixed ratio control

$$k_T - k_{Ts} = 0, \tag{5.46}$$

where the variables with subscript are predefined constants.

In the above, we introduce the basic concept of HVDC by deriving its basic equations (5.37)–(5.39).

5.3 Power Flow Calculation of AC/DC Interconnected Systems

When there are HVDC systems, the nonlinear algebra equations representing the whole system will have variables related to the DC system and the additional equations for the DC part. Power flow calculation cannot be directly carried out as stated in Chap. 2. Since the methods introduced in Chap. 2 are mature and are widely used in the industry, the most common power flow calculations of AC/DC interconnected systems are based on those approaches. There are mainly two ways: integrated iteration and alternating iteration.

The integrated iteration is based on the Newton algorithm in the form of polar coordinates. It is an integrated iteration process to solve for AC bus voltages, phase angles together with DC voltages, currents, converter transformer ratios, converter power factors, and control angles. This method has good convergence, and can achieve reliable solutions for different network topologies, parameters, and various DC control schemes. This method is also called the unified method.

The alternating iteration is a simplified integrated method. During iterations, it solves AC equations and DC equations separately. When solving AC, use the known real power and reactive power to represent the DC system. While solving DC, model the AC system as a constant voltage on the converter AC bus. Below we determine the per unit normalization for converters, build the mathematical model for power flow calculations of AC/DC interconnected systems, and introduce the detail process of integrated and alternating iterations.

5.3.1 Converter Basic Equations in the per Unit System

In power flow calculations, the DC system should use the per unit normalization as for the AC system. We will transfer the converter basic equations (5.37)–(5.39) into the per unit system so that they can be used with AC per unit equations. There are many ways to define the base values of various DC quantities, giving rise to different appearances of per unit basic equations. The base variables in the DC system are labeled with the dB subscript. The base variables should satisfy the same kind of relationship as the actual variable

$$\left.\begin{aligned} V_{\mathrm{dB}} &= R_{\mathrm{dB}} I_{\mathrm{dB}} \\ P_{\mathrm{dB}} &= V_{\mathrm{dB}} I_{\mathrm{dB}} \end{aligned}\right\}. \tag{5.47}$$

We can set two of the four base variables in the above so the other two are derived from the above equation. The base variables of AC quantities in the primary sides of converter transformers use the B subscript. Taking into consideration the coordination with AC systems, we define

$$P_{\mathrm{dB}} = S_{\mathrm{B}} = \sqrt{3} V_{\mathrm{B}} I_{\mathrm{B}}. \tag{5.48}$$

To make the converter basic equations have a concise form, we make

$$V_{\mathrm{dB}} = \frac{3\sqrt{2}}{\pi} n_{\mathrm{t}} k_{\mathrm{TB}} V_{\mathrm{B}}, \tag{5.49}$$

where k_{TB} is the converter transformer base ratio, that is, rated ratio.

We can derive the base values of DC current and resistor from (5.47).

$$I_{\mathrm{dB}} = \frac{P_{\mathrm{dB}}}{V_{\mathrm{dB}}} = \frac{\pi}{\sqrt{6} n_{\mathrm{t}} k_{\mathrm{TB}}} I_{\mathrm{B}} \tag{5.50}$$

$$\left.\begin{aligned} R_{\mathrm{dB}} &= \frac{V_{\mathrm{dB}}}{I_{\mathrm{dB}}} = \frac{3}{\pi} n_{\mathrm{t}} X_{\mathrm{cB}} \\ X_{\mathrm{cB}} &= \frac{6}{\pi} n_{\mathrm{t}} k_{\mathrm{TB}}^2 Z_{\mathrm{B}} \end{aligned}\right\}. \tag{5.51}$$

Equations (5.48)–(5.51) are the per unit system for converters. Dividing both sides of (5.37)–(5.39) by the corresponding base values and using "*" subscript to represent the per unit variables, we have

$$V_{d^*} = k_{T^*} V_{t^*} \cos\theta_d - X_{c^*} I_{d^*}, \tag{5.52}$$

$$V_{d^*} = k_\gamma k_{T^*} V_{t^*} \cos\varphi, \tag{5.53}$$

$$I_{t^*} = k_\gamma k_{T^*} I_{d^*}. \tag{5.54}$$

We introduce a constant X_{c^*} to simplify (5.52). Note that $X_{c^*} = X_c/X_{cB} \neq X_c/R_{dB}$. Variable θ_d and φ are the angles so they do not have base values or per unit values. The above three equations form the per unit converter basic equations. Below we are going to use per unit values in our analysis. For simplicity, we will ignore the per unit subscript "*".

5.3.2 Power Flow Equations

We identify a bus as a DC bus or a AC bus based on whether it is connected with a converter transformer or not. The bus connecting to the primary side of a converter transformer is called a DC bus. The bus having voltage V_t in Fig. 5.15 is a DC bus. Apparently, a bus without connection to a converter transformer is a pure AC bus. Assume that the total bus number in a system is n, and the number of converters is n_c. The number of DC buses is the same as the number of converters. The number of pure AC buses is $n_a = n - n_c$. For the simplicity of description, the numbering sequence of system buses is: the first n_a are AC buses and the next n_c are DC buses.

The basic principle of building power flow models for AC/DC interconnected systems is as follows. First use the extraction and injection power, $P_{tdc} + jQ_{tdc}$ of converter transformers at DC buses to represent converter transformers and the DC system behind them (refer to Fig. 5.15). Thus, the system topology does not have any converter transformers and associated DC systems, and becomes an AC network. Then we build the system bus admittance matrix using the methods introduced in Chap. 1. We can now establish the bus power equations as stated in Chap. 2. Note that the system may become several separate systems due to the removal of DC systems. As shown in Fig. 5.1, suppose that there is no other connection between system 1 and 2 other than the DC connection. The removal of DC system generates two separated AC system networks. Note that this does not mean that the two systems are de-coupled with each other. Their coupling is represented by the DC power at their DC buses. They are combined into one system when forming the bus admittance matrix. The above process applies to the case of two systems having different frequencies. As for power flow solution, frequency affects only network parameters and will not appear in bus power equations. Here the reference direction of rectifier DC voltages is from anode to cathode, and the opposite is true for inverters. The reference direction of DC current is always from valve positive pole to its negative pole.

1. Node power equations

Power equations are the same as (2.13) for AC buses

$$\left.\begin{aligned} \Delta P_i &= P_{is} - V_i \sum_{j\in i} V_j(G_{ij}\cos\theta_{ij} + B_{ij}\sin\theta_{ij}) = 0 \\ \Delta Q_i &= Q_{is} - V_i \sum_{j\in i} V_j(G_{ij}\sin\theta_{ij} - B_{ij}\cos\theta_{ij}) = 0 \end{aligned}\right\} \quad i = 1,2,\ldots,n_{\mathrm{a}}. \tag{5.55}$$

Note that j in the above equation can be an AC bus as well as a DC bus.
For a DC bus, suppose that converter transformer numbered k connects to bus i.
The complex power extracted from the bus is

$$P_{i\mathrm{dc}} + jQ_{i\mathrm{dc}} = V_i I_i(\cos\varphi_k + j\sin\varphi_k).$$

Substituting (5.54) into the above yields

$$P_{i\mathrm{dc}} + jQ_{i\mathrm{dc}} = k_\gamma k_{\mathrm{T}k} V_i I_{\mathrm{d}k}(\cos\varphi_k + j\sin\varphi_k). \tag{5.56}$$

We have assumed that there are ideal filters on both AC and DC sides so that harmonic power is zero. Neglecting the converter power loss, AC fundamental frequency power equals DC power. We have another expression of the extracted power

$$\left.\begin{aligned} P_{i\mathrm{dc}} &= V_{\mathrm{d}k} I_{\mathrm{d}k} \\ Q_{i\mathrm{dc}} &= V_{\mathrm{d}k} I_{\mathrm{d}k}\ \mathrm{tg}\ \varphi_k \end{aligned}\right\}. \tag{5.57}$$

The above two expressions are equivalent. AC bus voltage does not appear in (5.57). For easy programming, we will use (5.57). The difference between the DC bus power equation and (5.55) is an additional item of DC power

$$\left.\begin{aligned} \Delta P_i &= P_{is} - V_i \sum_{j\in i} V_j(G_{ij}\cos\theta_{ij} + B_{ij}\sin\theta_{ij}) \pm V_{\mathrm{d}k} I_{\mathrm{d}k} = 0 \\ \Delta Q_i &= Q_{is} - V_i \sum_{j\in i} V_j(G_{ij}\sin\theta_{ij} - B_{ij}\cos\theta_{ij}) \pm V_{\mathrm{d}k} I_{\mathrm{d}k}\ \mathrm{tg}\ \varphi_k = 0 \end{aligned}\right\}$$

$$i = n_{\mathrm{a}} + k, \quad k = 1,2,\ldots,n_{\mathrm{c}}, \tag{5.58}$$

where positive and negative signs correspond to inverters and rectifiers, respectively. Equations (5.55) and (5.58) make up the whole system power equations. Comparing with the conventional power equation (2.13), the difference is that DC bus power (5.58) has new variables $V_{\mathrm{d}k}$, $I_{\mathrm{d}k}$, and φ_k. Thus, the number of unknown variables is more than the number of equations. To make the power flow solvable, we need to add the following new equations.

2. Converter basic equations

Based on (5.52) and (5.58), for converter k we have

$$\Delta d_{1k} = V_{dk} - k_{Tk} V_{n_a+k} \cos\theta_{dk} + X_{ck} I_{dk} = 0 \quad k = 1, 2, \ldots, n_c, \tag{5.59}$$

$$\Delta d_{2k} = V_{dk} - k_\gamma k_{Tk} V_{n_a+k} \cos\varphi_k = 0 \quad k = 1, 2, \ldots, n_c. \tag{5.60}$$

3. DC network equations

DC network equations are the mathematical model of DC lines that describes the relationship between DC voltage and current. For multiterminal DC systems, pay attention to the reference directions of DC voltage and current. Eliminating the intermediate buses, we have

$$\Delta d_{3k} = \pm I_{dk} - \sum_{j=1}^{n_c} g_{dkj} V_{dj} = 0 \quad k = 1, 2, \ldots, n_c, \tag{5.61}$$

where g_{dkj} is the element of bus admittance matrix G_d after eliminating intermediate buses. The DC voltage and current in the above equation are converter output voltage and current, since the intermediate buses are removed. The positive and negative signs correspond to rectifiers and inverters, respectively.
For a simple two-terminal DC system as shown in Fig. 5.1, based on (5.1) and $I_{d1} = I_{d2}$ we have its DC network equations as

$$\begin{bmatrix} I_{d1} \\ -I_{d2} \end{bmatrix} = \begin{bmatrix} 1/R & -1/R \\ -1/R & 1/R \end{bmatrix} \begin{bmatrix} V_{d1} \\ V_{d2} \end{bmatrix}. \tag{5.62}$$

Note that DC line resistor R approaches zero when the electric distance between the two converters in a two-terminal system is very short (e.g., the back-to-back DC system connecting two systems having different frequencies). Neglecting this resistor, (5.62) becomes

$$\left.\begin{aligned} V_{d1} &= V_{d2} \\ I_{d1} &= I_{d2} \end{aligned}\right\}.$$

For easy programming, we will have a small enough R and still use (5.62) as the DC network equation.

4. Control equations

We introduce two new variables in the above three supplemental (5.59)–(5.61): converter transformer ratio k_{Tk} and converter control angle θ_{dk}. Based on the given control scheme, (5.42)–(5.46) can provide the values of the two variables to make the number of variables equal to the number of equations. To make the programs generally applicable, people usually add the two supplemental control equations. Note that the two converter variables must be independent. For example, the rectifier in a two-terminal DC system as shown in Fig. 5.1 uses fix current and fix voltage control scheme. As indicated in (5.62), the inverter current and voltage are determined by the rectifier control. Hence we should select other variables for the inverter. Generally, we use fixed ratio and fixed current control for rectifiers, and fixed ratio and fixed control angle for inverters. We have the common control equations as follows:

$$\Delta d_{4k} = d_{4k}(I_{\mathrm{d}k}, V_{\mathrm{d}k}, \cos\theta_{\mathrm{d}k}, k_{\mathrm{T}k}) = 0 \quad k = 1, 2, \ldots, n_{\mathrm{c}}, \tag{5.63}$$

$$\Delta d_{5k} = d_{5k}(I_{\mathrm{d}k}, V_{\mathrm{d}k}, \cos\theta_{\mathrm{d}k}, k_{\mathrm{T}k}) = 0 \quad k = 1, 2, \ldots, n_{\mathrm{c}}. \tag{5.64}$$

As seen in (5.42)–(5.46), the four variables, $I_{\mathrm{d}k}$, $V_{\mathrm{d}k}$, $k_{\mathrm{T}k}$, and $\cos\theta_{\mathrm{d}k}$, do not all appear in the real control equations; also $\theta_{\mathrm{d}k}$ appears as $\cos\theta_{\mathrm{d}k}$ in the above. To reduce the nonlinearity of equations (5.59), (5.63), and (5.64), we use $\cos\theta_{\mathrm{d}k}$ as the unknown variable instead of $\theta_{\mathrm{d}k}$.

The bus power equations (5.55), (5.58), supplemental equations (5.59)–(5.61), (5.63), and (5.64) make up the entire system equations. Comparing with the conventional power flow equations, AC/DC interconnected systems need to compute bus voltages and phase angles of all n buses, as well as DC voltage, DC current, control angle, converter transformer ratio, and converter power factor. For each converter, there will be five additional unknown variables and five additional equations.

Below we introduce the integrated iteration method, which uses similar basic principles as references [111, 112] while the per-unit system and approximations are different.

5.3.3 Jacobian Matrix of Power Flow Equations

From the mathematical point of view, the power flow equations established above have no difference in nature to the conventional power flow equations (2.13). They are both nonlinear multivariable algebraic equations. Thus, the Newton formula to solve (2.13) in Chap. 2 can be used to solve power flows for AC/DC interconnected systems as an extension of the existing methods. The extended equations include converter equations, DC network equations, and control equations. The extended variables are

$$\mathbf{X} = \begin{bmatrix} \mathbf{V}_{\mathrm{d}}^{\mathrm{T}} & \mathbf{I}_{\mathrm{d}}^{\mathrm{T}} & \mathbf{K}_{\mathrm{T}}^{\mathrm{T}} & \mathbf{W}_{\mathrm{d}}^{\mathrm{T}} & \mathbf{\Phi}^{\mathrm{T}} \end{bmatrix}^{\mathrm{T}},$$

where

$$\mathbf{I}_{\mathrm{d}} = \begin{bmatrix} I_{\mathrm{d}1} & I_{\mathrm{d}2} & \cdots & I_{\mathrm{d}n_{\mathrm{c}}} \end{bmatrix}^{\mathrm{T}},$$
$$\mathbf{K}_{\mathrm{T}} = \begin{bmatrix} k_{\mathrm{T}1} & k_{\mathrm{T}2} & \cdots & k_{\mathrm{T}n_{\mathrm{c}}} \end{bmatrix}^{\mathrm{T}},$$
$$\mathbf{W} = \begin{bmatrix} w_1 & w_2 & \cdots & w_{n_{\mathrm{c}}} \end{bmatrix}^{\mathrm{T}} = \begin{bmatrix} \cos\,\theta_{\mathrm{d}1} & \cos\,\theta_{\mathrm{d}2} & \cdots & \cos\,\theta_{\mathrm{d}n_{\mathrm{c}}} \end{bmatrix}^{\mathrm{T}},$$
$$\mathbf{\Phi} = \begin{bmatrix} \varphi_1 & \varphi_2 & \cdots & \varphi_{n_{\mathrm{c}}} \end{bmatrix}^{\mathrm{T}}.$$

We use notations ΔP_{a} and ΔP_{t} to represent power mismatches of pure AC buses and DC buses, respectively. Correspondingly, bus voltages and phase angles will use the same subscript. The mismatch equations for power flow of AC/DC interconnected systems corresponding to (2.40) are

$$\begin{bmatrix} \Delta\mathbf{P}_{\mathrm{a}} \\ \Delta\mathbf{P}_{\mathrm{t}} \\ \Delta\mathbf{Q}_{\mathrm{a}} \\ \Delta\mathbf{Q}_{\mathrm{t}} \\ \hline \Delta\mathbf{d}_1 \\ \Delta\mathbf{d}_2 \\ \Delta\mathbf{d}_3 \\ \Delta\mathbf{d}_4 \\ \Delta\mathbf{d}_5 \end{bmatrix} = \left[\begin{array}{cccc|ccccc} \mathbf{H}_{\mathrm{aa}} & \mathbf{H}_{\mathrm{at}} & \mathbf{N}_{\mathrm{aa}} & \mathbf{N}_{\mathrm{at}} & 0 & 0 & 0 & 0 & 0 \\ \mathbf{H}_{\mathrm{ta}} & \mathbf{H}_{\mathrm{tt}} & \mathbf{N}_{\mathrm{ta}} & \mathbf{N}_{\mathrm{tt}} & \mathbf{A}_{21} & \mathbf{A}_{22} & 0 & 0 & 0 \\ \mathbf{J}_{\mathrm{aa}} & \mathbf{J}_{\mathrm{at}} & \mathbf{L}_{\mathrm{aa}} & \mathbf{L}_{\mathrm{at}} & 0 & 0 & 0 & 0 & 0 \\ \mathbf{J}_{\mathrm{ta}} & \mathbf{J}_{\mathrm{tt}} & \mathbf{L}_{\mathrm{ta}} & \mathbf{L}_{\mathrm{tt}} & \mathbf{A}_{41} & \mathbf{A}_{42}0 & 0 & 0 & \mathbf{A}_{45} \\ \hline 0 & 0 & 0 & \mathbf{C}_{14} & \mathbf{F}_{11} & \mathbf{F}_{12} & \mathbf{F}_{13} & \mathbf{F}_{14} & 0 \\ 0 & 0 & 0 & \mathbf{C}_{24} & \mathbf{F}_{21} & 0 & \mathbf{F}_{23} & 0 & \mathbf{F}_{25} \\ 0 & 0 & 0 & 0 & \mathbf{F}_{31} & \mathbf{F}_{32} & 0 & 0 & 0 \\ 0 & 0 & 0 & 0 & \mathbf{F}_{41} & \mathbf{F}_{42} & \mathbf{F}_{43} & \mathbf{F}_{44} & \mathbf{F}_{45} \\ 0 & 0 & 0 & 0 & \mathbf{F}_{51} & \mathbf{F}_{52} & \mathbf{F}_{53} & \mathbf{F}_{54} & \mathbf{F}_{55} \end{array}\right] \begin{bmatrix} \Delta\boldsymbol{\theta}_{\mathrm{a}} \\ \Delta\boldsymbol{\theta}_{\mathrm{t}} \\ \Delta\mathbf{V}_{\mathrm{a}}/\mathbf{V}_{\mathrm{a}} \\ \Delta\mathbf{V}_{\mathrm{t}}/\mathbf{V}_{\mathrm{t}} \\ \hline \Delta\mathbf{V}_{\mathrm{d}} \\ \Delta\mathbf{I}_{\mathrm{d}} \\ \Delta\mathbf{K}_{\mathrm{T}} \\ \Delta\mathbf{W} \\ \Delta\mathbf{\Phi} \end{bmatrix}, \quad (5.65)$$

where

$$\Delta\mathbf{P}_{\mathrm{a}} = \begin{bmatrix} \Delta P_1 & \Delta P_2 & \ldots & \Delta P_{n_{\mathrm{a}}} \end{bmatrix}^{\mathrm{T}},$$
$$\Delta\mathbf{P}_{\mathrm{t}} = \begin{bmatrix} \Delta P_{n_{\mathrm{a}}+1} & \Delta P_{n_{\mathrm{a}}+2} & \ldots & \Delta P_{n_{\mathrm{a}}+n_{\mathrm{c}}} \end{bmatrix}^{\mathrm{T}},$$

$(\Delta\mathbf{Q}_{\mathrm{a}}, \Delta\mathbf{Q}_{\mathrm{t}})$, $(\Delta\boldsymbol{\theta}_{\mathrm{a}}, \Delta\boldsymbol{\theta}_{\mathrm{t}})$, and $(\Delta\mathbf{V}_{\mathrm{a}}/\mathbf{V}_{\mathrm{a}}, \Delta\mathbf{V}_{\mathrm{t}}/\mathbf{V}_{\mathrm{t}})$ have the similar expression structure as $(\Delta\mathbf{P}_{\mathrm{a}}, \Delta\mathbf{P}_{\mathrm{t}})$. While

$$\Delta\mathbf{d}_{\mathrm{m}} = \begin{bmatrix} \Delta d_{\mathrm{m}1} & \Delta d_{\mathrm{m}2} & \ldots & \Delta d_{\mathrm{m}n_{\mathrm{c}}} \end{bmatrix}^{\mathrm{T}} \quad m = 1, 2, 3, 4, 5.$$

It is not difficult to find the order of the above sub-matrices. It is clear that the order of the coefficient matrix (power flow Jacobian matrix of AC/DC interconnected systems) expands by $5n_{\mathrm{c}}$ comparing to conventional power flow calculations. We use sub-matrices **A**, **C**, and **F** to represent the extended parts of the coefficient matrix.

Below we are going to identify the actual expression of Jacobian matrix elements.

Since the DC power in bus power (5.58) takes the form of (5.57), the bus voltage magnitudes and phase angles do not appear in the equation. The part of the Jacobian matrix corresponding to conventional power flow is not changed by the introduction of DC transmission systems. The formation method of sub-matrices **H**, **N**, **J**, and **L** in (5.65) is the same as in (2.50). We can directly use expression in (2.41)–(2.49). Designate **E** as unit matrix of n_{c} order. From the established power flow calculation equations of AC/DC interconnected systems and the definition of a Jacobian matrix, it is not difficult to obtain

$$\mathbf{A}_{21} = \frac{\partial \Delta \mathbf{P}_{\mathrm{t}}}{\partial \mathbf{V}_{\mathrm{d}}} = \mathrm{diag}[\pm I_{\mathrm{d}k}], \tag{5.66}$$

$$\mathbf{A}_{22} = \frac{\partial \Delta \mathbf{P}_{\mathrm{t}}}{\partial \mathbf{I}_{\mathrm{d}}} = \mathrm{diag}[\pm V_{\mathrm{d}k}], \tag{5.67}$$

$$\mathbf{A}_{41} = \frac{\partial \Delta \mathbf{Q}_{\mathrm{t}}}{\partial \mathbf{V}_{\mathrm{d}}} = \mathrm{diag}[\pm I_{\mathrm{d}k}\ \mathrm{tg}\,\varphi_k], \tag{5.68}$$

$$\mathbf{A}_{42} = \frac{\partial \Delta \mathbf{Q}_{\mathrm{t}}}{\partial \mathbf{I}_{\mathrm{d}}} = \mathrm{diag}[\pm V_{\mathrm{d}k}\ \mathrm{tg}\,\varphi_k], \tag{5.69}$$

$$\mathbf{A}_{45} = \frac{\partial \Delta \mathbf{Q}_{\mathrm{t}}}{\partial \mathbf{\Phi}} = -\mathrm{diag}\left[V_{\mathrm{d}k} I_{\mathrm{d}k} \sec^2 \varphi_k\right], \tag{5.70}$$

$$\mathbf{F}_{11} = \frac{\partial \Delta \mathbf{d}_1}{\partial \mathbf{V}_{\mathrm{d}}} = \mathbf{E}, \tag{5.71}$$

$$\mathbf{F}_{21} = \frac{\partial \Delta \mathbf{d}_2}{\partial \mathbf{V}_{\mathrm{d}}} = \mathbf{E}, \tag{5.72}$$

$$\mathbf{F}_{31} = \frac{\partial \Delta \mathbf{d}_3}{\partial \mathbf{V}_{\mathrm{d}}} = -\mathbf{G}_{\mathrm{d}}, \tag{5.73}$$

$$\mathbf{F}_{12} = \frac{\partial \Delta \mathbf{d}_1}{\partial \mathbf{I}_{\mathrm{d}}} = \mathrm{diag}[X_{\mathrm{c}k}], \tag{5.74}$$

$$\mathbf{F}_{32} = \frac{\partial \Delta \mathbf{d}_3}{\partial \mathbf{I}_{\mathrm{d}}} = \mathbf{E}, \tag{5.75}$$

$$\mathbf{F}_{13} = \frac{\partial \Delta \mathbf{d}_1}{\partial \mathbf{K}_{\mathrm{T}}} = -\mathrm{diag}[V_{\mathrm{n}_a+k} w_k], \tag{5.76}$$

$$\mathbf{F}_{23} = \frac{\partial \Delta \mathbf{d}_2}{\partial \mathbf{K}_{\mathrm{T}}} = -\mathrm{diag}\left[k_\gamma V_{n_{\mathrm{a}}+k} \cos \varphi_k\right], \tag{5.77}$$

$$\mathbf{F}_{14} = \frac{\partial \Delta \mathbf{d}_1}{\partial \mathbf{W}} = -\mathrm{diag}[k_{\mathrm{T}k} V_{n_{\mathrm{a}}+k}], \tag{5.78}$$

$$\mathbf{F}_{25} = \frac{\partial \Delta \mathbf{d}_2}{\partial \mathbf{\Phi}} = \mathrm{diag}\left[k_\gamma k_{\mathrm{T}k} V_{n_{\mathrm{a}}+k} \sin \varphi_k\right], \tag{5.79}$$

$$\mathbf{C}_{14} = \frac{\partial \Delta \mathbf{d}_1}{\partial \mathbf{V}_t} \mathbf{V}_t = -\text{diag}\left[k_{Tk} V_{n_a+k} w_k\right], \tag{5.80}$$

$$\mathbf{C}_{24} = \frac{\partial \Delta \mathbf{d}_2}{\partial \mathbf{V}_t} \mathbf{V}_t = -\text{diag}\left[k_\gamma k_{Tk} V_{n_a+k} \cos \varphi_k\right]. \tag{5.81}$$

When all of the DC buses are PQ buses, the above sub-matrices are all n_c-order square matrices. All of them except $\mathbf{F}_{31}$ are diagonal. When a DC bus is a PV bus, the voltage magnitude on the bus is given. We should remove the corresponding columns in $\mathbf{C}_{14}$ and $\mathbf{C}_{24}$ and corresponding rows in $\mathbf{A}_{41}$ and $\mathbf{A}_{42}$. Sub-matrices $\mathbf{F}_{41}$–$\mathbf{F}_{45}$ and $\mathbf{F}_{51}$–$\mathbf{F}_{55}$ are related to the control schemes of the converters. Since the controlS (5.42)–(5.46) have very simple expressions, their sub-matrices are quite sparse. Besides, the corresponding rows in (5.73) for inverters should be multiplied by a negative sign, as shown in (5.62), due to the fact that we use the convention of load representation as the reference direction of inverter DC voltages and currents.

5.3.4 Integrated Iteration Formula of AC/DC Interconnected Systems

The power flow of interconnected systems can be solved by using the calculation process of the Newton algorithm for the conventional power flow. We can give the initial values of the n-bus voltage magnitudes and phase angles based on the flat start principle. The specialty here is to estimate the initial values for the variables associated with the extended equations and the control constraints.

Based on the estimated power or given power, we can use the converter basic equations to obtain the initial values of the extended variables. During the evaluation, we directly convert a variable into a constant if it is given in the converter control scheme. The initial values of bus voltages are 1.0 or the given voltages if the buses are PV buses. The power factors of converters are initialized to 0.9. The general evaluation process is as follows. If V_{Tk} and k_{Tk} are both unknown, take k_{Tk} as 1.0 and compute V_{Tk} from (5.60). If one of them is known, compute the other. If both are known, compute $\cos \varphi_k$ as power factor initial instead of assuming 0.9. In (5.59), if I_{dk} is unknown, compute it from (5.57) using estimated DC power P_{idc}. After I_{dk} is known, compute $\cos \theta_{dk}$ as the initial of ω_k from (5.59). If the resulted ω_k is greater than 1.0, take 1.0 as initial ω_k.

Operational constraints on extended variables can be similarly handled to variable limits in conventional power flow calculations. We set the extended variables to their limits if they go beyond their limits. Note that the minimum constraint on control angles corresponds to the conversion $\omega_k = \cos \theta_{dk}$, i.e., less than arccos $\theta_{dk\,\min}$. DC voltages and currents have maximum constraints, and transformer ratios have both upper and lower limits. Also, the transformer ratios are discrete variables but are treated as continuous variables in the above calculations.

We have discussed the basic principles of the Newton algorithm in power flow calculation of AC/DC interconnected systems. Although the Newton algorithm has good convergence, it requires repeated updates of the Jacobian matrix during iterations, resulting in low calculation efficiency. In a conventional power flow, the $P-Q$ decoupled method can increase the overall speed of calculation. With regard to the characteristics of the Jacobian matrix in AC/DC interconnected system, it is not difficult to consider applying a similar approximation to AC/DC interconnected systems to increase calculation speed. There are many simplification methods and iteration processes with the same basic concept. Below we will introduce one of them. To balance the convergence and calculation speed, we are going to take only the approximation conditions in the conventional $P-Q$ decoupled method.

Using the approximation conditions of the $P-Q$ decoupled method, (5.65) can be simplified by applying (2.81) and (2.82) in the form of the three low-order equations below

$$\Delta\mathbf{P}/\mathbf{V} = \mathbf{B}'\mathbf{V}\Delta\boldsymbol{\theta} + \mathbf{A}_1\Delta\mathbf{X}, \tag{5.82}$$

$$\Delta\mathbf{Q}/\mathbf{V} = \mathbf{B}''\Delta\mathbf{V} + \mathbf{A}_2\Delta\mathbf{X}, \tag{5.83}$$

$$\Delta\mathbf{D} = \mathbf{C}_2'\Delta\mathbf{V} + \mathbf{F}\Delta\mathbf{X}, \tag{5.84}$$

where

$$\Delta\mathbf{X} = \begin{bmatrix} \Delta\mathbf{V}_\mathrm{d}^\mathrm{T} & \Delta\mathbf{I}_\mathrm{d}^\mathrm{T} & \Delta\mathbf{K}_\mathrm{T}^\mathrm{T} & \Delta\mathbf{W}^\mathrm{T} & \Delta\boldsymbol{\Phi}^\mathrm{T} \end{bmatrix}^\mathrm{T},$$
$$\Delta\mathbf{D} = \begin{bmatrix} \Delta\mathbf{d}_1^\mathrm{T} & \Delta\mathbf{d}_2^\mathrm{T} & \Delta\mathbf{d}_3^\mathrm{T} & \Delta\mathbf{d}_4^\mathrm{T} & \Delta\mathbf{d}_5^\mathrm{T} \end{bmatrix}^\mathrm{T},$$

$$\mathbf{A}_1 = \begin{bmatrix} 0 & 0 & 0 & 0 & 0 \\ \mathbf{A}_{21} & \mathbf{A}_{22} & 0 & 0 & 0 \end{bmatrix}, \tag{5.85}$$

$$\mathbf{A}_2 = \begin{bmatrix} 0 & 0 & 0 & 0 & 0 \\ \mathbf{A}_{41} & \mathbf{A}_{42} & 0 & 0 & \mathbf{A}_{45} \end{bmatrix}, \tag{5.86}$$

$$\mathbf{C}_2' = \begin{bmatrix} 0 & 0 & 0 & 0 & 0 \\ \mathbf{C}_{14}'^\mathrm{T} & \mathbf{C}_{24}'^\mathrm{T} & 0 & 0 & 0 \end{bmatrix}^\mathrm{T}. \tag{5.87}$$

Note that in (5.84), voltage mismatch ΔV_i is not divided by V_i as in (5.65). The partial derivative corresponding to (5.80) and (5.81) is not multiplied by V_i. Thus

$$\mathbf{C}_{14}' = \frac{\partial\Delta\mathbf{d}_1}{\partial\mathbf{V}_\mathrm{t}} = -\mathrm{diag}[k_{\mathrm{T}k}w_k], \tag{5.88}$$

$$\mathbf{C}'_{24} = \frac{\partial \Delta \mathbf{d}_2}{\partial \mathbf{V}_t} = -\text{diag}\left[k_\gamma k_{\text{T}k} \cos \varphi_k\right], \tag{5.89}$$

From (5.83) we have

$$\Delta \mathbf{V} = \mathbf{B}''^{-1}[\Delta \mathbf{Q}/\mathbf{V} - \mathbf{A}_2 \Delta \mathbf{X}]. \tag{5.90}$$

Substituting (5.90) into (5.84) yields

$$\Delta \mathbf{D} - \mathbf{C}'_2 \mathbf{B}''^{-1} \Delta \mathbf{Q}/\mathbf{V} = \left[\mathbf{F} - \mathbf{C}'_2 \mathbf{B}''^{-1} \mathbf{A}_2\right] \Delta \mathbf{X} \tag{5.91}$$

Solving ΔX from (5.91) and substituting into (5.90) and (5.82) can obtain the values of $\Delta \mathbf{V}$ and $\mathbf{V}\Delta\boldsymbol{\theta}$. Let

$$\mathbf{B}''^{-1} \Delta \mathbf{Q}/\mathbf{V} = \mathbf{y}_q, \tag{5.92}$$

$$\mathbf{B}''^{-1} \mathbf{A}_2 = \mathbf{Y}_\mathbf{A}. \tag{5.93}$$

The solution process of (5.82)–(5.84) is

$$\mathbf{B}'' \mathbf{y}_q = \Delta \mathbf{Q}/\mathbf{V}, \tag{5.94}$$

$$\mathbf{B}'' \mathbf{Y}_\mathbf{A} = \mathbf{A}_2, \tag{5.95}$$

$$\Delta \mathbf{D} - \mathbf{C}'_2 \mathbf{y}_q = \left[\mathbf{F} - \mathbf{C}'_2 \mathbf{Y}_\mathbf{A}\right] \Delta \mathbf{X}, \tag{5.96}$$

$$\Delta \mathbf{V} = \mathbf{y}_q - \mathbf{Y}_\mathbf{A} \Delta \mathbf{X}, \tag{5.97}$$

$$\mathbf{B}' \mathbf{V} \Delta \theta = \Delta \mathbf{P}/\mathbf{V} - \mathbf{A}_1 \Delta \mathbf{X}. \tag{5.98}$$

Figure 5.16 shows the calculation process.

From the analysis in 2.4 we know matrices $\mathbf{B}''$ and $\mathbf{B}'$ are constant matrices. They only need to be generated once before iterating and only require LU decomposition once:

1. Solve vector $\mathbf{y}_q$ from (5.94). This step is consistent with the conventional *P–Q* decoupled method and computation.
2. Solve matrix $\mathbf{Y}_\mathbf{A}$ from (5.95). Since $\mathbf{A}_2$ changes in each iteration, it is built by (5.68)–(5.70). Apparently matrix A_2 consists of $3n_c$ sparse vectors with one nonzero element in each vector and $2n_c$ zero vectors because sub-matrices $\mathbf{A}_{31}$, $\mathbf{A}_{31}$, and $\mathbf{A}_{35}$ are all zero matrices. Thus, the corresponding sub-matrices $\mathbf{Y}_{\mathbf{A}13}$, $\mathbf{Y}_{\mathbf{A}23}$, $\mathbf{Y}_{\mathbf{A}14}$, and $\mathbf{Y}_{\mathbf{A}24}$ of $\mathbf{Y}_\mathbf{A}$ are all zero matrices (refer to (5.99)). The formation of $\mathbf{Y}_\mathbf{A}$ can be quickly accomplished by the sparse matrix method discussed in Chap. 2. As we know, the fast forward substitution of a sparse vector requires

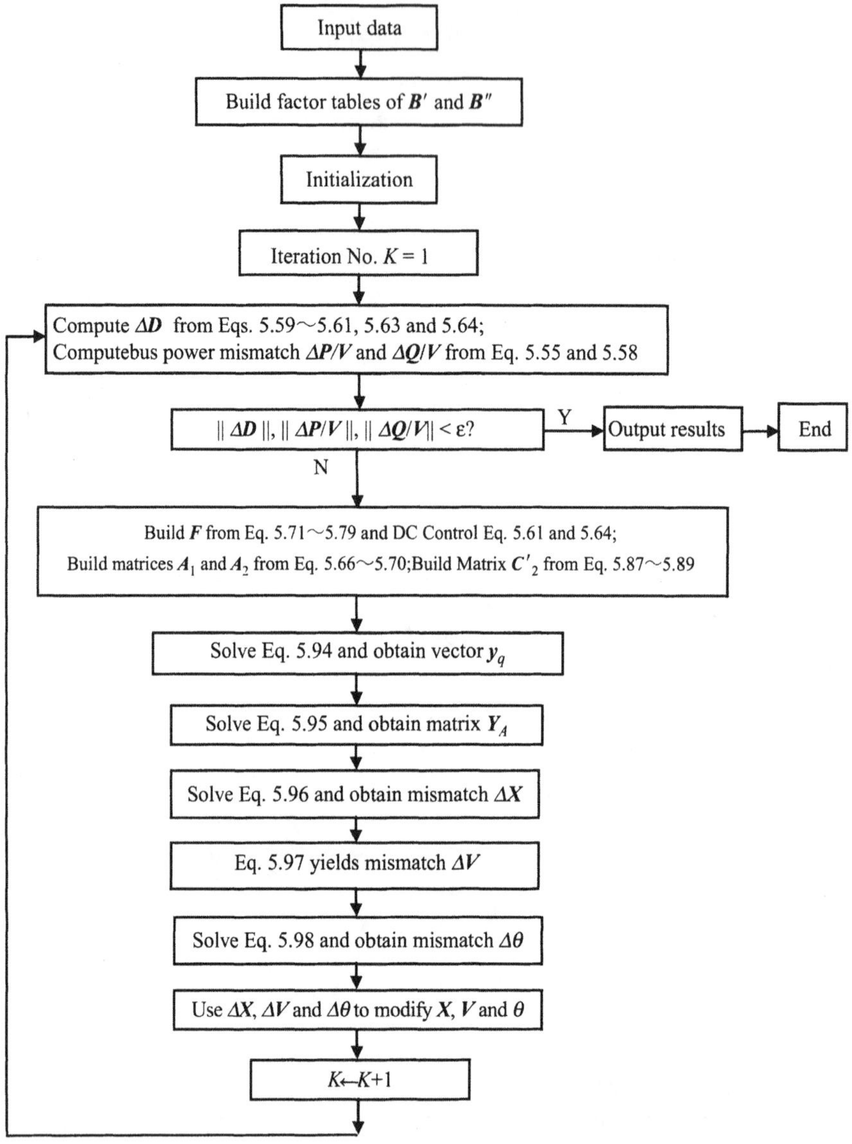

Fig. 5.16 Flowchart of PQ decoupled method

$(n - l + 1)(n - l + 2)/2$ times of multiplication if only the lth element of sparse n-order vector is nonzero. To save computation time, we should make l as large as possible. For example, the fast forward substitution of $n$cth column of $\mathbf{A}_2$, corresponding to $l = n$, needs only one multiplication according to (5.68) and (5.86). Hence it is necessary to put the bus power equations of pure AC buses at the beginning. This is to designate higher numbers to DC buses so that the only nonzero elements in $\mathbf{A}_2$ stay near the bottom of column vectors. When n_c is not large, the optimization of bus numbering should focus on the computation reduction of forming the factor table.

3. Establish the $5n_c$-order coefficient matrix $\mathbf{F} - \mathbf{C}_2' \mathbf{Y}_A$ of (5.96). We know $\mathbf{C}_2'$ is highly sparse from (5.87)–(5.89). Partitioning $\boldsymbol{Y}_A$ and multiplying by $\mathbf{C}_2'$ yields

$$\mathbf{C}_2'\mathbf{Y}_A = \begin{bmatrix} 0 & 0 & 0 & 0 & 0 \\ \mathbf{C}_{14}'^{T} & \mathbf{C}_{24}'^{T} & 0 & 0 & 0 \end{bmatrix}^{T} \begin{bmatrix} \mathbf{Y}_{A11} & \mathbf{Y}_{A12} & 0 & 0 & \mathbf{Y}_{A15} \\ \mathbf{Y}_{A21} & \mathbf{Y}_{A22} & 0 & 0 & \mathbf{Y}_{A25} \end{bmatrix}.$$

The last three columns of the above equation are zeros. The first two columns are

$$\begin{bmatrix} \mathbf{C}_{14}'\mathbf{Y}_{A21} & \mathbf{C}_{14}'\mathbf{Y}_{A22} & 0 & 0 & \mathbf{C}_{14}'\mathbf{Y}_{A25} \\ \mathbf{C}_{24}'\mathbf{Y}_{A21} & \mathbf{C}_{24}'\mathbf{Y}_{A22} & 0 & 0 & \mathbf{C}_{24}'\mathbf{Y}_{A25} \end{bmatrix}, \tag{5.99}$$

where the orders of the sub-matrices in the first and second rows of $\mathbf{Y}_A$ are $n_a \times n_c$ and $n_c \times n_c$ matrices. Note that $\mathbf{C}_{14}'$ and $\mathbf{C}_{24}'$ are both n_c-order diagonal matrices, thus

$$\left.\begin{aligned} \left(\mathbf{C}_{i4}'\mathbf{Y}_{A2j}\right)_{lm} &= \left(\mathbf{C}_{i4}'\right)_{ll}\left(\mathbf{Y}_{A2j}\right)_{lm} \\ i &= 1,2, \\ j &= 1,2,5, \\ l &= 1,2,\ldots,n_c, \\ m &= 1,2,\ldots,n_c. \end{aligned}\right\} \tag{5.100}$$

The computation of vector $\mathbf{C}_2' \mathbf{y}_q$ is as follows

$$\mathbf{C}_2'\mathbf{y}_q = \begin{bmatrix} 0 & \mathbf{C}_{14}' \\ 0 & \mathbf{C}_{24}' \\ 0 & 0 \\ 0 & 0 \\ 0 & 0 \end{bmatrix} \begin{bmatrix} \mathbf{y}_{q1} \\ \mathbf{y}_{q2} \end{bmatrix} = \begin{bmatrix} \mathbf{C}_{14}'\mathbf{y}_{q2} \\ \mathbf{C}_{24}'\mathbf{y}_{q2} \\ 0 \\ 0 \\ 0 \end{bmatrix}, \tag{5.101}$$

$$\left.\begin{aligned} \left(\mathbf{C}_{i4}'\mathbf{y}_{q2}\right)_l &= \left(\mathbf{C}_{i4}'\right)_{ll}\left(\mathbf{y}_{q2}\right)_l, \\ i &= 1,2, \\ l &= 1,2,\ldots,n_c. \end{aligned}\right\} \tag{5.102}$$

We can see that $2n_c$ among the $5n_c$ elements of $\mathbf{C}_2' \mathbf{y}_q$ are nonzero.

4. Building the vectors $\mathbf{A}_1\Delta\mathbf{X}$ in (5.98). From (5.85) we have

$$\begin{aligned} \mathbf{A}_1\Delta\mathbf{X} &= \begin{bmatrix} 0 & 0 & 0 & 0 & 0 \\ \mathbf{A}_{21} & \mathbf{A}_{22} & 0 & 0 & 0 \end{bmatrix} \left[\Delta\mathbf{V}_d^T \;\; \Delta\mathbf{I}_d^T \;\; \Delta\mathbf{K}_T^T \;\; \Delta\mathbf{W}^T \;\; \Delta\boldsymbol{\Phi}^T\right]^T \\ &= \begin{bmatrix} 0 \\ A_{21}\Delta\mathbf{V}_d + \mathbf{A}_{22}\Delta\mathbf{I}_d \end{bmatrix}. \end{aligned} \tag{5.103}$$

Equations (5.66) and (5.67) yield

$$(\mathbf{A}_{21}\Delta\mathbf{V}_{\mathrm{d}} + \mathbf{A}_{22}\Delta\mathbf{I}_{\mathrm{d}})_k = -(I_{\mathrm{d}k}\Delta V_{\mathrm{d}k} + V_{\mathrm{d}k}\Delta I_{\mathrm{d}k}), \quad k = 1, 2, \ldots, n_{\mathrm{c}}. \tag{5.104}$$

We have introduced the integrated iteration power flow calculations of interconnected systems.

5.3.5 Alternating Iteration for AC/DC Interconnected Systems

The alternating iteration method is a further simplification to the *P–Q* decoupled method in the integrated iteration power flow calculations. Based on the basic converter (5.52) and (5.53), the impact of AC systems on DC systems relies on the primary voltages V_{t} of converter transformers. If the AC voltages V_{t} of all converters in a multiterminal DC systems are known, the DC system will have (5.59)–(5.61), (5.63), and (5.64), a total of $5n_{\mathrm{c}}$ equations and $5n_{\mathrm{c}}$ unknown variables. We can obtain the $5n_{\mathrm{c}}$ unknown DC variables by solving only the DC system equations. The power taken out of, or injected into, AC systems, $P_{idc} + jQ_{idc}$ from converter transformers, represents the impact of DC systems on AC systems. If the power withdrawn from or injected into AC systems is known, power flow calculations of AC systems are not affected by DC systems. The ideal process is to designate the primary voltages of n_{c} converter transformers

$$\mathbf{V}_{\mathrm{t}}^{(0)} = \begin{bmatrix} V_{n_{\mathrm{a}}+1}^{(0)} & V_{n_{\mathrm{a}}+2}^{(0)} & \cdots & V_{n_{\mathrm{a}}+n_{\mathrm{c}}}^{(0)} \end{bmatrix}.$$

Obtain solution of DC variables $\mathbf{X}^{(0)}$. Substituting $\mathbf{X}^{(0)}$ into (5.57) yields the power of all converters $\mathbf{P}_{\mathrm{dc}}{}^{(0)}$ and $\mathbf{Q}_{\mathrm{dc}}{}^{(0)}$. Using converter power in AC system equations for conventional power flow calculations gives rise to convergent solution $\mathbf{V}^{(1)} = \begin{bmatrix} \mathbf{V}_{\mathrm{a}}^{(1)} & \mathbf{V}_{\mathrm{t}}^{(1)} \end{bmatrix}$. Ideally the calculation completes if $\mathbf{V}_{\mathrm{t}}^{(1)}$ equals $\mathbf{V}_{\mathrm{t}}^{(0)}$. Generally $\mathbf{V}_{\mathrm{t}}^{(1)}$ is not the same as $\mathbf{V}_{\mathrm{t}}^{(0)}$. The calculation is an iteration process.

Based on the above, AC and DC system equations are separately solved in alternating iterations. When solving AC system equations, we use the known power at the DC buses to represent the corresponding DC systems. While solving DC system equations, we model AC systems as constant voltages at the AC buses of converters. At each iteration, the solution of the AC systems provides the converter AC bus voltages for the next DC iteration; the solution of DC systems in turn produces the equivalent real and reactive power of converters for the further AC iteration. The iteration goes on and on until convergence is achieved. We must point out that the convergence of this method is mathematically related to the Gauss–Seidel iteration. In fact, the alternating iteration is not a complete Gauss–Seidel iteration. For the AC system equations in the alternating iteration method, we usually use the Newton algorithm or *P–Q* decoupled algorithm. For DC system equations, the Newton algorithm is the most common approach [114]. The Gauss–Seidel algorithm applies only to the coupling between AC and DC equations.

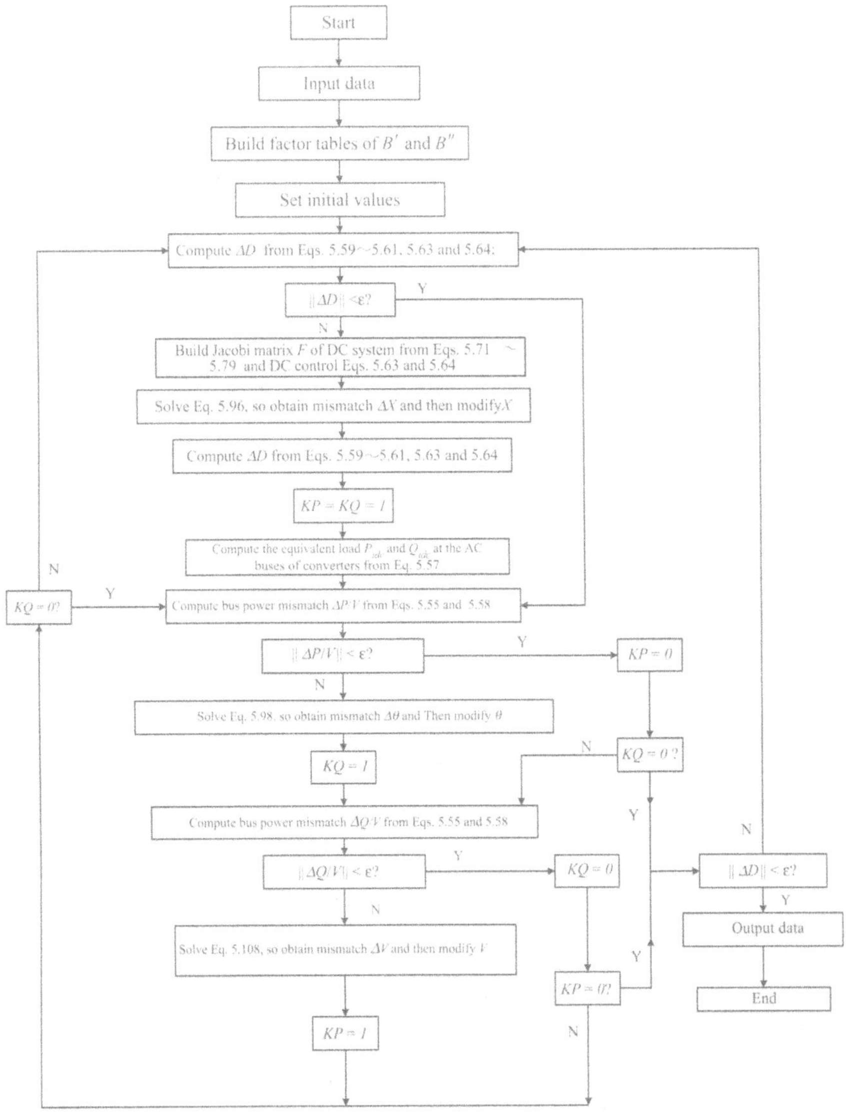

Fig. 5.17 Flow chart of alternating iteration method

With the above assumption, DC powers in the AC bus power (5.58) and AC voltages in the DC system (5.59), (5.61), (5.63), and (5.64) are also known constants. Sub-matrices **A** and **C** in (5.65) are all zero matrices to decouple AC and DC systems. Going one step further, using PQ decoupling method in AC equations results in the solution of the following three groups of equations for AC/DC interconnected power flow

$$\Delta\mathbf{D} = \mathbf{F}\Delta\mathbf{X}, \tag{5.105}$$

$$\Delta\mathbf{P}/\mathbf{V} = \mathbf{B}'\mathbf{V}\Delta\boldsymbol{\theta}, \tag{5.106}$$

$$\Delta\mathbf{Q}/\mathbf{V} = \mathbf{B}''\Delta\mathbf{V}. \tag{5.107}$$

Comparing the above three groups of equations with (5.82) and (5.84), we can find out that alternating iterations are equivalent to ignoring sub-matrices **A** and $\mathbf{C}_2'$ in integrated iterations. Figure 5.17 shows the flow chart of alternating iterations for AC/DC interconnected systems. ε in the figure represents the convergence tolerance. The whole system solution is the one that makes bus real power equations, bus reactive power equations and DC system equations all convergent. In the iteration process, DC variable **X** determines converter real and reactive powers. Any correction in DC variables forces reconfirmation of convergence with the new **X**, no matter whether the previous solution is convergent or not. The magnitudes of AC bus voltages, not their phase angles, affect the DC equations. The modification of AC bus voltages requires confirmation of convergence under the new **V** no matter whether the previous DC and real power equations were converged or not. Similarly, phase angle correction requires renewed convergence confirmation of reactive power.

Some DC variables can be outside of their ranges in iterations. There are various approaches to process over-limit scenarios. Take the converter transformer ratio k_T as an example. There is a voltage control terminal in a multiterminal DC system. If k_T is over its high limit, we can lower the given voltage at its voltage control terminal, and vice versa. If a control angle α or μ is less than its α_{min} or μ_{min}, we can change the terminal control scheme to fixed control angle to force it to its limit or another value, and release the previous given variable in the control equations.

There are some methods to solve DC (5.105) other than the above Newton algorithm. Below we introduce a simple and effective approach.

A practical operation control scheme for multiterminal DC systems is to designate one terminal to enforce voltage control; that is, to use fixed DC voltage as its control strategy. To be general, assume that converter n_c is voltage control terminal with DC voltage setting of V_{ds}. Fixed current and fixed power control apply to other terminals. As we pointed out before, the control angles of all converters should be as small as possible to reduce reactive power consumption. Based on this principle, suppose that the voltage control terminal of a DC system operates at the minimum control angle. Its control (5.63) and (5.64) become

$$\left.\begin{aligned} V_d &= V_{ds} \\ \theta_d &= \theta_{d\,min} \end{aligned}\right\}. \tag{5.108}$$

The control equations of nonvoltage control terminals are

$$I_{dk} = I_{dks} \text{ or } V_{dk}I_{dk} = P_{dks} \quad k = 1, 2, \ldots, n_c - 1. \tag{5.109}$$

Besides fixed current and fixed power control, we have the following additional equations to minimize the control angles

$$\cos\theta_{dk} = k_\theta \cos\theta_{dk\min} \quad k = 1,2,\ldots,n_c - 1, \tag{5.110}$$

where coefficient k_θ is given beforhand. We can obtain control angle θ_d with known k_θ. The above is a fixed control angle scheme in nature. Usually k_θ is set to 0.97. Taking (5.110) into (5.59) gives rise to converter equations of nonvoltage control terminals

$$\Delta d_{1k} = V_{dk} - k_{\theta k} k_{Tk} V_{n_a+k} \cos\theta_{dk\min} + X_{ck} I_{dk} = 0 \quad k = 1,2,\ldots,n_c - 1. \tag{5.111}$$

Equations (5.110) and (5.111) make up the converter control equations of nonvoltage control terminals.

For DC network equations, Gauss–Seidel iteration can be conveniently used due to the usually small n_c. We build bus resistance matrix of DC systems for easier iteration. Note that DC voltage in the DC network (5.61) is the voltage across the two poles of the converter. Thus the admittance matrix $\mathbf{G}_d$ is an indefinite admittance matrix and is singular. Taking bus n_c as DC voltage reference bus in the DC network, it is not difficult to build bus resistance matrix of the DC network from $\mathbf{G}_d$. Thus

$$V_{dk} = V_{ds} + \sum_{j=1}^{n_c-1} r_{kj} I_{dj} \quad k = 1,2,\ldots,n_c - 1, \tag{5.112}$$

where γ_{kj} is the element of bus resistance matrix. Substituting (5.109) into (5.112) we only have n_c-1 unknown DC voltages.

Below we list the steps of interconnected system power flow calculations:

1. Set all DC voltage initials to V_{ds}. Use Gauss–Seidel iteration to solve (5.112) for convergent $\mathbf{V}_d$.
2. Find DC current $I_d = P_{ds}/V_d$ of fixed power control converters based on known $\mathbf{V}_d$.
3. For each converter, compute secondary AC voltages $k_T V_t$ of their converter transformers. Calculation steps are as follows: For voltage control terminal, $k_T V_t$ can be found from (5.52) as V_d, I_d, and θ_d are known from (5.108) and the above step. For nonvoltage control terminals, $k_T V_t$ can be calculated from (5.111).
4. Compute converter power factor $\cos\varphi$ from (5.53).
5. Compute converter real and reactive power from (5.57).
6. Carry out AC system power flow calculations to find a convergent solution using converter power from Step 5.

7. Compute each converter transformer ratio k_T based on the primary voltage V_t in Step 6 and the secondary voltage $k_T V_t$ of converter transformers in Step 3.
8. If the ratios obtained are within their limits, calculation finishes. Otherwise go to Step 9.
9. Adjust the DC voltage at the voltage control terminal as follows. Select the maximum over-limit ratio k_{Tworst} among all over-limit ratios. Suppose its upper and lower limits are k_{Tmax} and k_{Tmin}. If $k_{Tworst} > k_{Tmax}$, take $V_{ds} \times k_{Tmax}/k_{Tworst}$ as new voltage control value; otherwise take $V_{ds} \times k_{Tmin}/k_{Tworst}$ as the new voltage control value. Return to Step 1.

The above steps are the main steps of this method. Its basic characteristics are the simplicity in theory and in programming. Comparing with Newton's iteration for DC equations, it saves considerable memory. When assuming the converter transformer ratio is a continuous variable and with no over-limits, DC and AC system power flow solution can be attained in one computation. With necessary amendment to the above method, it can be applied to power flow calculations for fixed control angle control, discrete transformer ratio changes, etc. In which case, iteration is required. The details can be found in [115].

In the above we have discussed the two major types of power flow calculation methods for AC/DC interconnected systems. Integrated iteration takes into consideration the complete coupling between AC and DC systems, and has good convergence for various conditions of network and system operation. The Jacobian matrix has a higher order than for pure AC systems. The approach requires more programming, uses more memory, and needs longer computation time. Alternating iteration can be accomplished by adding DC modules to the existing power flow programs, due to its separated solution of AC and DC equations. It is easier to take into consideration the constraints on DC variables and the adjustment of operation modes. However, the convergence of alternating iteration is not as good as integrated iteration. The computational practice indicates that its convergence is good when the AC system is strong. If the AC system is weak, its convergence deteriorates, requiring more iterations or even becoming nonconvergent. This is the shortcoming of the alternating iteration method. The strength of AC systems is related to the rated capacity of converters. Taking the converter rated power P_{dcN} as the base, the reciprocal of per unit equivalent reactance of AC system, as viewed from the AC bus of converters, is called the short-circuit ratio (SCR). The larger the SCR, the higher is the system strength. A weak AC system (SCR less than 3) has a larger equivalent reactance, making the AC bus voltage of the converter very sensitive to variation of the reactive power injection. Alternating iteration separates the solution of AC and DC equations, assuming constant V_t and Q_{tdc} at the boundary between AC and DC systems to neglect their coupling. If the AC system is weak, the variation of Q_{tdc} can bring potential change to V_t. This results in computational oscillation between Q_{tdc} and V_t in alternating iterations and worsening convergence [116]. There have been some improved calculation methods [117] for alternating iteration to make it applicable to weak AC systems. We are not going to discuss them here for brevity.

5.4 HVDC Dynamic Mathematical Models

We have introduced the steady-state models of HVDC systems in the previous section. The HVDC transients are quite complicated. The main causes of the complexity are as follows. (1) The firing pulses of bridge valves are triggered at discrete time points. In transients, the firing angle is regulated by the controller to make the corresponding time unevenly distributed. Thus the firing angle is a discrete variable with regards to computation. (2) We assume that AC systems are symmetric in steady-state analysis. From the steady-state analysis of converter valves we know that the valve on/off states are closely related to the commutating voltages, the time of firing, and the magnitudes of commutation angles. When firing angles or commutation angles are too large, commutation may fail. In transient states, AC systems are actually unsymmetrical. Some valves could have negative valve voltage and could not been turned on when firing pulses occur if commutation voltages are severe unsymmetrical. For HVDC systems under transient states, we need to establish derivative equations to take into consideration the variations of commutation voltages and firing angles as well as other exceptional conditions. The solution of these derivative equations reveals the time of valve state changes. (3) We should consider the distributed characteristics of DC lines for long distance transmission. Under such circumstances, the variations of voltage and current on DC lines become wave processes.

Due to the above conditions, we need to solve ordinary differential equations and partial differential equations with both continuous and discrete variables to calculate accurately the transients of HVDC transmission systems. From the mathematical point of view, it is not difficult to solve these equations. Many previous works [118–120] used detailed mathematical models resulting in huge computational requirements. We should simplify the transient DC models as much as possible without losing engineering accuracy. In general we can take a simple DC model for stability analysis, if AC systems are relatively strong; otherwise a detailed DC model is required. The general assumptions that we make in deriving DC steady-state models still apply to most analysis of power system stability. Thus we can use the steady-state mathematical models of converters (5.52), (5.54) as their dynamic models. Here we are going to introduce the mathematical models of control systems.

The controllers in HVDC systems consist of electronic circuits. Their basic working principles are as follows: receiving control inputs, sending outputs to phase-control circuits, and pulse generation device to set converter firing angles in order to control converter operation. Different control signals and different control strategies result in different controller structures and control characteristics, as well as the dynamics of DC systems or even the whole power system. To achieve better operation characteristics, the adjustments of rectifiers and inverters should be coordinated. As stated before, the basic control mode is fixed current or fixed power for rectifiers, and fixed voltage or fixed extinction angle for inverters. The transformer ratio adjustment is slow and is a discrete variable. The ratio is not changed

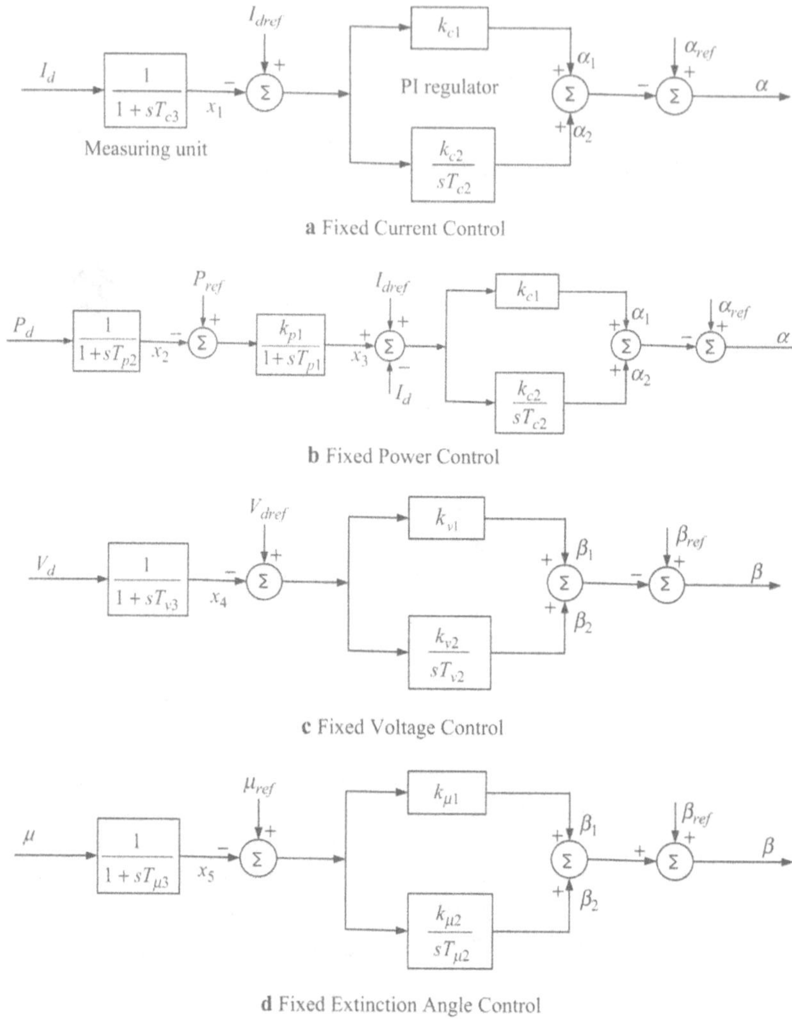

Fig. 5.18 Transfer function block diagrams of HVDC control systems

very often and is an ancillary means to optimize the converter operation point. Figure 5.18 shows the transfer function of the four basic control modes.

The transfer function of fixed current control is shown in Fig. 5.18a. It compares I_{d}, the output of DC current, and given current I_{dref}. The difference is amplified and goes through a proportional plus integral process. Then the signal is passed to the phase-shift control circuits to change the converter firing angle and to enforce the fixed current function.

The transfer function of fixed power control is shown in Fig. 5.18b. HVDC systems are usually required to transport power as planned. Fixed power is a basic control method. When the variations of AC voltages on both terminals are not large, using fixed current and fixed extinction angle can actually achieve fixed power control. When taking into consideration AC voltage fluctuations, using fixed current and fixed voltage can obtain exact fixed power control. These two control methods are to determine the DC current setting based on the given power and DC operation voltage at the control terminal. However DC voltage is related to DC current, so it is very difficult to set DC current beforehand. To overcome this problem, special control devices are set up for fixed power control. Fixed current control has high response speed, is capable of quickly constraining overcurrent to prevent converter overload, and is easy to set up. Power control devices are usually based on fixed current control and receive additional inputs rather than directly acting on phase control circuits. In the diagram, the DC power is compared with its target value. The difference is amplified and sent to the input of the fixed current controller. This works by changing the current setting of the fixed current control dynamically.

The transfer functions of fixed voltage and fixed extinction angle controls are shown in Fig. 5.18c,d. They share the same structure as Fig. 5.18a with different parameters. We need to point out that extinction angles cannot be directly measured. They are indirectly obtained by measuring the time interval between valve voltage and current zero-crossing points.

Although we do not show the quantity limitation block in the above diagrams, attention has to be paid to the constraints on various physical variables. There are minimum firing angle constraints for rectifier fixed current control, minimum extinction angle constraints for inverter fixed voltage control, etc.

We need to notice that the controllers here are all for DC internal adjustments. DC systems can be used to affect AC system operation through these DC internal adjustments. The inputs of controllers may include AC system operation parameters, line power, the velocity of some generators, system frequency, and so on. This kind of control is the integrated control of AC/DC systems, also called external adjustments. The control strategy and control signals in these cases are an important field of power system research.

5.5 Basic Principles and Mathematical Models of FACTS

After the introduction of the FACTS concept, many FACTS devices have been proposed. We can classify them into three groups based on the maturity of the technology. The first group has been applied in the power industry, such as static VAr compensators (SVR), thyristor controlled series capacitor (TCSC), and static synchronous compensators (STATCOM). The second group has industrial sample machines and is still under investigation, such as unified power flow controller (UPFC). The third group has only a theoretical design without any industrial application, such as static synchronous series compensator (SSSC), thyristor controlled phase shifting transformer (TCPST). We will introduce their basic principles and mathematical models in this section. The power flow calculation for systems having these devices will be discussed in the next section.

FACTS devices can be classified based on their connection types as series, shunt, and combined types. SVC and STATCOM are shunt type. TCSC and SSSC are series type. TCPST and UPFC are combined type. Designed by US Electrical Power Research Institute (EPRI), manufactured by Westinghouse, and installed at AEP power system in USA for industrial testing operation, UPFC is the most powerful FACTS device proposed as of today. Its control strategy is presently under further research.

5.5.1 Basic Principle and Mathematical Model of SVC

A common practice of system voltage adjustment is shunt reactive power compensation. The synchronous condenser was historically an important tool of shunt reactive power compensation. Since it is a rotating machine, its operation and maintenance are quite complicated. New synchronous condensers are now seldom installed. The static shunt reactive power compensation, as opposed to the rotating synchronous condenser, has wide industrial application due to its low cost and simple operation and maintenance. Conventional static shunt reactive power compensation is to install capacitors, reactors, or their combination, at the compensated buses to inject or extract reactive power from the system. Mechanical switches are used to put the shunt capacitor/reactors into or out of operation. There are three disadvantages in this type of compensation. First, their adjustment is discrete. Second, their control actions are slow and cannot meet system dynamic requirements. Third, they have negative voltage characteristics. When system voltages drop (rise), the reactive power injection of shunt capacitors decreases (increases). However, they are widely applied in power systems due to their economic advantages and easy maintenance.

Modern SVR with FACTS technology integrate power electronic elements into conventional static shunt reactive power compensation devices to achieve fast and continuously smooth adjustment. Ideal SVCs can maintain nearly constant voltages at the compensated buses. The good steady and dynamic characteristics render them widely applicable. Their basic elements are thyristor controlled reactors (TCRs) and thyristor switched capacitors. It is not difficult to understand other types of SVCs if we know the working principles of these two. Figure 5.19 shows their basic diagrams. To save cost, most SVCs connect to systems through step-down transformers. The valve control of the SVC produces harmonics. Filters are installed with SVCs to reduce harmonic contamination. They are capacitive as regards to fundamental frequency and inject reactive power into systems. Figure 5.20a, b shows TCR and TSC branches. Below we will analyze the control theory of TCR and TSC.

TCR branch consists of reactors connected with two back-to-back thyristors as control elements. The system voltage on the branch is sinusoidal and shown in Fig. 5.21a. The valve delayed firing angle is $\alpha \in [\pi/2, \pi]$. The firing time is

$$\omega t = \alpha + k\pi \quad k = 0, 1, 2, \ldots.$$

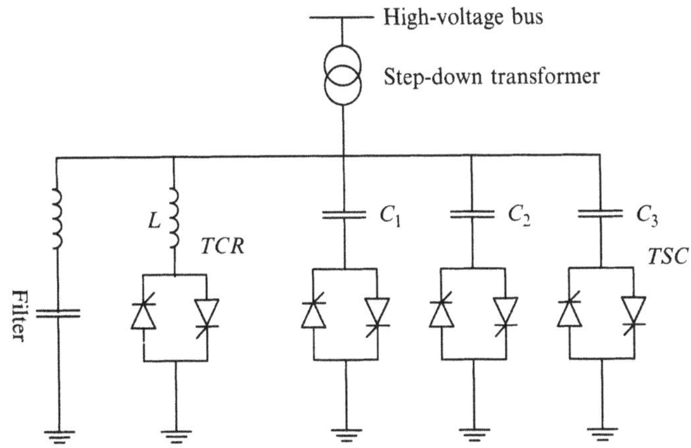

Fig. 5.19 SVC basic diagram

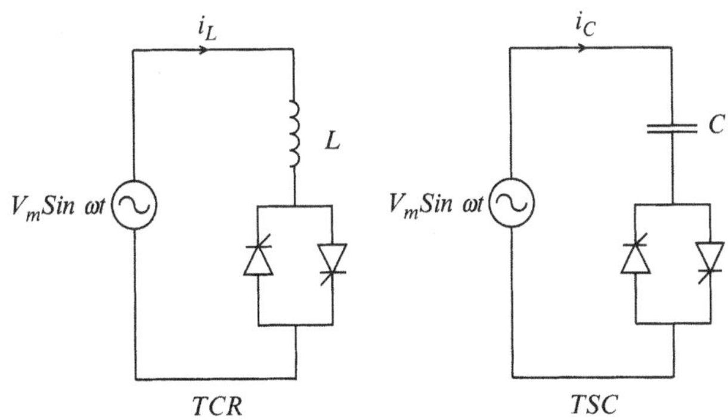

Fig. 5.20 TCR and TSC branches

Apparently the inductor current is zero when the two valves are off. When the valve conducts, neglecting the resistance in the reactor, the inductor current is

$$L\frac{\mathrm{d}i_{\mathrm{L}}}{\mathrm{d}t} = V_{\mathrm{m}} \sin \omega t, \tag{5.113}$$

where L is the inductance of the reactor, V_{m} is the magnitude of the system voltage. Its general solution is

$$i_{\mathrm{L}} = K - \frac{V_{\mathrm{m}}}{\omega L} \cos \omega t, \tag{5.114}$$

where K is the integral constant. Since the inductor current is zero at firing, the above equation yields

$$i_{\mathrm{L}} = K - \frac{V_{\mathrm{m}}}{\omega L} \cos(\alpha + k\pi) = 0.$$

Substituting the solution of K into (5.114) gives rise to the inductor current.

$$i_L = \frac{V_m}{\omega L}[\cos(\alpha + k\pi) - \cos\omega t] \quad k = 0, 1, 2, \ldots. \tag{5.115}$$

Based on the above equation, inductor current returns to zero at $\omega t = (k + 2)\pi - \alpha$. Thus the valve conducting period is

$$\omega t \in [k\pi + \alpha, (k + 2)\pi - \alpha] \quad k = 0, 1, 2, \ldots.$$

The waveform of inductor current is shown in Fig. 5.21b. The width of a single ripple of inductor current is

$$(k + 2)\pi - \alpha - (k\pi + \alpha) = 2(\pi - \alpha) = 2\beta.$$

$\beta = \pi - \alpha$ is called the conducting angle.

To make sure that there is always one valve conducting at any moment, we should have

$$(k + 2)\pi - \alpha = (k + 1)\pi + \alpha, \quad k = 0, 1, 2, \ldots.$$

One valve should conduct the moment and another one is turned off, so $\alpha = \pi/2$. This operation mode corresponds to connecting the shunt reactor directly to the system. From the waveforms we can see that the valve conducting period decreases from π to zero as the firing angle rises from $\pi/2$ to π. Now the two valves are turned off at all times, corresponding to reactors out of service. Besides when α is less than

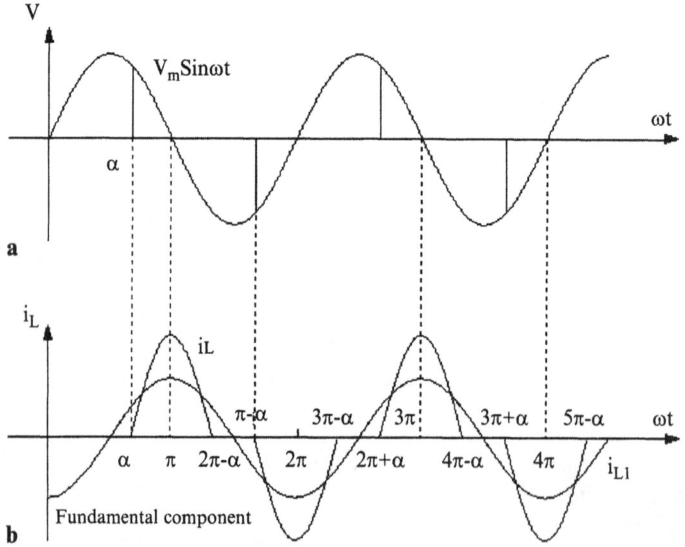

Fig. 5.21 (a) TCR voltage waveforms (b) TCR current waveforms

$\pi/2$, the moment at which the current of a conducting valve returns to zero is later than the firing moment of the off valve as

$$(k+2)\pi - \alpha > (k+1)\pi + \alpha.$$

In this case, the conducting valve has not been turned off when the other valve receives a firing pulse. The off valve cannot be triggered on due to zero valve voltage. One of the two valves is off at any moment. Thus the main component of inductor current is DC. The normal operating ranges of TCR firing angles are $\alpha \in [\pi/2, \pi]$.

Based on (5.115) and the waveforms, the current passing through the reactor is irregular and no longer sinusoidal due to valve control. The adjustment of firing angles changes the current peak values and conducting periods. Applying Fourier analysis to the current yields the magnitude of the fundamental frequency component

$$I_{L1} = \frac{2}{\pi} \int_{\alpha}^{2\pi-\alpha} \frac{V_m}{\omega L}(\cos\alpha - \cos\theta)\cos\theta d\theta = \frac{V_m}{\pi\omega L}[2(\alpha-\pi) - \sin 2\alpha].$$

And the instantaneous value of fundamental frequency component is

$$i_{L1} = I_{L1}\cos\omega t = \frac{V_m}{\pi\omega L}(2\beta - \sin 2\beta)\sin(\omega t - \pi/2). \tag{5.116}$$

The equivalent fundamental frequency reactance of the TCR branch is

$$X_L(\beta) = \frac{\pi\omega L}{2\beta - \sin 2\beta} \quad \beta \in \left[0, \frac{\pi}{2}\right]. \tag{5.117}$$

Thus the TCR equivalent reactance of fundamental frequency components is the function of conducting angle β or the firing angle α. The control of firing angle α can smoothly adjust the equivalent shunt reactance. The reactive power consumed by TCR is

$$Q_L = \dot{V}\dot{I}_{L1}^{*} = \frac{V^2}{X_L(\beta)} = \frac{2\beta - \sin 2\beta}{\pi\omega L}V^2. \tag{5.118}$$

As shown in Fig. 5.20b, the TSC branch consists of a capacitor connected in series with two thyristors connected in parallel and in opposite directions. The TSC source voltage is the same as TCR. Its waveforms are in Fig. 5.21a. The TSC creates two operating states for the capacitors through valve control: shunt capacitors in service or out of service. Stopping the firing can simply put the capacitor out of service. Note that the natural switch-off from conduction happens when the capacitor current is zero and its voltage at the peak of source voltage. Neglecting the capacitor leakage current, capacitor voltage maintains the peak value if firing stops after the natural switch-off. We need to pay attention to the timing of putting the capacitor into service. The principle is to reduce the impulse current in capacitors at the moment of in-service operation. We should use the correct valve based on the sign of the capacitor initial voltage, and put the capacitor into service at the moment when source voltage equals capacitor initial voltage. So the transient component of capacitor current is zero when put into service. After capacitors are in service, we need $\alpha = \pi/2$ to keep one valve conducting at all times. Ideally the capacitor voltage is the peak of source voltage. Using $\alpha = \pi/2$ makes no transients for the in-service operation. In reality, the source voltage and the capacitor initial voltage cannot be exactly the same. There is a small inductor in the TSC branch to reduce the possible impulse current. From the above analysis, we can see that the main difference between TSC and mechanically switched capacitors (MSC) is the fast control of in-service or out-of-service operation by valves in TSC. TSC dynamic characteristics can meet system control demands.

The reactive power injection of the capacitors is

$$Q_{\mathrm{C}} = \omega C V^2, \tag{5.119}$$

where C is the capacitance of the capacitor. From (5.118) and (5.119) we have the reactive power injection from the SVC is

$$Q_{\mathrm{SVC}} = Q_{\mathrm{C}} - Q_{\mathrm{L}} = \left(\omega C - \frac{2\beta - \sin 2\beta}{\pi \omega L}\right) V^2. \tag{5.120}$$

The SVC reactive power injection can be smoothly adjusted when $\beta \in [0, \pi/2]$. To expand the regulation ranges of SVC, we can have many TSC branches in one SVC, based on the compensation requirements. Figure 5.19 shows an SVC with three TSCs. When all three TSCs are in service, the C in (5.120) is $C_1 + C_2 + C_3$. To guarantee a continuous adjustment, the TCR capacity should be slightly larger than a group of TSCs, that is, $\omega C_1 < 1/\omega L$.

Based on (5.120), the equivalent reactance of SVC is

$$X_{\mathrm{SVC}} = -\left(\omega C - \frac{2\beta - \sin 2\beta}{\pi \omega L}\right)^{-1} = \frac{\pi \omega L}{2\beta - \sin 2\beta - \pi \omega^2 L C}. \tag{5.121}$$

The SVC equivalent voltage–current characteristics are the combination of TCR and TSC. As β increases from zero to $\pi/2$, X_{SVC} will change from capacitive maximum to inductive maximum. Generally, the control signal of SVC is derived from the voltage of the bus to which they are connected. Figure 5.22 shows that as the voltage V changes, the SVC equivalent reactance varies with β.

In Fig. 5.22, there is a straight line going through the original corresponding to every β. The slope of the straight line is X_{SVC}. Suppose that the system voltage

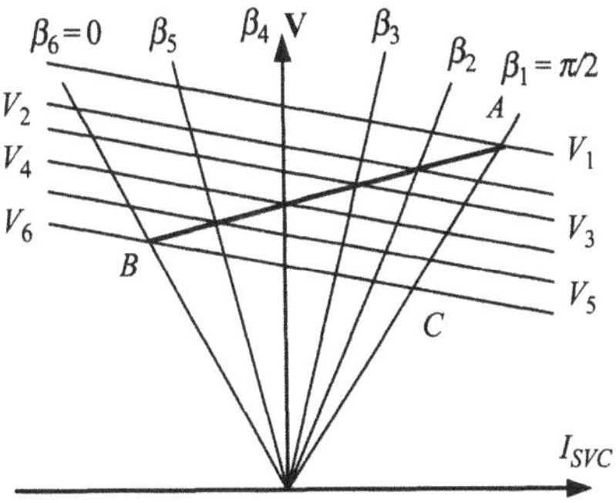

Fig. 5.22 Equivalent reactance variation with β as voltage changes

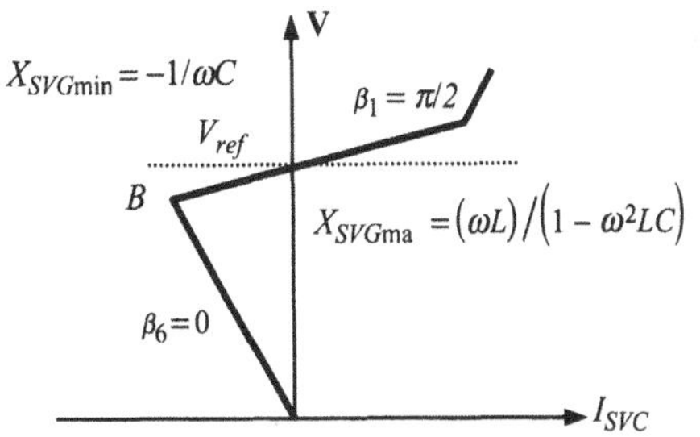

Fig. 5.23 Voltage–current characteristic

characteristic is V_1. The control scheme is to make the TCR conducting angle $\beta_1 = \pi/2$, corresponding to maximum equivalent inductive reactance. The SVC operating point is the crossover point A between system voltage characteristic V_1 and the straight line β_1. With system voltage characteristic V_2 and TCR conduction angle $\beta_2 < \beta_1$, X_{SVC} decreases and the SVC operating point shifts accordingly. Until system voltage characteristic is V_6 and conduction angle $\beta_6 = 0$, SVC equivalent reactance is maximum capacitive with operating point B. Apparently, voltage at B is higher than at C. When voltage changes between V_1 and V_6, the adjustment of β puts voltage under control. All the operating points constitute the straight line AB. The slope of AB and the crossover point with voltage axis V_{ref} is determined by the control scheme of β. From voltage control point of view, the slope of AB is zero at best, without steady-state error. To maintain the control stability, SVC should have a small steady-state error and the slope of AB is around 0.05. Taking into consideration the steady-state control scheme, the SVC voltage–current characteristics are shown in Fig. 5.23.

When system voltage varies within the SVC control range, SVC can be seen as a synchronous condenser having source voltage of V_{ref} and internal reactance of X_e.

$$V = V_{ref} + X_e I_{SVC}, \tag{5.122}$$

where X_e is the slope of the straight line AB in Fig. 5.23, V and I_{SVC} are the SVC terminal voltage and current. When system voltage is out of the SVC control range, SVC becomes a fixed reactor, X_{SVCmin} or X_{SVCmax}.

SVC is considered as a variable shunt reactor in system stability and control analysis. SVC controller determines its admittance. Reference [122] provided the controller block diagram.

We have introduced SVC basic principles. Special attention needs to be paid in industrial applications of SVC to capacity settings of reactors and capacitors, control strategy, flexibility of adjustments, protection, elimination of harmonics, etc. For example, in practical operation of an SVC, the range of the control angle is slightly less than [π/2, π] to make sure that valves can be triggered on and turned off securely.

5.5.2 *Basic Principle and Mathematical Model of STATCOM*

A STATCOM is also called an advanced static Var generator (ASVG). Its function is basically same as SVC with wider operation ranges and faster responses. As stated before, the control element of SVC is a thyristor, a semi-controllable element that can only be turned off when valve current crosses zero. STATCOM is made of fully controllable elements. Gyugyi et al. [123] presented the basic principles of using gate turn off thyristors (GTOs) to build a STATCOM. As yet, there have been several samples STATCOM operated in real systems [124–126].

The basic connection of a STATCOM is shown in Fig. 5.24. Its control element is the fully controlled valve (GTO). The ideal GTO switch characteristic is that the valve is turned on under positive valve voltage with positive control current on its gate; valve is turned off with negative control current on its gate. Valve resistor is zero when it conducts, and is infinity when it is turned off. A GTO can manage the switch-off by gate control in comparison with the thyristor where switch-off is only possible at current zero-crossing. STATCOM in Fig. 5.24 is a voltage type self-commutation full-bridge inverter according to power electronic theory.

The capacitor DC voltage acts as an ideal DC voltage source to support the inverter. The regular diode connected in the opposite direction and parallel with the GTO is a path for continuous current, providing route for the feedback energy from the AC side. The inverter normal operation is to transfer the DC voltage into AC voltage having controllable magnitude and phase angle at the same frequency as the AC system. The sum of instantaneous power of a symmetric three-phase system is a

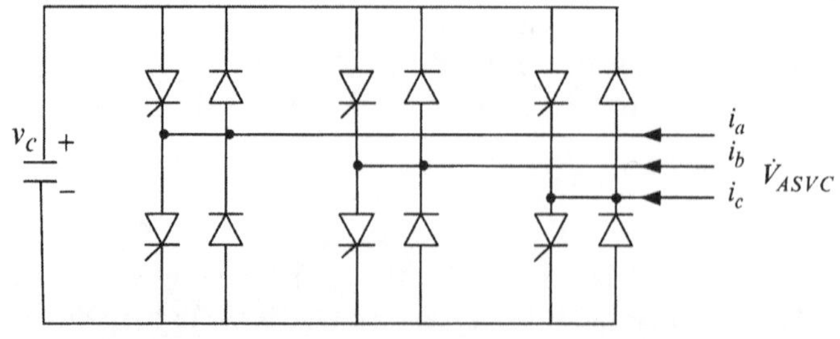

Fig. 5.24 Circuit of STATCOM

constant. Thus the reactive power exchanges periodically within phases instead of between source and load. There is no need to have an energy storage element on the DC side, if considering the inverter as a load. However, the interacting power among harmonics produces a small amount of reactive power exchange between the inverter and the system. The capacitor on the inverter DC side will provide both DC voltage and energy storage. The electrical energy stored in the capacitor is

$$W = \frac{1}{2}CV_{\mathrm{C}}^2.$$

If the above energy is not considered as energy support for AC systems during power system dynamic events, the value of capacitor C can be small while the reactive capacity provided by the STATCOM is much more than the stored energy. We will see later that the maximum reactive power capacity of a STATCOM depends on the inverter capacity. The STATCOM does not need large size reactors and capacitors as the SVC does.

Generally, there are three output voltage control modes for voltage-type inverters: phase-shift adjustment, pulse-width modulation, and direct DC source voltage control. The DC voltage of STATCOM is the charged voltage of the capacitor, not a DC source. So phase-shift adjustment and pulse-width modulation, instead of direct DC source voltage control, are usually used in STATCOMs. For brevity, we are not going to discuss the inverter working principles in detail. The width of the output voltage square waves θ is controlled by the GTO gates (the magnitude of voltage square waves is the DC voltage on the capacitor). By Fourier analysis, we have the fundamental frequency voltage on the AC side

$$V_{\mathrm{ASVG}} = KV_{\mathrm{C}} \sin\frac{\theta}{2}, \tag{5.123}$$

where K is a constant related to inverter structure; V_{C} is the capacitor DC voltage; θ is the control variable.

STATCOM connection to the systems is shown in Fig. 5.25. It must connect to systems through reactors or transformers because the use of voltage-bridge circuits. The connection reactor is needed to link the two unequal voltage sources, STATCOM and AC system. Its other function is to suppress the high-order harmonics in the current. Its inductance does not need to be large. The reactor in the figure is the transformer equivalent leakage reactance or the connection reactor. The resistor is the equivalent copper loss of the transformer or STATCOM loss. STATCOM is represented as an ideal synchronous condenser. Using the system voltage as the

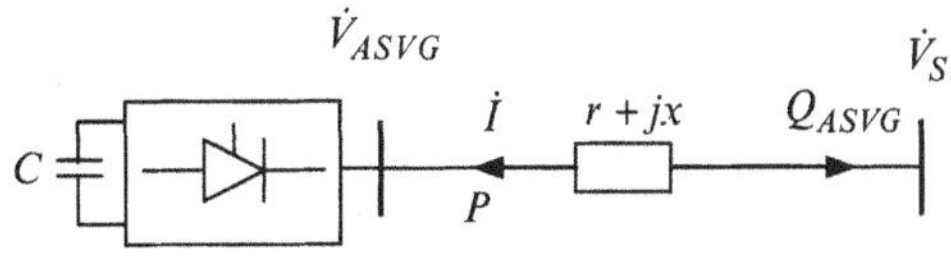

Fig. 5.25 STATCOM connection to systems

reference vector, the fundamental frequency component of the inverter output voltage is V_{ASVG} and lagged phase angle is δ. With $y = 1/\sqrt{r^2 + x^2}$, $\alpha =$ arctg r/x, we have the real power consumed by the inverter as

$$P = V_s V_{ASVG} y \sin(\delta + \alpha) - V_{ASVG}^2 y \sin \alpha. \tag{5.124}$$

The reactive power injection from STATCOM is

$$\begin{aligned} Q_{ASVG} &= \mathrm{Im}\left[-\dot{V}_s \dot{I}^*\right] = \mathrm{Im}\left[V_s \frac{V_{ASVG} \angle \delta - V_s}{r - jx}\right] \\ &= V_s V_{ASVG} y \cos(\delta - \alpha) - V_s^2 y \cos \alpha. \end{aligned} \tag{5.125}$$

In steady state, the inverter neither consumes nor generates real power. Based on (5.124), making P zero yields

$$V_{ASVG} = V_s \frac{\sin(\delta + \alpha)}{\sin \alpha}. \tag{5.126}$$

Taking (5.126) into (5.125) and (5.123) yields

$$Q_{ASVG} = \frac{V_s^2}{2r} \sin 2\delta, \tag{5.127}$$

$$V_C = \frac{V_s \sin(\delta + \alpha)}{K \sin \alpha \sin(\theta/2)}. \tag{5.128}$$

From the above two equations, we know that the adjustment of phase angle δ while maintaining constant pulse width θ can change the output reactive power as well as the capacitor voltage. The simultaneous adjustment of θ and δ can maintain capacitor voltage and change the reactive power output. The vector diagram of STATCOM steady-state operation is shown in Fig. 5.26. We use the equivalent resistance r to represent the inverter real power loss so that the inverter model neither consumes nor generates real power. In the diagram, compensation current $\dot{I}$ is perpendicular to inverter output voltage $\dot{V}_{ASVG}$. The inverter injects reactive power into the system when $\dot{I}$ leads $\dot{V}_{ASVG}$. Otherwise it consumes reactive power. While an SVC changes its equivalent inductance through adjusting the timing of its connection to the system, the STATCOM controls the magnitude and phase-angle of its output voltage.

As shown in the vector diagram, the reactive power provided by STATCOM is

$$Q_{ASVG} = \pm I V_s \cos \delta. \tag{5.129}$$

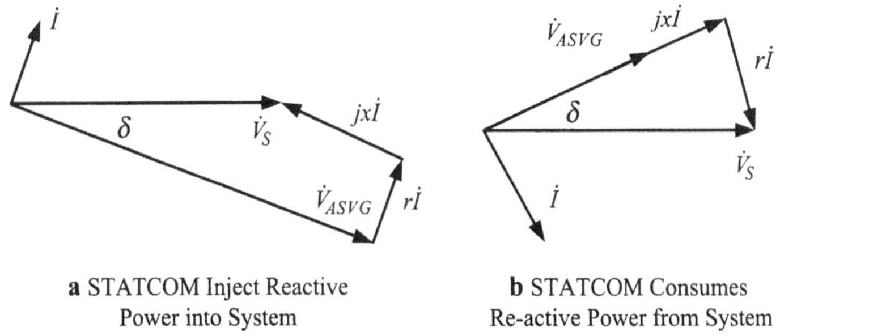

Fig. 5.26 STATCOM steady-state vector diagram

Fig. 5.27 STATCOM voltage adjustment

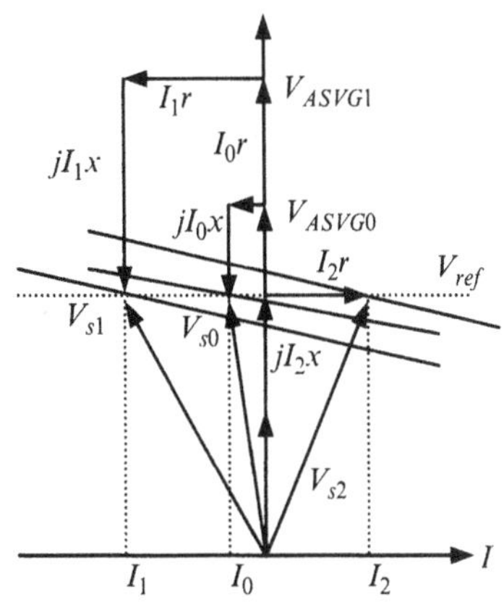

Note that δ is the angle by which vector $\dot{V}_{\mathrm{ASVG}}$ lags $\dot{V}_{\mathrm{s}}$. The positive sign corresponds to a greater than zero δ; the negative sign to a less than zero δ. Substituting the above into (5.127) yields the magnitude of compensation current as

$$I = \mp \frac{V_{\mathrm{s}}}{r} \sin\delta. \tag{5.130}$$

The phase angle of the compensation current is $\pm(\pi/2-\delta)$ as shown in Fig. 5.26. The real and reactive power components of the compensation current are

$$I_{\mathrm{P}} = I \cos\left(\frac{\pi}{2} - \delta\right) = \frac{V_{\mathrm{s}}}{r} \sin^2\delta = \frac{V_{\mathrm{s}}}{2r}(1 - \cos 2\delta), \tag{5.131}$$

$$I_{\mathrm{Q}} = \mp I \sin\left(\frac{\pi}{2} - \delta\right) = \frac{V_{\mathrm{s}}}{2r} \sin 2\delta. \tag{5.132}$$

Figure 5.27 shows the system voltage adjusted by STATCOM. Voltage V_{s0} is the voltage setting value V_{ref} under STATCOM output voltage of V_{ASVG0} and compensation current of I_0. When system operating conditions vary and the bus voltage

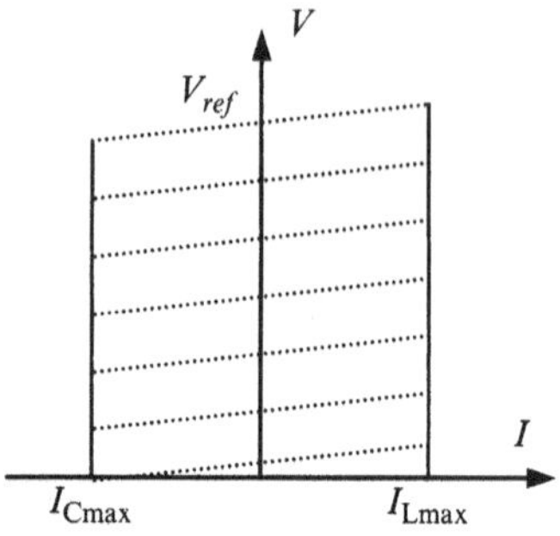

Fig. 5.28 STATCOM volt–ampere characteristics

decreases, STATCOM increases δ to inject more reactive power. The compensation current is I_1 while the voltage is maintained as V_{ref}. The STATCOM keeps system voltage constant through the adjustment of its control parameters. A practical STATCOM usually implements bus voltage mismatch control. From the above analysis, the operation characteristics of STATCOM are shown in Fig. 5.28, and approach rectangular. The constraints of maximum voltage and current are determined by the STATCOM capacity. Voltage setting is determined by the control scheme. Comparing with SVC inverse triangular operational characteristics, STATCOM has wider operation ranges.

We must notice that only one of the two control variables of the STATCOM is independent. The adjustment of δ will change both the magnitude and phase angle of the compensation current. The control variable θ is constrained by (5.128). As δ changes, θ should vary accordingly to maintain a constant capacitor voltage. The range of δ variation is very limited. When a STATCOM consumes reactive power from the system, $\dot{V}_s$ lags $\dot{V}_{ASVG}$ by δ. We can see from the vector diagram in Fig. 5.26b that δ is always less than α. The equivalent resistance r is much less than equivalent reactance x so that α is very small. When a STATCOM injects reactive power into the system, $\dot{V}_s$ leads $\dot{V}_{ASVG}$ by δ. As seen in (5.130) a small r makes δ less constrained by compensation current. Hence (5.132) indicates an approximately linear relationship between reactive compensation current and δ. To neglect resistance r for approximate analysis, setting α and δ to zero in (5.124) and (5.125) yields

$$P = 0, \quad Q_{ASVG} = V_s \frac{V_{ASVG} - V_s}{x}.$$

Now the free control variable of STATCOM is θ, and V_{ASVG} is determined by (5.123). If V_{ASVG} is greater than V_S, the STATCOM injects reactive power into the system, otherwise it consumes reactive power.

A STATCOM can be represented as a shunt connected, controllable current source as noted in (5.130) for power system stability and control analysis. The magnitude and phase angle are determined by the STATCOM controller.

5.5.3 Basic Principle and Mathematical Model of TCSC

TCSC can rapidly and continuously change the equivalent reactance of the compensated line, to maintain a constant power flow on the line within certain operating conditions. In system transients, the TCSC can increase system stability through its fast variation of line reactance. The earliest TCSC was first put into operation in USA in 1991. TCSC have many different structures. One of its basic formations is shown in Fig. 5.29.

Figure 5.29 shows a fixed capacitor and a parallel connected TCR. Its control element is the thyristor. We have seen TCR utilization in the above analysis of SVC. Since SVC is shunt-connected, the voltage on TCR is considered to be sinusoidal. However, the current flowing through TCR is irregular due to valve control, as shown in Fig. 5.21b. The TCR in a TCSC operates in different conditions as compared to those in an SVC. Note that the TCSC is series connected in the transmission line. The current flowing through the TCSC, the line current, is sinusoidal, due to harmonic filtering requirements and to physical operating constraints. Hence the irregular current in the TCR due to valve control will generate a nonsinusoidal capacitor voltage. This is the main difference between the two. Below we introduce the TCSC equivalent reactance at fundamental frequency to understand its working and control principles.

The reference directions of various physical variables are shown in Fig. 5.29. The line current is sinusoidal with the waveforms shown in Fig. 5.30a.

$$i = I_{\mathrm{m}} \sin \omega t \tag{5.133}$$

Suppose that the circuits are in steady state. When the valve conducts, we have the following equations based on circuit theory

$$i = i_{\mathrm{L}} + i_{\mathrm{C}}, \quad v = L\frac{\mathrm{d}i_{\mathrm{L}}}{\mathrm{d}t}, \quad i_{\mathrm{C}} = C\frac{\mathrm{d}v}{\mathrm{d}t}. \tag{5.134}$$

From the above we have

$$i_{\mathrm{L}} + LC\frac{\mathrm{d}^2 i_{\mathrm{L}}}{\mathrm{d}t^2} = I_{\mathrm{m}} \sin \omega t. \tag{5.135}$$

This is a nonhomogeneous differential equation of the inductor current. Its particular solution is the steady-state solution of the second-order circuit as follows:

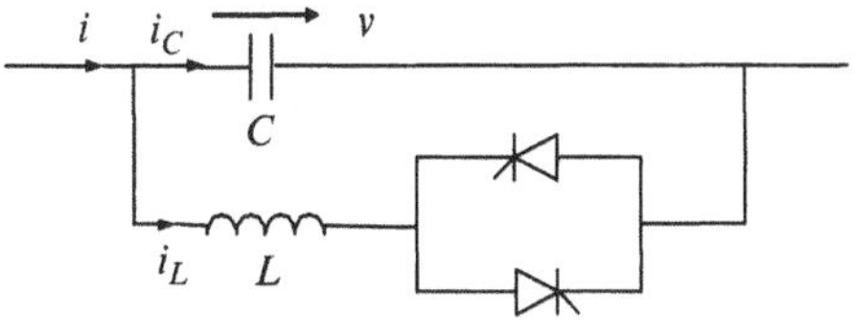

Fig. 5.29 Basic structure of TCSC

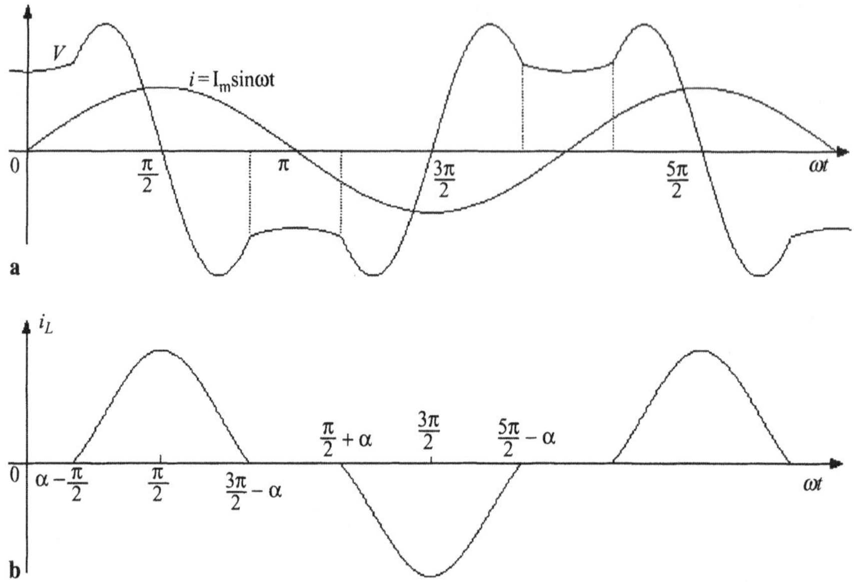

Fig. 5.30 (**a**) TCSC line current and capacitor voltage waveforms (**b**) inductor current waveforms

$$\left.\begin{array}{l} i_{\mathrm{L}}^{\mathrm{s}} = D\sin\omega t, \quad D = \dfrac{\lambda^2}{\lambda^2 - 1} I_{\mathrm{m}} \\ \lambda = \omega_0/\omega, \quad \omega_0 = 1/\sqrt{LC} \end{array}\right\}. \tag{5.136}$$

The complementary solution to homogeneous equation is

$$i_{\mathrm{L}}^{\mathrm{f}} = A\cos\omega_0 t + B\sin\omega_0 t, \tag{5.137}$$

where A and B are the undetermined coefficients. The general solution to (5.135) is

$$i_{\mathrm{L}} = A\cos\omega_0 t + B\sin\omega_0 t + D\sin\omega t. \tag{5.138}$$

Denote α as the firing angle and assume its value in $[\pi/2, \pi]$. A is the electrical angle from capacitor voltage crossing zero to the time of firing. Under steady state, the waveform of inductor current is symmetric to the time point of capacitor voltage crossing zero. The capacitor voltages at the moments of valve turning on and off are equal in magnitude and opposite in direction. Supposing that capacitor voltage magnitude is V_0 when valves turn on and off, the corresponding electrical angles are

$$\theta_k = \alpha - \frac{\pi}{2} + k\pi, \quad \delta_k = \frac{3\pi}{2} - \alpha + k\pi, \quad k = 0, 1, 2, \ldots. \tag{5.139}$$

Based on the initial conditions of inductor current and capacitor voltage: inductor current of zero and capacitor voltage of V_0 at turning on (refer to Fig. 5.30a), we have the following equations

$$A\cos\lambda\theta_k + B\sin\lambda\theta_k + D\sin\theta_k = 0, \tag{5.140}$$

$$L(-\omega_0 A\sin\lambda\theta_k + \omega_0 B\cos\lambda\theta_k + \omega D\cos\theta_k) = (-1)^k V_0. \tag{5.141}$$

The capacitor voltage is V_0 at the turning off time of the valve

$$L(-\omega_0 A\sin\lambda\delta_k + \omega_0 B\cos\lambda\delta_k + \omega D\cos\delta_k) = (-1)^{k+1} V_0. \tag{5.142}$$

The solution to the above three equations yields

$$A = -D\frac{\sin\theta_k}{\cos\lambda\beta}\cos\left(\lambda\frac{2k+1}{2}\pi\right), \tag{5.143}$$

$$B = -D\frac{\sin\theta_k}{\cos\lambda\beta}\sin\left(\lambda\frac{2k+1}{2}\pi\right), \tag{5.144}$$

$$V_0 = DL(\omega\sin\alpha + \omega_0\cos\alpha\,\mathrm{tg}\,\lambda\beta), \tag{5.145}$$

where $\beta = \pi - \alpha$, is called the conducting angle having a value within $[0, \pi/2]$. Substituting A and B in (5.138) yields the inductor current when the valve conducts as

$$i_L = D\left[\sin\omega t + (-1)^k\frac{\cos\alpha}{\cos\lambda\beta}\cos\lambda\left(\omega t - \frac{\pi}{2} - k\pi\right)\right]. \tag{5.146}$$

We can obtain the capacitor voltage from (5.134) as

$$v = DL\left[\omega\cos\omega t - (-1)^k\frac{\cos\alpha}{\cos\lambda\beta}\omega_0\sin\lambda\left(\omega t - \frac{\pi}{2} - k\pi\right)\right]. \tag{5.147}$$

The conducting period is

$$\omega t \in \left[\left(\alpha - \frac{\pi}{2} + k\pi\right), \left(\frac{3\pi}{2} - \alpha + k\pi\right)\right], \quad k = 0, 1, 2, \ldots.$$

The capacitor current $i_C = i + (-i_L)$. There are two components in capacitor current, one is the line current; the other has the same magnitude and opposite direction to the inductor current.

We have assumed that the firing angle is within $[\pi/2, \pi]$. The reason for such an assumption is the same as for the TCR in SVC. As seen from the waveforms of inductor current, to make one valve conduct at any time we have

$$\frac{3\pi}{2} - \alpha + k\pi = \frac{\pi}{2} + \alpha + k\pi.$$

There is one valve turned on when the other is turned off, so $\alpha = \pi/2$. The inductor current, as indicated in (5.146), is

$$i_L = D \sin \omega t.$$

This is the inductor current when the inductor connects directly with the capacitor in parallel. We usually call this bypass mode. When α increases from $\pi/2$ to π, the valve conducting period decreases from π to zero, corresponding to the turning-off of two valves at any moment. This is as if the inductor is not in operation, called off mode. Besides, if α is less than $\pi/2$, the time at which the current of the conducting valve crosses zero is later than the firing time of the other nonconducting valve. Thus

$$\frac{3\pi}{2} - \alpha + k\pi > \alpha - \frac{\pi}{2} + k\pi + \pi.$$

In this case the nonconducting valve cannot be triggered on with a zero voltage across it at the time of firing since the conducting valve is not turned off. Thus one of the valves is always nonconducting at any time, making DC current the main component in inductor current.

Under normal operation, the firing angle of TCR in TCSC has an operating range of $[\pi/2, \pi]$.

When both valves are turned off

$$\omega t \in \left[\left(\frac{\pi}{2} - \alpha + k\pi\right), \left(\alpha - \frac{\pi}{2} + k\pi\right)\right], \quad k = 0, 1, 2, \ldots.$$

The inductor current is zero while the capacitor current is the line current. The capacitor voltage is

$$C\frac{\mathrm{d}v}{\mathrm{d}t} = I_m \sin \omega t.$$

The solution is

$$v = K - \frac{I_m}{\omega C} \cos \omega t. \tag{5.148}$$

We know from (5.147) that the absolute value of capacitor voltage when the valve conducts is V_0. Take $\omega t = \alpha - \pi/2 + k\pi$ as the moment of valve turns on in the above, so

$$(-1)^k V_0 = K - \frac{I_m}{\omega C}\cos\left(\alpha - \frac{\pi}{2} + k\pi\right).$$

Obtaining integral constant K and substituting into (5.148) yields capacitor voltage as

$$v = (-1)^k\left[\frac{I_m}{\omega C}\sin\alpha + V_0\right] - \frac{I_m}{\omega C}\cos\omega t. \tag{5.149}$$

Equations (5.147) and (5.149) give the capacitor voltages when valves are turned on and off, respectively. Apparently, the capacitor voltage is not sinusoidal when $\alpha \neq \pi/2$. Figure 5.30a, b shows the waveforms of capacitor voltage and inductor current.

The Fourier analysis of nonsinusoidal capacitor voltage provides the fundamental frequency component

$$\begin{aligned} V_1 =& \frac{2}{\pi}\int_0^{\alpha-\pi/2}\left(\frac{I_m}{\omega C}\sin\alpha + V_0 - \frac{I_m}{\omega C}\cos\theta\right)\cos\theta\,\mathrm{d}\theta \\ &+ \frac{2}{\pi}\int_{\alpha-\pi/2}^{3\pi/2-\alpha} DL\left[\omega\cos\omega t - \frac{\cos\alpha}{\cos\lambda\beta}\omega_0\sin\lambda\left(\omega t - \frac{\pi}{2}\right)\right]\cos\theta\,\mathrm{d}\theta \,. \\ &+ \frac{2}{\pi}\int_{3\pi/2-\alpha}^{\pi}\left(-V_0 - \frac{I_m}{\omega C}\sin\alpha - \frac{I_m}{\omega C}\cos\theta\right)\cos\theta\,\mathrm{d}\theta \end{aligned} \tag{5.150}$$

The integral of the first item above equals the integral of the third. The sum of the two is

$$F_1 + F_3 = \frac{4}{\pi}\left[V_0\cos\alpha - \frac{I_m}{4\omega C}(2\alpha - \pi + \sin 2\alpha)\right]. \tag{5.151}$$

Taking into account (5.136), the integral of the second item is

$$F_2 = \frac{2}{\pi}DL\left[\omega\beta + \frac{\omega}{2}\sin 2\alpha - \frac{2\omega_0}{\lambda^2 - 1}(\lambda\,\mathrm{tg}\,\alpha + \mathrm{tg}\,\lambda\beta)\cos^2\alpha\right]. \tag{5.152}$$

Substituting (5.145) into (5.151) and rearranging yields the fundamental frequency reactance of TCSC

$$X_{TCSC} = \frac{V_1}{I_m} = \frac{F_1 + F_3 + F_2}{I_m} = K_\beta X_C, \tag{5.153}$$

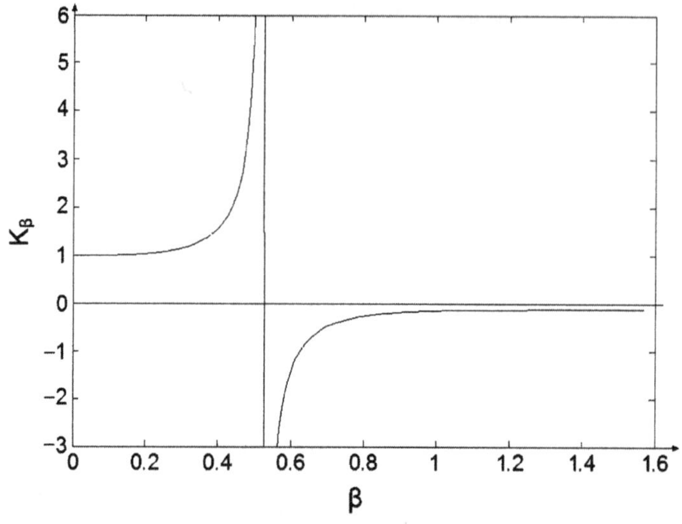

Fig. 5.31 $K_\beta - \beta$ curve

where

$$X_C = -1/\omega C, \tag{5.154}$$

$$K_\beta = 1 + \frac{2}{\pi}\frac{\lambda^2}{\lambda^2 - 1}\left[\frac{2\cos^2\beta}{\lambda^2 - 1}(\lambda \mathrm{tg}\,\lambda\beta - \mathrm{tg}\,\beta) - \beta - \frac{\sin 2\beta}{2}\right]. \tag{5.155}$$

As shown in (5.155), the adjustment of the firing angle changes the reactance XTCSC that is series connected in the line, rendering a controllable equivalent line reactance. The valve control scheme is predefined. TCSC ideal dynamic responses can allow the transmission line capacity to reach its thermal limit.

TCSC usually has $\omega L < 1/\omega C$ and λ^2 around 7 to reduce its cost. Figure 5.31 shows the $K_\beta - \beta$ curve at $\lambda = 3$. When $\beta \in [0, \pi/2\lambda]$, K_β is greater than zero and TCSC is capacitive. When $\beta \in [\pi/2\lambda, \pi/2]$, K_β is less than zero and TCSC is inductive. In the off mode, $\beta = 0$, $K_\beta = 1$. In by-pass mode, $\beta \to \pi/2$, $K_\beta \to 1/(1 - \lambda^2)$. When $\beta \to \pi/2\lambda$, $K_\beta \to \infty$ due to $\mathrm{tg}\lambda\beta \to \infty$, corresponding to parallel LC resonance. To prevent TCSC resonance over voltage, β is prohibited from being operated near $\pi/2\lambda$.

TCSC shown in Fig. 5.29 is a single module. A practical TCSC usually consists of many modules connected in series. Each module has its independent firing angle. The firing angle combination of different modules gives the TCSC equivalent reactance a wider range of variation and smoother adjustment. To protect the TCSC from damage due to overvoltages and overcurrents, there are various protection devices installed and corresponding operation constraints [127].

For power system stability and control analysis, a TCSC can be represented as a variable reactor series connected in the transmission line. The reactance is determined by the TCSC controller.

5.5.4 Basic Principle and Mathematical Model of SSSC

TCSC is a series compensation device using semi-controllable power electronic elements. There are many types of series compensation with fully controllable elements. Here we are going to introduce the SSSC built with GTO voltage-type inverters. The STATCOM discussed before uses voltage-type inverters.

Connected to systems in parallel via reactors or transformers; the SSSC employs voltage-type inverters connected in series in a transmission line through transformers. Neglecting the line-ground branches, the basic connection is shown in Fig. 5.32, where $r + jx$ is the line impedance. Note that the inverter is different to STATCOM as a DC source may be present on the DC side. With a DC source, SSSC can provide reactive power compensation as well as real power compensation to AC systems. When an SSSC only supplies or consumes reactive power, the capacity of its DC source can be small or even zero (the SSSC loss being provided by the AC system).

We know (from the introduction of the STATCOM) that the magnitude and phase angle of inverter output AC voltages are controllable. Hence we can consider the voltage of an SSSC, connected in series on a line, as an approximately ideal voltage source, as shown in Fig. 5.33a. Denote V_{SSSC} the voltage magnitude of the ideal voltage source and δ the voltage leading phase angle regarding voltage at bus l. The vector diagram is shown in Fig. 5.33b, where φ is the leading phase angle of voltage at bus l with regards to line current. Apparently

Fig. 5.32 SSSC basic connection

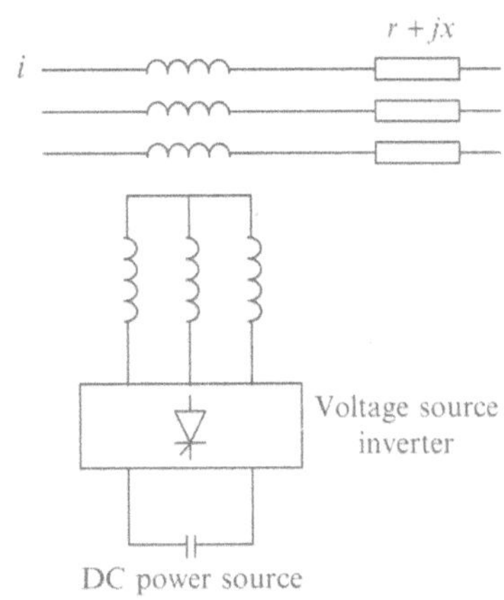

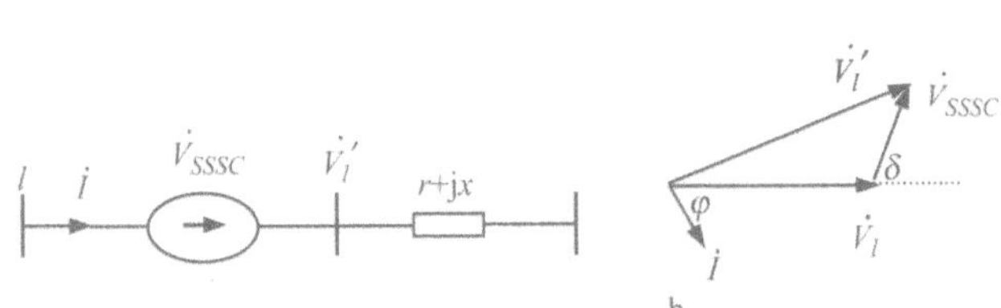

Fig. 5.33 (**a**) SSSC equivalent circuit (**b**) SSSC vector diagram

$$\dot{V}_1' = \dot{V}_1 + \dot{V}_{\text{SSSC}}. \tag{5.156}$$

For pure reactive power compensation, the inverter vector $\dot{V}_{\text{SSSC}}$ is perpendicular to line current $\dot{I}$

$$\delta + \varphi = \pm\pi/2. \tag{5.157}$$

In this way, SSSC corresponds to a reactor connected in series on a transmission line, denoting X_{SSSC} its equivalent reactance, so

$$\begin{aligned} \dot{V}_1 - \dot{V}_1' &= jX_{\text{SSSC}}\dot{I} = -\dot{V}_{\text{SSSC}}, \\ X_{\text{SSSC}} &= \mp V_{\text{SSSC}}/I. \end{aligned} \tag{5.158}$$

When $\dot{V}_{\text{SSSC}}$ leads $\dot{I}$, it is capacitive with negative sign; otherwise it is inductive with positive sign. Note that in the above equation, V_{SSSC} is not related to line current and is controlled by the inverter. Hence the adjustment of V_{SSSC} can change the equivalent reactance. In system analysis, once X_{SSSC} is given, the line current can be determined by

$$\dot{I} = \frac{\dot{V}_1 - \dot{V}_{\text{m}}}{r + j(x + X_{\text{SSSC}})}. \tag{5.159}$$

Thus

$$\left.\begin{aligned} V_{\text{SSSC}} &= I|X_{\text{SSSC}}| \\ \delta &= \pm\frac{\pi}{2} - \varphi \end{aligned}\right\}. \tag{5.160}$$

When X_{SSSC} is less than zero, use positive sign; otherwise use negative sign.

Generally, the source branch in Fig. 5.33a can be represented as a current source and impedance connected in parallel as shown in Fig. 5.34a by Norton's theorem. The current source is

$$\dot{I}_{\text{c}} = \dot{V}_{\text{SSSC}}/(r + jx). \tag{5.161}$$

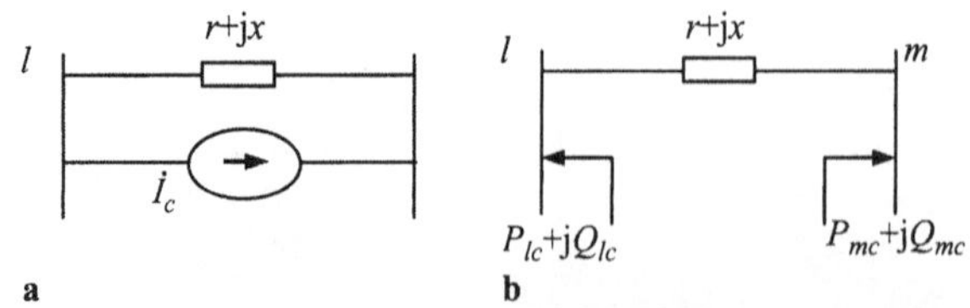

Fig. 5.34 **(a)** Use equivalent current source **(b)** use equivalent power injection

In power system analysis, the bus power injection is used in most cases, further simplifying Fig. 5.34a, b.

As indicated in (5.161)

$$\left.\begin{aligned} P_{\mathrm{lc}} + jQ_{\mathrm{lc}} &= -\dot{V}_{\mathrm{l}}\dot{I}_{\mathrm{c}}^{*} = -\dot{V}_{\mathrm{l}}\left(\frac{\dot{V}_{\mathrm{SSSC}}}{r+jx}\right)^{*} \\ P_{\mathrm{mc}} + jQ_{\mathrm{mc}} &= \dot{V}_{\mathrm{m}}\dot{I}_{\mathrm{c}}^{*} = \dot{V}_{\mathrm{m}}\left(\frac{\dot{V}_{\mathrm{SSSC}}}{r+jx}\right)^{*} \end{aligned}\right\}.$$

Note that the phase angle of $\dot{V}_{\mathrm{SSSC}}$ is $\theta_{\mathrm{l}} + \delta$, so

$$\left.\begin{aligned} P_{\mathrm{lc}} &= V_{\mathrm{l}}V_{\mathrm{SSSC}}(b\sin\delta - g\cos\delta) \\ Q_{\mathrm{lc}} &= V_{\mathrm{l}}V_{\mathrm{SSSC}}(g\sin\delta + b\cos\delta) \end{aligned}\right\}, \tag{5.162}$$

$$\left.\begin{aligned} P_{\mathrm{mc}} &= V_{\mathrm{m}}V_{\mathrm{SSSC}}[g\cos(\theta_{\mathrm{lm}}+\delta) - b\sin(\theta_{\mathrm{lm}}+\delta)] \\ Q_{\mathrm{mc}} &= -V_{\mathrm{m}}V_{\mathrm{SSSC}}[b\cos(\theta_{\mathrm{lm}}+\delta) + g\sin(\theta_{\mathrm{lm}}+\delta)] \end{aligned}\right\}, \tag{5.163}$$

where

$$\left.\begin{aligned} g &= \frac{r}{r^2+x^2} \\ b &= \frac{-x}{r^2+x^2} \\ \theta_{\mathrm{lm}} &= \theta_{\mathrm{l}} - \theta_{\mathrm{lm}} \end{aligned}\right\}. \tag{5.164}$$

The power generated by SSSC is

$$P_{\mathrm{SSSC}} + jQ_{\mathrm{SSSC}} = \dot{V}_{\mathrm{SSSC}}\dot{I}^{*} = \dot{V}_{\mathrm{SSSC}}\left(\frac{\dot{V}_{\mathrm{SSSC}} + \dot{V}_{\mathrm{l}} - \dot{V}_{\mathrm{m}}}{r+jx}\right)^{*},$$

$$P_{\mathrm{SSSC}} = gV_{\mathrm{SSSC}}^{2} + gV_{\mathrm{SSSC}}[V_{\mathrm{l}}\cos\delta - V_{\mathrm{m}}\cos(\theta_{\mathrm{lm}}+\delta)] + bV_{\mathrm{SSSC}}[V_{\mathrm{l}}\sin\delta - V_{\mathrm{m}}\sin(\theta_{\mathrm{lm}}+\delta)], \tag{5.165}$$

$$\begin{aligned} Q_{\mathrm{SSSC}} = &- bV_{\mathrm{SSSC}}^{2} + gV_{\mathrm{SSSC}}[V_{\mathrm{l}}\sin\delta - V_{\mathrm{m}}\sin(\theta_{\mathrm{lm}}+\delta)] \\ &- bV_{\mathrm{SSSC}}[V_{\mathrm{l}}\cos\delta - V_{\mathrm{m}}\cos(\theta_{\mathrm{lm}}+\delta)]. \end{aligned} \tag{5.166}$$

Apparently P_{SSSC} is zero for pure reactive power compensation. Neglecting line resistance, δ satisfies the following for pure reactive power compensation

$$V_{\mathrm{l}}\sin\delta = V_{\mathrm{m}}\sin(\theta_{\mathrm{lm}}+\delta). \tag{5.167}$$

In (5.166), the reactive power compensated by SSSC is not related to the line current directly since the adjustment of V_{SSSC} is not related to the line current.

The power flows from bus l to bus m is:

$$P_{lm} + jQ_{lm} = \dot{V}_l \dot{I}^* = \dot{V}_l \left(\frac{\dot{V}_{SSSC} + \dot{V}_l - \dot{V}_m}{r + jx} \right)^*,$$

$$\left. \begin{array}{l} P_{lm} = gV_l^2 + gV_l[V_{SSSC} \cos \delta - V_m \cos \theta_{lm}] - bV_l[V_{SSSC} \sin \delta + V_m \sin \theta_{lm}] \\ Q_{lm} = -bV_l^2 - gV_l[V_{SSSC} \sin \delta + V_m \sin \theta_{lm}] - bV_l[V_{SSSC} \cos \delta - V_m \cos \theta_{lm}] \end{array} \right\}. \tag{5.168}$$

So the power on the line is controlled by two parameters. For pure reactive power compensation, SSSC has only one independent control variable due to the constraint of (5.167) and has one control objective.

In power system stability and control analysis, SSSC can also be represented as a voltage source connected in series in the line. The controller determines the magnitude and phase angle. The voltage vector is always perpendicular to line current for pure reactive power compensation.

5.5.5 Basic Principle and Mathematical Model of TCPST

Thyristor controlled phase shifting transformer is abridged as TCPST. Phase shifters using mechanical switches to change the tap positions have been utilized in power systems for a long time. It is also called a series voltage booster. Since the response speeds of mechanical switches are slow in tap changing, this type of shifter can only be used in power system steady-state adjustment. Furthermore, the short operational life is a major drawback of this type of shifter. Substituting mechanical switches with thyristors can provide the phase shifter with faster responses and wider application. There are many types of implementation [130, 131]. We are going to use a relative simple type to introduce the working principles and mathematical models.

Figure 5.35 shows a basic connection of TCPST.

Phase shifters consist of parallel transformer (ET), series transformer (BT), and switches. Parallel and series transformers are also called excitation transformer and boosting transformer. Figure 5.35 shows only phase *c* of the secondary side of parallel transformer and secondary and the primary side of the series transformer. The other two phases have the same structure. Switch S is made up of a pair of thyristors connected in parallel in opposite directions, having the same working principles as discussed in TCSC. S_1–S_5 can only have one conducting and all others are turned off under all circumstances. We can see that the ratio of the parallel transformer varies with the conducting conditions of S_1–S_4. When S_1–S_4 are all turned off, S_5 must conduct to short-circuit the primary of the series transformer.

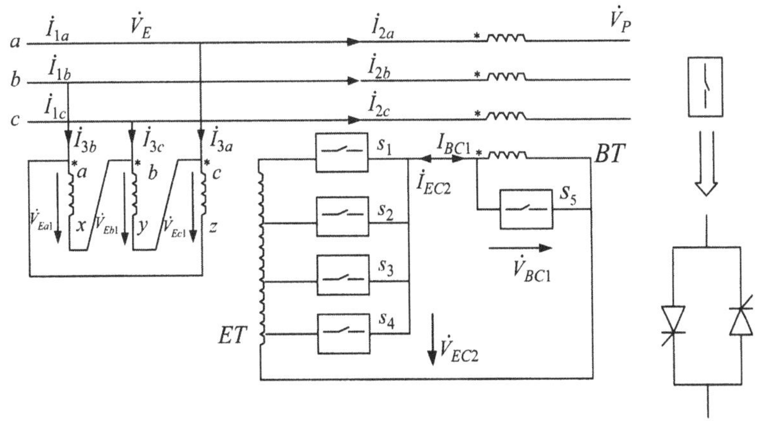

Fig. 5.35 TCPST basic configuration

This is to prevent series connection of the series transformer excitation reactance into the transmission line.

Notice the relationship between the primary voltage of the parallel transformer and the line phase voltage. Phase a, b, and c on the primary of the parallel transformer correspond to phase b, c, and a of the line voltage, respectively. Since the parallel transformer has Δ connection, the relationship between the primary voltages of the parallel transformer and the line phase voltages are

$$\left.\begin{aligned} \dot{V}_{\mathrm{E}a1} &= \dot{V}_{\mathrm{E}b} - \dot{V}_{\mathrm{E}c} \\ \dot{V}_{\mathrm{E}b1} &= \dot{V}_{\mathrm{E}c} - \dot{V}_{\mathrm{E}a} \\ \dot{V}_{\mathrm{E}c1} &= \dot{V}_{\mathrm{E}a} - \dot{V}_{\mathrm{E}b} \end{aligned}\right\}. \tag{5.169}$$

Supposing that the ratios of parallel and series transformers are k_{E} and k_{B}, respectively, and neglecting the voltage loss of transformers, the phase voltages on the secondary of the parallel transformer have the following relationship with the line phase voltages

$$\left.\begin{aligned} \dot{V}_{\mathrm{E}a2} &= k_{\mathrm{E}}\left(\dot{V}_{\mathrm{E}b} - \dot{V}_{\mathrm{E}c}\right)/\sqrt{3} = jk_{\mathrm{E}}\dot{V}_{\mathrm{E}a} \\ \dot{V}_{\mathrm{E}b2} &= k_{\mathrm{E}}\left(\dot{V}_{\mathrm{E}c} - \dot{V}_{\mathrm{E}a}\right)/\sqrt{3} = jk_{\mathrm{E}}\dot{V}_{\mathrm{E}b} \\ \dot{V}_{\mathrm{E}c2} &= k_{\mathrm{E}}\left(\dot{V}_{\mathrm{E}a} - \dot{V}_{\mathrm{E}b}\right)/\sqrt{3} = jk_{\mathrm{E}}\dot{V}_{\mathrm{E}c} \end{aligned}\right\}. \tag{5.170}$$

The phase voltages on the secondary of the series transformer are

$$\left.\begin{aligned} \dot{V}_{\mathrm{B}a2} &= k_{\mathrm{B}}\dot{V}_{\mathrm{B}a1} = k_{\mathrm{B}}\dot{V}_{\mathrm{E}a2} = jk_{\mathrm{B}}k_{\mathrm{E}}\dot{V}_{\mathrm{E}a} \\ \dot{V}_{\mathrm{B}b2} &= k_{\mathrm{B}}\dot{V}_{\mathrm{B}b1} = k_{\mathrm{B}}\dot{V}_{\mathrm{E}b2} = jk_{\mathrm{B}}k_{\mathrm{E}}\dot{V}_{\mathrm{E}b} \\ \dot{V}_{\mathrm{B}c2} &= k_{\mathrm{B}}\dot{V}_{\mathrm{B}c1} = k_{\mathrm{B}}\dot{V}_{\mathrm{E}c2} = jk_{\mathrm{B}}k_{\mathrm{E}}\dot{V}_{\mathrm{E}c} \end{aligned}\right\}. \tag{5.171}$$

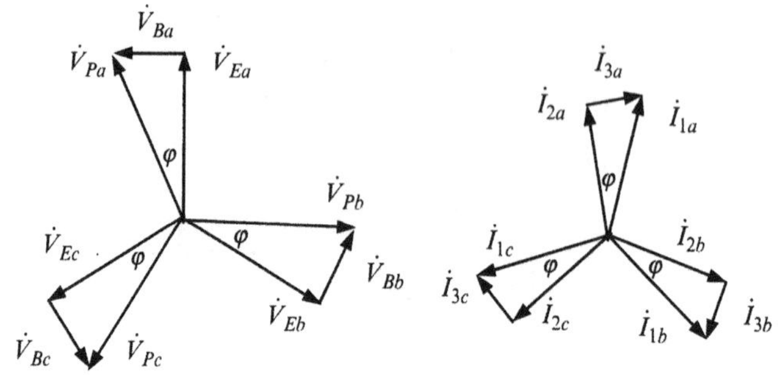

Fig. 5.36 Phase shifter vector diagram

Using a single phase expression to replace the above three-phase we have

$$\dot{V}_{\mathrm{B}} = jk_{\mathrm{B}}k_{\mathrm{E}}\dot{V}_{\mathrm{E}}, \tag{5.172}$$

where $\dot{V}_{\mathrm{E}}$ and $\dot{V}_{\mathrm{B}}$ are input voltage of the parallel transformer and output voltage of the series transformer. Similarly we can obtain the expression for the currents

$$\dot{I}_3 = -jk_{\mathrm{B}}k_{\mathrm{E}}\dot{I}_2. \tag{5.173}$$

The vector diagrams are shown in Fig. 5.36.

From (5.172), (5.173), and Fig. 5.35 we can obtain

$$\dot{V}_{\mathrm{P}} = \dot{V}_{\mathrm{E}} + \dot{V}_{\mathrm{B}} = (1 + jk_{\mathrm{B}}k_{\mathrm{E}})\dot{V}_{\mathrm{E}}, \tag{5.174}$$

$$\dot{I}_1 = \dot{I}_2 + \dot{I}_3 = (1 - jk_{\mathrm{B}}k_{\mathrm{E}})\dot{I}_2. \tag{5.175}$$

Hence we can represent phase shifter as a transformer having complex ratio as follows:

$$\left.\begin{aligned} \dot{K}_{\mathrm{P}} &= \frac{\dot{V}_{\mathrm{P}}}{\dot{V}_{\mathrm{E}}} = 1 + jk_{\mathrm{B}}k_{\mathrm{E}} = K_{\mathrm{P}}\angle\varphi \\ \varphi &= \mathrm{tg}^{-1}k_{\mathrm{B}}k_{\mathrm{E}} \\ K_{\mathrm{P}} &= \sqrt{1 + (k_{\mathrm{B}}k_{\mathrm{E}})^2} = \sec\varphi \end{aligned}\right\}. \tag{5.176}$$

Since the ratio of the parallel transformer, k_{E} is related to the on/off state of switches S_1–S_5, we can change φ by controlling switch states. Apparently φ is a discrete variable. Note that the product of k_{E} and k_{B} is much less than 1. V_{P} is a little larger than V_{E}. The main function of phase shifters is to change the phase angle φ of V_{E}.

Based on (5.174) and (5.175), we can use the phase shifter equivalent circuits shown in Fig. 5.37 for power system stability and control analysis.

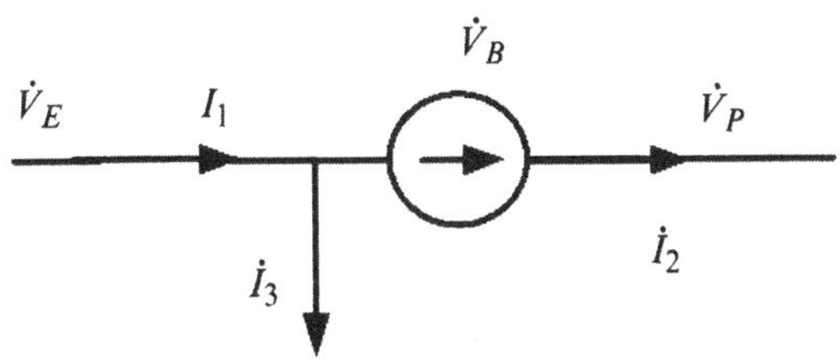

Fig. 5.37 Phase shifter equivalent circuit

The above phase shifter is also called a quadrature boosting transformer (QBT) since its output voltage $\dot{V}_B$ is always perpendicular to $\dot{V}_E$. A new type of phase shifter has been proposed to use a series voltage source $\dot{V}_B$, and its voltage magnitude and phase angle can both be continuously adjusted to provide easier implementation. Generally, this type of phase shifting transformer (PST) has the mathematical model shown in Fig. 5.37. The control variables are voltage $\dot{V}_B$ magnitude and phase angle. Note that the phase shifter is an inactive element. Neglecting its loss, phase shifter output complex power equals its input complex power. So

$$\dot{V}_B\dot{I}_2^* = \dot{V}_E\dot{I}_3^*. \tag{5.177}$$

The power generated from the series transformer is consumed by the shunt current source. Thus

$$\frac{\dot{V}_B}{\dot{V}_E} = \frac{\dot{I}_3^*}{\dot{I}_B^*} = k\angle\varphi,$$

where $k = \left|\frac{\dot{V}_B}{\dot{V}_E}\right|$.

The adjustment of $\dot{V}_B$ magnitude and phase angle can control k and φ. Both phase angle and magnitude of $\dot{V}_P$ are controllable, distinguishing this from QBT. There are two independent control variables. This type of phase shifter is similar to a UPFC and we are not going to discuss it in detail.

5.5.6 Basic Principle and Mathematical Model of UPFC

The FACTS devices that we discussed above manipulate only one of the three parameters affecting power transmission. TCSC and SSC compensate a line parameter. SVC and STATCOM control a bus voltage magnitude. TCPST adjusts bus voltage phase angle. The UPFC [132] is a combination of the above FACTS devices and can adjust the three parameters at the same time. In June 1998, the first UPFC was put into trial operation at AEP in the United States. Its application and control strategy are still under investigation. The basic structure of UPFC is shown in Fig. 5.38.

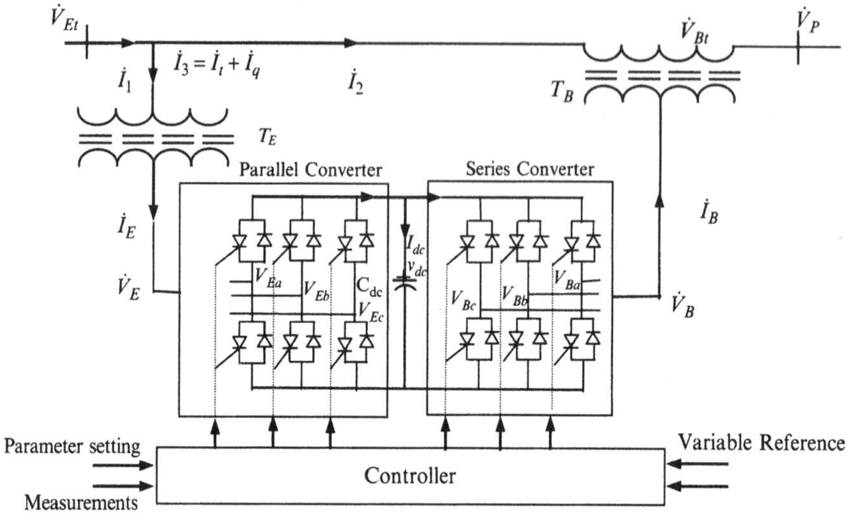

Fig. 5.38 UPFC basic configuration

UPFC is like a combination of SATCOM and SSSC. The two GTO voltage inverters share a capacitor to couple the STATCOM and the SSSC.

Nabavi-Niaki and Iravani [133] presented a dynamic model of UPFC at fundamental frequency and for symmetrical operation. The converters utilize sinusoid pulse width modulation (SPWM). Here we are going to introduce this model. The control variables of SPWM are modulation ratio and phase angle of sinusoidal control signal. As shown in Fig. 5.38, we can separate UPFC into an AC part and a DC part using transformers as the delimiters. The output voltages of the two converters are

$$\dot{V}_{\mathrm{E}} = \frac{1}{2\sqrt{2}} m_{\mathrm{E}} v_{\mathrm{dc}} \angle \delta_{\mathrm{E}} \quad m_{\mathrm{E}} \in [0, 1], \tag{5.178}$$

$$\dot{V}_{\mathrm{B}} = \frac{1}{2\sqrt{2}} m_{\mathrm{B}} v_{\mathrm{dc}} \angle \delta_{\mathrm{B}} \quad m_{\mathrm{B}} \in [0, 1], \tag{5.179}$$

where m_{E} and m_{B} are parallel and series converter modulation ratios; δ_{E} and δ_{B} are the phase angles of sinusoidal control signals; v_{dc} is the instantaneous voltage on the DC capacitor. It is not difficult to understand that there is the following relation between the variation rate of electrical energy on the capacitor and the real power of the converter

$$C_{\mathrm{dc}} v_{\mathrm{dc}} \frac{\mathrm{d}v_{\mathrm{dc}}}{\mathrm{d}t} = \mathrm{Re}\left[\dot{V}_{\mathrm{E}} \dot{I}_{\mathrm{E}}^{*} - \dot{V}_{\mathrm{B}} \dot{I}_{\mathrm{B}}^{*}\right], \tag{5.180}$$

where $\dot{I}_{\mathrm{E}}$ and $\dot{I}_{\mathrm{B}}$ are AC currents on the parallel and series converters. In power system stability analysis, we use a sub-steady-state model for power networks.

Correspondingly, the AC currents of the converters and the voltages on AC side have the following relation using the reference direction in Fig. 5.38

$$(r_{\mathrm{E}} + j\omega l_{\mathrm{E}})\dot{I}_{\mathrm{E}} = \dot{V}_{\mathrm{Et}} - \dot{V}_{\mathrm{E}}, \tag{5.181}$$

$$(r_{\mathrm{B}} + j\omega l_{\mathrm{B}})\dot{I}_{\mathrm{B}} = \dot{V}_{\mathrm{B}} - \dot{V}_{\mathrm{Bt}}, \tag{5.182}$$

where impedances Z_{B} and Z_{E} are equivalent impedances of parallel and series transformers and the converter losses; $\dot{V}_{\mathrm{Et}}$ and $\dot{V}_{\mathrm{Bt}}$ are the voltages transferred from UPFC terminal to converter side. Below we will convert (5.178)–(5.182) into the per unit system. To determine the voltage base, we generally assume that the converter output voltages are rated values when v_{dc} reaches its rated value v_{dcN} and modulation ratios approach 1. In the design of UPFC physical parameters, we have

$$\left.\begin{aligned} V'_{\mathrm{EN}} &= T_{\mathrm{E}} V_{\mathrm{EN}} = T_{\mathrm{E}}\left(\frac{\sqrt{2}}{4} \times 1 \times v_{\mathrm{dcN}}\right) = k_{\mathrm{E}} V_{\mathrm{N}} \\ V'_{\mathrm{BN}} &= T_{\mathrm{B}} V_{\mathrm{BN}} = T_{\mathrm{B}}\left(\frac{\sqrt{2}}{4} \times 1 \times v_{\mathrm{dcN}}\right) = k_{\mathrm{B}} V_{\mathrm{N}} \end{aligned}\right\}, \tag{5.183}$$

where T_{E} and T_{B} are parallel and series transformer ratios; V_{N} is the AC network rated voltage; V'_{EN} and V'_{BN} are the AC side voltages transformed from rated converter output voltages; k_{E} and k_{B} are the two parameters of UPFC. Due to voltage static security constraints, k_{E} and k_{B} must not be too large, for instance, 1.2 and 0.3, respectively. As seen from the above two equations, ratios of parallel and series transformers are different.

Having V_{N} as the voltage base of the AC network, we use the following voltage bases for converters to work with the network per unit system while considering (5.183):

$$V_{\mathrm{EB}} = \frac{V_{\mathrm{N}}}{T_{\mathrm{E}}} = \frac{v_{\mathrm{dcN}}}{2\sqrt{2}k_{\mathrm{E}}}, \tag{5.184}$$

$$V_{\mathrm{BB}} = \frac{V_{\mathrm{N}}}{T_{\mathrm{B}}} = \frac{v_{\mathrm{dcN}}}{2\sqrt{2}k_{\mathrm{B}}}. \tag{5.185}$$

We now have the corresponding current base and impedance base of converters from the above voltage base. DC voltage base is v_{dcN}.

The expressions of (5.178) and (5.179) in the per unit system are

$$\dot{V}_{\mathrm{E}^*} = k_{\mathrm{E}} m_{\mathrm{E}} v_{\mathrm{dc}^*} \angle \delta_{\mathrm{E}}, \quad m_{\mathrm{E}} \in [0, 1], \tag{5.186}$$

$$\dot{V}_{\mathrm{B}^*} = k_{\mathrm{B}} m_{\mathrm{B}} v_{\mathrm{dc}^*} \angle \delta_{\mathrm{B}}, \quad m_{\mathrm{B}} \in [0, 1]. \tag{5.187}$$

In the per unit system, (5.181) and (5.182) are unchanged. On dividing the two sides of (5.180) by the power base leaves, the right side is unchanged and the left side is

$$\frac{C_{\mathrm{dc}}v_{\mathrm{dc}}}{S_{\mathrm{B}}}\frac{\mathrm{d}v_{\mathrm{dc}}}{\mathrm{d}t}=\frac{2}{S_{\mathrm{B}}}\times\left(\frac{1}{2}C_{\mathrm{dc}}v_{\mathrm{dcN}}^{2}\right)\left(\frac{v_{\mathrm{dc}}}{v_{\mathrm{dcN}}}\right)\left(\frac{\mathrm{d}v_{\mathrm{dc}}/v_{\mathrm{dcN}}}{\mathrm{d}t}\right),$$
$$\frac{C_{\mathrm{dc}}v_{\mathrm{dc}}}{S_{\mathrm{B}}}\frac{\mathrm{d}v_{\mathrm{dc}}}{\mathrm{d}t}=\frac{2}{S_{\mathrm{B}}}\times\left(\frac{1}{2}C_{\mathrm{dc}}v_{\mathrm{dcN}}^{2}\right)\left(\frac{v_{\mathrm{dc}}}{v_{\mathrm{dcN}}}\right)\left(\frac{\mathrm{d}v_{\mathrm{dc}}/v_{\mathrm{dcN}}}{\mathrm{d}t}\right)=v_{\mathrm{dc}^*}T_{\mathrm{u}}\frac{\mathrm{d}v_{\mathrm{dc}^*}}{\mathrm{d}t}$$

where T_{u} is the UPFC time constant with the following value

$$T_{\mathrm{u}}=\frac{2W}{S_{\mathrm{B}}}=\frac{2}{S_{\mathrm{B}}}\times\frac{1}{2}C_{\mathrm{dc}}v_{\mathrm{dcN}}^{2}. \tag{5.188}$$

The time constant of a UPFC is related to the rated electrical energy stored in the DC capacitor. Equation (5.180) in the per unit system becomes

$$v_{\mathrm{dc}^*}T_{\mathrm{u}}\frac{\mathrm{d}v_{\mathrm{dc}^*}}{\mathrm{d}t}=\mathrm{Re}\left[\dot{V}_{\mathrm{E}*}\dot{I}_{\mathrm{E}}^{*}-\dot{V}_{\mathrm{B}}\dot{I}_{\mathrm{B}}^{*}\right]. \tag{5.189}$$

For the convenience of expression, we remove the subscripts for per unit system. In power system stability analysis, we need to use the two algebraic equations together with network equations. Substituting (5.186) and (5.187) into (5.181), (5.182), and (5.189), and separating real and imaginary parts we have

$$\begin{bmatrix} r_{\mathrm{E}} & -x_{\mathrm{E}} \\ x_{\mathrm{E}} & r_{\mathrm{E}} \end{bmatrix}\begin{bmatrix} I_{\mathrm{E}x} \\ I_{\mathrm{E}y} \end{bmatrix}=\begin{bmatrix} V_{\mathrm{Et}x} \\ V_{\mathrm{Et}y} \end{bmatrix}-\begin{bmatrix} k_{\mathrm{E}}m_{\mathrm{E}}v_{\mathrm{dc}}\cos\delta_{\mathrm{E}} \\ k_{\mathrm{E}}m_{\mathrm{E}}v_{\mathrm{dc}}\sin\delta_{\mathrm{E}} \end{bmatrix}, \tag{5.190}$$

$$\begin{bmatrix} r_{\mathrm{B}} & -x_{\mathrm{B}} \\ x_{\mathrm{B}} & r_{\mathrm{B}} \end{bmatrix}\begin{bmatrix} I_{\mathrm{B}x} \\ I_{\mathrm{B}y} \end{bmatrix}=\begin{bmatrix} k_{\mathrm{B}}m_{\mathrm{B}}v_{\mathrm{dc}}\cos\delta_{\mathrm{B}} \\ k_{\mathrm{B}}m_{\mathrm{B}}v_{\mathrm{dc}}\sin\delta_{\mathrm{B}} \end{bmatrix}-\begin{bmatrix} V_{\mathrm{Bt}x} \\ V_{\mathrm{Bt}y} \end{bmatrix}, \tag{5.191}$$

$$\begin{aligned} T_{\mathrm{u}}\frac{\mathrm{d}v_{\mathrm{dc}}}{\mathrm{d}t}&=k_{\mathrm{E}}m_{\mathrm{E}}\left(I_{\mathrm{E}x}\cos\delta_{\mathrm{E}}+I_{\mathrm{E}y}\sin\delta_{\mathrm{E}}\right)\\ &\quad-\mathrm{k}_{\mathrm{B}}m_{\mathrm{B}}\left(I_{\mathrm{B}x}\cos\delta_{\mathrm{B}}+I_{\mathrm{B}y}\sin\delta_{\mathrm{B}y}\right). \end{aligned} \tag{5.192}$$

Equations (5.190)–(5.192) constitute UPFC dynamic models in the per unit system.

In steady-state operation, UPFC is an inactive device and has constant capacitor voltage, so

$$\mathrm{Re}\left[\dot{V}_{\mathrm{E}}\dot{I}_{\mathrm{E}}^{*}-\dot{V}_{\mathrm{B}}\dot{I}_{\mathrm{B}}^{*}\right]=0. \tag{5.193}$$

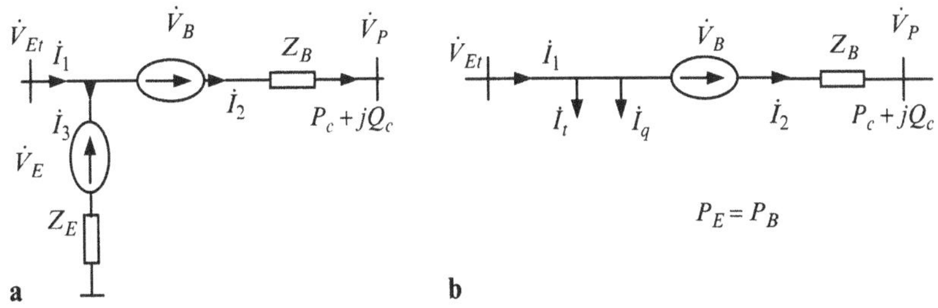

Fig. 5.39 Equivalent circuit of UPFC

Hence, UPFC can be represented as two branches having impedance, connected in series with ideal voltage sources, as shown in Fig. 5.39. $\dot{V}_B$ and $\dot{V}_E$ are adjusted by GTO gate control signals from the parallel and series converters.

Parallel branch current $\dot{I}_3$ can be separated into two components $\dot{I}_t$ and $\dot{I}_q$ as shown in Fig. 5.39.

$$\dot{I}_3 = \frac{\dot{V}_{Et} - \dot{V}_E}{Z_E} = \dot{I}_t + \dot{I}_q, \tag{5.194}$$

where $\dot{I}_t$ and $\dot{I}_q$ components are in phase and perpendicular to bus voltage $\dot{V}_{Et}$. The parallel branch power is

$$P_E = \mathrm{Re}\left(\dot{V}_{Et}\dot{I}_3^*\right) = \dot{V}_{Et}\dot{I}_t^* = \pm V_{Et}I_t, \tag{5.195}$$

$$jQ_E = j\mathrm{Im}\left(\dot{V}_{Et}\dot{I}_3^*\right) = \dot{V}_{Et}\dot{I}_q^* = \pm jV_{Et}I_q. \tag{5.196}$$

In (5.195), we use a negative sign when current is in opposite phase with voltage. In (5.196), we use a positive sign when current leads voltage; otherwise a use negative sign. For the parallel branch, the magnitude and phase of $\dot{V}_E$ determine the magnitudes of I_t and I_q from (5.194). We can see from the above two equations that I_q is the reactive power component of the parallel branch to provide parallel reactive power compensation as in the STATCOM; I_t is the real power component to consume or inject real power into the AC system. This is to maintain constant DC voltage V_{dc} and make the phase of the series voltage source $\dot{V}_B$ to be 360° controllable. The power generated from the series voltage source is

$$S_B = P_B + jQ_B = \dot{V}_B\dot{I}_2^*. \tag{5.197}$$

If we control the phase of $\dot{V}_B$ to make it perpendicular to line current, the function of the series voltage source is like the SSSC series compensation. Generally, the phase and magnitude of $\dot{V}_B$ are fully controllable. TCPST and SSSC do not have this kind of capability. Such a function of UPFC comes from the fact that the real power in

(5.197) is provided by the parallel branch. The control of two voltage sources under the conditions of (5.193) means that the real power generated or consumed by the series voltage source equals to that consumed or generated by the parallel voltage source. Apparently in this case, the electric field energy stored in the DC capacitor does not change and DC voltage is constant. This is the steady state of the UPFC. On the basis of the above analysis, the relationships for the variables in Fig. 5.39 under steady-state conditions are given:

$$\dot{V}_{\mathrm{p}} = \dot{V}_{\mathrm{Et}} + \dot{V}_{\mathrm{B}} - \dot{I}_2 Z_{\mathrm{B}}, \tag{5.198}$$

$$\dot{I}_2 = \dot{I}_1 - \dot{I}_{\mathrm{t}} - \dot{I}_{\mathrm{q}}, \tag{5.199}$$

$$\dot{I}_{\mathrm{t}} + \dot{I}_{\mathrm{q}} = \frac{\dot{V}_{\mathrm{Et}} - \dot{V}_{\mathrm{E}}}{Z_{\mathrm{E}}}, \tag{5.200}$$

$$I_{\mathrm{t}} = \left|\mathrm{Re}\left(\dot{V}_{\mathrm{B}}\dot{I}_2^*\right)/V_{\mathrm{Et}}\right|, \tag{5.201}$$

$$\arg\left(\dot{I}_{\mathrm{t}}\right) = \begin{cases} \arg\left(\dot{V}_{\mathrm{Et}}\right) + 0 & \mathrm{Re}\left(\dot{V}_{\mathrm{B}}\dot{I}_2^*\right) \le 0 \\ \arg\left(\dot{V}_{\mathrm{Et}}\right) + \pi & \mathrm{Re}\left(\dot{V}_{\mathrm{B}}\dot{I}_2^*\right) > 0 \end{cases}, \tag{5.202}$$

$$\arg\left(\dot{I}_{\mathrm{q}}\right) = \arg\left(\dot{V}_{\mathrm{Et}}\right) \pm \pi/2, \tag{5.203}$$

where arg represents the phase angle of the vector. Equations (5.201) and (5.202) correspond to (5.193). Although the magnitudes and phases of voltage sources $\dot{V}_{\mathrm{B}}$ and $\dot{V}_{\mathrm{E}}$ can be continuously adjusted, the constraint of (5.193) reduces the number of independent variables from four to three

$$\left.\begin{array}{l} 0 \le V_{\mathrm{B}} \le V_{\mathrm{B\,max}} \\ 0 \le \varphi_{\mathrm{B}} \le 2\pi \\ 0 \le I_{\mathrm{q}} \le I_{\mathrm{q\,max}} \end{array}\right\}, \tag{5.204}$$

where φ_{B} is the phase angle of parallel voltage source; V_{Bmax} and I_{qmax} are constants related to UPFC rated capacity. The phase vector diagram for UPFC steady-state operation is shown in Fig. 5.40.

For the convenience of analysis, Z_{B} is ignored in the vector diagram. The series connected $\dot{V}_{\mathrm{B}}$ changes the bus voltage from $\dot{V}_{\mathrm{Et}}$ to $\dot{V}_{\mathrm{p}}$. The change of $\dot{V}_{\mathrm{B}}$ makes $\dot{V}_{\mathrm{p}}$ vary within the circle centered at $\dot{V}_{\mathrm{Et}}$ to control the real and reactive power on the line directly. Note that the compensation of $\dot{I}_{\mathrm{q}}$ makes the magnitude of $\dot{V}_{\mathrm{Et}}$ controllable by the UPFC. One UPFC has three independent control variables to manipulate three operation variables P_{c}, Q_{c}, and V_{Et}. The steady-state equivalent circuit of UPFC can also be represented as shown in Fig. 5.39b.

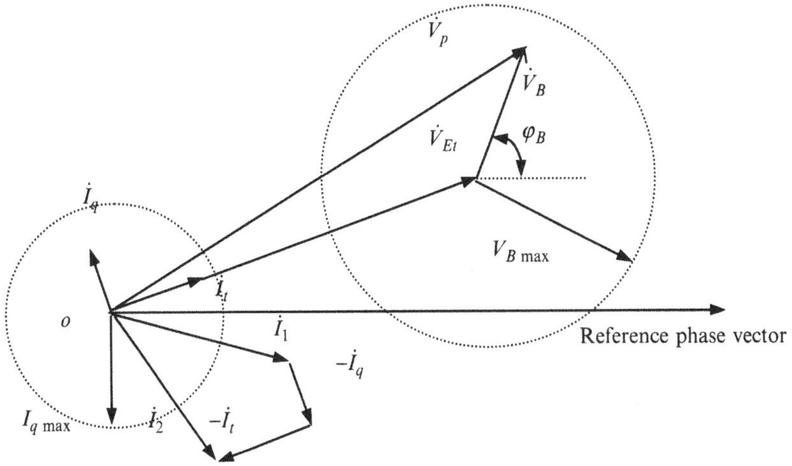

Fig. 5.40 Phase vector diagram for UPFC

We know from the previous analysis of STATCOM and SSSC that both of them require DC voltages V_C to be constant in steady-state operation. The converter AC voltage V_{ASVG} of a STATCOM is perpendicular to the AC current flowing out of the system; the V_{SSSC} of SSSC is perpendicular to the AC current on the transmission line. V_{ASVG} is to satisfy (5.126) and V_{SSSC} to satisfy (5.160). Their phase angles cannot be freely adjusted. Although UPFC still needs to maintain a constant DC voltage, the coupling between the two converters through the DC capacitor allows the real power consumed by the STATCOM to be sent back through the SSSC or vice versa. The magnitude and phase angle of series transformer output $\dot{V}_B$ can then be freely adjusted. The parallel transformer can provide not only reactive power compensation but also the real power transfer between the system and the series transformer. The functional difference between UPFC and PST is due to (5.193) and (5.177). I_q in UPFC is a free variable. For PST, the real power and reactive power taken from the system by the parallel branch are injected into the system by the series branch due to the constraint (5.177). Hence UPFC has STATCOM function while PST does not.

Thinking and Problem Solving

1. What are the factors that limit power transmission distance and capacity?
2. What are the advantages and disadvantages of AC transmission?
3. Discuss the advantages and disadvantages of DC transmission and the applications for which DC transmission is more suitable.
4. What are the characteristics of other new power transmission modes being studied at present?
5. What is the free load flow?

6. Why should flexible electrical power systems be introduced?
7. Can we contemplate $I_d < 0$ when $V_{2d} > V_{1d}$ in (5.1)? Why?
8. Why can DC transmission lines only transmit active power, but the converters absorb reactive power from the AC system?
9. What is the physical significance of phase-shifter resistance R_γ in (5.21)?
10. Distinguish trigger delay angle, phase-shifter angle, extinguish angle, trigger lead angle, and extinguish lead angle, paying attention to their operating areas.
11. Discuss the steady load flow control method of DC transmission.
12. Compare load flow calculation models with and without DC transmission lines.
13. Give an appropriate value L, C, and draw the curve of SVC equivalent reactance X_{SVC}–β denoted in (5.121).
14. Draw V–β curve according to (5.121), in which $V \in [0.9, 1.1]$, when per unit value $V_{ref} = 1.05$, $X_e = 0.05$ in (5.122).
15. Draw the equivalent circuit diagram when S_5 is also tripped with S_1–S_4 tripping in Fig. 5.35.
16. In steady state, UPFC can be regarded equivalently as two voltage source converters (VSC), with voltage amplitude values V_B and V_E, respectively, and phase angles δ_B and δ_E, respectively. Analyze why UPFC can only control three operational variables (active power P_c, reactive power Q_c, and nodal V_{Et} on a transmission line) in the steady state?
17. Discuss the capacitive and inductive value range of X_{TCSC} when conduction angle β should avoid the resonance region according to (5.153)–(5.155). Then discuss whether line transmission active power $P_c = P_{lp}$ controlled by TCSC can vary continuously.

Chapter 6
Mathematical Model of Synchronous Generator and Load

6.1 Introduction

The continuous increase of power system complexity and installation of more and more new equipment in power systems has demanded better methods for power system analysis, planning, and control. At present, analysis of modern power systems is generally based on digital computers. Hence, establishment of a mathematical model, describing the physical processes of a power system, is the foundation for the analysis and investigation of various power system problems. Correct and accurate computation for power system analysis requires a correct and accurate mathematical model of the power system.

Transient processes of the power system are very fast. This is why power system operation heavily relies on the applications of automatic control. With the installation of many different automatic control devices, the operation of which largely depends on the application of electronic and computing technology, modern power system operation has reached a very high level of automation. For such large-scale and complex systems, the mathematical description is nonlinear and high dimensional, consisting of a large number of nonlinear equations. Hence it is both appropriate and practical that the analysis and computation of such a system ought to start from simple local devices and be completed finally for the complex overall system. Therefore, in modeling of large-scale and complex power systems, these systems are first decomposed into independent basic components, such as synchronous generators, transformers, transmission lines, governors and automatic voltage regulators (AVR), etc. Then those components are modeled separately according to circuit theory or other related principles. Models of those components are building-bricks to construct the mathematical model of whole power systems.

For the study of different problems on the same system, different models are required. Mathematically, a power system is a nonlinear dynamic system. When the steady-state operation of a dynamic system is studied, the mathematical description of the system is in the form of algebraic equations. Differential equations (sometimes partial differential equations) give a mathematical description of system

X.-F. Wang et al., *Modern Power Systems Analysis.*
doi: 10.1007/978-0-387-72853-7,

dynamics. For the study of some specific problems, model parameters could be time-variable and variables may not be continuous. In addition, to meet the requirement of different computing accuracy, different models could be used. Obviously, a mathematical model for qualitative analysis could be simpler than that for quantitative analysis. Computing accuracy and speed are always two conflicting factors which need to be considered carefully when a power system model is established. The more accurate the computation is, the more the computing work and hence the longer the computing time. On the other hand, to sacrifice some computing accuracy will be compensated with high computing speed, which has been a common practice in modeling power systems and developing computing algorithms. In this aspect, the effort has been to develop a mathematical model of a power system and the associated solution methods, such that the need for both computing accuracy and computing speed is met. Often, the result is a compromise between those two requirements based on available computing tools.

There are two major issues in mathematical modeling. The first one is to describe a subject under investigation mathematically in the form of equations. There are two methods to establish those mathematical equations. The first method, the analytical method, is to derive those mathematical descriptions by using special knowledge and theory about the subject; the second is to identify them by carrying out experiments or using data obtained from its operation. That is the method of system identification in control theory. The second major issue in mathematical modeling is to obtain parameters of the mathematical description of the subject. No matter whether the plant is described by algebraic or differential equations, various parameters in those equations need to be obtained. Generally, for simple components of the subject, model parameters can be derived from design parameters according to certain physical (such as mechanical or electrical) principles. For example, four parameters of an overhead line, i.e., resistance, inductance, capacitance, and conductance to earth, can be obtained by applying electromagnetic theory to the way the line is arranged in space, the materials of the line, and the natural environment where the line is located. That is a typical analytical method. However, for complex components or systems, usually there is certain difference between the actual parameters and design parameters. A typical example is the generator parameter which could be affected by variations of power system operating conditions, saturation, and a series of complex conversion processes among mechanical, electrical, magnetic, and thermal energy. Therefore, in addition to the method of theoretical derivation, there is another important way to obtain model parameters of complex components and systems. This is the method of parameter estimation which is one of the methods in system identification. Parameter estimation and system identification is a research field which will not be discussed in this book.

In Chap. 1, the mathematical model of a power network has been introduced. Mathematical models of HVDC and FACTS are discussed in Chap. 5. Hence, in this chapter the focus is the introduction to mathematical models of generator and load, including the mathematical models of synchronous generator, excitation systems, and governing systems.

6.2 Mathematical Model of Synchronous Generator

The dynamics of a synchronous generator is the basis for the study of the dynamic behavior of the power system. In the history of developing the mathematical model of a synchronous generator, two milestones are the establishment of two-reaction theory in 1920s [146, 147] and the proposal of Park's transformation [148]. Under the ideally assumed conditions and by using two-reaction principle, Park derived the basic mathematical equations of a synchronous generator in *dq0* coordinate system. Since then, mathematical models of synchronous generators have been based on Park's contribution with further major development regarding the number of equivalent windings to model the generator rotor winding, different assumptions about when a synchronous generator should be described by transient or subtransient parameters, different ways to describe magnetic saturation, etc. Details of all these points above can be found in [149–152]. In this section, we shall focus on those mathematical descriptions of the synchronous generator which have been widely used. Readers should note that in other references, different symbols, defined positive directions of physical variables, form of transformation matrix, and selection of base values may be used.

From the structure of a synchronous generator we know that on the rotor, the field winding is a physical winding; while damping windings may just be electrically equivalent windings. For a salient-pole generator, damping windings represent the damping function of damping rods distributed on the rotor. While for a round rotor generator, they simulate the damping function produced by the eddy current inside the whole rotor. Since they are just equivalent windings, the damping function can be represented by a single or multiple damping windings. In theory, the more the equivalent damping windings, the more accurate the representation can be. However, if more equivalent damping windings are used, there could be two problems. The first is the increase of the order of differential equations in the mathematical model, adding computational burden for their solution. The second problem is that it is more difficult to obtain the relative electrical parameters accurately. Hence in the commonly used mathematical model of a synchronous generator, the number of equivalent damping windings is usually not more than three. Since the damping rods on the rotor of a salient-pole generator are more like real windings than the whole rotor of a round rotor generator and the magnetic circuit of the salient-pole generator is different in d and q directions, the damping function of the salient-pole generator is usually represented by two damping windings, one in the direction of direct axis (d), denoted as D damping winding and another in that of the quadrature axis (q), denoted as Q damping winding; For the round rotor generator, in addition to D and Q damping windings, one more equivalent damping winding in the quadrature direction (g winding) is used. Q and g winding represents the weaker and stronger eddy current effect, respectively.

According to the theory of electric machines, the ideal assumptions about the synchronous generators are that the magnetic circuits are symmetrical, saturation is negligible, and flux waveforms have sinusoidal space distribution. In the following, we shall first derive the mathematical model of an ideal synchronous generator with *D*, *g*, and *Q* damping winding, followed by introduction of a method considering magnetic saturation effects.

6.2.1 Basic Mathematical Equations of Synchronous Generator

6.2.1.1 Three-Phase Mathematical Equations

Figure 6.1a, b shows the structure of a synchronous generator and winding circuit diagram. We consider the general case of a salient-pole generator with *D*, *g*, *Q* three damping windings and treat a round rotor generator as a special case since it has only *D*, *Q* two damping windings. In the figures, the defined positive direction of voltage, current, and magnetic flux is related to the three-phase armature windings *abc*, field winding *f* and damping winding *D*, *g*, *Q*. It must be pointed out that the positive direction of magnetic flux related to the three-phase armature windings is opposite to that induced by the armature current of each winding in the positive direction; while magnetic flux associated with rotor windings is defined in the same direction as that induced by the current in each winding in the positive direction; *q*-axis leads *d*-axis by 90° in the rotational direction of the generator rotor. In addition, the positive direction of all flux axes is chosen to be those of the corresponding magnetic flux.

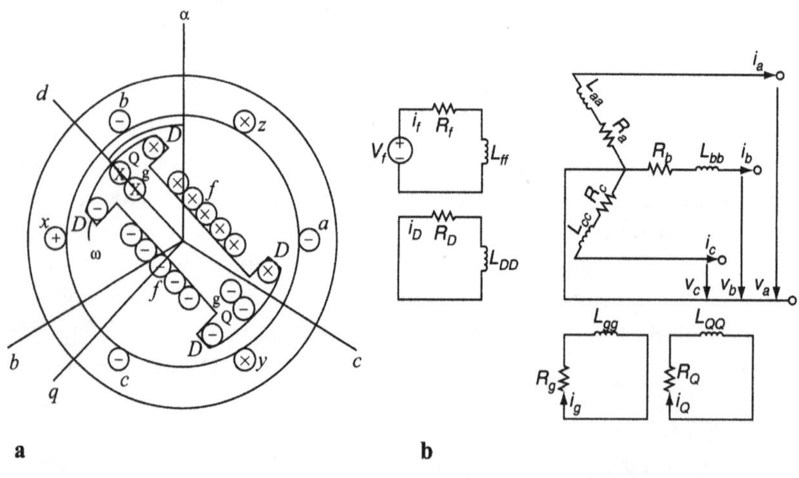

a **b**

Fig. 6.1 Structure of synchronous generator and winding circuit. (**a**) Structure of synchronous generator (**b**) winding circuit

From Fig. 6.1b, the following voltage equation for all the windings can be obtained

$$\begin{bmatrix} v_a \\ v_b \\ v_c \\ -- \\ v_f \\ 0 \\ 0 \\ 0 \end{bmatrix} = \left[\begin{array}{ccc|cccc} R_a & 0 & 0 & 0 & 0 & 0 & 0 \\ 0 & R_a & 0 & 0 & 0 & 0 & 0 \\ 0 & 0 & R_a & 0 & 0 & 0 & 0 \\ \hline 0 & 0 & 0 & R_f & 0 & 0 & 0 \\ 0 & 0 & 0 & 0 & R_D & 0 & 0 \\ 0 & 0 & 0 & 0 & 0 & R_g & 0 \\ 0 & 0 & 0 & 0 & 0 & 0 & R_Q \end{array}\right] \begin{bmatrix} -i_a \\ -i_b \\ -i_c \\ -- \\ i_f \\ i_D \\ i_g \\ i_Q \end{bmatrix} + p \begin{bmatrix} \varphi_a \\ \varphi_b \\ \varphi_c \\ -- \\ \varphi_f \\ \varphi_D \\ \varphi_g \\ \varphi_Q \end{bmatrix}, \tag{6.1}$$

where $p = \frac{d}{dt}$ denotes the differentiation operator.

For an ideal synchronous generator, magnetic saturation effects can be ignored. Hence magnetic flux linkage of each winding can be written in the form of self-inductance and mutual inductance as shown by the following flux linkage equation

$$\begin{bmatrix} \varphi_a \\ \varphi_b \\ \varphi_c \\ -- \\ \varphi_f \\ \varphi_D \\ \varphi_g \\ \varphi_Q \end{bmatrix} = \left[\begin{array}{ccc|cccc} L_{aa} & M_{ab} & M_{ac} & M_{af} & M_{aD} & M_{ag} & M_{aQ} \\ M_{ba} & L_{bb} & M_{bc} & M_{bf} & M_{bD} & M_{bg} & M_{bQ} \\ M_{ca} & M_{cb} & L_{cc} & M_{cf} & M_{cD} & M_{cg} & M_{cQ} \\ \hline M_{fa} & M_{fb} & M_{fc} & L_{ff} & M_{fD} & M_{fg} & M_{fQ} \\ M_{Da} & M_{Db} & M_{Dc} & M_{Df} & L_{DD} & M_{Dg} & M_{DQ} \\ M_{ga} & M_{gb} & M_{gc} & M_{gf} & M_{gD} & L_{gg} & M_{gQ} \\ M_{Qa} & M_{Qb} & M_{Qc} & M_{Qf} & M_{QD} & M_{Qg} & L_{QQ} \end{array}\right] \begin{bmatrix} -i_a \\ -i_b \\ -i_c \\ -- \\ i_f \\ i_D \\ i_g \\ i_Q \end{bmatrix}. \tag{6.2}$$

From circuit theory we know that the above coefficient matrix is symmetrical. From Fig. 6.1a we can see that due to the rotor rotation, the reluctance of the magnetic circuit of some windings changes periodically with the variation of rotor position. Hence the self-inductance and mutual inductance of those windings are a function of rotor position. According to the assumptions of an ideal synchronous generator, both the magnetomotive force (mmf) induced by armature current and mutual flux between armature windings and rotor windings have sinusoidal space distribution. Rotor position can be described by the angle between d-axis and flux axis of phase a armature winding $\theta = \theta_0 + \omega t$. Hence the self-inductance and mutual inductance of each winding can be expressed as follows [153].

1. Self-inductance and mutual inductance of armature windings

$$\left.\begin{aligned} L_{aa} &= l_0 + l_2 \cos 2\theta \\ L_{bb} &= l_0 + l_2 \cos 2(\theta - 2\pi/3) \\ L_{cc} &= l_0 + l_2 \cos 2(\theta + 2\pi/3) \end{aligned}\right\}, \tag{6.3}$$

$$\left.\begin{aligned} M_{ab} &= -[m_0 + m_2 \cos 2(\theta + \pi/6)] \\ M_{bc} &= -[m_0 + m_2 \cos 2(\theta - \pi/2)] \\ M_{ca} &= -[m_0 + m_2 \cos 2(\theta + 5\pi/6)] \end{aligned}\right\}. \tag{6.4}$$

Under the assumptions of an ideal synchronous generator, it can be proved that $l_2 = m_2$. Furthermore, for a round rotor generator, reluctance of magnetic circuits related to the self-inductance and mutual inductance of armature windings does not vary with rotor rotation, and we have $l_2 = m_2 = 0$. Hence those self-inductance and mutual inductance above are constant.

2. Mutual inductance between armature and rotor windings

$$\left.\begin{aligned} M_{af} &= m_{af}\cos\theta \\ M_{bf} &= m_{af}\cos(\theta - 2\pi/3) \\ M_{cf} &= m_{af}\cos(\theta + 2\pi/3) \end{aligned}\right\},\quad \left.\begin{aligned} M_{aD} &= m_{aD}\cos\theta \\ M_{bD} &= m_{aD}\cos(\theta - 2\pi/3) \\ M_{cD} &= m_{aD}\cos(\theta + 2\pi/3) \end{aligned}\right\}, \tag{6.5}$$

$$\left.\begin{aligned} M_{ag} &= -m_{ag}\sin\theta \\ M_{bg} &= -m_{ag}\sin(\theta - 2\pi/3) \\ M_{cg} &= -m_{ag}\sin(\theta + 2\pi/3) \end{aligned}\right\},\quad \left.\begin{aligned} M_{ag} &= -m_{aQ}\sin\theta \\ M_{bg} &= -m_{aQ}\sin(\theta - 2\pi/3) \\ M_{cg} &= -m_{aQ}\sin(\theta + 2\pi/3) \end{aligned}\right\}. \tag{6.6}$$

3. Self-inductance and mutual inductance of rotor windings

   Since rotor windings rotate with the generator rotor, for salient-pole or round rotor generator, reluctance of magnetic circuits does not vary with the change of rotor position. Hence self-inductance and mutual inductance of rotor windings are constant. *D*, *f* winding on direct axis (*d*) is vertical to *g*, *Q* winding on quadrature axis (*q*). Hence mutual inductance between them is zero, that is

$$M_{fg} = M_{fQ} = M_{Dg} = M_{DQ} = 0. \tag{6.7}$$

6.2.1.2 Basic Equations in *dq0* Coordinate

From the discussion above we know that the self-inductance and mutual inductance of generator windings are not constant and some of them vary with the position of the generator rotor. Hence (6.1) and (6.2) are time-variant differential equations which are difficult to solve. To transfer these into time-invariant differential equations, some of methods of coordinate transformation have been proposed, among which the *dq0* transformation proposed by Park [148] has been most widely used. In *dq0* coordinate, flux linkage equations become time invariant. Hence the mathematical model of a synchronous generator is presented as a group of time-invariant differential equations. In the following, we shall discuss the details of Park's transformation.

Park's transformation converts three-phase flux linkage, armature current, and voltage into *d*, *q*, *0* components in the *dq0* coordinates, through an equivalent coordinate transformation. It can be written as

$$\begin{bmatrix} A_d \\ A_q \\ A_0 \end{bmatrix} = \frac{2}{3}\begin{bmatrix} \cos\theta & \cos(\theta - 2\pi/3) & \cos(\theta + 2\pi/3) \\ -\sin\theta & -\sin(\theta - 2\pi/3) & -\sin(\theta + 2\pi/3) \\ 1/2 & 1/2 & 1/2 \end{bmatrix}\begin{bmatrix} A_a \\ A_b \\ A_c \end{bmatrix}. \tag{6.8}$$

For simplicity of expression, the equation above can be written in the compact form as follows

$$\mathbf{A}_{dq0} = \mathbf{P}\mathbf{A}_{abc}. \tag{6.9}$$

The inverse Park's transformation is

$$\begin{bmatrix} A_a \\ A_b \\ A_c \end{bmatrix} = \begin{bmatrix} \cos\theta & -\sin\theta & 1 \\ \cos(\theta - 2\pi/3) & -\sin(\theta - 2\pi/3) & 1 \\ \cos(\theta + 2\pi/3) & -\sin(\theta + 2\pi/3) & 1 \end{bmatrix} \begin{bmatrix} A_d \\ A_q \\ A_0 \end{bmatrix} \tag{6.10}$$

$$\text{or } \mathbf{A}_{abc} = \mathbf{P}^{-1}\mathbf{A}_{dq0}. \tag{6.11}$$

In (6.8)–(6.11), symbol $\mathbf{A}$ represents current, voltage, or flux linkage, i.e.,

$$\mathbf{i}_{dq0} = \mathbf{P}\mathbf{i}_{abc}, \mathbf{v}_{dq0} = \mathbf{P}\mathbf{v}_{abc}, \mathbf{\Psi}_{dq0} = \mathbf{P}\mathbf{\Psi}_{abc}, \tag{6.12}$$

$$\mathbf{i}_{abc} = \mathbf{P}^{-1}\mathbf{i}_{dq0}, \mathbf{v}_{abc} = \mathbf{P}^{-1}\mathbf{v}_{dq0}, \mathbf{\Psi}_{abc} = \mathbf{P}^{-1}\mathbf{\Psi}_{dq0}. \tag{6.13}$$

Applying the transformations of (6.12) and (6.13) as well as (6.3)–(6.7), (6.1) and (6.2) can be converted into the following equations in *dq0* coordinates

$$\begin{bmatrix} v_d \\ v_q \\ v_0 \\ \hline v_f \\ 0 \\ 0 \\ 0 \end{bmatrix} = \left[\begin{array}{ccc|cccc} R_a & 0 & 0 & 0 & 0 & 0 & 0 \\ 0 & R_a & 0 & 0 & 0 & 0 & 0 \\ 0 & 0 & R_a & 0 & 0 & 0 & 0 \\ \hline 0 & 0 & 0 & R_f & 0 & 0 & 0 \\ 0 & 0 & 0 & 0 & R_D & 0 & 0 \\ 0 & 0 & 0 & 0 & 0 & R_g & 0 \\ 0 & 0 & 0 & 0 & 0 & 0 & R_Q \end{array}\right] \begin{bmatrix} -i_d \\ -i_q \\ -i_0 \\ \hline i_f \\ i_D \\ i_g \\ i_Q \end{bmatrix} + p \begin{bmatrix} \varphi_d \\ \varphi_q \\ \varphi_0 \\ \hline \varphi_f \\ \varphi_D \\ \varphi_g \\ \varphi_Q \end{bmatrix} - \begin{bmatrix} \omega\varphi_q \\ -\omega\varphi_d \\ 0 \\ \hline 0 \\ 0 \\ 0 \\ 0 \end{bmatrix}, \tag{6.14}$$

$$\begin{bmatrix} \varphi_d \\ \varphi_q \\ \varphi_0 \\ \hline \varphi_f \\ \varphi_D \\ \varphi_g \\ \varphi_Q \end{bmatrix} = \left[\begin{array}{ccc|cccc} L_d & 0 & 0 & m_{af} & m_{aD} & 0 & 0 \\ 0 & L_q & 0 & 0 & 0 & m_{ag} & m_{aQ} \\ 0 & 0 & L_0 & 0 & 0 & 0 & 0 \\ \hline 3m_{af}/2 & 0 & 0 & L_f & m_{fD} & 0 & 0 \\ 3m_{aD}/2 & 0 & 0 & m_{fD} & L_D & 0 & 0 \\ 0 & 3m_{ag}/2 & 0 & 0 & 0 & L_g & m_{gQ} \\ 0 & 3m_{aQ}/2 & 0 & 0 & 0 & m_{gQ} & L_Q \end{array}\right] \begin{bmatrix} -i_d \\ -i_q \\ -i_0 \\ \hline i_f \\ i_D \\ i_g \\ i_Q \end{bmatrix}, \tag{6.15}$$

where

$$\begin{aligned} &L_d = l_0 + m_0 + 3l_2/2, L_q = l_0 + m_0 + 3l_2/2, L_0 = l_0 - 2m_0, \\ &L_f = L_{ff}, L_D = L_{DD}, L_g = L_{gg}, \\ &L_Q = L_{QQ}, m_{gQ} = M_{fD}, m_{gQ} = M_{gQ} \end{aligned} \tag{6.16}$$

and $\omega = \dfrac{\mathrm{d}\theta}{\mathrm{d}t}$ is the angular speed of the synchronous generator.

Park's transformation, in fact, replaces three-phase armature windings by their three structurally equivalent windings – d winding, q winding, and 0 winding. The difference is that the magnetic flux axis of three-phase armature windings is stationary in space; while that of $dq0$ windings rotates in space at rotor speed. The positive direction of magnetic flux axis of d winding and q winding is as same as that of d- and q-axis of generator rotor, respectively, to describe the behavior of electrical variables in the direction of d- and q-axis; while 0 winding represents the zero-sequence component in the three-phase armature current, voltage and flux linkage. L_d, L_q, and L_0 in (6.16) is the self-inductance of equivalent d, q, and 0 winding, corresponding to d, q, and 0 synchronous reactance, respectively. From (6.16) we can see that the coefficient matrix in (6.15) is a constant matrix. Hence the mathematical model of (6.14) of synchronous generator has been transformed into a set of time-invariant differential equations.

Equation (6.14) indicates that the phase voltage of the synchronous generator consists of three parts. The first part is the voltage drop across the resistance of armature windings; the second is the EMF induced from the variation of flux linking the armature windings, which is usually called the transformer voltage of a synchronous generator; the third part is the EMF due to the rotation of the synchronous generator which is termed speed voltage. The value of speed voltage is much greater than that of transformer voltage. The coefficient matrix in (6.15) is nonsymmetrical, i.e., the mutual inductance between windings on generator rotor and d, q, and 0 winding is not reciprocal. That is caused by the transformation. If the current in rotor windings is multiplied by 3/2, or an orthogonal transformation matrix is adopted, these mutual inductance will become reciprocal. From (6.16) we can see that for a salient-pole generator $L_d > L_q$ and round rotor generator $L_d = L_q$ because $l_2 = 0$. This difference makes it applicable to represent round rotor generators by the mathematical model of the salient-pole generator.

According to the reference direction of current and voltage given in Fig. 6.1b, the total output power from the three-phase armature windings is

$$p_o = v_a i_a + v_b i_b + v_c i_c = \mathbf{v}_{abc}^{\mathrm{T}} \mathbf{i}_{abc}. \tag{6.17}$$

Applying Park's transformation to the equation above, from (6.13) we can obtain the output power from armature windings in $dq0$ coordinates to be

$$p_o = (\mathbf{P}^{-1}\mathbf{v}_{dq0})^{\mathrm{T}}(\mathbf{P}^{-1}\mathbf{i}_{dq0}) = \frac{3}{2}(v_d i_d + v_q i_q + 2v_0 i_0). \tag{6.18}$$

6.2.1.3 Per Unit Equations of the Synchronous Generator

The per unit system is commonly used in power system analysis and calculation due to its many advantages. Parameters of synchronous generator are also usually given in per unit. Hence we need to convert the mathematical model of synchronous generator of (6.14) and (6.15) using actual values of various variables into the per unit equations where variables are described by per unit values. When we introduce the per unit equations used in HVDC in 4.3.1, we have mentioned the principle that, in a per unit system, the base values of different physical variables must have the same relationship that they have when using actual values. Hence in a per unit system, some base values are defined by users and others are derived from the physical relationships among variables. Obviously, the difference in defining those base values by users will lead to different per unit systems. This book will adopt a widely used per unit system – "unit excitation voltage/unit stator voltage" base value system. Subscript B is still used to denote base values of various physical variables and "*" to represent per unit variables.

Firstly we define the base value for generator speed to be the synchronous angular speed ω_s. Because $\omega t = \theta$ and θ is dimensionless (without base value), $\omega_B t_B = 1$ which can lead to the base value for time t. Hence

$$\left.\begin{aligned} \omega_B &= \omega_s \\ t_B &= 1/\omega_s \end{aligned}\right\}. \tag{6.19}$$

On the generator stator side, we define the magnitude of armature current and voltage as their base values. From the definition, we can derive the base values for power, impedance, and flux linkage as follows

$$S_B = 3 \times \frac{V_B}{\sqrt{2}} \times \frac{I_B}{\sqrt{2}} = \frac{3}{2} V_B I_B, \tag{6.20}$$

$$Z_B = \frac{V_B}{I_B} = \frac{3}{2} \times \frac{V_B^2}{S_B}, \tag{6.21}$$

$$\varphi_B = \frac{Z_B}{\omega_B} I_B = Z_B I_B t_B = V_B t_B. \tag{6.22}$$

In a per unit system, there should be only one base value for power. Hence for f, D, g, and Q, four rotor windings, we have

$$V_{fB} I_{fB} = V_{DB} I_{DB} = V_{gB} I_{gB} = V_{QB} I_{QB} = \frac{3}{2} V_B I_B = S_B. \tag{6.23}$$

Due to the constraint of above equation, we can only define one base value between current and voltage, for each rotor winding, and then derive the other. After the base values for the voltage and current of rotor windings are obtained, base values for impedance and flux linkage can be found from the following equations

$$\begin{aligned} &Z_{fB} = V_{fB}/I_{fB}, Z_{DB} = V_{DB}/I_{DB}, Z_{gB} = V_{gB}/I_{gB}, Z_{QB} = V_{QB}/I_{QB}, \\ &\varphi_{fB} = V_{fB} t_B, \varphi_{DB} = V_{DB} t_B, \quad \varphi_{gB} = V_{gB} t_B, \varphi_{QB} = V_{QB} t_B. \end{aligned} \tag{6.24}$$

With base values introduced, we can convert the mathematical equations of (6.14) and (6.15) into the description in per unit as follows.

Dividing both sides of each voltage equations of (6.14) by corresponding base voltage value and using the relationships among various base values as given in (6.19)–(6.24), we can obtain

$$\begin{bmatrix} v_{d*} \\ v_{q*} \\ v_{0*} \\ \hline v_{f*} \\ 0 \\ 0 \\ 0 \end{bmatrix} = \left[\begin{array}{ccc|cccc} R_{a*} & 0 & 0 & 0 & 0 & 0 & 0 \\ 0 & R_{a*} & 0 & 0 & 0 & 0 & 0 \\ 0 & 0 & R_{a*} & 0 & 0 & 0 & 0 \\ \hline 0 & 0 & 0 & R_{f*} & 0 & 0 & 0 \\ 0 & 0 & 0 & 0 & R_{D*} & 0 & 0 \\ 0 & 0 & 0 & 0 & 0 & R_{g*} & 0 \\ 0 & 0 & 0 & 0 & 0 & 0 & R_{Q*} \end{array}\right] \begin{bmatrix} -i_{d*} \\ -i_{q*} \\ -i_{0*} \\ \hline i_{f*} \\ i_{D*} \\ i_{g*} \\ i_{Q*} \end{bmatrix} + p_* \begin{bmatrix} \varphi_{d*} \\ \varphi_{q*} \\ \varphi_{0*} \\ \hline \varphi_{f*} \\ \varphi_{D*} \\ \varphi_{g*} \\ \varphi_{Q*} \end{bmatrix} - \begin{bmatrix} \omega_* \varphi_{q*} \\ -\omega_* \varphi_{d*} \\ 0 \\ \hline 0 \\ 0 \\ 0 \\ 0 \end{bmatrix}, \tag{6.25}$$

where p_* is differentiation operator in per unit:

$$\begin{aligned} p_* &= \frac{p}{\omega_B} = t_B \times \frac{d}{dt} = \frac{d}{dt_*}, \\ R_{a*} &= R_a / Z_B \\ R_{f*} &= \frac{R_f}{Z_{fB}} = \frac{2}{3} \times \frac{R_f}{Z_B} \times \left(\frac{I_{fB}}{I_B}\right)^2 \\ R_{D*} &= \frac{R_D}{Z_{DB}} = \frac{2}{3} \times \frac{R_D}{Z_B} \times \left(\frac{I_{DB}}{I_B}\right)^2. \\ R_{g*} &= \frac{R_g}{Z_{gB}} = \frac{2}{3} \times \frac{R_g}{Z_B} \times \left(\frac{I_{gB}}{I_B}\right)^2 \\ R_{Q*} &= \frac{R_Q}{Z_{QB}} = \frac{2}{3} \times \frac{R_Q}{Z_B} \times \left(\frac{I_{QB}}{I_B}\right)^2 \end{aligned} \tag{6.26}$$

Using the similar procedure for (6.15), we can obtain

$$\begin{bmatrix} \varphi_{d*} \\ \varphi_{q*} \\ \varphi_{0*} \\ \hline \varphi_{f*} \\ \varphi_{D*} \\ \varphi_{g*} \\ \varphi_{Q*} \end{bmatrix} = \left[\begin{array}{ccc|cccc} X_{d*} & 0 & 0 & X_{af*} & X_{aD*} & 0 & 0 \\ 0 & X_{q*} & 0 & 0 & 0 & X_{ag*} & X_{aQ*} \\ 0 & 0 & X_{0*} & 0 & 0 & 0 & 0 \\ \hline X_{af*} & 0 & 0 & X_{f*} & X_{fD*} & 0 & 0 \\ X_{aD*} & 0 & 0 & X_{fD*} & X_{D*} & 0 & 0 \\ 0 & X_{ag*} & 0 & 0 & 0 & X_{g*} & X_{gQ*} \\ 0 & X_{aQ*} & 0 & 0 & 0 & X_{gQ*} & X_{Q*} \end{array}\right] \begin{bmatrix} -i_{d*} \\ -i_{q*} \\ -i_{0*} \\ \hline i_{f*} \\ i_{D*} \\ i_{g*} \\ i_{Q*} \end{bmatrix}, \tag{6.27}$$

where

$$\left.\begin{aligned} X_{d^*} &= \omega_{\mathrm{B}} L_d / Z_{\mathrm{B}} \\ X_{q^*} &= \omega_{\mathrm{B}} L_q / Z_{\mathrm{B}} \\ X_{0^*} &= \omega_{\mathrm{B}} L_0 / Z_{\mathrm{B}} \\ X_{f^*} &= \frac{\omega_{\mathrm{B}} L_f}{Z_{f\mathrm{B}}} = \frac{2}{3} \times \frac{\omega_{\mathrm{B}} L_f}{Z_{\mathrm{B}}} \times \left(\frac{I_{f\mathrm{B}}}{I_{\mathrm{B}}}\right)^2 \\ X_{D^*} &= \frac{\omega_{\mathrm{B}} L_D}{Z_{D\mathrm{B}}} = \frac{2}{3} \times \frac{\omega_{\mathrm{B}} L_D}{Z_{\mathrm{B}}} \times \left(\frac{I_{D\mathrm{B}}}{I_{\mathrm{B}}}\right)^2 \\ X_{g^*} &= \frac{\omega_{\mathrm{B}} L_g}{Z_{g\mathrm{B}}} = \frac{2}{3} \times \frac{\omega_{\mathrm{B}} L_g}{Z_{\mathrm{B}}} \times \left(\frac{I_{g\mathrm{B}}}{I_{\mathrm{B}}}\right)^2 \\ X_{Q^*} &= \frac{\omega_{\mathrm{B}} L_Q}{Z_{Q\mathrm{B}}} = \frac{2}{3} \times \frac{\omega_{\mathrm{B}} L_Q}{Z_{\mathrm{B}}} \times \left(\frac{I_{Q\mathrm{B}}}{I_{\mathrm{B}}}\right)^2 \end{aligned}\right\} \tag{6.28a}$$

$$\begin{aligned} X_{af^*} &= \frac{\omega_{\mathrm{B}} m_{af}}{Z_{\mathrm{B}}} \left(\frac{I_{f\mathrm{B}}}{I_{\mathrm{B}}}\right), & X_{aD^*} &= \frac{\omega_{\mathrm{B}} m_{aD}}{Z_{\mathrm{B}}} \left(\frac{I_{D\mathrm{B}}}{I_{\mathrm{B}}}\right) \\ X_{ag^*} &= \frac{\omega_{\mathrm{B}} m_{ag}}{Z_{\mathrm{B}}} \left(\frac{I_{g\mathrm{B}}}{I_{\mathrm{B}}}\right), & X_{aQ^*} &= \frac{\omega_{\mathrm{B}} m_{aQ}}{Z_{\mathrm{B}}} \left(\frac{I_{Q\mathrm{B}}}{I_{\mathrm{B}}}\right) \\ X_{fD^*} &= \frac{2}{3} \times \frac{\omega_{\mathrm{B}} m_{fD}}{Z_{\mathrm{B}}} \left(\frac{I_{f\mathrm{B}} I_{D\mathrm{B}}}{I_{\mathrm{B}}^2}\right), & X_{gQ^*} &= \frac{2}{3} \times \frac{\omega_{\mathrm{B}} m_{gQ}}{Z_{\mathrm{B}}} \left(\frac{I_{g\mathrm{B}} I_{Q\mathrm{B}}}{I_{\mathrm{B}}^2}\right) \end{aligned} \tag{6.28b}$$

In addition, dividing both sides of (6.18) by S_{B}, from (6.20) we can obtain the output power from synchronous generator in per unit to be

$$p_{o^*} = v_{d^*} i_{d^*} + v_{q^*} i_{q^*} + 2 v_{0^*} i_{0^*}. \tag{6.29}$$

We should note that the per unit equations of the synchronous generator of (6.25) have the similar form to those using actual values of (6.14). However, in the per unit equations, the coefficient matrix in flux linkage equation of (6.27) is symmetrical, i.e., the mutual inductance between stator and rotor windings becomes reciprocal. Furthermore, with proper choice of base values for inductance, we can make the per unit values of inductance to be equal to that of reactance. Hence the coefficient matrix in the flux linkage equation can also be expressed by use of per unit reactance.

6.2.2 Mathematical Equations of Synchronous Generator Using Machine Parameters

For simplicity of expression, in the following discussion we shall use per unit system and omit the subscript "*" to express per unit variables.

In the mathematical equations of the synchronous generator of (6.25) and (6.27), a total of 18 parameters are presented in (6.26) and (6.28). We regard those 18 parameters as basic parameters of a synchronous generator which are decided by physical design and materials used. Strictly speaking, for two generators of the same type and same model, the parameters will not be exactly the same. Usually it is extremely difficult to obtain the values of those parameters through analytical calculation. Therefore, in practice we convert those 18 basic parameters of a synchronous generator into a group of 11 steady-state, transient, and subtransient parameters. These 11 parameters are called machine parameters and can be obtained directly from machine experiments. They are resistance of stator winding (R_a), q- and d-axis synchronous reactance (X_d, X_q), transient reactance (X'_d, X'_q), and subtransient reactance (X''_d, X''_q) as well as the four time constants ($T'_{d0}, T'_{q0}, T''_{d0}, T''_{q0}$). Because machine parameters are fewer than the basic parameters, certain assumptions are needed for the conversion between these two sets of parameters.

Firstly, from the basic (6.25) and (6.27) of a synchronous generator we can see that the magnetic field in space generated by zero-sequence component, i_0, is zero and hence it has no impact on any electrical variables associated with generator rotor. Therefore, the zero-sequence equation in (6.25) and (6.27) and the parameter X_0 can be ignored. Equation (6.25) now becomes

$$\begin{bmatrix} v_d \\ v_f \\ 0 \end{bmatrix} = \begin{bmatrix} R_a & 0 & 0 \\ 0 & R_f & 0 \\ 0 & 0 & R_D \end{bmatrix} \begin{bmatrix} -i_d \\ i_f \\ i_D \end{bmatrix} + p \begin{bmatrix} \varphi_d \\ \varphi_f \\ \varphi_D \end{bmatrix} - \begin{bmatrix} \omega\varphi_q \\ 0 \\ 0 \end{bmatrix}, \tag{6.30}$$

$$\begin{bmatrix} v_q \\ 0 \\ 0 \end{bmatrix} = \begin{bmatrix} R_a & 0 & 0 \\ 0 & R_g & 0 \\ 0 & 0 & R_Q \end{bmatrix} \begin{bmatrix} -i_q \\ i_g \\ i_Q \end{bmatrix} + p \begin{bmatrix} \varphi_q \\ \varphi_g \\ \varphi_Q \end{bmatrix} + \begin{bmatrix} \omega\varphi_d \\ 0 \\ 0 \end{bmatrix}. \tag{6.31}$$

Equation (6.27) can be written as

$$\begin{bmatrix} \varphi_d \\ \varphi_f \\ \varphi_D \end{bmatrix} = \begin{bmatrix} X_d & X_{af} & X_{aD} \\ X_{af} & X_f & X_{fD} \\ X_{aD} & X_{fD} & X_D \end{bmatrix} \begin{bmatrix} -i_d \\ i_f \\ i_D \end{bmatrix}, \tag{6.32}$$

$$\begin{bmatrix} \varphi_q \\ \varphi_g \\ \varphi_Q \end{bmatrix} = \begin{bmatrix} X_q & X_{ag} & X_{aQ} \\ X_{ag} & X_g & X_{gQ} \\ X_{aQ} & X_{gQ} & X_Q \end{bmatrix} \begin{bmatrix} -i_q \\ i_g \\ i_Q \end{bmatrix}. \tag{6.33}$$

We can assume that there exist relationships as shown in the following (6.34) among basic parameters in (6.32) and (6.33) [154]

$$\left.\begin{aligned} X_{af}X_D &= X_{aD}X_{fD} \\ X_{ag}X_Q &= X_{aQ}X_{gQ} \end{aligned}\right\}. \tag{6.34}$$

d-axis machine parameters are related to basic parameters as follows:

1. The definition of d-axis synchronous reactance X_d is that when f and D winding are open-circuited and there exists only the d-axis component of current in the armature winding, the measured armature reactance is X_d. From the definition we know that in (6.32) when $i_f = i_D = 0$, we have

$$\varphi_d = -X_d i_d,$$

i.e., basic parameter X_d is machine parameter X_d.

2. d-axis transient reactance X'_d is defined such that when f winding is short-circuited, D winding open-circuited and only a d-axis component of current suddenly flows through the armature winding, the measured armature reactance is X'_d. From the definition we know that with D winding being open-circuited, $i_D = 0$; and with f winding being short-circuited, at the moment of sudden flow of current through the armature winding, $\phi_f = 0$. Hence in (6.32) we have

$$\left.\begin{aligned} \varphi_d &= -X_d i_d + X_{af} i_f \\ \varphi_f &= -X_{af} i_d + X_f i_f = 0 \end{aligned}\right\}.$$

Canceling i_f in the above equations we obtain

$$\varphi_d = -\left(X_d - \frac{X_{af}^2}{X_f}\right) i_d.$$

Therefore

$$X'_d = \frac{\varphi_d}{-i_d} = X_d - \frac{X_{af}^2}{X_f}. \tag{6.35}$$

3. The definition of d-axis subtransient reactance X''_d is that when f and D winding are short-circuited and only a d component of current suddenly flows through the armature winding, the measured armature reactance is X''_d. According to the definition, with $\phi_f = \phi_D = 0$ in (6.32), we have

$$\left.\begin{aligned} \varphi_d &= -X_d i_d + X_{af} i_f + X_{aD} i_D \\ \varphi_f &= -X_{af} i_d + X_f i_f + X_{fD} i_D = 0 \\ \varphi_D &= -X_{ad} i_d + X_{fD} i_f + X_D i_D = 0 \end{aligned}\right\}.$$

By canceling i_f and i_D in the above equation, we obtain

$$\varphi_d = -\left(X_d - \frac{X_D X_{af}^2 - 2X_{af} X_{fD} X_{aD} + X_f X_{aD}^2}{X_D X_f - X_{fD}^2}\right) i_d.$$

That is

$$X_d'' = \frac{\varphi_d}{-i_d} = X_d - \frac{X_D X_{af}^2 - 2X_{af}X_{fD}X_{aD} + X_f X_{aD}^2}{X_D X_f - X_{fD}^2}. \tag{6.36}$$

From the first equation, on the previous assumption of (6.34), we can find X_{fD}. By substituting it into (6.36) we have

$$X_d'' = X_d - \frac{X_{aD}^2}{X_D}. \tag{6.37}$$

4. The definition of d-axis open-circuit transient time constant is the decaying time constant of i_f when d and D winding are open-circuited. This means that in (6.30) and (6.32), we have $i_d = i_D = 0$, $\phi_d = \phi_D = 0$. Hence

$$\left.\begin{aligned} v_f &= R_f i_f + p\varphi_f \\ \varphi_f &= X_f i_f \end{aligned}\right\}.$$

In per unit we have $X_f = L_f$. From the equation above we can obtain

$$v_f = R_f i_f + L_f \frac{\mathrm{d}i_f}{\mathrm{d}t}.$$

Hence

$$T_{d0}' = L_f/R_f = X_f/R_f. \tag{6.38}$$

In fact, when d and D winding are open-circuited, f winding becomes an isolated winding. Hence the decaying time constant of the winding current is the time constant of f winding itself.

5. d-axis open-circuit subtransient time constant T_{d0}'' is defined to be the decaying time constant of D winding when d winding is open-circuited and f winding short-circuited. From the definition we have $i_d = 0$, $v_f = 0$ in (6.30) and (6.32). Hence

$$\left.\begin{aligned} R_f i_f + p\varphi_f &= 0 \\ R_D i_D + p\varphi_D &= 0 \\ \varphi_f &= X_f i_f + X_{fD} i_D \\ \varphi_D &= X_{fD} i_f + X_D i_D \end{aligned}\right\}.$$

That is

$$\begin{bmatrix} X_f & X_{fD} \\ X_{fD} & X_D \end{bmatrix} p \begin{bmatrix} i_f \\ i_D \end{bmatrix} = \begin{bmatrix} -R_f & 0 \\ 0 & -R_D \end{bmatrix} \begin{bmatrix} i_f \\ i_D \end{bmatrix}.$$

This is obviously a second-order electrical circuit and hence there are two time constants. Because usually R_f is very small we can assume $R_f = 0$. By canceling i_f in the above equation we have

$$\left(X_D - \frac{X_{fD}^2}{X_f}\right) p i_D = -R_D i_D.$$

Hence

$$T''_{d0} = \left(X_D - \frac{X_{fD}^2}{X_f}\right) \Big/ R_D. \tag{6.39}$$

So far we have established the relationship between five d-axis machine parameters and basic parameters. In the similar way, from the definition of various q-axis machine parameters, q-axis voltage equation of (6.31), flux linkage equation of (6.33), and the assumption of (6.34), we can also obtain the relationship between five q-axis machine parameters and basic parameters. In total, the relationship between 11 machine parameters and 18 basic parameters can be listed as follows (on the left side of equations are the machine parameters and the right side the basic parameters).

$$R_a = R_a, X_d = X_d, X_q = X_q, \tag{6.40a}$$

$$\left.\begin{aligned} X'_d &= X_d - X_{af}^2/X_f \\ X'_q &= X_q - X_{ag}^2/X_g \end{aligned}\right\}, \tag{6.40b}$$

$$\left.\begin{aligned} X''_d &= X_d - X_{aD}^2/X_D \\ X''_q &= X_q - X_{aQ}^2/X_Q \end{aligned}\right\}, \tag{6.40c}$$

$$\left.\begin{aligned} T'_{d0} &= X_f/R_f \\ T'_{q0} &= X_g/R_g \end{aligned}\right\}, \tag{6.40d}$$

$$\left.\begin{aligned} T''_{d0} &= \left(X_D - X_{fD}^2/X_f\right)/R_D \\ T''_{q0} &= \left(X_Q - X_{gQ}^2/X_g\right)/R_Q \end{aligned}\right\}. \tag{6.40e}$$

Eleven machine parameters can be obtained through experiment. We should point out that the relationship between machine and basic parameters of (6.40) depends on the assumption given in (6.34). Different assumption may be made that will lead to different relationship between the machine and basic parameters such as that given in; while different relationships will result in different mathematical equations of a synchronous generator represented by using machine parameters. However, values of machine parameters are only affected by their definitions, irrelevant of the initial assumptions.

In the following, we will establish the mathematical equations of synchronous generators represented by machine parameters. To do so, we first introduce the no-load voltage that is proportional to the current of various rotor windings, and transient and subtransient excitation voltages that are proportional to the flux linkage of rotor windings as follows.

No-load voltage:

$$\left.\begin{aligned} e_{q1} &= X_{af} i_f \\ e_{d1} &= -X_{ag} i_g \\ e_{q2} &= X_{aD} i_D \\ e_{d2} &= -X_{aQ} i_Q \end{aligned}\right\}. \tag{6.41}$$

Transient and subtransient voltage:

$$\left.\begin{aligned} e'_q &= (X_{af}/X_f)\varphi_f \\ e'_d &= -(X_{ag}/X_g)\varphi_g \\ e''_q &= (X_{aD}/X_D)\varphi_D \\ e''_d &= -(X_{aQ}/X_Q)\varphi_Q \end{aligned}\right\}. \tag{6.42}$$

In the mathematical equations of a synchronous generator represented by basic parameters of (6.30)–(6.33), we can express current and flux linkage of all rotor windings by the associated voltage defined in (6.40)–(6.42). By using the relationship between basic and machine parameters of (6.40) and the assumption of (6.34), we will obtain the following mathematical equations of a synchronous generator represented by machine parameters.

Flux linkage equation of armature windings

$$\left.\begin{aligned} \varphi_d &= -X_d i_d + e_{q1} + e_{q2} \\ \varphi_q &= -X_q i_q - e_{d1} - e_{d2} \end{aligned}\right\}. \tag{6.43}$$

Flux linkage equation of rotor windings

$$\left.\begin{aligned} e'_q &= -(X_d - X'_d) i_d + e_{q1} + \frac{X_d - X'_d}{X_d - X''_d} e_{q2} \\ e''_q &= -(X_d - X''_d) i_d + e_{q1} + e_{q2} \\ e'_d &= (X_q - X'_q) i_q + e_{d1} + \frac{X_q - X'_q}{X_q - X''_q} e_{d2} \\ e''_d &= (X_q - X''_q) i_q + e_{d1} + e_{d2} \end{aligned}\right\}. \tag{6.44}$$

Voltage equation of armature windings

$$\left.\begin{aligned} v_d &= \rho\varphi_d - \omega\varphi_q - R_a i_d \\ v_q &= \rho\varphi_q + \omega\varphi_d - R_a i_q \end{aligned}\right\}. \tag{6.45}$$

Voltage equation of rotor windings

$$\left.\begin{aligned} T'_{d0}pe'_q &= E_{fq} - e_{q1} \\ T''_{d0}pe''_q &= -\frac{X'_d - X''_d}{X_d - X''_d} e_{q2} \\ T'_{q0}pe'_d &= -e_{d1} \\ T''_{q0}pe''_d &= -\frac{X'_q - X''_q}{X_q - X''_q} e_{d2} \end{aligned}\right\}, \tag{6.46}$$

where

$$E_{fq} = \frac{X_{af}}{R_f} u_f. \tag{6.47}$$

E_{fq} is the voltage across the armature winding when synchronous generator is connected to no load at the steady-state operation. In fact, v_f/R_f is an imaginary field current due to v_f at steady state. During the transient process, it is not equal to the actual i_f. From the definition of (6.41) we can see that the product of this steady-state field current and X_{af} gives the no-load voltage. Hence E_{fq} is called the imaginary voltage.

To express e_{q1}, e_{q2}, e_{d1}, and e_{d2} directly from the flux linkage equation of rotor windings of (6.44), we have

$$\left.\begin{aligned} e_{q1} &= \frac{X_d - X''_d}{X'_d - X''_d} e'_q - \frac{X_d - X'_d}{X'_d - X''_d} e''_q \\ e_{q2} &= -\frac{X_d - X''_d}{X'_d - X''_d} e'_q + \frac{X_d - X''_d}{X'_d - X''_d} e''_q + (X_d - X''_d) i_d \\ e_{d1} &= \frac{X_q - X''_q}{X'_q - X''_q} e'_d - \frac{X_q - X'_q}{X'_q - X''_q} e''_d \\ e_{d2} &= -\frac{X_q - X''_q}{X'_q - X''_q} e'_d + \frac{X_q - X''_q}{X'_q - X''_q} e''_d - (X_q - X'') i_q \end{aligned}\right\}. \tag{6.48}$$

Substituting (6.48) into (6.43) and (6.46) we can obtain the flux linkage equation of the armature windings

$$\left.\begin{aligned} \varphi_d &= e''_q - X''_d i_d \\ \varphi_q &= -e''_d - X''_q i_q \end{aligned}\right\} \tag{6.49}$$

and the voltage equation of rotor windings

$$\left.\begin{aligned} T'_{d0}pe'_q &= -\frac{X_d - X''_d}{X'_d - X''_d}e'_q + \frac{X_d - X'_d}{X'_d - X''_d}e''_q + E_{fq} \\ T''_{d0}pe''_q &= e'_q - e''_q - (X'_d - X''_d)i_d \\ T'_{q0}pe'_d &= -\frac{X_q - X''_q}{X'_q - X''_q}e'_d + \frac{X_q - X'_q}{X'_q - X''_q}e''_d \\ T''_{q0}pe''_d &= e'_d - e''_d + (X'_q - X''_q)i_q \end{aligned}\right\}. \tag{6.50}$$

In (6.47) we still have two basic parameters X_{af} and R_f. To avoid these two parameters in the expression, we need to choose proper base values such that in per unit system we have $X_{af} = R_f$ and hence $E_{fq} = v_f$. This choice of base values is usually called "unit excitation voltage/unit stator voltage" per unit system. The details are as follows.

As we have introduced previously, S_B is decided by the choice of base values on the generator stator side. For each winding on the rotor, we have to choose a base value for either voltage or current and derive the other. In the "unit excitation voltage/unit stator voltage" per unit system, we first choose the base value for the voltage of the field winding V_{fB} and then derive the base value for field current I_{fB} from (6.23). We choose V_{fB} such that when synchronous generator operates at steady state, is subject to no load and rotates at synchronous speed, the voltage of stator winding is equal to the base value of stator voltage. Obviously, V_{fB} can be gained by experiment. From the above definition about V_{fB}, in (6.14) and (6.15) we only have $i_f \neq 0$, we have

$$\left.\begin{aligned} v_d &= 0 \\ v_q &= \omega_B m_{af} i_f = V_B \\ v_f &= R_f i_f = V_{fB} \end{aligned}\right\}.$$

So we can obtain

$$V_{fB} = \frac{R_f}{\omega_B m_{af}} V_B.$$

Because $Z_{fB} = V_{fB}/I_{fB}$, we have

$$R_{f*} = \frac{R_f}{Z_{fB}} = R_f I_{fB} \times \frac{\omega_B m_{af}}{R_f V_B} = \frac{\omega_B m_{af}}{Z_B}\left(\frac{I_{fB}}{I_B}\right).$$

Comparing the above equation with X_{af*} in (6.28), we can see $R_{f*} = X_{f*}$. Hence in per unit

$$E_{fq} = \frac{X_{af}}{R_f} v_f = v_f. \tag{6.51}$$

Up to this point, we have established the mathematical model of synchronous generator represented by 11 machine parameters that consists of the voltage equation of armature windings (6.45), flux linkage equation of armature windings (6.49), and voltage equation of rotor windings (6.50). We should point out that this model only needs the specific choice of base value for field winding. Base value for the voltage or current of damping windings can be selected according to (6.23). Besides, voltage of field winding v_f is affected by excitation control and hence E_{fq} in (6.50) will be discussed further in Sect. 6.3.

6.2.3 Simplified Mathematical Model of Synchronous Generator

In the above discussion, we established the mathematical model of synchronous generator where four rotor windings, *f*, *g*, *D*, and *Q*, are used. From (6.50) we can see that the electromagnetic transient of rotor windings is depicted by four differential equations. In a modern power system, there could be over 1,000 generators in synchronous operation. Higher-order differential equations could result in numerical difficulty in power system analysis and calculation. Therefore, in practice the mathematical model of a synchronous generator is often simplified according to requirements of computing accuracy, and only for those generators that we are particularly concerned about are higher-order models used. The simplification can be classified according to how to ignore certain rotor windings, leading to three rotor-winding model, two rotor-winding model, nondamping-winding model and constant e'_q model (classical model). All these models can be derived from the full four rotor winding model of a synchronous generator.

1. *Three rotor winding model (f, D, Q)*. For a salient-pole generator, usually we only consider one equivalent damping winding Q on q-axis and ignore the existence of g winding. This means that in the four rotor winding model, $i_g = \phi_g = 0$. Hence in (6.41), $e_{d1} = 0$ and in (6.42), $e'_d = 0$ and $X'_q = X_q$. The voltage equations of rotor windings are reduced to an order three model

$$\left.\begin{aligned} T'_{d0}pe'_q &= -\frac{X_d - X''_d}{X'_d - X''_d}e'_q + \frac{X_d - X'_d}{X'_d - X''_d}e''_q + E_{fq} \\ T''_{d0}pe''_q &= e'_q - e''_q - (X'_d - X''_d)i_d \\ T''_{q0}pe''_d &= -e''_d + (X'_q - X''_q)i_q \end{aligned}\right\}. \tag{6.52}$$

   There is no change in the voltage and flux linkage equation of the armature windings.

2. *Two winding model (f, g or double-axis model)*. We only consider one damping winding g on q-axis and ignore D, Q damping winding. This is the same as the assumption $i_D = i_Q = \phi_D = \phi_Q = 0$ in four winding rotor model. Hence in (6.41), we have $e_{q2} = e_{d2} = 0$ and in (6.42), $e''_q = e''_d = 0$. The flux linkage equation of armature windings becomes

$$\left.\begin{aligned} \varphi_d &= e'_q - X'_d i_d \\ \varphi_q &= -e'_d - X'_q i_q \end{aligned}\right\}. \tag{6.53}$$

Voltage equation of rotor windings is reduced to a second-order model

$$\left.\begin{aligned} T'_{d0}\rho e'_q &= -e'_q - (X_d - X'_d)i_d + E_{fq} \\ T'_{q0}\rho e'_d &= -e'_d + (X_q - X'_q)i_q \end{aligned}\right\}. \tag{6.54}$$

There is no change in the voltage equation of armature windings.

3. *Nondamping winding model (f, or variable e'_q model).* Ignoring damping windings, we have $i_D = i_Q = i_g = \phi_D = \phi_Q = \phi_g = 0$. Hence in (6.41), $e_{d1} = e_{q2} = e_{d2} = 0$ and in (6.42), $e'_d = e''_q = e''_d = 0$. The flux linkage equation of armature windings becomes

$$\left.\begin{aligned} \varphi_d &= e'_q - X'_d i_d \\ \varphi_q &= -X_q i_q \end{aligned}\right\}. \tag{6.55}$$

Voltage equation of rotor winding is reduced to a first-order model

$$T'_{d0}\rho e'_q = -e'_q - (X_d - X'_d)i_d + E_{fq}. \tag{6.56}$$

There is no change in the voltage equation of armature windings.

4. *Constant e'_q model.* We neglect damping windings and transient of field winding. Also we consider the right-hand side of (6.56) to be zero due to the control function of AVR, i.e.,

$$e'_q \equiv (X'_d - X_d)i_d + E_{fq} = \text{constant}.$$

Thus the mathematical model of synchronous generator is comprised of only the voltage and flux linkage equation of armature windings of (6.45) and (6.55). There is no differential equation for the rotor windings. Constant e'_q model usually is used when the rotor motion equation of the synchronous generator is described by electrical torque.

5. *Classical model.* This is to use $X'_d = X'_q$ to further simplify the expression of output electrical power of a synchronous generator.

The discussion above is the simplification of depicting rotor windings to reduce the order of the mathematical model of a synchronous generator. On the other hand, in the analysis of power system steady-state operation, the voltage equation of armature windings can be simplified in the following two ways:

1. Ignoring the electromagnetic transient of armature windings. This is to neglect the induced voltage due to the variations of φ_{d} and φ_{q} in the voltage equation of armature windings. Thus the voltage equation of armature winding becomes

$$\left.\begin{aligned} v_d &= -\omega\varphi_q - R_a i_d \\ v_q &= \omega\varphi_d - R_a i_q \end{aligned}\right\}. \tag{6.57}$$

For power system stability studies the above simplification is very important. From the flux linkage equation of armature winding (6.49) we can see that the differential of flux linkage of armature windings with respect to time will involve that of armature current with respect to time. Because the armature windings of a synchronous generator are connected to a transmission network that is formed by a certain topology of resistance, inductance, and capacitance, the differentiation of armature current with respect to time will require the description of the network by differential equations. This will greatly increase the order of the mathematical model of whole power system. In addition, if the electromagnetic transients of armature windings and network are not ignored, the armature current of the synchronous generator will contain high-frequency components. Under this circumstance, we must take very small integration time steps to achieve the required computing accuracy in the numerical solution of power system mathematical equations. For a modern large power system, increase of the order of its mathematical model and decrease of the required integration time step would add a heavy computational burden such that normal calculation would become impossible. In fact, compared to the electromechanical process of the synchronous generator, the electromagnetic transient behavior of the power network is sufficiently fast that it can be ignored as far as its influence on power system stability analysis and computation is concerned. From (6.57) we can see that when the electromagnetic transient of armature windings of synchronous generator is neglected, its voltage equations become algebraic equations, i.e., those depicting steady-state operation of the synchronous generator.
2. In the voltage equation of armature windings, we consider the rotor speed of synchronous generator ω always to be the synchronous speed. This does not mean that during the transient, the rotor speed of synchronous generator does not change. It is because the range of ω is small due to the existence of various control functions in generator operation. Hence in the voltage equation of the armature winding, the numerical variation caused by the small change of ω is very small and hence can be ignored. This simplification does not result in great saving in computation. However, it has been shown that taking $\omega = 1$ in the voltage equation of the armature winding of synchronous generator can partly correct the computational errors caused by ignoring the

electromagnetic transient [153]. Therefore, the voltage equation of armature winding becomes

$$\left.\begin{aligned} v_d &= -\varphi_q - R_a i_d \\ v_q &= \varphi_d - R_a i_q \end{aligned}\right\}. \tag{6.58}$$

6.2.4 Steady-State Equations and Phasor Diagram

Mathematically, transient analysis of power systems is to solve a group of differential equations depicting power system transient behavior. Usually the steady-state operating point is the initial condition to solve the differential equations. In the following, we will derive the formula to calculate the initial conditions from the steady-state equations of the synchronous generator.

In steady-state operation, the generator rotates at synchronous speed, all electrical variables are balanced and the current of damping windings is zero. Current i_d, i_q, i_f, and e_{q1} associated with i_f as well as flux linkage of all windings are constant. In the following, we shall use capital letters to denote various steady-state electric variables.

1. Steady-state equations represented by synchronous reactance

   From (6.43) we have

$$\left.\begin{aligned} \Phi_d &= -X_d I_d + E_{q1} \\ \Phi_q &= -X_q I_q \end{aligned}\right\}. \tag{6.59}$$

   At steady state

$$E_{q1} = X_{af} I_f = X_{af} \frac{V_f}{R_f} = E_{fq}.$$

   Substituting (6.59) into (6.58) we obtain

$$\left.\begin{aligned} E_{fq} &= V_q + R_a I_q + X_d I_d \\ 0 &= V_d + R_a I_d - X_q I_q \end{aligned}\right\}. \tag{6.60}$$

   After load flow calculation, we have had terminal voltage $\dot{V}_t$ and current $\dot{I}_t$ of the synchronous generator in x–y coordinate. To obtain V_d, V_q, I_d, and I_q in d–q coordinate, we need to find the connection between these two coordinate systems, i.e., to find the angle between them. For this purpose, we multiply the first equation of (6.60) by j and add it to the second equation

$$\mathrm{j}E_{fq} - \mathrm{j}(X_d - X_q)I_d = \dot{V}_t + (R_a + \mathrm{j}X_q)\dot{I}_t.$$

We can define an imaginary voltage $\dot{E}_Q$ according to the above equation to be

$$\dot{E}_Q = \dot{V}_t + (R_a + jX_q)\dot{I}_t. \tag{6.61}$$

Because $\dot{E}_Q$ and jE_{fq} are in the same phase, from phasor diagram (Fig. 6.2a), we can see that the angle between $\dot{E}_Q$ and x, δ, is that between d–q and x–y coordinate. Hence from (6.61) we can find δ and obtain the transformation between two coordinate systems as follows

$$\begin{bmatrix} A_d \\ A_q \end{bmatrix} = \begin{bmatrix} \sin\delta & -\cos\delta \\ \cos\delta & \sin\delta \end{bmatrix} \begin{bmatrix} A_x \\ A_y \end{bmatrix}, \tag{6.62}$$

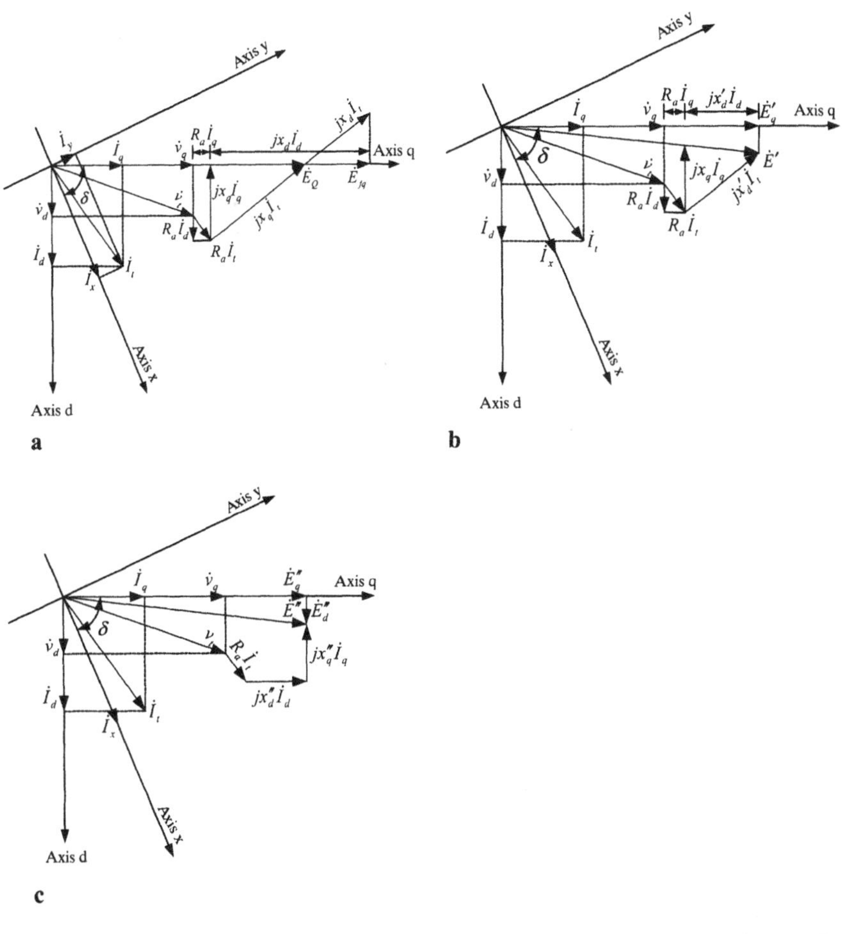

Fig. 6.2 Phasor diagram of steady-state operation of synchronous generator (**a**) when synchronous reactance is used (**b**) when transient reactance is used (**c**) when subtransient reactance is used

$$\begin{bmatrix} A_x \\ A_y \end{bmatrix} = \begin{bmatrix} \sin\delta & \cos\delta \\ -\cos\delta & \sin\delta \end{bmatrix} \begin{bmatrix} A_d \\ A_q \end{bmatrix}, \tag{6.63}$$

where A denotes current, voltage, flux linkage, and various EMF. After V_d, V_q, I_d, and I_q are found, from (6.60) we can calculate the initial value of v_f, $V_f = E_{fq}$.

2. Steady-state equations represented by transient reactance

   From the first and third equation in (6.44) we have

$$\left.\begin{aligned} E'_q &= -(X_d - X'_d)I_d + E_{q1} \\ E'_d &= (X_q - X'_q)I_q \end{aligned}\right\}.$$

Noting $E_{q1} = E_{fq}$ at steady state and canceling X_d and X_q by substituting the first and second equation in (6.60) into the first and second above equation, we have

$$\left.\begin{aligned} E'_q &= V_q + R_a I_q + X'_d I_d \\ E'_d &= V_d + R_a I_d - X'_q I_q \end{aligned}\right\}. \tag{6.64}$$

3. Steady-state equations represented by subtransient reactance

   From the second and fourth equation of (6.44), we can have

$$\left.\begin{aligned} E''_q &= -(X_d - X''_d)I_d + E_{q1} \\ E''_d &= (X_q - X''_q)I_q \end{aligned}\right\}.$$

Taking the similar procedure, from (6.64) we can obtain

$$\left.\begin{aligned} E''_q &= V_q + R_a I_q + X''_d I_d \\ E''_d &= V_d + R_a I_d - X''_q I_q \end{aligned}\right\}. \tag{6.65}$$

Equations (6.60), (6.64), and (6.65) comprise steady-state equations of a synchronous generator adopting the four rotor winding model. From those three equations we can calculate the initial values of five state variables, v_f, e'_q, e'_d, e''_d, and e''_q. Phasor diagrams related to those three equations are shown in Fig. 6.2.

When a simplified model of the synchronous generator is used, we can calculate required initial values of state variables directly from the above steady-state equations of the four rotor winding model. For example, when the damping windings are ignored, we have

$$\left.\begin{aligned} E_{fq} &= V_q + R_a I_q + X_d I_d \\ 0 &= V_d + R_a I_d - X_q I_q \\ E'_q &= V_q + R_a I_q + X'_d I_d \end{aligned}\right\}.$$

6.2.5 Mathematical Equations Considering Effect of Saturation

In the above discussion, we have established mathematical equations of the synchronous generator under the condition that the magnetic circuit of machine is unsaturated. In practice, to save materials, the design and manufacture of synchronous generator usually makes the iron core of both stator and rotor slightly saturated when operating at rated conditions. At some particular operating conditions, with the increase of flux density, saturation would become very obvious and serious. In system planning and operation analysis, errors caused by ignoring saturation are small. However, in certain applications, such as in transient stability analysis, with detailed model of AVR and its limiters included, the effect of machine saturation can greatly affect the accuracy of analysis and calculation.

Study on the effect of saturation started as early as about 60 years ago. The mathematical model of a synchronous generator will become extremely complicated if machine saturation is modeled in great detail. This is because the extent of saturation of a magnetic circuit is closely related to the total mmf in the machine air gap. It is required to combine *d*- and *q*-axis mmf to air-gap total mmf and then to find the corresponding magnetic flux and linkage from the saturation curve. Even though air-gap total mmf has a strict sinusoidal distribution in space, mmf varies in different positions. Thus saturation at various positions in space is different, which will cause distortion of the flux wave in the air gap. Therefore, in practice, considering the simplicity of model used, effectiveness of parameters and accuracy of computation, proper approximation is applied to take account of the effect of machine saturation [155–158]. In the following, we shall introduce a method commonly used in stability analysis [156]. The assumptions to apply the method are:

1. The effect of saturation is simply considered on *d*- and *q*-axis separately. The difference of magnetic reluctance in *d*- and *q*-axis magnetic circuits is only caused by that of length of air gap in the direction of *d*- and *q*-axis.
2. On a same axis, the extent of saturation depends on the Potier voltage behind Potier reactance X_{p}. The higher the Potier voltage, the more serious the saturation. Potier voltage on *d*- and *q*-axis is given by the following equation

$$\left.\begin{aligned} v_{d\mathrm{p}} &= v_d + R_a i_d - X_{\mathrm{p}} i_q \\ v_{q\mathrm{p}} &= v_q + R_a i_q + X_{\mathrm{p}} i_d \end{aligned}\right\}. \tag{6.66}$$

   In addition, the extent of saturation of voltage and flux linkages of armature and rotor windings is approximately considered to be same on the same axis.
3. The distortion of the distribution wave of air-gap flux does not affect the self-inductance and mutual inductance of various windings and the unsaturated values of winding reactance.

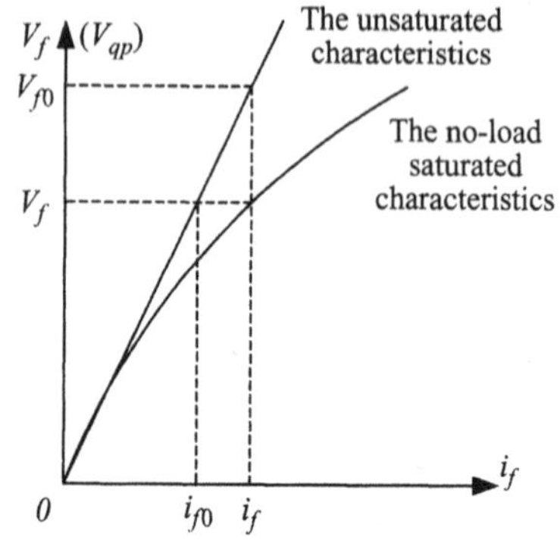

Fig. 6.3 No-load saturation characteristic of synchronous generator

The extent of saturation is described by saturation factor. For d-axis, saturation factor S_d can be calculated from the saturation characteristic of machine in no-load operation. This is because v_{qp} is equivalent to the voltage of q winding induced from the resultant d-axis air-gap flux. From Fig. 6.3 of no-load saturation characteristic of synchronous generator, we can find the unsaturated value of v_{qp0} from a certain value of v_{qp}. Hence we can define S_d to be

$$S_d = f(v_{qp}) = \frac{v_{qp0}}{v_{qp}} - 1. \tag{6.67}$$

Obviously, the bigger the value of S_d is, the more saturated is the synchronous generator. Zero S_d indicates the case of no saturation. For q-axis, the saturation characteristic is difficult to obtain through experiment. Hence from the first assumption above, the saturation factor S_q is also determined by using the no-load saturation characteristic of synchronous generator, using the following equation

$$S_q = \frac{X_q}{X_d} f(v_{dp}). \tag{6.68}$$

To calculate the saturation factor, one commonly used method is to approximately fit the no-load saturation characteristic curve of Fig. 6.3 by an analytical function, such as

$$i_f = aV_t + bV_t^n.$$

Obviously when $b = 0$, we have the characteristic curve without saturation

$$i_{f0} = aV_t.$$

Hence according to triangle similarity of Fig. 6.3, we have

$$S_d = \frac{v_{qp0}}{v_{qp}} - 1 = \frac{v_{qp0} - v_{qp}}{v_{qp}} = \frac{i_f - i_{f0}}{i_{f0}} = \frac{av_{qp} + bv_{qp}^n - av_{qp}}{av_{qp}} = \frac{b}{a} v_{qp}^{n-1}.$$

That is

$$S_d = cv_{qp}^{n-1}, \tag{6.69}$$

where $c = b/a$. Similarly, from (6.68) we have

$$S_q = c\frac{X_q}{X_d}v_{dp}^{n-1}. \tag{6.70}$$

In the following, we shall discuss the voltage equations of the field winding, voltage and flux linkage equations of armature windings of a synchronous generator, taking account of the saturation effect. From the derivation of voltage equations of field winding of (6.50) without considering saturation effect, we can see that on the right-hand side of the equations we have the voltage drop across the equivalent resistance of the field winding caused by field current and the external voltage applied on the rotor windings (i.e., the excitation voltage V_f). Hence there should no problem of saturation about this part in the equations. Hence when we consider the saturation effect, we shall still use unsaturated values for those on the right-hand side of the equations. On the other hand, on the left side of the equations, we have the induced voltage by variations of flux linkage with time. Hence when saturation effect is taken into account, for those terms on the left-hand side of the equations we should use their saturated values associated with actual flux linkage. According to the previous assumption (2) and (6.67), we know that on d-axis, the ratio of unsaturated value to the saturated of each voltage and flux linkage is $(1 + S_d)$. Similarly, on q-axis, the ratio is $(1 + S_q)$. Therefore, when we consider the saturation effect, the voltage equations of the field winding of a synchronous generator are

$$\left.\begin{aligned}
T'_{d0}pe'_{qs} &= -\frac{X_d - X''_d}{X'_d - X''_d}(1 + S_d)e'_{qs} + \frac{X_d - X'_d}{X'_d - X''_d}(1 + S_d)e''_{qs} + E_{fq}\\
T''_{d0}pe''_{qs} &= (1 + S_d)e'_{qs} - (1 + S_d)e''_{qs} - (X'_d - X''_d)i_d\\
T'_{q0}pe'_{ds} &= -\frac{X_q - X''_q}{X'_q - X''_q}(1 + S_q)e'_{ds} + \frac{X_q - X'_q}{X'_q - X''_q}(1 + S_q)e''_{ds}\\
T''_{q0}pe''_{ds} &= (1 + S_q)e'_{ds} - (1 + S_q)e''_{ds} + (X'_q - X''_q)i_q
\end{aligned}\right\}, \tag{6.71}$$

where subscript s denotes the saturated value of each voltage.

Taking saturation effects into account, we have the flux linkage equations of armature windings of (6.49) becoming

$$\left.\begin{aligned}
(1 + S_q)\varphi_{qs} &= -(1 + S_q)e''_{ds} - X''_q i_q\\
(1 + S_d)\varphi_{ds} &= (1 + S_d)e''_{qs} - X''_d i_d
\end{aligned}\right\}. \tag{6.72}$$

When we do not consider the saturation effect, from the voltage equations of armature windings of synchronous generator (6.58) and the definition of Potier voltage of (6.66), we can obtain the relationship between the Potier voltage and flux linkage of armature windings to be

$$\left.\begin{aligned} v_{dp0} &= -\varphi_q - X_p i_q \\ v_{qp0} &= \varphi_d + X_p i_d \end{aligned}\right\}. \tag{6.73}$$

According to the relationship between saturated and unsaturated value, we can have

$$\left.\begin{aligned} (1+S_q)v_{dp} &= -(1+S_q)\varphi_{qs} - X_p i_q \\ (1+S_d)v_{qp} &= (1+S_d)\varphi_{ds} + X_p i_d \end{aligned}\right\}.$$

Substituting (6.72) into the above equation we can establish the relationship between the Potier voltage and the EMF with saturation considered, to be

$$\left.\begin{aligned} v_{dp} &= e''_{ds} + \frac{X''_q - X_p}{1+S_q} i_q \\ v_{qp} &= e''_{qs} - \frac{X''_d - X_p}{1+S_d} i_d \end{aligned}\right\}. \tag{6.74}$$

Substituting the above equation into the defining equation of the Potier voltage, we have the voltage equations of armature windings with saturation being considered, to be

$$\left.\begin{aligned} v_d &= e''_{ds} - R_a i_d + \left(\frac{X''_q - X_p}{1+S_q} + X_p\right) i_q \\ v_q &= e''_{qs} - R_a i_q - \left(\frac{X''_d - X_p}{1+S_d} + X_p\right) i_d \end{aligned}\right\}. \tag{6.75}$$

Equations (6.66), (6.67), (6.71), and (6.75) form the mathematical model of synchronous generator with machine saturation being taken into account. From the model it would be straightforward to derive the steady-state equations of a synchronous generator. In practice, we often assume that stator leakage flux does not saturate. Hence we can use X_σ as the Potier reactance X_p.

6.2.6 *Rotor Motion Equation of Synchronous Generator*

6.2.6.1 Rotor Motion Equation of Stiff Rotor

If we consider the prime mover and generator rotor to be a single mass, the rotor motion equation of the whole generation unit is

$$\left.\begin{aligned}\frac{\mathrm{d}\delta}{\mathrm{d}t} &= (\omega_* - 1)\omega_\mathrm{s}\\ T_J\frac{\mathrm{d}\omega_*}{\mathrm{d}t} &= T_{\mathrm{m}^*} - T_{\mathrm{e}^*}\end{aligned}\right\}, \tag{6.76}$$

where $T_J = 2W_\mathrm{k}/S_\mathrm{B}$, δ is the electrical angle between q-axis of generator rotor and a reference axis x that rotates at synchronous speed. This angle is a dimensionless number and can be measured in radians (rad), T_J is the moment of inertia of generation unit measured in seconds (s), W_k the rotating kinetic energy of the rotor rotating at synchronous speed and measured in Joules (J), S_B the base value of generation capacity in V A; T_{m^*} and T_{e^*} are the output mechanical torque of prime mover and the electromagnetic torque of the generator in per unit; their base value is $S_\mathrm{B}/\Omega_\mathrm{s}$ measured in radian/second (rad s^{-1}), where Ω_s is the mechanical synchronous speed of rotor. The positive direction of T_{m^*} and T_{e^*} is taken to be as same as and opposite to that of rotation of the rotor, respectively. In some references, the mechanical inertia is represented by $H = W_\mathrm{k}/S_\mathrm{B}$. Obviously, $T_J = 2H$. In addition, we ought to note the following two issues:

1. Since the product of torque and speed is the power and $\Omega/\Omega_\mathrm{s} = \omega/\omega_\mathrm{s} = \omega_*$, in per unit we can have

$$\left.\begin{aligned}P_{\mathrm{m}^*} &= T_{\mathrm{m}^*}\omega_*\\ P_{\mathrm{e}^*} &= T_{\mathrm{e}^*}\omega_*\end{aligned}\right\}, \tag{6.77}$$

   where P_{m^*} is the output mechanical power from the prime mover and P_{e^*} the electromagnetic power of the synchronous generator. As pointed out before, various functions of power system stability control result in a small change of ω_*. Hence in order to save computational time, sometimes we can just simply take ω_* to be 1. Thus in per unit, torque is equal to power.
2. Rotor rotation is always subject to air resistance and friction between bearing and shaft. This results in a damping torque to rotor motion. Often we assume that this damping torque is approximately proportional to rotor speed and represent it by the product of a damping coefficient D and speed ω_*.

Considering what has been discussed above, when time is also represented in per unit, the rotor motion equation becomes

$$\left.\begin{aligned}\frac{\mathrm{d}\delta}{\mathrm{d}t_*} &= \omega_* - 1\\ T_{J^*}\frac{\mathrm{d}\omega_*}{\mathrm{d}t_*} &= -D\omega_* + P_{\mathrm{m}^*} - P_{\mathrm{e}^*}\end{aligned}\right\}. \tag{6.78}$$

We would point out that the mechanical torque and power involved in the above rotor motion equation are subject to the control of the governing system of generation unit. Hence the appearance of mechanical torque and power will lead to the establishment of equations for the governing system. This will be discussed in Sect. 6.4.

In (6.78), we consider a combined rotor of generator and prime mover to be a single lumped mass. This consideration will usually bring about no obvious errors when carrying out transient stability analysis. However, when the subsynchronous resonance of power systems is studied, we cannot ignore the existence of rotor shaft elasticity, since large steam-turbine generation units often consist of multiple stage turbines and their shafts can be as long as several tens of meters. In this case, we can consider the exciter, generator rotor, and each turbine section to be separate lumped masses. Thus elasticity of the whole shaft system can be treated as torsional springs between each mass. Therefore, with elasticity being taken into account, rotation speed of each mass could be different during a transient process, resulting in difference in relative angular position of each mass. The motion equation of each mass forms the motion equation of the whole shaft system. Detailed discussion can be found in [159].

6.2.6.2 Electromagnetic Torque and Power of Synchronous Generator

In the rotor motion equation of (6.78), mechanical torque (or power) from prime mover and electromagnetic torque (or power) of synchronous generator are introduced. The former is included in the mathematical model of the prime mover and governing system of the generation unit, which will be discussed in Sect. 6.4. Here we shall introduce the computing model of electromagnetic torque and power. Electromagnetic torque represents the function of force applied on the rotor from the mutual electric and magnetic interactions between stator and rotor of the synchronous generator. Theoretical proof has been provided that electromagnetic torque is equal to the partial differentiation of total magnetic field energy stored in various windings to rotor angle [148], i.e.,

$$T_{\mathrm{e}} = \frac{\partial W_F}{\partial \theta}, \tag{6.79}$$

where θ is the angle between d-axis and a-axis of armature winding (see Fig. 6.1a) and W_F is the total magnetic energy stored in three-phase armature windings and rotor windings, which can be represented as

$$W_F = -\frac{1}{2}(\varphi_a i_a + \varphi_b i_b + \varphi_c i_c) + \frac{1}{2}(\varphi_f i_f + \varphi_D i_D + \varphi_g i_g + \varphi_Q i_Q). \tag{6.80}$$

Because the reference positive direction of armature current is opposite to that of associated flux linkage, we have a negative sign in the above equation. From the base value for torque $T_{\mathrm{B}} = S_{\mathrm{B}}/\Omega_{\mathrm{B}}$ and (6.2)–(6.7), we can obtain

$$T_{e^*} = \varphi_{d^*} i_{q^*} - \varphi_{q^*} i_{d^*}. \tag{6.81}$$

The above equation shows that electromagnetic torque is independent of zero-sequence components, because they do not couple with rotor windings. In addition, although the above equation has been established from the four winding model, it is applicable to other higher or lower winding models with only slight differences in derivation.

When the four rotor winding model is used, substituting the flux linkage equation of (6.49) into the above equation, we can obtain the expression of electromagnetic torque to be

$$T_{e^*} = e''_{d^*} i_{d^*} + e''_{q^*} i_{q^*} - (X''_{d^*} - X''_{q^*}) i_{d^*} i_{q^*}. \tag{6.82}$$

From the above expression and (6.77), we can directly establish the expression of electromagnetic power where state variables ω_*, e''_d, and e''_q are included. This will bring about a heavy computing burden in the solution. Hence to solve this problem, we can substitute the voltage equation of armature windings of (6.45) into (6.81) and use (6.77) to obtain

$$p_{e^*} = u_{d^*} i_{d^*} + u_{q^*} i_{q^*} + R_{a^*}\left(i_{d^*}^2 + i_{q^*}^2\right) - i_{d^*} p_* \varphi_{d^*} - i_{q^*} p_* \varphi_{q^*}, \tag{6.83}$$

where $R_{a^*}\left(i_{d^*}^2 + i_{q^*}^2\right)$ is the copper loss of the armature windings. When the transient of armature windings is ignored, from the comparison of the above equation with the output power expression of the synchronous generator in (6.29), we can see that the electromagnetic power of the generator is the sum of generator output power and copper loss of generator armature windings. Finally we would like to mention here that (6.83) is also applicable to cases when other types of rotor winding model is used and/or machine saturation is considered.

6.3 Mathematical Model of Generator Excitation Systems

In (6.50) we introduced variable E_{fq} in the per unit system as "unit excitation voltage/unit stator voltage," that is equal to voltage v_f applied to the field winding. Hence we need to establish the mathematical model of generator excitation systems.

The basic function of a generator excitation system is to provide the generator field winding with appropriate DC current to generate a magnetic field in the distributed space of the generator armature windings. In earlier times, the excitation system regulated the excitation voltage through manual control, to maintain the required terminal voltage of the generator and reactive power supply from the generator. More recently, various types of excitation and AVR were proposed and used. In the 1960s, the proposal and application of power system stabilizers (PSS) further enhanced the role played by excitation control systems to improve power system stability. With the advancement of control theory and computer control technology, further new types of excitation regulators have been proposed. Their control tasks have been extended from simple terminal voltage regulation of the generator to multiple excitation control functions. Feedback signals used have developed from a single deviation of generator terminal voltage to the superimposition of various signals on the voltage deviation, based on factors such as electromagnetic power, electrical angular speed, system frequency, armature current, and deviation of excitation current or voltage and their combinations. The control strategy started with simple proportional control and has been enhanced by applying proportional–integral–differential (PID) control, multivariable linear system control schemes, self-tuning control, adaptive control, fuzzy control, and nonlinear control. In recent years, digital excitation controllers based on microprocessors or microcomputers have been developed and installed. In the near future, research into, and innovative applications of, excitation control will involve the development of digital excitation control systems realized by microcomputers and using modern control theory. Relatively accurate analysis of power system dynamics must be supported by mathematical models of the excitation system. Development and design of new types of excitation controller need to establish mathematical models for simulation to check if the dynamic performance is satisfactory. In this section, we shall only introduce the mathematical models of widely used excitation systems and the design principle of excitation regulators will not be discussed. Also we shall not discuss the newer type of excitation controllers, such as linear optimal excitation controller (LOEC), nonlinear optimal excitation controller (NOEC), because they are still at the stage of further theoretical research and testing. Figure 6.4 shows the construction of a general excitation system.

The exciter provides field current to the field winding of the generator. The regulator controls the field current. The measurement unit for generator terminal voltage and load compensation measures generator terminal voltage $\dot{V}_t$ and compensates for the load current of generator $\dot{I}_t$, respectively. The auxiliary control signals are sent through the auxiliary controller. One of the most widely used

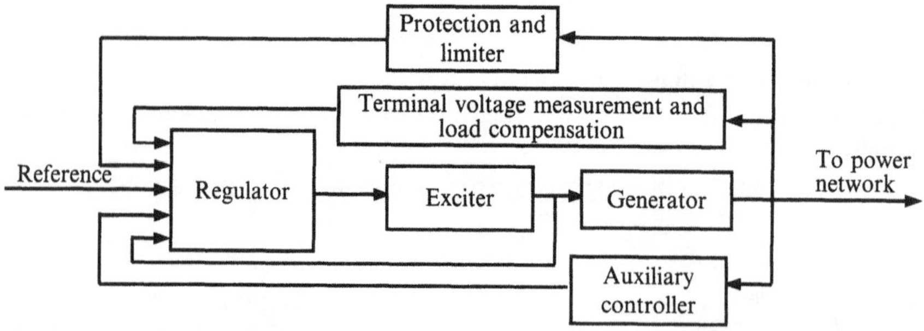

Fig. 6.4 Excitation system of generator

auxiliary controllers is PSS. Protection and limiter are incorporated to ensure the generator's operation within various allowed constraints.

In Sect. 6.2, we have discussed the mathematical model of a synchronous generator. In the following we shall introduce the mathematical models of excitation systems of generators for power system stability analysis, as shown in Fig. 6.4, block by block. These models are applicable to power system operation when system frequency deviation is within 5% and system oscillation frequency is below 3 Hz. Generally speaking, for the study of SSR or other problems of shaft torsional oscillations, these models would not be precise enough.

6.3.1 *Mathematical Model of Exciter*

According to the different means of providing excitation power sources, exciters can be classified into three types: DC exciter systems, AC exciter systems, and static excitation systems. The two former types are also called rotational excitation systems. In the following, we shall introduce each of the three types of exciter.

6.3.1.1 Mathematical Model of DC Exciter

Due to the high cost of maintenance, DC exciters have not been used in recently built large generation units. However, in some power systems, we can still see DC exciters in operation. Hence it is necessary to introduce their mathematical model. We shall introduce the establishment of a mathematical model of the general case of a DC exciter that has both self-excitation and separate excitation. Figure 6.5 shows the configuration of the DC exciter.

In Fig. 6.5, E represents armature of the exciter; R_{ef} and L_{ef}, R_{sf} and L_{sf} is the resistance and self-inductance of the self-excited and separately excited windings, respectively; i_{ef}, i_{sf}, and i_{cf} are the currents of the self-excitation, separate excitation,

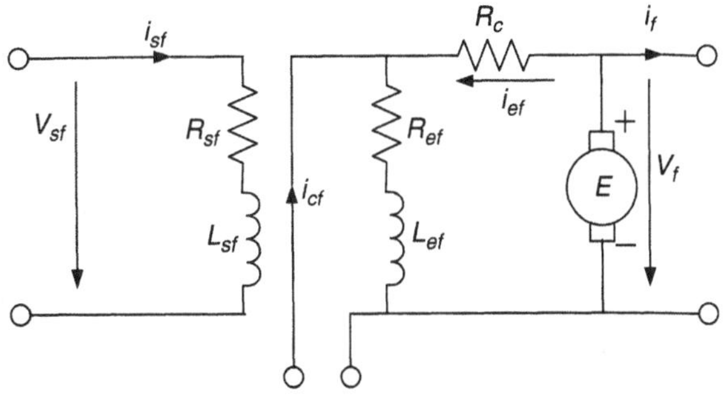

Fig. 6.5 Configuration of a DC exciter

and compound excitation, respectively; v_{sf} is the voltage externally applied on the separately excited winding; and R_c is a variable regulating resistor. For simplicity of analysis, we assume that self-excited and separately excited windings have the same number of turns, or number of turns and parameters of the separately excited winding have been transferred to the side of the self-excited winding. Hence we can obtain the following voltage equations and flux linkage equations (without considering magnetic saturation).

$$\left.\begin{aligned} v_f &= R_c i_{ef} + R_{ef}(i_{cf} + i_{ef}) + p\varphi_{ef} \\ v_{sf} &= R_{sf} i_{sf} + p\varphi_{sf} \end{aligned}\right\}, \tag{6.84}$$

$$\left.\begin{aligned} \varphi_{ef} &= L_{ef}(i_{cf} + i_{ef}) + M_{es} i_{sf} \\ \varphi_{sf} &= M_{es}(i_{cf} + i_{ef}) + L_{sf} i_{sf} \end{aligned}\right\}. \tag{6.85}$$

In the above flux linkage equations, we can approximately consider that the self-excited winding and the separately excited winding are coupled completely. Hence leakage reactance of each winding is zero and unsaturated self-inductance and all mutual inductance have the same value. From (6.85) we can have

$$\varphi_{L0} = \varphi_{ef} = \varphi_{sf} = L i_{f\Sigma}, \tag{6.86}$$

where

$$\left.\begin{aligned} L &= L_{ef} = L_{sf} = M_{es} \\ i_{f\Sigma} &= i_{cf} + i_{ef} + i_{sf} \end{aligned}\right\}. \tag{6.87}$$

φ_{L0} is the flux linkage of the self-excited winding and separately excited, winding without considering saturation, $i_{f\Sigma}$ is the total excitation current provided by the DC exciter.

If the saturation effect is considered, the relationship between the actual flux linkage φ_L and the total excitation current provided by DC exciter $i_{f\Sigma}$ is determined according to Fig. 6.6a, which shows the saturation characteristic curve of the DC exciter. Similarly to (6.67), we define the saturation factor of the DC exciter to be

$$S_E = \frac{\varphi_{L0}}{\varphi_L} - 1 = \frac{i_{f\Sigma}}{i_{f\Sigma 0}} - 1. \tag{6.88}$$

As shown in Fig. 6.6, in (6.88), $i_{f\Sigma 0}$ is the total excitation current required to generate φ_L without considering saturation. The value of S_E represents the level of saturation of the DC exciter, describing the relationship between saturated flux linkage φ_L and unsaturated flux linkage φ_{L0}. It is usually obtained from the load characteristic curve of the exciter. Figure 6.6b shows that because the load of the exciter is fixed, i.e., when the influence of excitation current of generator i_f on the

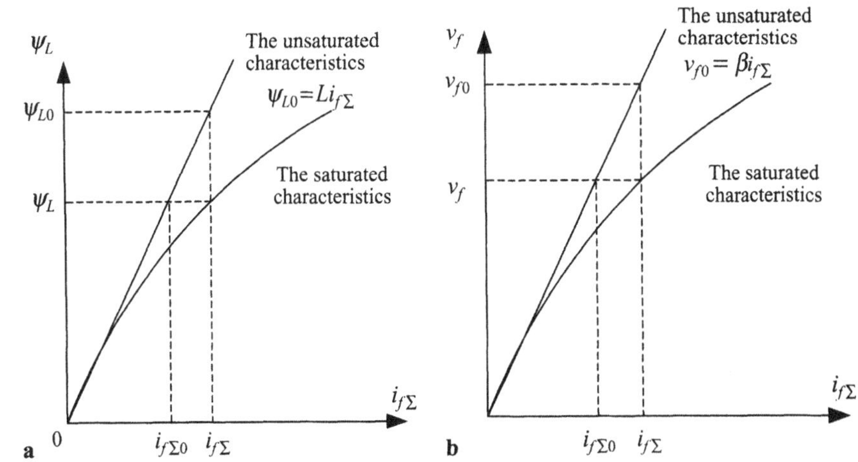

Fig. 6.6 Saturation characteristic curve of DC exciter (**a**) Relationship between flux linkage and excitation current (**b**) Load characteristic curve

armature voltage of the exciter during transients is ignored, the output voltage of exciter is approximately proportional to its internal EMF. If the variation of speed is neglected, flux linkage φ_{L} is proportional to voltage v_f.

Hence the unsaturated characteristic in Fig. 6.6b can be expressed as

$$v_{f0} = \beta i_{f\Sigma}. \tag{6.89}$$

β is the slope of the unsaturated load characteristic curve of exciter, measured in Ohms. From the equation above and (6.86) we can obtain

$$\varphi_{\mathrm{L0}} = \frac{L}{\beta} v_{f0}.$$

Because flux linkage φ_{L} is proportional to voltage v_f, the equation above can be extended to be

$$\varphi_{\mathrm{L}} = \frac{L}{\beta} v_f. \tag{6.90}$$

Dividing both sides of the first equation, (6.84), by $R_c + R_{ef}$, the second equation by R_{sf} and adding these two equations, as well as using (6.86), (6.87), and (6.90) we can obtain

$$\frac{v_f}{R_c + R_{ef}} + \frac{v_{sf}}{R_{sf}} = i_{f\Sigma} - \frac{R_c}{R_c + R_{ef}} i_{cf} + \frac{1}{\beta}\left(\frac{L}{R_c + R_{ef}} + \frac{L}{R_{sf}}\right) p v_f. \tag{6.91}$$

From (6.90), (6.88), and (6.89) we have

$$v_f = \frac{\beta}{L}\varphi_{\mathrm{L}} = \frac{\beta}{L}\frac{\varphi_{\mathrm{L0}}}{1+S_{\mathrm{E}}} = \frac{\beta}{L}\frac{Li_{f\Sigma}}{1+S_{\mathrm{E}}} = \frac{\beta i_{f\Sigma}}{1+S_{\mathrm{E}}}.$$

Substituting the above equation into (6.91) and canceling variable $i_{f\Sigma}$, we can obtain

$$\left[S_{\mathrm{E}} + \left(1 - \frac{\beta}{R_c + R_{ef}}\right) + (T_{ef} + T_{sf})\rho\right] v_f = \frac{\beta}{R_{sf}} v_{sf} + \frac{\beta R_c}{R_c + R_{ef}} i_{cf}, \tag{6.92}$$

where

$$\left.\begin{aligned} T_{ef} &= L/(R_c + R_{ef}) \\ T_{sf} &= L/R_{sf} \end{aligned}\right\}, \tag{6.93}$$

where T_{ef} and T_{sf} are the time constants of self-excited and separately excited windings, respectively (measured in seconds). Equation (6.92) gives the relationship between input v_{sf}, i_{cf}, and output v_f of the exciter using physical units. In order to combine it with the mathematical model of generator in per unit, established in Sect. 6.2, we need to convert (6.92) into per unit form. Here we should use the same base value $V_{f\mathrm{B}}$ that has been chosen in Sect. 6.2 for v_f. To decide the base value for v_{sf} and i_{cf}, we divide both sides of (6.92) by $V_{f\mathrm{B}}$. Then we can see that when base voltage for the excitation current and voltage of the separately excited winding of the exciter are chosen according to the following equation, the equation in per unit is in the most simple form.

$$\left.\begin{aligned} I_{f\Sigma\mathrm{B}} &= V_{f\mathrm{B}}/\beta \\ V_{sf\mathrm{B}} &= R_{sf}V_{f\mathrm{B}}/\beta \end{aligned}\right\}. \tag{6.94}$$

Hence (6.92) in per unit becomes

$$(S_{\mathrm{E}} + K_{\mathrm{E}} + T_{\mathrm{E}}\rho)v_{f*} = v_{sf*} + K_{cf}i_{cf*}, \tag{6.95}$$

where

$$\left.\begin{aligned} K_{\mathrm{E}} &= 1 - \beta/(R_c + R_{ef}) \\ T_{\mathrm{E}} &= T_{ef} + T_{sf} \\ K_{cf} &= R_c/(R_c + R_{ef}) \end{aligned}\right\}. \tag{6.96}$$

K_{E}, T_{E}, and K_{cf} are termed self-excitation factor, time constant, and gain of compound excitation, respectively. By changing variable resistance R_c, these three parameters can be adjusted properly. Equation (6.95) is the mathematical

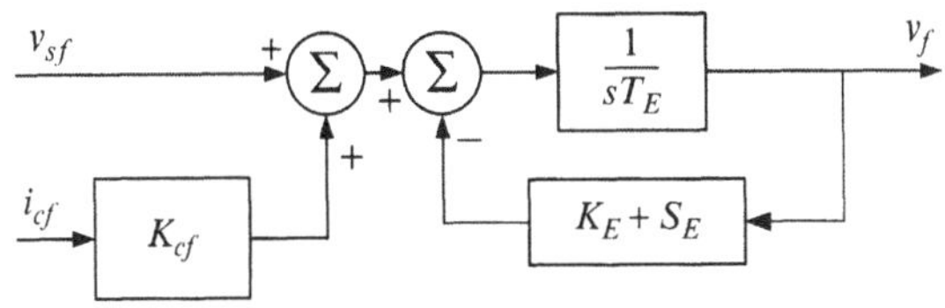

Fig. 6.7 Block diagram of DC exciter

model of the exciter shown in Fig. 6.5. Figure 6.7 is its block diagram where the per unit subscript * has been omitted.

Using the same method that has been adopted to consider the saturation effect in synchronous generators, in Sect. 6.2, we can obtain the relationship between the saturation factor S_E and output voltage of the DC exciter. To match the saturated load characteristic of the exciter by an approximate function, we can derive the following equation, as we have done (6.69)

$$S_E = a_E v_f^{n_E - 1} / b_E. \tag{6.97}$$

1. The case without separately excited winding is equivalent to $R_{sf} = \infty$, $v_{sf} = 0$. Hence from (6.93) and (6.96) we have $T_E = T_{ef}$.
2. The case with only a separately excited winding is equivalent to $R_c = \infty$, $i_{cf} = 0$. Hence from (6.93) and (6.96) we have $T_E = T_{sf}$ and $K_E = 1$.

6.3.1.2 Mathematical Model of AC Exciter

An AC exciter uses a synchronous machine (alternator), usually rotating on the shaft of the generator. AC output from the armature winding of the exciter is rectified through a three-phase noncontrollable, or controllable, bridge rectifier to supply current to the field winding of the generator. There are two types of rectifiers, stationary rectifiers and rotating rectifiers, and two methods of excitation: self-excitation and separate excitation. Hence there are different combinations of types of rectifiers and means of excitation. In the following, we shall first discuss the mathematical model of the exciter and then that of the rectifiers.

The majority of AC exciters use separate excitation. In this case, we can use the mathematical model of a synchronous generator, established in Sect. 6.2, to represent the AC exciter. However, the load of an AC exciter is the field winding of the generator and its operating conditions are much simpler than the generator's. Hence to reduce the effort in analysis and calculation, we need not describe an AC exciter in such detail as we have done for a generator. There are several methods to simplify the mathematical model of a synchronous generator to derive a mathematical model of an AC exciter. Here we shall introduce one simple and commonly used method as follows.

Because the load of the exciter is the field winding of the generator, the armature current of the exciter is almost purely inductive. Hence the q component of armature current of exciter is approximately zero. In the mathematical model of a synchronous generator without damping windings considered, ignoring (6.55) and (6.58), we can obtain the voltage equation of the armature winding of the exciter to be

$$\left.\begin{aligned} v_d &= 0 \\ v_q &= \varphi_d = e'_q - X'_d i_d \end{aligned}\right\}. \tag{6.98}$$

In the above equation, we can further ignore the influence of stator current of the exciter on stator voltage. This leads to stator voltage being equal to the transient voltage. In (6.56), due to the adoption of the base value system of "unit excitation/unit stator voltage", E_{fq} is equal to field voltage. Using the same assumption, denoting the exciter's field voltage by v_R, stator voltage by v_{E}, stator current by i_{E}, using subscript E to denote synchronous reactance, transient reactance, and various time constants of the exciter and following a similar procedure as for deriving (6.56), we can establish the mathematical model of the exciter without considering saturation as follows

$$\left.\begin{aligned} T_{\mathrm{E}} p v_{\mathrm{E}} &= v_R - e_{q\mathrm{E}} \\ e_{q\mathrm{E}} &= v_{\mathrm{E}} + (X_{d\mathrm{E}} - X'_{d\mathrm{E}}) i_{\mathrm{E}} \end{aligned}\right\}. \tag{6.99}$$

With saturation being considered, similarly to the procedure for deriving (6.71), we can obtain

$$(1 + S_{\mathrm{E}}) v_{\mathrm{E}} = e_{q\mathrm{E}} - (X_{d\mathrm{E}} - X'_{d\mathrm{E}}) i_{\mathrm{E}}, \tag{6.100}$$

where we use the same method to gain the saturation factor of exciter S_{E} as we have done for a DC exciter, i.e., to fit the saturation curve of exciter by the approximate function

$$S_{\mathrm{E}} = a_{\mathrm{E}} v_{\mathrm{E}}^{n_{\mathrm{E}}-1} / b_{\mathrm{E}}. \tag{6.101}$$

We should note that stator voltage v_{E} and current i_{E} only enter the field winding of the generator after rectification. The relationship between v_{E} and v_f will be established later in the mathematical model of rectifier. Here we shall derive the connection between i_{E} and i_f first.

When the exciter supplies the field winding of the generator through a three-phase noncontrollable bridge rectifier, output current from the rectifier i_f is the field current of the generator that is approximately proportional to the input current of the rectifier, i.e., armature current of exciter i_{E}. Hence replacing $(X_{d\mathrm{E}} - X'_{d\mathrm{E}}) i_{\mathrm{E}}$ in (6.100) by $K_D i_f$, we can describe this relationship as

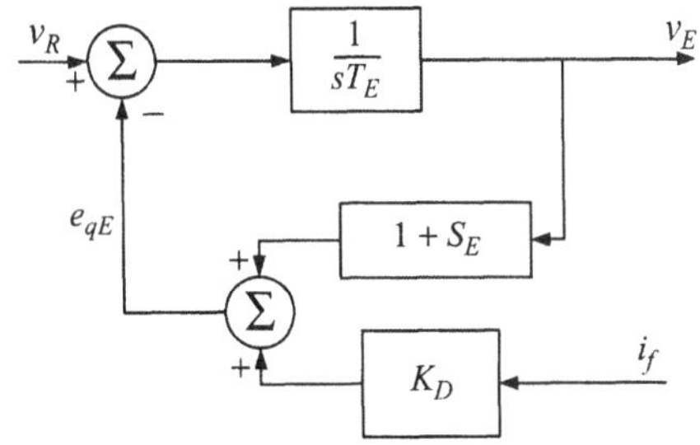

Fig. 6.8 Block diagram of transfer function of separately excited AC exciter

$$\left.\begin{aligned} T_{\mathrm{E}}\rho v_{\mathrm{E}} &= v_R - e_{q\mathrm{E}} \\ e_{q\mathrm{E}} &= (1+S_{\mathrm{E}})v_{\mathrm{E}} + K_D i_f \end{aligned}\right\}. \tag{6.102}$$

From this, the mathematical model of a separately excited AC exciter, using a three-phase noncontrollable bridge rectifier, can be expressed by the block diagram of Fig. 6.8.

For self-excited exciter, replacing $(1 + S_{\mathrm{E}})$ in (6.102) and Fig. 6.8 by $(K_{\mathrm{E}} + S_{\mathrm{E}})$ we can obtain its mathematical model [160], where K_{E} is self-excitation factor and $K_{\mathrm{E}} < 1$.

Because an AC exciter is connected to the field winding of the synchronous generator through a rectifier, base values of its armature and field voltage and current must not only satisfy various rules used when the mathematical model of synchronous generator is established in Sect. 6.2, but also be related to the mathematical model of the rectifier. This will be discussed in Sect. 6.3.1.3.

6.3.1.3 Mathematical Model of Power Rectifier

An AC exciter usually supplies excitation to the generator through a three-phase noncontrollable or controllable rectifying circuit. In the following, we shall introduce a mathematical model of a noncontrollable rectifier. The input to the rectifier is the stator voltage of the AC exciter v_{E}, the output voltage and current are the field voltage and current of the synchronous generator, respectively.

It is very complicated to accurately model the transient response of a rectifier. Engineering practice also suggests that a transient rectifier model is unnecessary. Consequently, a so-called quasisteady-state mathematical model is usually adopted. That is, although during a transient v_{E}, v_f, and i_f satisfy the transient equations of the rectifier, for their instantaneous values in numerical solutions we approximate them as satisfying a steady-state equation. In this way, the transient process is approximated as a series of continuous steady-state processes.

A rectifier has three operational modes according to the value of its commutating angle, γ, being less than, equal to, or greater than 60°. When γ is less than 60° and harmonics are ignored, the steady-state equation of the rectifier using actual values of variables is

$$V_f = \frac{3\sqrt{2}}{\pi}V_{\mathrm{E}} - \frac{3X_\gamma}{\pi}I_f, \tag{6.103}$$

where V_E is the effective value of stator line voltage of the AC exciter, X_γ is the commutating reactance of the rectifier (that is often taken to be the subtransient reactance or negative-sequence reactance of the exciter). Comparing with (4.37), we can see that the equation above in fact treats the three-phase uncontrollable bridge rectifier as for the case of a six-pulse rectifier in HVDC when its firing angle α is zero. In the above equation, $3X_\gamma I_f/\pi$ reflects commutating voltage drop.

To connect with the mathematical model of a generator, we need to convert (6.103) into per unit form. Hence, we divide both sides of the equation by the base value of field voltage of the generator V_{fB}

$$V_{f^*} = F_{EX} V_{E^*}, \tag{6.104}$$

where

$$V_{E^*} = \frac{3\sqrt{2}V_E}{\pi V_{fB}}, \tag{6.105}$$

$$\begin{aligned} F_{EX} &= 1 - I_N/\sqrt{3} \\ I_N &= K_C I_{f^*}/V_{E^*} \\ K_C &= X_\gamma/(\sqrt{3}\pi Z_{fB}) \end{aligned} \tag{6.106}$$

and K_C is a constant. We should point out that commutating angle γ is not included in (6.106). In fact, when γ is less that 60°, I_N is in the range of (0–0.433). It can be proved that when γ is equal to or greater that 60°, (5.104) can still be used as the mathematical model of the rectifier. However, in this case, the relationship between F_{EX} and I_N has changed. When I_N is between zero and 1, F_{EX} is given by the following equation.

$$F_{EX} = \begin{cases} 1 - I_N/\sqrt{3} & 0 \le I_N < 0.443 \\ \sqrt{0.75 - I_N^2} & 0.443 \le I_N \le 0.75\,. \\ \sqrt{3}(1 - I_N) & 0.75 < I_N < 1 \end{cases} \tag{6.107}$$

To use the above model, I_N must be nonnegative and less that 1. If for some reason, the value of I_N is greater than 1, F_{EX} should be set to zero. From (6.104), (6.106), and (6.107), we can obtain the block diagram of the transfer function of the rectifier, and the relationship of F_{EX} and I_N as shown in Fig. 6.9, where the subscript of per unit has been omitted as before.

In the following, we shall discuss the base value of various variables in the mathematical model of an AC exciter. Because we have established the mathematical model of an exciter directly by using that of a synchronous generator, choice of per unit system of the exciter should be kept consistent with that of the synchronous generator. From (6.105) and noting that V_E is the effective value of stator line voltage of the exciter, obviously the base value of stator line voltage of the exciter, V_{LEB}, should be

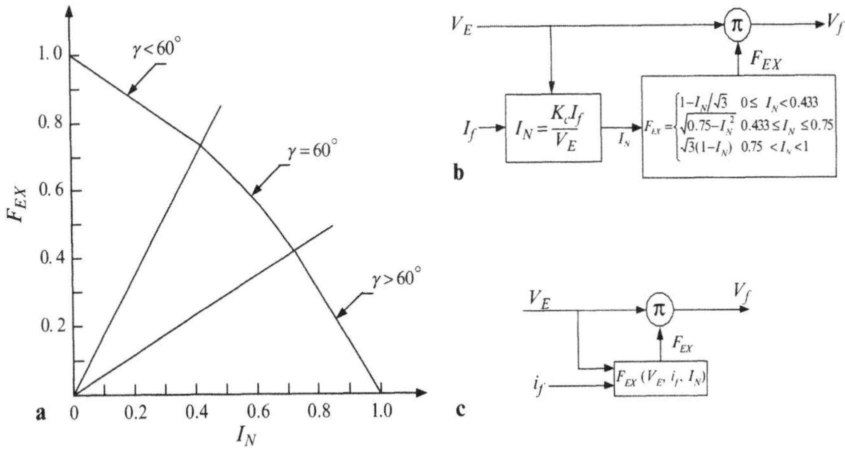

Fig. 6.9 Mathematical model of power rectifier (**a**) F_{EX}–I_N relationship curve (**b**) block diagram of transfer function (**c**) simplified representation of block diagram of transfer function

$$V_{\mathrm{LEB}} = \frac{\pi}{3\sqrt{2}} V_{f\mathrm{B}}.$$

From the relationship between line and phase voltage as well as effective and maximum value, we can obtain the base value of the maximum value of stator phase voltage of the exciter to be

$$V_{\mathrm{EB}} = \frac{\sqrt{2}}{\sqrt{3}} V_{\mathrm{LEB}} = \frac{\pi}{3\sqrt{3}} V_{f\mathrm{B}}. \tag{6.108a}$$

From (6.20), we can give the base value of the maximum value of stator phase current of the AC exciter to be

$$I_{\mathrm{EB}} = \frac{2\sqrt{3} S_{\mathrm{B}}}{\pi V_{f\mathrm{B}}}. \tag{6.108b}$$

The base value of field voltage of the exciter $V_{R\mathrm{B}}$ should be decided through experiment according to the principle of "unit excitation voltage/unit stator voltage." Base value of the field current of exciter is obtained by (6.23) to be

$$I_{\mathrm{RB}} = S_{\mathrm{B}}/V_{\mathrm{RB}}. \tag{6.109}$$

In (6.102), $K_D i_f$ represents the load effect of the exciter. From previous derivation of (6.102) we have known that i_f is approximately proportional to i_{E}, that is, $i_f = k i_{\mathrm{E}}$

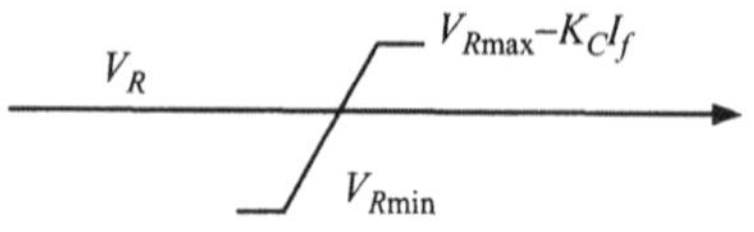

Fig. 6.10 Mathematical model of AC exciter when generator excitation is provided through controllable rectifier

where k is a coefficient of proportionality. From the second term on the right-hand side of (6.100), we have

$$(X_{d\text{E}^*} - X'_{d\text{E}^*})i_{\text{E}^*} = \frac{X_{d\text{E}} - X'_{d\text{E}}}{Z_{\text{EB}}} \times \frac{i_{\text{E}}}{I_{\text{EB}}} = \frac{X_{d\text{E}} - X'_{d\text{E}}}{Z_{\text{EB}}} \times \frac{i_f}{kI_{\text{EB}}} = \frac{X_{d\text{E}} - X'_{d\text{E}}}{kZ_{\text{EB}}} \times \frac{I_{f\text{B}}}{I_{\text{EB}}} i_{f^*}.$$

Hence, K_D in (6.102) is given by the following equation as

$$K_D = -\frac{X_{d\text{E}} - X'_{d\text{E}}}{kV_{\text{EB}}} I_{f\text{B}}. \tag{6.110}$$

When an AC exciter supplies generator excitation by use of a controllable rectifier, the AC exciter itself often is excited in the form of self-excitation. The voltage regulator of the exciter controls the firing angle of its rectifiers to maintain an approximately constant output voltage. In this case, the mathematical model of the exciter is simplified. In addition, in this case, terminal voltage of the AC exciter is set at a high level. This results in relatively small commutating voltage drop for the controllable commutating bridge. At normal operating conditions, commutating voltage drop can be ignored. Only under automatic field-forcing or reduction, is commutating voltage drop represented as $K_C I_f$ in the upper limit of output voltage. Hence when an AC exciter provides generator excitation through controllable rectifiers, the mathematical model of the AC exciter is a unit with bidirectional limiters, as shown in Fig. 6.10. We shall discuss input–output relationships of various limiters later.

6.3.1.4 Mathematical Model of Stationary Exciter

A stationary exciter takes terminal voltage or terminal current plus voltage as the power source of excitation for the generator. The former is called a self-excited potential-source system; the latter is a self-excited compound-source system. In a self-excited potential-source system, generator terminal voltage is reduced via an exciter transformer to supply generator excitation through a controllable rectifier. Firing angle of the controllable rectifier is set by a regulator. The block diagram of the transfer function of the self-excited potential-source system is shown in Fig. 6.11. We can see that it is very similar to that of the AC exciter of Fig. 6.10. This is because the only physical difference between the two is that of the excitation power

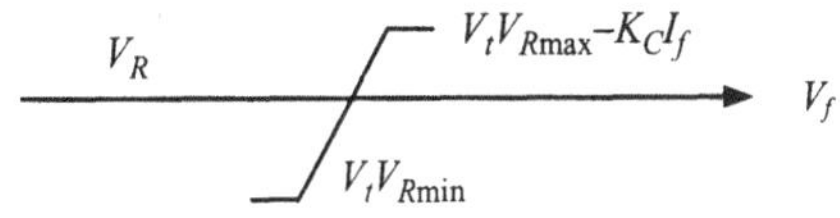

Fig. 6.11 Mathematical model of stationary exciter

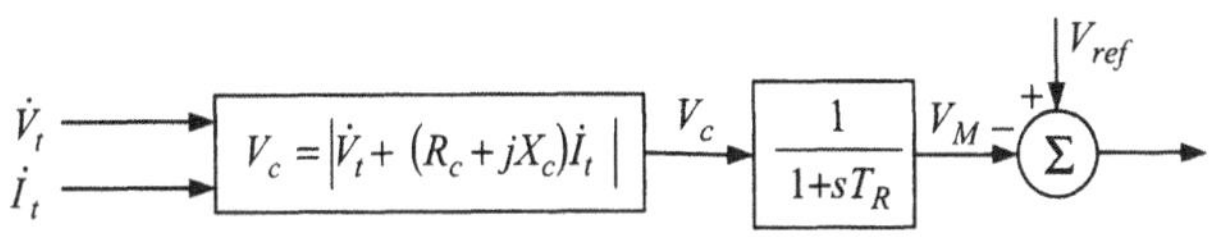

Fig. 6.12 Voltage measurement and load compensation

source. This difference is demonstrated in the output limiters of the two excitation systems. In self-excited potential-source system, the excitation power source is the generator itself. Hence upper and lower limit of output voltage is related to the terminal voltage of generator V_t to be $V_tV_{R\max} - K_CI_f$ and $V_tV_{R\min}$, respectively. $V_{R\max}$ and $V_{R\min}$, respectively, are the maximum and minimum value of no-load voltage of the rectifier when $V_t = 1$.

In a self-excited compound-source system, the power source of the controllable rectifier is supplied by an exciter voltage transformer and current transformer. Measured voltage and current can be accumulated before or after rectification in the form of parallel or series addition. This results in many different types and here we shall not introduce their mathematical models. Details can be found in [160].

6.3.2 Voltage Measurement and Load Compensation Unit

The function of an AVR is to maintain generator terminal voltage at an ideal level. The voltage measurement unit takes generator terminal voltage $\dot{V}_t$ to convert it into a DC signal through stages of voltage reduction, rectification and filtering, etc. The voltage measurement and conversion unit can be described by a first-order simple lag, as shown by the block diagram transfer function of Fig. 6.12. The function of load compensation is to compensate the load current of generator $\dot{I}_t$ so as to maintain the required constant voltage at a controlled voltage point in steady-state operation when the load changes. $R_C + jX_C$ represents the impedance between the controlled voltage point and generator terminal. When R_C and X_C are positive, the controlled voltage point is inside the generator; otherwise, outside the generator. In addition, automatic distribution of reactive load among electrically close generators is related to the voltage droop of the generator; while the voltage droop of the generator is realized by adjusting parameter R_C and X_C. For simplicity, often R_C is ignored and set to zero. When X_C is greater than zero, we have positive droop; that is, the larger the load current, the higher the terminal voltage. On the other hand, when X_C is less than zero we have negative droop. Terminal voltage decreases with the increase of load current. In the case without compensation, parameters R_C, and X_C are zero. Voltage measurement and load compensation units may have different time constants. For simplicity, usually we just use a single time constant T_R for their description. T_R is called the time constant of the measurement unit and usually is less than 60 ms. For many systems, it is very close to zero. Hence in computation, we often take it to be zero. Its output voltage V_M is compared against reference voltage V_{ref}. After amplification, the error signal is used as the control signal of the excitation system of the generator. Although the reference voltage V_{ref} is set artificially, it reflects the ideal value for the controlled voltage point of the generator and must satisfy the initial steady-state operating conditions of the whole power system.

6.3.3 Limiters

In the mathematical model of an excitation system, due to functional limitations, physical limits, or the existence of saturation, the output of certain units is subject to limitations, which we represent via limiters. There are two types of limiters, windup limiters and nonwindup limiters. Limiters often appear in integral units, the first-order simple lag and lead-lag units. Figure 6.13a, b show block diagrams of those two types of limiters. In the following, we shall discuss the example of an integral unit and its input–output relationship. We leave those of windup and nonwindup first-order simple lag and lead-lag units for readers to establish.

Equation of an integral unit is $dv/dt = u$. The limiting function of the two types of limiters is different. For a windup limiter, if variable v is greater than lower limit B and lower than upper limit A, output variable y is v; If v is greater or equal to upper limit A, output variable y is constrained to be upper limit A; If v is less or equal to lower limit B, output variable y is constrained to be lower limit B. We should note that variable v is not constrained and only the next output variable y is. If v is beyond limitation, output variable y is constrained to be the value of the upper or lower limit. For a nonwindup limiter, output variable y is directly constrained between upper and lower limit. If y is within the limits, input–output relationship is $dv/dt = u$. If y is equal to upper limit and tends to increase with time, i.e., $dy/dt > 0$, input–output relationship is $dy/dt = 0$ and y takes the value of upper limit A. If y is equal to lower limit and tends to decrease, i.e., $dy/dt < 0$, the relationship is $dy/dt = 0$ and y takes value of lower limit B. When the output variable y takes the value of the upper or lower limit, once input variable u changes sign, y enters within

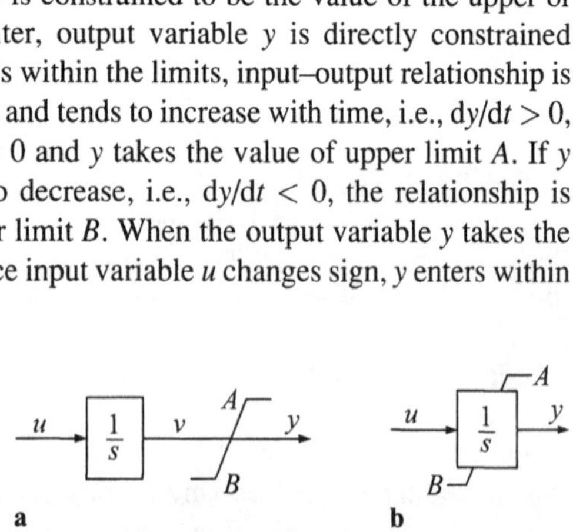

Fig. 6.13 Limiters (**a**) integral unit with windup limiter (**b**) integral unit with nonwindup limiter

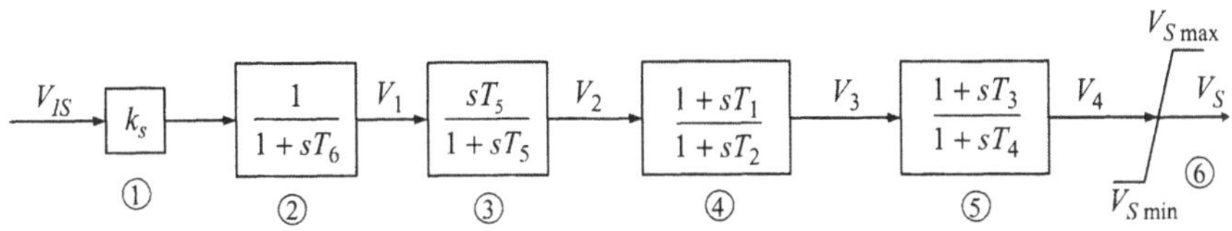

Fig. 6.14 Block diagram of transfer function of power system stabilizer

the limits. However, for a windup limiter, only when variable v returns within the limits, so does the output variable.

6.3.4 *Mathematical Model of Power System Stabilizer*

Power system stabilizer (PSS) is a widely used auxiliary regulator in excitation control. Its function is to suppress power system low-frequency oscillations or increase system damping. Its basic principle is to provide the AVR with an auxiliary control signal to make the generator produce an electrical torque in phase with the deviation of rotor speed. Details about the PSS principle, parameter setting, and installation locations can be found in [153]. There are several forms of PSS. Here we give a commonly used block diagram of a PSS transfer function as shown in Fig. 6.14.

In Fig. 6.14, block① is the gain of PSS; ② is the measurement unit with a time constant T_6 (usually very small and can be ignored); ③ is a wash-out unit or low-frequency filter to block steady-state input signal to disable PSS at steady-state operation. T_5 usually is as large as about 5 s. ④; and ⑤; are two lead-lag networks. PSS should consist of at least one lead-lag network. ⑥; is a limiter. Input signal to PSS, V_{IS}, usually is generator speed, terminal voltage, power, system frequency, or combination of some of them. Output signal V_s is superimposed on the AVR input signal. For PSS to play an effective role, its installing location must be selected and parameters be set properly.

6.3.5 *Mathematical Model of Excitation Systems*

The function of an AVR is to treat and amplify the input control signal to generate a suitable excitation control signal. The AVR usually consists of power amplifier, excitation system stabilizer, and limiters. In the following, we shall introduce mathematical models of different excitation systems. In each block diagram shown below, basic input signal V_C is the output from voltage measurement and load compensation unit of Fig. 6.12 and V_s is an auxiliary regulation signal of the AVR, such as the output signal from a PSS.

6.3.5.1 Excitation System with DC Exciter

With different types of AVR being used, there are three types of DC excitation systems: controllable phase compound regulator, compound excitation plus load compensation, and thyristor-controlled regulator. The former two DC excitation systems are usually used for small generation units (100 MW or below) and have been gradually passing out of use. The block diagram of an excitation system adopting controllable phase compound regulator is shown in Fig. 6.15, where $\dot{V}_t$ and $\dot{I}_t$ are the terminal voltage and current of generator, respectively. Block ①; represents phase compound excitation, blocks ②; and ③; are load compensation and measurement unit, ④; is composite amplifying unit, ⑤; is limiter with input signal being compound excitation current of exciter, ⑥ and ⑦ are units of the DC exciter. To improve performance of the excitation system, a soft negative feedback unit ⑧ is often used to provide a series adjustment to field voltage of the generator. Control parameters that can be set are K_V, K_I, R_C, X_C, K_E, K_A, T_A, K_F, and T_F. For the excitation system adopting compound excitation plus load compensation, the block diagram can still be used except that block ①; needs to be replaced by a simple amplifier of I_t.

Figure 6.16 shows the block diagram of a DC excitation system using thyristor-controlled regulator. T_B and T_C are time constants of the excitation regulator itself. They are usually very small and considered to be zero. Time constant and gain of composite amplifying unit is T_A and K_A. Due to the saturation of the amplifier and limitation of power output, the block for the amplifier has a nonwindup limiter. V_F

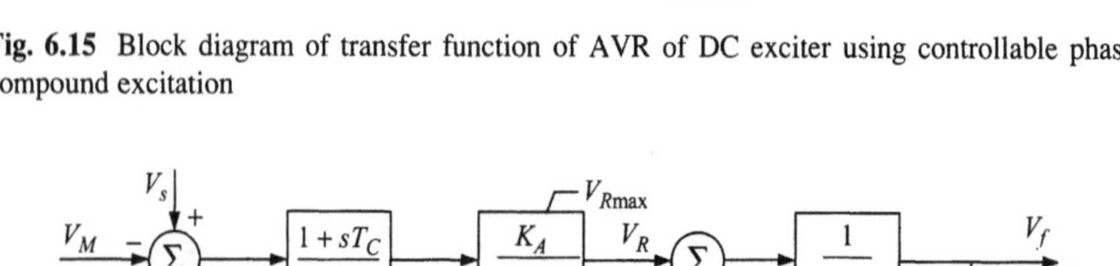

Fig. 6.15 Block diagram of transfer function of AVR of DC exciter using controllable phase compound excitation

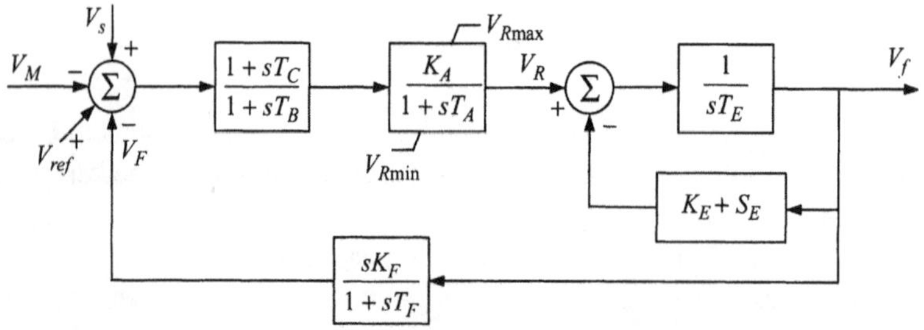

Fig. 6.16 Block diagram of transfer function of a DC excitation system

is the output of the soft negative feedback unit of excitation voltage to improve the dynamic performance of whole excitation system. V_R is the excitation voltage of the DC exciter. Parameters to be set for the normal operation of excitation control are R_C, X_C, K_E, K_A, T_A, K_F, and T_F.

6.3.5.2 Excitation System with AC Exciter

Excitation system with an AC exciter is widely used for 100 MW or above generation units. Most excitation systems with AC exciter adopt uncontrollable power rectifier. They can be classified into two groups: stationary rectifier excitation systems and rotating rectifier excitation systems. Here we introduce one type of AC excitation system as shown by the block diagram of Fig. 6.17. Introduction to other types of AC excitation systems can be found in [160]. In Fig. 6.17, parameters T_B, T_C, K_A, T_A, K_F, and T_F describe three blocks belonging to the excitation regulator similar to that in Fig. 6.16. The input signal to the series regulation unit is the no-load voltage $e_{q\mathrm{E}}$ of the AC exciter (6.102). Another kind of arrangement is to use the field voltage of the generator V_f as the feedback input signal. Field current I_f is also an input signal of the excitation regulator and constant K_D represents the equivalent load effect of the AC exciter. In Fig. 6.17, the exciter is separately excited. When self-excitation is used, we need to replace the block $1 + S_\mathrm{E}$ by $k_\mathrm{E} + S_\mathrm{E}$, where k_E and S_E are the self-excitation coefficient and saturation factor of the AC exciter, respectively. Because the input to rectifier requires V_E not to be negative, in the block of the exciter the integral unit represented by T_E has a single-directional windup limiter to prevent V_E from becoming negative. Parameters to be set for the normal operation of excitation control are R_C, X_C, K_E, K_A, T_A, K_F, and T_F.

The block diagram of an AC excitation system adopting a controllable rectifier to supply generator excitation is shown in Fig. 6.18. The rectifier is controlled by an independent voltage regulator and hence its output is kept approximately constant. Therefore, the mathematical model of an AC exciter and controllable rectifier is shown in Fig. 6.10. In Fig. 6.18, this has been combined with an equivalent

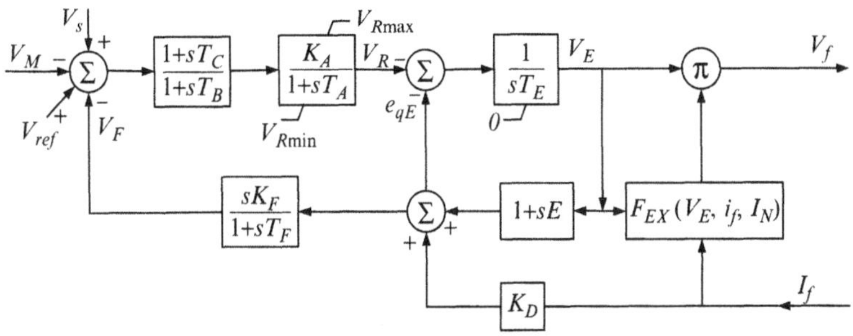

Fig. 6.17 Block diagram of transfer function of excitation system with AC exciter adopting uncontrolled power rectifier

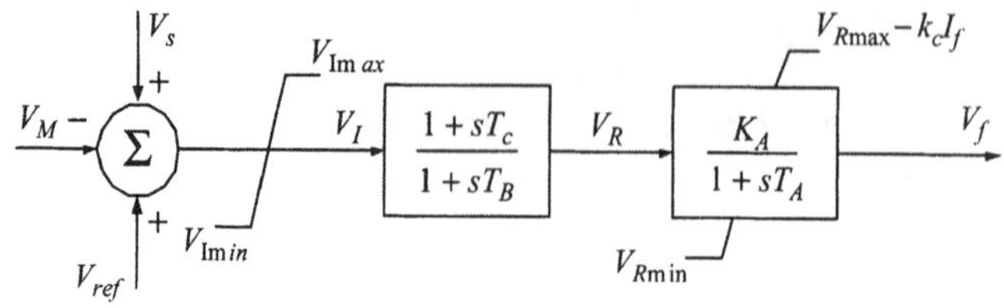

Fig. 6.18 Block diagram of transfer function of excitation system with AC exciter adopting controllable power rectifier

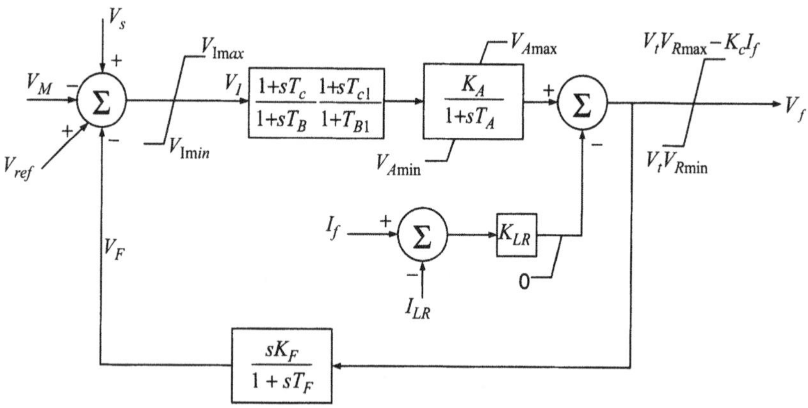

Fig. 6.19 Block diagram of transfer function of self-excited potential-source system

composite amplifying unit, where time constant T_A and gain K_A depict the dynamic performance of the controllable rectifier and its regulator. To improve system dynamic performance, this type of excitation system usually adopts a series regulator instead of a shunt regulator. The time constants of the series regulator are T_B and T_C. We should point out that the load of the controllable rectifier is limited to ensure I_N between 0 and 0.433 (6.107). Load effect of the excitation system is reflected in the upper limit of the bidirectional limiter. Parameters to be set for the normal operation of excitation control are R_C, X_C, K_A, T_A, T_C, and T_B. Here, because an independent AC exciter is used, the values of upper and lower limits of the bidirectional nonwindup limiter are not connected to the terminal voltage of the generator.

6.3.5.3 Stationary Excitation System

Figure 6.19 shows the block diagram of a self-excited potential-source system and controllable rectifier described by a bi-directional limiter. As has been introduced before, the power into stationary excitation is from the generator terminal. Hence the value of upper and lower limit is related to the terminal voltage of the generator. This type pf excitation system can provide very high automatic field forcing voltages. To avoid overloading of generator field and rectifier, the field current of generator I_f is constrained by K_{LR} and I_{LR} in the diagram. Proportional unit K_{LR} has a windup lower limit. To avoid this unit we can simply set K_{LR} to zero. K_A and T_A are the system composite equivalent gain and time constant, respectively. Both series regulation and shunt regulation are displayed in the diagram. Usually only one of them is used. Hence when series regulation is used, we can set K_F to zero. Or when shunt regulation is used, we just set time constants T_B and T_C to zero. Time constants T_{BI} and T_{CI} are for the increase of system dynamic gain. Usually we have $T_{CI} > T_{BI}$. To simplify the model, this unit can be ignored by setting both of these time constant to zero. Here we should point out that the block diagram of Fig. 6.19 can represent the excitation system adopting full-wave controllable rectifying bridge. When a half-wave controllable rectifying bridge is used, we can simply set the lower limit of the bidirectional limiter at the system output to zero. Parameters to be set for the normal operation of excitation control are R_C, X_C, K_A, T_A, K_F, T_F, T_C, T_B, K_{LR}, and I_{LR}. In [160], we can find more about the mathematical model and block diagrams of other types of stationary excitation system.

6.4 Mathematical Model of Prime Mover and Governing System

Variable P_{m} in the rotor movement equation of the generator (6.78) is the mechanical power output from the prime mover. P_{m} is related to the operating condition of the prime mover and controlled by a governing system. Excluding wind, sun, and wave power generation, there are two types of prime mover used for large-scale power generation, hydraulic turbines, and steam turbines. The hydraulic turbine (or steam turbine) converts hydraulic energy (or steam thermal energy) into rotating kinetic energy of the prime mover which is then converted into electric power by the generator. Obviously, the amount of power being converted is associated with the opening position of the wicket gate of a hydraulic turbine and steam valve of a steam turbine. Because the generator rotor is driven by the prime mover and rotates on the same shaft with the prime mover, if we assume that the generator output power is fixed, when the opening position increases, the generator will accelerate; and conversely it decelerates. Therefore, regulation of the gate or valve position will change the output power from the prime mover to control generator speed. Hence it is easy to see that the main control signal to the opening position should be generator speed. From the rotor movement equation (6.78) we can see that when a power system is subject to a disturbance at steady-state operation, electric power output from the generator changes. This change destroys the balance between electric power output from the generator and mechanical power input to the generator from the prime mover, leading to variation of the generator speed. Change of generator speed results in a response of the governing system to adjust the opening position of the wicket gate (of a hydraulic turbine) or steam valve (of a steam turbine). The disturbance causes the system to engage in a complex transient process of mechanical, magnetic, and electrical interactions. Therefore, when the function of the governing system is considered, resulting in the variable P_m, we need to establish a mathematical model of the prime mover and the governing system in order to quantitatively analyze electromechanical transients in power systems.

6.4.1 *Mathematical Model of Hydroturbine and Governing System*

6.4.1.1 Mathematical Model of Hydraulic Turbine

Dynamics of hydraulic turbines are closely related to those of water flow through a penstock, whereas the characteristics of water flow through a penstock are affected by many factors, such as water inertia, water compressibility, and pipe wall elasticity in the penstock. For example, due to water inertia inside a penstock, change of water flow inside a hydraulic turbine lags the opening position change of the wicket gate. When the opening position of the wicket gate increases suddenly, water volume at the wicket gate increases. However, due to the water inertia, speed of water flow at other points inside the pipe cannot increase immediately. This results in input water pressure of the hydraulic turbine decreasing instead of increasing for a short of period of time after the change, leading to a decrease of input power of the hydraulic turbine instead of an increase. On the other hand, when the opening position of the wicket gate decreases suddenly, input water pressure and input power will increase temporarily and then decrease. This phenomenon is usually called the water hammer effect. Furthermore, for the movement of a compressible fluid inside an elastic pipe, the change of water flow volume and pressure at each point inside the pipe is a wave movement, quite similar to the wave process of transmission lines with evenly distributed parameters. A detailed derivation of the mathematical model of input water pressure on the turbine with wave effects considered requires extensive application of fluid mechanics. This is only necessary for the case with a long pressure pipe. In the following, we shall establish a mathematical model of a hydraulic turbine, useful for the analysis of power system stability, with the wave effect of water flow ignored. That is, to assume that the pressure pipe is inelastic, and water is not compressible. Additionally we shall only consider an ideal hydraulic turbine, i.e., (1) neglecting the mechanical power loss caused by the resistance against water flow from the penstock wall; (2) power output of the hydraulic turbine being proportional to the product of net water head and water flow volume; and (3) speed of water flow being proportional to the product of the opening position of the wicket gate and square root of the stationary water head. Hence we can obtain the hydraulic equations as follows:

$$U = K_U \mu \sqrt{H}, \tag{6.111}$$

$$P_m = K_P H U, \tag{6.112}$$

$$\frac{dU}{dt} = -\frac{g}{L}(H - H_0), \tag{6.113}$$

where U is the water velocity; K_U the proportional constant; H the net water head of hydraulic turbine; μ the opening position of wicket gate; P_m the mechanical power output of hydraulic turbine; K_P the proportional constant; g the gravity acceleration constant; L the length of penstock; and H_0 is the steady-state value of H.

Taking the initial value of various variables as their base value, the above hydraulic equations can be converted into the following per unit form (subscript * is omitted as before)

$$U = \mu\sqrt{H}, \tag{6.114}$$

$$P_m = HU, \tag{6.115}$$

$$\frac{dU}{dt} = -\frac{1}{T_w s}(H - 1), \tag{6.116}$$

where

$$T_w = LU_0/(gH_0) \tag{6.117}$$

T_w is the time constant of equivalent water hammer effect and physically it is the time required for water head H_0 to accelerate water flow in penstock from a stationary state to the flowing speed U_0. We ought to point out that this time constant is affected by U_0, i.e., related to the load condition of the hydraulic turbine. The heavier the load is, the higher the time constant. Usually under full load condition, T_w is set by the manufacturer between 0.5 and 4 s.

Assuming that at initial steady state, the operating point of the hydraulic turbine shifts slightly due to small disturbances from the load, the above hydraulic equations can be linearized at the initial steady-state operating point and after Laplace transformation they become

$$\Delta U = \sqrt{H_0}\Delta\mu + \frac{1}{2}\frac{\mu_0}{\sqrt{H_0}}\Delta H, \tag{6.118}$$

$$\Delta P = H_0 \Delta U + U_0 \Delta H, \tag{6.119}$$

$$T_w s \Delta U = -\Delta H. \tag{6.120}$$

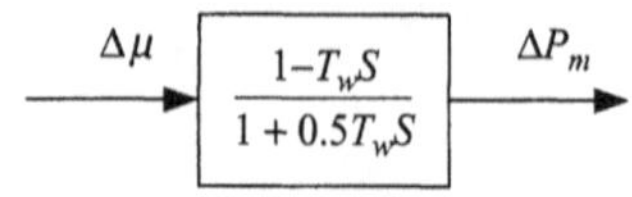

Fig. 6.20 Transfer function of classical model of hydraulic turbine

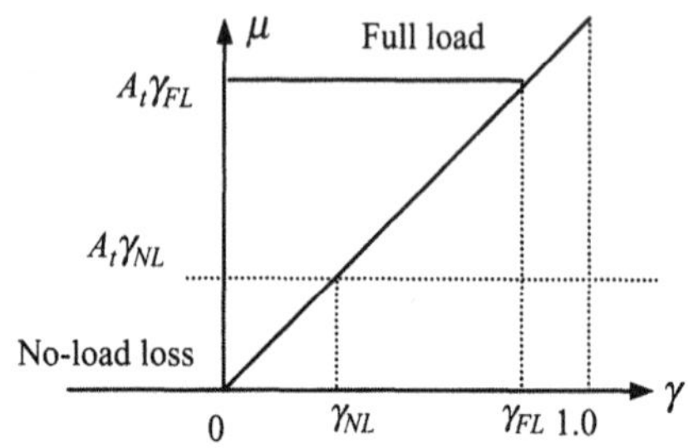

Fig. 6.21 Relationships between actual and ideal opening position

Eliminating variables ΔH and ΔU in the above three equations we can obtain (per unit value of H_0 is 1)

$$\Delta P_{\mathrm{m}} = \frac{1 - T_{\mathrm{w}}s}{1 + 0.5T_{\mathrm{w}}s}\Delta\mu. \tag{6.121}$$

The model above is called the classical model of a hydraulic turbine. Its transfer function block diagram is shown in Fig. 6.20.

In the analysis of power system stability, the above classical model of a hydraulic turbine is used. From the assumptions used to derive the model we know that the classical model is applicable to cases with relatively small variations of load. When load changes over a wide range, the model may cause a large computational error [161]. In the following, we shall establish a nonlinear model of a hydraulic turbine. Basic assumptions will be the same as those for deriving the classical model except that mechanical power loss and dead zone are taken into account. Opening position of wicket gate μ in (6.111) is that of an ideal wicket gate with the dead zone of the hydraulic turbine caused by factors such as friction being ignored, i.e., it is assumed that when μ changes from 0 to 1, operation of the hydraulic turbine goes from no load to full load. With mechanical power loss being considered, position change of the wicket gate from closing to opening, initially has to overcome stationary friction forces in the hydraulic turbine without causing the turbine to start rotation immediately. Hence we need to replace the ideal opening position μ by the actual one γ. From Fig. 6.21, we can see their relationship to be

$$\mu = A_{\mathrm{t}}\gamma, \tag{6.122}$$

where

$$A_{\mathrm{t}} = \frac{1}{\gamma_{\mathrm{FL}} - \gamma_{\mathrm{NL}}}. \tag{6.123}$$

When the actual opening position is γ_{NL}, the hydraulic turbine is still at no-load. When it is γ_{FL}, the hydraulic turbine operates at full load. With power loss being considered, hydraulic equations (6.112) become

$$P_m = K_P HU - P_L, \tag{6.124}$$

$$P_L = K_P U_{NL} H, \tag{6.125}$$

where P_L is no-load loss of the hydraulic turbine; U_{NL} critical water speed when the hydraulic turbine changes from stationary to rotating. Obviously that is when the actual opening position is γ_{NL}. Taking the rated parameters of the hydraulic turbine as the corresponding base value, we can convert (6.111), (6.113), (6.124), and (6.125) into the following per unit form

$$U = \mu\sqrt{H}, \tag{6.126}$$

$$P_m = (U - U_{NL})H, \tag{6.127}$$

$$\frac{dU}{dt} = -\frac{1}{T_W}(H - H_0), \tag{6.128}$$

$$\text{where } T_W = \frac{LU_B}{gH_B}. \tag{6.129}$$

T_W is the time constant of the equivalent water hammer effect at rated load. From (6.117) we can see that the relationship of the time constant between any load condition and at rated load is

$$T_w = \frac{U_0 H_B}{U_B H_0} T_W. \tag{6.130}$$

In (6.127), base value of power is the rated power of the hydraulic turbine. To connect it to the mathematical model of the generator, we can convert the base value of power to the rated power of generator S_B

$$P_m = P_r(U - U_{NL})H, \tag{6.131}$$

$$P_r = P_B/S_B. \tag{6.132}$$

Rewriting (6.126)

$$H = \left(\frac{U}{\mu}\right)^2. \tag{6.133}$$

From (6.122) and (6.133) we can eliminate H in (6.128) and (6.131) to obtain

$$\frac{\mathrm{d}U}{\mathrm{d}t} = -\frac{1}{T_W}\left(\left(\frac{U}{A_t\gamma}\right)^2 - H_0\right), \tag{6.134}$$

$$P_m = P_r(U - U_{NL})\left(\frac{U}{A_t\gamma}\right)^2. \tag{6.135}$$

The two equations above are the nonlinear model of a hydraulic turbine. From the physical meaning of U_{NL} we know that when the actual opening position of wicket gate γ is γ_{NL}, acceleration of water flow is zero. From (6.134) we have

$$\left.\frac{\mathrm{d}U}{\mathrm{d}t}\right|_{U=U_{NL}} = -\frac{1}{T_W}\left(\left(\frac{U_{NL}}{A_t\gamma_{NL}}\right)^2 - H_0\right) = 0, \quad U_{NL} = A_t\gamma_{NL}\sqrt{H_0}. \tag{6.136}$$

Normally H_0 is 1 and hence U_{NL} is a constant. From (6.122), (6.133), (6.128), and (6.131) we can show the nonlinear model of the hydraulic turbine in Fig. 6.22.

6.4.1.2 Mathematical Model of Governing System of Hydraulic Turbine

Modern generation units usually use an electrical-hydraulic governing system. However, the principle of mechanical hydraulic governing system is easier to illustrate. Hence we shall take it as representative to establish the mathematical model of a governing system of a hydraulic turbine. Figure 6.23 shows the configuration of a governing system using a centrifugal pendulum (fly-ball governor). In the following we shall present the equation of motion of each component of the governing system, where variables are in per unit and their positive direction is indicated in Fig. 6.23. Compressibility of hydraulic oil will be neglected:

1. *Equation of the centrifugal pendulum.* The function of the centrifugal pendulum (fly balls) is to measure generator speed. Relative ring position of the fly balls is denoted by η. When generator speed increases, the fly balls move away from each other due to the increase of centrifugal force and consequently η decreases. On the other hand, when generator speed decreases, the fly balls come closer because of the decrease of centrifugal force and hence η increases. Ignoring the

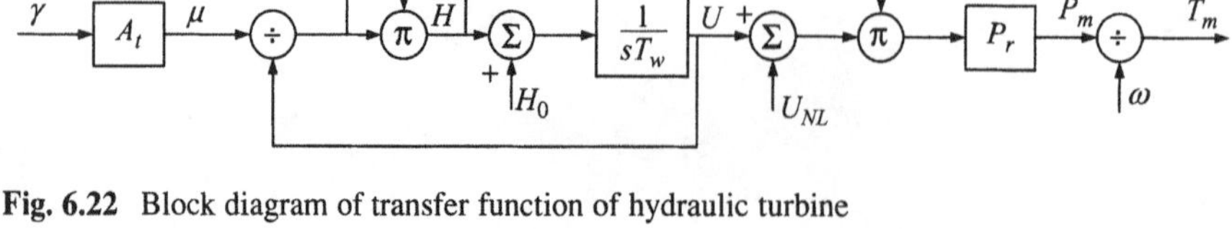

Fig. 6.22 Block diagram of transfer function of hydraulic turbine

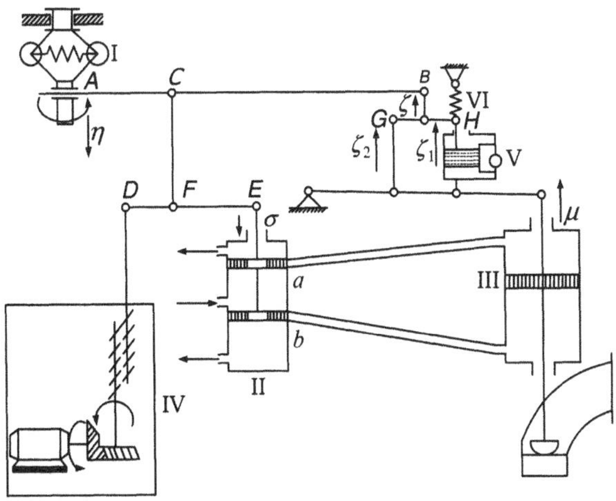

Fig. 6.23 Illustration of governing system of a centrifugal pendulum

mass of the fly balls and damping of motion, η is approximately proportional to the deviation of generator speed with a proportional coefficient k_δ, that is

$$\eta = k_\delta(\omega_0 - \omega). \tag{6.137}$$

2. *Equation of pilot valve*. If the servomotor does not function (point D in Fig. 6.23 is fixed) and the inertia of the pilot valve is ignored, the relationship between position of pilot valve, σ, and that of point B, ζ, is

$$\sigma = \eta - \varsigma. \tag{6.138}$$

3. *Equation of relay valve*. Position of relay valve, μ, is the integral of the position of the pilot valve, σ, with respect to time, i.e., position speed of relay valve is proportional to position of pilot valve, σ. The proportional constant is called the time constant of the relay. Hence

$$T_s p\mu = \sigma. \tag{6.139}$$

4. *Feedback equation*. From Fig. 6.23, we can see that when η increases, σ increases accordingly and μ also increases. Increase of η and μ results in an increasing ζ, leading σ to decrease and μ to decrease accordingly. Hence ζ is a position variable exhibiting feedback from μ. There are two parts in ζ, ζ_1, and ζ_2. ζ_1 is a soft feedback due to the existence of a spring and dashpot; while ζ_2 is proportional to μ and hence it is a hard feedback. We have

$$\varsigma = \varsigma_1 + \varsigma_2 = \frac{k_\beta T_i s}{1 + T_i s}\mu + k_\alpha \mu, \tag{6.140}$$

where $k_\alpha = \alpha/\delta$, $k_\beta = \beta/\delta$, $\delta = 1/k_\delta$. k_β and T_i are the gain and time constant of the soft feedback, respectively; k_σ the gain of hard feedback; δ the sensitivity of measuring component; β coefficient of soft feedback; and α is the droop coefficient. Due to the inertia, water flow cannot follow the change of opening position of the wicket gate quickly. Hence when speed deviation of the generator changes fast, the governor needs a strong negative feedback from the opening position μ to slow down the change of opening position of wicket gate such that water flow and output power from the hydroturbine can follow that of μ. For slow variations of generator speed in steady-state operation, the governor needs to respond promptly. Hence the gain of the negative feedback should take a small value. From (6.140) we can see that dynamic gain of the whole negative feedback unit is high. When $t = 0$, it is the summation of k_β and k_α. The time constant of the soft negative feedback is large, usually between 0.5 and 5 s. At steady state, the steady-state gain of the soft negative feedback is zero and the gain of the whole negative feedback is only that of the hard feedback k_α. This provides the generator with a certain droop coefficient at steady-state operation, such that a generator speed decrease will increase generator output. The droop characteristic ensures stable load sharing among multiple generation units in parallel steady-state operation, to realize the function of primary frequency control. k_α and k_β are often about 0.04 and 0.4, respectively.

Opening position of both pilot valve and wicket gate have certain limitations imposed. In addition, due to the existence of mechanical friction and gap, there exists a certain dead zone of the governing function. Hence in the mathematical model, there are associated limiters and a nonlinear unit representing the dead zone. From (6.137) to (6.140) we can obtain the transfer function block diagram of the governing system of a hydraulic turbine as shown in Fig. 6.24.

The function of an electrical-hydraulic governing system of a hydraulic turbine is quite similar to that of the mechanical hydraulic system introduced above, but more simple and flexible as far as the regulation of basic parameters is concerned. A mathematical model of an electrical-hydraulic governing system of the hydraulic turbine adopting PID control can be found in [162, 163].

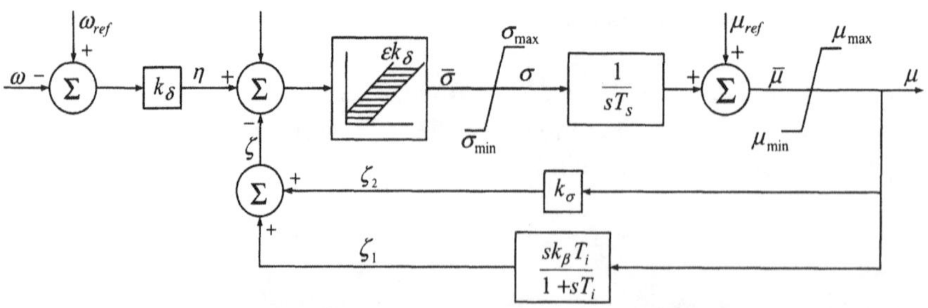

Fig. 6.24 Block diagram of governing system of centrifugal pendulum (fly ball) of a hydraulic turbine

6.4.2 Mathematical Model of Steam Turbine and Governing System

6.4.2.1 Mathematical Model of Steam Turbine

Dynamics of steam turbines are mainly related to the volume effect of steam. In the following, we shall first derive the time constant for the general steam volume effect. As shown in Fig. 6.25, volume of the vessel is V (m^3) and input and output steam mass flow rates are Q_{in} and Q_{out} (kg s^{-1}), respectively.

We have

$$\frac{dW}{dt} = V\frac{d\rho}{dt} = Q_{in} - Q_{out}, \tag{6.141}$$

where W is the weight of steam in the vessel (kg) and ρ is the density of steam (kg m^{-3}).

Assuming that the output steam flow is proportional to steam pressure in the vessel, we have

$$Q_{out} = \frac{Q_N}{P_N}P, \tag{6.142}$$

where P is steam pressure in the vessel (kPa), P_N the rated steam pressure in the vessel (kPa), and Q_N is the rated output of steam out of the vessel (kg s^{-1}).

With the steam temperature in the vessel being constant, we have

$$\frac{d\rho}{dt} = \frac{dP}{dt}\frac{\partial\rho}{\partial P}, \tag{6.143}$$

where the rate of change of steam density with pressure, at a given temperature, $\partial\rho/\partial P$, can be obtained from steam parameter tables and is a constant. From (6.141) to (6.143) and after Laplace transformation, we have

$$Q_{out} = \frac{1}{1 + sT_V}Q_{in}, \tag{6.144}$$

$$\text{where } T_V = \frac{P_N}{Q_N}V\frac{\partial\rho}{\partial P}. \tag{6.145}$$

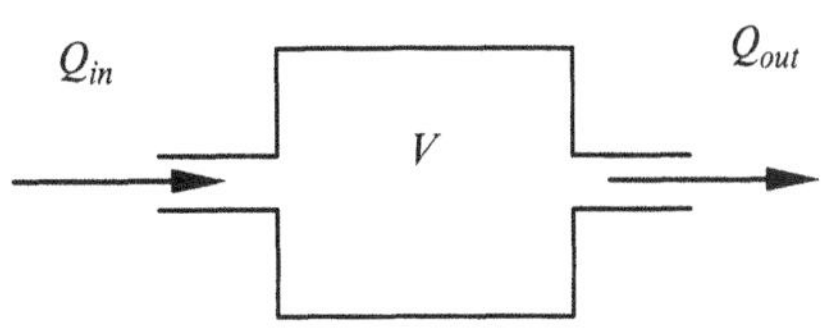

Fig. 6.25 Steam vessel

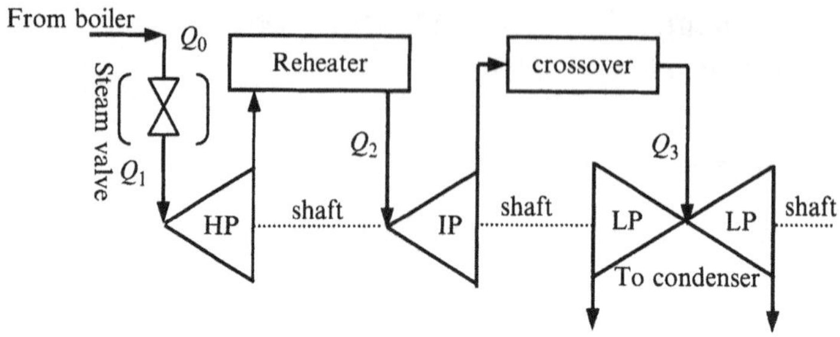

Fig. 6.26 Illustration of a multistage steam turbine

T_V is called the time constant of steam volume effect. From (6.145) we can see that the bigger the volume of the vessel, the higher is the time constant of volume effect. From (6.144) we can see that when the input steam flow increases (or decreases) suddenly, the output steam flow will not increases (or decreases) immediately because the pressure inside the vessel cannot increase (or decrease) instantly. Change of output steam flow lags that of input steam flow. This is the steam volume effect phenomena.

There are many types of configuration of steam turbines. Modern steam turbine units consist of multiple-stage steam turbines to drive a single generator. According to the difference in rated operating steam pressure, multiple-stage turbines can be classified as high pressure (HP), intermediate pressure (IP), and low pressure (LP) turbines.

Medium and small steam turbine units may have only a one-stage turbine. To increase thermal efficiency, modern steam turbine units usually have an intermediate reheater (RH). Figure 6.26 shows the configuration of a steam turbine with reheater.

From Fig. 6.26 we can see that the high-pressure high-temperature steam from the boiler enters the HP stage through a main valve and steam chest. We should note the existence of a certain volume of steam in the pipe and chest from the main valve to the nozzle of the HP stage. Exhaust steam from the HP stage is sent into the reheater section to raise temperature before entering the IP stage. Similarly we ought to note that there exists a certain volume of steam between the output point of the HP stage and input point of IP stage. Exhaust steam from IP stage enters the LP stage through crossover that also has a certain volume. Volume effects of the three volumes mentioned above can be described by time constant T_{CH}, T_{RH}, and T_{CO}, respectively. Usually T_{CH} is between 0.2 and 0.3 s, time constant of reheater T_{RH} is large, between 5 and 10 s, and T_{CO} is about 0.5 s.

Output mechanical torque of the steam turbine is proportional to the steam flow at the nozzle. In addition, we assume that input steam flow to the HP stage is approximately proportional to the opening position of the main steam valve μ. We denote the proportionality coefficient of mechanical power of HP, IP, and LP stages to be F_{HP}, F_{IP}, F_{LP}. Usually F_{HP}, F_{IP}, F_{LP} is 0.3, 0.3, 0.4 and their summation is one.

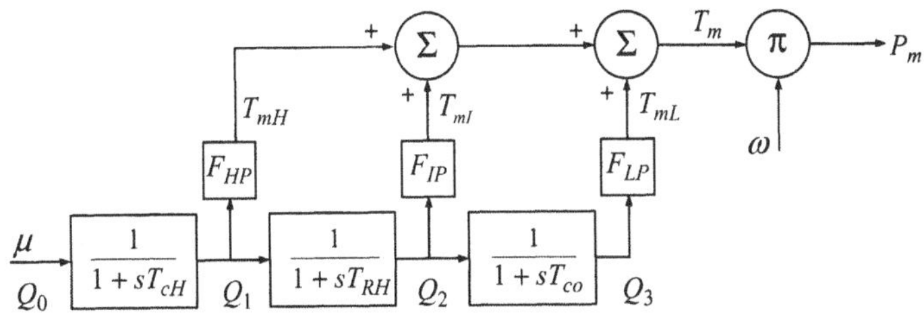

Fig. 6.27 Block diagram of transfer function of a multi-stage steam turbine

From the analysis above and taking proper base values for per unit expressions, we can obtain the mathematical model of the steam turbine in per unit to be

$$\left.\begin{aligned} Q_1 &= \frac{1}{1+T_{\mathrm{CH}}s}Q_0 \\ Q_2 &= \frac{1}{1+T_{\mathrm{RH}}s}Q_1 \\ Q_3 &= \frac{1}{1+T_{\mathrm{CO}}s}Q_2 \end{aligned}\right\}, \tag{6.146}$$

$$\left.\begin{aligned} \mu &= Q_0 \\ T_{\mathrm{mH}} &= F_{\mathrm{HP}}Q_1 \\ T_{\mathrm{mI}} &= F_{\mathrm{IP}}Q_2 \\ T_{\mathrm{mL}} &= F_{\mathrm{LP}}Q_3 \\ T_{\mathrm{m}} &= T_{\mathrm{mH}} + T_{\mathrm{mI}} + T_{\mathrm{mL}} \\ F_{\mathrm{HP}} &+ F_{\mathrm{IP}} + F_{\mathrm{LP}} = 1 \end{aligned}\right\}, \tag{6.147}$$

where T_{mH}, T_{mI}, T_{mL} is the output mechanical torque of HP, IP, and LP turbines, respectively, flows Q_0–Q_3 are shown in Fig. 6.26. Block diagram of the transfer function of the mathematical model above is shown in Fig. 6.27. Other types of mathematical models of steam turbines can be found in [164, 165].

6.4.2.2 Mathematical Model of Governing System of Steam Turbine

Basic functions of the governing system of a steam turbine include normal primary frequency control, secondary frequency control, over-speed control, over-speed generation shedding and generation starting and stopping control in normal operation, as well as auxiliary steam pressure control. Normal primary frequency control and secondary frequency control of the steam turbine is quite similar to those of a hydroturbine. Primary frequency control provides a droop around 4–5% to ensure stable load sharing among parallel generation units. Secondary frequency control is achieved through adjusting the load reference. In modern steam turbine units, usually there are more control valves in addition to the main valve shown in Fig. 6.26. For example, in a steam turbine unit equipped with a reheater there is a stop valve behind the RH stage. When over-speeding generation requires an emergency reduction of output power from the steam turbine, it would not be enough to just turn off the main valve, because the steam volume of the RH stage is very large. Under this circumstance, usually the main valve and stop valve must be turned off simultaneously. Primary frequency control and secondary frequency control function only by adjusting the main valve shown in Fig. 6.26. Usually in the study of power system stability, only the control of the main valve is considered and that of other valves is ignored. However, if emergency stop and generation shedding are used as the method for stability control, the control function of other valves needs to be taken into account. In this book, we shall only introduce the control model of the main valve. The control model of other valves can be found in [164, 165].

Governing systems of steam turbines can be classified into three types, mechanical hydraulic, electrohydraulic, and power-frequency electrohydraulic. The principle of mechanical hydraulic governing system is the same as that of a centrifugal pendulum governor introduced previously, except that the governing system of a steam turbine does not need the soft feedback unit and only uses hard feedback. The coefficient of hard feedback is 1. Hence a mechanical hydraulic governing system of a steam turbine can be shown by the transfer function block diagram of Fig. 6.28, where the simple lag with time constant T_1 represents the pilot valve in the governor. The value of T_1 usually is small and hence this unit can be ignored. In an electrohydraulic governor, the low power output unit in the mechanical hydraulic governor, i.e., the part from speed measurement to servomotor, is realized by an electronic circuit. Compared to the mechanical hydraulic governor, an electrohydraulic governor is of better applicability and flexibility, with quicker responds speed. In order to obtain better performance and linear response, the feedback channel from steam flow (or steam pressure at the first stage in the HP turbine) and valve position of the servomotor is introduced in the electrical-hydraulic governor. The transfer function block diagram is shown in Fig. 6.29.

The transfer function block diagram of the power-frequency electrohydraulic governor is shown in Fig. 6.30. By comparing frequency and power signals with the given reference, an error signal is obtained and then amplified. A PID controller conditions the amplified signal. Its output electrical signal is converted into a hydraulic signal by an electrical-hydraulic converter to drive a relay and servomotor to regulate the main valve of the steam turbine. In Fig. 6.30, k_P, k_I and k_D are the gains of proportional, integral, and differential units, respectively; T_{EL} the time constant of the electrical-hydraulic converter; and T_s is the time constant of the relay.

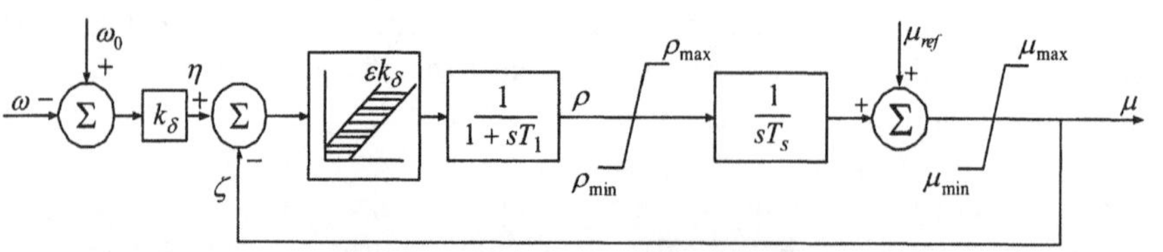

Fig. 6.28 Block diagram of mechanical hydraulic governing system of steam turbine

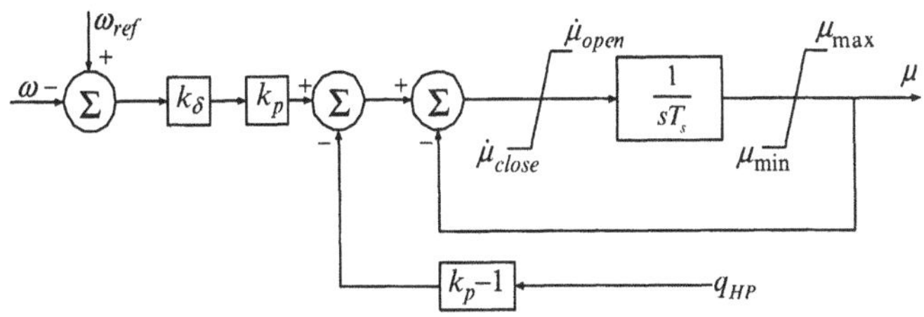

Fig. 6.29 Block diagram of transfer function of electrohydraulic governing system

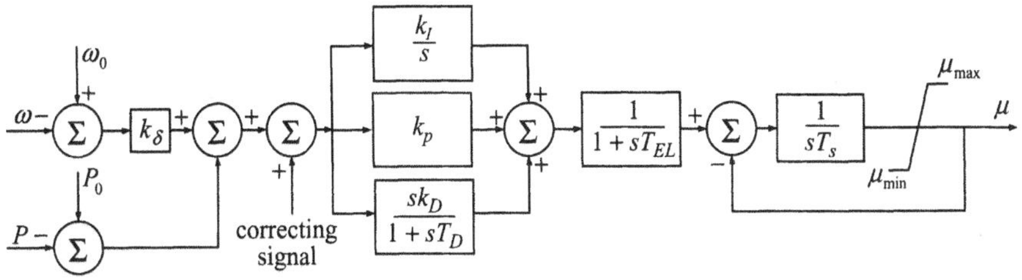

Fig. 6.30 Transfer function of power-frequency electro hydraulic governor

In the recent 20 years, digital governing systems for steam turbines have been developed, in which the operating unit of the main valve of the steam turbine is connected to a digital controller via a digital-analogue hybrid unit. The control function is realized by software. A digital governor provides more flexible and universal functions than an electrical-hydraulic governor. Response speed is enhanced greatly with the time constant being about 0.03 s. More details about digital governors can be found in [167].

We ought to point out that (6.147) is equivalent to ignoring the transients of the thermodynamic system. If thermodynamics is considered, obviously Q_0 will be determined by a mathematical model describing the thermodynamic system. As far as the time scale for the computation of power system stability (usually for 5 s following a disturbance) is concerned, the time constant of the thermodynamic system is very large. Hence the thermodynamic system can be considered as operating in steady state. However, for long-term power system stability analysis, involving system dynamics for several minutes after a disturbance, the dynamics of the thermodynamic systems, such as the boiler, will play an important role. Mathematical modeling of the thermodynamic system is still a research subject at the moment.

6.5 Mathematical Model of Load

Load is an important part in a power system. To study power system behavior in various operational states, we need to establish a mathematical model of system load. It is not difficult to establish a mathematical model of certain power-consuming equipment in the power system. However, it is neither necessary nor possible mathematically to describe each of hundreds and thousands of loads in detail. Hence in this section, power system load refers to all electrical equipment connected at a common node in the power system. It includes not only various end-users of power-consuming equipment but also under-load tap changing transformer, distribution network, various kinds of reactive power compensators, voltage regulation units, and even some small generators, etc. The relationship between active and reactive power absorbed, by all those mentioned above, at the node and the node voltage and system frequency constitutes the mathematical model of nodal load. Obviously, for different types of node, such as residential, commercial, industrial, and rural, the composition of load is quite different. Besides, for the same node, during different time periods, such as different seasons in a year, different days in a week, and different hours in a day, the composition of nodal load can vary. Due to the variety, randomness, and time variance of load, it is an extremely difficult problem to establish an accurate load model. A large number of studies have demonstrated that the conclusions from power system analysis are greatly affected by whether the mathematical model of the load has been established properly or not. From the point of view of system operation analysis and control, improper mathematical modeling of the load will result in analytical conclusions being poorly matched with practical results, either being too conservative – leading to inefficient utilization of the system, or too optimistic – causing hidden risks to system operation. An even more difficult problem is that at the moment there is no way to know if a certain load model is always conservative or optimistic under any disturbance. The importance and complexity of establishing mathematical models of the load has become a special research field, resulting in a large number of studies over many years [168–170].

There are many methods for the establishment of mathematical models of load, but these can be classified into two groups: "method of theoretical aggregation" [170] and "method of identification aggregation" [171]. In theoretical aggregation, nodal load is considered to be the combination of various individual users. Firstly those users are electrically categorized and average characteristics of each category are determined. Then a statistical percentage of each category of users is worked out and finally the total load model is aggregated. The method of identification aggregation uses collected field data. After a proper structure for the load model is chosen, the model parameters are identified by using field data. The two methods have their own merits and disadvantages. The former is simple and easy to use, but its accuracy is not satisfactory. The latter can produce more accurate mathematical models by treating and analyzing field data using modern identification theory. However, it is still difficult to obtain an accurate dynamic model of load because voltage and frequency of the real power system cannot vary over a large range. Therefore, power system load modeling remains a research topic to be pursued and no fully matured method is available.

There are quite a few methods to classify power system load models. With regard to whether the load model can describe load dynamics, a model is categorized as either static or dynamic. Obviously, a static load model is a set of algebraic equations, while a dynamic model includes differential equations. Other classifications include: linear load model or nonlinear load model and voltage-related model or frequency-related model. Conventionally, we consider load models related to both voltage and frequency to be frequency-related models. According to the way that the model is established, we have derived-model or input–output model. A derived model has clear physical meaning and can easily be understood. It is usually adopted when few types of load are considered. Nonderived models only concern the mathematical relationship between load input and output. Due to the limitation of space, in this section, we shall only introduce several commonly used types of load.

The simplest load model is to use a constant impedance to represent the load. That is, to assume that the equivalent impedance of the load does not change during system transients and its value is determined by the node voltage and power absorbed by the load at steady state before the occurrence of a disturbance. This load model is rather rough. However, due to its simplicity, it is still widely used when requirements on computational accuracy are not high.

6.5.1 Static Load Model

The static characteristic of load is the relationship between node voltage or frequency and power absorbed by the load, when voltage or frequency varies slowly. The usual forms of static load model are as follows.

1. *Static load voltage or frequency characteristic described by a polynomial.* Without considering variations of frequency, the relationship between node voltage and power absorbed by load is taken to be

$$\left.\begin{aligned} P_{\mathrm{L}} &= P_{\mathrm{L0}}\left[a_P\left(\frac{V_{\mathrm{L}}}{V_{\mathrm{L0}}}\right)^2 + b_P\left(\frac{V_{\mathrm{L}}}{V_{\mathrm{L0}}}\right) + c_P\right] = P_{\mathrm{L0}}\left(a_P V_{\mathrm{L*}}^2 + b_P V_{\mathrm{L*}} + c_P\right) \\ Q_{\mathrm{L}} &= Q_{\mathrm{L0}}\left[a_Q\left(\frac{V_{\mathrm{L}}}{V_{\mathrm{L0}}}\right)^2 + b_Q\left(\frac{V_{\mathrm{L}}}{V_{\mathrm{L0}}}\right) + c_Q\right] = Q_{\mathrm{L0}}\left(a_Q V_{\mathrm{L*}}^2 + b_Q V_{\mathrm{L*}} + c_Q\right) \end{aligned}\right\} \tag{6.148}$$

where P_{L0}, Q_{L0}, and V_{L0} are the active, reactive power absorbed by the load and node voltage before the occurrence of a disturbance. Parameters, a_P, b_P, c_P, a_Q, b_Q, and c_Q have different values for different nodes and satisfy

$$\left.\begin{aligned} a_P + b_P + c_P &= 1 \\ a_Q + b_Q + c_Q &= 1 \end{aligned}\right\}. \tag{6.149}$$

From (6.148) we can see that this model is in fact equivalent to representing the load in three parts. Coefficient a, b, and c represent the percentage of constant impedance (Z), constant current (I), and constant power (P) in the total nodal load, respectively. Hence this type of load model is also called a *ZIP* model.

Because system frequency does not vary much during transients, static frequency characteristics of load can be represented linearly. Without considering variation of node voltage, the relationship between node power and system frequency is

$$\left.\begin{aligned} P_{\mathrm{L}} &= P_{\mathrm{L0}}\left(1 + k_P \frac{f - f_0}{f_0}\right) \\ Q_{\mathrm{L}} &= Q_{\mathrm{L0}}\left(1 + k_Q \frac{f - f_0}{f_0}\right) \end{aligned}\right\}, \tag{6.150}$$

where P_{L0}, Q_{L0}, and f_0 are the active, reactive power absorbed by load and system frequency before the occurrence of a disturbance, respectively. Parameters k_P and k_Q have different values at different nodes and their physical meaning is the differential of node power to variation of system frequency at steady state, that is

$$\left.\begin{aligned} k_P &= \frac{f_0}{P_{\mathrm{L0}}} \frac{\mathrm{d}P_{\mathrm{L}}}{\mathrm{d}f}\bigg|_{f=f_0} = \frac{\mathrm{d}P_{L^*}}{\mathrm{d}f_*}\bigg|_{f=f_0} \\ k_Q &= \frac{f_0}{Q_{\mathrm{L0}}} \frac{\mathrm{d}Q_{\mathrm{L}}}{\mathrm{d}f}\bigg|_{f=f_0} = \frac{\mathrm{d}Q_{\mathrm{L}^*}}{\mathrm{d}f_*}\bigg|_{f=f_0} \end{aligned}\right\}. \tag{6.151}$$

With variation of voltage and frequency being taken into account, the mathematical model of load is the product of the two per unit model expressions above, that is

$$\left.\begin{aligned} P_{\mathrm{L}^*} &= \left(a_P V_{\mathrm{L}^*}^2 + b_P V_{\mathrm{L}^*} + c_P\right)\left(1 + k_P \Delta f_*\right) \\ Q_{\mathrm{L}^*} &= \left(a_Q V_{\mathrm{L}^*}^2 + b_Q V_{\mathrm{L}^*} + c_Q\right)\left(1 + k_Q \Delta f_*\right) \end{aligned}\right\}. \tag{6.152}$$

We would like to point out here that in statistical computation, various base values must be converted to maintain consistency with system base values.

2. *Static load voltage characteristics expressed by exponentials.* Without considering variation of frequency, static load voltage characteristics can be described by the following exponential form

$$\left.\begin{aligned} P_{\mathrm{L}} &= P_{\mathrm{L0}}\left(\frac{V_{\mathrm{L}}}{V_{\mathrm{L0}}}\right)^{\alpha} \\ Q_{\mathrm{L}} &= Q_{\mathrm{L0}}\left(\frac{V_{\mathrm{L}}}{V_{\mathrm{L0}}}\right)^{\beta} \end{aligned}\right\}. \tag{6.153}$$

For composite load, power α usually is between 0.5 and 1.8, β changes significantly between different nodes, typically between 1.5 and 6.

With the effect of frequency change being taken into account, we have

$$\left.\begin{aligned}\frac{P_{\mathrm{L}}}{P_{\mathrm{L0}}} &= \left(\frac{V_{\mathrm{L}}}{V_{\mathrm{L0}}}\right)^{\alpha}\left(1+k_P\frac{f-f_0}{f_0}\right)\\ \frac{Q_{\mathrm{L}}}{Q_{\mathrm{L0}}} &= \left(\frac{V_{\mathrm{L}}}{V_{\mathrm{L0}}}\right)^{\beta}\left(1+k_Q\frac{f-f_0}{f_0}\right)\end{aligned}\right\}. \tag{6.154}$$

Although static load models are widely used in routine computation of power system stability due to their simplicity, computational errors could be very high when the magnitude of node voltage involved in the computation varies over a wide range. For example, discharge lighting load takes over 20% of commercial load. When the per unit voltage value reaches as low as 0.7 p.u., the light goes off and the load consumes zero power. When the voltage recovers, the light goes on after a short delay. Some induction motors are equipped with low voltage protection. When the voltage decreases below a certain level, the motor will be disconnected from the network. Also, due to transformer saturation at higher voltages, reactive power absorbed is very sensitive to changes in the magnitude of nodal voltage. All these factors make static load models inapplicable when nodal voltage varies over a large range. A common method to cope with this problem is to use different model parameters in different voltage ranges or to use a simple constant impedance load when the node voltage is below 0.3–0.7 p.u. Other algebraic forms of static load model can be found in [170].

6.5.2 *Dynamic Load Model*

When node voltage changes quickly over a large range, adoption of purely static load models will bring about excessive computational errors; especially in the study of voltage stability (or load stability) where high accuracy is required in load modeling. Many studies using different types of load model have shown that at sensitive nodes, dynamic load models should be used [172–175]. In computational practice, those nodal loads are considered to consist of two parts: static and dynamic. Although there are many different types of industrial load, induction motors takes the largest share. Hence, load dynamics are mainly determined by the transient behavior of an induction motor. In the following, we shall introduce mathematical models of induction motors of two types: a model considering only mechanical transients and a more detailed model including both electromechanical transients and mechanical transients. Induction motors of large and small capacity have obviously different dynamics. For small capacity motors, only mechanical transients need to be considered [168].

6.5.2.1 Dynamic Load Model Considering Mechanical Transients of an Induction Motor

In this type of model, electromechanical transients of an induction motor are ignored, with only the mechanical transient being taken into consideration. From machine theory we know that the dynamics of an induction motor can be described by the equivalent circuit of an induction motor as in Fig. 6.31, where X_1 and X_2 are leakage reactance of armature and field windings, respectively; X_μ the mutual impedance between armature and field windings; R_2/s the equivalent rotor resistance. If system frequency and motor speed are denoted by ω and ω_m, respectively, machine slip speed $s = (\omega - \omega_m)/\omega = 1 - \omega_{m*}$ should satisfy the following equation of motion of the rotor

$$T_{JM}\frac{ds}{dt} = T_{mM} - T_{eM}, \tag{6.155}$$

where T_{JM} is the equivalent moment of inertia of the machine rotor and mechanical load and T_{mM} and T_{eM} are the mechanical torque of load and machine electrical torque, respectively. Derivation of above equation is the same as that used to derive the rotor motion equation for a synchronous generator, noting its reference positive direction of torque is just opposite to that for the synchronous generator. From the above equation we can see that when load torque is greater than electrical torque, slip speed of the induction motor increases, i.e., motor speed decreases. Ignoring electromechanical transients, electrical torque of an induction motor can be expressed to be

$$T_{eM} = \frac{2T_{eM\,max}}{\dfrac{s}{s_{cr}} + \dfrac{s_{cr}}{s}}\left(\frac{V_L}{V_{LN}}\right)^2, \tag{6.156}$$

where $T_{eM\,max}$ is the maximal electrical torque of the induction motor at rated voltage and s_{cr} is the critical slip speed for steady-state stability of the induction motor. For a certain induction motor, $T_{eM\,max}$ and s_{cr} are constant when change of frequency is not considered. V_L and V_{LN} are the terminal voltage and rated voltage of the induction motor, respectively. Mechanical torque of an induction motor is related to the characteristics of the mechanical load and often a function of motor speed. Traditionally it is given as

$$T_{mM} = k[\alpha + (1 - \alpha)(1 - s)^{p_m}], \tag{6.157}$$

where α is the portion of mechanical load torque that is independent of motor speed, p_m the exponent associated with the characteristic of the mechanical load, and k is

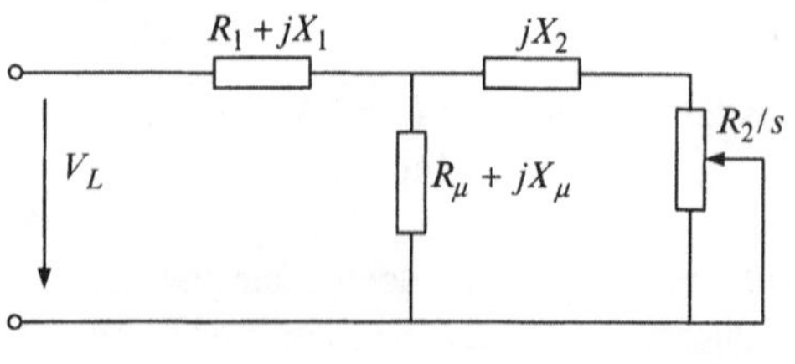

Fig. 6.31 Equivalent circuit of induction motor

the percentage of motor load. To achieve better flexibility and wider applicability of computation, currently mechanical torque is expressed as the summation of polynomial and exponential forms [168]

$$\frac{T_{\mathrm{mM}}}{T_{\mathrm{mM0}}} = a_{\mathrm{m}}\left(\frac{\omega_{\mathrm{m}}}{\omega_{\mathrm{m0}}}\right)^2 + b_{\mathrm{m}}\frac{\omega_{\mathrm{m}}}{\omega_0} + c_{\mathrm{m}} + d_{\mathrm{m}}\left(\frac{\omega_{\mathrm{m}}}{\omega_{\mathrm{m0}}}\right)^{\gamma}, \tag{6.158}$$

where T_{mM0} and ω_{m0} are mechanical torque and motor speed before occurrence of disturbance. a_{m}, b_{m}, c_{m}, d_{m}, and γ are the characteristic parameters of mechanical torque. Parameter c_{m} is calculated from the following equation

$$c_{\mathrm{m}} = 1 - (a_{\mathrm{m}} + b_{\mathrm{m}} + d_{\mathrm{m}}). \tag{6.159}$$

From Fig. 6.31, we can obtain the equivalent impedance of an induction motor to be

$$Z_{\mathrm{M}} = R_1 + \mathrm{j}X_1 + \frac{(R_\mu + \mathrm{j}X_\mu)(R_2/s + \mathrm{j}X_2)}{(R_\mu + R_2/s) + \mathrm{j}(X_\mu + X_2)}, \tag{6.160}$$

where Z_{M} is a function of motor slip speed. Rotor motion equation of an induction motor (6.155), electrical torque ignoring electromechanical transients (6.156), load mechanical torque (6.157), (6.158), or (6.159), and equivalent impedance (6.160) form the mathematical model of an induction motor load with electromechanical transients neglected. Input variables to the model are node voltage and system frequency. Output variable is the equivalent impedance. Hence when V_{L} and ω, as functions of time, are known, s can be found by solving the above equations to obtain the equivalent impedance Z_{M} at any time.

As pointed out previously, nodal load includes all electrical equipment connected at the node. Because so many types of electrical equipment may be connected, the dynamics of nodal load are very complicated. In the following, we shall introduce a method of simplifying nodal load by use of the classical model of an induction motor. The key issue in the simplification is to obtain the equivalent impedance of nodal load at any time.

Step 1. We separate the total power $P_{\mathrm{L(0)}}$ and $Q_{\mathrm{L(0)}}$ absorbed by the nodal load, in steady-state operation, into two parts. One part is expressed by a static load model with power $P_{\mathrm{LS(0)}}$ and $Q_{\mathrm{LS(0)}}$. The corresponding equivalent impedance is denoted as $Z_{\mathrm{LS(0)}} = V_{\mathrm{L(0)}}^2/[P_{\mathrm{LS(0)}} - jQ_{\mathrm{LS(0)}}]$. Another part is modeled by an induction motor with only mechanical transients considered. The power of this part is denoted as $P_{\mathrm{LM(0)}}$ and $Q_{\mathrm{LM(0)}}$ with corresponding equivalent impedance $Z_{\mathrm{LM(0)}} = V_{\mathrm{L(0)}}^2/[P_{\mathrm{LM(0)}} - \mathrm{j}Q_{\mathrm{LM(0)}}]$. Equivalent impedance of nodal load is $Z_{\mathrm{L(0)}} = Z_{\mathrm{LS(0)}} + Z_{\mathrm{LM(0)}}$.

Step 2. It is approximated that all equipment connected at the node, with their dynamics being considered, is a certain typical induction motor. Model parameters of the typical motor are $s_{(0)}$, T_{JM}, $T_{\mathrm{eM\,max}}$, s_{cr}, R_1, X_1, R_2, X_2, R_μ, X_μ and k, α, p_{m} or a_{m}, b_{m}, d_{m}, γ. From (6.160) we can find the steady-state equivalent impedance of the typical motor $Z_{\mathrm{M(0)}}$. Obviously, steady-state equivalent impedance of the typical motor does not have to be equal to the steady-state equivalent impedance of an equivalent motor.

Step 3. In a system transient, node voltage and system frequency vary with time. By using some numerical methods to solve system equations and rotor motion equation of the typical motor (details about the numerical method are introduced in Chaps. 7 and 8), we can obtain the slip speed $s_{(t)}$ of the typical motor, nodal voltage magnitude $V_{L(t)}$ and system frequency $\omega_{(t)}$ at time t. From (6.160) we then can calculate the equivalent impedance of the typical motor $Z_{M(t)}$ at time t. From the static load model we can find the equivalent impedance of static load $Z_{LS(t)}$.

Step 4. We suppose that at any time, the ratio between the equivalent impedance of equivalent motor and equivalent impedance of typical motor is a constant. Hence at any time t, the equivalent impedance of the equivalent motor is

$$Z_{LM(t)} = (c_r + jc_i)Z_{M(t)}, \tag{6.161}$$

where the proportionality constant can be found from steady-state conditions

$$c_r + jc_i = Z_{LM(0)}/Z_{M(0)}. \tag{6.162}$$

Finally we obtain the equivalent impedance of nodal load at time t to be

$$Z_{L(t)} \doteq Z_{LS(t)} + Z_{LM(t)}. \tag{6.163}$$

6.5.2.2 Load Dynamic Model with Electromechanical Transients of Induction Motors Considered

Compared to the model introduced above, this model considers electromechanical transients of the field winding of induction motors. Similar to the case of a synchronous generator, because the transient of the armature winding is very fast, we do not consider the electromechanical transient of armature windings for an induction motor either. Details about deriving the mathematical model of an induction motor with electromechanical transients of the field winding being taken into account can be found in [153]. In the following, we shall give a simple derivation method by use of the mathematical model of a synchronous generator established in Sect. 6.2.

In fact, as far as the transient equation of the machine is concerned, an induction motor can be considered to be a synchronous generator being completely symmetrical in the two directions of d- and q-axes. Hence in some algorithms of power system transient stability analysis, modeling of induction motors and synchronous generators is treated in the same way. When an induction motor is considered individually; for simplicity, the subtransient process of a synchronous generator is ignored. In the mathematical model, the f winding has the same structure as that of the g winding but is short-circuited. Under these conditions, in equations of the synchronous generator ((6.43)–(6.46)), letting $X_d = X_q = X$, $X'_d = X'_q = X$, $e_{q2} = e_{d2} = e''_q = e''_d = 0$, $p\varphi_d = p\varphi_q = 0$, $T'_{d0} = T'_{q0}$, $\omega = 1 - s$, $R_a = R_1$, we have per unit equations of an induction motor to be

$$\left.\begin{aligned} v_q &= (1-s)(e'_q - X'i_d) - R_1 i_q \\ v_d &= (1-s)(e'_d + X'i_q) - R_1 i_d \\ T'_{d0}pe'_q &= -e'_q - (X - X')i_d \\ T'_{d0}pe'_d &= -e'_d + (X - X')i_q \end{aligned}\right\}, \tag{6.164}$$

where machine parameters, X, X', and T'_{d0} can be derived from Fig. 6.31. Because d- and q-axis are completely symmetrical and the structure of f and g windings is the same, in (6.32) and (6.33) we have

$$X_{af} = X_{ag} = X_\mu. \tag{6.165}$$

Hence according to the definition of synchronous reactance, we have the following equation for the stator side,

$$X = X_d = X_q = X_1 + X_\mu. \tag{6.166}$$

Similarly on the rotor side, we have

$$X_f = X_g = X_2 + X_\mu. \tag{6.167}$$

Substituting (6.166) and (6.167) into (6.40b), we can obtain

$$X' = X'_d = X'_q = X_1 + \frac{X_2 X_\mu}{X_2 + X_\mu}. \tag{6.168}$$

We denote the resistance in (6.30) and (6.31) $R_f = R_g$ as R_2. Substituting (6.167) into (6.40b) we have

$$T'_{d0} = T'_{q0} = (X_2 + X_\mu)/R_2. \tag{6.169}$$

Equation (6.164) can be simplified by converting it in d–q coordinates from (6.62) to system unified x–y coordinates. Differentiation of (6.62) to per unit time can result in

$$p\begin{bmatrix} A_d \\ A_q \end{bmatrix} = \begin{bmatrix} \sin\delta & -\cos\delta \\ \cos\delta & \sin\delta \end{bmatrix} p\begin{bmatrix} A_x \\ A_y \end{bmatrix} + \begin{bmatrix} \cos\delta & \sin\delta \\ -\sin\delta & \cos\delta \end{bmatrix}\begin{bmatrix} A_x \\ A_y \end{bmatrix} p\delta. \tag{6.170}$$

From the geometrical meaning of α and (6.78) we know $p\delta = -s$. Hence in x–y coordinates (6.164) becomes

$$\left.\begin{aligned} v_x &= (1-s)e'_x + (1-s)X'i_y - R_1 i_x \\ v_y &= (1-s)e'_y - (1-s)X'i_x - R_1 i_y \end{aligned}\right\}, \tag{6.171}$$

$$\left.\begin{aligned} T'_{d0}pe'_x &= T'_{d0}se'_y - e'_x + (X-X')i_y \\ T'_{d0}pe'_y &= -T'_{d0}se'_x - e'_y - (X-X')i_x \end{aligned}\right\}. \tag{6.172}$$

At quasisteady state, multiplying the second equation in (6.171) and (6.172) by j and adding to the first equation, we have

$$\dot{V}_{\mathrm{L}} = (1-s)\dot{E}'_{\mathrm{M}} - [R_1 + \mathrm{j}(1-s)X']\dot{I}_{\mathrm{M}}, \tag{6.173}$$

$$T'_{d0}p\dot{E}'_{\mathrm{M}} = -(1 + \mathrm{j}\,sT'_{d0})\dot{E}'_{\mathrm{M}} - \mathrm{j}(X-X')\dot{I}_{\mathrm{M}}, \tag{6.174}$$

where $\dot{V}_{\mathrm{L}} = V_x + \mathrm{j}V_y$, $\dot{I}_{\mathrm{M}} = I_x + \mathrm{j}I_y$, $\dot{E}'_{\mathrm{M}} = E'_x + \mathrm{j}E'_y$.

However, with subtransient process ignored, the mathematical model of a synchronous generator cannot be converted into the form of (6.173) and (6.174) if d- and q-axis are not symmetrical.

Treating an induction motor as a synchronous generator and from (6.81), (6.43), and (6.44), we can obtain the electrical torque of an induction motor to be

$$T_{\mathrm{eM}} = -(e'_q i_q + e'_d i_d) = -(e'_x i_x + e'_y i_y), \tag{6.175}$$

where the negative sign is because the positive reference direction of electrical torque of an induction motor is just opposite to that of a synchronous generator. Because a generator model is used, the reference direction of current is going out of, instead of into the induction motor. Therefore, the mathematical model of an induction motor considering electromagnetic transients consists of (6.155), (6.173)–(6.175) and the load mechanical torque of (6.157) or (6.158).

For nodal composite load, we can use the same method adopted previously with mechanical transients being considered. For the typical motor, $p\dot{E}'_{\mathrm{M}} = 0$ in steady-state operation, from (6.173) and (6.174) we can find $\dot{I}_{\mathrm{M(0)}}, \dot{E}'_{\mathrm{M(0)}}$. Hence the equivalent impedance of the typical motor at steady state is $Z_{\mathrm{M(0)}} = -\dot{V}_{\mathrm{L(0)}}/\dot{I}_{\mathrm{M(0)}}$. Equivalent impedance of the equivalent motor can be calculated from nodal voltage and load power at steady state. Thus the ratio between equivalent impedance of typical and equivalent motors can be computed from (6.162). During transients, solving the combined equation describing the typical motor and system we can obtain $\dot{I}_{\mathrm{M}(t)}$, $\dot{V}_{\mathrm{L}(t)}$, and $Z_{\mathrm{M}(t)}$. Hence the equivalent impedance of equivalent motor and composite load can be calculated from (6.161) and (6.163). During transients, variation of slip speed has little effect on armature voltage, numerically. It can be ignored in a simple computation and hence in the armature voltage equation of the motor, (6.173), s is taken to be a constant 0. Typical parameters of induction motors can be found in [168, 176].

There are other forms of load dynamic model. For some special loads with large capacity, such as large rolling machines, electric-arc furnaces, electric trains, large units of temperature control, and synchronous motors in pumping or energy storage power plants, etc., the model needs to be established individually. For long-term stability analysis, transformer saturation, adjustment of under-load tap changing transformers, voltage regulation arising from reactive compensators, and the action of low-frequency low-voltage load-shedding equipment, etc., ought to be represented within the load models. Overall, load modeling is still a developing subject.

Thinking and Problem Solving

1. How is the relationship between the electrical quantities in stator and rotor of a synchronous generator set up?
2. Does the mutual inductance between stator winding and rotor winding vary with time, according to whether the generator is round-rotor or salient-pole?
3. What is the function of *Park* conversion?
4. In the state equation of (6.1), each winding flux linkage is a state variable. Considering the motion equation of the rotor, the electrical rotational speed ω of generator is also a state variable. Discuss the nonlinearity of the generator model according to this formula.
5. Discuss the physical significance of the right-hand three items in (6.14), and thereby explain the electrical mechanism of power output of a generator.
6. What are the usual applications of round-rotor generators and salient-pole generators in electrical power systems, and why?
7. The form of synchronous generator model will be influenced by such factors as the choosing of positive direction of magnetic axis, the suppositions taken during converting original parameters into rotor parameters, selection method of base values, and so on. By consulting other books, compare the common and different points of synchronous machine models with the forms that are introduced in this book.
8. By consulting books on synchronous generator experiments, find out about and describe the methods that can be used to empirically determine the parameters of a synchronous generator.
9. During a transient in an electrical power system, the electromagnetic transient process in the electrical network is much faster than the rotor flux dynamics of the generator, so in the synchronous generator model that is used to analyze the electromechanical transient process, the time derivative of stator winding flux linkage is taken to be zero. Analyze the effect of this approximation on calculation quantity.

10. There are three kinds of coordinates used to describe the electrical quantities of a generator. These are the electrical quantities in three-phase static coordinates a, b, c; three-axis rotating coordinates $d, q, 0$; and complex plane $x - y$. Discuss the relationship among these three kinds of description.
11. Given one salient-pole synchronous generator, its terminal voltage $\dot{U}_t = 1.0$, and unit power output $P + jQ = 1.0 + j0.1$. The parameters of the generator unit are $X_d = 1.0$, $X_q = 0.6$, $X'_d = 0.3, X'_q = 0.2, X''_d = 0.15, X''_q = 0.1$. If the stator resistance is neglected, calculate the emfs E'_q, E'_d, E''_q, and E''_q of this generator.
12. During the formulation of a synchronous generator model, in which formulae can the electrical rotational speed ω be considered approximately as invariable, and in which formulae can the electrical rotational speed ω not be considered approximately invariable? Why?
13. Discuss the effect of excitation current on the operating state of a synchronous generator, according to (6.50) and (6.51).
14. Discuss the working mechanism of the Washout link in PSS (in Fig. 6.14).
15. Discuss the necessity and difficulty of building steady and dynamical synthetic load models.

Chapter 7
Power System Transient Stability Analysis

7.1 Introduction

The mechanical–electrical transient of a power system that has experienced a large disturbance can evolve into two different situations. In the first situation, the relative rotor angles among generators exhibit swing (or oscillatory) behavior, but the magnitude of oscillation decays asymptotically; the relative motions among generators eventually disappear, thus the system migrates into a new stable state, and generators remain in synchronous operation. The power system is said to be transiently stable. In another situation, the relative motions of some generator rotors continue to grow during the mechanical–electrical transient, and the relative rotor angles increase, resulting in the loss of synchronism of these generators. The system is said to be transiently unstable. When a generator loses synchronization with the remaining generators in the system, its rotor speed will be above or below what is required to produce a voltage at system frequency, and the slip motion between the rotating stator magnetic field (relative to system frequency) and rotor magnetic field causes generator power output, current and voltage to oscillate with very high magnitudes, making some generators and loads trip and, in the worst case, causing the system to split or collapse.

A necessary condition that a power system maintains normal operation is the synchronous operation of all generators. Therefore, analyzing the stability of a power system after a large disturbance is equivalent to analyzing the ability of generators to maintain synchronous operation after the system experiences a large disturbance, this is called power system transient stability analysis.

The aforementioned power system transient stability analysis typically involves the short-term (within some 10 s) dynamic behavior of a system, nevertheless, sometimes we have to study system midterm (10 s to several minutes) and long-term (several minutes to tens of minutes) dynamic behavior, this would be termed power system midterm and long-term stability analysis.

Midterm and long-term stability mainly concerns the dynamic response of a power system that experiences a severe disturbance. A severe disturbance can cause system voltages, frequency, and power flows to undergo drastic changes; therefore, it is meaningful to look into certain slow dynamics, control, and protection

X.-F. Wang et al., *Modern Power Systems Analysis.*
doi: 10.1007/978-0-387-72853-7,

performance that are not addressed in a short-term transient stability analysis. The response time of devices that affect voltage and frequency can be from a few seconds (such as the response time of generator control and protection devices) to several minutes (such as the response time of a prime mover system and on-load tap changing regulators, etc.)

A long-term stability analysis focuses on the slow phenomenon with long duration that occur after a large disturbance has happened, and the significant mismatch between active/reactive power generation and consumption. The phenomena of concern include: boiler dynamics, water gate and water-pipe dynamics of hydraulic turbines, automatic generation control (AGC), control and protection of power plant and transmission system, transformer saturation, abnormal frequency effects of load and network, and so forth. When performing long-term stability analysis, one is often concerned about the responses of a system under extremely severe disturbances that are not taken into consideration in system design. After the occurrence of an extremely severe disturbance, a power system can undergo cascading faults and can be split into several isolated parts. The question a stability study has to answer is whether or not each isolated part can reach acceptable stable operation after any load-shedding occurs.

Midterm response refers to response whose timeframe is between that of short-term response and long-term response. Midterm stability study investigates the synchronous power oscillations among generators, including some slow phenomena and possibly large voltage and frequency deviations [177].

Large disturbances are severe threatens to power system operation, but in reality they cannot be avoided. The consequence of losing stability after a power system experiences a disturbance is in general very serious, it can even be a disaster. In fact the various large disturbances, such as short circuit, tripping or committing of large capacity generator, load, or important transmission facility, appear as probabilistic events, therefore when designing and scheduling a power system, one always ensures that the system can maintain stable operation under a set of reasonably specified credible contingencies, rather than requiring that the system can sustain the impact of any disturbance. Because every country has their own stability requirements, the selection of credible contingencies can be based on different standards.

To check if a power system can maintain stable operation under credible contingencies, one needs to perform transient stability analysis. When the system under study is not stable, efficient measures that can improve system stability need to be sought. When a system experiences extremely severe stability problems, fault analysis is required to find the weak points in the system, and develop corresponding strategies.

In power system stability analysis, the mathematical models of system components not only directly relate to the analysis results, but also have a significant effect on the complexity of the analysis. Therefore, if appropriate mathematical models for each system component are developed, stability analysis can be made simple and accurate. This is a crucial step in stability analysis.

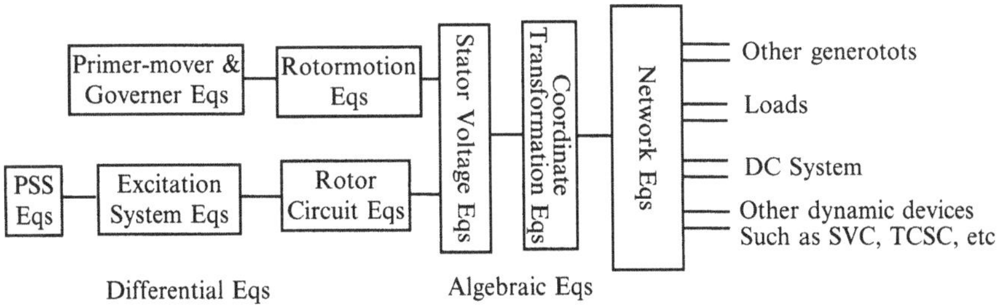

Fig. 7.1 Conceptual framework of mathematical models for stability studies

Figure 7.1 conceptualizes the mathematical model of all system components for power system stability studies. From the figure one can see that the mathematical model consists of the models of synchronous machines and the associated excitation systems, prime mover and speed-governing system, electrical load, and other dynamic devices and electrical network. Apparently, all the dynamic components of the system are independent; it is the electrical network that connects them with each other.

Mathematically, the complete system model can be described as a set of differential-algebraic equations as follows:

$$\frac{\mathrm{d}\mathbf{x}}{\mathrm{d}t} = \mathbf{f}(\mathbf{x}, \mathbf{y}), \tag{7.1}$$

$$\mathbf{0} = \mathbf{g}(\mathbf{x}, \mathbf{y}). \tag{7.2}$$

This chapter first introduces the composition of the component models for transient stability analysis and the numerical solution algorithms for differential-algebraic equations, then describes the mathematical relationship between dynamic components and electrical network, followed by an exposition of how to model network switches and faults. The chapter also presents in detail the solution algorithms for simple transient stability analysis and for analysis of systems with FACTS devices represented by full mathematical models.

7.2 Numerical Methods for Transient Stability Analysis

Power system transient stability analysis can be viewed as an initial value problem of differential-algebraic equations. In this section we first introduce the numerical methods for ordinary differential equations, and then discuss the numerical methods for differential-algebraic equations. We provide a general procedure for transient stability analysis at the end of the section.

7.2.1 *Numerical Methods for Ordinary Differential Equations*

7.2.1.1 Fundamental Concept

Consider the following first-order differential equation:

$$\frac{dx}{dt} = f(t, x), \quad x(t_0) = x_0. \tag{7.3}$$

In general, the function f in the above equation is a nonlinear function of x and t. In many real world situations, f is not an explicit function of time t, therefore the above equation reduces to

$$\frac{dx}{dt} = f(x), \quad x(t_0) = x_0. \tag{7.4}$$

In power system stability analysis, the right-hand side of all the differential equations does not contain explicitly the time variable t.

When f in (7.4) is a linear function of x, one can easily obtain the closed-form solution of the differential equation. For example, consider the following differential equation:

$$\frac{dx}{dt} = x. \tag{7.5}$$

The closed-form solution is given as

$$x = A\,e^t, \tag{7.6}$$

where A is a constant. Equation (7.6) represents a family of integral curves.

Given an initial condition in the form of $x(t_0) = x_0$, one can determine a solution curve. For instance, if $x(0) = 1$, then from (7.6) the integral constant can be found as $A = 1$, thus the solution curve is as follows:

$$x = e^t. \tag{7.7}$$

The differential equations of real world engineering problems appear to be more complex, the right-hand sides of the equations are typically not integrable, therefore closed-form solutions, like (7.6), of such differential equations cannot be obtained. To solve these differential equations, one must rely on numerical methods. The idea of numerical methods is to employ a certain integral formula to solve for the approximate value of x_n at each instant in the time series $t_n = t_0 + nh$, $n = 1,2, \ldots$ (here h is the step size) in a step-by-step fashion, starting from the initial state ($t = t_0$, $x = x_0$). This method of solving differential equation is called step-by-step integration.

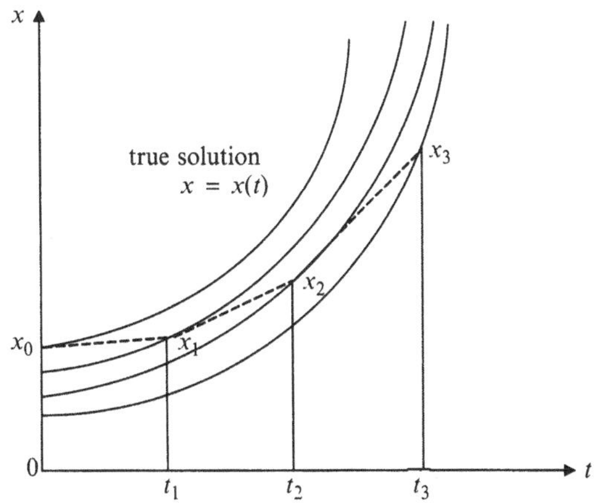

Fig. 7.2 Illustration of Euler's method for solving differential equations

In the following, we illustrate the basic idea of step-by-step integration using Euler's method as an example.

Suppose that the exact solution of the first-order differential (7.3) at $t_0 = 0$, $x(t_0) = x_0$ is as follows:

$$x = x(t). \tag{7.8}$$

The graph of the function, that is, the integral curve of the differential (7.3) passing through the point $(0, x_0)$ is depicted in Fig. 7.2 .

Euler's method is also called the Euler's tangent method or Euler's polygon method. The idea of the method is to approximate the integral curve by an Euler's polygon, the slope of each Euler's polygon is obtained by solving for (7.3) with the initial value of the Euler's polygon as input. Specifically, the computational steps are as follows:

For the first segment, the slope of the integral curve at point $(0, x_0)$ is

$$\left.\frac{\mathrm{d}x}{\mathrm{d}t}\right|_0 = f(x_0, 0).$$

Replacing the first segment with a straight line which has a slope of $\left.\frac{\mathrm{d}x}{\mathrm{d}t}\right|_0$, one can find the incremental of x at $t_1 = h$ (h is the step size) as follows:

$$\Delta x_1 = \left.\frac{\mathrm{d}x}{\mathrm{d}t}\right|_0 h.$$

Therefore the approximation of x at $t_1 = h$ should be

$$x_1 = x_0 + \Delta x_1 = x_0 + \left.\frac{\mathrm{d}x}{\mathrm{d}t}\right|_0 h.$$

For the second segment, the integral curve will be approximated by another straight line segment, the slope of which can be obtained by substituting the initial value of the segment (that is, the starting point of the segment (t_1, x_1)) into (7.3):

$$\left.\frac{\mathrm{d}x}{\mathrm{d}t}\right|_1 = f(x_1, t_1).$$

An approximate value of x at $t_2 = 2h$ can be found based on

$$x_2 = x_1 + \left.\frac{\mathrm{d}x}{\mathrm{d}t}\right|_1 h$$

as illustrated in Fig. 7.2. The above procedure can be repeated to find an approximate value of x_3 at t_3 and so forth. In general, the recursive formula for computing an approximate value of the $n + 1$ point is as follows:

$$x_{n+1} = x_n + \left.\frac{\mathrm{d}x}{\mathrm{d}t}\right|_n h, \quad n = 0, 1, 2, \ldots. \tag{7.9}$$

Now we turn to analyzing the error introduced by this recursive formula which is used to compute (t_{n+1}, x_{n+1}) from (t_n, x_n). To do so, expand the integral function (7.8) at (t_n, x_n) using Taylor's formula as follows:

$$x_{n+1} = x_n + x'_n h + x''_n \frac{h^2}{2!} + \cdots + x^{(r)}_{\xi_n} \frac{h^r}{r!}, \tag{7.10}$$

where $x'_n, x''_n, \ldots$ are the first-order, second-order,... derivatives of the integral function with regard to variable t. The symbol ξ_n represents a number in the interval [t_n, t_{n+1}], and $x^{(r)}_{\xi_n}$ is the residual of the Taylor's series. When $r = 2$, (7.10) becomes

$$x_{n+1} = x_n + x'_n h + x''_{\xi'_n} \frac{h^2}{2!} \tag{7.11}$$

or in an alternative form

$$x_{n+1} = x_n + \left.\frac{\mathrm{d}x}{\mathrm{d}t}\right|_n h + \left.\frac{\mathrm{d}^2x}{\mathrm{d}t^2}\right|_{\xi'_n} \frac{h^2}{2!}. \tag{7.12}$$

Here the symbol ξ'_n still represents a number in interval [t_n, t_{n+1}] and in general $\xi'_n \neq \xi_n$.

Obviously, Euler's recursive (7.9) can be obtained after neglecting the residual term $\left.\frac{\mathrm{d}^2x}{\mathrm{d}t^2}\right|_{\xi'_n} \frac{h^2}{2!}$ in (7.12).

Therefore when computing the function value at point $n+1$ from that at n, the error introduced by the approximation is

$$E_{n+1} = \left.\frac{\mathrm{d}^2x}{\mathrm{d}t^2}\right|_{\xi'_n} \frac{h^2}{2!}. \tag{7.13}$$

Suppose that within the computing interval $[0,t_m]$, the maximum value of $\frac{\mathrm{d}^2x}{\mathrm{d}t^2} = f'(x,t)$ is M, then the error E_{n+1} should satisfy

$$E_{n+1} \leq \frac{M}{2}h^2, \tag{7.14}$$

where M is independent of the choice of step size h. The errors in (7.13) and (7.14) are due to the approximation made when computing the function value at point $n+1$ from that at n, it is called local truncation error. The truncation error of Euler's formula is in proportion to h^2, and often expressed as of order h^2 or $0(h^2)$.

It should be noted that before obtaining x_{n+1}, x_n is solved using the same recursive formula, therefore x_n itself also contains error. As a matter of fact, when computing x_{n+1} based on (7.9), one should take into account the impact of the error of x_n, in addition to the impact of the local truncation error associated with neglecting residual term. This error is called global truncation error or simply put truncation error. Consequently the error introduced by the inaccuracy of Euler's formula is larger than the local truncation error expressed in (7.13) and (7.14). It can be proved that the global truncation error of Euler's formula is in proportion to h, in other words, it is $0(h)$.

Based on the above discussion, a smaller step size h should be selected to reduce the computational error of the Euler's formula. But it is false to assert that the smaller the step size h is, the smaller the error would be.

In the aforementioned discussion, we did not take into consideration the round-off error of the computer. When a small step size h is used, the computational effort adversely increases; thus, the impact of rounding errors increases, as illustrated in Fig. 7.3. In the figure, $h_{\min}$ is the step size associated with the minimum error, therefore we cannot merely rely on reducing the step size to reduce computational error. If higher computational precision is desired, a better computational algorithm has to be used.

In the above calculations, when computing the function value at t_{n+1}, only the function value x_n at the previous point t_n is required, this algorithm is called a single-step algorithm. The algorithms to be presented in this section belong to this category. There are multistep or multivalue algorithms which are more accurate. These algorithms require the information of previous steps (t_n, x_n), (t_{n+1}, x_{n+1}), $\ldots$, (t_{n-k+1}, x_{n-k+1}) when solving for the value x_{n+1} corresponding to time t_{n+1}.

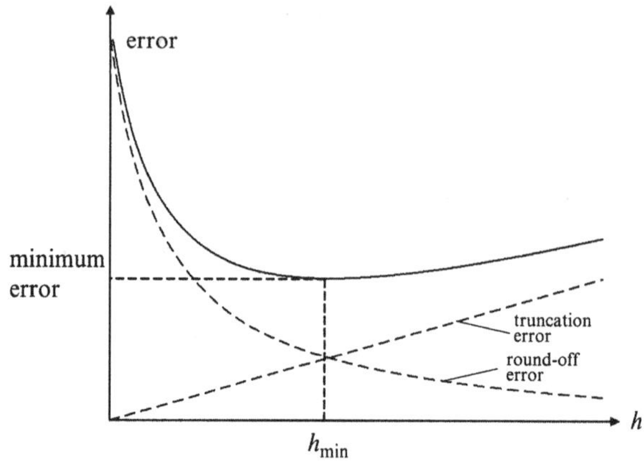

Fig. 7.3 Relationship between error and step size

7.2.1.2 Modified Euler's Method

The large error of Euler's method comes from the fact that the derivative $\frac{dx}{dt}\big|_n = f(x_n, t_n)$ of the starting point of an Euler's polygon is used for the entire segment $[t_n, t_{n+1}]$. In other words, the slope of each Euler's polygon is entirely determined by the starting point of the polygon. If the slope of an Euler's segment is replaced with the average of slopes of starting point and end point, we should expect improved solution precision. This is the basic idea of the modified Euler's method.

For first-order differential equation (7.3), let the initial value is given as $t_0 = 0$, $x(t_0) = x_0$, in what follows we introduce the computational steps of the modified Euler's method.

To find out the function value x_1 at $t_1 = h$, first compute an approximate value of x_1 using Euler's method:

$$x_1^{(0)} = x_0 + \frac{dx}{dt}\bigg|_0 h, \tag{7.15}$$

where $\frac{dx}{dt}\big|_0 = f(x_0, t_0)$.

When $x_1^{(0)}$ is obtained based on (7.15), substitute $t_1, x_1^{(0)}$ into (7.3) to solve for the derivative at the end point of the segment:

$$\frac{dx}{dt}\bigg|_1^{(0)} = f(x_1^{(0)}, t_1).$$

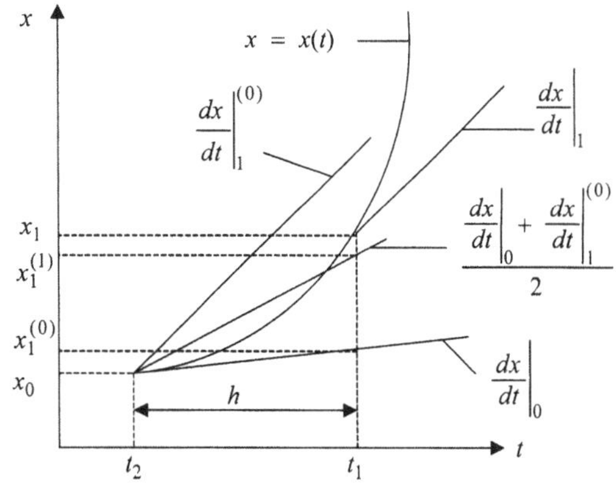

Fig. 7.4 Geometrical explanation of the modified Euler's method

Now the average of $\frac{dx}{dt}\Big|_0$ and $\frac{dx}{dt}\Big|_1^{(0)}$ can be used to calculate an improved solution of x_1 as follows:

$$x_1^{(1)} = x_0 + \frac{\left.\frac{dx}{dt}\right|_0 + \left.\frac{dx}{dt}\right|_1^{(0)}}{2} h. \tag{7.16}$$

The solution $x_1^{(1)}$ computed this way better approximates the true solution x_1 than does $x_1^{(0)}$ which is computed using a standard Euler's method. Figure 7.4 provides a geometrical explanation.

To compute (t_{n+1}, x_{n+1}) from (t_n, x_n), the following general formula can be used

$$\left.\begin{aligned} \left.\frac{dx}{dt}\right|_n &= f(x_n, t_n) \\ x_{n+1}^{(0)} &= x_n + \left.\frac{dx}{dt}\right|_n h \\ \left.\frac{dx}{dt}\right|_{n+1}^{(0)} &= f(x_{n+1}^{(0)}, t_{n+1}) \\ x_{n+1} = x_{n+1}^{(1)} &= x_n + \frac{\left.\frac{dx}{dt}\right|_n + \left.\frac{dx}{dt}\right|_{n+1}^{(0)}}{2} h \end{aligned}\right\}. \tag{7.17}$$

Eliminate x_n in (7.17), the fourth formula of (7.17) can be modified as

$$x_{n+1} = x_{n+1}^{(0)} + \left.\frac{dx}{dt}\right|_{n+1}^{a} h, \tag{7.18}$$

where

$$\left.\frac{dx}{dt}\right|_{n+1}^{a} = \frac{1}{2}\left(\left.\frac{dx}{dt}\right|_{n+1}^{(0)} - \left.\frac{dx}{dt}\right|_{n}\right).$$

As such, the general formula of the modified Euler's method can be summarized as follows:

$$\left.\begin{aligned} \left.\frac{dx}{dt}\right|_{n} &= f(x_n, t_n) \\ x_{n+1}^{(0)} &= x_n + \left.\frac{dx}{dt}\right|_{n} h \\ \left.\frac{dx}{dt}\right|_{n+1}^{a} &= \frac{1}{2}\left(f(x_{n+1}^{(0)}, t_{n+1}) - \left.\frac{dx}{dt}\right|_{n}\right) \\ x_{n+1} = x_{n+1}^{(1)} &= x_{n+1}^{(0)} + \left.\frac{dx}{dt}\right|_{n+1}^{a} h \end{aligned}\right\}. \tag{7.19}$$

When solving for x_{n+1} based on (7.19), because it takes the same form as that of $x_{n+1}^{(0)}$, the computer code can be simplified. In addition, x_n need not be saved after $x_{n+1}^{(0)}$ is obtained, thus computer memory can be saved.

In what follows we discuss the local truncation error of modified Euler's method. To do so, recall the Taylor's expansion formula of equation (7.10):

$$x_{n+1} = x_n + x_n' h + x_n'' \frac{h^2}{2!} + x_{\xi_n''}''' \frac{h^3}{3!}, \tag{7.20}$$

where $x_{\xi_n''}''' \frac{h^3}{3!}$ is the residual term of the Taylor's expansion.

The fourth equation of the modified Euler's method (7.17) can be re-cast as

$$x_{n+1}^{(1)} = x_n + x_n' \frac{h}{2} + \frac{h}{2} f(x_{n+1}^{(0)}, t_{n+1}).$$

Substituting the first formula in (7.17) into the above equation, one obtains

$$x_{n+1}^{(1)} = x_n + x_n' \frac{h}{2} + \frac{h}{2} f(x_n + x_n' h, t_n + h). \tag{7.21}$$

Expand the third term in the right-hand side of the above equation using Taylor's formula,

$$\frac{h}{2} f(x_n + x_n' h, t_n + h) = \frac{h}{2} f(x_n, t_n) + \frac{h^2}{2} \left.\frac{\partial f}{\partial x}\right|_{n} x_n' + \frac{h^2}{2} \left.\frac{\partial f}{\partial t}\right|_{n} + 0(h^3).$$

Since

$$x''_n = \left.\frac{\partial f}{\partial x}\right|_n x'_n + \left.\frac{\partial f}{\partial t}\right|_n,$$

therefore

$$\frac{h}{2}f(x_n + x'_n h, t_n + h) = \frac{h}{2}x'_n + \frac{h^2}{2}x''_n + 0(h^3),$$

substituting the above formula into (7.21), it follows

$$x_{n+1}^{(1)} = x_n + x'_n h + x''_n \frac{h^2}{2} + 0(h^3), \tag{7.22}$$

subtracting the above formula from (7.20), we have

$$E_{n+1} = x_{n+1} - x_{n+1}^{(1)} = x'''_{\xi''_n}\frac{h^3}{3!} - 0(h^3).$$

The above equation shows that the local truncation error of the modified Euler's method is $0(h^3)$. By the same token, it can be proved that the global truncation error of the modified Euler's method is $0(h^2)$.

[Example 7.1] Solve the following differential equation by the modified Euler's method

$$\frac{dx}{dt} = x - \frac{2t}{x},$$

where the initial values are $t_0 = 0$ and $x_0 = 1$.

[Solution] Taking 0.2 as step length, the computational results are summarized in the following table:

n	t_n	x_n	$\left.\frac{dx}{dt}\right\|_n$	$x_{n+1}^{(0)}$	t_{n+1}	$\left.\frac{dx}{dt}\right\|_{n+1}^{0}$	$\frac{\left.\frac{dx}{dt}\right\|_n + \left.\frac{dx}{dt}\right\|_{n+1}^{(0)}}{2}$	x_n
0	0	1	1	1.2	0.2	0.8667	0.9333	1.18667
1	0.2	1.18667	0.84959	1.35658	0.4	0.7669	0.8083	1.34832
2	0.4	1.34832	0.75499	1.49932	0.6	0.6990	0.7270	1.49372
3	0.6	1.49837	0.69036	1.63179	0.8	0.6513	0.6708	1.62788
4	0.8	2.62790	0.64500	1.75690	1.0	0.6185	0.6318	1.75430
5	1.0	1.75430						

The true solution of this differential equation is

$$x = \sqrt{2t + 1}.$$

When $t = 1, x = 1.73205$, therefore the error is equal to

$$|1.73205 - 1.7543| = 0.02225.$$

The modified Euler's method can also be employed to solve first-order differential equations. For instance, for the following differential equations:

$$\left.\begin{aligned} \frac{dx}{dt} &= f_1(x, y, t) \\ \frac{dy}{dt} &= f_2(x, y, t) \end{aligned}\right\}. \tag{7.23}$$

Let the initial values be t_0, x_0, y_0, when step length h is determined, for the first segment, one can compute the approximate value of the true solution as follows:

$$\begin{aligned} x_1^{(0)} &= x_0 + \left.\frac{dx}{dt}\right|_0 h \\ y_1^{(0)} &= y_0 + \left.\frac{dy}{dt}\right|_0 h \end{aligned},$$

where

$$\begin{aligned} \left.\frac{dx}{dt}\right|_0 &= f_1(x_0, y_0, t_0) \\ \left.\frac{dy}{dt}\right|_0 &= f_2(x_0, y_0, t_0) \end{aligned}.$$

From $t_1 = h, x_1^{(0)}, y_1^{(0)}$, we have

$$\begin{aligned} \left.\frac{dx}{dt}\right|_1^{(0)} &= f_1(x_1^{(0)}, y_1^{(0)}, t_1) \\ \left.\frac{dy}{dt}\right|_1^{(0)} &= f_2(x_1^{(0)}, y_1^{(0)}, t_1) \end{aligned},$$

thus the true solution of the differential equation at t should be

$$x_1 = x_0 + \frac{\left.\frac{dx}{dt}\right|_0 + \left.\frac{dx}{dt}\right|_1^{(0)}}{2} h = x_1^{(0)} + \left.\frac{dx}{dt}\right|_1^a h,$$

$$y_1 = y_0 + \frac{\left.\frac{dy}{dt}\right|_0 + \left.\frac{dy}{dt}\right|_1^{(0)}}{2} h = y_1^{(0)} + \left.\frac{dy}{dt}\right|_1^a h,$$

where

$$\left.\frac{dx}{dt}\right|_1^a = \frac{1}{2}\left(\left.\frac{dx}{dt}\right|_1^{(0)} - \left.\frac{dx}{dt}\right|_0\right)$$

$$\left.\frac{dy}{dt}\right|_1^a = \frac{1}{2}\left(\left.\frac{dy}{dt}\right|_1^{(0)} - \left.\frac{dy}{dt}\right|_0\right)$$

and so forth.

From (7.17), it can be concluded that the modified Euler's method applied to one segment requires computational effort, that is, two times of that of the Euler's method. On the other hand, if the same step length is used, the modified Euler's method provides more accurate calculation results than the Euler's method. As discussed before, the truncation error of the modified Euler's method is $0(h^2)$, while the Euler's method is $0(h)$. Figure 7.5 illustrates that, when the tolerance is equal to ε_1, the difference between the required step length of the modified Euler's method h_1' and the Euler's step length h_1 is small. Under such circumstance, the computational effort required by the modified Euler's method is larger than that of the Euler's method. When the tolerance is equal to ε_2, the required step length of the modified Euler's method h_2' is significantly larger than the Euler's step length h_2. Obviously, if $h_2' > 2h_2$, then the total computational effort of the modified Euler's method is smaller than that of the Euler's method.

7.2.1.3 Runge–Kutta Method

The modified Euler's method is based on the observation that x_{n+1} can be estimated using the derivatives or slopes of two points in the interval $[t_n, t_{n+1}]$, and since the Taylor's series of the integral function is approximated by the first three terms, the local truncation error is $0(h^3)$. This has motivated the following question: is it possible to estimate x_{n+1} using the derivatives of more points in the interval $[t_n, t_{n+1}]$,

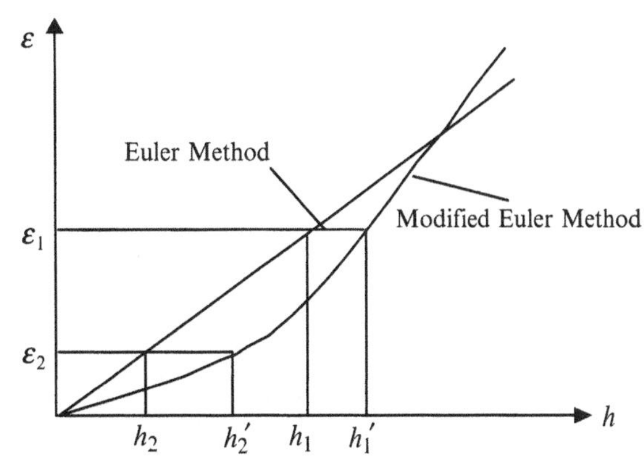

Fig. 7.5 Comparison between the modified Euler's method and Euler's method

such that more terms of the Taylor's series can be included? The answer to this question is positive. The well-known Runge–Kutta method is built upon this idea. The most popular Runge–Kutta method is the fourth-order method. In this method, x_{n+1} is estimated using the derivatives of four points in the interval $[t_n, t_{n+1}]$, thus the first five terms of Taylor's series are included in the approximation:

$$x_{n+1} = x_n + x'_n h + x''_n \frac{h^2}{2!} + x_n^{(3)} \frac{h^3}{3!} + x_n^{(4)} \frac{h^4}{4!} + 0(h^5).$$

The local truncation error of the method is $0(h^5)$, and the global truncation error is $0(h^4)$.

For differential (7.3), the following Runge–Kutta formula should be used:

$$\left.\begin{aligned} x_{n+1} &= x_n + \frac{1}{6}(k_1 + 2k_2 + 2k_3 + k_4) \\ k_1 &= hf(x_n, t_n) \\ k_2 &= hf\left(x_n + \frac{k_1}{2}, t_n + \frac{h}{2}\right) \\ k_3 &= hf\left(x_n + \frac{k_2}{2}, t_n + \frac{h}{2}\right) \\ k_4 &= hf(x_n + k_3, t_n + h) \end{aligned}\right\} \tag{7.24}$$

to solve for $x_1, x_2, x_3, \ldots$.

[Example 7.2] Solve the first-order differential equation in Example 7.1 using the Runge–Kutta method

[Solution] Let the step length $h = 0.2$, the computational steps are described in the following table:

t_n	x_n	k_1	$t_n + \frac{h}{2}$	$xn + \frac{k_1}{2}$	k_2	$t_n + \frac{h}{2}$	$xn + \frac{k_2}{2}$	k_3	$t_n + h$	$x_n + k_3$	k_4
0	1	0.2	0.1	1.1	0.1836364	0.1	1.0918182	0.1817274	0.2	1.181727	0.1686478
0.2	1.1832292	0.1698342	0.3	1.267746	0.1588930	0.3	1.262676	0.1574990	0.4	1.340728	0.1488074
0.4	1.3416668	0.1490788	0.5	1.416026	0.1420188	0.5	1.412676	0.1409600	0.6	1.482627	0.1346506
0.6	1.483281	0.1348528	0.7	1.550707	0.1295786	0.7	1.548070	0.1287436	0.8	1.612025	0.1238970
0.8	1.612513	0.1240546	0.9	1.674541	0.1199240	0.9	1.672475	0.1192452	1.0	1.731759	0.1153728
1	1.732141										

The above table shows that, based on the Runge–Kutta method, the value of x at $t = 1$ is $x = 1.732141$. Comparing this result with the true solution, the error is equal to

$$|1.73205 - 1.732141| = 0.00009,$$

which is a much better result in comparison with the result obtained in Example 7.1.

The Runge–Kutta method can also be used to solve first-order differential equations. As an example, the differential equation (7.23) can be solved using the following recursive formula:

$$
\begin{aligned}
x_{n+1} &= x_n + \frac{1}{6}(k_1 + 2k_2 + 2k_3 + k_4), \\
y_{n+1} &= y_n + \frac{1}{6}(l_1 + 2l_2 + 2l_3 + l_4),
\end{aligned}
$$

where

$$
\left.\begin{aligned}
k_1 &= hf_1(x_n, y_n, t_n) \\
k_2 &= hf_1\left(x_n + \frac{k_1}{2}, y_n + \frac{l_1}{2}, t_n + \frac{h}{2}\right) \\
k_3 &= hf_1\left(x_n + \frac{k_2}{2}, y_n + \frac{l_2}{2}, t_n + \frac{h}{2}\right) \\
k_4 &= hf_1(x_n + k_3, y_n + l_3, t_n + h)
\end{aligned}\right\},
$$

$$
\left.\begin{aligned}
l_1 &= hf_2(x_n, y_n, t_n) \\
l_2 &= hf_2\left(x_n + \frac{k_1}{2}, y_n + \frac{l_1}{2}, t_n + \frac{h}{2}\right) \\
l_3 &= hf_2\left(x_n + \frac{k_2}{2}, y_n + \frac{l_2}{2}, t_n + \frac{h}{2}\right) \\
l_4 &= hf_2(x_n + k_3, y_n + l_3, t_n + h)
\end{aligned}\right\}.
$$

Although the Runge–Kutta method has the advantage of higher precision, it requires larger computational effort which is four times that required by the Euler's method. The trend is that multiple-step methods, which require less computational effort, are replacing Runge–Kutta methods when higher computational accuracy is required. Runge–Kutta methods are typically used as auxiliary methods only to initiate multiple-step methods in the first few steps.

7.2.1.4 Implicit Integration Methods

Explicit and implicit methods are the major categories of solution methods for differential equations. The methods described in the previous sections belong to the category of explicit methods. From (7.9), (7.17), and (7.24), one can see that the right-hand sides of the formulas are known quantities; therefore, the value of the end point x_{n+1} can be directly computed using those recursive formulas. In contrast, an implicit method does not work with recursive equations, it first converts differential equations into difference equations, then solves for the value x_{n+1} using the methods of difference equations.

Let us first introduce the method of the trapezoidal rule.

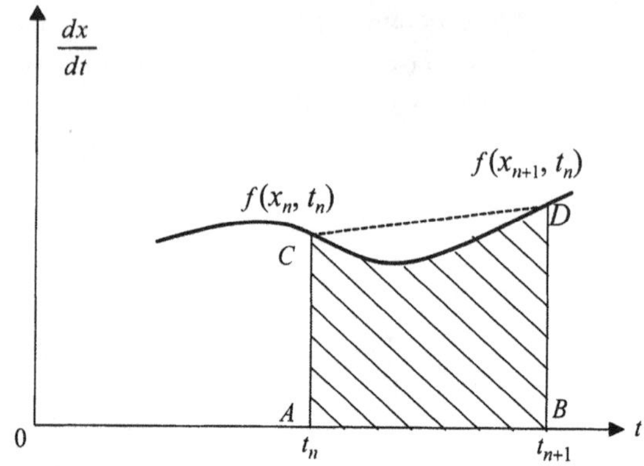

Fig. 7.6 Geometrical illustration of trapezoidal rule

When x_n at t_n is known, the function value x_{n+1} at time $t_{n+1} = t_n + h$ of the differential equation (7.3) can be solved using the following formula:

$$x_{n+1} = x_n + \int_{t_n}^{t_{n+1}} f(x, t)\mathrm{d}t. \tag{7.25}$$

The solution of the definite integral of the above equation is equal to the area of the shaded region in Fig. 7.6. Observe that if the step size h is sufficiently small, the graph of the function $f(x, t)$ between t_n and t_{n+1} can be approximated by a straight line as illustrated in the figure. Apparently the area of the shaded region is equal to the area of the trapezoid ABCD. Equation (7.25) can thus be reformulated as

$$x_{n+1} = x_n + \frac{h}{2}[f(x_n, t_n) + f(x_{n+1}, t_{n+1})]. \tag{7.26}$$

This is the difference equation of the trapezoidal rule.

Obviously, one cannot rely on certain recursive formula to compute x_{n+1} because the right-hand side of (7.26) also includes unknown x_{n+1}. Equation (7.26) has to solve as an algebraic equation to find x_{n+1}.

Generally speaking, the idea of implicit methods is to transform a numerical initial value problem of differential equations into a sequence of algebraic equation problems. For example, given starting point t_0 and x_0, according to (7.26) the difference equation for the first step should be

$$x_1 = x_0 + \frac{h}{2}[f(x_0, t_0) + f(x_1, t_0 + h)],$$

where the only unknown variable is x_1, which can be solved for using the methods for solving algebraic equation. Given t_1 and x_1, based on (7.26), the difference formula for the next step should be

$$x_2 = x_1 + \frac{h}{2}[f(x_1, t_1) + f(x_2, t_1 + h)]$$

from which x_2 can be computed, and so forth.

If $f(x_n, t_n)$ and $f(x_{n+1}, t_{n+1})$ are viewed as the slopes of the integral curve at the starting point and terminating point of the interval $[t_n, t_{n+1}]$, then it is reasonable to term the implicit trapezoidal rule as an implicit modified Euler's method. In other words, difference equation (7.26) can be viewed as the solution formula of the implicit modified Euler's method. In fact, the idea of implicit methods are applicable not only to the modified Euler's method, but also to the previously mentioned Euler's method, Runge–Kutta method, and multistep methods. For example, the recursive formula of the Euler's method (7.9) can be rewritten as

$$x_{n+1} = x_n + x'_{n+1}h = x_n + f(x_{n+1}, t_{n+1})h. \quad (7.27)$$

Changing the derivate value x'_n of the starting point of the interval $[t_n, t_{n+1}]$ to x'_{n+1}, one obtains the implicit Euler's method. Equation (7.27) is the difference formula of the implicit Euler's method.

The difference equations (7.26) and (7.27) can be nonlinear as a result of the nonlinearity of the function $f(x, t)$ in (7.3). Therefore the algorithms for implicit methods are more complex than those of explicit methods.

It is not difficult to find out that the truncation error of implicit trapezoidal rule is introduced by the approximation of replacing the trapezoid with the shaded area (see Fig. 7.6). Using the same arguments as before, one can prove that the local truncation error of difference equation (7.26) is $0(h^3)$.

The advantage of implicit methods over explicit methods is that a larger step size can be used. This issue involves the numerical stability of numerical initial value problems; readers are referred to relevant references. Here we illustrate using a simple example.

Suppose we have the following differential equation:

$$\frac{dx}{dt} = -100x. \quad (7.28)$$

The initial values are $t = 0$ and $x_0 = 1$.

For the above differential equation, the true solution is $x = e^{-100t}$.

This is an exponential function, as depicted in Fig. 7.7.

When the step length is equal to $h = 0.025$, the numerical solution using the Euler's method is as follows:

Steps	t_n	x_n	x'_n	$x'_n h$
0	0.000	1	−100	−2.5
1	0.025	−1.5	150	3.75
2	0.050	2.25	−225	−5.625
3	0.075	−3.375		

Observe that the function value oscillates as time increases, and the magnitude of the oscillation increases, as illustrated in the dotted line in Fig. 7.7. Mathematically,

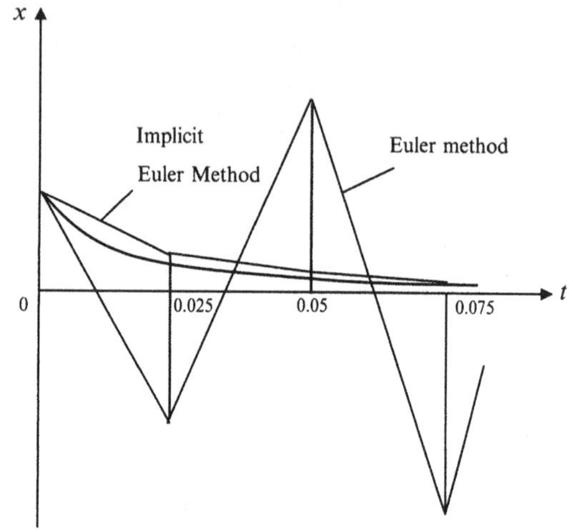

Fig. 7.7 Illustration of solutions obtained using different methods

this indicates that the numerical solution obtained using the Euler's method is not stable.

This situation can be avoided if the implicit Euler's method is used. Let us first transform (7.28) into a difference equation as follows:

$$x_{n+1} = x_n + x'_{n+1}h = x_n - 100x_{n+1}h.$$

Thus

$$x_{n+1} = \frac{x_n}{1 + 100h}.$$

When $h = 0.025$,

$$x_{n+1} = \frac{x_n}{3.5}.$$

One obtains the following calculation results:

Steps	t_n	x_n
0	0.000	1
1	0.025	1/3.5
2	0.050	$(1/3.5)^2$
3	0.075	$(1/3.5)^3$

The function value in the above table decays as time increases, see Fig. 7.7.

To explain the relationship between step size and numerical solutions, we rewrite differential equation (7.28) into more general form:

$$\frac{dx}{dt} = -\frac{x}{T}, \tag{7.29}$$

where the constant T has the unit of time, which is termed the time constant.

Substituting (7.29) into the Euler's equation (7.9), we have

$$x_{n+1} = x_n\left(1 - \frac{h}{T}\right).$$

Therefore

$$x_{n+1} = x_0\left(1 - \frac{h}{T}\right)^{n+1}. \tag{7.30}$$

Obviously in order for x to be a monotonically decaying function, the right-hand side of (7.30) has to meet the following condition:

$$0 < 1 - \frac{h}{T} < 1.$$

Thus the step size should be selected such that

$$h < T. \tag{7.31}$$

For (7.28), the step size must satisfy the inequality $h < 0.01$ to find a stable solution using the explicit method.

The difference equation of (7.29) in the setting of the implicit method is

$$x_{n+1} = x_n - \frac{h}{T}x_{n+1}.$$

After simple manipulation:

$$x_{n+1} = \left(\frac{1}{1 + h/T}\right)x_n.$$

Therefore we have

$$x_{n+1} = x_0\left(\frac{1}{1 + h/T}\right)^{n+1}. \tag{7.32}$$

From the above formula we conclude that, using an implicit Euler's method, any choice of step size can meet the requirement that the function x is monotonically decaying.

In general, the choice of step size in an explicit method is restricted by the time constants in the differential equations under study. If this restriction is violated, one can expect false computational results. An implicit method allows a larger step size and therefore is not subject to such a restriction.

7.2.1.5 Choosing a Numerical Method for Ordinary Differential Equations

As discussed before, numerical methods for ordinary differential equation fall into the category of explicit or implicit methods and single-step or multistep methods. In an explicit method, the integration formula can be directly applied to solve each differential equation, therefore the method requires less computational effort but is numerically less stable; in an implicit method, each differential equation is implicitly integrated and solved as a set of algebraic equations, obviously the method is more complicated but guarantees higher numerical stability. A single-step method requires only the information of the previous step. Therefore it is self-starting and superior when the right-hand side function of the differential equation contains any discontinuity. A multistep method requires the information of previous steps, thus in theory it is more efficient, but it requires a restart when the right-hand side function of the differential equation exhibits a discontinuity. To determine which numerical solution method to use, at least the following three aspects should be investigated:

(1) *Accuracy of the method*: There are two classes of errors involved in the numerical solution of differential equations; these are round-off error and truncation error. The occurrence of round-off errors is due to the inability of computers to represent real numbers perfectly; thus to reduce this type of error, the only means is to use computing with a higher degree of precision.

The difference between the true solution and computed solution is mainly caused by truncation error, which determines the required precision of the solution method. The magnitude of truncation error is dependent on the specific integration formula. Obviously, the higher the order of the method, the higher is the precision offered, for the same step size. Furthermore, local truncation errors are accumulated as integration proceeds, therefore the difference between the true solution at a certain instant and the computed solution is called accumulated truncation error.

(2) *Numerical stability of the method*: Roughly speaking, the numerical stability of an integration method is related to the propagation of numerical errors in the integration process. In a numerically unstable method, errors are accumulated and increase step by step, and in the end they distort the true solution. In a numerically stable method, errors do not increase, in fact, they can even decrease sometimes.

There are a variety of definitions of numerical stability, which will not be discussed here. To make a comparison, we define two broad classes of methods: those "very stable" and those "very unstable." In general, implicit methods have better stability properties than their explicit counterparts.

(3) *Adaptability to stiff differential equations*: Stiffness of differential equations is a property that corresponds to ill-conditioning in algebraic equations. In general, if the difference between the largest time constant and the smallest time constant in a differential equation is large, the differential equation is considered stiff (or ill-conditioned). More precisely, stiffness is measured by the largest eigenvalue divided by the smallest eigenvalue.

For a stiff problem, to ensure that the truncation error is below a numerically safe threshold, a numerical method that is relatively unstable would require a very small step size to capture the fast components in the system response. To obtain solutions with the same degree of accuracy, a numerically stable integration method allows the use of a larger step size, because it tolerates higher errors.

With the exception of simple classical models, the differential equation models for power system transient stability analysis are typically stiff, and the degree of stiffness increases as the complexity of the synchronous machine model increases, as a result of the diversity of time constants. The algebraic equations in transient stability analysis are also often poorly conditioned; this is particularly true when loads are not represented as impedances. There are also discontinuities in the differential-algebraic equations, and regulator limits in the models of generating units, etc.

Of the wide variety of solution methods for initial value problems of differential equations, very few are suitable for power system applications. This is mainly due to the characteristics of the differential equation models for stability studies. To choose a specific numerical solution method, one should take into account the requirements for computational speed and precision, numerical stability, adaptability to stiffness, and modeling flexibility (in dealing with discontinuity and limit actions). Research into these aspects has been extensively reported, with many successful applications. For the time being, the explicit methods, such as the Euler's method, modified Euler's method, and Runge–Kutta method, together with the implicit trapezoidal rule, have all been used with a degree of success. These methods are available in many production-grade commercial packages.

7.2.2 Numerical Methods for Differential-Algebraic Equations

In power system transient stability analysis, what needs to be done is to obtain the solution of a set of simultaneous differential-algebraic equations. The key here is to deal with the interfacing between differential equation (7.1) and algebraic equation (7.2). Either an alternate solution method or a simultaneous solution method, to be described below, can be used for this purpose.

7.2.2.1 Alternate Solution Method

In an alternate solution method, a numerical integration algorithm is applied to the differential equations first, to compute **x**, and then the algebraic equations are solved to obtain the solution of **y**. Clearly, the solution algorithms for integration and for algebraic equation solution can be independent. In general **x** and **y** can be solved for, alternately in a specified manner. In the setting of alternate methods, an explicit method for the differential equations is quite different from an implicit method. The following describes two examples which demonstrate how to solve for $\mathbf{x}_{(t+\Delta t)}$ and $\mathbf{y}_{(t+\Delta t)}$ at $t+\Delta t$, given $\mathbf{x}_{(t)}$ and $\mathbf{y}_{(t)}$. (By convention, for power system stability analysis, a variable with subscript (t) represents the calculated value at t, and Δt means step size.)

Based on the Runge–Kutta formula in (7.24), we have the following computational procedure:

(1) Compute $\mathbf{k}_1 = \Delta t\mathbf{f}(\mathbf{x}_{(t)}, \mathbf{y}_{(t)})$

(2) Compute $\mathbf{x}_1 = \mathbf{x}_{(t)} + \frac{1}{2}\mathbf{k}_1$ and solve algebraic equation $\mathbf{0} = \mathbf{g}(\mathbf{x}_1, \mathbf{y}_1)$ to get $\mathbf{y}_1$, then compute $\mathbf{k}_2 = \Delta t\mathbf{f}(\mathbf{x}_1, \mathbf{y}_1)$

(3) Compute $\mathbf{x}_2 = \mathbf{x}_{(t)} + \frac{1}{2}\mathbf{k}_2$, then solve algebraic equation $\mathbf{0} = \mathbf{g}(\mathbf{x}_2, \mathbf{y}_2)$ to get $\mathbf{y}_2$, then compute $\mathbf{k}_3 = \Delta t\mathbf{f}(\mathbf{x}_2, \mathbf{y}_2)$

(4) Compute $\mathbf{x}_3 = \mathbf{x}_{(t)} + \mathbf{k}_3$, then solve algebraic equation $\mathbf{0} = \mathbf{g}(\mathbf{x}_3, \mathbf{y}_3)$ to get $\mathbf{y}_3$, then compute $\mathbf{k}_4 = \Delta t\mathbf{f}(\mathbf{x}_3, \mathbf{y}_3)$

(5) Finally compute $\mathbf{x}_{(t+\Delta t)} = \mathbf{x}_{(t)} + \frac{1}{6}(\mathbf{k}_1 + 2\mathbf{k}_2 + 2\mathbf{k}_3 + \mathbf{k}_4)$ and solve the corresponding algebraic $\mathbf{0} = \mathbf{g}(\mathbf{x}_{(t+\Delta t)}, \mathbf{y}_{(t+\Delta t)})$ for $\mathbf{y}_{(t+\Delta t)}$

To solve the differential equations according to the implicit trapezoidal rule (7.26), one needs to solve the following algebraic equations:

$$\mathbf{x}_{(t+\Delta t)} = \mathbf{x}_{(t)} + \frac{\Delta t}{2}\left[\mathbf{f}(\mathbf{x}_{(t+\Delta t)}, \mathbf{y}_{(t+\Delta t)}) + \mathbf{f}(\mathbf{x}_{(t)}, \mathbf{y}_{(t)})\right], \tag{7.33}$$

$$\mathbf{0} = \mathbf{g}(\mathbf{x}_{(t+\Delta t)}, \mathbf{y}_{(t+\Delta t)}). \tag{7.34}$$

The alternate solution steps for the above nonlinear algebraic equations are as follows:

(1) Given initial estimate $\mathbf{y}_{(t+\Delta t)}^{[0]}$ of $\mathbf{y}_{(t+\Delta t)}$, solve for the estimated value $\mathbf{x}_{(t+\Delta t)}^{[0]}$ of $\mathbf{x}_{(t+\Delta t)}$ based on (7.33); that is, solve the following equation:

$$\mathbf{x}_{(t+\Delta t)}^{[0]} = \mathbf{x}_{(t)} + \frac{\Delta t}{2}\left[\mathbf{f}(\mathbf{x}_{(t+\Delta t)}^{[0]}, \mathbf{y}_{(t+\Delta t)}^{[0]}) + \mathbf{f}(\mathbf{x}_{(t)}, \mathbf{y}_{(t)})\right]$$

(2) Compute the improved solution $\mathbf{y}^{[1]}_{(t+\Delta t)}$ of $\mathbf{y}_{(t+\Delta t)}$ using $\mathbf{x}^{[0]}_{(t+\Delta t)}$ and (7.34), that is, solve the following equation for $\mathbf{y}^{[1]}_{(t+\Delta t)}$:

$$\mathbf{0} = \mathbf{g}(\mathbf{x}^{[0]}_{(t+\Delta t)}, \mathbf{y}^{[1]}_{(t+\Delta t)})$$

(3) Replace $\mathbf{y}^{[0]}_{(t+\Delta t)}$ with $\mathbf{y}^{[1]}_{(t+\Delta t)}$, return to step (1). This procedure continues until convergence is reached

To obtain better starting values for the purpose of reducing the number of iterations, $\mathbf{y}^{[0]}_{(t+\Delta t)}$ can be set to the value of the previous step, it can also be calculated by extrapolation using the values of previous steps. From the above iteration sequence, it is not difficult to find that the solutions $\mathbf{x}_{(t+\Delta t)}$ and $\mathbf{y}_{(t+\Delta t)}$ need not be consistent unless the number of iterations is unlimited. That is, $\mathbf{x}_{(t+\Delta t)}$ and $\mathbf{y}_{(t+\Delta t)}$ cannot satisfy (7.33) and (7.34) to the same degree of precision, this error is termed interfacing error. Obviously, the only means to reduce interface error is to increase the number of iterations, which necessarily increases the computational burden.

7.2.2.2 Simultaneous Solution Method

Simultaneous solution methods typically use implicit methods. The basic steps are as follows. First transform the differential equations into algebraic equations according to an implicit integration formula. The transformed equations together with the original algebraic equations form a set of simultaneous algebraic equations. By solving this set of simultaneous algebraic equations, one obtains the desired result. Evidently this solution method does not have the problem of interfacing error. If the implicit trapezoidal rule is applied, the simultaneous method is equivalent to solving (7.33) and (7.34) simultaneously. Normally the simultaneous algebraic equations are solved using the Newton method. To improve computational efficiency, the sparsity of the Jacobian matrix should be fully exploited.

7.2.3 General Procedure for Transient Stability Analysis

The basic idea of transient stability study is to compute the dynamic response of a power system after a disturbance occurs, and judge if the system is stable based on the dynamic response.

In fact, the differential-algebraic equation (7.1) and (7.2) for modeling the dynamic behavior of a power system should be nonautonomous and discontinuous because of the nature of the power system. The composition and/or form of the differential-algebraic equations can change during a transient stability study period depending on the occurrence of faults or switching. A power system can be subject to a variety of faults or switching; for example, short circuits, removal of transmission facilities, action of relays and auto-reclosing, forced connection/disconnection of series capacitors and braking resistors, etc. Under these circumstances, the topology and/or parameters of a power network change; as a result, the algebraic equations must be modified accordingly. As further examples, the tripping of a generator, the activation of a high-speed excitor, and the fast control of a valve can all change the model/parameters of a generator, thus the algebraic equations underlying the models have to be correctly adjusted. Aside from faults and switching, the existence of hard limits in regulator models also results in discontinuity in the differential-algebraic equations for stability study.

The faults or switching occurring at various times introduce discontinuities into the differential-algebraic equations for stability study, causing the operating variable $\mathbf{y}(t)$ to exhibit jump behavior. However, a fundamental fact about differential equations is that the solution of a differential equation depends continuously on initial values. This ensures that the state variable $\mathbf{x}(t)$ changes in a continuous pattern. As a matter of the fact, the entire transient stability study period can be divided into multiple time intervals identified by the times at which faults or switch changes occur. Within each interval, the structures of the functions $\mathbf{f}$ and $\mathbf{g}$ remain constant, and the differential-algebraic equations for each interval are autonomous. At the end of an interval (instant t_{-0}) and the beginning of next interval (instant t_{+0}), the structure and parameters of (7.1) and (7.2) should be appropriately modified to reflect the fault or switch operation. Since $\mathbf{x}(t_{+0}) = \mathbf{x}(t_{-0})$, the modified network equations can be solved viewing $\mathbf{x}(t_{+0})$ as a known parameter. Similarly, $\mathbf{y}(t_{+0})$ can be computed. Having $\mathbf{x}(t_{+0})$ and $\mathbf{y}(t_{+0})$ for the new interval, the solution methods introduced in Sect. 7.2.2 can be applied to solve the differential-algebraic equations.

Normally the time at which the system under study suffers a major disturbance is set to the starting time (i.e., $t = 0$ s). In the process of solving the differential-algebraic equations, the stability of the system is determined based on certain criteria evaluated from the system operating state. The basic computational procedure for a transient stability study is illustrated in Fig. 7.8.

Before starting the procedure, one should compute the prefault operating state of the system under study, using a load flow program. More specifically, one uses a load flow program to calculate the voltage and power injection at each node, and compute the operating variables $\mathbf{y}_{(0)}$, then compute the initial values of state variables $\mathbf{x}_{(0)}$, as illustrated in blocks ① and ②.

In block 3, differential equations for each system component are formed, and the corresponding network equations are set up according to the specified solution algorithm. It should be noted that the network model used for transient stability study is different from that used in regular load flow study. The former should take into account the effect of generators and loads. Later in this chapter we will revisit this issue.

The formal transient stability calculation starts in block 4. In existing production-grade code, integration step size Δt is fixed to be a constant. If we assume that the computation reaches the point t, then $\mathbf{x}_{(t)}$ and $\mathbf{y}_{(t)}$ are known quantities. Before

Fig. 7.8 Basic procedure for transient stability study

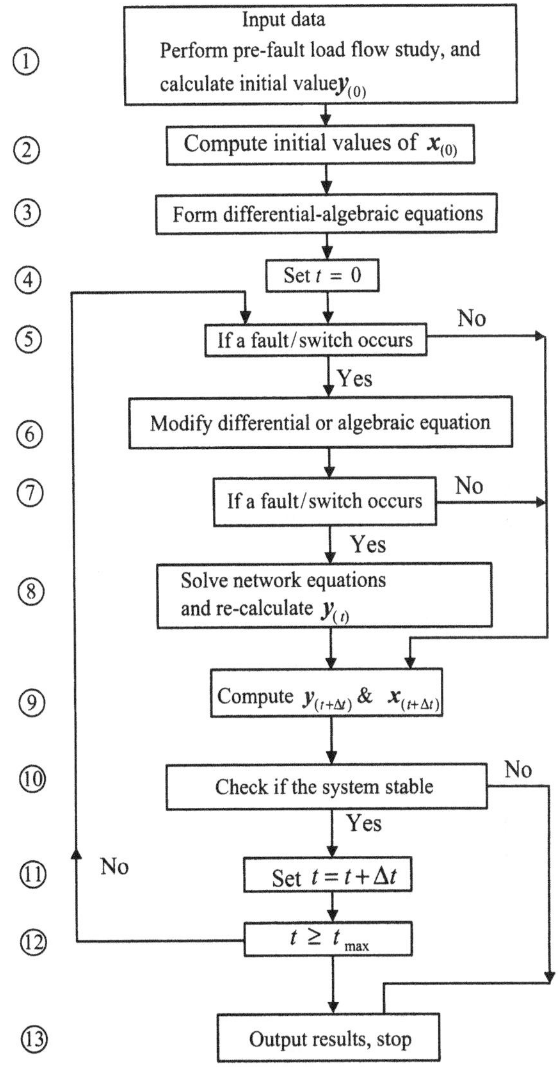

proceeding to compute $\mathbf{x}_{(t+\Delta t)}$ and $\mathbf{y}_{(t+\Delta t)}$, one should check if there is a fault or switch change at instant t. If the answer to the question is yes, then the differential-algebraic equations should be appropriately modified, as shown in blocks 5 and 6 in the figure. Furthermore, if the fault or switching occurs in the network, system operation variables $\mathbf{y}_{(t)}$ can exhibit jump behavior therefore the network equation must be resolved to get the postfault or postswitching system variable $\mathbf{y}_{(t+0)}$, as in blocks 7 and 8 in the figure. Because the state variables do not have jump behavior, the prefault $\mathbf{x}_{(t)}$ and postfault $\mathbf{x}_{(t+0)}$ are identical.

Block 9 is for solving the differential-algebraic equations, that is to solve for $\mathbf{x}_{(t+\Delta t)}$ and $\mathbf{y}_{(t+\Delta t)}$ given the known quantities $\mathbf{x}_{(t)}$ and $\mathbf{y}_{(t)}$ using an alternating method or a simultaneous method. Then in block 10 the stability is checked against certain criteria (e.g., if the maximum relative rotor angle among generators exceed

180°, the system is considered unstable). If the system becomes unstable, then the calculation results can be output, and the procedure stops (block 13); otherwise, increase the time by Δt, via block 11, to proceed to the next iteration. The calculation is repeated until the prespecified termination time t_{max} is reached (block 13).

The choice of t_{max} is related to the problem under study. If only the first swing is of interest, $t_{\text{max}} = 1 - 1.5$ s. Transient stability computations of this type allow for a number of simplifications. For instance, the governor responses of generators can be neglected, thus the outputs of prime movers are viewed as constants. The role of excitation systems is approximated as keeping generator voltage sources constant during the transient stability study period. The stability analysis under these assumptions will be described in detail in Sect. 7.4. For a large-scale interconnected power system, the process of loss of stability slowly develops. Typically the simulation needs to be performed for several, even a dozen, seconds in order to obtain meaning results. Under such circumstances, detailed component models should be assumed. For example, excitation systems, prime movers and speed-governing systems, HVDC systems, and other controls must be correctly modeled. This is the topic of Sect. 7.5.

Most importantly, it is worth noting that a commercial-grade transient stability study code must fulfill the following basic requirements:

(1) The code provides sufficiently accurate results: the errors of generator rotor angles during the entire study period should be less than a few percent.
(2) The algorithm is reliable: the numerical stability of the solution algorithm and the convergence performance of the iteration process should meet minimum requirements.
(3) The code requires sufficiently low memory space: this allows one to perform large system stability studies using a computer with limited memory space.
(4) The code should be flexible and provide easy maintenance: this allows a user to perform customized studies using different component models.

The development of commercial code must make a trade-off among computational speed, precision, reliability, memory, and flexibility.

7.3 Network Mathematical Model for Transient Stability Analysis

Similar to load flow and short circuit studies, the node voltage model of the network for transient stability study can be represented in vector form as follows:

$$\mathbf{YV} = \mathbf{I}, \tag{7.35}$$

where $\mathbf{I}$ and $\mathbf{V}$ denote network node current injection and node voltage in vector form, respectively, $\mathbf{Y}$ denotes the node admittance matrix. Network equation (7.35) is linear, and the admittance matrix $\mathbf{Y}$ is determined by network topology and parameters only.

In subsequent analysis, the network equation is cast in real-number formulation:

$$\begin{bmatrix} \begin{bmatrix} G_{11} & -B_{11} \\ B_{11} & G_{11} \end{bmatrix} & \cdots & \begin{bmatrix} G_{1i} & -B_{1i} \\ B_{1i} & G_{1i} \end{bmatrix} & \cdots & \begin{bmatrix} G_{1n} & -B_{1n} \\ B_{1n} & G_{1n} \end{bmatrix} \\ \vdots & & \vdots & & \vdots \\ \begin{bmatrix} G_{i1} & -B_{i1} \\ B_{i1} & G_{i1} \end{bmatrix} & \cdots & \begin{bmatrix} G_{ii} & -B_{ii} \\ B_{ii} & G_{ii} \end{bmatrix} & \cdots & \begin{bmatrix} G_{in} & -B_{in} \\ B_{in} & G_{in} \end{bmatrix} \\ \vdots & & \vdots & & \vdots \\ \begin{bmatrix} G_{n1} & -B_{n1} \\ B_{n1} & G_{n1} \end{bmatrix} & \cdots & \begin{bmatrix} G_{ni} & -B_{ni} \\ B_{ni} & G_{ni} \end{bmatrix} & \cdots & \begin{bmatrix} G_{nn} & -B_{nn} \\ B_{nn} & G_{nn} \end{bmatrix} \end{bmatrix} \begin{bmatrix} \begin{bmatrix} V_{x1} \\ V_{y1} \end{bmatrix} \\ \vdots \\ \begin{bmatrix} V_{xi} \\ V_{yi} \end{bmatrix} \\ \vdots \\ \begin{bmatrix} V_{xn} \\ V_{yn} \end{bmatrix} \end{bmatrix} = \begin{bmatrix} \begin{bmatrix} I_{x1} \\ I_{y1} \end{bmatrix} \\ \vdots \\ \begin{bmatrix} I_{xi} \\ I_{yi} \end{bmatrix} \\ \vdots \\ \begin{bmatrix} I_{xn} \\ I_{yn} \end{bmatrix} \end{bmatrix}, \quad (7.36)$$

where n denotes the number of nodes, G_{ij} and B_{ij} denote the real and imaginary parts of the elements Y_{ij} of the network admittance matrix, respectively, I_{xi}, I_{yi}, V_{xi}, and V_{yi} denote the real and imaginary parts of node current injection and node voltage, respectively.

In an interconnected power system, dynamic devices which seem independent from each other are interconnected via the transmission network. At any instant during a transient stability study period, current injections of these dynamic devices follow their operating characteristics, the currents in the entire network follow the fundamental Kirchhoff's law. The former is determined by the algebraic equations of the dynamic device, while the later is reflected in the network equations. Therefore, to solve the network equations, the algebraic equations of dynamic devices have to be set up so that the network equations and device equations can be solved as a set of simultaneous equations. Generally speaking, the current injections introduced from dynamic devices are the state variables, and they are functions of corresponding node voltages. Deriving the formula of these functions is the main objective of this section.

Faults or network switching occurring during a stability study period can cause changes to the network topology and parameters. Particularly, if the fault or switching is unsymmetrical in the setting of the three-phase system, the topology and parameters of the network equations are dependent on positive sequence network as well as negative sequence and zero sequence network. Thus the network equations under various fault or switching conditions, during a transient stability study period, have to be carefully laid out.

7.3.1 The Relationship Between Network and Dynamic Devices

7.3.1.1 The Relationship Between Generators and the Network

For the various synchronous machine models discussed in Sects. 6.2.2 and 6.2.3, the stator voltage equation under $d - q$ coordinates can be unified as follows:

$$\begin{bmatrix} V_d \\ V_q \end{bmatrix} = \begin{bmatrix} \bar{E}_d \\ \bar{E}_q \end{bmatrix} - \begin{bmatrix} R_a & -\bar{X}_q \\ \bar{X}_d & R_a \end{bmatrix} \begin{bmatrix} I_d \\ I_q \end{bmatrix}, \quad (7.37)$$

Table 7.1 Machine model and parameters

Models	Parameters			
	$\bar{E}_d$	$\bar{E}_q$	$\bar{X}_d$	$\bar{X}_q$
Varying E'_q, E''_q, E'_d, and E''_d or varying E'_q, E''_q, and E''_d	E''_d	E''_q	X''_d	X''_q
Varying E'_q and E'_d	E'_d	E'_q	X'_d	X'_q
Varying E'_q or constant $E'_q = C$	0	E'_q	X'_d	X_q
$E' = E'_q = C, X_q = X'_d$	0	E'_q	X'_d	X'_d

where $\bar{E}_d$, $\bar{E}_q$, $\bar{X}_d$, and $\bar{X}_q$ denote the d-axis and the q-axis voltage and impedance, respectively. Depending upon the model used to represent the machine, their values can be determined by comparing (7.37) with the original stator voltage equation, as summarized in Table 7.1.

Apply the coordinate transformation (6.62) to (7.37), the stator voltage equation under $x - y$ coordinate is obtained as

$$\begin{bmatrix} \sin\delta & -\cos\delta \\ \cos\delta & \sin\delta \end{bmatrix}\begin{bmatrix} V_x \\ V_y \end{bmatrix} = \begin{bmatrix} \bar{E}_d \\ \bar{E}_q \end{bmatrix} - \begin{bmatrix} R_a & -\bar{X}_q \\ \bar{X}_d & R_a \end{bmatrix}\begin{bmatrix} \sin\delta & -\cos\delta \\ \cos\delta & \sin\delta \end{bmatrix}\begin{bmatrix} I_x \\ I_y \end{bmatrix}. \quad (7.38)$$

After simple manipulations, the generator current injections can be derived from (7.38) as follows:

$$\begin{bmatrix} I_x \\ I_y \end{bmatrix} = \begin{bmatrix} g_x & b_x \\ b_y & g_y \end{bmatrix}\begin{bmatrix} \bar{E}_d \\ \bar{E}_q \end{bmatrix} - \begin{bmatrix} G_x & B_x \\ B_y & G_y \end{bmatrix}\begin{bmatrix} V_x \\ V_y \end{bmatrix}, \quad (7.39)$$

where

$$\left.\begin{aligned} g_x &= \frac{R_a \sin\delta - \bar{X}_d \cos\delta}{R_a^2 + \bar{X}_d\bar{X}_q} & b_x &= \frac{R_a \cos\delta + \bar{X}_q \sin\delta}{R_a^2 + \bar{X}_d\bar{X}_q} \\ b_y &= \frac{-R_a \cos\delta - \bar{X}_d \sin\delta}{R_a^2 + \bar{X}_d\bar{X}_q} & g_y &= \frac{R_{ag} \sin\delta - \bar{X}_{qg} \cos\delta}{R_a^2 + \bar{X}_d\bar{X}_q} \\ G_x &= \frac{R_a - (\bar{X}_d - \bar{X}_q)\sin\delta\cos\delta}{R_a^2 + \bar{X}_d\bar{X}_q} & B_x &= \frac{\bar{X}_d \cos^2\delta + \bar{X}_q \sin^2\delta}{R_a^2 + \bar{X}_d\bar{X}_q} \\ B_y &= \frac{-\bar{X}_d \sin^2\delta - \bar{X}_q \cos^2\delta}{R_a^2 + \bar{X}_d\bar{X}_q} & G_y &= \frac{R_a + (\bar{X}_d - \bar{X}_q)\sin\delta\cos\delta}{R_a^2 + \bar{X}_d\bar{X}_q} \end{aligned}\right\}. \quad (7.40)$$

Substituting the current injection formula derived from (7.39) into network equation (7.36), and applying some simple manipulations, one can conclude that the interconnection of a generator is equivalent to a current injection at the corresponding node:

$$\begin{bmatrix} I'_x \\ I'_y \end{bmatrix} = \begin{bmatrix} g_x & b_x \\ b_y & g_y \end{bmatrix}\begin{bmatrix} \bar{E}_d \\ \bar{E}_q \end{bmatrix}.$$

This current is termed generator pseudocurrent. Furthermore, the corresponding block of the admittance matrix of the network should be added to by a matrix as follows:

$$\begin{bmatrix} G_x & B_x \\ B_y & G_y \end{bmatrix}.$$

It is not difficult to see that, after connecting a generator into the system, the network equations during the stability study period are still linear, however, the generator pseudocurrents and the corresponding admittance matrix are functions of the generator variables $\bar{E}_d$, $\bar{E}_q$, and δ. Thus these linear equations are time varying.

If simpler synchronous machine models are used in the study, the network equations can be simplified too. These simplified equations appear as n-order equations in the complex plane. Unless there is a fault or switch change, the network equations remain unaltered. Thus during the study period, the coefficient matrix of the network equations needs to be refactorized using triangular factorization only when there is a fault or switch change. In what follows we discuss the network model associated with two simplified machine models.

If the effect of damper windings is not considered, the varying E'_q or $E'_q = C$ model for synchronous machines in Table 7.1 should be applied. In this case, (7.39) can be reformulated as

$$\begin{bmatrix} I_x \\ I_y \end{bmatrix} = \begin{bmatrix} \dfrac{R_a - \dfrac{X'_d - X_q}{2}\sin 2\delta}{R_a^2 + X'_d X_q} & \dfrac{\dfrac{X'_d + X_q}{2} + \dfrac{X'_d - X_q}{2}\cos 2\delta}{R_a^2 + X'_d X_q} \\ \dfrac{-\dfrac{X'_d + X_q}{2} + \dfrac{X'_d - X_q}{2}\cos 2\delta}{R_a^2 + X'_d X_q} & \dfrac{R_a + \dfrac{X'_d - X_q}{2}\sin 2\delta}{R_a^2 + X'_d X_q} \end{bmatrix} \times \begin{bmatrix} E'_q \cos\delta - V_x \\ E'_q \sin\delta - V_y \end{bmatrix}. \tag{7.41}$$

From the above, one obtains the formula of generator current into node i represented in the complex domain:

$$\dot{I}_i = \dot{I}'_i - Y'_i \dot{V}_i, \tag{7.42}$$

where

$$Y'_i = \frac{R_{ai} - j\frac{1}{2}(X'_{di} + X_{qi})}{R_{ai}^2 + X'_{di} X_{qi}}, \tag{7.43}$$

$$\left.\begin{aligned} \dot{I}'_i &= \frac{R_{ai} - jX_{qi}}{R_{ai}^2 + X'_{di} X_{qi}} \dot{E}'_{qi} - j\frac{\frac{1}{2}(X'_{di} - X_{qi})}{R_{ai}^2 + X'_{di} X_{qi}} e^{j2\delta_i} \hat{V}_i \\ \dot{E}'_{qi} &= E'_{qi} e^{j\delta_i} \end{aligned}\right\}. \tag{7.44}$$

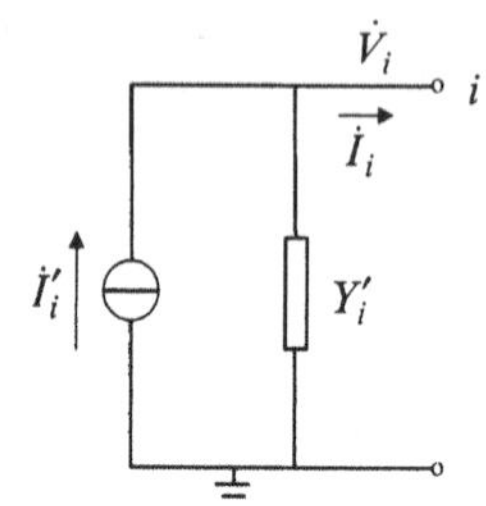

Fig. 7.9 Generator equivalent circuit when damper winding is not considered

The concept underlying (7.42) can be explained using the circuit model illustrated in Fig. 7.9, where Y'_i is called generator pseudoadmittance and is dependent only on generator parameters. The generator pseudoadmittance can be incorporated into the network admittance matrix; $\dot{I}'_i$ is the generator pseudocurrent injection which is related to generator terminal voltage. The network equations are now nonlinear, thus can only be solved using an iterative procedure. As one example, assume an initial value of voltage $\dot{V}_i$, compute $\dot{I}'_i$ based on (7.44), then solve the network equations for an improved solution of $\dot{V}_i$, taking $\dot{I}'_i$ as current injection. This procedure is repeated until convergence is reached. In normal computational steps, the iteration converges within 2–3 steps; while under fault or switching conditions, it may take a few more steps to obtain a converged solution [196].

If synchronous machines are represented by classical models, the effects of damper windings and salient poles are neglected; in addition, the transient voltages E' of generators behind X'_d are assumed to be constants during the stability study period. This situation is shown in Table 7.1, where $E' = E'_q = C$ and $X_q = X'_d$. Correspondingly, from (7.42)–(7.44), it follows that

$$Y'_i = \frac{1}{R_{ai} + jX'_{di}}, \tag{7.45}$$

$$\left.\begin{aligned} \dot{I}'_i &= \frac{1}{R_{ai} + jX'_{di}}\dot{E}'_i \\ \dot{E}'_i &= E'_i e^{j\delta_i} \end{aligned}\right\}. \tag{7.46}$$

Obviously generator pseudocurrent $\dot{I}'_i$ is independent of generator terminal voltage $\dot{V}_i$; thus, once pseudoadmittance Y'_i is incorporated into the network admittance matrix, the network equations can be solved by direct Gauss elimination since $\dot{I}'_i$ is a known quantity.

7.3.1.2 Relationship Between Loads and Network

Depending upon the characteristics of loads, the ways loads are treated are different:

1. If loads are represented by constant impedance models, the constant impedances can be incorporated into the network admittance matrix.

2. If loads are modeled as dynamic devices and only the mechanical–electrical interactions of induction motors in synthesized loads are taken into consideration, loads are still modeled as impedances. However, these impedances are not constant during the stability study period, but vary as the slip-speeds of the induction motors vary. Therefore the impedances representing induction motor loads must be updated, given the slips of the induction motors in each step of the transient stability computation. This means that the diagonal elements of the network admittance matrix are varying in the calculation. The network admittance matrix has to be refactorized in each step of the transient stability calculation when solving the network equations.
3. Again, if loads are modeled as dynamic devices and only the mechanical–electrical interactions of induction motors in synthesized loads are taken into consideration, they can be represented using the Norton equivalent circuit described in Sect. 5.5.2, as illustrated in Fig. 7.10. That is, the load impedances $R + jX$ and $K_M\,(r_1 + jx')$ are incorporated into the network; thus, the loads become simple current sources. This treatment is similar to the way generators connected to the network are treated.

   In the above load representations, the underlying networks are linear.
4. If loads are modeled based on steady-state voltage characteristics, the corresponding node current injections are nonlinear functions of node voltage; as a result, the network equations are nonlinear. According to (6.148) and (6.153), the steady-state voltage characteristics of loads have two formulations, these are the second-order polynomial formulation and exponential formulation:

$$\left.\begin{aligned} P_i &= P_{i(0)}\left[a_P\left(\frac{V_i}{V_{i(0)}}\right)^2 + b_P\left(\frac{V_i}{V_{i(0)}}\right) + c_P\right] \\ Q_i &= Q_{i(0)}\left[a_Q\left(\frac{V_i}{V_{i(0)}}\right)^2 + b_Q\left(\frac{V_i}{V_{i(0)}}\right) + c_Q\right] \end{aligned}\right\} \left.\begin{aligned} P_i &= P_{i(0)}\left(\frac{V_i}{V_{i(0)}}\right)^m \\ Q_i &= Q_{i(0)}\left(\frac{V_i}{V_{i(0)}}\right)^n \end{aligned}\right\}. \tag{7.47}$$

Note that the active and reactive powers in the above equations are the loads absorbed from the network.

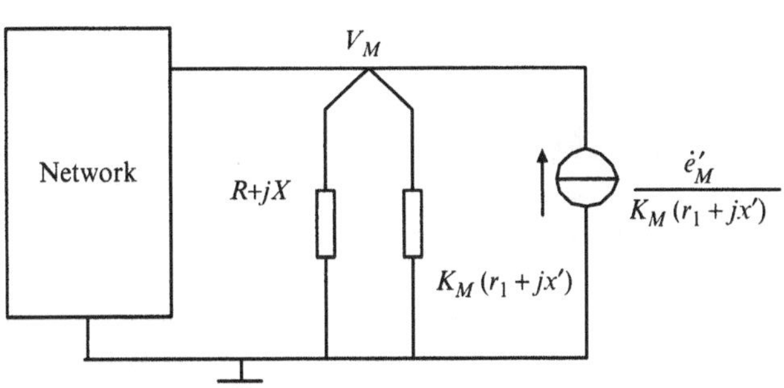

Fig. 7.10 Load representation

Node voltage, current injection, and power injection are connected by the following relationship:

$$-P_i - jQ_i = \dot{V}_i\hat{\dot{I}}_i = (V_{xi} + jV_{yi})(I_{xi} - jI_{yi})$$

from which it is easy to find the relationships between load current injections and node voltages.

If loads are represented by second-order polynomial forms, the load current injections are found to be

$$\left.\begin{aligned}
I_{xi} &= -\frac{P_{i(0)}a_P V_{xi} + Q_{i(0)}a_Q V_{yi}}{V_{i(0)}^2} - \frac{P_{i(0)}b_P V_{xi} + Q_{i(0)}b_Q V_{yi}}{V_{i(0)}\sqrt{V_{xi}^2 + V_{yi}^2}} - \frac{P_{i(0)}c_P V_{xi} + Q_{i(0)}c_Q V_{yi}}{V_{xi}^2 + V_{yi}^2} \\
I_{yi} &= \frac{Q_{i(0)}a_Q V_{xi} - P_{i(0)}a_P V_{yi}}{V_{i(0)}^2} + \frac{Q_{i(0)}b_Q V_{xi} - P_{i(0)}b_P V_{yi}}{V_{i(0)}\sqrt{V_{xi}^2 + V_{yi}^2}} + \frac{Q_{i(0)}c_Q V_{xi} - P_{i(0)}c_P V_{yi}}{V_{xi}^2 + V_{yi}^2}
\end{aligned}\right\}, \tag{7.48}$$

where the terms proportional to the square of voltages can be incorporated into the network admittance matrix as constant admittances, thus (7.48) is reduced to the last two terms only.

If loads are represented by exponential functions, the load current injections are found to be

$$\left.\begin{aligned}
I_{xi} &= -\frac{P_{i(0)}V_i^{m-2}V_{xi}}{V_{(0)}^m} - \frac{Q_{i(0)}V_i^{n-2}V_{yi}}{V_{(0)}^n} \\
I_{yi} &= \frac{Q_{i(0)}V_i^{n-2}V_{xi}}{V_{(0)}^n} - \frac{P_{i(0)}V_i^{m-2}V_{yi}}{V_{(0)}^m}
\end{aligned}\right\}. \tag{7.49}$$

7.3.1.3 The Relationship Between FACTS Devices and the Network

Here we will only describe the relationship between SVC/TCSC and the network; also the relationship between the other FACTS devices and the network can be derived following the same concept.

1. **SVC:** In general an SVC is connected to a high-voltage bus of the network through a transformer (let the index of this bus be i). Thus the shunt susceptance of the device is equal to

$$j\frac{B_{\mathrm{SVC}}}{1 - X_{\mathrm{T}}B_{\mathrm{SVC}}}.$$

From the relationship between nodal voltage $\dot{V}_i$ and current injection $\dot{I}_i$ it is not difficult to find the real and imaginary parts of the current injection as follows:

$$\left.\begin{aligned} I_{xi} &= \frac{B_{\mathrm{SVC}}}{1 - X_{\mathrm{T}}B_{\mathrm{SVC}}} V_{yi} \\ I_{yi} &= -\frac{B_{\mathrm{SVC}}}{1 - X_{\mathrm{T}}B_{\mathrm{SVC}}} V_{xi} \end{aligned}\right\}, \tag{7.50}$$

where X_{T} is the impedance of the transformer, B_{SVC} is the equivalent susceptance of the SVC, V_{xi} and V_{yi} are the real and imaginary parts of the voltage of the high-voltage bus.

2. **TCSC:** Regardless of the place where the TCSC is connected in series in a line, it is always possible to put two nodes around the TCSC, let the nodes be i and j. As a matter of the fact, the role a TCSC plays is equivalent to two current sources having the same magnitude but opposite directions at node i and j, the current injections are easily derived as

$$\left.\begin{aligned} I_{xi} &= -I_{xj} = B_{\mathrm{TCSC}}(V_{yi} - V_{yj}) \\ I_{yj} &= -I_{yi} = B_{\mathrm{TCSC}}(V_{xi} - V_{xj}) \end{aligned}\right\}, \tag{7.51}$$

where B_{TCSC} is the equivalent susceptance of the TCSC, V_{xi}, V_{yi}, V_{xj}, and V_{yj} are the real and imaginary parts of the voltages of the two nodes.

7.3.1.4 The Relationship Between Two-Terminal HVDC and the Network

Let variables with subscript "d" denote quantities on the DC side, and subscripts "R" and "I" denote rectifier and inverter sides (they have the same meaning in subsequent text), respectively. From (5.52)–(5.54) and (5.57) (where $k_\gamma \approx 1$), the steady-state equations of the rectifier are as follows:

$$\left.\begin{aligned} V_{\mathrm{dR}} &= k_{\mathrm{R}} V_{\mathrm{R}} \cos\alpha - X_{\mathrm{cR}} I_{\mathrm{dR}} \\ V_{\mathrm{dR}} &= k_{\mathrm{R}} V_{\mathrm{R}} \cos\varphi_{\mathrm{R}} \\ I_{\mathrm{R}} &= k_{\mathrm{R}} I_{\mathrm{dR}} \\ P_{\mathrm{R}} &= V_{\mathrm{dR}} I_{\mathrm{dR}} = \sqrt{3} V_{\mathrm{R}} I_{\mathrm{R}} \cos\varphi_{\mathrm{R}} \\ Q_{\mathrm{R}} &= P_{\mathrm{R}} \mathrm{tg}\,\varphi_{\mathrm{R}} \end{aligned}\right\} \tag{7.52}$$

and the inverter side steady-state equations:

$$\left.\begin{aligned} V_{\mathrm{dI}} &= k_{\mathrm{I}} V_{\mathrm{I}} \cos\beta + X_{\mathrm{cI}} I_{\mathrm{dI}} \\ V_{\mathrm{dI}} &= k_{\mathrm{I}} V_{\mathrm{I}} \cos\varphi_{\mathrm{I}} \\ I_{\mathrm{I}} &= k_{\mathrm{I}} I_{\mathrm{dI}} \\ P_{\mathrm{I}} &= V_{\mathrm{dI}} I_{\mathrm{dI}} = \sqrt{3} V_{\mathrm{I}} I_{\mathrm{I}} \cos\varphi_{\mathrm{I}} \\ Q_{\mathrm{I}} &= P_{\mathrm{I}} \mathrm{tg}\,\varphi_{\mathrm{I}} \end{aligned}\right\}. \tag{7.53}$$

Based on (7.52) and (7.53), the power injections into the AC system by the DC system can denoted by functions of I_{d}, α, β, $V_{x\mathrm{R}}$, $V_{y\mathrm{R}}$, $V_{x\mathrm{I}}$, and $V_{y\mathrm{I}}$.

The power injection into the AC bus from the rectifier is given by

$$\left.\begin{aligned} \overline{P}_{\mathrm{R}} &= -P_{\mathrm{R}} = -V_{\mathrm{dR}} I_{\mathrm{dR}} = X_{\mathrm{cR}} I_{\mathrm{dR}}^2 - k_{\mathrm{R}} I_{\mathrm{dR}} \sqrt{V_{x\mathrm{R}}^2 + V_{y\mathrm{R}}^2} \cos\alpha \\ \overline{Q}_{\mathrm{R}} &= -Q_{\mathrm{R}} = \overline{P}_{\mathrm{R}} \frac{\sqrt{k_{\mathrm{R}}^2 V_{\mathrm{R}}^2 - V_{\mathrm{dR}}^2}}{V_{dR}} = -I_{\mathrm{dR}} \sqrt{k_{\mathrm{R}}^2 V_{\mathrm{R}}^2 - V_{\mathrm{dR}}^2} \\ &= -I_{\mathrm{dR}} \sqrt{k_{\mathrm{R}}^2 (V_{x\mathrm{R}}^2 + V_{y\mathrm{R}}^2) \sin^2\alpha + 2 k_{\mathrm{R}} X_{\mathrm{cR}} I_{\mathrm{dR}} \sqrt{V_{x\mathrm{R}}^2 + V_{y\mathrm{R}}^2} \cos\alpha - X_{\mathrm{cR}}^2 I_{\mathrm{dR}}^2} \end{aligned}\right\} \tag{7.54}$$

and the power injection from the inverter is

$$\left.\begin{aligned} \overline{P}_{\mathrm{I}} &= P_{\mathrm{I}} = V_{\mathrm{dI}} I_{\mathrm{dI}} = X_{\mathrm{cI}} I_{\mathrm{dI}}^2 + k_{\mathrm{I}} I_{\mathrm{dI}} \sqrt{V_{x\mathrm{I}}^2 + V_{y\mathrm{I}}^2} \cos\beta \\ \overline{Q}_{\mathrm{I}} &= Q_{\mathrm{I}} = P_{\mathrm{I}} \frac{\sqrt{k_{\mathrm{I}}^2 V_{\mathrm{I}}^2 - V_{\mathrm{dI}}^2}}{V_{\mathrm{dI}}} = I_{\mathrm{dI}} \sqrt{k_{\mathrm{I}}^2 V_{\mathrm{I}}^2 - V_{\mathrm{dI}}^2} \\ &= I_{\mathrm{dI}} \sqrt{k_{\mathrm{I}}^2 (V_{x\mathrm{I}}^2 + V_{y\mathrm{I}}^2) \sin^2\beta - 2 k_{\mathrm{I}} X_{\mathrm{cI}} I_{\mathrm{dI}} \sqrt{V_{x\mathrm{I}}^2 + V_{y\mathrm{I}}^2} \cos\beta - X_{\mathrm{cI}}^2 I_{\mathrm{dI}}^2} \end{aligned}\right\}. \tag{7.55}$$

Thus the current injections into the AC system from the rectifier and inverter are obtained as

$$\left.\begin{aligned} I_{x\mathrm{R}} &= \frac{\overline{P}_{\mathrm{R}} V_{x\mathrm{R}} + \overline{Q}_{\mathrm{R}} V_{y\mathrm{R}}}{V_{x\mathrm{R}}^2 + V_{y\mathrm{R}}^2}, \quad I_{y\mathrm{R}} = \frac{\overline{P}_{\mathrm{R}} V_{y\mathrm{R}} - \overline{Q}_{\mathrm{R}} V_{x\mathrm{R}}}{V_{xR}^2 + V_{y\mathrm{R}}^2} \\ I_{x\mathrm{I}} &= \frac{\overline{P}_{I} V_{xI} + \overline{Q}_{\mathrm{I}} V_{y\mathrm{I}}}{V_{x\mathrm{I}}^2 + V_{y\mathrm{I}}^2}, \quad I_{y\mathrm{I}} = \frac{\overline{P}_{\mathrm{I}} V_{y\mathrm{I}} - \overline{Q}_{\mathrm{I}} V_{x\mathrm{I}}}{V_{x\mathrm{I}}^2 + V_{y\mathrm{I}}^2} \end{aligned}\right\}. \tag{7.56}$$

Substituting (7.54) and (7.55) into (7.56), and eliminating variables $\overline{P}_{\mathrm{R}}$, $\overline{Q}_{\mathrm{R}}$, $\overline{P}_{\mathrm{I}}$, and $\overline{Q}_{\mathrm{I}}$, the current injections $I_{x\mathrm{R}}$ and $I_{y\mathrm{R}}$ become functions of I_{dR}, α, $V_{x\mathrm{R}}$ and $V_{y\mathrm{R}}$, and $I_{x\mathrm{I}}$ and $I_{y\mathrm{I}}$ are functions of variables I_{dI}, , $\beta V_{x\mathrm{I}}$, and $V_{y\mathrm{I}}$.

7.3.2 Modeling Network Switching and Faults

When a fault or switch change is applied to a network, the network admittance matrix needs to be correctly modified. If the fault or switch is three-phase symmetrical, for example, a three-phase short circuit, the removal of three phases of a device, the forced connection of a series capacitor, the introduction or removal of a braking resistor, etc., the modification to the admittance matrix is straightforward because such a fault or switching operation results in a parameter change in a shunt branch or series branch of the network.

Most of short circuits and device removals are unsymmetrical, and thus have to be analyzed using a symmetrical components method. In addition to dealing with the positive sequence network of the power system under study, one has to consider the negative sequence and zero sequence networks. On the other hand, in stability studies we are mostly interested in the quantities of the positive sequence network, paying little if any attention to the quantities of the negative and zero sequence networks. The effects of the negative and zero sequence networks can be modeled using an equivalent impedance viewed from the positive sequence network.

When analyzing unsymmetrical problem using the concept of symmetrical components, phase A is often taken as the reference, the boundary conditions of various types of short circuit or open-conductor are expressed in terms of the sequence quantities of phase A. When a short circuit or an open-conductor occurs, the phase that exhibits different behavior compared with the other two phases is called the special phase. For instances, the special phase in a single-line-to-ground fault is the phase connected to ground; the special phase in a double-line-to-ground or line-to-line fault is the phase that is not faulted. The special phase in a single-line-open-conductor is the phase that is open, and the special phase in a double-line-open-conductor is the phase that is intact. When the special phase of a short circuit or open-conductor is phase A, the three sequence networks can be directly connected to form the so-called composite sequence network according to certain boundary conditions. This is equivalent to connecting supplementary impedance to the faulted terminals of the positive sequence network. The size of the supplementary impedance depends on the type of fault, as illustrated in Tables 7.2 and 7.3. Here the "faulted terminals" mean, in a short circuit the terminals between faulted bus and ground and in an open-conductor fault the two nodes resulting from the open-conductor. The network admittance matrices under these circumstances can be easily formed.

Table 7.2 Supplemental impedances of short circuits

Type of short circuit	Supplemental impedance
Single-line-to-ground	$Z_\Sigma^{(2)} + Z_\Sigma^{(0)}$
Double-line-to-ground	$Z_\Sigma^{(2)} Z_\Sigma^{(0)} / (Z_\Sigma^{(2)} + Z_\Sigma^{(0)})$
Line-to-line	$Z_\Sigma^{(2)}$

$Z_\Sigma^{(2)}$ is the self-impedance of the short circuit in negative sequence network, $Z_\Sigma^{(0)}$ is the self-impedance of the short circuit in zero sequence network

Table 7.3 Supplemental impedances of open-conductor

Type of open-conductor	Supplemental impedance
Single-line-open-conductor	$Z^{(2)}Z^{(0)}/(Z^{(2)}+Z^{(0)})$
Double-line-open-conductor	$Z^{(2)}+Z^{(0)}$

$Z^{(2)}$ is the equivalent impedance of the open-conductor terminals in the negative sequence network, $Z^{(0)}$ is the equivalent impedance of the open-conductor terminals in the zero sequence network

Table 7.4 The ratios of ideal transformers

Special phase	Sequence		
	Zero	Positive	Negative
A	1	1	1
B	1	a^2	a
C	1	a	a^2

When the special phase is not phase A, there is a complex operator $a = e^{j120^\circ}$ in the boundary conditions, therefore the three sequence networks cannot be directly connected to form a combined sequence network. However, we can connect the three sequence networks via three ideal transformers with ratios $1{:}n'^{(0)}$, $1{:}n'^{(1)}$, and $1{:}n'^{(2)}$ in the zero, positive, and negative sequence networks. The two sides of these transformers have the same voltage/current ratios thus the transformers introduce no losses. For different special phases, these transformers have different ratios in different sequence networks, as described in Table 7.4.

After introducing ideal transformers, the various types of unsymmetrical short circuit and open-conductor can be classified into two categories: series and shunt (or parallel) faults based on the topology of the three sequence networks. The faults belonging to the series category include single-line-to-ground, double-line-open-conductor, and single-line-to-ground of a series capacitor. The boundary conditions of these faults are as follows: the sum of three sequence voltages is zero, and the sequence currents are identical in the nonstandard ratio side of the transformer. The faults belonging to the shunt category include double-line-to-ground, single-line-open-conductor, and double-line short circuit of capacitors. The boundary conditions of this class of faults are as follows: in the nonstandard ratio side of the transformer, the sum of sequence currents is equal to zero, and the sequence voltages are equal.

When simultaneous short circuits or open-conductors occur, and they occur in different phases, the method for handling single faults can still be applied to modify the admittance matrix of the positive sequence network, but now the concept of supplementary impedance is generalized to that of synthesized impedance matrix. In what follows, we introduce the basic concept of synthesized impedance matrix using single-line-to-ground and single-line-open-conductor faults as examples.

Suppose a single-line-to-ground fault occurs at bus k (let this be fault 1), and a single-line-open-conductor occurs between buses i and j (let this be fault 2), and the two faults occur in different phases. By the boundary conditions of the three sequence components at the place where a fault occurs, the combined sequence

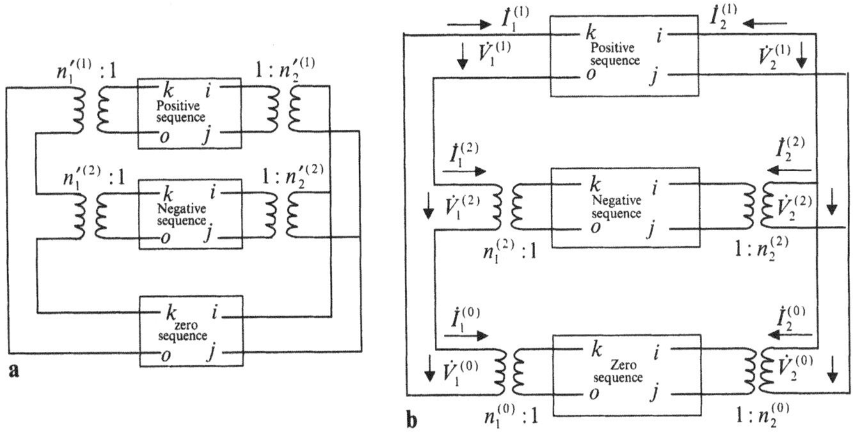

Fig. 7.11 Combined sequences of two simultaneous faults

network can be obtained as in Fig. 7.11a. In the figure, $n_1'^{(1)}$, $n_2'^{(1)}$, $n_1'^{(2)}$, and $n_2'^{(2)}$ are the ratios of ideal transformers, the specific values of them depending on the special phases. For ease of mathematical manipulation, let us recast the combined sequence network in Fig. 7.11a as that in Fig. 7.11b. It is not difficult to see that the ratios in the two figures obey the following relationships:

$$n_1^{(2)} = n_1'^{(2)}/n_1'^{(1)}, \quad n_2^{(2)} = n_2'^{(2)}/n_2'^{(1)}, \quad n_1^{(0)} = 1/n_1'^{(1)}, \quad n_2^{(0)} = 1/n_2'^{(1)}.$$

In the following, we derive the impedance matrix $\mathbf{Z}_\mathrm{f}$ viewed from the fault buses of the positive sequence network into the negative and zero sequence network based on the combined sequence network. We call $\mathbf{Z}_\mathrm{f}$ the synthesized impedance matrix of simultaneous faults.

In Fig. 7.11b, the single-line-to-ground part on the left forms a loop circuit, let the loop current be $\dot{I}_1^{(1)}$, and the single-line-open-conductor part on the right forms two independent loop circuits, let the currents in these circuits be $\dot{I}_2^{(1)}$ and $\dot{I}_2^{(0)}$. Therefore the currents $\dot{I}_1^{(2)}$, $\dot{I}_2^{(2)}$, $\dot{I}_1^{(0)}$, and $\dot{I}_2^{(0)}$ of the faulted buses in the negative and zero sequence networks can be obtained in terms of these loop currents as follows:

$$\mathbf{I}_\mathrm{S} = \mathbf{C}\mathbf{I}_\mathrm{L}, \tag{7.57}$$

where $\mathbf{C}$ the coincidence matrix is dependent on fault conditions. The definitions of the symbols are

$$\mathbf{I}_\mathrm{S} = \begin{bmatrix} \dot{I}_1^{(2)} \\ \dot{I}_2^{(2)} \\ \dot{I}_1^{(0)} \\ \dot{I}_2^{(0)} \end{bmatrix}, \quad \mathbf{I}_\mathrm{L} = \begin{bmatrix} \dot{I}_1^{(1)} \\ \dot{I}_2^{(1)} \\ \dot{I}_2^{(0)} \end{bmatrix}, \quad \mathbf{C} = \begin{bmatrix} 1 & 0 & 0 \\ 0 & -1 & -1 \\ 1 & 0 & 0 \\ 0 & 0 & 1 \end{bmatrix}. \tag{7.58}$$

Based on loop voltage equations, the relationship among the voltages of the faulted buses in each sequence can be obtained as

$$\mathbf{V}_{\mathrm{L}} = \mathbf{C}^{\mathrm{T}}\mathbf{V}_{\mathrm{S}}, \tag{7.59}$$

where $\mathbf{C}^{\mathrm{T}}$ is the transpose of matrix $\mathbf{C}$ and

$$\mathbf{V}_{\mathrm{L}} = \begin{bmatrix} \dot{V}_{ok}^{(1)} \\ \dot{V}_{ji}^{(1)} \\ 0 \end{bmatrix}, \quad \mathbf{V}_{\mathrm{S}} = \begin{bmatrix} \dot{V}_{1}^{(2)} \\ \dot{V}_{2}^{(2)} \\ \dot{V}_{1}^{(0)} \\ \dot{V}_{2}^{(0)} \end{bmatrix}. \tag{7.60}$$

From the transformer nonstandard ratio side point of view, the relationship among the currents and voltages of negative and zero sequence networks is expressed as

$$\begin{bmatrix} \dot{V}_{1}^{(2)} \\ \dot{V}_{2}^{(2)} \end{bmatrix} = \begin{bmatrix} Z_{11}^{(2)} & Z_{12}^{(2)} \\ Z_{21}^{(2)} & Z_{22}^{(2)} \end{bmatrix} \begin{bmatrix} \dot{I}_{1}^{(2)} \\ \dot{I}_{2}^{(2)} \end{bmatrix}, \tag{7.61}$$

$$\begin{bmatrix} \dot{V}_{1}^{(0)} \\ \dot{V}_{2}^{(0)} \end{bmatrix} = \begin{bmatrix} Z_{11}^{(0)} & Z_{12}^{(0)} \\ Z_{21}^{(0)} & Z_{22}^{(0)} \end{bmatrix} \begin{bmatrix} \dot{I}_{k}^{(0)} \\ \dot{I}_{i}^{(0)} \end{bmatrix}. \tag{7.62}$$

Because of the existence of ideal transformers in negative and zero sequence networks, the impedance matrices in (7.61) and (7.62) are unsymmetrical, in general. The computation of the elements of these matrices will be introduced in subsequent sections. Let us incorporate (7.61) and (7.62) into a single equation:

$$\begin{bmatrix} \dot{V}_{1}^{(2)} \\ \dot{V}_{2}^{(2)} \\ \dot{V}_{1}^{(0)} \\ \dot{V}_{2}^{(0)} \end{bmatrix} = \begin{bmatrix} Z_{11}^{(2)} & Z_{12}^{(2)} & 0 & 0 \\ Z_{21}^{(2)} & Z_{22}^{(2)} & 0 & 0 \\ 0 & 0 & Z_{11}^{(0)} & Z_{12}^{(0)} \\ 0 & 0 & Z_{21}^{(0)} & Z_{22}^{(0)} \end{bmatrix} \begin{bmatrix} \dot{I}_{1}^{(2)} \\ \dot{I}_{2}^{(2)} \\ \dot{I}_{1}^{(0)} \\ \dot{I}_{2}^{(0)} \end{bmatrix} \tag{7.63}$$

or in compact form:

$$\mathbf{V}_{\mathrm{S}} = \mathbf{Z}\mathbf{I}_{\mathrm{S}}. \tag{7.64}$$

Making use of the matrix $\mathbf{Z}$ and coincidence matrix $\mathbf{C}$, the relationship between positive sequence voltage and current can be derived. To this end, substituting (7.64) and (7.57) into (7.59), the relationship between faulted loop voltage and current is found to be

$$\mathbf{V}_{\mathrm{L}} = \mathbf{Z}_{\mathrm{L}}\mathbf{I}_{\mathrm{L}}, \tag{7.65}$$

where $\mathbf{Z}_{\mathrm{L}}$ is termed the loop impedance matrix, defined by

$$\mathbf{Z}_{\mathrm{L}} = \mathbf{C}^{\mathrm{T}}\mathbf{Z}\mathbf{C}. \tag{7.66}$$

In this example,

$$\mathbf{Z}_{\mathrm{L}} = \begin{bmatrix} 1 & 0 & 1 & 0 \\ 0 & -1 & 0 & 0 \\ 0 & -1 & 0 & 1 \end{bmatrix} \begin{bmatrix} Z_{11}^{(2)} & Z_{12}^{(2)} & 0 & 0 \\ Z_{21}^{(2)} & Z_{22}^{(2)} & 0 & 0 \\ 0 & 0 & Z_{11}^{(0)} & Z_{12}^{(0)} \\ 0 & 0 & Z_{21}^{(0)} & Z_{22}^{(0)} \end{bmatrix} \begin{bmatrix} 1 & 0 & 0 \\ 0 & -1 & -1 \\ 1 & 0 & 0 \\ 0 & 0 & 1 \end{bmatrix}$$

$$= \begin{bmatrix} Z'_{11} & Z'_{12} & Z'_{13} \\ Z'_{21} & Z'_{22} & Z'_{23} \\ Z'_{31} & Z'_{32} & Z'_{33} \end{bmatrix}. \tag{7.67}$$

Eliminating current $\dot{I}_2^{(0)}$ in (7.65), it follows:

$$\begin{bmatrix} \dot{V}_{ok}^{(1)} \\ \dot{V}_{ji}^{(1)} \end{bmatrix} = \begin{bmatrix} Z_{11} & Z_{12} \\ Z_{21} & Z_{22} \end{bmatrix} \begin{bmatrix} \dot{I}_1^{(1)} \\ \dot{I}_2^{(1)} \end{bmatrix}, \tag{7.68}$$

where the elements Z_{mn} (m and n can be equal to 1 or 2) of the impedance matrix are computed based on:

$$Z_{mn} = Z'_{mn} - \frac{Z'_{m3}Z'_{3n}}{Z'_{33}}, \tag{7.69}$$

(7.68) is rewritten in compact form as

$$\mathbf{V}_{\mathrm{f}} = \mathbf{Z}_{\mathrm{f}}\mathbf{I}_{\mathrm{f}}. \tag{7.70}$$

Finally, the impedance matrix $\mathbf{Z}_{\mathrm{f}}$, viewed from the faulted buses of the positive sequence network into the negative and zero sequence networks, is obtained.

Equation (7.70) can also be expressed in the form of synthesized admittance matrix as follows:

$$\mathbf{I}_{\mathrm{f}} = \mathbf{Y}_{\mathrm{f}}\mathbf{V}_{\mathrm{f}}, \tag{7.71}$$

where $\mathbf{Y}_{\mathrm{f}} = \mathbf{Z}_{\mathrm{f}}^{-1}$. Once $\mathbf{Y}_{\mathrm{f}}$ is determined, the elements of the matrix can be incorporated into the correct position of the admittance matrix of the positive sequence network. In this example, notice that

$$\dot{V}_{ok}^{(1)} = -\dot{V}_k^{(1)}, \quad \dot{V}_{ji}^{(1)} = \dot{V}_j^{(1)} - \dot{V}_i^{(1)}, \quad \dot{I}_k^{(1)} = -\dot{I}_1^{(1)}, \quad \dot{I}_i^{(1)} = -\dot{I}_2^{(1)}, \quad \dot{I}_j^{(1)} = \dot{I}_2^{(1)}. \tag{7.72}$$

The above relationships together with (7.71) give us the relationship among the voltages and currents at node k, i and j in the positive sequence network:

$$\begin{bmatrix} \dot{I}_k^{(1)} \\ \dot{I}_i^{(1)} \\ \dot{I}_j^{(1)} \end{bmatrix} = \begin{bmatrix} Y_{11} & Y_{12} & -Y_{12} \\ Y_{21} & Y_{22} & -Y_{22} \\ -Y_{21} & -Y_{22} & Y_{22} \end{bmatrix} \begin{bmatrix} \dot{V}_k^{(1)} \\ \dot{V}_i^{(1)} \\ \dot{V}_j^{(1)} \end{bmatrix}. \tag{7.73}$$

In summary, the calculation of synthesized impedance matrix includes the following steps:

1. Form the impedance matrix of the faulted buses of the negative and zero sequence network (refer to (7.63))
2. By use of the coincidence matrix that represents the boundary conditions of simultaneous faults, form the loop impedance matrix $\mathbf{Z}_{\mathrm{L}}$ (refer to (7.66) and (7.67))
3. Eliminate the closed circuit from the synthesized impedance matrix $\mathbf{Z}_{\mathrm{f}}$ (refer to (7.68) and (7.69))

In what follows we describe the above steps in detail.

(1) *Forming the impedance matrices of the faulted buses of the negative and zero sequence networks*: In a transient stability study, the admittance matrices of each sequence network should be formed first, followed by calculation of the triangular factors for these matrices. In this way the impedance matrices of the faulted buses of each sequence network can be easily obtained given the fault information.

For the negative sequence network, observing Fig. 7.11b it is not difficult to see that, if one injects unity current into the nonstandard ratio node k of the ideal transformer with zero current injections to the other nodes, that is, $\dot{I}_k^{(2)} = 1$ and $\dot{I}_m^{(2)} = 0$ (m is a node other than node k), then solve the equation of the negative sequence network including the ideal transformer for voltages $\dot{V}_k^{(2)}$ and $\dot{V}_{ij}^{(2)} = \dot{V}_i^{(2)} - \dot{V}_j^{(2)}$. These quantities are the desired quantities for the first column $Z_{11}^{(2)}$ and $Z_{21}^{(2)}$ of the impedance matrix in (7.71).

More specifically, injecting unity current into node k of the nonstandard transformer is equivalent to injecting into node k of negative sequence network a current $\dot{I}_k'^{(2)} = \hat{n}_1^{(2)}$, thus after performing sparse forward substitution and backward substitution on the admittance matrix of the negative sequence network, voltages $\dot{V}_k'^{(2)}$ and $\dot{V}_{ij}'^{(2)} = \dot{V}_i'^{(2)} - \dot{V}_j'^{(2)}$ are obtained; in addition, we have $\dot{V}_k^{(2)} = n_1^{(2)}\dot{V}_k'^{(2)}$ and $\dot{V}_{ij}^{(2)} = n_2^{(2)}\dot{V}_{ij}'^{(2)}$.

By the same token, injecting into nodes i and j of the negative sequence network currents $+\hat{n}_2^{(2)}$ and $-\hat{n}_2^{(2)}$, and performing sparse forward substitution and backward substitution, one obtains voltages $\dot{V}_k'^{(2)}$ and $\dot{V}_{ij}'^{(2)}z = \dot{V}_i'^{(2)} - \dot{V}_j'^{(2)}$, furthermore, $\dot{V}_k^{(2)} = n_1^{(2)}\dot{V}_k'^{(2)}$, $\dot{V}_{ij}^{(2)} = n_2^{(2)}\dot{V}_{ij}'^{(2)}$, the quantities which we seek for the elements $Z_{12}^{(2)}$ and $Z_{22}^{(2)}$ of the second column in the impedance matrix in (7.61).

The same principle applies to compute the elements of the impedance matrix in (7.62).

(2) *Forming loop impedance matrix from the coincidence matrix*: As discussed before, the combined sequence network of a series fault is formed by putting the three sequence networks together in series, therefore there is only one independent loop circuit. The combined sequence network of a shunt fault is formed by putting together the three sequence networks in parallel, resulting in two independent loop circuits. Besides, a line-to-line fault is viewed as a special shunt fault.

From (7.57) and (7.59), the coincidence matrix $\mathbf{C}$ expresses the relationship between the loop current of the boundary circuit of the combined sequence network and the current of the faulted buses of the negative and zero sequence networks. Thus the number of rows of the coincidence matrix equals the dimension of $\mathbf{I}_\mathrm{S}$, that is, two times the number of simultaneous faults (when a line-to-line fault occurs, an empty faulted bus in the zero sequence network is designated). The number of columns of the coincidence matrix equals the dimension of $\mathbf{I}_\mathrm{L}$. A series fault occupies one column in the coincidence matrix as illustrated below:

$$[\underbrace{0 \;\cdots\; 0 \;\; 1 \;\; 0 \;\cdots\; 0}_{\substack{\text{corresponding to the current of the}\\ \text{faulted bus in negative sequence network}}}\;\underbrace{0 \;\cdots\; 0 \;\; 1 \;\; 0 \;\ldots\; 0}_{\substack{\text{corresponding to the current of the}\\ \text{faulted bus in zero sequence network}}}]^\mathrm{T},$$

where the column number of the nonzero is equal to the index number of the fault among all faults. A shunt fault occupies two columns in the coincidence matrix as illustrated below:

$$\underbrace{\begin{bmatrix} 0 & \cdots & 0 & -1 & 0 & \cdots & 0 \\ 0 & \cdots & 0 & -1 & 0 & \cdots & 0 \end{bmatrix}}_{\substack{\text{corresponding to the current of the}\\ \text{faulted bus in negative sequence network}}}\underbrace{\begin{bmatrix} 0 & \cdots & \cdots & \cdots & \cdots & \cdots & 0 \\ 0 & \cdots & 0 & 1 & 0 & \cdots & 0 \end{bmatrix}^\mathrm{T}}_{\substack{\text{corresponding to the current of the}\\ \text{faulted bus in zero sequence network}}},$$

where the first column contains the information on how the negative sequence network is connected with the positive sequence network, while the second column describe the connectivity between the zero sequence and negative sequence networks. The column number of the nonzero corresponds to the index number of the fault among all faults. For a line-to-line fault, the coincidence matrix has only the first column because there is no circuit connection between negative and zero sequence networks.

Based on the above principle, we can easily find the coincidence matrix that represents the boundary conditions of arbitrarily complex simultaneous faults given the types of the faults. For example, if three faults simultaneously occur, and the faults are, in order, single-line-to-ground, single-line-open-conductor, and line-to-line, then the coincidence matrix is as follows:

$$\mathbf{C} = \left[\begin{array}{c|cc|c} 1 & 0 & 0 & 0 \\ 0 & -1 & -1 & 0 \\ 0 & 0 & 0 & -1 \\ \hline 1 & 0 & 0 & 0 \\ 0 & 0 & 1 & 0 \\ 0 & 0 & 0 & 0 \\ \text{single-line-to-ground} & \multicolumn{2}{c|}{\text{single-line-open-conductor}} & \text{line-to-line} \end{array}\right] \begin{array}{l} \left.\right\} \text{negative sequence part} \\ \left.\right\} \text{zero sequence part} \end{array}.$$

With the coincidence matrix that describes the boundary conditions of complex simultaneous faults, the loop impedance matrix of the combined sequence network can be obtained using (7.66) and (7.67). The manipulations on these matrices can be accomplished by simple addition and subtraction operations.

(3) *Eliminating the closed circuit of shunt faults to form synthesized impedance matrix*: The order of the loop impedance matrix equals the number of independent loop currents in the combined sequence network. To seek the synthesized impedance matrix, the currents of zero and negative sequence network must be eliminated (refer to (7.68) and (7.69)).

7.4 Transient Stability Analysis with Simplified Model

For a regional power system, the duration of losing synchronous stability is very short, typically a simulation study of the first swing (1–1.5 s) after a disturbance is applied suffices to judge whether or not the system can maintain synchronous operation. In stability studies like this, the effects of speed-governing systems can be neglected, thus the output of prime movers can be assumed to be constant, the reason is that the inertias of the prime movers are sufficient to keep the outputs of the prime movers constant; besides, because the time constants of the excitor windings are relatively large, their flux linkages do not change drastically in a short range of time, as a result the effect of the excitation system can be modeled as keeping generator transient voltages E_q' or E' constant. In other words, the free current components of excitor windings are compensated by the regulation of excitation systems, thus the flux linkages ψ_{f} of excitor windings remain constant. Correspondingly, the effects of damper windings are also ignored.

The simplified models for transient stability analysis are widely used in power system operation and planning. Specific applications include feasibility studies on system topologies and operating schedules, computation of maximum transfer capabilities, calculation of critical clearing times, and investigations into the effects of stability controls, etc.

Using different models for generators, loads, and network, one can build codes for various simplified stability analyses. Which portfolio of models to use depends on the fundamental characteristics of the problem under study. To explain the principles and procedures of simplified transient stability analysis, the subsequent sections assume the following mathematical models and solution algorithms have been applied to the transient stability analysis procedure:

Generators: Generator transient voltage E_q' remains constant
Loads: Small loads are modeled as constant impedances, while larger loads are modeled as motors with mechanical–electrical interactions
Network: Modeled with admittance matrix

The differential equations are solved by the modified Euler's method while the network equations are solved by Gauss elimination method.

The overall procedure for a transient stability calculation is still as described in Fig. 7.8. The computer implementation of the calculation is provided below.

7.4.1 Computing Initial Values

Before starting the numerical integration, the initial values of the differential equations should be calculated based on the prefault operating state obtained by performing a load flow study. In a simplified transient stability study, the calculation of initial values include prefault generator transient voltages, rotor angles, the output of prime movers, and the slips and equivalent admittances of motors representing loads, etc. These parameters do not change discontinuously at the instant immediately after the fault is applied. In what follows the initial value variables are marked with subscripts (0).

First we describe how to calculate the initial values of generators.

From a load flow study the generator terminal voltages before the disturbance and the generator powers are given by $\dot{V}_{(0)} = V_{x(0)} + jV_{y(0)}$ and $S_{(0)} = P_{(0)} + jQ_{(0)}$. Furthermore, the generator currents injected into the network are computed by

$$\dot{I}_{(0)} = I_{x(0)} + jI_{y(0)} = \frac{\hat{S}_{(0)}}{\hat{\dot{V}}_{(0)}}. \tag{7.74}$$

Thus by (6.61), one can find the pseudovoltage $\dot{E}_{Q(0)}$ as

$$\dot{E}_{Q(0)} = E_{Qx(0)} + jE_{Qy(0)} = \dot{V}_{(0)} + (R_a + jX_q)\dot{I}_{(0)}. \tag{7.75}$$

Subsequently, the generator rotor angles are calculated by

$$\delta_{(0)} = \operatorname{arctg}(E_{Qy(0)}/E_{Qx(0)}). \tag{7.76}$$

Under steady-state operation, generators rotate at synchronous speed, therefore:

$$\omega_{(0)} = 1. \tag{7.77}$$

Using coordinate transformation formula (6.62), the d, q components of generator stator voltages and currents are given by

$$\begin{aligned}\begin{bmatrix} V_{d(0)} \\ V_{q(0)} \end{bmatrix} &= \begin{bmatrix} \sin\delta_{(0)} & -\cos\delta_{(0)} \\ \cos\delta_{(0)} & \sin\delta_{(0)} \end{bmatrix} \begin{bmatrix} V_{x(0)} \\ V_{y(0)} \end{bmatrix} \begin{bmatrix} I_{d(0)} \\ I_{q(0)} \end{bmatrix} \\ &= \begin{bmatrix} \sin\delta_{(0)} & -\cos\delta_{(0)} \\ \cos\delta_{(0)} & \sin\delta_{(0)} \end{bmatrix} \begin{bmatrix} I_{x(0)} \\ I_{y(0)} \end{bmatrix}.\end{aligned} \tag{7.78}$$

Now based on (6.64), the values of transient voltages are obtained as

$$E'_{q(0)} = V_{q(0)} + R_a I_{q(0)} + X'_d I_{d(0)}. \tag{7.79}$$

In addition, the electrical powers $P_{e(0)}$ of generators under steady-state operation are equal to the mechanical powers of the prime movers $P_{m(0)}$, that is,

$$P_{m(0)} = P_{e(0)} = P_{(0)} + (I_{x(0)}^2 + I_{y(0)}^2)R_a. \tag{7.80}$$

The calculation of the initial values of loads is simple.

The prefault node voltages $\dot{V}_{(0)}$ and powers $S_{(0)}$ consumed by loads are obtained from a load flow study, therefore the equivalent admittances of loads are computed by

$$Y_{(0)} = \frac{\hat{S}_{(0)}}{V_{(0)}^2}. \tag{7.81}$$

When loads are modeled as constant impedances, the corresponding equivalent admittances remain constant in the study period, and thus can be incorporated into the network admittance matrix as discussed earlier. For loads representing motors with mechanical–electrical interactions, since the slips of motors do not jump at the instant of disturbance, the equivalent admittances of loads do not change. In other words, the equivalent admittances of loads after the disturbance are identical to those of loads under normal steady-state operation.

7.4.2 Solving Network Equations with Gauss Elimination Method

In this solution method, the network equations are represented in the domain of real numbers, as in (7.36). Before starting the simulation, the loads represented by constant impedances should be incorporated into the network to obtain the network with constant impedance loads, this set of network equations remains constant during the simulation period.

Suppose a motor load is connected at node j. In the transient period the motor slip s_j is time varying, and given the s_j at an instant, the actual impedance of the motor load can be calculated based on (6.160):

$$Z_{Mj} = \left\{ R_1 + jX_1 + \frac{(R_\mu + jX_\mu)(R_2/s_j + jX_2)}{(R_\mu + jX_\mu) + (R_2/s_j + jX_2)} \right\} \frac{Z_{Mj(0)}}{Z_{M(0)}}, \tag{7.82}$$

where $Z_{Mj(0)}$ and $Z_{M(0)}$ are the equivalent impedance of all the motors under normal operation and the equivalent impedance of a typical motor. The admittance associated with the actual impedance can be rewritten as

$$Y_{Mj} = \frac{1}{Z_{Mj}} = G_{Mj} + jB_{Mj}. \tag{7.83}$$

Now suppose a generator is located at node i of the network. When the generator is represented by a varying E'_q model, with reference to Table 7.1, in (7.39) let $\overline{E}_{di} = 0$, $\overline{E}_{qi} = E'_{qi}$, $\overline{X}_{di} = X'_{di}$, and $\overline{X}_{qi} = X_{qi}$, the formula for generator current is as follows:

$$\begin{bmatrix} I_{xi} \\ I_{yi} \end{bmatrix} = \begin{bmatrix} b_{xi} \\ g_{yi} \end{bmatrix} E'_{qi} - \begin{bmatrix} G_{xi} & B_{xi} \\ B_{yi} & G_{yi} \end{bmatrix} \begin{bmatrix} V_{xi} \\ V_{yi} \end{bmatrix}, \tag{7.84}$$

where the elements can be rewritten, based (7.40), as follows:

$$\left.\begin{aligned} b_{xi} &= \frac{R_{ai}\cos\delta_i + X_{qi}\sin\delta_i}{R_{ai}^2 + X'_{di}X_{qi}}, & g_{yi} &= \frac{R_{ai}\sin\delta_i - X_{qi}\cos\delta_i}{R_{ai}^2 + X'_{di}X_{qi}} \\ G_{xi} &= \frac{R_{ai} - (X'_{di} - X_{qi})\sin\delta_i\cos\delta_i}{R_{ai}^2 + X'_{di}X_{qi}}, & B_{xi} &= \frac{X'_{di}\cos^2\delta_i + X_{qi}\sin^2\delta_i}{R_{ai}^2 + X'_{di}X_{qi}} \\ B_{yi} &= \frac{-X'_{di}\sin^2\delta_i - X_{qi}\cos^2\delta_i}{R_{ai}^2 + X'_{di}X_{qi}}, & G_{yi} &= \frac{R_{ai} + (X'_{di} - X_{qi})\sin\delta_i\cos\delta_i}{R_{ai}^2 + X'_{di}X_{qi}} \end{aligned}\right\}. \tag{7.85}$$

Substitute the generator current representations (7.84) into the network equations with constant impedance loads, and do the same for the equivalent admittance [(7.83)] of motors, we obtain the new set of network equations. Obviously, the new network equations are just modifications of the original network equations: the diagonal elements of the admittance matrix are modified, and there are nonzero pseudocurrents in elements of the current vector associated with generators, the current injections of other nodes are zero, that is:

The ith diagonal block of the admittance matrix is changed to

$$\begin{bmatrix} G_{xi} + G_{ii} & B_{xi} - B_{ii} \\ B_{yi} + B_{ii} & G_{yi} + G_{ii} \end{bmatrix} \tag{7.86}$$

and the jth diagonal block changes to

$$\begin{bmatrix} G_{Mj} + G_{jj} & -B_{Mj} - B_{jj} \\ B_{Mj} + B_{jj} & G_{Mj} + G_{jj} \end{bmatrix}. \tag{7.87}$$

The pseudocurrent injections at generator nodes are given by

$$\begin{bmatrix} I'_{xi} \\ I'_{yi} \end{bmatrix} = \begin{bmatrix} b_{xi} \\ g_{yi} \end{bmatrix} E'_{qi}. \tag{7.88}$$

Now the linear equations obtained in each integration step can be solved by Gauss elimination or the triangular factorization method. This gives us the real and imaginary part V_x and V_y of the network voltages for this step. Finally, based on (7.84), the generator currents I_x and I_y can be found.

7.4.3 Solving Differential Equations by Modified Euler's Method

In a transient stability analysis using simplified models, the differential equations comprise the motion (6.76) of generator rotors and the motion (6.155) of motor rotors representing loads:

$$\left.\begin{aligned} \frac{\mathrm{d}\delta_i}{\mathrm{d}t} &= \omega_s(\omega_i - 1) \\ \frac{\mathrm{d}\omega_i}{\mathrm{d}t} &= \frac{1}{T_{Ji}}(P_{mi} - P_{ei}) \\ \frac{\mathrm{d}s_j}{\mathrm{d}t} &= \frac{1}{T_{JMi}}(M_{mMj} - M_{eMj}) \end{aligned}\right\}. \tag{7.89}$$

Suppose the simulation of mechanical–electrical interactions has been completed up to time t, now let us discuss how to calculate the system states for time $t + \Delta t$. Before calculating system states for the next step, whether or not there is a fault or switch operation at time t should be checked first. If the answer is no, then one proceeds to compute the states of the next step, given the states of time t; otherwise, one has to calculate the postswitch or postfault network operating parameters first, and then continue the calculation for the next step. The computational procedure for solving differential equations based on the modified Euler's method is as follows:

(1) Given generator $\delta_{i(t)}$ and motor $s_{j(t)}$ at time t, compute system voltages $V_{x(t)}$ and $V_{y(t)}$, and generator currents $I_{xi(t)}$ and $I_{yi(t)}$ based on the method described in Sect. 7.4.2.

(2) Based on (7.89), compute the derivatives for time t:

$$\left.\begin{aligned}
\left.\frac{d\delta_i}{dt}\right|_t &= \omega_s(\omega_{i(t)} - 1) \\
\left.\frac{d\omega_i}{dt}\right|_t &= \frac{1}{T_{Ji}}(P_{mi} - P_{ei(t)}) \\
\left.\frac{ds_j}{dt}\right|_t &= \frac{1}{T_{JMi}}(M_{mMj(t)} - M_{eMj(t)})
\end{aligned}\right\}, \tag{7.90}$$

where generator power $P_{ei(t)}$ is calculated by

$$P_{ei(t)} = (V_{ix(t)}I_{ix(t)} + V_{iy(t)}I_{iy(t)}) + (I_{ix(t)}^2 + I_{iy(t)}^2)R_{ai}. \tag{7.91}$$

The mechanical torque $T_{mMj(t)}$ of generators and electrical torque $T_{eMj(t)}$ of motors are computed based on (6.157) and (6.156) as follows:

$$\left.\begin{aligned}
M_{mMj(t)} &= k[\alpha + (1-\alpha)(1 - s_{j(t)})^2] \\
M_{eMj(t)} &= \frac{2M_{eM\max}}{\dfrac{s_{j(t)}}{s_{crj}} + \dfrac{s_{crj}}{s_{j(t)}}} \frac{V_{jx(t)}^2 + V_{jy(t)}^2}{V_{jx(0)}^2 + V_{jy(0)}^2}
\end{aligned}\right\}, \tag{7.92}$$

in which $V_{jx(0)}$ and $V_{jy(0)}$ denote the real and imaginary parts of prefault node voltage at node j.

(3) Compute an initial estimate of state variables for time $t + \Delta t$:

$$\left.\begin{aligned}
\delta_{i(t+\Delta t)}^{[0]} &= \delta_{i(t)} + \left.\frac{d\delta_i}{dt}\right|_t \Delta t \\
\omega_{i(t+\Delta t)}^{[0]} &= \omega_{i(t)} + \left.\frac{d\omega_i}{dt}\right|_t \Delta t \\
s_{j(t+\Delta t)}^{[0]} &= s_{j(t)} + \left.\frac{ds_j}{dt}\right|_t \Delta t
\end{aligned}\right\}. \tag{7.93}$$

(4) Similar to step (1), given generator $\delta_{i(t+\Delta t)}^{[0]}$ and motor $s_{j(t+\Delta t)}^{[0]}$, compute system node voltages $V_{x(t+\Delta t)}^{[0]}$ and $V_{y(t+\Delta t)}^{[0]}$, generator currents $I_{xi(t+\Delta t)}^{[0]}$ and $I_{yi(t+\Delta t)}^{[0]}$ based on the method of Sect. 7.4.2.

(5) Similar to step (2), compute the estimated derivatives $\left.\frac{d\delta_i^{[0]}}{dt}\right|_{t+\Delta t}$, $\left.\frac{d\omega_i^{[0]}}{dt}\right|_{t+\Delta t}$, and $\left.\frac{ds_j^{[0]}}{dt}\right|_{t+\Delta t}$ for step $t + \Delta t$. To this end, one should replace $\omega_{i(t)}$, $P_{ei(t)}$, $M_{mMj(t)}$, and $M_{eMj(t)}$ in (7.92) with $\omega_{i(t+\Delta t)}^{[0]}$, $P_{ei(t+\Delta t)}^{[0]}$, $M_{mMj(t+\Delta t)}^{[0]}$, and $M_{eMj(t+\Delta t)}^{[0]}$. To compute them, one should also replace $V_{ix(t)}$, $V_{iy(t)}$, $I_{ix(t)}$,

$I_{iy(t)}$, $s_{j(t)}$, $V_{jx(t)}$, and $V_{jy(t)}$ with $V^{[0]}_{ix(t+\Delta t)}$, $V^{[0]}_{iy(t+\Delta t)}$, $I^{[0]}_{ix(t+\Delta t)}$, $I_{iy(t)}$, $s_{j(t)}$, $V^{[0]}_{jx(t+\Delta t)}$, and $V^{[0]}_{jy(t+\Delta t)}$.

(6) Finally, compute the variable values for step $t + \Delta t$, that is:

$$\left.\begin{aligned} \delta_{i(t+\Delta t)} &= \delta_{i(t)} + \frac{\Delta t}{2}\left[\left.\frac{\mathrm{d}\delta_i}{\mathrm{d}t}\right|_t + \left.\frac{\mathrm{d}\delta_i^{[0]}}{\mathrm{d}t}\right|_{t+\Delta t}\right] \\ \omega_{i(t+\Delta t)} &= \omega_{i(t)} + \frac{\Delta t}{2}\left[\left.\frac{\mathrm{d}\omega_i}{\mathrm{d}t}\right|_t + \left.\frac{\mathrm{d}\omega_i^{[0]}}{\mathrm{d}t}\right|_{t+\Delta t}\right] \\ s_{j(t+\Delta t)} &= s_{j(t)} + \frac{\Delta t}{2}\left[\left.\frac{\mathrm{d}s_j}{\mathrm{d}t}\right|_t + \left.\frac{\mathrm{d}s_j^{[0]}}{\mathrm{d}t}\right|_{t+\Delta t}\right] \end{aligned}\right\}. \tag{7.94}$$

[Example 7.3] Consider the 9-bus system in Fig. 7.12 [178]. This system consists of three generators, three loads, and nine branches. The generator and branch parameters are listed in Tables 7.5 and 7.6, respectively. The system load flow under normal operation is illustrated in Table 7.7, and the system frequency is 60 Hz.

[Solution] A stability analysis based on the simplified system model will be described below. The disturbances are as follows: at time zero a three-line-to-ground fault occurs in line 5–7 at the node 7 side, the fault is cleared five cycles (about 0.08333 s) later by the removal of line 5–7.

Generators are modeled as constant E_q', loads are modeled as impedances, the network is modeled by admittance matrix, the differential equations are solved by the modified Euler's method, and the network equations are solved by a direct method.

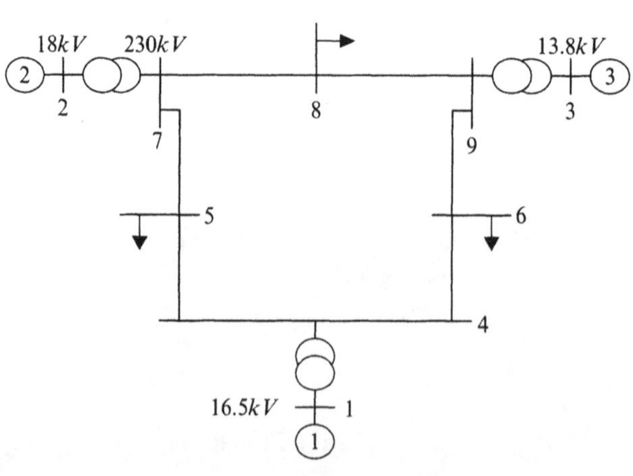

Fig. 7.12 Single-line diagram of 9-bus system

Table 7.5 Branch data

From-end bus	To-end bus	Resistance (in per unit)	Reactance (in per unit)	Half of the admittance (in per unit)	Non-standard ratio of transformer
4	5	0.010	0.085	0.088	
4	6	0.017	0.092	0.079	
5	7	0.032	0.161	0.153	
6	9	0.039	0.170	0.179	
7	8	0.0085	0.072	0.0745	
8	9	0.0119	0.1008	0.1045	
1	4	0.0	0.0576		1.0
2	7	0.0	0.0625		1.0
3	9	0.0	0.0586		1.0

Table 7.6 Generator data

Generator	Bus	T_J	R_a	X_d	X'_d	X_q	X'_q	T'_{d0}	T'_{q0}	D
1	1	47.28	0.0	0.1460	0.0608	0.0969	0.0969	8.96		0.0
2	2	12.80	0.0	0.8958	0.1198	0.8645	0.1969	6.00	0.535	0.0
3	3	6.02	0.0	1.3125	0.1813	1.2578	0.2500	5.89	0.600	0.0

The units for all time constants are "seconds," the units of all damping coefficients D, resistances and impedances are in "per unit"

Table 7.7 Load flow under normal system operation

Bus	Voltage		Generator		Load	
	Magnitude	Phase angle (degree)	Active power	Reactive power	Active power	Reactive power
1	1.040	0.0000	0.7164	0.2705		
2	1.0250	9.2800	1.6300	0.0665		
3	1.0250	4.6648	0.8500	−0.1086		
4	1.0258	−2.2168				
5	0.9956	−3.9888			1.2500	0.5000
6	1.0127	−3.6874			0.9000	0.3000
7	1.0258	3.7197				
8	1.0159	0.7275			1.0000	0.3500
9	1.0324	1.9667				

Based on the general procedure described in Fig. 7.8 and the method described in the previous section, the transient stability analysis can be summarized below:

1. *Initial value computation*: Compute the equivalent shunt admittances of loads according to (7.81), and the results are as follows:

   Load (node 5): $1.26099 - j0.50440$
   Load (node 6): $0.87765 - j0.29255$
   Load (node 8): $0.96898 - j0.33914$

   Then compute, based on (7.74)–(7.80), generator transient voltage E'_q, initial rotor angle $\delta_{(0)}$, and mechanical power $P_{m(0)}$. The results are in Table 7.8. The

Table 7.8 E'_q, $\delta_{(0)}$, and $P_{m(0)}$ of generators

Generator	Neglecting the effect of salient poles		With the effect of salient poles		$P_{m(0)}$
	E'_q	$\delta_{(0)}$	E'_q	$\delta_{(0)}$	
1	1.05664	2.27165	1.05636	3.58572	0.71641
2	1.05020	19.73159	0.78817	61.09844	1.63000
3	1.01697	13.16641	0.76786	54.13662	0.85000

initial values of generator rotor angles are set to $\omega_{1(0)} = \omega_{2(0)} = \omega_{3(0)} = 1$. In the calculations to be described below, the effect of generator salient poles is neglected, which is to say $E'_q = C$ and $X_q = X'_d$. This is the classical model of generators.

2. *The fault-on system and post-fault system model*: In the fault-on system, a shunt branch with zero impedance is connected at node 7, to model this shunt branch, the diagonal element Y_{77} of the admittance matrix $\mathbf{Y}$ is set to a very high value (say 10^{20}). The admittance matrix of the fault-on system is $\mathbf{Y}_F$.

   In the postfault network, branch 5–7 is removed. Because the contribution of line 5–7 to admittance matrix is equal to:

$$
\begin{array}{rc}
 & \begin{array}{ccccc} \qquad & \qquad 5 \qquad & \qquad & \qquad 7 \qquad & \qquad \end{array} \\
\mathbf{Y}_{l(5-7)} = \begin{array}{c} \\ 5 \\ \\ 7 \\ \\ \end{array} & \begin{bmatrix} \vdots & & \vdots & & \vdots \\ \cdots & \dfrac{1}{r+jx}+jb & \cdots & -\dfrac{1}{r+jx} & \cdots \\ \vdots & \vdots & \vdots & \vdots & \vdots \\ \cdots & -\dfrac{1}{r+jx} & \cdots & \dfrac{1}{r+jx}+jb & \cdots \\ & \vdots & \vdots & \vdots & \end{bmatrix}
\end{array},
$$

   where $r = 0.032$, $x = 0.161$, and $b = 0.153$. Thus the postfault admittance matrix is $\mathbf{Y}_P = \mathbf{Y} - \mathbf{Y}_{l(5-7)}$.

3. *Integrating the differential-algebraic equations*: We will only compute the transient duration from the instant the fault occurs to time equals 2 s. Thus the system for the duration 0–2 s is divided into two autonomous systems: that is, the fault-on system for duration 0–0.08333 s, and the postfault system for duration 0.08333–2 s. The step size for numerical integration is 0.001 s. Table 7.9 lists the rotor angles $\delta(t)$ and relative maximum rotor angles, with and without consideration of the effect of salient pole. The later is also depicted in Fig. 7.13.

   From Fig. 7.13, observe that the system is stable, whether or not the salient pole effect is taken into consideration. When the salient pole effect is considered, the maximum relative rotor angle is $\delta_{21} = 151.48396°$ ($t = 0.80133$s). When the

Table 7.9 Generator $\delta(t)$ and relative maximum rotor angles

	Without consideration of salient pole effects			With consideration of salient pole effects			$\bar{\delta}_2 - \bar{\delta}_1$	$\tilde{\delta}_2 - \tilde{\delta}_1$
	$\bar{\delta}_1$	$\bar{\delta}_2$	$\bar{\delta}_3$	$\tilde{\delta}_1$	$\tilde{\delta}_2$	$\tilde{\delta}_3$		
0.00000	2.27165	19.73159	13.16641	3.58572	61.09844	54.13662	17.45994	57.51272
0.04200	2.28779	22.15764	14.63856	3.69016	63.52449	55.31617	19.86985	59.83433
0.08333	2.34848	29.28237	18.86248	4.00270	70.64922	58.74084	26.93389	66.64652
0.13333	2.40803	41.21540	25.92757	4.48409	82.69359	64.79619	38.80737	78.20951
0.18333	2.58251	53.48395	33.68320	5.09318	95.37520	72.05804	50.90144	90.28202
0.23333	3.19401	65.30378	41.94182	6.11234	108.03487	80.45556	62.10977	101.92253
0.28333	4.52397	76.12258	50.42907	7.80288	120.26523	89.78211	71.59862	112.46235
0.33333	6.79447	85.62591	58.80933	10.37892	131.92171	99.74929	78.83143	121.54280
0.38333	10.16415	93.68682	66.72892	13.99684	143.06625	110.07065	83.52267	129.06941
0.43333	14.73304	100.29635	73.85628	18.75788	153.88636	120.53143	85.56332	135.12847
0.48333	20.54980	105.50168	79.91656	24.71833	164.62086	131.02170	84.95188	139.90254
0.53333	27.61528	109.36539	84.73139	31.90242	175.50595	141.53491	81.75011	143.60353
0.58333	35.87927	111.95248	88.27281	40.31440	186.74309	152.14627	76.07321	146.42869
0.63333	45.23062	113.34805	90.72587	49.94781	198.48469	162.98542	68.11743	148.53688
0.68333	55.48563	113.70536	92.52996	60.79180	210.83172	174.21138	58.21973	150.03992
0.73333	66.38413	113.31385	94.35019	72.83490	223.83807	185.99198	46.92972	151.00317
0.78333	77.60323	112.66232	96.94304	86.06717	237.51766	198.48689	35.05909	151.45049
0.83333	88.79177	112.45974	100.94170	100.48127	251.85143	211.83293	23.66797	151.37016
0.88333	99.61769	113.57983	106.66335	116.07313	266.79269	226.13087	13.96214	150.71956
0.93333	109.81335	116.92253	114.05272	132.84229	282.27022	241.43372	7.10918	149.42793
0.98333	119.20630	123.22495	122.79557	150.79173	298.18912	257.73701	4.01865	147.39740
1.03333	127.73265	132.88223	132.52906	169.92694	314.43044	274.97226	5.14958	144.50351
1.08333	135.43743	145.83696	143.03225	190.25360	330.85111	293.00522	10.39953	140.59752
1.13333	142.46732	161.57457	154.30791	211.77290	347.28707	311.64166	19.10725	135.51417
1.18333	149.05619	179.23461	166.52345	234.47339	363.56384	330.64456	30.17842	129.09045
1.23333	155.50082	197.81063	179.84443	258.31838	379.52044	349.76706	42.30981	121.20206

(*continued*)

Table 7.9 (continued)

	Without consideration of salient pole effects			With consideration of salient pole effects			$\bar{\delta}_2 - \bar{\delta}_1$	$\tilde{\delta}_2 - \tilde{\delta}_1$
	$\bar{\delta}_1$	$\bar{\delta}_2$	$\bar{\delta}_3$	$\tilde{\delta}_1$	$\tilde{\delta}_2$	$\tilde{\delta}_3$		
1.28333	162.12683	216.36810	194.25664	283.22935	395.05209	368.80340	54.24126	111.82274
1.33333	169.25129	234.19924	209.48217	309.06874	410.17107	387.65294	64.94795	101.10233
1.38333	177.15096	250.87411	225.02567	335.62983	425.07106	406.38036	73.72315	89.44123
1.43333	186.04298	266.20363	240.30096	362.64368	440.16296	425.24462	80.16065	77.51928
1.48333	196.07909	280.15828	254.76080	389.80754	456.04799	444.67239	84.07919	66.24045
1.53333	207.34945	292.78226	267.98613	416.82762	473.42157	465.17298	85.43281	56.59395
1.58333	219.88920	304.12763	279.73658	443.45984	492.94327	487.22074	84.23842	49.48343
1.63333	233.68131	314.22084	289.98411	469.53546	515.12417	511.14408	80.53953	45.58871
1.68333	248.65233	323.06766	298.94296	494.97070	540.25876	537.05069	74.41533	45.28806
1.73333	264.66308	330.69730	307.08305	519.76777	568.39552	564.80111	66.03422	48.62775
1.78333	281.50116	337.23926	315.08928	544.01356	599.32930	594.03622	55.73810	55.31574
1.83333	298.88496	343.01311	323.73003	567.87632	632.61227	624.26498	44.12814	64.73595
1.88333	316.48551	348.59530	333.64692	591.59440	667.60021	655.00155	32.10979	76.00581
1.93333	333.96386	354.81755	345.15535	615.45111	703.55468	685.90126	20.85369	88.10357
1.98333	351.01413	362.66346	358.17917	639.73684	739.78895	716.82750	11.64932	100.05211

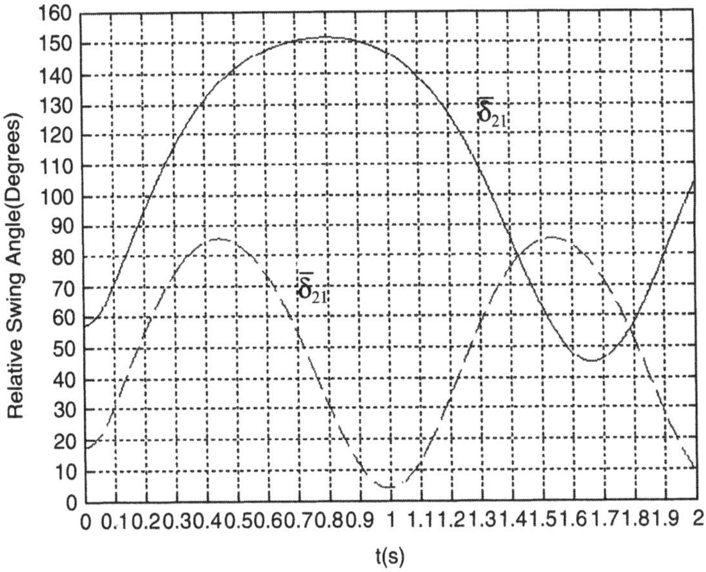

Fig. 7.13 Relative rotor angles as functions of time

salient pole effect is not considered, the maximum relative rotor angle is $\delta_{21} = 85.65788°$ ($t = 0.44633$ s), and the angle of the second swing $\delta_{21} = 85.43378°$ ($t = 1.53433$ s), which is smaller than that of the first swing.

Lastly, the critical clearing times, with and without consideration of salient pole effects, are calculated. It turns out that the critical clearing time under the former circumstance is between 0.162 and 0.163 s, and it is between 0.085 and 0.086 s under the second condition. The swing curves under these circumstances are provided in Figs. 7.14 and 7.15, respectively.

7.4.4 Numerical Integration Methods for Transient Stability Analysis Under Classical Model

In a modern energy management system (EMS), to assess system security, transient stability under various prespecified contingencies is predicted online within a limited amount of time. Because the number of contingencies is quite large, to meet the requirement of online assessment, each transient stability analysis must be completed rapidly. Obviously the traditional integration methods for transient stability analysis are no longer suitable because of the limitation of speed, fast methods customized for online applications need to be developed.

Dynamic security assessment is one of the "hot" areas in the power system stability field. As early as 1983, IEEE established a transient stability analysis working group, the responsibility of which was to lead and review the research in

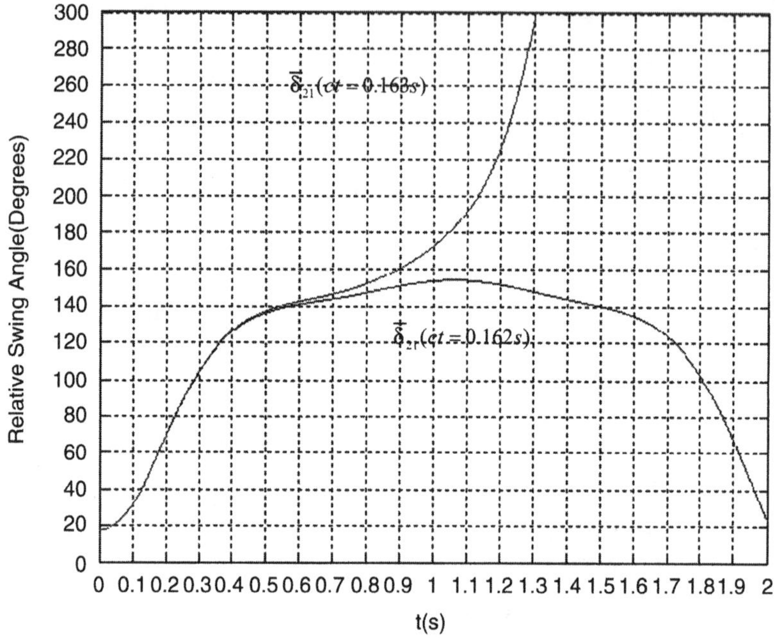

Fig. 7.14 Relative rotor angle near critical clearing time (salient pole effects not considered)

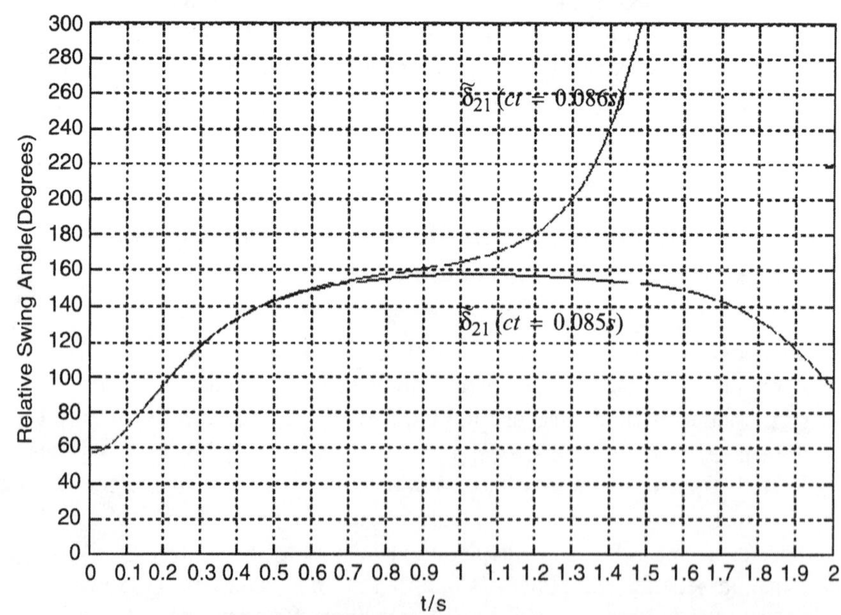

Fig. 7.15 Relative rotor angle near critical clearing time (salient pole effects considered)

this area. A dynamic security assessment method must be extra-rapid, especially in an online environment. Though the requirement on computational precision can be relaxed to some extent, reliability and robustness are still required. Currently, the methods for improving the speed of online dynamic security assessment include no more than: first, simplifying the mathematical model for stability analysis; and second, developing rapid algorithms for stability analysis. In what follows we introduce a rapid algorithm for transient stability analysis based on a classical model.

1. *The classical model of power system stability*: The "classical model" places the following assumptions on the mathematical model:

   (1) Assume that the generator mechanical power remains constant during the transient stability period, and neglect the effects of damper windings
   (2) Assume that generator transient voltage E' does not change during the transient stability study period, and furthermore the phase angle of this voltage is equal to the rotor angle
   (3) Loads are modeled as constant impedances

   With the above assumptions, the motion equation of the ith generator is obtained as

$$\left.\begin{aligned}\frac{\mathrm{d}\delta_i}{\mathrm{d}t} &= \omega_s(\omega - 1)\\ \frac{\mathrm{d}\omega_i}{\mathrm{d}t} &= \frac{1}{T_{Ji}}(P_{mi} - P_{ei})\end{aligned}\right\} i = 1, 2, \ldots, m. \tag{7.95}$$

   From a load flow study one finds the transient voltage:

$$\dot{E}_i' = E_i' \angle \delta_{i(0)} = \dot{V}_{i(0)} + (R_{ai} + jX_{di}')\frac{P_{i(0)} - jQ_{i(0)}}{\hat{\dot{V}}_{i(0)}}. \tag{7.96}$$

   And the normal operation conditions give

$$\omega_{i(0)} = 1. \tag{7.97}$$

   Based on (7.42), (7.45), and (7.46) in Sect. 7.3.1, which describe the generator–network relationship, we can incorporate the pseudoadmittances of generators (see (7.45)) and load equivalent admittances (see (7.81)) into the network. The diagonal elements of the admittance matrix of the network (7.35) should have the generator pseudoadmittances or load equivalent admittances added. The right-hand side vector, as given by (7.46), contains nonzeros in rows corresponding to generator nodes only, the other rows are zero.

The expression for electromagnetic power of generators is easily obtained as

$$P_{ei} = \mathrm{Re}\left(\dot{E}'_i \frac{\hat{E}'_i - \hat{V}_i}{R_{ai} - jX'_{di}}\right). \tag{7.98}$$

2. *Solving for the network equations*: First perform triangular factorization on the admittance matrix $\mathbf{Y}$ (a symmetrical matrix):

$$\mathbf{Y} = \mathbf{U}^{\mathrm{T}}\mathbf{D}\mathbf{U}, \tag{7.99}$$

where $\mathbf{U}$ is an upper triangular matrix, $\mathbf{D}$ is a matrix with nonzeros on the diagonal only. After performing the following forward substitution and backward substitution, one obtains the voltage:

$$\mathbf{F} = \mathbf{D}^{-1}\mathbf{U}^{-\mathrm{T}}\mathbf{I}, \tag{7.100}$$

$$\mathbf{V} = \mathbf{U}^{-1}\mathbf{F}. \tag{7.101}$$

Vector $\mathbf{I}$ in the above equations is a sparse vector. To compute the electromagnetic powers of generators, it is only necessary to know the generator voltages. Thus the unknown vector $\mathbf{V}$ is also sparse. Therefore, the network equations can be solved using rapid forward and backward substitutions. The experience of medium size computations demonstrates that, some 1/3 computational effort can be saved if a sparse vector method is used to perform rapid forward and back substitution.

When solving the network equations with a sparse vector method, the majority of the time is spent in factorizing the admittance matrix. In dynamic security assessment, the admittance matrices for fault-on and postfault networks under different contingency scenarios are different. If these admittance matrices are factored every time, it would take a large amount of computer time. However, in general, the fault-on and postfault network admittance matrices differ from the prefault network admittance matrix only in a few places. This allows the utilization of compensation methods for solving network equations. The idea of a compensation method is to avoid matrix refactorizations, thus computational burden can be greatly relieved.

Consider network equation:

$$(\mathbf{Y} + \Delta\mathbf{Y})\mathbf{V} = \mathbf{I}, \tag{7.102}$$

where $\mathbf{Y}$ is the prefault network admittance matrix, and $\Delta\mathbf{Y}$ is the adjustment to $\mathbf{Y}$ due to network switching or a fault and can be represented as

$$\Delta\mathbf{Y} = \mathbf{M}\delta\mathbf{y}\mathbf{M}^{\mathrm{T}}, \tag{7.103}$$

where $\delta\mathbf{y}$ is a $(q \times q)$ matrix, including the information for the adjustment to $\mathbf{Y}$, q is in general of order 1 or 2, and $\mathbf{M}$ is a $(n \times q)$ coincidence matrix related to the specific fault or switch.

By the matrix inversion lemma, (7.102) and (7.103) become

$$\mathbf{V} = (\mathbf{Y}^{-1} - \mathbf{Y}^{-1}\mathbf{M}\mathbf{C}\mathbf{M}^{\mathrm{T}}\mathbf{Y}^{-1})\mathbf{I}, \tag{7.104}$$

where the $(q \times q)$ matrix $\mathbf{C}$ is equal to

$$\mathbf{C} = [(\delta\mathbf{y})^{-1} + \mathbf{Z}]^{-1}, \tag{7.105}$$

while the $(q \times q)$ matrix $\mathbf{Z}$ is

$$\mathbf{Z} = \mathbf{M}^{\mathrm{T}}\mathbf{Y}^{-1}\mathbf{M}. \tag{7.106}$$

Thus according to (7.104), taking into account (7.99), the computational steps for solving the network (7.102) using compensation are as follows:

Preparatory calculation:

$$\left.\begin{aligned} &(1) \quad \mathbf{W} = \mathbf{U}^{-\mathrm{T}}\mathbf{M} \\ &(2) \quad \tilde{\mathbf{W}} = \mathbf{D}^{-1}\mathbf{W} \\ &(3) \quad \mathbf{Z} = \tilde{\mathbf{W}}^{\mathrm{T}}\mathbf{W} \\ &(4) \quad \mathbf{C} = [(\delta\mathbf{y})^{-1} + \mathbf{Z}]^{-1} \end{aligned}\right\}. \tag{7.107}$$

Solving the network equations:

$$\left.\begin{aligned} &(1) \quad \tilde{\mathbf{F}} = \mathbf{U}^{-\mathrm{T}}\mathbf{I} \\ &(2) \quad \Delta\mathbf{F} = -\mathbf{W}\mathbf{C}\tilde{\mathbf{F}}^{\mathrm{T}}\tilde{\mathbf{F}} \\ &(3) \quad \mathbf{F} = \tilde{\mathbf{F}} + \Delta\mathbf{F} \\ &(4) \quad \mathbf{V} = \mathbf{U}^{-1}\mathbf{D}^{-1}\mathbf{F} \end{aligned}\right\}. \tag{7.108}$$

The forward and backward substitutions in (7.107) and (7.108) are all completed using a sparse vector method.

3. *A numerical integration algorithm for second-order conservative systems*: The differential equations in (7.95) can be rewritten in the following compact form:

$$\frac{d^2\boldsymbol{\delta}}{dt^2} = \mathbf{f}(\delta), \tag{7.109}$$

where

$$\left.\begin{aligned} \boldsymbol{\delta} &= [\delta_1, \ldots, \delta_m]^T \\ \mathbf{f}(\boldsymbol{\delta}) &= [f_1(\boldsymbol{\delta}), \ldots, f_m(\boldsymbol{\delta})]^T \\ f_i(\boldsymbol{\delta}) &= \frac{\omega_s}{T_{Ji}}(P_{mi} - P_{ei}) \end{aligned}\right\}. \tag{7.110}$$

The right-hand side functions of the differential equations in (7.109) do not contain arguments with first-order derivatives, the equations are thus termed a second-order conservative system. Compared with solving two first-order equations, the equations can be solved by direct differencing which results in an efficiency one level higher. Consider the Stormer and Numerov integration formula [186]:

$$\boldsymbol{\delta}_{k+2} = 2\boldsymbol{\delta}_{k+1} - \boldsymbol{\delta}_k + h^2\mathbf{f}(\boldsymbol{\delta}_{k+1}), \tag{7.111}$$

$$\boldsymbol{\delta}_{k+2} = 2\boldsymbol{\delta}_{k+1} - \boldsymbol{\delta}_k + \frac{h^2}{12}[\mathbf{f}(\boldsymbol{\delta}_{k+2}) + 10\mathbf{f}(\boldsymbol{\delta}_{k+1}) + \mathbf{f}(\boldsymbol{\delta}_k)]. \tag{7.112}$$

Equation (7.111) is an explicit second-order method, while (7.112) is an implicit forth-order method. To solve the differential (7.109) based on (7.111) requires a smaller step size because of the poor numerical stability. To solve (7.109) based on (7.112) allows a larger step size because the method has higher order and has a larger region of absolute stability. However this method still takes a large amount of computational effort because it involves solving a set of nonlinear simultaneous equations. On the other hand, if good initial estimates $\boldsymbol{\delta}_{k+2}^{[0]}$ are provided when solving (7.112), the convergence can be speeded up. This suggests the application of a predictor–corrector method for solving (7.109); specifically, the explicit method (7.111) is adopted for the predictor, while the implicit method (7.112) is adopted for the corrector.

Let P and C represent the application of one predictor and one corrector, E represent computing function $\mathbf{f}(\boldsymbol{\delta})$ once, the pair of predictor–corrector is formed as PECE. More concretely, one computes $\boldsymbol{\delta}_{k+2}^{[0]}$ based on the predictor, and calculates $\mathbf{f}_{k+2}^{[0]} = \mathbf{f}(\boldsymbol{\delta}_{k+2}^{[0]})$, substitute the result into the corrector to obtain $\boldsymbol{\delta}_{k+2}^{[1]}$, and finally compute $\mathbf{f}_{k+2}^{[1]} = \mathbf{f}(\boldsymbol{\delta}_{k+2}^{[1]})$.

The above method falls into the category of multistep methods, the procedure can be started using the following special fourth-order Runge–Kutta formula [185]:

$$\left.\begin{aligned}
\boldsymbol{\delta}_{k+1} &= \boldsymbol{\delta}_k + h\boldsymbol{\delta}'_k + \frac{h^2}{6}(\mathbf{k}_1 + 2\mathbf{k}_2) \\
\boldsymbol{\delta}'_{k+1} &= \boldsymbol{\delta}'_k + \frac{h}{6}(\mathbf{k}_1 + 4\mathbf{k}_2 + \mathbf{k}_3) \\
\mathbf{k}_1 &= \mathbf{f}(\boldsymbol{\delta}_k) \\
\mathbf{k}_2 &= \mathbf{f}\left(\boldsymbol{\delta}_k + \frac{h}{2}\boldsymbol{\delta}'_k + \frac{h^2}{8}\mathbf{k}_1\right) \\
\mathbf{k}_3 &= \mathbf{f}\left(\boldsymbol{\delta}_k + h\boldsymbol{\delta}'_k + \frac{h^2}{2}\mathbf{k}_2\right)
\end{aligned}\right\}. \tag{7.113}$$

The classical model for transient stability analysis applies to "first swing" (about 1.5 s after the disturbance). This model is free from the stiffness problem and therefore permits the use of larger step size (0.1–0.2 s).

7.5 Transient Stability Analysis with FACTS Devices

To study in detail the transient stability of a large scale interconnected power system experiencing various large disturbances and to analyze the effects of control devices on system stability, often for the purpose of seeking mechanisms for improving stability, a detailed component model for transient stability analysis is required.

As the technology of HVDC develops, HVDC systems are widely used in long-distance transmission, under-sea cable transmission and system interconnection. The technology of flexible AC transmission (FACTS), matured only in recent years, is also receiving much acceptance from the industry. FACTS devices not only help to improve system steady-state performance, they also improve the dynamic performance of power systems to a certain degree, as a result system transfer capabilities are enlarged considerably. The dynamic performance of a power system is also affected by generator prime movers and speed-governing systems, excitation systems, PSSs, and other control devices. A power system with increasing scale, and increasing installations of dynamic devices, exhibits complex behavior after it experiences a disturbance. The mechanical–electrical interaction of such a system lasts longer, and the duration of oscillation of the system before loss of stability occurs can be as long as several seconds to a dozen seconds.

This section introduces the basic transient stability analysis method for large-scale interconnected power systems with many dynamic devices which are modeled in detail. It should be noted that the material presented does not address the detailed implementation of a commercial code, rather it concerns the basic principles.

The mathematical models for the dynamic devices are as follows: synchronous machines which are modeled by a sixth-order model with varying E'_q, E''_q, E'_d, E''_d, and rotor variables; hydroprime mover and their speed-governing system; excitation systems with thyristor-based DC excitors, PSSs with generator speed deviation as input; two-terminal HVDC; SVC and TCSC of the FACTS family; constant impedance loads or loads with second-order voltage characteristics. If a different model other than those described above is used for a component, the same principle applies.

The above large scale dynamic system is a typical stiff system because of the existence of dynamic devices with drastically different time constants. To solve such systems with an explicit numerical method, a very small step size has to be assumed because the stability region is relatively small. The implicit trapezoidal rule is a second-order algorithm with the left half plane being the stability region, therefore it allows for the use of a larger step size. In early commercial codes explicit methods such as forth-order Runge–Kutta method were quite popular. Because of their better numerical properties, adaptability to stiffness, and the introduction of fast control schemes with small time constants, the second-order trapezoidal rule has become almost an industry standard since the 1970s. Many commercial grade codes, for example, the transient stability analysis package developed by Bonneville power administration (BPA), the power system analysis software package, are based on this method. In a typical transient stability analysis, the trapezoidal rule with constant step size, between 0.01 and 0.02 s (or even longer), is assumed. The difference and algebraic equations are solved by a simultaneous method or a sequential method.

In the large-scale transient stability analysis procedure to be presented below, the implicit trapezoidal rule is used to solve the differential equations, while a Newton method is used to solve the simultaneous difference-algebraic equations of the detailed system model.

7.5.1 Initial Values and Difference Equations of Generators

7.5.1.1. Generators

The mathematical model of a synchronous machine comprises rotor motion equations, rotor electromagnetic equations, etc., together with stator voltage equations and the expressions for electromagnetic powers. Based on (6.1)–(6.4), these equations can be rewritten as follows:

Rotor motion equations:

$$\left.\begin{aligned} \frac{\mathrm{d}\delta}{\mathrm{d}t} &= \omega_s(\omega - 1) \\ \frac{\mathrm{d}\omega}{\mathrm{d}t} &= \frac{1}{T_J}(P_m - P_e - D\omega) \end{aligned}\right\}. \tag{7.114}$$

Rotor electromagnetic equations:

$$\left.\begin{aligned}\frac{dE'_q}{dt} &= \frac{1}{T'_{d0}}[E_{fq} - k_d E'_q + (k_d - 1)E''_q]\\ \frac{dE''_q}{dt} &= \frac{1}{T''_{d0}}[E'_q - E''_q - (X'_d - X''_d)I_d]\\ \frac{dE'_d}{dt} &= \frac{1}{T'_{q0}}[-k_q E'_d + (k_q - 1)E''_d]\\ \frac{dE''_d}{dt} &= \frac{1}{T''_{q0}}[E'_d - E''_d + (X'_q - X''_q)I_q]\end{aligned}\right\}, \tag{7.115}$$

where $k_d = \dfrac{X_d - X''_d}{X'_d - X''_d}$ and $k_q = \dfrac{X_q - X''_q}{X'_q - X''_q}$

Stator voltage equations:

$$\left.\begin{aligned}V_d &= E''_d - R_a I_d + X''_q I_q\\ V_q &= E''_q - X''_d I_d - R_a I_q\end{aligned}\right\}. \tag{7.116}$$

The electrical power is equal to the output power plus stator copper loss:

$$P_e = P_{\text{out}} + |\dot{I}|^2 R_a = V_x I_x + V_y I_y + (I_x^2 + I_y^2)R_a. \tag{7.117}$$

Given a load flow solution, some of the initial values of generators can be computed based on (7.74)–(7.78). Note that the current flows in damper windings under steady-state operation are equal to zero, based on (6.60), (6.64), and (6.65), the initial values of generator no-load synchronous voltages, transient voltages, and sub-transient voltages can be easily obtained as

$$E_{fq(0)} = V_{q(0)} + R_a I_{q(0)} + X_d I_{d(0)}, \tag{7.118}$$

$$\left.\begin{aligned}E'_{q(0)} &= V_{q(0)} + R_a I_{q(0)} + X'_d I_{d(0)}\\ E'_{d(0)} &= V_{d(0)} + R_a I_{d(0)} - X'_q I_{q(0)}\end{aligned}\right\}, \tag{7.119}$$

$$\left.\begin{aligned}E''_{q(0)} &= V_{q(0)} + R_a I_{q(0)} + X''_d I_{d(0)}\\ E''_{d(0)} &= V_{d(0)} + R_a I_{d(0)} - X''_q I_{q(0)}\end{aligned}\right\}. \tag{7.120}$$

Besides, the electrical power $P_{e(0)}$ of generators under steady-state operation can be computed directly from (7.117):

$$P_{e(0)} = P_{(0)} + (I_{x(0)}^2 + I_{y(0)}^2)R_a. \tag{7.121}$$

Set $\frac{d\omega}{dt} = 0$ in (7.114), the prime mover outputs $P_{m(0)}$ are equal to

$$P_{m(0)} = P_{e(0)} + D. \tag{7.122}$$

To solve the difference equations we first apply the trapezoidal rule for the rotor motion (7.114),

$$\delta_{(t+\Delta t)} = \delta_{(t)} + \frac{\omega_s \Delta t}{2}(\omega_{(t+\Delta t)} + \omega_{(t)} - 2), \tag{7.123}$$

$$\omega_{(t+\Delta t)} = \omega_{(t)} + \frac{\Delta t}{2T_J}(P_{m(t+\Delta t)} - P_{e(t+\Delta t)} - D\omega_{(t+\Delta t)} + P_{m(t)} - P_{e(t)} - D\omega_{(t)}). \tag{7.124}$$

From (7.124) one obtains the expression for $\omega_{(t+\Delta t)}$, substituting this into (7.123), it follows that:

$$\delta_{(t+\Delta t)} = \alpha_J(P_{m(t+\Delta t)} - P_{e(t+\Delta t)}) + \delta_0, \tag{7.125}$$

where

$$\alpha_J = \frac{\omega_s (\Delta t)^2}{4T_J + 2D\Delta t}, \tag{7.126}$$

$$\delta_0 = \delta_{(t)} + \alpha_J\left(P_{m(t)} - P_{e(t)} + \frac{4T_J}{\Delta t}\omega_{(t)}\right) - \omega_s \Delta t. \tag{7.127}$$

In (7.126), α_J is a function of step size Δt and some other constants. If a fixed step size is assumed, it becomes a constant. As for δ_0 in (7.127), it is a constant only in difference equation (7.125), it takes different values in each computational step.

After $\delta_{(t+\Delta t)}$ is found, $\omega_{(t+\Delta t)}$ is calculated based on (7.123):

$$\omega_{(t+\Delta t)} = \frac{2}{\omega_s \Delta t}(\delta_{(t+\Delta t)} - \delta_{(t)}) - \omega_{(t)} + 2. \tag{7.128}$$

Now applying the trapezoidal rule to the electromagnetic equation (7.115), it follows:

$$\left.\begin{aligned} E'_{q(t+\Delta t)} &= E'_{q(t)} + \frac{\Delta t}{2T'_{d0}}[E_{fq(t+\Delta t)} - k_d E'_{q(t+\Delta t)} + (k_d - 1)E''_{q(t+\Delta t)} \\ &\quad + E_{fq(t)} - k_d E'_{q(t)} + (k_d - 1)E''_{q(t)}] \\ E''_{q(t+\Delta t)} &= E''_{q(t)} + \frac{\Delta t}{2T''_{d0}}[E'_{q(t+\Delta t)} - E''_{q(t+\Delta t)} - (X'_d - X''_d)I_{d(t+\Delta t)} \\ &\quad + E'_{q(t)} - E''_{q(t)} - (X'_d - X''_d)I_{d(t)}] \end{aligned}\right\}, \tag{7.129}$$

$$\left.\begin{aligned} E'_{d(t+\Delta t)} &= E'_{d(t)} + \frac{\Delta t}{2T'_{q0}}[-k_q E'_{d(t+\Delta t)} + (k_q-1)E''_{d(t+\Delta t)} \\ &\quad - k_q E'_{d(t)} + (k_q-1)E''_{d(t)}] \\ E''_{d(t+\Delta t)} &= E''_{d(t)} + \frac{\Delta t}{2T''_{q0}}[E'_{d(t+\Delta t)} - E''_{d(t+\Delta t)} + (X'_q - X''_q)I_{q(t+\Delta t)} \\ &\quad + E'_{d(t)} - E''_{d(t)} + (X'_q - X''_q)I_{q(t)}] \end{aligned}\right\}. \tag{7.130}$$

Eliminating variables $E'_{q(t+\Delta t)}$ and $E'_{d(t+\Delta t)}$ in (7.129) and (7.130), we have

$$E''_{q(t+\Delta t)} = -\alpha''_d(X'_d - X''_d)I_{d(t+\Delta t)} + \alpha''_d\alpha_{d1}E_{fq(t+\Delta t)} + E''_{q0}, \tag{7.131}$$

$$E''_{d(t+\Delta t)} = \alpha''_q(X'_q - X''_q)I_{q(t+\Delta t)} + E''_{d0}, \tag{7.132}$$

where

$$\left.\begin{aligned} E''_{q0} &= \alpha''_d\Big\{\alpha_{d1}E_{fq(t)} - (X'_d - X''_d)I_{d(t)} + 2(1-k_d\alpha_{d1})E'_{q(t)} \\ &\quad + \left[\alpha_{d1}(k_d-1) + \frac{1}{\alpha_{d2}} - 2\right]E''_{q(t)}\Big\} \\ E''_{d0} &= \alpha''_q\Big\{(X'_q - X''_q)I_{q(t)} + 2(1-k_q\alpha_{q1})E'_{d(t)} \\ &\quad + \left[\alpha_{q1}(k_q-1) + \frac{1}{\alpha_{q2}} - 2\right]E''_{d(t)}\Big\} \end{aligned}\right\}, \tag{7.133}$$

$$\left.\begin{aligned} &\alpha_{d1} = \frac{\Delta t}{2T'_{d0} + k_d\Delta t}, \quad \alpha_{q1} = \frac{\Delta t}{2T'_{q0} + k_q\Delta t} \\ &\alpha_{d2} = \frac{\Delta t}{2T''_{d0} + \Delta t}, \quad \alpha_{d2} = \frac{\Delta t}{2T''_{d0} + \Delta t} \\ &\alpha''_d = [\alpha_{d1}(1-k_d) + 1/\alpha_{d2}]^{-1}, \quad \alpha''_q = [\alpha_{q1}(1-k_q) + 1/\alpha_{q2}]^{-1} \end{aligned}\right\}. \tag{7.134}$$

The coefficients α_{d1}, α_{d2}, α''_d, α_{q1}, α_{q2}, and α''_q in (7.134) are all constants if a fixed step size Δt is assumed, while in (7.133), E''_{q0} and E''_{d0} are known quantities at step t, although they take different values in each step.

After $E''_{q(t+\Delta t)}$ and $E''_{d(t+\Delta t)}$ are calculated, now based on (7.129) and (7.130), $E'_{q(t+\Delta t)}$ and $E'_{d(t+\Delta t)}$ can be obtained by

$$\left.\begin{aligned} E'_{q(t+\Delta t)} &= \alpha_{d1}\left[\frac{2T'_{do} - k_d\Delta t}{\Delta t}E'_{q(t)} + E_{fq(t+\Delta t)} + E_{fq(t)} + (k_d-1)(E''_{q(t+\Delta t)} + E''_{q(t)})\right] \\ E'_{d(t+\Delta t)} &= \alpha_{q1}\left[\frac{2T'_{qo} - k_q\Delta t}{\Delta t}E'_{d(t)} + (k_q-1)(E''_{d(t+\Delta t)} + E''_{d(t)})\right] \end{aligned}\right\}. \tag{7.135}$$

7.5.1.2 Exitation System and PSS

Taking the DC exitor with thyristor-based regulator, illustrated in Fig. 6.16, as an example, let us derive the differential-algebraic equations based on the transfer function diagram. We will neglect the effects of R_{C}, and equivalent time constants T_{B} and T_{C} of the analog regulator. Under the "one per unit exitation voltage/one per stator voltage" system, by (6.51) it follows that $V_f = E_{fq}$.

The measurement and filter system

$$\frac{\mathrm{d}V_M}{\mathrm{d}t} = \frac{1}{T_{\mathrm{R}}}(V_{\mathrm{C}} - V_M), \quad V_{\mathrm{C}} = \left|\dot{V} + jX_{\mathrm{C}}\dot{I}\right| \tag{7.136}$$

The transient droop feedback:

$$\frac{\mathrm{d}(K_{\mathrm{F}}E_{fq} - T_{\mathrm{F}}V_{\mathrm{F}})}{\mathrm{d}t} = V_{\mathrm{F}} \tag{7.137}$$

The amplifier:

$$\left.\begin{array}{l} f = \dfrac{1}{T_{\mathrm{A}}}[K_{\mathrm{A}}(V_{\mathrm{REF}} + V_{\mathrm{S}} - V_M - V_{\mathrm{F}}) - V_{\mathrm{R}}] \\ \text{if } V_{\mathrm{R}} = V_{\mathrm{RMAX}} \quad \text{and} \quad f > 0, \quad \dfrac{\mathrm{d}V_{\mathrm{R}}}{\mathrm{d}t} = 0, \quad V_{\mathrm{R}} = V_{\mathrm{RMAX}} \\ \text{if } V_{\mathrm{RMIN}} < V_{\mathrm{R}} < V_{\mathrm{RMAX}}, \quad \dfrac{\mathrm{d}V_{\mathrm{R}}}{\mathrm{d}t} = f \\ \text{if } V_{\mathrm{R}} = V_{\mathrm{RMIN}} \quad \text{and} \quad f < 0, \quad \dfrac{\mathrm{d}V_{\mathrm{R}}}{\mathrm{d}t} = 0, \quad V_{\mathrm{R}} = V_{\mathrm{RMIN}} \end{array}\right\} \tag{7.138}$$

The exiter

$$\frac{\mathrm{d}E_{fq}}{\mathrm{d}t} = \frac{1}{T_{\mathrm{E}}}[V_{\mathrm{R}} - (K_{\mathrm{E}} + S_{\mathrm{E}})E_{fq}], \tag{7.139}$$

where the saturation coefficient S_{E} is modeled as an exponential function according to (6.101). Under one per unit excitation voltage/one per unit stator voltage system, (6.101) is simplified to

$$S_{\mathrm{E}} = C_{\mathrm{E}}E_{fq}^{N_{\mathrm{E}}-1}. \tag{7.140}$$

The saturation function can also be piece-wise linearized as follows:

$$S_{\mathrm{E}}E_{fq} = K_1E_{fq} - K_2. \tag{7.141}$$

From Fig. 6.14 we have the PSS equations:

$$\left.\begin{aligned} \frac{\mathrm{d}V_1}{\mathrm{d}t} &= \frac{1}{T_6}(K_S V_{IS} - V_1) \\ \frac{\mathrm{d}(V_1 - V_2)}{\mathrm{d}t} &= \frac{1}{T_5}V_2 \\ \frac{\mathrm{d}(T_1 V_2 - T_2 V_3)}{\mathrm{d}t} &= V_3 - V_2 \\ \frac{\mathrm{d}(T_3 V_3 - T_4 V_4)}{\mathrm{d}t} &= V_4 - V_3 \end{aligned}\right\}. \tag{7.142}$$

The limits of PSS output are

$$\left.\begin{aligned} &\text{If } V_4 \geq V_{\mathrm{Smax}}, \quad V_S = V_{\mathrm{Smax}} \\ &\text{If } V_{\mathrm{Smin}} < V_4 < V_{\mathrm{Smax}}, \quad V_S = V_4 \\ &\text{If } V_4 \leq V_{\mathrm{Smin}}, \quad V_S = V_{\mathrm{Smin}} \end{aligned}\right\}. \tag{7.143}$$

The initial values of excitation system variables can be found by setting, in the transfer function diagram, $s = 0$, or alternatively setting the left-hand side of the differential equations of the excitation system to zero. The effects of limiters can in general be ignored since the variables with limiters under normal operation do not in general exceed their corresponding limits. In the following, we describe how to compute the initial values of the excitation system mentioned above, the other excitation systems can be dealt with likewise.

Setting the left-hand side of (7.139) to zero, one obtains the initial value for the amplifier

$$V_{R(0)} = (S_{E(0)} + K_E)E_{fq(0)}, \tag{7.144}$$

where the saturation coefficient is calculated based on (7.140), that is,

$$S_{E(0)} = C_E E_{fq(0)}^{N_E - 1}.$$

Setting the left-hand side of (7.136), (7.137), and (7.138) to zero, it follows:

$$\left.\begin{aligned} &V_{F(0)} = 0, \quad V_{M(0)} = \left|\dot{V}_{(0)} + jX_C\dot{I}_{(0)}\right| \\ &V_{REF} = V_{M(0)} + \frac{V_{R(0)}}{K_A} \end{aligned}\right\}. \tag{7.145}$$

Setting the left-hand side of (7.142) to zero, and taking into account the relationship expressed in (7.142), we have the initial value of PSS:

$$\left.\begin{aligned} &V_{S(0)} = V_{4(0)} = V_{3(0)} = V_{2(0)} = 0 \\ &V_{1(0)} = K_S V_{IS(0)} = 0 \end{aligned}\right\}, \tag{7.146}$$

where V_{IS} is equal to zero since it often takes the form of speed, or change of active power.

Applying the trapezoidal rule to (7.136), we have the difference equations of measurement and filter systems:

$$V_{M(t+\Delta t)} = \alpha_{\mathrm{R}} V_{\mathrm{C}(t+\Delta t)} + V_{M0}, \tag{7.147}$$

in which

$$\alpha_{\mathrm{R}} = \frac{\Delta t}{2T_{\mathrm{R}} + \Delta t}, \tag{7.148}$$

$$V_{M0} = \alpha_{\mathrm{R}} V_{\mathrm{C}(t)} + \frac{2T_{\mathrm{R}} - \Delta t}{2T_{\mathrm{R}} + \Delta t} V_{M(t)}, \tag{7.149}$$

$$\left.\begin{aligned} V_{\mathrm{C}(t+\Delta t)} &= \left|\dot{V}_{(t+\Delta t)} + jX_{\mathrm{C}}\dot{I}_{(t+\Delta t)}\right| \\ V_{\mathrm{C}(t)} &= \left|\dot{V}_{(t)} + jX_{\mathrm{C}}\dot{I}_{(t)}\right| \end{aligned}\right\}. \tag{7.150}$$

Applying the trapezoidal rule to (7.137), we have

$$V_{\mathrm{F}(t+\Delta t)} = \alpha_{\mathrm{F}} E_{fq(t+\Delta t)} + V_{\mathrm{F0}}, \tag{7.151}$$

where

$$\alpha_{\mathrm{F}} = \frac{2K_{\mathrm{F}}}{2T_{\mathrm{F}} + \Delta t}, \tag{7.152}$$

$$V_{\mathrm{F0}} = \frac{2T_{\mathrm{F}} - \Delta t}{2T_{\mathrm{F}} + \Delta t} V_{\mathrm{F}(t)} - \alpha_{\mathrm{F}} E_{fq(t)}. \tag{7.153}$$

When limiters are not taken into consideration, applying the trapezoidal rule to (7.138), we have the difference equation:

$$V_{\mathrm{R}(t+\Delta t)} = \alpha_{\mathrm{A}}\left(V_{\mathrm{S}(t+\Delta t)} - V_{M(t+\Delta t)} - V_{\mathrm{F}(t+\Delta t)}\right) + V_{\mathrm{R0}}, \tag{7.154}$$

where

$$\alpha_{\mathrm{A}} = \frac{K_{\mathrm{A}}\Delta t}{2T_{\mathrm{A}} + \Delta t}, \tag{7.155}$$

$$V_{\mathrm{R0}} = \alpha_{\mathrm{A}}\left(2V_{\mathrm{REF}} + V_{\mathrm{S}(t)} - V_{M(t)} - V_{\mathrm{F}(t)}\right) + \frac{2T_{\mathrm{A}} - \Delta t}{2T_{\mathrm{A}} + \Delta t} V_{\mathrm{R}(t)}. \tag{7.156}$$

Substituting (6.141) into (6.139), and applying trapezoidal rule, we have the difference equations of the excitor:

$$E_{fq(t+\Delta t)} = \alpha_{\mathrm{E}} V_{\mathrm{R}(t+\Delta t)} + V_{\mathrm{E0}}, \tag{7.157}$$

where

$$\alpha_{\mathrm{E}} = \frac{\Delta t}{2T_{\mathrm{E}} + (K_{\mathrm{E}} + K_1)\Delta t}, \tag{7.158}$$

$$V_{E0} = \alpha_E[V_{R(t)} - 2(K_E + K_1)E_{fq(t)} + 2K_2] + E_{fq(t)}. \tag{7.159}$$

Applying the trapezoidal rule to (7.142), it follows:

$$\left.\begin{aligned} V_{1(t+\Delta t)} &= \alpha_1 V_{IS(t+\Delta t)} + V_{10} \\ V_{2(t+\Delta t)} &= \alpha_2 V_{1(t+\Delta t)} + V_{20} \\ V_{3(t+\Delta t)} &= \alpha_3 V_{2(t+\Delta t)} + V_{30} \\ V_{4(t+\Delta t)} &= \alpha_4 V_{3(t+\Delta t)} + V_{40} \end{aligned}\right\}, \tag{7.160}$$

in the above formula

$$\alpha_1 = \frac{K_S \Delta t}{2T_6 + \Delta t}, \quad \alpha_2 = \frac{2T_5}{2T_5 + \Delta t}, \quad \alpha_3 = \frac{2T_1 + \Delta t}{2T_2 + \Delta t}, \quad \alpha_4 = \frac{2T_3 + \Delta t}{2T_4 + \Delta t}, \tag{7.161}$$

$$\left.\begin{aligned} V_{10} &= \alpha_1 V_{IS(t)} + \frac{2T_6 - \Delta t}{2T_6 + \Delta t} V_{1(t)} \\ V_{20} &= \frac{2T_5 - \Delta t}{2T_5 + \Delta t} V_{2(t)} - \alpha_2 V_{1(t)} \\ V_{30} &= \frac{2T_2 - \Delta t}{2T_2 + \Delta t} V_{3(t)} - \frac{2T_1 - \Delta t}{2T_2 + \Delta t} V_{2(t)} \\ V_{40} &= \frac{2T_4 - \Delta t}{2T_4 + \Delta t} V_{4(t)} - \frac{2T_3 - \Delta t}{2T_4 + \Delta t} V_{3(t)} \end{aligned}\right\}. \tag{7.162}$$

Eliminating the intermediate variables $V_{1(t+\Delta t)}$, $V_{2(t+\Delta t)}$, and $V_{3(t+\Delta t)}$ in (7.160), it follows:

$$V_{4(t+\Delta t)} = \alpha_4\alpha_3\alpha_2\alpha_1 V_{IS(t+\Delta t)} + V_{40} + \alpha_4[V_{30} + \alpha_3(V_{20} + \alpha_2 V_{10})]. \tag{7.163}$$

If the input of the PSS is set to $V_{IS} = \omega - \omega_s$, apparently $V_{IS(t)} = \omega_{(t)} - \omega_s$. Substituting $V_{IS(t+\Delta t)} = \omega_{(t=\Delta t)} - \omega_s$ into (7.163), and making use of (7.128) to eliminate variable $\omega_{(t+\Delta t)}$, we have

$$V_{4(t+\Delta t)} = \alpha_S \delta_{(t+\Delta t)} + V_{S0}, \tag{7.164}$$

where

$$\alpha_S = \frac{2\alpha_4\alpha_3\alpha_2\alpha_1}{\omega_s \Delta t}, \tag{7.165}$$

$$V_{S0} = V_{40} + \alpha_4[V_{30} + \alpha_3(V_{20} + \alpha_2 V_{10})] - \alpha_S \delta_{(t)} + \alpha_4\alpha_3\alpha_2\alpha_1(2 - \omega_s - \omega_{(t)}). \tag{7.166}$$

If the limits of PSS outputs are not considered, obviously we get

$$V_{S(t+\Delta t)} = V_{4(t+\Delta t)}. \tag{7.167}$$

If PSS takes other forms of input signals, following the same derivations, we should be able to find the corresponding expressions.

Eliminating the intermediate variables $V_{4(t+\Delta t)}$, $V_{S(t+\Delta t)}$, $V_{M(t+\Delta t)}$, $V_{F(t+\Delta t)}$, and $V_{R(t+\Delta t)}$ in (7.164), (7.167), (7.147), (7.151), (7.154), and (7.157), the difference equations of the excitation system without taking into account the affects of limiters are obtained as

$$E_{fq(t+\Delta t)} = \beta_1 \delta_{(t+\Delta t)} - \beta_2 \left| \dot{V}_{(t+\Delta t)} + jX_C \dot{I}_{(t+\Delta t)} \right| + E_{fq0}, \tag{7.168}$$

where

$$\beta_1 = \frac{\alpha_E \alpha_A \alpha_S}{1 + \alpha_E \alpha_A \alpha_F}, \quad \beta_2 = \frac{\alpha_E \alpha_A \alpha_R}{1 + \alpha_E \alpha_A \alpha_F}, \tag{7.169}$$

$$E_{fq0} = \frac{V_{E0} + \alpha_E [V_{R0} + \alpha_A (V_{S0} - V_{M0} - V_{F0})]}{1 + \alpha_E \alpha_A \alpha_F}. \tag{7.170}$$

7.5.1.3 The Prime Movers and Their Speed-Governing Systems

Taking the hydrogenerator and its speed-governing system illustrated in Fig. 6.24 as an example, based on the transfer function we have the differential-algebraic equations:

The acentric flyball

$$\eta = K_\delta (\omega_{REF} - \omega) \tag{7.171}$$

The valve

The dead zones are

$$\left.\begin{array}{ll} \text{If } -\dfrac{\varepsilon K_\delta}{2} < \eta - \xi < \dfrac{\varepsilon K_\delta}{2}, & \overline{\sigma} = 0 \\ \text{If } \eta - \xi \geq \dfrac{\varepsilon K_\delta}{2}, & \overline{\sigma} = \eta - \xi - \dfrac{\varepsilon K_\delta}{2} \\ \text{If } \eta - \xi \leq -\dfrac{\varepsilon K_\delta}{2}, & \overline{\sigma} = \eta - \xi + \dfrac{\varepsilon K_\delta}{2} \end{array}\right\}. \tag{7.172}$$

The limits of value position are

$$\left.\begin{array}{ll} \text{If } \sigma_{MIN} < \overline{\sigma} < \sigma_{MAX}, & \sigma = \overline{\sigma} \\ \text{If } \overline{\sigma} \geq \sigma_{MAX}, & \sigma = \sigma_{MAX} \\ \text{If } \overline{\sigma} \leq \sigma_{MAX}, & \sigma = \sigma_{MIN} \end{array}\right\}. \tag{7.173}$$

The servo system

$$\frac{d\overline{\mu}}{dt} = \frac{\sigma}{T_S} \tag{7.174}$$

The limits of the valve

$$\left.\begin{array}{l}\text{If } \mu_{\text{MIN}} < \overline{\mu} < \mu_{\text{MAX}}, \quad \mu = \overline{\mu} \\ \text{If } \overline{\mu} \geq \mu_{\text{MAX}}, \quad \mu = \mu_{\text{MAX}} \\ \text{If } \overline{\mu} \leq \mu_{\text{MAX}}, \quad \mu = \mu_{\text{MIN}}\end{array}\right\} \tag{7.175}$$

The feedback system

$$\frac{\mathrm{d}[\xi - (K_\beta + K_i)\mu]}{\mathrm{d}t} = \frac{1}{T_i}(K_i\mu - \xi) \tag{7.176}$$

The hydrogenerator

$$\frac{\mathrm{d}(P_m + 2K_{mH}\mu)}{\mathrm{d}t} = \frac{2}{T_\omega}(K_{mH}\mu - P_m), \tag{7.177}$$

where parameter K_{mH} is defined as follows:

$$K_{mH} = \frac{P_{\mathrm{H}}(MW)}{S_{\mathrm{B}}(MVA)}. \tag{7.178}$$

In general, the parameters of a prime mover and its speed-governing system are provided in the per unit system with the nominal capability of the generator being the base. With the introduction of parameters K_{mH}, P_m, and P_e all expressed in per unit system with system base S_{B}.

Similar to the calculation of excitation system initial values, the initial values of prime mover and speed-governing systems can be found by setting, in the transfer functions, $s = 0$, or alternatively by setting the left-hand side of the differential equations to zero. Again the dead zones of measurement systems and various limiters need not be considered in general. Setting the left-hand side of (7.177), (7.176), and (7.174) to zero, and making use of the linear relationships in (7.171), (7.172), (7.173), and (7.175), together with (7.77), we have the initial values of each state variable:

$$\left.\begin{array}{l}\mu_{(0)} = \overline{\mu}_{(0)} = \dfrac{P_{m(0)}}{K_{mH}}, \quad \eta_{(0)} = \xi_{(0)} = K_i\mu_{(0)}, \quad \sigma_{(0)} = \overline{\sigma}_{(0)} = 0, \\ \omega_{\text{REF}} = \omega_{(0)} + \dfrac{\xi_{(0)}}{K_\delta} = 1 + \dfrac{\xi_{(0)}}{K_\delta}.\end{array}\right\} \tag{7.179}$$

Based on (7.171), the equation corresponding to instant $t + \Delta t$ for the acentric flyball is as follows:

$$\eta_{(t+\Delta t)} = K_\delta(\omega_{\text{REF}} - \omega_{(t+\Delta t)}). \tag{7.180}$$

Neglecting the measurement dead zone, based on (7.172), it follows:

$$\overline{\sigma}_{(t+\Delta t)} = \eta_{(t+\Delta t)} - \xi_{(t+\Delta t)}. \tag{7.181}$$

Also neglecting the limits on valve position, and based on (7.173), obviously we have:

$$\sigma_{(t+\Delta t)} = \overline{\sigma}_{(t+\Delta t)}. \tag{7.182}$$

Applying the trapezoidal rule to (7.174), we obtain the following difference equation:

$$\overline{\mu}_{(t+\Delta t)} = \alpha_{\mathrm{S}}\sigma_{(t+\Delta t)} + \mu_0, \tag{7.183}$$

where

$$\alpha_{\mathrm{S}} = \frac{\Delta t}{2T_{\mathrm{S}}}, \tag{7.184}$$

$$\mu_0 = \alpha_{\mathrm{S}}\sigma_{(t)} + \overline{\mu}_{(t)}. \tag{7.185}$$

Neglect again the limit on valve position, based on (7.175), we have

$$\mu_{(t+\Delta t)} = \overline{\mu}_{(t+\Delta t)}. \tag{7.186}$$

Applying the trapezoidal rule to (7.176), we have the difference equation of the feedback block:

$$\xi_{(t+\Delta t)} = \alpha_i\mu_{(t+\Delta t)} + \xi_0, \tag{7.187}$$

where

$$\alpha_i = K_i + \frac{2T_iK_\beta}{2T_i + \Delta t}, \tag{7.188}$$

$$\xi_0 = \frac{2T_i - \Delta t}{2T_i + \Delta t}[\xi_{(t)} - K_i\mu_{(t)}] - \frac{2T_iK_\beta}{2T_i + \Delta t}\mu_{(t)}. \tag{7.189}$$

Applying the trapezoidal rule to (7.177), the difference equations of hydrogenerators are obtained, as follows:

$$P_{m(t+\Delta t)} = -\alpha_{\mathrm{H}}\mu_{(t+\Delta t)} + P_0, \tag{7.190}$$

where

$$\alpha_{\mathrm{H}} = \frac{K_{mH}(2T_\omega - \Delta t)}{T_\omega + \Delta t}, \tag{7.191}$$

$$P_0 = \frac{T_\omega - \Delta t}{T_\omega + \Delta t} P_{m(t)} + \frac{K_{mH}(2T_\omega + \Delta t)}{T_\omega + \Delta t} \mu_{(t)}. \tag{7.192}$$

Eliminating the intermediate variables $\eta_{(t+\Delta t)}$, $\overline{\sigma}_{(t+\Delta t)}$, $\sigma_{(t+\Delta t)}$, $\overline{\mu}_{(t+\Delta t)}$, $\mu_{(t+\Delta t)}$, and $\xi_{(t+\Delta t)}$ in (7.180), (7.182), (7.183), (7.186), (7.187), and (7.190), and eliminating variable $\omega_{(t+\Delta t)}$ based on (7.128), we find the difference equations for step $t+\Delta t$ of hydrogenerators and their speed-governing systems, without consideration of limiters:

$$P_{m(t+\Delta t)} = \beta_3 \delta_{(t+\Delta t)} + P_{m0}, \tag{7.193}$$

in which

$$\beta_3 = \frac{2\alpha_{\mathrm{H}}\alpha_{\mathrm{S}}K_\delta}{(1 + \alpha_{\mathrm{S}}\alpha_i)\omega_s \Delta t}, \tag{7.194}$$

$$P_{m0} = P_0 - \beta_3 \delta_{(t)} + \frac{\alpha_{\mathrm{H}}[\alpha_{\mathrm{S}} K_\delta (2 - \omega_{\mathrm{REF}} - \omega_{(t)}) + \alpha_{\mathrm{S}}\xi_0 - \mu_0]}{1 + \alpha_{\mathrm{S}}\alpha_i}. \tag{7.195}$$

Finally, substituting (7.117) of $P_{e(t+\Delta t)}$ and (7.193) of $P_{m(t+\Delta t)}$ into (7.125), and substituting difference equation (7.168) of $E_{fq(t+\Delta t)}$ into (7.131), together with (7.132) we obtain the difference equations of generators for step $t + \Delta t$. Let us transform the state currents under *d–q* coordinates into those under *x–y* coordinates, for notational simplicity, neglecting the subscripts $(t + \Delta t)$, it follows:

$$\left.\begin{aligned}
&(1 - \alpha_J\beta_3)\delta + \alpha_J[V_x I_x + V_y I_y + R_a(I_x^2 + I_y^2)] - \alpha_J P_{m0} - \delta_0 = 0\\
&E''_q + \alpha''_d(X'_d - X''_d)(I_x \sin\delta - I_y \cos\delta) - \alpha''_d \alpha_{d1}\beta_1\delta\\
&\quad + \alpha''_d\alpha_{d1}\beta_2\sqrt{(V_x - X_{\mathrm{C}}I_y)^2 + (V_y + X_{\mathrm{C}}I_x)^2} - \alpha''_d\alpha_{d1}E_{fq0} - E''_{q0} = 0\\
&E''_d - \alpha''_q(X'_q - X''_q)(I_x\cos\delta + I_y\sin\delta) - E''_{d0} = 0
\end{aligned}\right\}. \tag{7.196}$$

The set of simultaneous (7.196) consists of three equations, the first reflects the mechanical motion of the generators, while the other two reflect the electromagnetic interactions. Based on (7.39), generator currents I_x and I_y are functions of V_x, V_y, δ, E''_q, and E''_d (refer to (7.258) for details), and therefore can be eliminated. The above set of simultaneous equations thus has three state variables δ, E''_q, and E''_d, plus two operating parameters V_x and V_y.

7.5.2 *Initial Values and Difference Equations of FACTS and HVDC*

7.5.2.1 SVC

Here we will focus on an SVC model comprising a fixed capacitor (FC) and a thyristor-controlled reactor (TCR). For ease of exposition, we will take a proportional regulator-based SVC as an example; its transfer function is illustrated in (7.16).

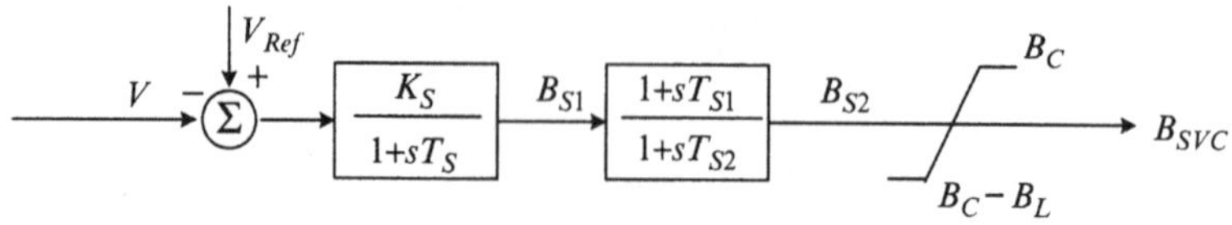

Fig. 7.16 A simple model of SVC

An SVC is generally connected to a high-voltage system via a transformer. The equivalent admittance of TCR is controlled by the firing angle α of a thyristor, thus the equivalent admittance B_{SVC} of the SVC is manipulated. This mechanism facilitates the control of voltage V given the input V_{REF}. The mathematical model of the SVC is obtained easily from Fig. 7.16 as

$$\left.\begin{aligned} \frac{\mathrm{d}B_{\mathrm{S1}}}{\mathrm{d}t} &= \frac{1}{T_{\mathrm{S}}}[K_{\mathrm{S}}(V_{\mathrm{REF}} - V) - B_{\mathrm{S1}}] \\ \frac{\mathrm{d}(T_{\mathrm{S2}}B_{\mathrm{S2}} - T_{\mathrm{S1}}B_{\mathrm{S1}})}{\mathrm{d}t} &= B_{\mathrm{S1}} - B_{\mathrm{S2}} \end{aligned}\right\}. \tag{7.197}$$

The limit on SVC output is

$$\left.\begin{aligned} &\text{If } B_{\mathrm{C}} - B_{\mathrm{L}} < B_{\mathrm{S2}} < B_{\mathrm{C}}, \quad B_{\mathrm{SVC}} = B_{\mathrm{S2}} \\ &\text{If } B_{\mathrm{S2}} \geq B_{\mathrm{C}}, \quad B_{\mathrm{SVC}} = B_{\mathrm{C}} \\ &\text{If } B_{\mathrm{S2}} \leq B_{\mathrm{C}} - B_{\mathrm{L}}, \quad B_{\mathrm{SVC}} = B_{\mathrm{C}} - B_{\mathrm{L}} \end{aligned}\right\}, \tag{7.198}$$

where $B_{\mathrm{C}} = \omega C$ is the susceptance of the fixed capacitor, $B_{\mathrm{L}} = 1/\omega L$ is the susceptance of the reactor, the output B_{SVC} is the equivalent susceptance of the SVC. The upper limit of the SVC corresponds to the point at which the thyristor is completely shut off, while the lower limit corresponds to the point at which the thyristor is like a lossless conductor. The position between the limits corresponds to a point at which the thyristor is partially closed.

Although an SVC is connected at the low-voltage side of a transformer, it can still be viewed as a reactive power source at the high-voltage side, intended to control the voltage at the high-voltage side bus of the transformer. Therefore, the high-voltage bus can be effectively set as a PV node in load flow studies ($P = 0$, $V = V^{\mathrm{SP}}$). From the result of a load flow study, one obtains $\dot{V}_{(0)} = V^{\mathrm{SP}}\angle\theta_{(0)}$ and the power injection from the SVC $S_{(0)} = jQ_{(0)}$. Let the reactance of the transformer be X_{T}, the power injected into the network from the SVC is given by

$$Q_{(0)} = \frac{V_{(0)}^2}{\dfrac{1}{B_{\mathrm{SVC}(0)}} - X_{\mathrm{T}}}. \tag{7.199}$$

Setting both sides of (7.197) to zero, and noticing the relationship in (7.198) and (7.199), we find the initial values of the SVC as

$$\left.\begin{aligned} B_{\text{SVC}(0)} &= B_{\text{S2}(0)} = B_{\text{S1}(0)} = \frac{1}{X_{\text{T}} + \frac{V_{(0)}^2}{Q_{(0)}}} \\ V_{\text{REF}} &= V^{\text{SP}} + \frac{B_{\text{SVC}(0)}}{K_{\text{S}}} \end{aligned}\right\}. \tag{7.200}$$

Applying the trapezoidal rule to the first (7.197), it follows:

$$B_{\text{S1}(t+\Delta t)} = -v_1 V_{(t+\Delta t)} + B_{\text{S10}}, \tag{7.201}$$

where

$$v_1 = \frac{K_{\text{S}}\Delta t}{2T_{\text{S}} + \Delta t}, \tag{7.202}$$

$$B_{\text{S10}} = v_1(2V_{\text{REF}} - V_{(t)}) + \frac{2T_{\text{S}} - \Delta t}{2T_{\text{S}} + \Delta t} B_{\text{S1}(t)}. \tag{7.203}$$

And applying the trapezoidal rule to the second equation in (7.197), and eliminating $B_{\text{S1}(t+\Delta t)}$ from (7.201), it follows:

$$B_{\text{S2}(t+\Delta t)} = B_{\text{SVCO}} - v_{\text{S}}\sqrt{V_{x(t+\Delta t)}^2 + V_{y(t+\Delta t)}^2}, \tag{7.204}$$

where

$$v_{\text{S}} = v_1 \frac{2T_{\text{S1}} + \Delta t}{2T_{\text{S2}} + \Delta t}, \tag{7.205}$$

$$B_{\text{SVCO}} = \frac{2T_{\text{S1}} + \Delta t}{2T_{\text{S2}} + \Delta t} B_{\text{S10}} + \frac{2T_{\text{S2}} - \Delta t}{2T_{\text{S2}} + \Delta t} B_{\text{S2}(t)} - \frac{2T_{\text{S1}} - \Delta t}{2T_{\text{S2}} + \Delta t} B_{\text{S1}(t)}. \tag{7.206}$$

If the limit of the SVC is ignored, then $B_{\text{S}(t+\Delta t)} = B_{\text{S2}(t+\Delta t)}$, thus:

$$B_{\text{SVC}(t+\Delta t)} = B_{\text{SVCO}} - v_{\text{S}}\sqrt{V_{x(t+\Delta t)}^2 + V_{y(t+\Delta t)}^2}. \tag{7.207}$$

7.5.2.2 TCSC

A thyristor-controlled series compensator (TCSC) is connected into a transmission line in series, it changes its equivalent admittance thus achieving the goal of controlling the equivalent admittance of the transmission line. Here we will only give the mathematical model of a TCSC composed of FC and TCR connected in parallel (similar to an SVC):

$$\left.\begin{aligned}\frac{dB_{T1}}{dt} &= \frac{1}{T_T}\left[K_T(P_{REF} - P_T) - B_{T1}\right]\\ \frac{d(T_{T2}B_{T2} - T_{T1}B_{T1})}{dt} &= B_{T1} - B_{T2}\end{aligned}\right\}, \tag{7.208}$$

where the input signal P_T is the power flowing through the line in which the TCSC is connected, the output B_{TCSC} is the equivalent susceptance of the TCSC.

The limits of the TCSC are

$$\left.\begin{aligned}&\text{If } B_{TCSC}^{MIN} < B_{T2} < B_{TCSC}^{MAX}, \quad B_{TCSC} = B_{T2}\\ &\text{If } B_{T2} \geq B_{TCSC}^{MAX}, \quad B_{TCSC} = B_{TCSC}^{MAX}\\ &\text{If } B_{T2} \leq B_{TCSC}^{MIN}, \quad B_{TCSC} = B_{TCSC}^{MIN}\end{aligned}\right\}, \tag{7.209}$$

where the specific values of B_{TCSC}^{MAX} and B_{TCSC}^{MIN} depend on the sizes of L and C. They can be computed based on (5.153)–(5.155).

As usual a load flow study provides $B_{TCSC(0)}$ and $P_{T(0)} = P^{SP}$, similar to computing the initial values of an SVC, we have

$$\left.\begin{aligned}B_{TCSC(0)} &= B_{T2(0)} = B_{T1(0)}\\ P_{REF} &= P_{T(0)} + \frac{B_{TCSC(0)}}{K_T}\end{aligned}\right\}. \tag{7.210}$$

If the measured value of P_T of TCSC flows from bus i to bus j, the expression of P_T is easily obtained as

$$P_T = B_{TCSC}(V_{xi}V_{yj} - V_{yi}V_{xj}). \tag{7.211}$$

Apply the trapezoidal rule to the first equation in (7.208), one obtains

$$B_{T1(t+\Delta t)} = -\zeta_1 P_{T(t+\Delta t)} + B_{T10}, \tag{7.212}$$

in which

$$\zeta_1 = \frac{K_T \Delta t}{2T_T + \Delta t}, \tag{7.213}$$

$$B_{T10} = \zeta_1(2P_{REF} - P_{T(t)}) + \frac{2T_T - \Delta t}{2T_T + \Delta t} B_{T1(t)}. \tag{7.214}$$

Now applying the trapezoidal rule to the second equation of (7.208), eliminating $B_{\mathrm{T1}(t+\Delta t)}$ and $P_{\mathrm{T}(t+\Delta t)}$ based on (7.212) and (7.211), we have

$$[1+\zeta_{\mathrm{T}}(V_{xi(t+\Delta t)}V_{yj(t+\Delta t)}-V_{yi(t+\Delta t)}V_{xj(t+\Delta t)})]B_{\mathrm{T2}(t+\Delta t)}-B_{\mathrm{TCSC0}}=0, \quad (7.215)$$

in which

$$\zeta_{\mathrm{T}}=\zeta_1\frac{2T_{\mathrm{T1}}+\Delta t}{2T_{\mathrm{T2}}+\Delta t}, \quad (7.216)$$

$$B_{\mathrm{TCSC0}}=\frac{2T_{\mathrm{T1}}+\Delta t}{2T_{\mathrm{T2}}+\Delta t}B_{\mathrm{T10}}+\frac{2T_{\mathrm{T2}}-\Delta t}{2T_{\mathrm{T2}}+\Delta t}B_{\mathrm{T2}(t)}-\frac{2T_{\mathrm{T1}}-\Delta t}{2T_{\mathrm{T2}}+\Delta t}B_{\mathrm{T1}(t)}. \quad (7.217)$$

If the output limits of the TCSC are neglected, obviously we have $B_{\mathrm{TCSC}(t+\Delta t)}=B_{\mathrm{T2}(t+\Delta t)}$, then

$$[1+\zeta_{\mathrm{T}}(V_{xi(t+\Delta t)}V_{yj(t+\Delta t)}-V_{yi(t+\Delta t)}V_{xj(t+\Delta t)})]B_{\mathrm{TCSC}(t+\Delta t)}-B_{\mathrm{TCSC0}}=0. \quad (7.218)$$

7.5.2.3 HVDC Systems

In stability studies, the network equations of the AC system appear in terms of positive sequence quantities, this places a fundament limitation on the model of an HVDC system. In particular, commutation failure in the HVDC system cannot be predicted. A commutation failure may be the result of a severe three-line-to-ground fault occurring close to the rectifier, an unsymmetrical fault on the AC side of the rectifier, or saturation of HVDC transformer operating during a transient period.

Earlier HVDC models included the dynamic characteristics of transmission lines and converter dynamics. In recent years, there is a trend toward adopting simpler models. Two models for an HVDC system are popular, these are, a simplified model and a steady-state model.

1. The simplified model

An HVDC system some distance away from the study area has little impact on the results of a stability study, and thus can be modeled using a simple model: the system is viewed as a pair of active and reactive power sources connected at the converter AC substation.

A more realistic model is termed the steady-state model. Based on (5.2), the DC line is modeled by the algebraic equation of a resistor:

$$V_{\mathrm{dR}}=V_{\mathrm{dI}}+R_{\mathrm{dc}}I_{\mathrm{d}}, \quad (7.219)$$

where R_{dc} denotes the resistance of the line.

Noticing that $I_{dR} = I_{dI} = I_d$, from (7.52) and (7.53), eliminating V_{dR} and V_{dI} in (7.219), it follows:

$$RI_d = k_R V_R \cos\alpha - k_I V_I \cos\beta, \tag{7.220}$$

where

$$R = R_{dc} + X_{cR} + X_{cI}. \tag{7.221}$$

The pole control action is assumed to be instantaneous; many of the control functions are represented in terms of their net effects, rather than actual characteristics of the hardware. This model appears in the form of an algebraic equation, the interaction between AC and DC system is similar to that in a load flow model.

2. Quasi-steady-state model

If the short circuit currents in any of the converters are relatively low, then the dynamics of DC system elements has a non-negligible impact on the AC system. As a result, a more detailed DC model is necessary for conducting a transient stability analysis.

In a quasi-steady-state model, the converter characteristics are still modeled by the equation governing the relationship between average DC values and the nominal values of fundamental frequency components. In this setting, the DC transmission line can adopt different models given different requirements on precision. The simplest DC line model is just that of a steady-state model, as in (7.220). A more detailed model is based on an *R–L* circuit:

$$L\frac{dI_d}{dt} + RI_d = k_R V_R \cos\alpha - k_I V_I \cos\beta, \tag{7.222}$$

where R is defined in (7.221), besides,

$$L = L_{dc} + L_R + L_I, \tag{7.223}$$

where L_{dc}, L_R, and L_I are the reactance of DC line, and the smoothing reactors.

For the control system, taking the control mode of constant current and constant voltage as an example, from the transfer function given in Fig. 5.18, we have the differential equations:

$$\left.\begin{aligned} \frac{dx_1}{dt} &= \frac{1}{T_{c3}}(I_d - x_1) \\ \frac{d(K_{c1}x_1 - \bar{\alpha})}{dt} &= \frac{K_{c2}}{T_{c2}}(I_{dREF} - x_1) \end{aligned}\right\}. \tag{7.224}$$

The limits on delayed ignition angle include

$$\left.\begin{array}{l}\text{If } \alpha_{\text{MIN}} < \bar{\alpha} < \alpha_{\text{MAX}}, \quad \alpha = \bar{\alpha} \\ \text{If } \bar{\alpha} \geq \alpha_{\text{MAX}}, \quad \alpha = \alpha_{\text{MAX}} \\ \text{If } \bar{\alpha} \leq \alpha_{\text{MIN}}, \quad \alpha = \alpha_{\text{MIN}}\end{array}\right\}, \tag{7.225}$$

$$\left.\begin{array}{l}\dfrac{\mathrm{d}x_4}{\mathrm{d}t} = \dfrac{1}{T_{\text{v3}}}(V_{\text{dI}} - x_4) \\ \dfrac{\mathrm{d}(K_{\text{v1}}x_4 - \bar{\beta})}{\mathrm{d}t} = \dfrac{K_{\text{v2}}}{T_{\text{v2}}}(V_{\text{dREF}} - x_4)\end{array}\right\}. \tag{7.226}$$

The limits on ignition advance angle include

$$\left.\begin{array}{l}\text{If } \beta_{\text{MIN}} < \bar{\beta} < \beta_{\text{MAX}}, \quad \beta = \bar{\beta} \\ \text{If } \bar{\beta} \geq \beta_{\text{MAX}}, \quad \beta = \beta_{\text{MAX}} \\ \text{If } \bar{\beta} \leq \beta_{\text{MIN}}, \quad \beta = \beta_{\text{MIN}}\end{array}\right\}. \tag{7.227}$$

When the rectifier is under constant current control, and the inverter is under constant voltage control, we have $I_{\text{d(0)}} = I_{\text{d}}^{\text{SP}}$ and $V_{\text{dI(0)}} = V_{\text{dI}}^{\text{SP}}$. From a load flow study we have $V_{\text{R(0)}}$ and $V_{\text{I(0)}}$. Based on (7.224)–(7.227), noticing the relationships in (7.219) and/or (7.222), and (7.52) and (7.53), it follows:

$$\left.\begin{array}{l}I_{\text{dREF}} = x_{1(0)} = I_{\text{d(0)}} \\ \alpha_{(0)} = \bar{\alpha}_{(0)} = \cos^{-1}\left(\dfrac{V_{\text{dI(0)}} + (R_{\text{dc}} + X_{\text{cR}})I_{\text{d(0)}}}{k_{\text{R}}V_{\text{R(0)}}}\right) \\ V_{\text{dREF}} = x_{4(0)} = V_{\text{dI(0)}} \\ \beta_{(0)} = \bar{\beta}_{(0)} = \cos^{-1}\left(\dfrac{V_{\text{dI(0)}} - X_{\text{cI}}I_{\text{d(0)}}}{k_{\text{I}}V_{\text{I(0)}}}\right)\end{array}\right\}. \tag{7.228}$$

Applying trapezoidal rule to (7.224), we find

$$x_{1(t+\Delta t)} = \gamma_1 I_{\text{d}(t+\Delta t)} + x_{10}, \tag{7.229}$$

where

$$\gamma_1 = \frac{\Delta t}{2T_{\text{c3}} + \Delta t}, \tag{7.230}$$

$$x_{10} = \gamma_1 I_{\text{d}(t)} + \frac{2T_{\text{c3}} - \Delta t}{2T_{\text{c3}} + \Delta t} x_{1(t)}. \tag{7.231}$$

Using the second formula in (7.224), and eliminating $x_{1(t+\Delta t)}$ by making use of (7.229), we have

$$\bar{\alpha}_{(t+\Delta t)} = \gamma_2 I_{d(t+\Delta t)} + \alpha_0, \tag{7.232}$$

where

$$\gamma_2 = \gamma_1\left(K_{c1} + \frac{K_{c2}\Delta t}{2T_{c2}}\right), \tag{7.233}$$

$$\alpha_0 = \left(K_{c1} + \frac{K_{c2}\Delta t}{2T_{c2}}\right)x_{10} + \bar{\alpha}_{(t)} - \frac{K_{c2}\Delta t}{T_{c2}}I_{dREF} - \left(K_{c1} - \frac{K_{c2}\Delta t}{2T_{c2}}\right)x_{1(t)}. \tag{7.234}$$

Neglecting the limits on ignition angle α, it is obvious that:

$$\alpha_{(t+\Delta t)} = \bar{\alpha}_{(t+\Delta t)}. \tag{7.235}$$

Applying the trapezoidal rule to the first formula of (7.226), it follows:

$$x_{4(t+\Delta t)} = \gamma_3 V_{dI(t+\Delta t)} + x_{40}, \tag{7.236}$$

where

$$\gamma_3 = \frac{\Delta t}{2T_{v3} + \Delta t}, \tag{7.237}$$

$$x_{40} = \gamma_3 V_{dI(t)} + \frac{2T_{v3} - \Delta t}{2T_{v3} + \Delta t}x_{4(t)}. \tag{7.238}$$

Applying the trapezoidal to the second formula in (7.226), making use of (7.236) to eliminate $x_{4(t+\Delta t)}$, and noticing the relationship in (7.53) allows $V_{dI(t+\Delta t)}$ to be eliminated, after simple manipulations, we have

$$\bar{\beta}_{(t+\Delta t)} = \gamma_4 V_{I(t+\Delta t)} \cos \beta_{(t+\Delta t)} + \gamma_5 I_{d(t+\Delta t)} + \beta_0, \tag{7.239}$$

in which

$$\gamma_4 = \frac{\gamma_3}{n_I}\left(K_{v1} + \frac{K_{v2}\Delta t}{2T_{v2}}\right), \quad \gamma_5 = \gamma_4 n_I R_{cI}, \tag{7.240}$$

$$\beta_0 = \left(K_{v1} + \frac{K_{v2}\Delta t}{2T_{v2}}\right)x_{40} + \bar{\beta}_{(t)} - \frac{K_{v2}\Delta t}{T_{v2}}V_{dREF} - \left(K_{v1} - \frac{K_{v2}\Delta t}{2T_{v2}}\right)x_{4(t)}. \tag{7.241}$$

Neglecting the limits on ignition angle β, it follows:

$$\beta_{(t+\Delta t)} = \bar{\beta}_{(t+\Delta t)}. \tag{7.242}$$

Under a quasi-steady-state model, different difference equations can be developed, with or without consideration for the transient duration of the DC transmission line.

If the transient duration of the DC line is not considered, the DC line is modeled based on (7.220), where I_d can be expressed as a function of α, β, V_{xR}, V_{yR}, V_{xI}, and V_{yI}:

$$I_d = \frac{k_R}{R}\sqrt{V_{xR}^2 + V_{yR}^2}\cos\alpha - \frac{k_I}{R}\sqrt{V_{xI}^2 + V_{yI}^2}\cos\beta. \tag{7.243}$$

Let us eliminate $\bar{\alpha}_{(t+\Delta t)}$ and $I_{d(t+\Delta t)}$ in (7.232), (7.235), and (7.243), we then obtain the difference equation of the rectifier under constant current control, when the limits on α are not considered:

$$\begin{aligned}\alpha_{(t+\Delta t)} &- \rho_1\sqrt{V_{xR(t+\Delta t)}^2 + V_{yR(t+\Delta t)}^2}\cos\alpha_{(t+\Delta t)} \\ &+ \rho_2\sqrt{V_{xI(t+\Delta t)}^2 + V_{yI(t+\Delta t)}^2}\cos\beta_{(t+\Delta t)} - \alpha_0 = 0,\end{aligned} \tag{7.244}$$

where

$$\rho_1 = \frac{k_R}{R}\gamma_2, \quad \rho_2 = \frac{k_I}{R}\gamma_2. \tag{7.245}$$

Similarly, eliminating variables $\bar{\beta}_{(t+\Delta t)}$ and $I_{d(t+\Delta t)}$ in (7.239), (7.242), and (7.243), we get the difference equation of the inverter under constant voltage control, when the limits on β are not considered:

$$\begin{aligned}\beta_{(t+\Delta t)} &- \rho_3\sqrt{V_{xR(t+\Delta t)}^2 + V_{yR(t+\Delta t)}^2}\cos\alpha_{(t+\Delta t)} \\ &- \rho_4\sqrt{V_{xI(t+\Delta t)}^2 + V_{yI(t+\Delta t)}^2}\cos\beta_{(t+\Delta t)} - \beta_0 = 0,\end{aligned} \tag{7.246}$$

where

$$\rho_3 = \frac{k_R}{R}\gamma_5, \quad \rho_4 = \gamma_4 - \frac{k_I}{R}\gamma_5. \tag{7.247}$$

If the transient response of the DC line is considered, the DC line is modeled by (7.222), applying the trapezoidal rule to this equation, it follows:

$$I_{d(t+\Delta t)} = \gamma_6 V_{R(t+\Delta t)}\cos\alpha_{(t+\Delta t)} - \gamma_7 V_{I(t+\Delta t)}\cos\beta_{(t+\Delta t)} + I_{d0}, \tag{7.248}$$

where

$$\gamma_6 = \frac{k_R \Delta t}{2L + R\Delta t}, \quad \gamma_7 = \gamma_6 \frac{k_I}{k_R}, \tag{7.249}$$

$$I_{d0} = \gamma_6 V_{R(t)} \cos \alpha_{(t)} - \gamma_7 V_{I(t)} \cos \beta_{(t)} + \frac{2L - R\Delta t}{2L + R\Delta t} I_{d(t)}. \tag{7.250}$$

Now eliminating $\bar{\alpha}_{(t+\Delta t)}$ and $I_{d(t+\Delta t)}$ in (7.232), (7.235), and (7.248), we find the difference equation of the rectifier, under constant current control, when the limits on α are not considered:

$$\begin{aligned} &\alpha_{(t+\Delta t)} - \rho_5 \sqrt{V_{xR(t+\Delta t)}^2 + V_{yR(t+\Delta t)}^2} \cos \alpha_{(t+\Delta t)} \\ &\quad + \rho_6 \sqrt{V_{xI(t+\Delta t)}^2 + V_{yI(t+\Delta t)}^2} \cos \beta_{(t+\Delta t)} - u_0 = 0, \end{aligned} \tag{7.251}$$

where

$$\rho_5 = \gamma_2 \gamma_6, \quad \rho_6 = \gamma_2 \gamma_7, \tag{7.252}$$

$$u_0 = \alpha_0 + \gamma_2 I_{d0}. \tag{7.253}$$

By the same token, eliminating $\bar{\beta}_{(t+\Delta t)}$ and $I_{d(t+\Delta t)}$ in (7.239), (7.242), and (7.248), we find the difference equation of the inverter, under constant current control, when the limits on β are not considered:

$$\begin{aligned} &\alpha_{(t+\Delta t)} - \rho_7 \sqrt{V_{xR(t+\Delta t)}^2 + V_{yR(t+\Delta t)}^2} \cos \alpha_{(t+\Delta t)} \\ &\quad + \rho_8 \sqrt{V_{xI(t+\Delta t)}^2 + V_{yI(t+\Delta t)}^2} \cos \beta_{(t+\Delta t)} - v_0 = 0, \end{aligned} \tag{7.254}$$

where

$$\rho_7 = \gamma_5 \gamma_6, \quad \rho_8 = \gamma_4 - \gamma_5 \gamma_7, \tag{7.255}$$

$$v_0 = \beta_0 + \gamma_5 I_{d0}. \tag{7.256}$$

7.5.3 Forming Network Equations

The network equations expressed in the domain of real numbers are provided in (7.36). In transient stability studies, the nodes in the network are divided into three classes: nodes connected in parallel with dynamic devices (including generator nodes, SVC nodes, and load nodes); nodes connected in series with dynamic devices (including the AC buses of an HVDC system, TCSC nodes, etc.); and the faulted nodes or nodes not connected with any device. Substitute the network current expressions of each dynamic device, illustrated in Sect. 7.3.1, into the network equations, and properly dealing with a fault or switch, as described in Sect. 7.3.2, we obtain the network equations ready for subsequent simulation.

7.5.3.1 Nodes Connected in Parallel with Dynamic Devices

If a dynamic device is connected at node i, then the network equation for this node is

$$\left.\begin{aligned} \Delta I_{xi} &= I_{xi} - \sum_{k \in i} (G_{ik}V_{xk} - B_{ik}V_{yk}) = 0 \\ \Delta I_{yi} &= I_{yi} - \sum_{k \in i} (G_{ik}V_{yk} + B_{ik}V_{xk}) = 0 \end{aligned}\right\}. \tag{7.257}$$

The expressions for currents I_{xi} and I_{yi} at node i depend on what device is connected.

(1) *Connected with a generator*: Note that a generators is represented using a varying E'_q, E''_q, E'_d, and E''_d model, therefore assigning the corresponding values to the elements in (7.40), based on Table 7.1, it turns out that the expression for the generator current (7.39) can be rewritten as

$$\left.\begin{aligned} I_{xi} = &\frac{1}{R_{ai}^2 + X''_{di}X''_{qi}} \Big\{ (R_{ai}\cos\delta_i + X''_{qi}\sin\delta_i)E''_{qi} + (R_{ai}\sin\delta_i - X''_{di}\cos\delta_i)E''_{di} \\ &- [R_{ai} - (X''_{di} - X''_{qi})\sin\delta_i\cos\delta_i]V_{xi} - (X''_{di}\cos^2\delta_i + X''_{qi}\sin^2\delta_i)V_{yi} \Big\} \\ I_{yi} = &\frac{1}{R_{ai}^2 + X''_{di}X''_{qi}} \Big\{ (R_{ai}\sin\delta_i - X''_{qi}\cos\delta_i)E''_{qi} - (R_{ai}\cos\delta_i + X''_{di}\sin\delta_i)E''_{di} \\ &+ (X''_{di}\sin^2\delta_i + X''_{qi}\cos^2\delta_i)V_{xi} - [R_{ai} + (X''_{di} - X''_{qi})\sin\delta_i\cos\delta_i]V_{yi} \Big\} \end{aligned}\right\}. \tag{7.258}$$

(2) *Connected with a load*: As described in Sect. 7.3.1, if the load is a constant impedance, it can be incorporated in the network. If the load is modeled as a second-order polynomial function, it is modeled as a current injection in (7.48), note that the constant impedance part of the load can also be incorporated into the network. The last two terms in (7.48) are treated as current injections. If the load is modeled as an exponential function of voltage, it can be viewed as a current injection (7.49).

(3) *Connected with an SVC*: The expression of the current injection of an SVC is given in (7.50).

7.5.3.2 Nodes Connected with Series Devices

If a dynamic device is connected in series between node i and node j, the network equations between node i and j are

$$\left.\begin{aligned} \Delta I_{xi} &= I_{xi} - \sum_{k \in i} (G_{ik} V_{xk} - B_{ik} V_{yk}) = 0 \\ \Delta I_{yi} &= I_{yi} - \sum_{k \in i} (G_{ik} V_{yk} + B_{ik} V_{xk}) = 0 \\ \Delta I_{xj} &= I_{xj} - \sum_{k \in j} (G_{jk} V_{xk} - B_{jk} V_{yk}) = 0 \\ \Delta I_{yj} &= I_{yj} - \sum_{k \in j} (G_{jk} V_{yk} + B_{jk} V_{xk}) = 0 \end{aligned}\right\}. \tag{7.259}$$

The currents I_{xi}, I_{yi}, I_{xj}, and I_{yj} take different forms, depending upon what device is connected between node i and j.

1. Series TCSC
   The expressions for currents I_{xi}, I_{yi}, I_{xj}, and I_{yj} are given by (7.51).
2. Series HVDC system
   If the AC–DC system is solved by a simultaneous approach, the expressions for DC currents I_{xi}, I_{yi}, I_{xj}, and I_{yj} injected into the AC nodes are given in (7.56). In the formula of (7.54) and (7.55), the DC current I_{d} can be replaced by (7.243) or (7.248), thus the current injections are functions of α, β, $V_{x\text{R}}$, $V_{y\text{R}}$, $V_{x\text{I}}$, and $V_{y\text{I}}$ only.

7.5.3.3 A Connection Node or Faulted Node

A connection node has zero current injection. As discussed before, any type of fault can be modeled by adjusting the admittance matrix of the positive sequence network based on the concept of the synthesized impedance matrix. Therefore, in the faulted node of the extended positive sequence network, there is no current injection. The faulted node is therefore a connection node. The network equation of a connection node or faulted node is

$$\left.\begin{aligned} \Delta I_{xf} &= 0 - \sum_{k \in f} (G_{fk} V_{xk} - B_{fk} V_{yk}) = 0 \\ \Delta I_{yf} &= 0 - \sum_{k \in f} (G_{fk} V_{yk} + B_{fk} V_{xk}) = 0 \end{aligned}\right\}. \tag{7.260}$$

7.5.4 Simultaneous Solution of Difference and Network Equations

All the equations of the system for step $t + \Delta t$ have been given, they include the network equations and the difference equations of each dynamic device. In this system of equations, the unknowns include the operating variables of the power system under study V_x and V_y; all the state variables of dynamic devices, for example, δ, E''_q, and E''_d of each generator, the B_{SVC} of each SVC, the B_{TCSC} of each TCSC, and the α and β of the HVDC system. Assume that there are n nodes, n_G generators, n_S SVCs, n_T TCSCs, n_D HVDC systems, then the number of unknowns is equal to $2n + 3n_G + n_S + n_T + 2n_D$, which is just equal to the number of equations. The system of equations is well defined.

Network equations (7.257), (7.259), (7.260), generator difference (7.196), SVC (7.207), TCSC (7.218), together with HVDC difference equations (7.244), (7.246) or (7.251), (7.254) comprise a set of nonlinear equations. The current injections and the difference equations are time varying, while the network equations assume the same structure, except for the steps at which a disturbance (either a fault or a switch operation) occurs. To compute the system states immediately after a disturbance, only the network equations need to be resolved. The state variables δ, E''_q, E''_d, B_{SVC}, B_{TCSC}, α, and β of the dynamic devices should take the values obtained before the disturbance. The set of nonlinear equations comprising the difference and network equations is solved in a recursive manner to provide the states of the study system at each integration step.

The above set of nonlinear equations is typically solved using a Newton method. Since the Newton method is already fairly familiar, the computational procedure of the method will only be briefly described below:

1. Set, for step $t + \Delta t$, the initial values of generator state variables δ, E''_q, E''_d, the initial values of SVC state variables B_{SVC}, the initial values of TCSC state variables B_{TCSC}, the initial values of HVDC state variables α, β, and the initial values of network voltage V_x and V_y. These initial values either can be set to the values at step t, or may be extrapolated from the values of the previous steps.
2. For the set of nonlinear equations comprising generator difference equations, SVC difference equations, TCSC difference equations, HVDC difference equations, and the network equations, compute the Jacobian matrix and mismatches given the initial values obtained in step (1), then solve the linear equations for the updates to the variables.
3. Check if the iteration has converged. If yes, stop; otherwise, return to step (2). The iteration continues until convergence is reached.
4. After the quantities of the state variables for $t + \Delta t$ are obtained, proceed to compute the values of the other dynamic variables according to the difference/algebraic equations derived in Sects. 7.5.1 and 7.5.2. These values will be useful for the computation of the next step. It should be noted that, the effects of limiters should be considered in this step.

Thinking and Problem Solving

1. What is meant by the transient stability of electrical power systems? What methods can be adopted to analyze it? How can we judge the transient stability of electrical power systems?
2. What are the consequences of loss of transient stability in a power system?
3. What suppositions are made within the transient stability analysis model of electrical power systems and what is the underlying theory?
4. What main measures exist to improve the transient stability of electrical power systems? What is the principle of each measure?
5. Give the method used to modify an admittance matrix when short-circuit failures at different locations and of different fault types occur on one transmission line in an electrical network, and list essential calculation formulas.
6. What aspects should be considered in choosing appropriate integration methods when numerical integration is used to analyze the transient stability of electrical power systems?
7. What kinds of initial value problems of differential equations belong to the class of "stiff" problems? What requirements are there for numerical integration methods to solve stiff problems?
8. What are the advantages and disadvantages of the alternating solution method and the simultaneous solution method in solving initial value problems of differential-algebraic equations?
9. How can we deal with limiters when a numerical integration method is used to find the time-domain solution of each state variable in an electrical power system?
10. Although there are many numerical integration methods, the implicit trapezoidal integration method obtains wide application in transient stability analysis of electrical power systems, why?
11. During dynamic security evaluation, it is required to carry out rapid transient stability analysis of electrical power systems under each contingency. What aspects can be considered to improve the speed of the transient stability analysis?
12. When analyzing the transient stability of electrical power systems, each generator can be represented by one of the following models: $E' = C$; $E_q' = C$; E_q' vary; E_q', E_d' vary; E_q', E_q'', E_d'' vary; E_q', E_q'', E_d', E_d'' vary. Explain the applicability of each model.
13. It can be seen from the numerical solution of the transient stability analysis of a real electrical power system that the current on an inductance and the voltage across the two terminals of a capacitance will change significantly at the second that failure occurs, which seems to not satisfy the law of electromagnetic induction. Why?
14. During transient stability calculation using the improved Euler's method, when considering the transient process of the excitation winding and the influence of excitation system, select a type of excitation system, and list the relevant formulas of the transient process calculation for one step.

Chapter 8
Small-Signal Stability Analysis of Power Systems

8.1 Introduction

Small-signal stability analysis is about power system stability when subject to small disturbances. If power system oscillations caused by small disturbances can be suppressed, such that the deviations of system state variables remain small for a long time, the power system is stable. On the contrary, if the magnitude of oscillations continues to increase or sustain indefinitely, the power system is unstable. Power system small-signal stability is affected by many factors, including initial operation conditions, strength of electrical connections among components in the power system, characteristics of various control devices, etc. Since it is inevitable that power system operation is subject to small disturbances, any power system that is unstable in terms of small-signal stability cannot operate in practice. In other words, a power system that is able to operate normally must first be stable in terms of small-signal stability. Hence, one of the principal tasks in power system analysis is to carry out small-signal stability analysis to assess the power system under the specified operating conditions.

The dynamic response of a power system subject to small disturbances can be studied by using the method introduced in Chap. 7 to determine system stability. However, when we use the method for power system small-signal stability analysis, in addition to slow computational speed, the weakness is that after a conclusion of instability is drawn, we cannot carry out any deeper investigation into the phenomenon and cause of system instability. The Lyapunov linearized method has provided a very useful tool for power system small-signal stability analysis. Based on the fruitful results of eigensolution analysis of linear systems, the Lyapunov linearized method has been widely used in power system small-signal stability analysis. In the following, we shall first introduce the basic mathematics of power system small-signal stability analysis.

The Lyapunov linearized method is closely related to the local stability of nonlinear systems. Intuitively speaking, movement of a nonlinear system over a small range should have similar properties to its linearized approximation.

X.-F. Wang et al., *Modern Power Systems Analysis.*
doi: 10.1007/978-0-387-72853-7,

Assume the nonlinear system described by

$$\frac{\mathrm{d}\mathbf{x}}{\mathrm{d}t} = f(\mathbf{x}).$$

Its Taylor expansion at the origin is

$$\frac{\mathrm{d}\Delta\mathbf{x}}{\mathrm{d}t} = \mathbf{A}\Delta\mathbf{x} + \mathbf{h}(\Delta\mathbf{x}), \tag{8.1}$$

where

$$\mathbf{A} = \left.\frac{\partial \mathbf{f}(\mathbf{x}_e + \Delta\mathbf{x})}{\partial \Delta\mathbf{x}}\right|_{\Delta\mathbf{x}=\mathbf{0}} = \left.\frac{\partial \mathbf{f}(\mathbf{x})}{\partial \mathbf{x}}\right|_{\mathbf{x}=\mathbf{x}_e}.$$

If in the neighborhood of $\Delta\mathbf{x} = \mathbf{0}$, $\mathbf{h}(\Delta\mathbf{x})$ is a high-order function of $\Delta\mathbf{x}$, we can use the stability of the following linear system

$$\frac{\mathrm{d}\Delta\mathbf{x}}{\mathrm{d}t} = \mathbf{A}\Delta\mathbf{x}. \tag{8.2}$$

To study the stability of the nonlinear system at point $\mathbf{x}_e$

(1) If the linearized system is asymptotically stable, i.e., all eigenvalues of $\mathbf{A}$ have negative real parts, the actual nonlinear system is asymptotically stable at the equilibrium point.
(2) If the linearized system is unstable, i.e., at least one of eigenvalues of $\mathbf{A}$ has a positive real part, the actual nonlinear system is unstable at the equilibrium point.
(3) If the linearized system is critically stable, i.e., real parts of all eigenvalues of $\mathbf{A}$ are nonpositive but the real part of at least one of them is zero, no conclusion can be drawn about the stability of the nonlinear system from its linearized approximation.

The basic principle of the Lyapunov linearized method is to draw conclusions about the local stability of the nonlinear system around the equilibrium point from the stability of its linear approximation.

When carrying out small-signal stability analysis of a power system, we always assume that the system at normal operation at equilibrium point $\mathbf{x} = \mathbf{x}_e$ or $\Delta\mathbf{x} = \mathbf{0}$ is disturbed instantly at the moment $t = t_0$ when system state moves from $\mathbf{0}$ to $\Delta\mathbf{x}(t_0)$. $\Delta\mathbf{x}(t_0)$ is the initial state of system free movement after disappearance of the disturbance. Because the disturbance is sufficiently small, $\Delta\mathbf{x}(t_0)$ is within a sufficiently small neighborhood of $\Delta\mathbf{x} = \mathbf{0}$. Thus in the neighborhood of $\Delta\mathbf{x} = \mathbf{0}$, $\mathbf{h}(\Delta\mathbf{x})$ is a high-order indefinitely small variable. Hence according to the Lyapunov linearized method, we can study the stability of the linearized system to investigate that of the actual nonlinear power system.

Linearizing the differential-algebraic dynamic description of a power system of (8.1) and (8.2) at steady-state operating point $(\mathbf{x}_{(0)}, \mathbf{y}_{(0)})$, we can obtain

$$\begin{bmatrix} \mathrm{d}\Delta\mathbf{x}/\mathrm{d}t \\ \mathbf{0} \end{bmatrix} = \begin{bmatrix} \tilde{\mathbf{A}} & \tilde{\mathbf{B}} \\ \tilde{\mathbf{C}} & \tilde{\mathbf{D}} \end{bmatrix} \begin{bmatrix} \Delta\mathbf{x} \\ \Delta\mathbf{y} \end{bmatrix}, \tag{8.3}$$

where

$$\tilde{\mathbf{A}} = \begin{bmatrix} \frac{\partial f_1}{\partial x_1} & \cdots & \frac{\partial f_1}{\partial x_n} \\ \vdots & & \vdots \\ \frac{\partial f_n}{\partial x_1} & \cdots & \frac{\partial f_n}{\partial x_n} \end{bmatrix}_{\substack{\mathbf{x}=\mathbf{x}_{(0)} \\ \mathbf{y}=\mathbf{y}_{(0)}}}, \quad \tilde{\mathbf{B}} = \begin{bmatrix} \frac{\partial f_1}{\partial y_1} & \cdots & \frac{\partial f_1}{\partial y_m} \\ \vdots & & \vdots \\ \frac{\partial f_n}{\partial y_1} & \cdots & \frac{\partial f_n}{\partial y_m} \end{bmatrix}_{\substack{\mathbf{x}=\mathbf{x}_{(0)} \\ \mathbf{y}=\mathbf{y}_{(0)}}},$$

$$\tilde{\mathbf{C}} = \begin{bmatrix} \frac{\partial g_1}{\partial x_1} & \cdots & \frac{\partial g_1}{\partial x_n} \\ \vdots & & \vdots \\ \frac{\partial g_m}{\partial x_1} & \cdots & \frac{\partial g_m}{\partial x_n} \end{bmatrix}_{\substack{\mathbf{x}=\mathbf{x}_{(0)} \\ \mathbf{y}=\mathbf{y}_{(0)}}}, \quad \tilde{\mathbf{D}} = \begin{bmatrix} \frac{\partial g_1}{\partial y_1} & \cdots & \frac{\partial g_1}{\partial y_m} \\ \vdots & & \vdots \\ \frac{\partial g_m}{\partial y_1} & \cdots & \frac{\partial g_m}{\partial y_m} \end{bmatrix}_{\substack{\mathbf{x}=\mathbf{x}_{(0)} \\ \mathbf{y}=\mathbf{y}_{(0)}}}.$$

$\mathbb{R}$ denotes the set of real numbers, $\mathbb{R}^n$ is the n-dimensional space of real vectors, $\mathbb{R}^{m\times n}$ is the set of m-row n-column real matrices. We define $\mathbb{R}^n$ to be $\mathbb{R}^{n\times 1}$, i.e., elements in $\mathbb{R}^n$ are column vectors. On the other hand, elements in $\mathbb{R}^{1\times n}$ are row vectors. Obviously in the above equation, $\tilde{\mathbf{A}} \in \mathbb{R}^{n\times n}$, $\tilde{\mathbf{B}} \in \mathbb{R}^{n\times m}$, $\tilde{\mathbf{C}} \in \mathbb{R}^{m\times n}$, $\tilde{\mathbf{D}} \in \mathbb{R}^{m\times m}$.

Deleting $\Delta\mathbf{y}$ in (8.3), we have

$$\frac{\mathrm{d}\Delta\mathbf{x}}{\mathrm{d}t} = \mathbf{A}\Delta\mathbf{x}, \tag{8.4}$$

where

$$\mathbf{A} = \tilde{\mathbf{A}} - \tilde{\mathbf{B}}\tilde{\mathbf{D}}^{-1}\tilde{\mathbf{C}}. \tag{8.5}$$

Matrix $\mathbf{A} \in \mathbb{R}^{n\times n}$ is often referred to as the state matrix or coefficient matrix.

Therefore, small-signal stability studies local characteristics of the power system, i.e., asymptotic stability of an equilibrium point before the system is disturbed. Obviously, the theoretical basis to study power system small-signal stability by using the Lyapunov linearized method is that the disturbance must be sufficiently small. When the power system is subject to any such disturbance, state variables of the transient system model vary over a very small range. Hence asymptotic stability of the linearized system can guarantee a certain type of asymptotic stability of the actual nonlinear system.

We know that when the power system is subject to a sufficiently small disturbance at steady-state operation, there can be two consequences. One is that the disturbance approaches zero with time (i.e., disturbed movement approaches the undisturbed movement and all eigenvalues of corresponding matrix **A** have negative real parts) in this case the system is asymptotically stable at steady-state operation. The disturbed system will eventually return to the steady-state operation before the occurrence of disturbance. Another possible consequence is that disturbance $\Delta\mathbf{x}$ increases indefinitely with time, no matter how small the disturbance is (i.e., the real part of at least one of the eigenvalues of **A** is positive). Obviously the system is then unstable at this steady-state operating point. For the operation of a real power system, study of the critically stable situation is not so important, except that we can see it as the limiting case of small-signal stability.

Here we need to point out that in our previous discussion about system stability, we assumed that the disturbance was instantaneous. That is, the system state moves instantly from $\Delta\mathbf{x} = \mathbf{0}$ to $\Delta\mathbf{x}(t_0)$, and the disturbance disappears when the movement happens. However, the same theory is applicable to the study of stability when the system is subject to a permanent disturbance, because we can consider this as a case of stability subject to an instantaneous disturbance but operating at a new equilibrium point.

Furthermore, for certain operating conditions in which the system is unstable in terms of small-signal stability or lacks damping, we can determine relationships between some controller parameters and the system eigenvalues (representing system stability) by using eigensolution analysis. In doing so, we can find certain ways to improve the power system small-signal stability. Hence small-signal stability analysis is a very important aspect of power system analysis and control.

Therefore, power system analysis for operation at a steady state and subject to small disturbances includes:

(1) Computation of steady-state values of various variables of the power system at a given steady-state operating condition,
(2) Linearization of the differential-algebraic description of power system nonlinear dynamics to obtain the linearized differential-algebraic equations,
(3) Formation of system state matrix **A** from the system linearized differential-algebraic equations to determine system stability by calculation of the eigenvalues of **A**.

In our above discussion, only the electromechanical oscillations between generators are considered in small-signal stability analysis. That is, we consider generators to be lumped rigid masses. However, the mechanical structure of real large-scale steam-turbine generation units is very complicated. It consists of several major lumped masses, such as turbine rotor, generator rotor, exciter rotor, etc. These lumped masses are connected by a rigid shaft of limited length. When generation units are disturbed, rotational speeds of the lumped masses are different during the system transient, due to elasticity between the lumped mass. This leads to torsional oscillations between each lumped mass. Because the inertia of each lumped mass is smaller than the total inertia of generation units, and taking into account the relevant stiffness, the frequency of torsional oscillations between each lumped mass is higher than that of electromechanical oscillations between generation units. Frequency of torsional oscillation is between about 10 and 50 Hz. This oscillation is often referred to as sub-synchronous oscillation (SSO).

When SSO occurs, there is an oscillating torsional torque between lumped masses connected by the common shaft of a generation unit. Fatigue accumulation due to repeated episodes of torsional oscillation on the shaft will reduce shaft operating life. If the torsional torque exceeds a certain limit, shaft cracking, even breaking, can happen. Occurrence of SSO is mainly related to excitation control, governing control, HVDC control, and interaction between transmission line and series compensation of the line. When carrying out torsional oscillation analysis, we need to first establish a mathematical model of the shaft system of the steam turbine and generator. In addition, because frequency of torsional oscillation is high, a quasi-steady-state model of various components cannot be used. Instead, the electromagnetic transients in the power system need to be considered. Detailed analysis on torsional oscillation is outside the scope of this book.

In this chapter, we first derive linearized models of various dynamic components in power systems, to establish the linearized equations of the whole system, in order to demonstrate the basic steps for computation of small-signal stability in power systems. Then, we will discuss the eigensolution problem in power system small-signal stability analysis and the analytical methods required to study power system oscillations.

8.2 Linearized Equations of Power System Dynamic Components

In power system small-signal stability analysis, we need to linearize various dynamic components in the power system. In linearization, limiters in control devices often need not be considered. This is because in normal steady-state operation, the values of state variables associated with control devices are within the range determined by the limiters. If disturbances are sufficiently small, variations of state variables will not go beyond these limitations. As far as dead zones associated with certain control devices are concerned, we generally consider the dead zone to be small and hence ignored. If the dead zone is large, we can simply consider that in this case the control device does not function.

8.2.1 Linearized Equation of Synchronous Generator

8.2.1.1 Linearized Equation of Each Part of a Synchronous Generator

(1) *Synchronous generator*: For a synchronous generator described by (7.114)–(7.116) at a given steady-state operating condition, steady-state values of

various variables are $\delta_{(0)}$, $\omega_{(0)}$, $E'_{q(0)}$, $E''_{q(0)}$, $E'_{d(0)}$, $E''_{d(0)}$, $I_{d(0)}$, $I_{q(0)}$, $V_{d(0)}$, $V_{q(0)}$, $P_{m(0)}$, $P_{e(0)}$, $E_{fq(0)}$ which can be calculated from (7.74–7.78) and (7.118–7.122). Linearizing each equation at these steady-state values, we obtain the linearized equation of a synchronous generator

$$\left.\begin{aligned}
\frac{\mathrm{d}\Delta\delta}{\mathrm{d}t} &= \omega_s\Delta\omega \\
\frac{\mathrm{d}\Delta\omega}{\mathrm{d}t} &= \frac{1}{T_J}\Big\{-D\Delta\omega - I_{q(0)}\Delta E''_q - I_{d(0)}\Delta E''_d + \Delta P_m - \left[E''_{d(0)} - \left(X''_d - X''_q\right)I_{q(0)}\right] \\
&\qquad \Delta I_d - \left[E''_{q(0)} - \left(X''_d - X''_q\right)I_{d(0)}\right]\Delta I_q\Big\} \\
\frac{\mathrm{d}\Delta E'_q}{\mathrm{d}t} &= \frac{1}{T'_{d0}}\left[-k_d\Delta E'_q + (k_d - 1)\Delta E''_q + \Delta E_{fq}\right] \\
\frac{\mathrm{d}\Delta E''_q}{\mathrm{d}t} &= \frac{1}{T''_{d0}}\left[\Delta E'_q - \Delta E''_q - (X'_d - X''_d)\Delta I_d\right] \\
\frac{\mathrm{d}\Delta E'_d}{\mathrm{d}t} &= \frac{1}{T'_{q0}}\left[-k_q\Delta E'_d + (k_q - 1)\Delta E''_d\right] \\
\frac{\mathrm{d}\Delta E''_d}{\mathrm{d}t} &= \frac{1}{T''_{q0}}\left[\Delta E'_d - \Delta E''_d + \left(X'_q - X''_q\right)\Delta I_q\right]
\end{aligned}\right\}, \tag{8.6}$$

$$\left.\begin{aligned}
\Delta V_d &= \Delta E''_d - R_a\Delta I_d + X''_q\Delta I_q \\
\Delta V_q &= \Delta E''_q - X''_d\Delta I_d - R_a\Delta I_q
\end{aligned}\right\}. \tag{8.7}$$

(2) *Excitation system*: Taking an excitation system consisting of a DC exciter with thyristor-controlled regulator as an example, we can derive the linearized equation of (7.136–7.140) as follows.

For measurement unit with $V_C = \left|\dot{V} + jX_C\dot{I}\right|$, from coordinate transformation of (6.63), *d*, *q* components of voltage and current at generator terminals can be represented as

$$\dot{V} = (V_d + jV_q)\mathrm{e}^{j(\delta-\pi/2)}, \quad \dot{I} = (I_d + jI_q)\mathrm{e}^{j(\delta-\pi/2)}. \tag{8.8}$$

Obviously we have

$$\begin{aligned}
V_C &= \left|[(V_d + jV_q) + jX_C(I_d + jI_q)]\mathrm{e}^{j(\delta-\pi/2)}\right| \\
&= \left|(V_d + jV_q) + jX_C(I_d + jI_q)\right| \\
&= \sqrt{(V_d - X_C I_q)^2 + (V_q + X_C I_d)^2}
\end{aligned}. \tag{8.9}$$

Linearizing the above equations at steady-state values, we can obtain

$$\Delta V_C = K_{cd}(\Delta V_d - X_C\Delta I_q) + K_{cq}(\Delta V_q + X_C\Delta I_d), \tag{8.10}$$

where

$$\left.\begin{aligned} K_{cd} &= (V_{d(0)} - X_C I_{q(0)})/V_{C(0)}, \quad K_{cq} = (V_{q(0)} + X_C I_{d(0)})/V_{C(0)} \\ V_{C(0)} &= \sqrt{(V_{d(0)} - X_C I_{q(0)})^2 + (V_{q(0)} + X_C I_{d(0)})^2} \end{aligned}\right\}. \tag{8.11}$$

Linearizing (7.136), substituting (8.10) into it to cancel ΔV_C, we can obtain the linearized equation of measuring and filtering unit to be

$$\frac{\mathrm{d}\Delta V_M}{\mathrm{d}t} = \frac{1}{T_R}(-\Delta V_M + K_{cq}X_C\Delta I_d - K_{cd}X_C\Delta I_q + K_{cd}\Delta V_d + K_{cq}\Delta V_q). \tag{8.12}$$

Representing the saturation characteristic of the exciter by (7.140) and linearizing (7.139) at steady-state operating point, we can have the linearized equation of the exciter to be

$$\frac{\mathrm{d}\Delta E_{fq}}{\mathrm{d}t} = \frac{1}{T_E}\left[-\left(K_E + n_E c_E E_{fq(0)}^{n_E-1}\right)\Delta E_{fq} + \Delta V_R\right]. \tag{8.13}$$

Finally, linearizing (7.137, 7.138) and rearranging them together with (8.12) and (8.13), we can obtain the linearized equation of the whole DC excitation system to be

$$\left.\begin{aligned} \frac{\mathrm{d}\Delta E_{fq}}{\mathrm{d}t} &= -\frac{K_E + n_E c_E E_{fq(0)}^{n_E-1}}{T_E}\Delta E_{fq} + \frac{1}{T_E}\Delta V_R \\ \frac{\mathrm{d}\Delta V_R}{\mathrm{d}t} &= -\frac{1}{T_A}\Delta V_R - \frac{K_A}{T_A}\Delta V_F - \frac{K_A}{T_A}\Delta V_M + \frac{K_A}{T_A}\Delta V_S \\ \frac{\mathrm{d}\Delta V_F}{\mathrm{d}t} &= -\frac{K_F\left(K_E + n_E c_E E_{fq(0)}^{n_E-1}\right)}{T_E T_F}\Delta E_{fq} + \frac{K_F}{T_E T_F}\Delta V_R - \frac{1}{T_F}\Delta V_F \\ \frac{\mathrm{d}\Delta V_M}{\mathrm{d}t} &= -\frac{1}{T_R}\Delta V_M + \frac{K_{cq}X_C}{T_R}\Delta I_d - \frac{K_{cd}X_C}{T_R}\Delta I_q + \frac{K_{cd}}{T_R}\Delta V_d + \frac{K_{cq}}{T_R}\Delta V_q \end{aligned}\right\}. \tag{8.14}$$

(3) *Power system stabilizer*: For a Power system stabilizer (PSS) of Fig. 6.14, from (7.142) and (7.143) we can establish the following linearized equations when input to PSS is the deviation of rotor speed, $V_{IS} = \omega - \omega_s$

$$\left.\begin{aligned} \frac{\mathrm{d}\Delta V_1}{\mathrm{d}t} &= \frac{K_S}{T_6}\Delta\omega - \frac{1}{T_6}\Delta V_1 \\ \frac{\mathrm{d}(\Delta V_1 - \Delta V_2)}{\mathrm{d}t} &= \frac{1}{T_5}\Delta V_2 \\ \frac{\mathrm{d}(T_1\Delta V_2 - T_2\Delta V_3)}{\mathrm{d}t} &= \Delta V_3 - \Delta V_2 \\ \frac{\mathrm{d}(T_3\Delta V_3 - T_4\Delta V_S)}{\mathrm{d}t} &= \Delta V_S - \Delta V_3 \end{aligned}\right\}. \tag{8.15}$$

After rearrangement, we can obtain linearized state equations of the PSS as follows

$$\left.\begin{aligned}
\frac{\mathrm{d}\Delta V_1}{\mathrm{d}t} &= \frac{K_S}{T_6}\Delta\omega - \frac{1}{T_6}\Delta V_1\\
\frac{\mathrm{d}\Delta V_2}{\mathrm{d}t} &= \frac{K_S}{T_6}\Delta\omega - \frac{1}{T_6}\Delta V_1 - \frac{1}{T_5}\Delta V_2\\
\frac{\mathrm{d}\Delta V_3}{\mathrm{d}t} &= \frac{K_S T_1}{T_2 T_6}\Delta\omega - \frac{T_1}{T_2 T_6}\Delta V_1 - \frac{T_1 - T_5}{T_2 T_5}\Delta V_2 - \frac{1}{T_2}\Delta V_3\\
\frac{\mathrm{d}\Delta V_S}{\mathrm{d}t} &= \frac{K_S T_1 T_3}{T_2 T_4 T_6}\Delta\omega - \frac{T_1 T_3}{T_2 T_4 T_6}\Delta V_1 - \frac{T_3(T_1 - T_5)}{T_2 T_4 T_5}\Delta V_2 - \frac{T_3 - T_2}{T_2 T_4}\Delta V_3 - \frac{1}{T_4}\Delta V_S
\end{aligned}\right\}. \tag{8.16}$$

(4) *Prime mover and governing system*: For the hydraulic turbine and its governing system, of Fig. 6.24, we can obtain its linearized equation from (7.171)–(7.177) to be

$$\left.\begin{aligned}
\frac{\mathrm{d}\Delta\mu}{\mathrm{d}t} &= -\frac{K_\delta}{T_S}\Delta\omega - \frac{1}{T_S}\Delta\xi\\
\frac{\mathrm{d}\Delta\xi}{\mathrm{d}t} &= -\frac{K_\delta(K_i + K_\beta)}{T_S}\Delta\omega + \frac{K_i}{T_i}\Delta\mu - \left(\frac{1}{T_i} + \frac{K_i + K_\beta}{T_S}\right)\Delta\xi\\
\frac{\mathrm{d}\Delta P_m}{\mathrm{d}t} &= \frac{2K_{mH}K_\delta}{T_S}\Delta\omega + \frac{2K_{mH}}{T_\omega}\Delta\mu + \frac{2K_{mH}}{T_S}\Delta\xi - \frac{2}{T_\omega}\Delta P_m
\end{aligned}\right\}. \tag{8.17}$$

8.2.1.2 Matrix Description of Linearized Equation of Synchronous Generator Unit and Coordinate Transformation

(1) *Matrix description of generation unit*: For a generation unit described by (8.6), (8.7), (8.9), (8.15) and (8.17), its state variables can be arranged to form the following vector:

$$\begin{aligned}
\Delta\mathbf{x}_g = [&\Delta\delta, \Delta\omega, \Delta E'_q, \Delta E''_q, \Delta E'_d, \Delta E''_d, \Delta E_{fq}, \Delta V_R,\\
&\Delta V_F, \Delta V_M, \Delta V_1, \Delta V_2, \Delta V_3, \Delta V_S, \Delta\mu, \Delta\xi, \Delta P_m]^{\mathrm{T}}.
\end{aligned} \tag{8.18}$$

We define

$$\Delta\mathbf{V}_{dqg} = [\Delta V_d, \Delta V_q]^{\mathrm{T}}, \quad \Delta\mathbf{I}_{dqg} = [\Delta I_d, \Delta I_q]^{\mathrm{T}}. \tag{8.19}$$

Linearized differential equations of each generation unit can be written in the following matrix form

$$\frac{\mathrm{d}\Delta\mathbf{x}_g}{\mathrm{d}t} = \bar{\mathbf{A}}_g\Delta\mathbf{x}_g + \bar{\mathbf{B}}_{Ig}\Delta\mathbf{I}_{dqg} + \bar{\mathbf{B}}_{Vg}\Delta\mathbf{V}_{dqg}. \tag{8.20}$$

Linearized equations of armature voltage equations can be arranged as

$$\Delta\mathbf{V}_{dqg} = \bar{\mathbf{P}}_g\Delta\mathbf{x}_g + \bar{\mathbf{Z}}_g\Delta\mathbf{I}_{dqg}. \tag{8.21}$$

In the two equations above, elements in the coefficient matrices $\bar{\mathbf{A}}_g, \bar{\mathbf{B}}_{Ig}, \bar{\mathbf{B}}_{Vg}, \bar{\mathbf{P}}_g, \bar{\mathbf{Z}}_g$, can be obtained easily by comparing (8.20) and (8.6), (8.9), (8.15), (8.17) and comparing (8.21) and (8.7) as follows:

$$
\bar{\mathbf{A}}_g = \left[\begin{array}{c|c|c|c|c|c|c|c|c|c|c|c|c|c|c|c|c}
 & \omega_s & & & & & & & & & & & & & & & \\
 & -\frac{D}{T_J} & & -\frac{I_{q(0)}}{T_J} & & -\frac{I_{d(0)}}{T_J} & & & & & & & & & & & \frac{1}{T_J} \\
 & & -\frac{k_d}{T'_{d0}} & \frac{k_d-1}{T'_{d0}} & & & \frac{1}{T'_{d0}} & & & & & & & & & & \\
 & & \frac{1}{T''_{d0}} & -\frac{1}{T''_{d0}} & & & & & & & & & & & & & \\
 & & & & -\frac{k_q}{T'_{q0}} & \frac{k_q-1}{T'_{q0}} & & & & & & & & & & & \\
 & & & & \frac{1}{T''_{q0}} & -\frac{1}{T''_{q0}} & & & & & & & & & & & \\
\hline
 & & & & & & -\frac{K'_E}{T_E} & \frac{1}{T_E} & & & & & & & & & \\
 & & & & & & & -\frac{1}{T_A} & -\frac{K_A}{T_A} & -\frac{K_A}{T_A} & & & & \frac{K_A}{T_A} & & & \\
 & & & & & & -\frac{K_F K'_E}{T_E T_F} & \frac{K_F}{T_E T_F} & -\frac{1}{T_F} & & & & & & & & \\
 & & & & & & & & & -\frac{1}{T_R} & & & & & & & \\
\hline
 & \frac{K_S}{T_6} & & & & & & & & & -\frac{1}{T_6} & & & & & & \\
 & \frac{K_S}{T_6} & & & & & & & & & -\frac{1}{T_6} & -\frac{1}{T_5} & & & & & \\
 & \frac{K_S T_1}{T_2 T_6} & & & & & & & & & -\frac{T_1}{T_2 T_6} & \frac{T_5 - T_1}{T_2 T_5} & -\frac{1}{T_2} & & & & \\
 & \frac{K_S T_1 T_3}{T_2 T_4 T_6} & & & & & & & & & -\frac{T_1 T_3}{T_2 T_4 T_6} & \frac{T_3 (T_5 - T_1)}{T_2 T_4 T_5} & \frac{T_2 - T_3}{T_2 T_4} & -\frac{1}{T_4} & & & \\
\hline
 & -\frac{K_\delta}{T_S} & & & & & & & & & & & & & & -\frac{1}{T_S} & \\
 & -\frac{K_\delta (K_\iota + K_\beta)}{T_S} & & & & & & & & & & & & & \frac{K_\iota}{T_\iota} & -\left(\frac{1}{T_\iota} + \frac{K_\iota + K_\beta}{T_S}\right) & \\
 & \frac{2K_{mH} K_\delta}{T_S} & & & & & & & & & & & & & \frac{2K_{mH}}{T_\omega} & \frac{2K_{mH}}{T_S} & -\frac{2}{T_\omega}
\end{array}\right]
$$

$$\bar{\mathbf{B}}_{Ig} = \left[\begin{array}{c|c}
 & \\
\dfrac{(X''_d - X''_q)I_{q(0)} - E''_{d(0)}}{T_J} & \dfrac{(X''_d - X''_q)I_{d(0)} - E''_{q(0)}}{T_J} \\
 & \\
\dfrac{X''_d - X'_d}{T''_{d0}} & \\
 & \\
 & \dfrac{X'_q - X''_q}{T''_{q0}} \\
\hline
 & \\
 & \\
 & \\
\dfrac{K_{cq}X_C}{T_R} & -\dfrac{K_{cd}X_C}{T_R} \\
\hline
 & \\
 & \\
 & \\
 & \\
\hline
 & \\
 & \\
 &
\end{array}\right], \quad
\bar{\mathbf{B}}_{Vg} = \left[\begin{array}{c|c}
 & \\
 & \\
 & \\
 & \\
 & \\
 & \\
\hline
 & \\
 & \\
 & \\
\dfrac{K_{cd}}{T_R} & \dfrac{K_{cq}}{T_R} \\
\hline
 & \\
 & \\
 & \\
 & \\
\hline
 & \\
 & \\
 &
\end{array}\right],$$

$$\bar{\mathbf{P}}_g = \left[\begin{array}{c|c|c|c|c|c|c|c|c|c|c|c|c|c|c|c|c}
 & & & & & 1 & & & & & & & & & & & \\
\hline
 & & & 1 & & & & & & & & & & & & &
\end{array}\right],$$

$$\bar{\mathbf{Z}}_g = \begin{bmatrix} -R_a & X''_q \\ -X''_d & -R_a \end{bmatrix}$$

If different models of synchronous generator, excitation system and governing system are adopted, using the same procedure we can always first derive the linearized equation of each unit and then represent them in the form of (8.20) and (8.21). Of course, in (8.18), the sequential arrangement of state variables can be different, which would lead to different coefficient matrices.

(2) *Coordinate transformation*: In (8.20) and (8.21), $\Delta\mathbf{V}_{dqg}$ and $\Delta\mathbf{I}_{dqg}$ are the deviation of d, q voltage and current components of each generator, respectively. Hence we must convert them into a representation in a unified x–y coordinate rotating at the same speed, so that they can then be connected to a common power network.

Coordinate transformation for the terminal voltages of the generator is given by (6.62)

$$\begin{bmatrix} V_d \\ V_q \end{bmatrix} = \begin{bmatrix} \sin\delta & -\cos\delta \\ \cos\delta & \sin\delta \end{bmatrix} \begin{bmatrix} V_x \\ V_y \end{bmatrix}. \tag{8.22}$$

Steady-state values $V_{d(0)}$, $V_{q(0)}$, $V_{x(0)}$, $V_{y(0)}$, and $\delta_{(0)}$ should also satisfy (8.22). That is

$$\begin{bmatrix} V_{d(0)} \\ V_{q(0)} \end{bmatrix} = \begin{bmatrix} \sin\delta_{(0)} & -\cos\delta_{(0)} \\ \cos\delta_{(0)} & \sin\delta_{(0)} \end{bmatrix} \begin{bmatrix} V_{x(0)} \\ V_{y(0)} \end{bmatrix}. \tag{8.23}$$

Linearizing (8.22) at steady-state values we have

$$\begin{bmatrix} \Delta V_d \\ \Delta V_q \end{bmatrix} = \begin{bmatrix} \sin\delta_{(0)} & -\cos\delta_{(0)} \\ \cos\delta_{(0)} & \sin\delta_{(0)} \end{bmatrix} \begin{bmatrix} \Delta V_x \\ \Delta V_y \end{bmatrix} + \begin{bmatrix} \cos\delta_{(0)} & \sin\delta_{(0)} \\ -\sin\delta_{(0)} & \cos\delta_{(0)} \end{bmatrix} \times \begin{bmatrix} V_{x(0)} \\ V_{y(0)} \end{bmatrix} \Delta\delta. \tag{8.24}$$

From (8.23) we can rewrite (8.24) as

$$\begin{bmatrix} \Delta V_d \\ \Delta V_q \end{bmatrix} = \begin{bmatrix} \sin\delta_{(0)} & -\cos\delta_{(0)} \\ \cos\delta_{(0)} & \sin\delta_{(0)} \end{bmatrix} \begin{bmatrix} \Delta V_x \\ \Delta V_y \end{bmatrix} + \begin{bmatrix} V_{q(0)} \\ -V_{d(0)} \end{bmatrix} \Delta\delta \tag{8.25}$$

that can be written simply as

$$\Delta\mathbf{V}_{dqg} = \mathbf{T}_{g(0)}\Delta\mathbf{V}_g + \mathbf{R}_{Vg}\Delta\mathbf{x}_g, \tag{8.26}$$

where

$$\Delta\mathbf{V}_g = \begin{bmatrix} \Delta V_x \\ \Delta V_y \end{bmatrix}, \quad \mathbf{R}_{Vg} = \begin{bmatrix} V_{q(0)} & 0 & \cdots & 0 \\ -V_{d(0)} & 0 & \cdots & 0 \end{bmatrix},$$

$$\mathbf{T}_{g(0)} = \begin{bmatrix} \sin\delta_{(0)} & -\cos\delta_{(0)} \\ \cos\delta_{(0)} & \sin\delta_{(0)} \end{bmatrix}.$$

Note that $\mathbf{T}_{g(0)}$ is an orthogonal matrix, satisfying

$$\mathbf{T}_{g(0)}^{-1} = \mathbf{T}_{g(0)}^{\mathrm{T}}. \tag{8.27}$$

Similarly, for generator current we can obtain

$$\Delta \mathbf{I}_{dqg} = \mathbf{T}_{g(0)}\Delta \mathbf{I}_g + \mathbf{R}_{Ig}\Delta \mathbf{x}_g, \tag{8.28}$$

where

$$\Delta \mathbf{I}_g = \begin{bmatrix} \Delta I_x \\ \Delta I_y \end{bmatrix}, \quad \mathbf{R}_{Ig} = \begin{bmatrix} I_{q(0)} & 0 & \cdots & 0 \\ -I_{d(0)} & 0 & \cdots & 0 \end{bmatrix}.$$

Substituting (8.26) and (8.28) into (8.21) to cancel $\Delta \mathbf{V}_{dqg}$ and $\Delta \mathbf{I}_{dqg}$, we have

$$\Delta \mathbf{I}_g = \mathbf{C}_g \Delta \mathbf{x}_g + \mathbf{D}_g \Delta \mathbf{V}_g, \tag{8.29}$$

where

$$\left.\begin{aligned} \mathbf{C}_g &= \mathbf{T}^{\mathrm{T}}_{g(0)}[\bar{\mathbf{Z}}_g^{-1}(\mathbf{R}_{Vg} - \bar{\mathbf{P}}_g) - \mathbf{R}_{Ig}] \\ \mathbf{D}_g &= \mathbf{T}^{\mathrm{T}}_{g(0)}\bar{\mathbf{Z}}_g^{-1}\mathbf{T}_{g(0)} \end{aligned}\right\}. \tag{8.30}$$

Substituting (8.26) and (8.28) into (8.20) to cancel $\Delta \mathbf{V}_{dqg}$ and $\Delta \mathbf{I}_{dqg}$ and using (8.29), (8.30) to cancel $\Delta \mathbf{I}_g$, we can obtain

$$\frac{\mathrm{d}\Delta \mathbf{x}_g}{\mathrm{d}t} = \mathbf{A}_g \Delta \mathbf{x}_g + \mathbf{B}_g \Delta \mathbf{V}_g, \tag{8.31}$$

where

$$\left.\begin{aligned} \mathbf{A}_g &= \bar{\mathbf{A}}_g + \bar{\mathbf{B}}_{Ig}\bar{\mathbf{Z}}_g^{-1}(\mathbf{R}_{Vg} - \bar{\mathbf{P}}_g) + \bar{\mathbf{B}}_{Vg}\mathbf{R}_{Vg} \\ \mathbf{B}_g &= (\bar{\mathbf{B}}_{Ig}\bar{\mathbf{Z}}_g^{-1} + \bar{\mathbf{B}}_{Vg})\mathbf{T}_{g(0)} \end{aligned}\right\}. \tag{8.32}$$

Equations (8.31) and (8.29) consist of linearized equations of every generator, in the form of the state equation and output equation for a general time-invariant linear system.

8.2.2 Linearized Equation of Load

In small-signal stability analysis, a static load model is usually adopted. If a certain amount of induction motor load needs to be considered, we can use procedures similar to those used to derive the linearized equations of a synchronous generator, to establish the linearized equations of an induction motor.

No matter which form is adopted to model the static voltage characteristics of load, deviation of injected current into the load has the following relationship to nodal voltage:

$$\Delta \mathbf{I}_l = \mathbf{Y}_l \Delta \mathbf{V}_l, \tag{8.33}$$

where

$$\Delta \mathbf{I}_l = \begin{bmatrix} \Delta I_x \\ \Delta I_y \end{bmatrix}, \quad \mathbf{Y}_l = \begin{bmatrix} G_{xx} & B_{xy} \\ -B_{yx} & G_{yy} \end{bmatrix}, \quad \Delta \mathbf{V}_l = \begin{bmatrix} \Delta V_x \\ \Delta V_y \end{bmatrix}. \tag{8.34}$$

The coefficients can be calculated from the following relationship between injected current and nodal voltage at the load node

$$\left.\begin{aligned} G_{xx} &= \left.\frac{\partial I_x}{\partial V_x}\right|_{\substack{V_x=V_{x(0)}\\ V_y=V_{y(0)}}}, \quad & B_{xy} &= \left.\frac{\partial I_x}{\partial V_y}\right|_{\substack{V_x=V_{x(0)}\\ V_y=V_{y(0)}}}, \\ B_{yx} &= -\left.\frac{\partial I_y}{\partial V_x}\right|_{\substack{V_x=V_{x(0)}\\ V_y=V_{y(0)}}}, \quad & G_{yy} &= \left.\frac{\partial I_y}{\partial V_y}\right|_{\substack{V_x=V_{x(0)}\\ V_y=V_{y(0)}}}. \end{aligned}\right. \tag{8.35}$$

If the static voltage characteristic of the load is modeled by a quadratic polynomial, we can use the relationship of (8.48) between injected current and node voltage and (8.35) to calculate relevant coefficients in (8.34) directly

$$\left.\begin{aligned} G_{xx} &= \frac{P_{(0)}V_{x(0)}^2(b_P+2c_P)+Q_{(0)}V_{x(0)}V_{y(0)}(b_Q+2c_Q)}{V_{(0)}^4}-\frac{P_{(0)}}{V_{(0)}^2} \\ B_{xy} &= \frac{Q_{(0)}V_{y(0)}^2(b_Q+2c_Q)+P_{(0)}V_{x(0)}V_{y(0)}(b_P+2c_P)}{V_{(0)}^4}-\frac{Q_{(0)}}{V_{(0)}^2} \\ B_{yx} &= \frac{Q_{(0)}V_{x(0)}^2(b_Q+2c_Q)-P_{(0)}V_{x(0)}V_{y(0)}(b_P+2c_P)}{V_{(0)}^4}-\frac{Q_{(0)}}{V_{(0)}^2} \\ G_{yy} &= \frac{P_{(0)}V_{y(0)}^2(b_P+2c_P)-Q_{(0)}V_{x(0)}V_{y(0)}(b_Q+2c_Q)}{V_{(0)}^4}-\frac{P_{(0)}}{V_{(0)}^2} \end{aligned}\right\}. \tag{8.36}$$

When an exponential function is used to model static voltage characteristics of the load, the relationship between load injected current and node voltage, of (7.49), can be used jointly with (8.35) to derive relevant coefficients in (8.34) directly as

$$\left.\begin{aligned} G_{xx} &= \frac{P_{(0)}}{V_{(0)}^2}\left((2-m)\frac{V_{x(0)}^2}{V_{(0)}^2}-1\right)+\frac{Q_{(0)}}{V_{(0)}^2}\left((2-n)\frac{V_{x(0)}V_{y(0)}}{V_{(0)}^2}\right) \\ B_{xy} &= \frac{Q_{(0)}}{V_{(0)}^2}\left((2-n)\frac{V_{y(0)}^2}{V_{(0)}^2}-1\right)+\frac{P_{(0)}}{V_{(0)}^2}\left((2-m)\frac{V_{x(0)}V_{y(0)}}{V_{(0)}^2}\right) \\ B_{yx} &= \frac{Q_{(0)}}{V_{(0)}^2}\left((2-n)\frac{V_{x(0)}^2}{V_{(0)}^2}-1\right)-\frac{P_{(0)}}{V_{(0)}^2}\left((2-m)\frac{V_{x(0)}V_{y(0)}}{V_{(0)}^2}\right) \\ G_{yy} &= \frac{P_{(0)}}{V_{(0)}^2}\left((2-m)\frac{V_{y(0)}^2}{V_{(0)}^2}-1\right)-\frac{Q_{(0)}}{V_{(0)}^2}\left((2-n)\frac{V_{x(0)}V_{y(0)}}{V_{(0)}^2}\right) \end{aligned}\right\}. \tag{8.37}$$

Especially, when there is not enough information about the static voltage characteristics of the load, a normally acceptable load model is to represent load active power by a constant current (i.e., taking $m = 1$) and load reactive power by a constant impedance (i.e., taking $n = 2$).

8.2.3 Linearized Equation of FACTS Components

1. SVC

From (7.197) and (7.198) we can obtain the following linearized equation directly

$$\left.\begin{aligned}\frac{\mathrm{d}\Delta B_{S1}}{\mathrm{d}t}&=-\frac{K_S}{T_S}\Delta V-\frac{1}{T_S}\Delta B_{S1}\\ \frac{\mathrm{d}(T_{S2}\Delta B_{\mathrm{SVC}}-T_{S1}\Delta B_{S1})}{\mathrm{d}t}&=\Delta B_{S1}-\Delta B_{\mathrm{SVC}}\end{aligned}\right\}. \tag{8.38}$$

Because $V^2=V_x^2+V_y^2$, after linearization we have

$$\Delta V=\frac{V_{x(0)}}{V_{(0)}}\Delta V_x+\frac{V_{y(0)}}{V_{(0)}}\Delta V_y. \tag{8.39}$$

Substituting the above equation into (8.38) and after rearrangement we obtain

$$\frac{\mathrm{d}\Delta\mathbf{x}_s}{\mathrm{d}t}=\mathbf{A}_s\Delta\mathbf{x}_s+\mathbf{B}_s\Delta\mathbf{V}_s, \tag{8.40}$$

where

$$\left.\begin{aligned}\Delta\mathbf{x}_s&=\begin{bmatrix}\Delta B_{S1}\\ \Delta B_{\mathrm{SVC}}\end{bmatrix},\quad \Delta\mathbf{V}_s=\begin{bmatrix}\Delta V_x\\ \Delta V_y\end{bmatrix}\\ \mathbf{A}_s&=\begin{bmatrix}-\frac{1}{T_S} & 0\\ \frac{T_S-T_{S1}}{T_ST_{S2}} & -\frac{1}{T_{S1}}\end{bmatrix},\quad \mathbf{B}_s=-\frac{K_S}{T_SV_{(0)}}\begin{bmatrix}V_{x(0)} & V_{y(0)}\\ \frac{T_{S1}}{T_{S2}}V_{x(0)} & \frac{T_{S1}}{T_{S2}}V_{y(0)}\end{bmatrix}\end{aligned}\right\}. \tag{8.41}$$

In addition, from (7.50) we can obtain the relationship of deviation between SVC injected current and nodal voltage to be

$$\Delta\mathbf{I}_s=\mathbf{C}_s\Delta\mathbf{x}_s+\mathbf{D}_s\Delta\mathbf{V}_s, \tag{8.42}$$

where

$$\left.\begin{aligned}\Delta\mathbf{I}_s&=\begin{bmatrix}\Delta I_x\\ \Delta I_y\end{bmatrix},\quad \Delta\mathbf{V}_s=\begin{bmatrix}\Delta V_x\\ \Delta V_y\end{bmatrix}\\ \mathbf{C}_s&=\frac{1}{(1-X_TB_{\mathrm{SVC}(0)})^2}\begin{bmatrix}0 & V_{y(0)}\\ 0 & -V_{x(0)}\end{bmatrix},\quad \mathbf{D}_s=\frac{B_{\mathrm{SVC}(0)}}{1-X_TB_{\mathrm{SVC}(0)}}\begin{bmatrix}0 & 1\\ -1 & 0\end{bmatrix}\end{aligned}\right\}. \tag{8.43}$$

Hence (8.40) and (8.42) form the linearized equation of the SVC.

2. TCSC

From (7.208) and (7.209) we can obtain the following linearized equation directly

$$\left.\begin{aligned}\frac{\mathrm{d}\Delta B_{T1}}{\mathrm{d}t} &= -\frac{K_T}{T_T}\Delta P_T - \frac{1}{T_T}\Delta B_{T1}\\ \frac{\mathrm{d}(T_{T2}\Delta B_{\mathrm{TCSC}} - T_{T1}\Delta B_{T1})}{\mathrm{d}t} &= \Delta B_{T1} - \Delta B_{\mathrm{TCSC}}\end{aligned}\right\}. \tag{8.44}$$

From (7.211) we have

$$\begin{aligned}\Delta P_T = {} & (V_{xi(0)}V_{yj(0)} - V_{yi(0)}V_{xj(0)})\Delta B_{\mathrm{TCSC}} + B_{\mathrm{TCSC}(0)}V_{yj(0)}\Delta V_{xi}\\ & - B_{\mathrm{TCSC}(0)}V_{xj(0)}\Delta V_{yi} - B_{\mathrm{TCSC}(0)}V_{yi(0)}\Delta V_{xj} + B_{\mathrm{TCSC}(0)}V_{xi(0)}\Delta V_{yj}\end{aligned}. \tag{8.45}$$

Substituting the above equation into (8.44) and after rearrangement we have

$$\frac{\mathrm{d}\Delta \mathbf{x}_t}{\mathrm{d}t} = \mathbf{A}_t\Delta\mathbf{x}_t + \mathbf{B}_t\Delta\mathbf{V}_t, \tag{8.46}$$

where

$$\left.\begin{aligned}\Delta\mathbf{x}_t &= \begin{bmatrix}\Delta B_{T1}\\ \Delta B_{\mathrm{TCSC}}\end{bmatrix}, \quad \Delta\mathbf{V}_t = \begin{bmatrix}\Delta V_{xi} & \Delta V_{yi} & \Delta V_{xj} & \Delta V_{yj}\end{bmatrix}^{\mathrm{T}}\\ \mathbf{A}_t &= \begin{bmatrix}-\dfrac{1}{T_T} & \dfrac{K_T}{T_T}(V_{yi(0)}V_{xj(0)} - V_{xi(0)}V_{yj(0)})\\ \dfrac{1}{T_{T2}} - \dfrac{T_{T1}}{T_{T2}}\dfrac{1}{T_T} & \dfrac{T_{T1}}{T_{T2}}\dfrac{K_T}{T_T}(V_{yi(0)}V_{xj(0)} - V_{xi(0)}V_{yj(0)}) - \dfrac{1}{T_{T2}}\end{bmatrix}\\ \mathbf{B}_t &= \frac{K_T B_{\mathrm{TCSC}(0)}}{T_T}\begin{bmatrix}-V_{yj(0)} & V_{xj(0)} & V_{yi(0)} & -V_{xi(0)}\\ -\dfrac{T_{T1}}{T_{T2}}V_{yj(0)} & \dfrac{T_{T1}}{T_{T2}}V_{xj(0)} & \dfrac{T_{T1}}{T_{T2}}V_{yi(0)} & -\dfrac{T_{T1}}{T_{T2}}V_{xi(0)}\end{bmatrix}\end{aligned}\right\}. \tag{8.47}$$

In addition, from (7.51) we can directly obtain the relationship of deviation in TCSC injected current and nodal voltage to be

$$\Delta\mathbf{I}_t = \mathbf{C}_t\Delta\mathbf{x}_t + \mathbf{D}_t\Delta\mathbf{V}_t, \tag{8.48}$$

where

$$\left.\begin{aligned}&\Delta\mathbf{I}_t = \begin{bmatrix}\Delta I_{xi} & \Delta I_{yi} & \Delta I_{xj} & \Delta I_{yj}\end{bmatrix}^{\mathrm{T}}\\ &\mathbf{C}_t = \begin{bmatrix}0 & V_{yi(0)} - V_{yj(0)}\\ 0 & V_{xj(0)} - V_{xi(0)}\\ 0 & V_{yj(0)} - V_{yi(0)}\\ 0 & V_{xi(0)} - V_{xj(0)}\end{bmatrix}, \quad \mathbf{D}_t = B_{\mathrm{TCSC}(0)}\begin{bmatrix}0 & 1 & 0 & -1\\ -1 & 0 & 1 & 0\\ 0 & -1 & 0 & 1\\ 1 & 0 & -1 & 0\end{bmatrix}\end{aligned}\right\}. \tag{8.49}$$

Thus (8.46) and (8.48) form the linearized equations of a TCSC.

8.2.4 Linearized Equation of HVDC Transmission System

When transient behavior of an HVDC transmission line is considered, the control equations of HVDC transmission line, rectifier, and inverter are given by (7.222), and (7.224)–(7.227). Canceling V_{dI} in (7.226) by using the first equation in (7.53) and ignoring the limitation on α and β, we can obtain the following linearized equation around steady state

$$\left.\begin{aligned}\frac{\mathrm{d}\Delta I_d}{\mathrm{d}t} &= -\frac{R}{L}\Delta I_d - \frac{k_R V_{R(0)}\sin\alpha_{(0)}}{L}\Delta\alpha + \frac{k_I V_{I(0)}\sin\beta_{(0)}}{L}\Delta\beta \\ &\quad + \frac{k_R\cos\alpha_{(0)}}{L}\Delta V_R - \frac{k_I\cos\beta_{(0)}}{L}\Delta V_I \\ \frac{\mathrm{d}\Delta x_1}{\mathrm{d}t} &= \frac{1}{T_{c3}}(\Delta I_d - \Delta x_1) \\ \frac{\mathrm{d}(K_{c1}\Delta x_1 - \Delta\alpha)}{\mathrm{d}t} &= -\frac{K_{c2}}{T_{c2}}\Delta x_1 \\ \frac{\mathrm{d}\Delta x_4}{\mathrm{d}t} &= \frac{X_{cI}}{T_{v3}}\Delta I_d - \frac{1}{T_{v3}}\Delta x_4 - \frac{k_I V_{I(0)}\sin\beta_{(0)}}{T_{v3}}\Delta\beta + \frac{k_I\cos\beta_{(0)}}{T_{v3}}\Delta V_I \\ \frac{\mathrm{d}(K_{v1}\Delta x_4 - \Delta\beta)}{\mathrm{d}t} &= -\frac{K_{v2}}{T_{v2}}\Delta x_4\end{aligned}\right\}. \tag{8.50}$$

Relationships between the magnitude of AC bus voltages of rectifier and inverter and their x, y components are $V_R^2 = V_{xR}^2 + V_{yR}^2$, $V_I^2 = V_{xI}^2 + V_{yI}^2$.

Linearizing the above equations, we have

$$\left.\begin{aligned}\Delta V_R &= \frac{V_{xR(0)}}{V_{R(0)}}\Delta V_{xR} + \frac{V_{yR(0)}}{V_{R(0)}}\Delta V_{yR} \\ \Delta V_I &= \frac{V_{xI(0)}}{V_{I(0)}}\Delta V_{xI} + \frac{V_{yI(0)}}{V_{I(0)}}\Delta V_{yI}\end{aligned}\right\}. \tag{8.51}$$

Substituting (8.51) into (8.50) to cancel ΔV_R and ΔV_I and after rearrangement we obtain

$$\frac{\mathrm{d}\Delta\mathbf{x}_d}{\mathrm{d}t} = \mathbf{A}_d\Delta\mathbf{x}_d + \mathbf{B}_d\Delta\mathbf{V}_d, \tag{8.52}$$

where

$$\left.\begin{aligned}\Delta\mathbf{x}_d &= [\Delta I_d \quad \Delta x_1 \quad \Delta x_4 \quad \Delta\alpha \quad \Delta\beta]^{\mathrm{T}} \\ \Delta\mathbf{V}_d &= \left[\Delta V_{xR} \quad \Delta V_{yR} \quad \Delta V_{xI} \quad \Delta V_{yI}\right]^{\mathrm{T}}\end{aligned}\right\}, \tag{8.53}$$

where coefficient matrices $\mathbf{A}_d$ and $\mathbf{B}_d$ can be easily obtained by comparing (8.52) and the original equation.

Algebraic equations of a two-terminal HVDC transmission system can be derived from relationships of power and current on the AC and DC sides of the converter. For the rectifier, the power relationship is

$$V_{xR}I_{xR} + V_{yR}I_{yR} = X_{cR}I_d^2 - k_R I_d V_R \cos\alpha. \tag{8.54}$$

Linearizing the above equation we have

$$\begin{aligned} V_{xR(0)}\Delta I_{xR} + V_{yR(0)}\Delta I_{yR} = &- I_{xR(0)}\Delta V_{xR} - I_{yR(0)}\Delta V_{yR} + 2X_{cR}I_{d(0)}\Delta I_d \\ &- k_R V_{R(0)} \cos\alpha_{(0)}\Delta I_d - k_R I_{d(0)} \cos\alpha_{(0)}\Delta V_R \, . \\ &+ k_R I_{d(0)} V_{R(0)} \sin\alpha_{(0)}\Delta\alpha \end{aligned} \tag{8.55}$$

In addition, from the third equation in (8.52) we have

$$I_R^2 = I_{xR}^2 + I_{yR}^2 = k_R^2 I_d^2. \tag{8.56}$$

The linearized form of the above equation is

$$I_{xR(0)}\Delta I_{xR} + I_{yR(0)}\Delta I_{yR} = k_R^2 I_{d(0)}\Delta I_d. \tag{8.57}$$

Substituting (8.51) into (8.55) to cancel ΔV_R and noting the reactive power injection into the AC system from the rectifier, $Q_{R(0)} = V_{yR(0)} I_{xR(0)} - V_{xR(0)} I_{yR(0)}$, is always nonzero, we can derive the deviation of node injected current from (8.55) and (8.57) and have the following matrix form

$$\Delta \mathbf{I}_R = \mathbf{C}_R \Delta \mathbf{x}_d + \mathbf{D}_R \Delta \mathbf{V}_R, \tag{8.58}$$

$$\left.\begin{aligned}
&\Delta \mathbf{I}_R = \begin{bmatrix} \Delta I_{xR} \\ \Delta I_{yR} \end{bmatrix}, \quad \Delta \mathbf{V}_R = \begin{bmatrix} \Delta V_{xR} \\ \Delta V_{yR} \end{bmatrix} \\
&\mathbf{C}_R = \frac{1}{V_{xR(0)}I_{yR(0)} - V_{yR(0)}I_{xR(0)}} \begin{bmatrix} C_{11} & 0 & 0 & C_{14} & 0 \\ C_{21} & 0 & 0 & C_{24} & 0 \end{bmatrix} \\
&C_{11} = 2X_{cR}I_{d(0)}I_{yR(0)} - k_R V_{R(0)} I_{yR(0)} \cos\alpha_{(0)} - k_R^2 I_{d(0)} V_{yR(0)} \\
&C_{14} = k_R V_{R(0)} I_{yR(0)} I_{d(0)} \sin\alpha_{(0)} \\
&C_{21} = -2X_{cR}I_{d(0)}I_{xR(0)} + k_R V_{R(0)} I_{xR(0)} \cos\alpha_{(0)} + k_R^2 I_{d(0)} V_{xR(0)} \\
&C_{24} = -k_R V_{R(0)} I_{xR(0)} I_{d(0)} \sin\alpha_{(0)} \\
&\mathbf{D}_R = \frac{1}{V_{xR(0)}I_{yR(0)} - V_{yR(0)}I_{xR(0)}} \begin{bmatrix} -D_{11} & -D_{12} \\ D_{21} & D_{22} \end{bmatrix} \\
&D_{11} = I_{yR(0)} \left[I_{xR(0)} + \frac{k_R V_{xR(0)} I_{d(0)} \cos\alpha_{(0)}}{V_{R(0)}} \right], \quad D_{12} = I_{yR(0)} \left[I_{yR(0)} + \frac{k_R V_{yR(0)} I_{d(0)} \cos\alpha_{(0)}}{V_{R(0)}} \right] \\
&D_{21} = I_{xR(0)} \left[I_{xR(0)} + \frac{k_R V_{xR(0)} I_{d(0)} \cos\alpha_{(0)}}{V_{R(0)}} \right], \quad D_{22} = I_{xR(0)} \left[I_{yR(0)} + \frac{k_R V_{yR(0)} I_{d(0)} \cos\alpha_{(0)}}{V_{R(0)}} \right]
\end{aligned}\right\} . \tag{8.59}$$

Power relationship of the inverter is

$$V_{xI}I_{xI} + V_{yI}I_{yI} = X_{cI}I_d^2 + k_I I_d V_I \cos\beta. \tag{8.60}$$

After linearization we have

$$\begin{aligned} V_{xI(0)}\Delta I_{xI} + V_{yI(0)}\Delta I_{yI} = & -I_{xI(0)}\Delta V_{xI} - I_{yI(0)}\Delta V_{yI} + 2X_{cI}I_{d(0)}\Delta I_d \\ & + k_I V_{I(0)} \cos\beta_{(0)}\Delta I_d + k_I I_{d(0)} \cos\beta_{(0)}\Delta V_I \\ & - k_I I_{d(0)} V_{I(0)} \sin\beta_{(0)}\Delta\beta. \end{aligned} \tag{8.61}$$

Similarly from the third equation in (8.53) and after linearization we have

$$I_{xI(0)}\Delta I_{xI} + I_{yI(0)}\Delta I_{yI} = k_I^2 I_{d(0)}\Delta I_d. \tag{8.62}$$

Again, substituting (8.51) into (8.61) to cancel ΔV_I, from (8.61) and (8.62) we obtain the following matrix form for current and voltage deviation

$$\Delta\mathbf{I}_I = \mathbf{C}_I\Delta\mathbf{x}_d + \mathbf{D}_I\Delta\mathbf{V}_I, \tag{8.63}$$

where

$$\Delta\mathbf{I}_I = \begin{bmatrix} \Delta I_{xI} \\ \Delta I_{yI} \end{bmatrix}, \quad \Delta\mathbf{V}_I = \begin{bmatrix} \Delta V_{xI} \\ \Delta V_{yI} \end{bmatrix}. \tag{8.64}$$

Equation (8.58) and (8.63) form the algebraic equation of the DC system

$$\Delta\mathbf{I}_d = \mathbf{C}_d\Delta\mathbf{x}_d + \mathbf{D}_d\Delta\mathbf{V}_d, \tag{8.65}$$

where

$$\Delta\mathbf{I}_d = \begin{bmatrix} \Delta\mathbf{I}_R \\ \Delta\mathbf{I}_I \end{bmatrix}, \quad \Delta\mathbf{V}_d = \begin{bmatrix} \Delta\mathbf{V}_R \\ \Delta\mathbf{V}_I \end{bmatrix}, \quad \mathbf{C}_d = \begin{bmatrix} \mathbf{C}_R \\ \mathbf{C}_I \end{bmatrix}, \quad \mathbf{D}_d = \begin{bmatrix} \mathbf{D}_R & \mathbf{0} \\ \mathbf{0} & \mathbf{D}_I \end{bmatrix}. \tag{8.66}$$

When different mathematical models of the DC system are used, we can follow similar procedures to derive linearized equations like (8.52) and (8.65).

8.3 Steps in Small-Signal Stability Analysis

8.3.1 Network Equation

For convenience of expression, we write the network equation of (8.36) in the form of block matrices. Noting that the network equation is itself linear, we can write the linear equation for the relationship between deviation of node injection current and node voltage in x–y coordinates directly, to be

$$\begin{bmatrix} \Delta\mathbf{I}_1 \\ \vdots \\ \Delta\mathbf{I}_i \\ \vdots \\ \Delta\mathbf{I}_n \end{bmatrix} = \begin{bmatrix} \mathbf{Y}_{11} & \cdots & \mathbf{Y}_{1i} & \cdots & \mathbf{Y}_{1n} \\ \vdots & & \vdots & & \vdots \\ \mathbf{Y}_{i1} & \cdots & \mathbf{Y}_{ii} & \cdots & \mathbf{Y}_{in} \\ \vdots & & \vdots & & \vdots \\ \mathbf{Y}_{n1} & \cdots & \mathbf{Y}_{ni} & \cdots & \mathbf{Y}_{nn} \end{bmatrix} \begin{bmatrix} \Delta\mathbf{V}_1 \\ \vdots \\ \Delta\mathbf{V}_i \\ \vdots \\ \Delta\mathbf{V}_n \end{bmatrix}, \tag{8.67}$$

where

$$\Delta\mathbf{I}_i = \begin{bmatrix} \Delta I_{xi} \\ \Delta I_{yi} \end{bmatrix}, \quad \Delta\mathbf{V}_i = \begin{bmatrix} \Delta V_{xi} \\ \Delta V_{yi} \end{bmatrix}, \quad \mathbf{Y}_{ij} = \begin{bmatrix} G_{ij} & -B_{ij} \\ B_{ij} & G_{ij} \end{bmatrix}, \tag{8.68}$$
$$i,j = 1,2,\ldots,n.$$

For load nodes, we can substitute the relationship between deviation of injected current and node voltage into the above equation to cancel the current deviation at the load node. Assuming load is connected at node i, then the network equation after canceling this node is just a simple correction to the original network equation of (8.67): current deviation at node i becomes zero, the ith diagonal block in the network admittance matrix changes to $\mathbf{Y}_{ii} - \mathbf{Y}_{li}$ and nothing more.

Without loss of generality, we assume the sequence of nodes in the network is: firstly each generator node, then each SVC node followed by two-terminal nodes of TCSC, then AC bus nodes of each HVDC transmission line (the node on the rectifier side first and inverter side second), and finally the remaining nodes. Canceling current deviation of all load nodes, we have the following block-matrix form of the network equation

$$\begin{bmatrix} \Delta\mathbf{I}_G \\ \Delta\mathbf{I}_S \\ \Delta\mathbf{I}_T \\ \Delta\mathbf{I}_D \\ \mathbf{0} \end{bmatrix} = \begin{bmatrix} \mathbf{Y}_{GG} & \mathbf{Y}_{GS} & \mathbf{Y}_{GT} & \mathbf{Y}_{GD} & \mathbf{Y}_{GL} \\ \mathbf{Y}_{SG} & \mathbf{Y}_{SS} & \mathbf{Y}_{ST} & \mathbf{Y}_{SD} & \mathbf{Y}_{SL} \\ \mathbf{Y}_{TG} & \mathbf{Y}_{TS} & \mathbf{Y}_{TT} & \mathbf{Y}_{TD} & \mathbf{Y}_{TL} \\ \mathbf{Y}_{DG} & \mathbf{Y}_{DS} & \mathbf{Y}_{DT} & \mathbf{Y}_{DD} & \mathbf{Y}_{DL} \\ \mathbf{Y}_{LG} & \mathbf{Y}_{LS} & \mathbf{Y}_{LT} & \mathbf{Y}_{LD} & \mathbf{Y}_{LL} \end{bmatrix} \begin{bmatrix} \Delta\mathbf{V}_G \\ \Delta\mathbf{V}_S \\ \Delta\mathbf{V}_T \\ \Delta\mathbf{V}_D \\ \Delta\mathbf{V}_L \end{bmatrix}, \tag{8.69}$$

where $\Delta\mathbf{I}_G$ and $\Delta\mathbf{V}_G$ are vectors consisting of deviation of injected current and node voltage of all generators, respectively; $\Delta\mathbf{I}_S$ and $\Delta\mathbf{V}_S$ vectors of deviation of node injection current and node voltage of all SVC nodes; $\Delta\mathbf{I}_T$ and $\Delta\mathbf{V}_T$ all TCSC nodes; $\Delta\mathbf{I}_D$ and $\Delta\mathbf{V}_D$ those at AC busbars of all converters; $\Delta\mathbf{V}_L$ associated with voltage of remaining nodes. All those vectors can be written as

$$\left.\begin{aligned} &\Delta\mathbf{I}_G = [\Delta\mathbf{I}_{g1} \quad \Delta\mathbf{I}_{g2} \quad \cdots]^{\mathrm{T}}, \quad \Delta\mathbf{V}_G = [\Delta\mathbf{V}_{g1} \quad \Delta\mathbf{V}_{g2} \quad \cdots]^{\mathrm{T}} \\ &\Delta\mathbf{I}_S = [\Delta\mathbf{I}_{s1} \quad \Delta\mathbf{I}_{s2} \quad \cdots]^{\mathrm{T}}, \quad \Delta\mathbf{V}_S = [\Delta\mathbf{V}_{s1} \quad \Delta\mathbf{V}_{s2} \quad \cdots]^{\mathrm{T}} \\ &\Delta\mathbf{I}_T = [\Delta\mathbf{I}_{t1} \quad \Delta\mathbf{I}_{t2} \quad \cdots]^{\mathrm{T}}, \quad \Delta\mathbf{V}_T = [\Delta\mathbf{V}_{t1} \quad \Delta\mathbf{V}_{t2} \quad \cdots]^{\mathrm{T}} \\ &\Delta\mathbf{I}_D = [\Delta\mathbf{I}_{d1} \quad \Delta\mathbf{I}_{d2} \quad \cdots]^{\mathrm{T}}, \quad \Delta\mathbf{V}_D = [\Delta\mathbf{V}_{d1} \quad \Delta\mathbf{V}_{d2} \quad \cdots]^{\mathrm{T}} \\ &\Delta\mathbf{V}_L = [\Delta\mathbf{V}_1 \quad \Delta\mathbf{V}_2 \quad \cdots]^{\mathrm{T}} \end{aligned}\right\}. \tag{8.70}$$

8.3.2 Linearized Differential Equations of Whole Power System

Equations of all generation units are formed from (8.31) and (8.29) to be

$$\frac{d\Delta \mathbf{x}_G}{dt} = \mathbf{A}_G \Delta \mathbf{x}_G + \mathbf{B}_G \Delta \mathbf{V}_G, \tag{8.71}$$

$$\Delta \mathbf{I}_G = \mathbf{C}_G \Delta \mathbf{x}_G + \mathbf{D}_G \Delta \mathbf{V}_G, \tag{8.72}$$

where

$$\left.\begin{array}{ll} \mathbf{A}_G = \text{diag}\{\mathbf{A}_{g1} \quad \mathbf{A}_{g2} \quad \cdots\}, & \mathbf{B}_G = \text{diag}\{\mathbf{B}_{g1} \quad \mathbf{B}_{g2} \quad \cdots\} \\ \mathbf{C}_G = \text{diag}\{\mathbf{C}_{g1} \quad \mathbf{C}_{g2} \quad \cdots\}, & \mathbf{D}_G = \text{diag}\{\mathbf{D}_{g1} \quad \mathbf{D}_{g2} \quad \cdots\} \end{array}\right\}. \tag{8.73}$$

Equations (8.40) and (8.42) of each SVC can form equations of all SVCs to be

$$\frac{d\Delta \mathbf{x}_S}{dt} = \mathbf{A}_S \Delta \mathbf{x}_S + \mathbf{B}_S \Delta \mathbf{V}_H, \tag{8.74}$$

$$\Delta \mathbf{I}_S = \mathbf{C}_S \Delta \mathbf{x}_S + \mathbf{D}_S \Delta \mathbf{V}_S, \tag{8.75}$$

where

$$\left.\begin{array}{ll} \mathbf{A}_S = \text{diag}\{\mathbf{A}_{s1} \quad \mathbf{A}_{s2} \quad \cdots\}, & \mathbf{B}_S = \text{diag}\{\mathbf{B}_{s1} \quad \mathbf{B}_{s2} \quad \cdots\} \\ \mathbf{C}_S = \text{diag}\{\mathbf{C}_{s1} \quad \mathbf{C}_{s2} \quad \cdots\}, & \mathbf{D}_S = \text{diag}\{\mathbf{D}_{s1} \quad \mathbf{D}_{s2} \quad \cdots\} \end{array}\right\}. \tag{8.76}$$

TCSC equations are formed from (8.46) and (8.48)

$$\left.\begin{array}{l} (\mathbf{A} - s_1\mathbf{I})\mathbf{v}_\mathbf{A} = (\lambda_\mathbf{A} - s_1)\mathbf{v}_\mathbf{A} \\ (\mathbf{A} - s_2\mathbf{I})\mathbf{v}_\mathbf{A} = (\lambda_\mathbf{A} - s_2)\mathbf{v}_\mathbf{A} \end{array}\right\}, \tag{8.77}$$

$$\Delta \mathbf{I}_T = \mathbf{C}_T \Delta \mathbf{x}_T + \mathbf{D}_T \Delta \mathbf{V}_T, \tag{8.78}$$

where

$$\left.\begin{array}{ll} \mathbf{A}_T = \text{diag}\{\mathbf{A}_{t1} \quad \mathbf{A}_{t2} \quad \cdots\}, & \mathbf{B}_T = \text{diag}\{\mathbf{B}_{t1} \quad \mathbf{B}_{t2} \quad \cdots\} \\ \mathbf{C}_T = \text{diag}\{\mathbf{C}_{t1} \quad \mathbf{C}_{t2} \quad \cdots\}, & \mathbf{D}_T = \text{diag}\{\mathbf{D}_{t1} \quad \mathbf{D}_{t2} \quad \cdots\} \end{array}\right\}. \tag{8.79}$$

All two-terminal HVDC transmission lines have the following equations

$$\frac{d\Delta \mathbf{x}_D}{dt} = \mathbf{A}_D \Delta \mathbf{x}_D + \mathbf{B}_D \Delta \mathbf{V}_D, \tag{8.80}$$

$$\Delta \mathbf{I}_D = \mathbf{C}_D \Delta \mathbf{x}_D + \mathbf{D}_D \Delta \mathbf{V}_D, \tag{8.81}$$

where

$$\left.\begin{array}{ll}\mathbf{A}_D = \mathrm{diag}\{\mathbf{A}_{d1} \quad \mathbf{A}_{d2} \quad \cdots\}, & \mathbf{B}_D = \mathrm{diag}\{\mathbf{B}_{d1} \quad \mathbf{B}_{d2} \quad \cdots\} \\ \mathbf{C}_D = \mathrm{diag}\{\mathbf{C}_{d1} \quad \mathbf{C}_{d2} \quad \cdots\}, & \mathbf{D}_D = \mathrm{diag}\{\mathbf{D}_{d1} \quad \mathbf{D}_{d2} \quad \cdots\}\end{array}\right\}. \tag{8.82}$$

Substituting (8.72), (8.75), (8.78), and (8.81) into (8.69) to cancel $\Delta\mathbf{I}_G$, $\Delta\mathbf{I}_S$, $\Delta\mathbf{I}_T$, and $\Delta\mathbf{I}_D$, together with (8.71), (8.74), (8.77), and (8.80), we can obtain the matrix formulations, as required by (8.3):

$$\left.\begin{array}{l}
\left.\begin{array}{l}\Delta\mathbf{x} = [\Delta\mathbf{x}_G \quad \Delta\mathbf{x}_S \quad \Delta\mathbf{x}_T \quad \Delta\mathbf{x}_D]^{\mathrm{T}} \\ \Delta\mathbf{y} = [\Delta\mathbf{V}_G \quad \Delta\mathbf{V}_S \quad \Delta\mathbf{V}_T \quad \Delta\mathbf{V}_D \quad \Delta\mathbf{V}_L]^{\mathrm{T}}\end{array}\right\} \\
\tilde{\mathbf{A}} = \begin{bmatrix} \mathbf{A}_G & 0 & 0 & 0 \\ 0 & \mathbf{A}_S & 0 & 0 \\ 0 & 0 & \mathbf{A}_T & 0 \\ 0 & 0 & 0 & \mathbf{A}_D \end{bmatrix} \\
\tilde{\mathbf{B}} = \begin{bmatrix} B_G & 0 & 0 & 0 & 0 \\ 0 & B_S & 0 & 0 & 0 \\ 0 & 0 & B_T & 0 & 0 \\ 0 & 0 & 0 & B_D & 0 \end{bmatrix}, \quad \tilde{\mathbf{C}} = \begin{bmatrix} -C_G & 0 & 0 & 0 \\ 0 & -C_S & 0 & 0 \\ 0 & 0 & -C_T & 0 \\ 0 & 0 & 0 & -C_D \\ 0 & 0 & 0 & 0 \end{bmatrix} \\
\tilde{\mathbf{D}} = \begin{bmatrix} \mathbf{Y}_{GG} - \mathbf{D}_G & \mathbf{Y}_{GS} & \mathbf{Y}_{GT} & \mathbf{Y}_{GD} & \mathbf{Y}_{GL} \\ \mathbf{Y}_{SG} & \mathbf{Y}_{SS} - \mathbf{D}_S & \mathbf{Y}_{ST} & \mathbf{Y}_{SD} & \mathbf{Y}_{SL} \\ \mathbf{Y}_{TG} & \mathbf{Y}_{TS} & \mathbf{Y}_{TT} - \mathbf{D}_T & \mathbf{Y}_{TD} & \mathbf{Y}_{TL} \\ \mathbf{Y}_{DG} & \mathbf{Y}_{DS} & \mathbf{Y}_{DT} & \mathbf{Y}_{DD} - \mathbf{D}_D & \mathbf{Y}_{DL} \\ \mathbf{Y}_{LG} & \mathbf{Y}_{LS} & \mathbf{Y}_{LT} & \mathbf{Y}_{LD} & \mathbf{Y}_{LL} \end{bmatrix}
\end{array}\right\}. \tag{8.83}$$

Obviously, $\tilde{\mathbf{A}}$, $\tilde{\mathbf{B}}$, and $\tilde{\mathbf{C}}$ are sparse block matrices, as is $\tilde{\mathbf{D}}$ which is also an admittance matrix. Using matrices $\tilde{\mathbf{A}}$, $\tilde{\mathbf{B}}$, $\tilde{\mathbf{C}}$ and $\tilde{\mathbf{D}}$, and from (8.5) we can obtain the system state matrix $\mathbf{A}$. By now, we have obtained the linearized equations of a power system at a steady-state operating point.

Finally, we would like to point out:

1. If this linearized system is asymptotically stable, i.e., the real part of all eigenvalues of matrix $\mathbf{A}$ are negative, the actual nonlinear system is asymptotically stable at this equilibrium point.
2. The method used to form matrix $\mathbf{A}$ is different in various commercial software packages. In the above, we only give one way to form it to introduce the principles and techniques used in forming matrix $\mathbf{A}$ [189, 190]. There are various alternative formats of matrices $\tilde{\mathbf{A}}$, $\tilde{\mathbf{B}}$, $\tilde{\mathbf{C}}$, and $\tilde{\mathbf{D}}$ in (8.83) that are related with the sequence order of state variables, format of network equations, algebraic equations of various dynamic components and ways to treat the network equations.

Different methods determine the complexity of, and flexibility in developing, the program, but do not change the resulting eigensolution.

3. In the formation of the above linearized equations, we have considered generation units, SVC, TCSC, two-terminal HVDC transmission lines. We can also treat other dynamic components in power systems in similar ways. For example, for dynamic components (such as induction motor loads) we can derive their linearized equation in the same way as treating generators; for multiterminal HVDC transmission lines, we can obtain the linearized equation as we have done in treating two-terminal HVDC transmission lines. We can then arrange the linearized equations into the equation of the whole power system.
4. Matrix **A**, as formed, must have a zero eigenvalue. A zero eigenvalue exists because the absolute angle of the generator rotors is not unique. In other words, there is a redundant rotor angle in a power system model. In fact, power distribution among generators is determined by the relative rotor angle of generators. If the absolute rotor angle of all generators is added to by a fixed value, the power distribution does not change at all. Hence this does not affect system stability. To eliminate the zero eigenvalue, we only need to choose the rotor angle of any particular generator as a reference and then use the relative rotor angle of other generators as the new state variable. In doing so, the dimension of state matrix **A** and the corresponding state variable vector is reduced by one.
5. In the case that all generator torques are not directly related to rotor speed, i.e., when there is no damping term in the swing equation and the governing effect is ignored; matrix **A** will have another zero eigenvalue. Similarly, to remove this zero eigenvalue, we only need to choose the rotor speed of any generator as a reference and use the relative rotor speed of the other generators as new state variables. Again, in doing so, the order of matrix **A** and the corresponding state variable vector is reduced by one.

Knowing the origin of the zero eigenvalues, we do not have to apply the treatment of (4) and (5) above, but simply eliminate the corresponding zero eigenvalues in our computational results. However, due to errors in load flow calculation and in the computation of eigenvalues, we should note that the theoretically zero eigenvalues will be computed as eigenvalues with very small magnitude.

8.3.3 Program Package for Small-Signal Stability Analysis

From what we have discussed previously, we can develop a program package for small-signal stability analysis of an AC/DC power system with FACTS devices such as SVC and TCSC installed. Basic steps in developing the stability analysis program are:

1. Load flow calculation at a given steady-state operating condition of the power system. This includes finding the voltage, current, and power at each node in the system.
2. Formation of the admittance matrix in (8.67).
3. Treatment of load. Load power and voltage at steady-state operation are known to be $P_{(0)}$, $Q_{(0)}$, $V_{x(0)}$, and $V_{y(0)}$. From parameters of the static voltage characteristics of load, we calculate matrix elements G_{xx}, B_{xy}, B_{yx}, and G_{yy} in (8.34) from

(8.36) or (8.37). These will be used to adjust diagonal blocks related to loads in the admittance matrix.

4. Establishment of linearized equations of dynamic components in the system. Firstly we calculate initial values of all variables of generators from (7.74)–(7.78) and (7.118)–(7.122). Then we can form matrices $\overline{\mathbf{A}}_g$, $\overline{\mathbf{B}}_{Ig}$, $\overline{\mathbf{B}}_{Vg}$, $\overline{\mathbf{P}}_g$, and $\overline{\mathbf{Z}}_g$ in (8.20) and (8.21) as well as matrices $\mathbf{T}_{g(0)}$, $\mathbf{R}_{Vg}$, and $\mathbf{R}_{Ig}$ in (8.26) and (8.28). Finally we calculate matrices from (8.30) and (8.32) to establish the linearized equation of generators. In a similar way we can obtain linearized equations of all dynamic components in the power system.
5. Formation of system state matrix $\mathbf{A}$ from (8.5). This is obtained by forming matrices $\tilde{\mathbf{A}}$, $\tilde{\mathbf{B}}$, $\tilde{\mathbf{C}}$, and $\tilde{\mathbf{D}}$ from (8.71)–(8.83).
6. Calculation of all eigenvalues of the state matrix $\mathbf{A}$ by using the QR method [187, 188]. The result of this calculation is used to determine system small-signal stability. The QR method to calculate all eigenvalues of matrix $\mathbf{A}$ will be introduced in Sect. 8.4.

[Example 8.1] Single-line diagram of the 9-node power system, line data, generator parameters, and load flow at a steady-state operating condition is given in Fig. 7.7, Tables 7.5–7.7, respectively. System frequency is 60 Hz.

All loads in the system are modeled by constant impedance. Generator 1 uses the classical model, generators 2 and 3 the double-axis model with self-excited potential-source excitation system. Parameters of the exciter are:

$$X_C = 0, \quad K_A = 200, \quad T_R = 0.03\,\mathrm{s}, \quad T_A = 0.02\,\mathrm{s}, \quad T_B = 10.0\,\mathrm{s}, \quad T_C = 1.0\,\mathrm{s}.$$

In addition, the damping coefficient of each generator D_i is 1.0.

[Solution] In the following, we will demonstrate the process of small-signal stability analysis for the example power system. For simplicity of expression, a blank in the matrix will represent either zero or a zero matrix:

(1) From load flow calculation, (7.74)–(7.78) and (7.118)–(7.122) we calculate initial values of all variables of generators shown in Table 8.1. Equivalent admittance of loads has been calculated in example 8.1 which is included in the power network model.
(2) Establishment of linearized equations of generators using the method introduced in Sect. 8.2.1.

Generator 1

We can calculate coefficient matrices in (8.20), (8.21), (8.26), and (8.28) as follows:

Table 8.1 Initial values of generator variables

	$\delta_{(0)}$	$V_{q(0)}$	$V_{d(0)}$	$I_{q(0)}$	$I_{d(0)}$	$E_{fq(0)}$	$E'_{q(0)}$	$E'_{d(0)}$
1	2.27165	1.03918	0.04122	0.67801	0.28716		1.05664	
2	61.09844	0.63361	0.80571	0.93199	1.29015	1.78932	0.78817	0.62220
3	54.13662	0.66607	0.77909	0.61941	0.56147	1.40299	0.76786	0.62424

$$\overline{\mathbf{A}}_1 = \begin{matrix} \Delta\delta_1 \\ \Delta\omega_1 \end{matrix} \begin{bmatrix} 0.0 & 376.99112 \\ 0.0 & -0.02115 \end{bmatrix}, \quad \overline{\mathbf{B}}_{I1} = \begin{bmatrix} 0.0 & 0.0 \\ 0.0 & -0.02235 \end{bmatrix}, \quad \overline{\mathbf{B}}_{V1} = \begin{bmatrix} 0.0 & 0.0 \\ 0.0 & 0.0 \end{bmatrix},$$

$$\overline{\mathbf{P}}_1 = \begin{bmatrix} 0.0 & 0.0 \\ 0.0 & 0.0 \end{bmatrix}, \quad \overline{\mathbf{Z}}_1 = \begin{bmatrix} 0.0 & 0.0608 \\ -0.0608 & 0.0 \end{bmatrix},$$

$$\mathbf{T}_{1(0)} = \begin{bmatrix} 0.03964 & -0.99921 \\ 0.99921 & 0.03964 \end{bmatrix}, \quad \boldsymbol{R}_{V1} = \begin{bmatrix} 1.03918 & 0.0 \\ -0.04122 & 0.0 \end{bmatrix},$$

$$\boldsymbol{R}_{I1} = \begin{bmatrix} 0.67801 & 0.0 \\ -0.28716 & 0.0 \end{bmatrix}.$$

From (8.32), (8.30), and the matrices above, we can obtain matrices in the linearized equation of the generator of (8.29) and (8.31) to be

$$\mathbf{A}_1 = \begin{bmatrix} 0.0 & 376.99112 \\ -0.38198 & -0.02115 \end{bmatrix}, \quad \mathbf{B}_1 = \begin{bmatrix} 0.0 & 0.0 \\ -0.01457 & 0.36729 \end{bmatrix},$$

$$\mathbf{C}_1 = \begin{bmatrix} 17.36532 & 0.0 \\ 0.68886 & 0.0 \end{bmatrix}, \quad \mathbf{D}_1 = \begin{bmatrix} 0.0 & -16.44737 \\ 16.44737 & 0.0 \end{bmatrix}.$$

Generator 2

In a similar way, we can obtain coefficient matrices of the linearized equation of generator 2 to be

$$\overline{\mathbf{A}}_2 = \begin{matrix} \Delta\delta_2 \\ \Delta\omega_2 \\ \Delta E'_{q2} \\ \Delta E'_{d2} \\ \Delta E_{fq2} \\ \Delta V_{R2} \\ \Delta V_{M2} \end{matrix} \begin{bmatrix} & 376.99112 & & & & & \\ & -0.07813 & -0.07281 & -0.10079 & & & \\ & & -0.16667 & & 0.16667 & & \\ & & & -1.86916 & & & \\ & & & & -0.10000 & -4.90000 & -1000.0 \\ & & & & & -50.0000 & -10000.0 \\ & & & & & & -33.33333 \end{bmatrix},$$

$$\overline{\boldsymbol{B}}_{I2} = \begin{bmatrix} & \\ -0.05422 & -0.06935 \\ -0.12933 & \\ & 1.24785 \\ & \\ & \\ & \end{bmatrix}, \quad \overline{\boldsymbol{B}}_{V2} = \begin{bmatrix} & \\ & \\ & \\ & \\ & \\ & \\ 26.20186 & 20.60518 \end{bmatrix},$$

$$\overline{\mathbf{P}}_2 = \left[\begin{array}{c:c:c:c:c:c:c} & & & 1.0 & & & \\ \hdashline & & 1.0 & & & & \end{array}\right], \quad \overline{\mathbf{Z}}_2 = \left[\begin{array}{c:c} 0.0000 & 0.1969 \\ \hdashline -0.1198 & 0.0000 \end{array}\right],$$

$$\mathbf{T}_{2(0)} = \left[\begin{array}{c:c} 0.87545 & -0.48331 \\ \hdashline 0.48331 & 0.87545 \end{array}\right],$$

$$\mathbf{R}_{V2} = \left[\begin{array}{c:c:c:c:c:c:c} 0.63361 & & & & & & \\ \hdashline -0.80571 & & & & & & \end{array}\right], \quad \boldsymbol{R}_{I2} = \left[\begin{array}{c:c:c:c:c:c:c} 0.93199 & & & & & & \\ \hdashline -1.29015 & & & & & & \end{array}\right],$$

$$\mathbf{A}_2 = \left[\begin{array}{c:c:c:c:c:c:c} & 376.99112 & & & & & \\ \hdashline -0.58783 & -0.07813 & -0.52542 & 0.25140 & & & \\ \hdashline -0.86982 & & -1.24624 & & 0.16667 & & \\ \hdashline 4.01549 & & & -8.20664 & & & \\ \hdashline & & & & -0.10000 & -4.90000 & -1000.0 \\ \hdashline & & & & & -50.0000 & -10000.0 \\ \hdashline & & & & & & -33.33333 \end{array}\right],$$

$$\mathbf{B}_2 = \left[\begin{array}{c:c} & \\ \hdashline -0.08958 & 0.56646 \\ \hdashline 0.52177 & 0.94512 \\ \hdashline 5.54816 & -3.06294 \\ \hdashline & \\ \hdashline & \\ \hdashline 32.89707 & 5.37531 \end{array}\right],$$

$$\mathbf{C}_2 = \left[\begin{array}{c:c:c:c:c:c:c} 7.25066 & & 7.30761 & -2.45458 & & & \\ \hdashline 1.14659 & & -4.03428 & -4.44617 & & & \end{array}\right],$$

$$\mathbf{D}_2 = \left[\begin{array}{c:c} -1.38295 & -7.58377 \\ \hdashline 5.84220 & 1.38295 \end{array}\right].$$

Generator 3

Similarly we have

$$\bar{\mathbf{A}}_3 = \begin{matrix} \Delta\delta_3 \\ \Delta\omega_3 \\ \Delta E'_{q3} \\ \Delta E'_{d3} \\ \Delta E_{fq3} \\ \Delta V_{R3} \\ \Delta V_{M3} \end{matrix} \begin{bmatrix} & 376.99112 & & & & & \\ & -0.16611 & -0.10289 & -0.09327 & & & \\ & & -0.16978 & & 0.16978 & & \\ & & & -1.66667 & & & \\ & & & & -0.10000 & -4.90000 & -1000.0 \\ & & & & & -50.0000 & -10000.0 \\ & & & & & & -33.33333 \end{bmatrix},$$

$$\bar{\boldsymbol{B}}_{I3} = \begin{bmatrix} & \\ -0.11076 & -0.13396 \\ -0.19205 & \\ & 1.67967 \\ & \\ & \\ & \end{bmatrix}, \quad \bar{\boldsymbol{B}}_{V3} = \begin{bmatrix} & \\ & \\ & \\ & \\ & \\ & \\ 25.33623 & 21.66071 \end{bmatrix},$$

$$\bar{\mathbf{P}}_3 = \begin{bmatrix} & & & 1.0 & & & \\ & & 1.0 & & & & \end{bmatrix}, \quad \bar{\mathbf{Z}}_3 = \begin{bmatrix} 0.0000 & 0.2500 \\ -0.1813 & 0.0000 \end{bmatrix},$$

$$\mathbf{T}_{3(0)} = \begin{bmatrix} 0.81042 & -0.58586 \\ 0.58586 & 0.81042 \end{bmatrix},$$

$$\boldsymbol{R}_{V3} = \begin{bmatrix} 0.66607 & & & & & & \\ -0.77907 & & & & & & \end{bmatrix}, \quad \boldsymbol{R}_{I3} = \begin{bmatrix} 0.61941 & & & & & & \\ -0.56147 & & & & & & \end{bmatrix},$$

$$\mathbf{A}_3 = \begin{bmatrix} & 376.99112 & & & & & \\ -0.83288 & -0.16611 & -0.71383 & 0.44257 & & & \\ -0.82530 & & -1.22910 & & 0.16978 & & \\ 4.47508 & & & -8.38533 & & & \\ & & & & -0.10000 & -4.90000 & -1000.0 \\ & & & & & -50.0000 & -10000.0 \\ & & & & & & -33.33333 \end{bmatrix},$$

$$\mathbf{B}_3 = \begin{bmatrix} & \\ -0.07633 & 0.80903 \\ 0.62061 & 0.85849 \\ 5.44492 & -3.93616 \\ & \\ & \\ 33.22292 & 2.71085 \end{bmatrix},$$

$$\mathbf{C}_3 = \begin{bmatrix} 4.87039 & & 4.47003 & -2.34342 & & & \\ 0.45951 & & -3.23141 & -3.24166 & & & \end{bmatrix},$$

$$\mathbf{D}_3 = \begin{bmatrix} -0.71964 & -4.99548 \\ 4.52023 & 0.71964 \end{bmatrix}.$$

(3) Linearized equations of the system

Obviously, matrices in (8.3) (see 8.83) are

$$\tilde{A} = A_G, \quad \tilde{B} = [\boldsymbol{B}_G \quad \boldsymbol{0}], \quad \tilde{C} = \begin{bmatrix} -\boldsymbol{C}_G \\ \boldsymbol{0} \end{bmatrix}, \quad \tilde{D} = \begin{bmatrix} \boldsymbol{Y}_{GG} - \boldsymbol{D}_G & \boldsymbol{Y}_{GL} \\ \boldsymbol{Y}_{LG} & \boldsymbol{Y}_{LL} \end{bmatrix},$$

$$\mathbf{A}_G = \begin{bmatrix} \mathbf{A}_1 & \mathbf{0} & \mathbf{0} \\ \mathbf{0} & \mathbf{A}_2 & \mathbf{0} \\ \mathbf{0} & \mathbf{0} & \mathbf{A}_3 \end{bmatrix}, \quad \mathbf{B}_G = \begin{bmatrix} \mathbf{B}_1 & \mathbf{0} & \mathbf{0} \\ \mathbf{0} & \mathbf{B}_2 & \mathbf{0} \\ \mathbf{0} & \mathbf{0} & \mathbf{B}_3 \end{bmatrix}, \quad \mathbf{C}_G = \begin{bmatrix} \mathbf{C}_1 & \mathbf{0} & \mathbf{0} \\ \mathbf{0} & \mathbf{C}_2 & \mathbf{0} \\ \mathbf{0} & \mathbf{0} & \mathbf{C}_3 \end{bmatrix},$$

$$\mathbf{D}_G = \begin{bmatrix} \mathbf{D}_1 & \mathbf{0} & \mathbf{0} \\ \mathbf{0} & \mathbf{D}_2 & \mathbf{0} \\ \mathbf{0} & \mathbf{0} & \mathbf{D}_3 \end{bmatrix},$$

$$\mathbf{Y}_{GG} - \mathbf{D}_G = \begin{bmatrix} & 33.80848 & & & & \\ -33.80848 & & & & & \\ & & 1.38295 & 23.58377 & & \\ & & -21.84220 & -1.38295 & & \\ & & & & 0.71964 & 22.06033 \\ & & & & -21.58508 & -0.71964 \end{bmatrix},$$

$\mathbf{Y}_{LL} =$

3.30738	39.30889	−1.36519	−11.60410	−1.94219	−10.51068						
−39.30889	3.30738	11.60410	−1.36519	10.51068	−1.94219						
−1.36519	−11.60410	**3.81379**	**17.84263**			−1.18760	−5.97513				
11.60410	−1.36519	**−17.84263**	**3.81379**			5.97513	−1.18760				
−1.94219	−10.51068			**4.10185**	**16.13348**					−1.28201	−5.58824
10.51068	−1.94219			**−16.13348**	**4.10185**					5.58824	−1.28201
		−1.18760	−5.97513			2.80473	35.44561	1.61712	−13.69798		
		5.97513	−1.18760			−35.44561	2.80473	13.69798	−1.61712		
						−1.61712	−13.69798	**3.74119**	**23.64239**	−1.15509	−9.78427
						13.69798	−1.61712	**−23.64239**	**3.74119**	9.78427	−1.15509
				−1.28201	−5.58824			−1.15509	−9.78427	2.43710	32.15386
				5.58824	−1.28201			9.78427	−1.15509	−32.15386	2.43710

,

$\mathbf{A} =$

$\Delta\delta_1$		376.99112														
$\Delta\omega_1$	−0.06244	−0.02115	0.03518		0.02485	−0.02020				0.02726		0.01775	−0.01913			
$\Delta\delta_2$				376.99112												
$\Delta\omega_2$	0.11924		−0.19905	−0.07813	−0.24571	0.03212				0.07982		0.03553	−0.06960			
$\Delta E'_{q2}$	0.20108		−0.32411		−0.52072	−0.04750	0.16667			0.12303		0.13428	−0.04154			
$\Delta E'_{d2}$	−0.62862		1.13669		0.44018	−4.61925				−0.50807		0.37374	0.93904			
ΔE_{fq2}							−0.10000	−4.90000	−1000.0							
ΔV_{R2}								−50.0	−10000.0							
ΔV_{M2}	1.23883		−1.48647		15.66751	13.92522			−33.33333	0.24764		4.10804	3.08961			
$\Delta\delta_3$											376.99112					
$\Delta\omega_3$	0.20654		0.17777		0.10937	−0.11476				−0.38432	−0.16611	−0.40080	0.14517			
$\Delta E'_{q3}$	0.22459		0.16072		0.21160	−0.01561				−0.38531		−0.62589	−0.04685	0.16978		
$\Delta E'_{d3}$	−0.96568		−1.06680		0.15877	1.32599				2.03248		0.43705	−4.99508			
ΔE_{fq3}														−0.10000	−4.90000	−1000.0
ΔV_{R3}															−50.0	−10000.0
ΔV_{M3}	0.95082		−0.73652		4.92556	4.68110				−0.21429		13.98229	11.82683			−33.3333

$$
\mathbf{Y}_{LG} = \begin{bmatrix}
 & -17.36111 & & & & \\
17.36111 & & & & & \\
 & & & & & \\
 & & & & & \\
 & & & & & \\
 & & & & & \\
 & & & -16.00000 & & \\
 & & 16.00000 & & & \\
 & & & & & \\
 & & & & & \\
 & & & & & -17.06485 \\
 & & & & 17.06485 &
\end{bmatrix},
$$

$$
\mathbf{Y}_{GL} = \begin{bmatrix}
 & -17.36111 & & & & & & & & & & \\
17.36111 & & & & & & & & & & & \\
 & & & & & & & -16.00000 & & & & \\
 & & & & & & 16.00000 & & & & & \\
 & & & & & & & & & & & -17.06485 \\
 & & & & & & & & & & 17.06485 &
\end{bmatrix}.
$$

From (8.5) we can obtain the state matrix

$$
\begin{aligned}
\mathbf{A} &= \mathbf{A}_G - [\mathbf{B}_G \quad \mathbf{0}] \begin{bmatrix} \mathbf{Y}_{GG} - \mathbf{D}_G & \mathbf{Y}_{GL} \\ \mathbf{Y}_{LG} & \mathbf{Y}_{LL} \end{bmatrix}^{-1} \begin{bmatrix} -\mathbf{C}_G \\ \mathbf{0} \end{bmatrix} \\
&= \mathbf{A}_G + \mathbf{B}_G [\mathbf{Y}_{GG} - \mathbf{D}_G - \mathbf{Y}_{GL} \mathbf{Y}_{LL}^{-1} \mathbf{Y}_{LG}]^{-1} \mathbf{C}_G.
\end{aligned}
$$

(4) Eigenvalues and eigenvectors of state matrix **A**

All eigenvalues of **A** are obtained by using the QR method as

$$
\begin{aligned}
&\lambda_1 = -53.05299, \quad \lambda_2 = -51.80217, \quad \lambda_3 = -30.41762, \quad \lambda_4 = -28.21401, \\
&\lambda_{5,6} = -0.75497 \mp j12.86370, \quad \lambda_{7,8} = -0.15154 \mp j8.67125, \\
&\lambda_9 = -5.58205, \quad \lambda_{10} = -3.72276, \\
&\lambda_{11,12} = -1.13701 \mp j0.91540, \quad \lambda_{13,14} = -0.48432 \mp j0.657417, \\
&\lambda_{15} = -0.04571, \quad \lambda_{16} = 0.00000.
\end{aligned}
$$

Obviously, except the zero eigenvalue we expect, the real part of the remaining eigenvalues is negative. Hence the power system is stable in terms of small-signal stability.

8.4 Eigenvalue Problem in Small-Signal Stability Analysis

Nonlinear system stability, when the system is subject to small disturbances, can be analyzed from the stability of its linearized system as determined by the eigenvalues of state matrix **A**. Hence, in the following, we shall introduce the method of eigensolution analysis for a state matrix **A**.

From the discussion above we can see that state matrix **A** is a real asymmetric matrix. Hence, in the following, all our discussion will be under the condition that $\mathbf{A} \in \mathbb{R}^{n\times n}$. We denote the set of complex numbers by $\mathbb{C}$, n-dimensional complex vector space (column vector) by $\mathbb{C}^n$, and set of all m-row n-column complex matrices by $\mathbb{C}^{m\times n}$. Operations of scalar multiplication, addition and multiplication of complex matrices are similar to those for real matrices. However, transposition of a complex matrix is taken as conjugate transposition (denoted by superscript H), i.e., $\mathbf{C} = \mathbf{A}^H \Rightarrow c_{ij} = \hat{a}_{ji}$. Dot product of n-dimensional vector x and y is $s = \mathbf{x}^H\mathbf{y} = \sum_{i=1}^{n} \hat{x}_i y_i$. In addition, unit vector (normalized vector) under norm p is a vector x satisfying $\|\mathbf{x}\|_p = 1$. For example, unit vectors x under 1-norm, 2-norm, and infinite norm, respectively, are

$$\left.\begin{aligned} |\mathbf{x}|_1 &= |x_1| + \cdots + |x_n| = 1 \\ |\mathbf{x}|_2 &= \sqrt{|x_1|^2 + \cdots + |x_n|^2} = \sqrt{\mathbf{x}^H\mathbf{x}} = 1 \\ |\mathbf{x}|_\infty &= \max_{1\le i\le n} |x_i| = 1 \end{aligned}\right\}. \tag{8.84}$$

The process to convert a vector to a unit vector is called normalization.

8.4.1 Characteristics of State Matrix Given by Its Eigensolution

8.4.1.1 Eigenvalue

For a scalar $\lambda \in \mathbb{C}$ and vector $\mathbf{v} \in \mathbb{C}^n$, if equation

$$\mathbf{A}\mathbf{v} = \lambda\mathbf{v} \tag{8.85}$$

has a nonsingular solution (i.e., $\mathbf{v} \neq \mathbf{0}$), λ is an eigenvalue of matrix **A**.

To calculate eigenvalues, (8.85) can be written as

$$(\mathbf{A} - \lambda\mathbf{I})\mathbf{v} = \mathbf{0}. \tag{8.86}$$

A sufficient and necessary condition for existence of a nonsingular solution of the equation is

$$\det(\mathbf{A} - \lambda\mathbf{I}) = 0. \tag{8.87}$$

Expansion of the determinant in the above equation gives the following polynomial equation

$$\alpha_0 + \alpha_1\lambda + \cdots + \alpha_{n-1}\lambda^{n-1} + (-1)^n\lambda^n = 0. \tag{8.88}$$

It is called the characteristic equation of matrix **A**. The polynomial on the left side of the above equation is called the characteristic polynomial. Because the coefficient of λ^n is nonzero, there are a total of n roots. The set of all roots is called the spectrum and is denoted by $\lambda(\mathbf{A})$. If $\lambda(\mathbf{A}) = \{\lambda_1, \cdots, \lambda_n\}$, we have

$$\det(\mathbf{A}) = \lambda_1\lambda_2\cdots\lambda_n.$$

In addition, if we define the trace of **A** to be

$$\mathrm{tr}(\mathbf{A}) = \sum_{i=1}^{n} a_{ii}.$$

Then $\mathrm{tr}(\mathbf{A}) = \lambda_1 + \lambda_2 + \cdots + \lambda_n$, can be proved.

Eigenvalues of a real asymmetric matrix can be real or complex numbers. Complex eigenvalues always appear in the form of conjugate pairs. Moreover, similar matrices have the same eigenvalues and transposition of a matrix does not change its eigenvalues.

8.4.1.2 Eigenvectors

For any eigenvalue λ_i, any nonzero vector $\mathbf{v}_i \in \mathbb{C}^n$ satisfying equation

$$\mathbf{A}\mathbf{v}_i = \lambda_i\mathbf{v}_i \quad i = 1, 2, \ldots, n \tag{8.89}$$

is called a right eigenvector of matrix **A** corresponding to eigenvalue λ_i. Since it is a homogenous equation, $k\mathbf{v}_i$ (k is a scalar) is also the solution of the equation to be a right eigenvector of matrix **A** corresponding to eigenvalue λ_i. In the following (unless explicitly stated otherwise) "eigenvector" refers to "right eigenvector." An eigenvector defines a one-dimensional subspace that remains invariable under the operation of left multiplication by matrix **A**.

Similarly, any nonzero vector $\mathbf{u}_i \in \mathbb{C}^n$ satisfying equation

$$\mathbf{A}^{\mathrm{T}}\mathbf{u}_i = \lambda_i\mathbf{u}_i \quad i = 1, 2, \ldots, n \tag{8.90}$$

is called a right eigenvector of matrix $\mathbf{A}^{\mathrm{T}}$ corresponding to eigenvalue λ_i. Taking transposition on both sides of equation, we have

$$\mathbf{u}_i^{\mathrm{T}}\mathbf{A} = \lambda_i\mathbf{u}_i^{\mathrm{T}}, \quad i = 1, 2, \ldots, n. \tag{8.91}$$

We call row vector $\mathbf{u}_i^{\mathrm{T}}$ the left eigenvector of matrix $\mathbf{A}$ corresponding to eigenvalue λ_i.

To express the eigensolution of matrix $\mathbf{A}$ clearly, we form a diagonal matrix Λ consisting of all eigenvalues of matrix $\mathbf{A}$, a matrix $\mathbf{X}_{\mathrm{R}}$ of all right eigenvectors arranged in columns, a matrix $\mathbf{X}_{\mathrm{L}}$ of all left eigenvectors in rows. That is

$$\left.\begin{aligned} \mathbf{\Lambda} &= \mathrm{diag}\{\lambda_1 \quad \lambda_2 \quad \cdots \quad \lambda_n\} \\ \mathbf{X}_{\mathrm{R}} &= [\mathbf{v}_1 \quad \mathbf{v}_2 \quad \cdots \quad \mathbf{v}_n] \\ \mathbf{X}_{\mathrm{L}} &= [\mathbf{u}_1 \quad \mathbf{u}_2 \quad \cdots \quad \mathbf{u}_n]^{\mathrm{T}} \end{aligned}\right\}. \tag{8.92}$$

These three n-dimensional square matrices are called modal matrices.

Using (8.92), (8.89), and (8.91) can be expressed in the following matrix form:

$$\left.\begin{aligned} \mathbf{A}\mathbf{X}_{\mathrm{R}} &= \mathbf{X}_{\mathrm{R}}\mathbf{\Lambda} \\ \mathbf{X}_{\mathrm{L}}\mathbf{A} &= \mathbf{\Lambda}\mathbf{X}_{\mathrm{L}} \end{aligned}\right\}. \tag{8.93}$$

Premultiplying the first equation above by $\mathbf{X}_{\mathrm{L}}$, and postmultiplying the second by $\mathbf{X}_{\mathrm{R}}$, we have

$$(\mathbf{X}_{\mathrm{L}}\mathbf{X}_{\mathrm{R}})\mathbf{\Lambda} = \mathbf{\Lambda}(\mathbf{X}_{\mathrm{L}}\mathbf{X}_{\mathrm{R}}) \tag{8.94}$$

or

$$\lambda_j \mathbf{u}_i^{\mathrm{T}}\mathbf{v}_j = \lambda_i \mathbf{u}_i^{\mathrm{T}}\mathbf{v}_j, \quad i,j = 1,2,\ldots,n.$$

Obviously, left and right eigenvectors corresponding to different eigenvalues are orthogonal; for the same eigenvalue their product is a nonzero number that can be converted to 1 after normalization of left and right eigenvectors. That is

$$\mathbf{u}_i^T \mathbf{v}_j = \begin{cases} 0 & i \neq j \\ 1 & i = j \end{cases}. \tag{8.95}$$

Please note that $\mathbf{u}_i^{\mathrm{T}}\mathbf{v}_j$ is not the normal inner product of two vectors. The matrix form of above equation is

$$\mathbf{X}_{\mathrm{L}}\mathbf{X}_{\mathrm{R}} = \mathbf{I}, \quad \mathbf{X}_{\mathrm{L}}^{-1} = \mathbf{X}_{\mathrm{R}}. \tag{8.96}$$

From (8.93) and (8.96) we have

$$\mathbf{X}_{\mathrm{R}}^{-1}\mathbf{A}\mathbf{X}_{\mathrm{R}} = \mathbf{\Lambda}. \tag{8.97}$$

8.4.1.3 Free Movement of Dynamic System

From the state equation, (8.4), we can see that the rate of change of every state variable is a linear combination of all state variables. Hence due to the coupling among state variables, it is difficult to clearly see the system movement.

To cancel the coupling among state variables, we introduce a new state variable vector $\mathbf{z}$. Its relationship with the original state variable vector $\Delta\mathbf{x}$ is defined to be

$$\Delta\mathbf{x} = \mathbf{X}_{\mathrm{R}}\mathbf{z}. \tag{8.98}$$

Substituting the above equation into (8.4) and using (8.14), the state equation can be written as

$$\frac{\mathrm{d}\mathbf{z}}{\mathrm{d}t} = \Lambda z. \tag{8.99}$$

The difference from the original state equation is that Λ is a diagonal matrix, while $\mathbf{A}$ usually is not. Equation (8.99) can be expressed as n decoupled first-order differential equations

$$\frac{\mathrm{d}z_i}{\mathrm{d}t} = \lambda_i z_i, \quad i = 1, 2, \ldots, n. \tag{8.100}$$

Its solution in the time domain is

$$z_i(t) = z_i(0)\mathrm{e}^{\lambda_i t}, \tag{8.101}$$

where initial values of z_i, $z_i(0)$ can be expressed from (8.98) by $\mathbf{u}_i^{\mathrm{T}}$ and $\Delta\mathbf{x}(0)$

$$z_i(0) = \mathbf{u}_i^{\mathrm{T}}\Delta\mathbf{x}(0). \tag{8.102}$$

Substituting (8.101) and (8.102) into the transformation of (8.98), we have the solution of the original state vector in the time domain to be

$$\Delta\mathbf{x} = \sum_{i=1}^{n} \mathbf{v}_i z_i(0)\mathrm{e}^{\lambda_i t},$$

where solution of the ith state variable in the time domain is

$$\begin{aligned} \Delta x_i(t) &= v_{i1}z_1(0)\mathrm{e}^{\lambda_1 t} + v_{i2}z_2(0)\mathrm{e}^{\lambda_2 t} + \cdots + v_{in}z_n(0)\mathrm{e}^{\lambda_n t}, \\ &= 1, 2, \ldots, n, \end{aligned} \tag{8.103}$$

where v_{ik} is the ith element of vector $\mathbf{v}_k$. The above equation is the time response of system free movement expressed by eigenvalues, left and right eigenvectors. Eigenvalue λ_i represents the ith mode of the system, with corresponding time characteristic $e^{\lambda_i t}$. Hence, time response of system free movement is the linear combination of n system modes. Therefore, system stability is determined by the eigenvalues:

(1) A real eigenvalue represents a nonoscillatory mode. A negative real eigenvalue is a decaying mode and the bigger its absolute value, the faster it decays. A positive real eigenvalue indicates nonperiodic instability. Eigenvectors, and $z(0)$, corresponding to real eigenvalues are real valued.
(2) Complex eigenvalues always appear in conjugate pairs, i.e.,

$$\lambda = \sigma \pm j\omega. \tag{8.104}$$

Each pair of complex eigenvalues represents an oscillation mode. Eigenvectors, and $z(0)$, corresponding to complex eigenvalues are complex valued. Hence

$$(a + jb)e^{(\sigma - j\omega)t} + (a - jb)e^{(\sigma + j\omega)t} = e^{\sigma t}(2a \cos \omega t + 2b \sin \omega t)$$

should exhibit as $e^{\sigma t} \sin(\omega t + \theta)$.

Obviously, the real part of the eigenvalue describes system oscillation damping and the imaginary part gives the frequency of oscillation. A negative real part is a decaying oscillation mode and positive an increasing oscillation mode. Oscillation frequency (Hz) is

$$f = \frac{\omega}{2\pi}. \tag{8.105}$$

Damping ratio is defined to be

$$\zeta = \frac{-\sigma}{\sqrt{\sigma^2 + \omega^2}}. \tag{8.106}$$

This determines the decay rate property of the oscillation magnitude.

8.4.2 *Modal Analysis of Linear Systems*

8.4.2.1 Mode and Eigenvector

From the discussion above we know that the relationship among system time response, vectors $\Delta\mathbf{x}$ and $\mathbf{z}$ are

$$\left.\begin{aligned} \Delta\mathbf{x}(t) &= \mathbf{X}_R\mathbf{z}(t) = [\mathbf{v}_1 \quad \mathbf{v}_2 \quad \cdots \quad \mathbf{v}_n]z(t) = \sum_{i=1}^{n} \mathbf{v}_i z_i(t) \\ \mathbf{z}(t) &= \mathbf{X}_L\Delta\mathbf{x}(t) = [\mathbf{u}_1 \quad \mathbf{u}_2 \quad \cdots \quad \mathbf{u}_n]^{\mathrm{T}}\Delta\mathbf{x}(t) \end{aligned}\right\}. \tag{8.107}$$

Variables Δx_1, Δx_2, ..., Δx_n are the original state variables depicting system dynamics. Variables $z_1, z_2, \cdots, z_n$ are state variables after transformation, each of which represents a mode of the system.

From the first equation of (8.107) we can see that right eigenvectors decide the form of exhibition of each mode, i.e., when a specific mode is excited, the relative activity of each state variable is described by the right eigenvector. For example, when the ith mode is excited, the kth element v_{ki} of right eigenvector $\mathbf{v}_i$ gives the level of influence of this mode on state variable x_k. Magnitude of each element in $\mathbf{v}_i$ represents the level of activity of each of the n state variables resulting from the ith mode; while the angle of each element represents the effect of the mode on the phase shift of each state variable.

From the second equation of (8.107) we can see that the left eigenvector $\mathbf{u}_i^{\mathrm{T}}$ represents the way the original state variables combine to effect the ith mode. Therefore, the kth element in the right eigenvector $\mathbf{v}_i$ measures the level of activity of state variable x_k in the ith mode; while the kth element of the left eigenvector $\mathbf{u}_i^{\mathrm{T}}$ weights the contribution of the exhibited activity to the ith mode.

8.4.2.2 Eigenvalue Sensitivity

Firstly we consider the sensitivity of an eigenvalue to each element a_{kj} in matrix $\mathbf{A}$ (the k-row, j-column element in $\mathbf{A}$). Taking partial derivatives to a_{kj} on both sides of (8.89), we have

$$\frac{\partial \mathbf{A}}{\partial a_{kj}}\mathbf{v}_i + \mathbf{A}\frac{\partial \mathbf{v}_i}{\partial a_{kj}} = \frac{\partial \lambda_i}{\partial a_{kj}}\mathbf{v}_i + \lambda_i \frac{\partial \mathbf{v}_i}{\partial a_{kj}}. \tag{8.108}$$

Premultiplying both sides of the above equation by $\mathbf{u}_i^{\mathrm{T}}$ and from (8.91) and (8.95) we can obtain

$$\frac{\partial \lambda_i}{\partial a_{kj}} = \mathbf{u}_i^{\mathrm{T}} \frac{\partial \mathbf{A}}{\partial a_{kj}}\mathbf{v}_i. \tag{8.109}$$

Obviously, in $\partial \mathbf{A}/\partial a_{kj}$ the kth-row, jth-column element is 1 and remaining elements are zero. Hence

$$\frac{\partial \lambda_i}{\partial a_{kj}} = u_{ki} v_{ji} \tag{8.110}$$

where, v_{ji} is the jth element in $\mathbf{v}_i$ and u_{ki} is the kth element in $\mathbf{u}_i$.

Assuming α is a scalar, $\mathbf{A}(\alpha)$ is an n-order square matrix with elements being $a_{kj}(\alpha)$ and for all k and j, $a_{kj}(\alpha)$ is a differentiable function, we have

$$\frac{\mathrm{d}\mathbf{A}(\alpha)}{\mathrm{d}\alpha} = \left(\frac{\mathrm{d}a_{kj}(\alpha)}{\mathrm{d}\alpha}\right). \tag{8.111}$$

Therefore, similarly we can find the eigenvalue sensitivity to scalar α to be

$$\frac{\partial \lambda_i}{\partial \alpha} = \mathbf{u}_i^{\mathrm{T}} \frac{\partial \mathbf{A}}{\partial \alpha} \mathbf{v}_i. \tag{8.112}$$

8.4.2.3 Participation Factor

To determine the relationship between state variables and system modes, we establish a so-called participation matrix **P** by combining right and left eigenvectors to measure the level of coupling between state variables and system modes.

$$\mathbf{P} = \begin{array}{c} \Delta x_1 \\ \\ \Delta x_k \\ \\ \Delta x_n \end{array} \overset{\begin{array}{ccccc} \lambda_1 & & \lambda_i & & \lambda_n \end{array}}{\begin{bmatrix} u_{11}v_{11} & \cdots & u_{1i}v_{1i} & \cdots & u_{1n}v_{1n} \\ \vdots & \ddots & \vdots & \ddots & \vdots \\ u_{k1}v_{k1} & \cdots & u_{ki}v_{ki} & \cdots & u_{kn}v_{kn} \\ \vdots & \ddots & \vdots & \ddots & \vdots \\ u_{n1}v_{n1} & \cdots & u_{ni}v_{ni} & \cdots & u_{nn}v_{nn} \end{bmatrix}}. \tag{8.113}$$

Element $p_{ki} = u_{ki}\, v_{ki}$ in matrix **P** is called a participation factor [193] that measures the level of participation of the ith mode and the kth state variable Δx_k with each other. The ith row of matrix **P**, $\mathbf{p}_i$, is the participation vector of the ith mode. Since v_{ki} measures the level of activity of Δx_k in the ith mode and u_{ki} weights the contribution of the activity to the mode, their product p_{ki} can measure the pure participation. The product of corresponding elements in left and right eigenvectors is a dimensionless result, independent of the dimensions selected for the eigenvectors.

Assuming $\Delta\mathbf{x}(0) = \mathbf{e}_k$, i.e., $\Delta x_k(0) = 1$ and $\Delta x_{j \neq k}(0) = 0$, from (8.102) we have $z_i(0) = u_{ki}$. From (8.103) we can obtain

$$\Delta x_k(t) = \sum_{i=1}^{n} v_{ki} u_{ki} \mathrm{e}^{\lambda_i t} = \sum_{i=1}^{n} p_{ki} \mathrm{e}^{\lambda_i t}. \tag{8.114}$$

This equation shows that the ith mode excited by initial value $\Delta x_k(0) = 1$ participates in response $\Delta x_k(t)$ with a participation coefficient p_{ki}. That is why it is called a participation factor.

For all modes or all state variables, it is easy to prove that

$$\sum_{i=1}^{n} p_{ki} = \sum_{k=1}^{n} p_{ki} = 1. \tag{8.115}$$

To set $t = 0$ in (8.114), we can easily obtain the summation of the kth-row elements of **P** to be 1. Summation of the ith-column elements of matrix **P** is equal to $\mathbf{u}_i^{\mathrm{T}}\mathbf{v}_i$

which is 1 according to (8.95). In addition, from (8.110) we can see that participation factor p_{ki} in fact is the sensitivity of eigenvalue λ_i to diagonal element a_{kk} of matrix **A**, i.e.,

$$p_{ki} = \frac{\partial \lambda_i}{\partial a_{kk}}. \tag{8.116}$$

8.4.3 *Computation of Eigenvalues*

8.4.3.1 QR Method

Among numerical methods to compute all the eigenvalues of a general matrix, the QR method is usually the first choice. It was proposed by J. G. F. Francis in 1962, and has advantages such as strong robustness and fast speed of convergence. It has been found to be the most effective method of eigensolution so far.

For a given $\mathbf{A} \in \mathbb{R}^{n\times n}$ and orthogonal matrix $\mathbf{Q}_0 \in \mathbb{R}^{n\times n}$, we have the following iteration:

$$\begin{aligned}
&\mathbf{A}_0 = \mathbf{Q}_0^{\mathrm{T}}\mathbf{A}\mathbf{Q}_0,\\
&k = 1, 2, \ldots,\\
&\mathbf{A}_{k-1} = \mathbf{Q}_k\mathbf{R}_k \quad \text{(QR decomposition)},\\
&\mathbf{A}_k = \mathbf{R}_k\mathbf{Q}_k,
\end{aligned} \tag{8.117}$$

where each $\mathbf{Q}_k \in \mathbb{R}^{n\times n}$ is an orthogonal matrix and $\mathbf{R}_k \in \mathbb{R}^{n\times n}$ upper triangular matrix. By an inductive approach we have

$$\mathbf{A}_k = (\mathbf{Q}_0\mathbf{Q}_1 \cdots \mathbf{Q}_k)^{\mathrm{T}}\mathbf{A}(\mathbf{Q}_0\mathbf{Q}_1 \cdots \mathbf{Q}_k). \tag{8.118}$$

Hence each $\mathbf{A}_k$ is similar to **A**. Because matrix **A** has complex eigenvalues, $\mathbf{A}_k$ will not converge to a strict "eigenvalue exposed" upper triangular matrix, but satisfy a computational real Schur decomposition.

An upper triangular matrix with diagonal elements being 1×1 blocks or 2×2 blocks is called an upper quasi-triangular matrix. Real Schur decomposition is a real operation to reduce a matrix to an upper quasi-triangular matrix. If $\mathbf{A} \in \mathbb{R}^{n\times n}$, there exists an orthogonal matrix $\mathbf{Q} \in \mathbb{R}^{n\times n}$ to lead to

$$\mathbf{Q}^{\mathrm{T}}\mathbf{A}\mathbf{Q} = \begin{bmatrix} \mathbf{R}_{11} & \mathbf{R}_{12} & \cdots & \mathbf{R}_{1m} \\ \mathbf{0} & \mathbf{R}_{22} & \cdots & \mathbf{R}_{2m} \\ \vdots & \vdots & \ddots & \vdots \\ \mathbf{0} & \mathbf{0} & \cdots & \mathbf{R}_{mm} \end{bmatrix}, \tag{8.119}$$

where $\mathbf{R}_{ii}$ is either a 1×1 matrix or 2×2 matrix. If it is 1×1, the element is an eigenvalue of matrix $\mathbf{A}$; if 2×2, the eigenvalues of $\mathbf{R}_{ii}$ are a pair of conjugate complex eigenvalue of $\mathbf{A}$.

To effectively complete a Schur decomposition, we select the initial orthogonal matrix of a similarity transformation as in (8.117) $\mathbf{Q}_0$ to make $\mathbf{A}_0$ become an upper Hessenberg matrix. Doing so, the computational complexity of one iteration is reduced from $O(n^3)$ to $O(n^2)$.

In an upper Hessenberg matrix, except for the sub-diagonal elements, those below the diagonal are zero. For example, in a 6×6 upper Hessenberg matrix, nonzero elements are distributed as shown below:

$$\begin{bmatrix} \times & \times & \times & \times & \times & \times \\ \times & \times & \times & \times & \times & \times \\ & \times & \times & \times & \times & \times \\ & & \times & \times & \times & \times \\ & & & \times & \times & \times \\ & & & & \times & \times \end{bmatrix}.$$

This form of matrix can be obtained by performing a series of Householder transformations. Since Householder transformation is a symmetrical orthogonal similarity transformation, the upper Hessenberg matrix obtained has the same eigenvalues as the original matrix.

Finally, if values of elements of $\mathbf{A}$ have large differences, implementation of the iterative method could result in large computational errors in eigenvalues. The level of sensitivity of eigenvalue computation to round off can be reduced by a balancing operation. Since usually errors of eigensolution from numerical computation are proportional to a Euclidean norm, the idea of the balancing operation is to make the norm of corresponding rows and columns as close as possible through similarity transformation. Thus, the total norm of the matrix is reduced without changing the eigenvalues of the matrix.

Implementation of the balancing operation is to determine the diagonal matrix $\mathbf{D}$ through $O(n^2)$ computation such that

$$\tilde{\mathbf{A}} = \mathbf{D}^{-1}\mathbf{A}\mathbf{D} = [\mathbf{c}_1, \mathbf{c}_2, \ldots, \mathbf{c}_n] = [\mathbf{r}_1, \mathbf{r}_2, \ldots, \mathbf{r}_n]^{\mathrm{T}} \tag{8.120}$$

with $\| \mathbf{r}_i \|_\infty \approx \| \mathbf{c}_i \|_\infty$, $i = 1,2, \ldots, n$. Diagonal matrix $\mathbf{D}$ is selected to have the form $\mathbf{D} = \mathrm{diag}\{b^{i_1}, b^{i_2}, \ldots, b^{i_n}\}$, where b is the floating-point base. Thus round off in computing $\tilde{\mathbf{A}}$ is avoided. After $\mathbf{A}$ goes through the balancing operation, computation of eigenvalues will become more accurate.

8.4.3.2 The Power Method

In practical applications, often we do not need to compute all eigenvalues of matrix $\mathbf{A}$, but only that with largest modulus (often called the dominant eigenvalue). The power method is a very effective iterative method to calculate the dominant eigenvalue.

Assuming that $\mathbf{A} \in \mathbb{C}^{n\times n}$ can be diagonalized and $\mathbf{X}^{-1}\,\mathbf{A}\,\mathbf{X} = \mathrm{diag}\,(\lambda_1, \lambda_2, \ldots, \lambda_n)$, where $\mathbf{X} = [\mathbf{x}_1, \mathbf{x}_2, \ldots, \mathbf{x}_n]$, $|\lambda_1| > |\lambda_2| \geq \cdots \geq |\lambda_n|$. For a given initial unit vector under the 2-norm $\mathbf{v}^{(0)} \in \mathbb{C}^n$, the power method generates the following series of vectors $\mathbf{v}^{(k)}$

$$\left.\begin{aligned} \mathbf{z}^{(k)} &= \mathbf{A}\mathbf{v}^{(k-1)} \\ \mathbf{v}^{(k)} &= \mathbf{z}^{(k)}/\left\|\mathbf{z}^{(k)}\right\|_2 \\ \lambda^{(k)} &= [\mathbf{v}^{(k)}]^H\mathbf{A}\mathbf{v}^{(k)} \end{aligned}\right\}, \quad k = 1, 2, \ldots. \tag{8.121}$$

Obviously, the series of vectors in the above iteration $\mathbf{v}^{(k)}$ are unit vectors under the 2-norm.

Because

$$\mathrm{dist}(\mathrm{span}\{\mathbf{v}^{(k)}\},\ \mathrm{span}\{\mathbf{x}_1\}) = O\left(\left|\frac{\lambda_2}{\lambda_1}\right|^k\right)$$

and

$$\left|\lambda_1 - \lambda^{(k)}\right| = O\left(\left|\frac{\lambda_2}{\lambda_1}\right|^k\right).$$

Obviously, only $\lambda_2/\lambda_1 < 1$, when $k \to \infty$, we have

$$\lambda^{(k)} \to \lambda_1, \quad \mathbf{v}^{(k)} \to \mathbf{x}_1. \tag{8.122}$$

The power method is of linear convergence and its applicability depends on the ratio $|\lambda_2|/|\lambda|_1$, which reflects the rate of convergence.

After the dominant eigenvalue of $\mathbf{A}$ is obtained by using the power method, we can compute the remaining eigenvalues through a deflation technique. There are many deflation methods but only a few of them are numerically stable. In the following, we shall introduce a deflation method based on similarity transformation.

Assuming λ_1 and $\mathbf{v}_1$ are known, we can find a Householder matrix $\mathbf{H}_1$ to satisfy $\mathbf{H}_1\,\mathbf{v}_1 = k_1\,\mathbf{e}_1$ and $k_1 \neq 0$. From $\mathbf{A}_1\,\mathbf{v}_1 = \lambda_1\,\mathbf{v}_1$, we have $\mathbf{H}_1\mathbf{A}_1(\mathbf{H}_1^{-1}\mathbf{H}_1)\mathbf{v}_1 = \lambda_1\mathbf{H}_1\mathbf{v}_1$. Obviously, $\mathbf{H}_1\mathbf{A}_1\mathbf{H}_1^{-1}\mathbf{e}_1 = \lambda_1\mathbf{e}_1$, that is, the first column of $\mathbf{H}_1\mathbf{A}_1\mathbf{H}_1^{-1}$ is $\lambda_1\ \mathbf{e}_1$. Denoting

$$\mathbf{A}_2 = \mathbf{H}_1\mathbf{A}_1\mathbf{H}_1^{-1} = \left[\begin{array}{c|c} \lambda_1 & \mathbf{b}_1^{\mathrm{T}} \\ \hline \mathbf{0} & \mathbf{B}_2 \end{array}\right], \tag{8.123}$$

where $\mathbf{B}_2$ is an $(n-1)$th-order square matrix that obviously has eigenvalues to be $\lambda_2, \ldots, \lambda_n$. Under the condition that $|\lambda_2| > |\lambda_3|$, we can use power method to compute the dominant eigenvalue of $\mathbf{B}_2$, λ_2, and corresponding eigenvector, $\mathbf{y}_2$, where

$\mathbf{B}_2 \mathbf{y}_2 = \lambda_2 \mathbf{y}_2$. Assuming $\mathbf{A}_2 \mathbf{z}_2 = \lambda_2 \mathbf{z}_2$ and to calculate $\mathbf{z}_2$, assuming α a constant to be found and $\mathbf{y}$ an $(n-1)$th dimensional vector, we have

$$\mathbf{z}_2 = \begin{bmatrix} \alpha \\ \mathbf{y} \end{bmatrix}, \quad \begin{cases} \lambda_1 \alpha + \mathbf{b}_1^{\mathrm{T}} \mathbf{y} = \lambda_2 \alpha \\ \mathbf{B}_2 \mathbf{y} = \lambda_2 \mathbf{y} \end{cases}.$$

Because $\lambda_1 \neq \lambda_2$, we can choose $\mathbf{y} = \mathbf{y}_2$, $\alpha = \frac{\mathbf{b}_1^{\mathrm{T}} \mathbf{y}}{\lambda_2 - \lambda_1}$, thus we can find $\mathbf{z}_2$. $\mathbf{v}_2 = \mathbf{H}_1^{-1} \mathbf{z}_2$ is the eigenvector of $\mathbf{A}$ corresponding to λ_2.

With application of the above method and Householder matrix, we have

$$\left.\begin{aligned} k_1 &= -\mathrm{sgn}(\mathbf{e}_1^{\mathrm{T}} \mathbf{v}_1) \|\mathbf{v}_1\|_2 \\ \beta &= \left[\|\mathbf{v}_1\|_2 \left(\|\mathbf{v}_1\|_2 + \|\mathbf{e}_1^T \mathbf{v}_1\| \right) \right]^{-1} \\ \mathbf{u} &= \mathbf{v}_1 - k_1 \mathbf{e}_1 \\ \mathbf{A}_2 &= \mathbf{H}_1 \mathbf{A}_1 \mathbf{H}_1^{-1} = (\mathbf{I} - \beta \mathbf{u}\mathbf{u}^{\mathrm{T}}) \mathbf{A}_1 (\mathbf{I} - \beta \mathbf{u}\mathbf{u}^{\mathrm{T}}) \end{aligned}\right\}. \tag{8.124}$$

After λ_2 and $\mathbf{v}_2$ are computed, we can continue to deflate $\mathbf{B}_2$ to calculate the rest of the eigenvalues and eigenvectors. In theory, if eigenvalues of $\mathbf{A}$ are arranged according to their modulus and those with higher values can be separated, we can use the above method to compute those eigenvalues. A drawback of the deflation method is that it changes elements of the original matrix, so that any sparsity in the matrix cannot be maintained during deflation.

Finally, we would like to point out that it is not so straightforward to compute the dominant eigenvalue and corresponding eigenvector by using the power method as introduced above, because we only discussed the case of a single dominant eigenvalue. In fact, λ_1 could be one of a set of multiple real eigenvalues or λ_1 and λ_2 could have the same modulus but are real eigenvalues with opposite sign, or λ_1 and λ_2 are a pair of conjugate complex eigenvalues. For those different cases, the power method will be slightly different. Details can be found in [187].

8.4.3.3 The Inverse Power Method

Eigenvalues of inverse matrix $\mathbf{A}^{-1}$ of a nonsingular matrix $\mathbf{A}$ are reciprocal values of the eigenvalues of $\mathbf{A}$. Hence the reciprocal of the dominant eigenvalue of $\mathbf{A}^{-1}$ is the eigenvalue of $\mathbf{A}$ with smallest modulus. Applying the power method on $\mathbf{A}^{-1}$ is called the inverse power method (or inverse iterative method) to compute the eigenvalue of the nonsingular matrix $\mathbf{A}$ with smallest modulus and corresponding eigenvector.

For a given initial unit vector under the 2-norm, $\mathbf{v}^{(0)} \in \mathbb{C}^n$, the inverse power method generates the following iterative series

$$\left.\begin{aligned} \mathbf{A}\mathbf{z}^{(k)} &= \mathbf{v}^{(k-1)} \\ \mathbf{v}^{(k)} &= \mathbf{z}^{(k)} / \|\mathbf{z}^{(k)}\|_2 \\ \lambda^{(k)} &= [\mathbf{v}^{(k)}]^H \mathbf{A} \mathbf{v}^{(k)} \end{aligned}\right\}, \quad k = 1, 2, \ldots. \tag{8.125}$$

When $k \to \infty$,

$$\lambda^{(k)} \to \frac{1}{\lambda_n}\mathbf{v}^{(k)} \to \mathbf{x}_n \tag{8.126}$$

Another more useful form of inverse power method is to apply the power method to matrix $(\mathbf{A} - \tau\mathbf{I})^{-1}$, where τ is a real or complex constant. For a given initial unit vector under the 2-norm, $\mathbf{v}^{(0)} \in \mathbb{C}^n$, the iterative process is as the following:

$$\left.\begin{array}{l} (\mathbf{A} - \tau\mathbf{I})\mathbf{z}^{(k)} = \mathbf{v}^{(k-1)} \\ \mathbf{v}^{(k)} = \mathbf{z}^{(k)}/\|\mathbf{z}^{(k)}\|_2 \\ \lambda^{(k)} = [\mathbf{v}^{(k)}]^H\mathbf{A}\mathbf{v}^{(k)} \end{array}\right\}, \quad k = 1, 2, \ldots. \tag{8.127}$$

When $k \to \infty$,

$$\left.\begin{array}{l} \lambda^{(k)} \to \dfrac{1}{\lambda_p - \tau}\tau + \dfrac{1}{\lambda^{(k)}} \to \lambda_p \\ \mathbf{v}^{(k)} \to \mathbf{x}_p \end{array}\right\}, \tag{8.128}$$

where λ_p is that closest to τ among all eigenvalues of $\mathbf{A}$ and $\mathbf{x}_p$ is the corresponding eigenvector. We need to explain (8.128) further as follows.

Because eigenvalues of nonsingular matrix $\mathbf{A} - \tau\mathbf{I}$ are $\lambda_j - \tau$ ($j = 1, 2, \ldots, n$), those corresponding to matrix $(\mathbf{A} - \tau\,\mathbf{I})^{-1}$ are $\frac{1}{\lambda_j - \tau}$ $(j = 1, 2, \ldots, n)$. Applying the power method to matrix $(\mathbf{A} - \tau\mathbf{I})^{-1}$, we obtain eigenvalue $\frac{1}{\lambda_p - \tau}$ with largest modulus that means $\lambda_p - \tau$ with smallest modulus. Hence λ_p is the closest to τ.

Hence if we need to compute the eigenvalue of matrix $\mathbf{A}$ with a value closest to number τ and corresponding eigenvector, we can use the inverse power method given by (8.127). Another application of the inverse power method is that with a known approximation τ of an eigenvalue of matrix $\mathbf{A}$, we can use the inverse power method to compute the corresponding eigenvector and improve the accuracy of computation of the eigenvalue.

Using (8.127), we can apply triangular decomposition on matrix $\mathbf{A} - \tau\mathbf{I}$,

$$\mathbf{A} - \tau\mathbf{I} = \mathbf{L}\mathbf{U},$$

where $\mathbf{L}$ is a unit lower triangular matrix and $\mathbf{U}$ upper triangular. Then equation of solution becomes

$$\mathbf{L}\mathbf{U}\mathbf{z}^{(k)} = \mathbf{v}^{(k-1)}.$$

8.4.4 Eigensolution of Sparse Matrix

In small-signal stability analysis, the dynamics of a power system are described by differential-algebraic equations of (8.3). From (8.83) we can see that $\tilde{A}$, $\tilde{B}$, $\tilde{C}$, and $\tilde{D}$ are all sparse matrices. When we obtain matrix $\mathbf{A}$ from (8.5) to compute its eigenvalues, we can find that matrix **A** has lost its sparsity almost completely. Since the implementation of the QR method cannot take advantage of matrix sparsity, it is not important whether **A** is sparse or not when we compute its eigenvalues. However, when other iterative methods, such as power method, inverse power method, and sub-space method (to be introduced later), are used to compute part of the eigenvalues of matrix **A**; if we can take full advantage of the sparsity of the original matrices to compute those eigenvalues directly from (8.3), computational efficiency will be greatly enhanced.

For an eigenvalue of **A**, λ, a nonzero vector $\mathbf{v} \in \mathbb{C}^n$ satisfying the following equation

$$\begin{bmatrix} \tilde{\mathbf{A}} & \tilde{\mathbf{B}} \\ \tilde{\mathbf{C}} & \tilde{\mathbf{D}} \end{bmatrix} \begin{bmatrix} \mathbf{v} \\ \mathbf{w} \end{bmatrix} = \lambda \begin{bmatrix} \mathbf{v} \\ \mathbf{0} \end{bmatrix} \tag{8.129}$$

is the right eigenvector of **A**, corresponding to this eigenvalue. The matrix on the left-hand side of above equation is called the augmented state matrix.

It is not difficult to prove the above conclusion. In fact, from (8.129) we have $\mathbf{w} = -\tilde{D}^{-1}\tilde{C}\mathbf{v}$. Canceling **w** and from (8.5), we can obtain

$$(\tilde{\mathbf{A}} - \tilde{\mathbf{B}}\tilde{\mathbf{D}}^{-1}\tilde{\mathbf{C}})\mathbf{v} = \mathbf{A}\mathbf{v} = \lambda\mathbf{v}. \tag{8.130}$$

Hence we can compute eigenvalues and eigenvectors of matrix **A** from the eigensolution of the augmented state matrix of (8.129) without destroying system sparsity. In the following, we shall introduce the sparse realization of the power method and the inverse power method. In addition, the sparse expression of eigenvalue sensitivity to scalar α will be presented.

8.4.4.1 Sparse Realization of Power Method of (8.121)

Since the relationship between $\mathbf{z}^{(k)}$ and $\mathbf{v}^{(k-1)}$ given by the equation

$$\begin{bmatrix} \mathbf{z}^{(k)} \\ \mathbf{0} \end{bmatrix} = \begin{bmatrix} \tilde{\mathbf{A}} & \tilde{\mathbf{B}} \\ \tilde{\mathbf{C}} & \tilde{\mathbf{D}} \end{bmatrix} \begin{bmatrix} \mathbf{v}^{(k-1)} \\ \mathbf{w}^{(k-1)} \end{bmatrix}$$

is equivalent to $\mathbf{z}^{(k)} = \mathbf{A}\mathbf{v}^{(k-1)}$, computation of the first equation in (8.121) can be replaced by the following equations:

$$\left.\begin{aligned} \tilde{\mathbf{D}}\mathbf{w}^{(k-1)} &= -\tilde{\mathbf{C}}\mathbf{v}^{(k-1)} \\ \mathbf{z}^{(k)} &= \tilde{\mathbf{A}}\mathbf{v}^{(k-1)} + \tilde{\mathbf{B}}^{(k-1)} \end{aligned}\right\}. \tag{8.131}$$

Before the iteration of (8.121) is implemented, we only apply sparse triangular decomposition on $\tilde{\mathbf{D}}$ once, i.e., $\tilde{\mathbf{D}}$. Hence computation of (8.131) in each iteration is only the multiplication of some sparse matrices and vectors and solution of two triangular equations.

8.4.4.2 Sparse Realization of Inverse Power Method of (8.127)

Since the relationship between $\mathbf{z}^{(k)}$ and $\mathbf{v}^{(k-1)}$ given by the equation

$$\begin{bmatrix} \tilde{\mathbf{A}} - \tau\mathbf{I} & \tilde{\mathbf{B}} \\ \tilde{\mathbf{C}} & \tilde{\mathbf{D}} \end{bmatrix} \begin{bmatrix} \mathbf{z}^{(k)} \\ \mathbf{w}^{(k)} \end{bmatrix} = \begin{bmatrix} \mathbf{v}^{(k-1)} \\ \mathbf{0} \end{bmatrix} \tag{8.132}$$

is equivalent to $(\mathbf{A} - \tau\mathbf{I})\mathbf{z}^{(k)} = \mathbf{v}^{(k-1)}$, solution of the first equation in (8.127) can be replaced by that of (8.132) to obtain vector $\mathbf{z}^{(k)}$.

For a given number τ, we first calculate $\tilde{\mathbf{D}}^* = \tilde{\mathbf{D}} - \tilde{\mathbf{C}}\,(\tilde{\mathbf{A}} - \tau\mathbf{I})^{-1}\tilde{\mathbf{B}}$ and apply sparse triangular decomposition $\tilde{\mathbf{D}}^* = \mathbf{LU}$. Noting $(\tilde{\mathbf{D}})$ is a diagonal block matrix (a diagonal block is from a dynamic component in power system), we can obtain $(\tilde{\mathbf{A}} - \tau\mathbf{I})^{-1}$ by calculating the inverse of diagonal block matrices directly. In addition, $\tilde{\mathbf{D}}^*$ and $\tilde{\mathbf{D}}$ have the same sparse structure (2×2 block sparse matrix). Hence solution of (8.132) can be summarized in the following steps:

(1) Calculate $\mathbf{w}^{(k)}$ from solution of the equation

$$\tilde{\mathbf{D}}^*\mathbf{w}^{(\mathbf{k})} = -\tilde{\mathbf{C}}(\tilde{\mathbf{A}} - \tau\mathbf{I})^{-1}\mathbf{v}^{(\mathbf{k}-1)}$$

(2) Calculate $\mathbf{z}^{(k)} = (\tilde{\mathbf{A}} - \tau\mathbf{I})^{-1}(\mathbf{v}^{(k-1)} - \tilde{\mathbf{B}}w^{(k)})$

8.4.4.3 Eigenvalue Sensitivity to Scalar α

Similar to (8.129), for a left eigenvector we have

$$\begin{bmatrix} \mathbf{u}^{\mathrm{T}} & \mathbf{y}^{\mathrm{T}} \end{bmatrix} \begin{bmatrix} \tilde{\mathbf{A}} & \tilde{\mathbf{B}} \\ \tilde{\mathbf{C}} & \tilde{\mathbf{D}} \end{bmatrix} = \lambda \begin{bmatrix} \mathbf{u}^{\mathrm{T}} & \mathbf{0} \end{bmatrix}. \tag{8.133}$$

Hence using a similar derivation, we can obtain

$$\frac{\partial \lambda_i}{\partial \alpha} = \begin{bmatrix} \mathbf{u}_i^{\mathrm{T}} & \mathbf{y}_i^{\mathrm{T}} \end{bmatrix} \begin{bmatrix} \dfrac{\partial \tilde{\mathbf{A}}}{\partial \alpha} & \dfrac{\partial \tilde{\mathbf{B}}}{\partial \alpha} \\ \dfrac{\partial \tilde{\mathbf{C}}}{\partial \alpha} & \dfrac{\partial \tilde{\mathbf{D}}}{\partial \alpha} \end{bmatrix} \begin{bmatrix} \mathbf{v}_i \\ \mathbf{w}_i \end{bmatrix} \tag{8.134}$$

8.4.5 Application of Eigenvalue Sensitivity Analysis

In analysis of power system operation and design of power system controllers, we often need to investigate the influence of certain parameters, such as the gain and time constant of a controller, on power system stability. This will help in the selection or setting of those parameters to stabilize the power system or to improve system stability.

Since system state matrix **A** is a function of a system parameter α, i.e., $\mathbf{A}(\alpha)$, any eigenvalue of matrix **A**, λ_i, is also a function of parameter α, i.e., $\lambda_i(\alpha)$, $i = 1, 2, \ldots, n$. When parameter α varies, $\lambda_i(\alpha)$ will accordingly change. Variation of $\lambda_i(\alpha)$ represents the influence of variation of parameter α on power system stability.

Assuming that parameter α changes from $\alpha_{(0)}$ to $\alpha_{(0)} + \Delta\alpha$, the corresponding change of system eigenvalue is from $\lambda_i(\alpha_{(0)})$ to $\lambda_i(\alpha_{(0)} = \Delta\alpha)$. Taylor expansion of $\lambda_i(\alpha_{(0)} + \Delta\alpha)$ at $\alpha_{(0)}$ is

$$\lambda_i(\alpha_{(0)} + \Delta\alpha) = \lambda_i(\alpha_{(0)}) + \left.\frac{\partial\lambda_i(\alpha)}{\partial\alpha}\right|_{\alpha=\alpha_{(0)}}\Delta\alpha + \left.\frac{\partial^2\lambda_i(\alpha)}{\partial\alpha^2}\right|_{\alpha=\alpha_{(0)}}(\Delta\alpha)^2 + \cdots.$$

When $\Delta\alpha$ is very small, change of λ_i can be approximately expressed as

$$\Delta\lambda_i(\alpha_{(0)}, \Delta\alpha) = \lambda_i(\alpha_{(0)} + \Delta\alpha) - \lambda_i(\alpha_{(0)}) = \left.\frac{\partial\lambda_i(\alpha)}{\partial\alpha}\right|_{\alpha=\alpha_{(0)}}\Delta\alpha, \tag{8.135}$$

where partial derivative $\partial\lambda_i/\partial\alpha$ is the first-order sensitivity of eigenvalue λ_i to parameter α, referred to simply as eigenvalue sensitivity. Hence if we can calculate $\partial\lambda_i/\partial\alpha$, $\Delta\alpha$ can be approximately determined from the required change of eigenvalue $\Delta\lambda_i$.

Calculation of the first-order sensitivity of eigenvalue λ_i to parameter α can be summarized as follows:

(1) Set $\alpha = \alpha_{(0)}$ to form state matrix $\mathbf{A}(\alpha_{(0)})$
(2) Calculate eigenvalue of $\mathbf{A}(\alpha_{(0)})$, λ_i, and corresponding left and right eigenvector $\mathbf{u}_i^H$ and $\mathbf{v}_i$ such that $\mathbf{u}_i^H\mathbf{v}_i = 1$
(3) Calculate $\left.\frac{\partial\mathbf{A}(\alpha)}{\partial\alpha}\right|_{\alpha=\alpha_{(0)}}$
(4) $\frac{\partial\lambda_i(\alpha)}{\partial\alpha} = \mathbf{u}_i^H\left.\frac{\partial\mathbf{A}(\alpha)}{\partial\alpha}\right|_{\alpha=\alpha_{(0)}}\mathbf{v}_i$

In the following, we shall give an example taking the gain K_S, time constant T_1, T_2, T_3, and T_4 of lead-lag network of PSS as parameter α to demonstrate the calculation of $\frac{\partial\mathbf{A}(\alpha)}{\partial\alpha}$. In the equations of generation unit g of (8.20) and (8.21), except $\overline{\mathbf{A}}_g, \mathbf{B}_{Ig}, \overline{\mathbf{B}}_{Vg}, \overline{\mathbf{P}}_g$, and $\overline{\mathbf{Z}}_g$ are independent of α. In addition, $\mathbf{R}_{Ig}$, $\mathbf{R}_{Vg}$, and $\mathbf{T}_{g(0)}$ are also independent of α. Hence from (8.30) and (8.32) we have

$$\frac{\partial \mathbf{B}_g}{\partial \alpha} = \frac{\partial \mathbf{C}_g}{\partial \alpha} = \frac{\partial \mathbf{D}_g}{\partial \alpha} = \mathbf{0},$$

where matrix $\frac{\partial \bar{\mathbf{A}}_g}{\partial \alpha}$ can be calculate from matrix $\bar{\mathbf{A}}_g$ in (8.20).

Obviously in the equation of the whole system, from (8.83) we can obtain

$$\frac{\partial \tilde{\mathbf{A}}}{\partial \alpha} = \begin{bmatrix} \frac{\partial \mathbf{A}_G}{\partial \alpha} & \mathbf{0} & \mathbf{0} & \mathbf{0} \\ \mathbf{0} & \mathbf{0} & \mathbf{0} & \mathbf{0} \\ \mathbf{0} & \mathbf{0} & \mathbf{0} & \mathbf{0} \\ \mathbf{0} & \mathbf{0} & \mathbf{0} & \mathbf{0} \end{bmatrix}, \quad \frac{\partial \tilde{\mathbf{B}}}{\partial \alpha} = \mathbf{0}, \quad \frac{\partial \tilde{\mathbf{C}}}{\partial \alpha} = \mathbf{0}, \quad \frac{\partial \tilde{\mathbf{D}}}{\partial \alpha} = \mathbf{0},$$

where $\partial \mathbf{A}_G / \partial \alpha$ can be calculated from (8.73). Moreover, from (8.5) and the equation above we have

$$\frac{\partial \mathbf{A}}{\partial \alpha} = \frac{\partial \tilde{\mathbf{A}}}{\partial \alpha}.$$

The partial derivative of matrix **A** to other parameters can be calculated similarly.

In the analysis of eigenvalue sensitivity, in addition to the eigenvalue sensitivity to parameters introduced above, eigenvalue sensitivity to power system operating conditions has been proposed. To enhance computational accuracy, second-order eigenvalue sensitivity has also been suggested, with some effective computational methods proposed. Details can be found in references [215–218].

8.5 Oscillation Analysis of Power Systems

A power system cannot operate without proper control. System operators can satisfy the predicted load demand through automatic generation control, and also through switching on, or off, various other controllable devices. Certain automatic control devices, such as the governor and AVR of a generator, HVDC control and FACTS control, etc., carry out the task of fast automatic regulation to maintain system frequency and voltage within required limits, when the power system is subject to disturbances.

Since the middle of the twentieth century, the power industry has found that interconnection of regional power systems can lead to more reliable and economical operation of power systems. This has resulted in the increasing scale of modern power systems. In the 1960s, the interconnection of two Northern American power systems suffered from increasing oscillations. Power system oscillations have subsequently been reported in many countries. Investigation into power system oscillations has revealed that when regional power networks are connected through long-distance transmission lines, the resulting weak coupling of large power generation centers implies weak damping of interarea power oscillations. Another cause of reduced, or even negative, damping of power system oscillation is the application of high-gain, fast-acting excitation systems. Electrical engineers have found that through the introduction of a supplementary control signal from PSS, system damping can be increased. Experience of Northern American power system interconnection has shown that application of PSS is very effective in damping power system oscillations. Increasing oscillations prevent power networks from exploiting interconnection. In some interconnected power systems, power exchange between interconnected networks has to be kept below a certain limit to avoid the occurrence of oscillations. This greatly reduces the value of interconnecting regional power networks. In some interconnected power systems, low-gain AVR have to be adopted to avoid the oscillation problem. Hence before the scheme of asynchronous interconnection via HVDC was proposed, further interconnection was abandoned in some power systems.

Since the 1940s, it has been known that excitation control can enhance the stability limit of a synchronous generator. Since in some cases, excitation control can successfully improve power system dynamic performance, in addition to the control being fast and efficient, electrical engineers have held high expectations of the function of excitation control. However, effectiveness of excitation control is not unlimited. Fast-acting excitation systems can improve synchronous torque to enhance system first swing stability. However, fact-acting excitation is often a negative feedback system with high gain that has little influence on oscillation damping after the first swing. Sometimes it could provide negative damping. When a power system exhibits negative oscillation damping, fast-acting excitation control (usually with high gain) often increases the negative damping to the detriment of system operating conditions.

In an m machine interconnected power system, there are a total of $m - 1$ electromechanical oscillation modes. From field records of real power system oscillation [229] and extensive experience from power system simulation, these oscillation modes can be classified, according to the area of coverage, into two types [189], local modes and interarea modes:

(1) Local modes only involve power swings of generation units in a power plant to the rest of the power system. Oscillation frequency usually is between 1 and 2 Hz.
(2) Interarea modes are power swings of a group of generators in an area to another group of generators in another area. This interarea oscillation often occurs between two or more generators in a weakly connected power system. Because the moment of inertia of the equivalent generator in each area is very large, the oscillation frequency of an interarea oscillation is lower than that of local-mode oscillation, being in the range of 0.1–0.7 Hz. When the oscillation is exhibited between two groups of generators, oscillation frequency is between about 0.1 and 0.3 Hz. When it is an oscillation among multiple groups of generators, oscillation frequency is about 0.4–0.7 Hz.

Since the frequency of those two types of oscillation is low, they are often called power system low-frequency oscillations. In addition to electromechanical oscillation modes, control modes and torsional oscillation modes may exist in a power system. Torsional modes have been previously introduced. Control modes are related to various control devices installed in the power system. Since regulation of control devices is fast and controllers have small time constant, frequency of control modes is usually high. Here we are only concerned about electromechanical oscillation modes. Analysis regarding control modes and torsional modes is out of the scope of this book.

Small disturbances can lead to power system low-frequency oscillations. If the oscillations of all modes are decaying, the power system is stable in terms of small-signal stability. However, in real power system operation, usually only where the damping ratio of electromechanical oscillation modes is greater than 0.05, is the power system operation acceptable. Of course this value is not fixed. With variations of system operating conditions and small changes of oscillation modes, lower damping ratios (such as 0.03) could also be acceptable.

It is apparent that small-signal instability of real power systems is mainly due to system oscillations caused by lack of damping. In 1969, Demello and Concordia [218] obtained conditions of power system small-signal stability with regard to the operation of a thyristor-controlled excitation system for a single-machine infinite-bus power system. These are certain requirements on the setting of AVR gain and the introduction of an auxiliary control signal of generator rotor deviation. Their work clearly revealed the cause of power system oscillation in the single-machine infinite-bus model and laid down a solid theoretical foundation for the design of PSSs. Based on their idea and principles proposed, researchers have attempted extensions into multimachine power systems for the analysis of local-mode oscillations, and further to interarea oscillations in interconnected power systems. However, we have to point out that some of the simple extensions are often found to be inappropriate.

A large-scale multimachine power system is a typical nonlinear dynamic system. Increasing oscillations caused by disturbances are dependant on many factors. Network topology and parameters, characteristics of dynamic components, system operating conditions, control strategies and parameters of various controllers all play an important role in system oscillations. It is a challenging task to clearly analyze the cause of power system electromechanical oscillations and to propose effective measure to overcome the problem. With the increasing demand of economics in a modern society, especially with the trend toward electricity markets, more and more load is required to be carried over existing power networks. However, economics and security of the power system are two conflicting requirements. When a power system operates under a light load condition before it is disturbed, damping windings of generators can provide adequate torque proportional to rotor speed. This damping can usually absorb the energy involved in system oscillations and thus the magnitude of oscillations decays continuously. The power system is stable in terms of small-signal stability. If the power system operates at heavy load conditions before it is disturbed, damping windings of generators cannot completely dissipate the energy involved in the system oscillations, so that the oscillations can grow continuously. The power system is unstable in terms of small-signal stability. Moreover, to increase the capability of power transmission and to improve system transient stability or other system performance aspects, large numbers of various types of controller are installed in the power system. Some of these may clash with the damping of system oscillations, due to improper control strategies or parameters, or mismatch among controller functions. This may again lead to unstable system oscillations. The purpose of oscillation analysis of power systems is to study key factors affecting oscillation modes, so that useful measures can be worked out to suppress oscillations effectively.

[Example 8.2] In Example 8.1, all the eigenvalues of the system state matrix have been calculated. In the following, we shall study system oscillation modes.

[Solution] Table 8.2 gives oscillation frequency and damping ratio of several oscillation modes; corresponding left and right eigenvectors and participation vectors are given in Tables 8.3 and 8.4.

In the following we will carry out modal analysis from the results in Tables 8.3 and 8.4, where all vectors have been normalized to unit vectors under the infinite norm. Firstly, we identify electromechanical oscillation modes from the participation vector of specified modes: if the component with largest modulus in a participation vector is related to generator speed, we identify that the mode is an electromechanical oscillation mode. Then we can observe the exhibition of modes from right eigenvectors: for those components in right eigenvectors related to generator speed, a group of components with similar modulus and directional phase identifies a group of coherent generators. Incoherent generators are associated with those components with opposite phase. The right eigenvector of a local mode is dominated by variables related to one or a group of closely located generators. Components of the right eigenvector of an interarea mode evenly distribute in all regions in a power system.

For oscillation mode $\lambda_{5,6}$, the element with largest modulus in its participation vector is related to $\Delta\omega_3$. Hence it is an electromechanical oscillation mode. Besides, in its right eigenvector, components associated with $\Delta\omega_1$, $\Delta\omega_2$ have small modulus (being 0.00018 and 0.00121, respectively) and similar phase (being $-170.62°$ and $-166.98°$, respectively); the component associated with $\Delta\omega_3$ has large modulus (0.00411) and opposite directional phase to those above (being 10.52°). Hence this mode will exhibit as an electromechanical oscillation between generator 1, 2 and generator 3. It is a local oscillation mode with oscillation frequency being 2.04732 Hz.

Table 8.2 Oscillation frequency and damping ratio of several oscillation modes

	$\lambda_{5,6}$	$\lambda_{7,8}$	$\lambda_{11,12}$	$\lambda_{13,14}$
f	2.04732	1.38007	0.14569	0.10463
ξ	0.05859	0.01747	0.77893	0.59313

Table 8.3 Left and right eigenvectors and participation vectors of oscillation modes

	λ_6						λ_8					
	$\mathbf{u}_6$		$\mathbf{v}_6$		$\mathbf{p}_6$		$\mathbf{u}_8$		$\mathbf{v}_8$		$\mathbf{p}_8$	
	Modulus	Phase	Modulus	Phase	Modulus	Phase	Modulus	Phase	Modulus	Phase	Modulus	Phase
$\Delta\delta_1$	0.01137	−90.60	0.00532	96.02	0.01470	−5.10	0.02300	90.86	0.04052	155.40	0.43955	2.47
$\Delta\omega_1$	0.33268	176.13	**0.00018**	**−170.62**	**0.01470**	−5.00	1.00000	0.00	**0.00093**	**−113.60**	**0.43957**	2.61
$\Delta\delta_2$	0.02281	−85.77	0.03527	99.66	0.19568	3.36	0.01871	−89.81	0.11332	−26.92	0.99988	−0.52
$\Delta\omega_2$	0.66769	−178.79	**0.00121**	**−166.98**	**0.19574**	3.71	0.81346	179.71	**0.00261**	**64.08**	**1.0000**	0.00
$\Delta E'_{q2}$	0.02124	−85.82	0.00156	−162.26	0.00808	101.41	0.02567	−77.72	0.00265	48.13	0.03212	86.62
$\Delta E'_{d2}$	0.00888	110.43	0.00688	36.04	0.01485	135.95	0.00413	42.44	0.01286	−91.43	0.02506	67.22
ΔE_{fq2}	0.00028	−178.73	0.03610	72.77	0.00241	−116.48	0.00049	−168.06	0.10048	−5.93	0.02337	−57.78
ΔV_{R2}	0.00003	−13.37	0.36141	76.77	0.00233	52.88	0.00005	2.07	1.00000	0.00	0.02253	118.28
ΔV_{M2}	0.00198	53.99	0.00184	−88.59	0.00089	−45.12	0.00248	71.83	0.00506	−170.13	0.00591	17.92
$\Delta\delta_3$	0.03416	92.62	0.12030	−82.84	0.99933	−0.74	0.00430	−86.22	0.06144	−23.80	0.12456	6.19
$\Delta\omega_3$	1.00000	0.00	**0.00411**	**10.52**	**1.0000**	0.00	0.18692	−176.13	**0.00141**	**67.20**	**0.12458**	7.29
$\Delta E'_{q3}$	0.03534	94.07	0.00411	−12.05	0.03536	71.50	0.00870	−68.23	0.00054	19.43	0.00222	67.42
$\Delta E'_{d3}$	0.01302	−78.12	0.02094	−152.38	0.06632	118.98	0.00257	−12.05	0.00513	−94.98	0.00621	9.19
ΔE_{fq3}	0.00047	1.15	0.09989	−4.01	0.01132	−13.38	0.00017	−158.57	0.07102	−6.82	0.00570	−49.18
ΔV_{R3}	0.00005	166.51	1.0000	0.00	0.01091	155.99	0.00002	11.56	0.70681	−0.89	0.00550	126.88
ΔV_{M3}	0.00336	−126.13	0.00509	−165.36	0.00416	57.99	0.00086	81.33	0.00358	−171.02	0.00144	26.52

Table 8.4 Left and right eigenvectors and participation vectors of oscillation modes

	λ_{12}						λ_{14}					
	$\mathbf{u}_{12}$		$\mathbf{v}_{12}$		$\mathbf{p}_{12}$		$\mathbf{u}_{14}$		$\mathbf{v}_{14}$		$\mathbf{p}_{14}$	
	Modulus	Phase	Modulus	Phase	Modulus	Phase	Modulus	Phase	Modulus	Phase	Modulus	Phase
$\Delta\delta_1$	0.00383	140.64	0.05071	85.09	0.04408	150.68	0.00049	−146.95	0.04062	106.90	0.00159	−74.61
$\Delta\omega_1$	1.00000	0.00	**0.00020**	**−133.74**	**0.04458**	151.21	0.23071	87.88	**0.00009**	**−126.72**	**0.00162**	−73.39
$\Delta\delta_2$	0.00240	−31.48	0.05334	73.44	0.02911	−33.09	0.00194	−49.58	0.04627	122.47	0.00715	38.33
$\Delta\omega_2$	0.64734	−170.64	**0.00021**	**−145.39**	**0.03036**	−31.08	0.94544	−171.29	**0.00010**	**−111.15**	**0.00756**	43.00
$\Delta E'_{q2}$	0.36458	34.69	0.00812	−178.71	0.67247	140.93	0.43208	38.37	0.01202	0.10	0.41422	3.91
$\Delta E'_{d2}$	0.15380	156.74	0.00066	−13.57	0.02317	68.12	0.08490	153.90	0.00528	171.90	0.03577	−68.77
ΔE_{fq2}	0.04393	−103.88	0.06692	−40.05	0.66740	141.02	0.09457	−81.94	0.05319	115.53	0.40108	−0.97
ΔV_{R2}	0.00440	75.05	1.00000	0.00	**1.00000**	0.00	0.00936	97.30	0.48473	−176.05	0.36171	−113.31
ΔV_{M2}	0.02583	171.93	0.00489	−178.93	0.02866	−82.04	0.04856	148.04	0.00240	4.71	0.00930	118.19
$\Delta\delta_3$	0.00149	−52.19	0.05286	75.02	0.01782	−52.22	0.00194	115.83	0.04028	63.91	0.00622	145.17
$\Delta\omega_3$	0.41946	171.13	**0.00021**	**−143.82**	**0.01949**	−47.74	1.00000	0.00	**0.00009**	**−169.72**	**0.00696**	155.72
$\Delta E'_{q3}$	0.16500	42.95	0.00685	−177.95	0.25677	149.95	0.50272	−141.75	0.02494	176.31	**1.00000**	0.00
$\Delta E'_{d3}$	0.10380	156.78	0.00099	−11.69	0.02343	70.04	0.14005	−30.18	0.01276	−12.20	0.14246	−76.95
ΔE_{fq3}	0.02025	−95.61	0.05590	−39.27	0.25702	150.07	0.11208	97.94	0.10972	−68.42	0.98069	−5.04
ΔV_{R3}	0.00203	83.32	0.83532	0.78	0.38511	9.05	0.01109	−82.82	1.00000	0.00	0.88442	−117.38
ΔV_{M3}	0.01191	−179.80	0.00408	−178.14	0.01104	−72.99	0.05756	−32.08	0.00495	−179.24	0.02273	114.12

Similarly, for $\lambda_{7,8}$, the element with largest modulus in its participation vector is related to $\Delta\omega_2$. Hence it is an electromechanical oscillation mode. In addition, in its right eigenvector, components associated with $\Delta\omega_2$, $\Delta\omega_3$ have relatively large modulus (being 0.00261 and 0.00141, respectively) and the same direction (phase being 64.08° and 67.20°, respectively); component associated with $\Delta\omega_1$ has small modulus (0.00093) and opposite direction. Hence this mode will exhibit as electromechanical oscillation between generator 1, 2 and generator 3. It is also a local oscillation mode with oscillation frequency being 1.38007 Hz. Though this mode is stable, the damping ratio (0.01747) is not sufficient, exhibiting poor dynamic performance as far as oscillation decay is concerned.

For mode $\lambda_{11,12}$, the element with largest modulus in its participation vector is related to ΔV_{R2}; for $\lambda_{13,14}$, element with largest modulus in its participation vector is related to $\Delta E'_{q3}$. Hence those modes are not electromechanical oscillation modes but control modes.

[Example 8.3] We take the 39-node 10-machine simplified New England system as an example to demonstrate the procedure of power system oscillation analysis [221]. In the power system, ten machines are at nodes 30–39 and the machine at node 39 is an equivalent generator. Generators at nodes 30–38 have fast static excitation systems installed.

[Solution] We obtain the system linearized equation by using the methods introduced above and then compute all eigenvalues of the system state matrix. Damping of nine modes associated with electromechanical oscillations is not sufficient and some of eigenvalues have positive real parts. For two eigenvalues $0.1022 \pm j7.215$ (mode 1) and 0.037 ± 4.301 (mode 9), components associated with generator speed in their right eigenvectors are given in Table 8.5.

From Table 8.5 we can see that in the eigenvector of the first mode, there are three components with large modulus (highlighted by bold figures), among these the direction of the first component (with phase being 0°) is opposite to that of the

Table 8.5 Components associated with generator speed in right eigenvectors of mode 1 and 2

Generator number	Mode 1		Mode 9	
	Modulus	Phase (degree)	Modulus	Phase (degree)
30	**1.0**	0.0	0.5574	−9.9
31	0.1408	−44.5	0.4757	−3.4
32	0.0797	241.9	0.5208	−5.5
33	0.1851	152.3	0.7601	−5.3
34	0.4777	−32.1	1.0	0.0
35	**0.7935**	170.2	0.7961	−5.7
36	**0.7797**	170.5	0.7977	−6.8
37	0.3468	10.1	0.5084	−12.4
38	0.1664	111.4	0.6694	−3.3
39	0.0170	191.3	**0.4052**	−179.6

other two (with phase being about 170°). This indicates that the mode mainly exhibits as an electromechanical oscillation between generator 30 and generators 35, 36; with oscillation frequency being $7.215/2\pi = 1.148$ Hz. This is a local oscillation mode. In the eigenvector of the second mode, except for the component associated with generator 39 with relatively small modulus, other components have similar values of modulus. Moreover, the first nine components have opposite direction (with phase being about 0°) to that of the last one (with phase being about −180°). This indicates that this mode exhibits mainly as an electromechanical oscillation between generators 30–38 and generator 39 (the equivalent generator that can be seen as a regional network). Hence this is an interarea oscillation with oscillation frequency being $4.301/2\pi = 0.685$ Hz.

In addition to identifying electromechanical oscillation modes, participation vectors can also be used to estimate the relative effects of generator controls on specified oscillation modes. For example, a component associated with rotor speed in a participation vector gives eigenvalue sensitivity to the variation of damping applied on the associated generator. If it is zero, this indicates that installation of PSS on the generator will have no impact in improving oscillation damping. If it is a large positive number, this shows that the associated generator is a good candidate place to install PSS, to effectively increase damping of the relevant oscillation mode. In Table 8.6, components associated with generator rotor speed are given. From Table 8.6 we can see that for the local mode, a component in the participation vector associated with generator 30 has the largest value, about equal to the sum of components associated with generators 35 and 36. Hence applying damping control at generator 30 should almost be equivalent to similar applications at both generators 35 and 36 simultaneously. For the interarea oscillation mode, a component associated with generator 39 has the largest modulus. However, generator 39 is an equivalent machine and damping control cannot be applied there. The sum of components associated with generators 30–38 is about equal to the component related to generator 39. This means that damping control applied at generators 30–38 will achieve a similar effect as that applied at generator 39. Moreover, we should note that although some generators have large participation factors, there will be little effect in applying damping control on those generators if their capacity is small. Applying damping control on generators with large capacity will be more effective than applying it on those with small capacity, as far as increased oscillation damping is concerned.

Table 8.6 Components in participation vector

Generator number	30	31	32	33	34	35	36	37	38	39
Mode 1	**1.0**	0.01	0.005	0.02	0.13	**0.42**	**0.43**	0.07	0.02	0.001
Mode 2	0.17	0.09	0.12	0.22	0.33	0.26	0.21	0.07	0.18	**1.0**

Values in the table are estimated from graphs presented in [221]

Thinking and Problem Solving

1. What are the purpose and significance of small-signal stability analysis for electrical power systems?
2. What are the basic principle and basic procedures of small-signal stability analysis?
3. What are the main methods to solve the eigenvalues of a linearized electrical power system? What are their advantages and disadvantages?
4. Why is the QR method not suitable for the eigenvalue analysis of large-scale electrical power systems?
5. What is the critical eigenvalue? What methods are there for calculating critical eigenvalues of large-scale electrical power systems? What advantages and disadvantages are there for each method?
6. How can we apply sparse matrix techniques to critical eigenvalue calculations for large-scale electrical power systems? Can the sparse matrix technique be used in the QR method?
7. How are the eigenvalue and corresponding left and right eigen vectors used to represent the modes of a linear system?
8. What is the participation factor? Why can the participation factor be used to represent both the observability and controllability of a system?
9. What are the main causes of increasing amplitude, low-frequency oscillation?
10. What are the major manifestations of low-frequency oscillation? Why is the oscillating frequency among local generators lower than that among generators in a plant, and the oscillating frequency among regional generators is lower than that among local generators?
11. What are the main measures to control low-frequency oscillation?

References

1. W.F. Tinney, I.W. Waiker, "Direct solutions of sparse network equation by optimal ordered triangular factorization," *Proceedings of IEEE* 55(11), 1801–1809, 1967
2. W.F. Tinney, Some examples of sparse matrix methods for power system problems, *Proceedings of Power Systems Computation Conference (PSCC)*, Rome, June 23–27, 1969
3. W.F. Tinney, V. Brandwajn, S.M. Chen, "Sparse vector methods," *IEEE Transactions on Power Apparatus and Systems,* 104, 295–301, 1985
4. G.W. Stagg, A.H. El-Abiad, *Computer Methods in Power Systems*, McGraw Hill, New York, 1968
5. R.G. Andreich, H.E. Brown, H.H. Happ, C.E. Person, "The piecewise solution of the impedance matrix load flow," *IEEE Transactions on Power Apparatus and Systems,* 87(10), 1877–1882, 1968
6. W.F. Tinney, C.E. Hart, "Power flow solution by Newton's method," *IEEE Transactions on Power Apparatus and Systems* 86(4), 1449–1460, 1967
7. W.F. Tinney, "Compensation methods for network solutions by optimal ordered triangular factorization," *IEEE Transactions on Power Apparatus and Systems*, 91(1), 123–127, 1972
8. B. Scott, O. Alsac, "Fast decoupled load flow," *IEEE Transactions on Power Apparatus and Systems*, 93(3), 859–869, 1974
9. R. Van Amerongen, "A general-purpose version of the fast decoupled load flow," *IEEE Transactions on Power Systems*, 4(2), 760–770, 1989
10. A. Monticeli, O.R. Savendra, "Fast decoupled load flow: Hypothesis, derivations, and testing," *IEEE Transactions on Power Systems,* 5(4), 1425–1431, 1990
11. L. Wang, X. Rong Li, "Robust fast decoupled power flow," *IEEE Transactions on Power Systems*, 15(1), 208–215, 2000
12. V.M. da Costa, N. Martins, J.L. Pereira, "Developments in the Newton Raphson power flow formulation based on current injections," *IEEE Transactions on Power Systems*, 14(4), 1320–1326, 1999
13. A. Semlyen, F. de Leon, "Quasi-Newton power flow using partial Jacobian updates," *IEEE Transactions on Power Systems*, 16(3), 332–339, 2001
14. V.H. Quintana, N. Muller, "Studies of load flow method in polar and rectangular coordinates," *Electric Power System Research*, 20(1), 225–235, 1991
15. R.P. Klump, T. J. Overbye, "Techniques for improving power flow convergence," *Proceedings of PES Summer Meeting*, Seattle, USA, vol. 1, July 2000
16. K.L. Lo, Y.J. Lin, W.H. Siew, "Fuzzy-logic method for adjustment of variable parameters in load flow calculation," *IEE Proceedings of Generation Transmission Distribution*, 146(3), 276–282, 1999
17. W.L. Chan, A.T.P. So, L.L. Lai, "Initial applications of complex artificial neural networks to load-flow analysis," *IEE Proceedings of Generation Transmission Distribution*, 147(6), 361–366, 2000

18. T. Nguyen, "Neural network load-flow," *IEE Proceedings of Generation Transmission Distribution*, 142(1), 51–58, 1995
19. K.P. Wong, A. Li, T.M.Y. Law, "Advanced constrained genetic algorithm load flow method," *Proceedings of Generation Transmission Distribution*, 146(6), 609–616, 1999
20. P.K. Mannava, L. Teeslink, A.R. Hasan, "Evaluation of efficiency of parallelization of power flow algorithms," *Proceedings of the 40th Midwest Symposium on Circuit and Systems, Sacramento, California, USA,* August 1997, pp. 127–130
21. N. balu, T. Bertram, A. Bose, et al, "On-Line power system security analysis," *Proceedings of the IEEE*, 80, 262–282, 1992
22. J. Carpentier, "Static security assessment and control: A short survey," IEEE/NTUA Athens Power Tech Conference on Planning, Operation and Control of Today's Electric Power Systems", Athens, Greece, September 5–8, 1993, pp. 1–9
23. B. Stott, O. Alsac, F.L. Alvarado, "Analytical and computational improvement in performance index raking algorithm for networks," *International Journal of Electrical Power and Energy Systems*, 7(3), 154–160, 1985
24. T.A. Mikolinas, B.F. Wollenberg, "An advanced contingency selection algorithm," *IEEE Transactions on Power Apparatus and Systems,* 100(2), 608–617, 1981
25. V. Brandwajn, "Efficient bounding method for linear contingency analysis," *IEEE Transactions on Power Systems*, 3(2), 726–733, 1988
26. R. Bacher, W.F. Tinney, "Faster local power flow solutions: The zero mismatch approach," *IEEE Transactions on Power Systems*, 4(4), 1345–1354, 1989
27. W.P. Luan, K.L. Lo, Y.X. Yu, "NN-based pattern recognition technique for power system security assessment," *Proceedings of the International Conference on Electric Utility Deregulation and Restructuring and Power Technologies*, London, April 2000
28. Task Force on Probabilistic Aspects of Reliability Criteria of the IEEE PES Reliability, Risk and Probability Applications Subcommittee, "Probabilistic security assessment for power system operations", IEEE PES General Meeting, Denver, USA, June 6–10, 2004
29. A.M. Leite da Silva, L.C. Resende, L.A.F. Manso, et al, "Well-being analysis for composite generation and transmission systems", IEEE Transactions on Power Systems, 19(4), 1763–1770, 2004
30. R. Billinton, R.N. Allan, *Reliability evaluation of Large Electric Power Systems*, Kluwer, Boston, 1988
31. R. Billinton, R. Allan, *Reliability Evaluation of Power Systems*, Plenum Press, New York, 1996
32. X. Wang, J. McDonald, *Modern Power System Planning*, McGraw-Hill, London, 1994, pp. 108–110
33. R. Billinton, W. Li, *Reliability Evaluation of Electric Power Systems using Monte Carlo Methods*, Plenum Press, New York, 1994
34. A.M. Leite de Silva, S.M.P. Ribeiro, V.L. Arienti, et al., "Probabilistic load flow techniques applied to power system expansion planning", *IEEE Transactions on Power Systems*, 5(4), 1047–1053, 1990
35. P. Zhang, S.T. Lee, "Probabilistic load flow computation using the method of combined cumulants and Gram-Charlier expansion", *IEEE Transactions on Power Systems*, 19(1), 676–682, 2004
36. Z. Hu, X. Wang, "A probabilistic load flow method considering branch outages," *IEEE Transactions on Power Systems*, 21(2), 507–514, 2006
37. K. Maurice, S. Alan, *The Advanced Theory of Statistics*, vol. 1, Macmillan, USA, 1977
38. J.C. Spall, "Estimation via Markov chain Monte Carlo," *IEEE Control System Magazine*, 23 (2), 35–45, 2003
39. R. Chen, J.S. Liu, X. Wang, "Convergence analyses and comparisons of Markov chain Monte Carlo algorithms in digital communications," *IEEE Transactions on Signal Processing*, 50(2), 255–270, 2002

40. Reliability Test System Task Force, "IEEE reliability test system," *IEEE Transactions on Power System*, 14(3), 1010–1020, 1999
41. R. Von Mises, *Mathematical Theory of Probability and Statistics*, Academic Press, New York, 1964
42. P.A. Jensen, J.W. Barnes, *Network-Flow Programming*, Wiley, New York, 1980
43. Y.K. Lin, "Reliability of a stochastic flow network with unreliable branches and nodes, under budget constraints," *IEEE Transactions on Reliability*, 53(3), 381–387, 2004
44. D. Maagee, A. Refsum, "RESIN, A desktop-computer program for finding cut set", *IEEE Transactions on Reliability*, 30(5), 407–410, 1981
45. K. Kobayashi, H. Yamamoto, "A new algorithm in enumerating all minimal paths in a sparse network", *Reliability Engineering and System Safety*, 65(1), 11–15, 1999
46. S. Ross, *Introduction to Probability Models*, Academic Press, NewYork, 2006
47. X. Wang, C. Pottle, "A concise frequency and duration approach to generating system reliability studies", IEEE Transactions on Power Amplifier Symposium, 102(8), 2521–2530, 1983
48. F.A. Rahimi, A. Vojdani, "Meet the emerging transmission market segments", IEEE Computer Application in Power, 12(1), 26–32, 1999
49. F.C. Schweppe, M.C. Caramanis, R.D. Tabors, et al, *Spot Pricing of Electricity*, Kluwer, Boston, 1988
50. J. Carpentier, "Contribution a'l'etude du dispatching economique," Bulletin de la Societe Francaise des Electricients, 3, 431–447, 1962
51. H.W. Dommel, W.F. Tinney, "Optimal power flow solutions," *IEEE Transactions on Power Apparatus and Systems*, 87(12), 1866–1876, 1968
52. A.M. Sasson, "Combined use of the parallel and fletcher-powell non-linear programming methods for optimal load flows," *IEEE Transactions on Power Apparatus and Systems*, 88 (10), 1530–1537, 1969
53. A.M. Sasson, "Decomposition technique applied to the non-linear programming load flow method," *IEEE Transactions on Power Apparatus and Systems*, 89(1), 78–82, 1970
54. R. Divi, H.K. Kesavan, "A shifted penalty function approach for optimal power flow," *IEEE Transactions on Power Apparatus and Systems*, 101(9), 3502–3512, 1982
55. S.N. Talukdar, T.C. Giras, V.K. Kalyan, "Decompositions for optimal power flows," *IEEE Transactions on Power Apparatus and Systems*, 102(12), 3877–3884, 1983
56. G.F. Reid, L. Hasdorf, "Economic dispatch using quadratic programming," *IEEE Transactions on Power Apparatus and Systems*, 92, 2015–2023, 1973
57. R.C. Burchett, H.H. Happ, D.R. Vierath, "Quadratically convergent optimal power flow," *IEEE Transactions on Power Apparatus and Systems*, 103, 3267–3276, 1984
58. D.W. Wells, "Method for economic secure loading of a power systems," *Proceedings of IEEE*, 115(8), 606–614, 1968
59. C.M. Shen, M.A. Laughton, "Power system load scheduling with security constraints using dual linear programming," *Proceedings of IEEE*, 117(1), 2117–2127, 1970
60. N. Nabona, L.L. Ferris, "Optimization of economic dispatch through quadratic and linear programming," *Proceedings of IEEE*, 120(5), 1973
61. Z. Yan, N.D. Xiang, B.M. Zhang, et al, "A hybrid decoupled approach to optimal power flow," *IEEE on Power Systems*, 11(2), 947–954, 1996
62. K.R. Frish, "Principles of linear programming: The double gradient form of the logarithmic potential method," Memorandum, Institute of Economics, University of Oslo, Oslo, Norway, 1954
63. P. Huard, "Resolution of mathematical programming with nonlinear constraints by the method of centers," *Nonlinear Programming*, 209–219, 1967
64. I.I. Dikin, "Iterative solution of problems of linear and quadratic programming," *Soviet Mathematics*, 8, 674–675, 1967
65. N. Karmarkar, "A new polynomial-time algorithm for linear programming," *Combinatorica*, 4, 373–395, 1984

66. P.E. Gill, W. Murray, M.A. Saunders, et al, "On the projected newton barrier methods for linear programming and an equivalence to karmarkar's projective method," *Mathematical Programming*, 36, 183–209, 1986
67. X. Guan, W.H. Edwin Liu, A.D. Papalexopoulos, "Application of a fuzzy set method in an optimal power flow," *Electric Power Systems Research*, 34(1), 11–18, 1995
68. Y.T. Hsiao, C.C. Liu, H.D. Chiang, et al, "A new approach for optimal VAR sources planning in large scale electric power systems," *IEEE Trunsactions on Power Systems*, 8(3), 988–996, 1993
69. K.R. Frisch, "The Logarithmic Potential Method for Convex Programming," Memorandum, Institute of Economics, University of Oslo, Norway, May, 1955
70. A.V. Fiacco, G.P. MoCormic, *Nonlinear Programming: Sequential Unconstrained Minimization Techniques*, Wiley, New York, 1968
71. H. Wei, H. Sasaki, J. Kubokawa, et al, "An interior point nonlinear programming for optimal power flow problems with a novel data structure," *IEEE Transactions on Power Systems*, 13 (3), 870–877, 1998
72. F.C. Schweppe, M.C. Caramanis, R.D. Tabors, et al, "*Spot Pricing of Electricity*," Kluwer, Boston, 1988
73. M.C. Carmanis, R.E. Bohn, F.C. Schweppe, "Optimal spot pricing: Practice and theory," *IEEE Transactions on Power Apparatus and Systems*, 101(9), 3234–3245, 1982
74. M.C. Carmanis, R.E. Bohn, F.C. Schweppe, "The costs of wheeling and optimal wheeling rates," *IEEE Transactions on Power Systems*, 1(1), 63–73, 1986
75. D. Ray, F. Alvarado, "Use of an engineering model for economic analysis in the electricity utility industry," *The Advanced Workshop on Regulation and Public Utility Economics*, Rutgers University, New Jersey, May 25–27, 1988
76. M.L. Baughman, S.N. Siddiqi, "Real time pricing of reactive power: theory and case study result," *IEEE Transactions on Power Systems*, 6(1), 23–29, 1991
77. S.N. Siddiqi, M.L. Baughman, "Reliability differentiated pricing of spinning reserve," *IEEE Transactions on Power Systems*, 10(3), 1211–1218, 1993
78. A. Zobian, M.D. llic, "Unbundling of transmission and ancillary services," *IEEE Transactions on Power systems*, 12(2), 539–558, 1997
79. M.L. Baughman, S.N. Siddiqi, J.M. Zarnikau, "Advanced pricing in electrical system. Part 1: Theory," *IEEE Transactions on Power Systems*, 12(1), 489–495, 1997
80. M.L. Baughman, S.N. Siddiqi, J.M. Zarnikau, "Advanced pricing in electrical system. Part 2: Implication," *IEEE Transactions on Power Systems*, 12(1), 496–502, 1997
81. K. Xie, Y.H. Song, J. Stonham, et al, "Decomposition model and interior point methods for optimal spot pricing of electricity in deregulation environments," *IEEE Transactions on Power Systems*, 15(1), 39–50, 2000
82. C.N. Yu, M.D. Ilic, "An algorithm for implementing transmission rights in competitive power industry," *IEEE Power Engineering Society Winter Meeting*, 3, 1708–1714, 2000
83. X. Wang, Y.H. Song, Q. Lu, et al, "Series FACTS devices in financial transmission rights auction for congestion management," *IEEE Power Engineering Review*, 21(11), 41–44, 2001
84. R. Bacher, H. Glavitsch, "Loss reduction by network switching," *IEEE Transactions on Power Systems*, 3(2), 447–454, 1988
85. R. Baldick, E. Kahn, "Contract paths, phase shifters and efficient electricity trade," *IEEE Power Engineering Society Winter Meeting*, 2, 968–974, 2000
86. S.Y. Ge, T.S. Chung, Y.K. Wong, "A new method to incorporate FACTS devices in optimal power flow," *Proceedings of International Conference on Energy Management and Power Delivery*, 1, 122–271, 1998
87. X. Wang, Y.H. Song, Q. Lu, "Primal-dual interior point linear programming optimal power flow for real-time congestion management," *IEEE Power Engineering Society Winter Meeting*, 3, 1643–1649, 2000
88. G. Hamoud, "Assessment of available transfer capability of transmission system," *IEEE Transactions on Power System*, 15(1), 27–32, 2000

89. X. Luo, A.D. Patton, C. Singh, "Real power transfer capability calculations using multi-layer feed-forward neutral networks," *IEEE Transactions on Power Systems*, 15(2), 903–908, 2000
90. M. Pavella, D. Ruiz-Vega, J. Giri, et al, "An integrated scheme for on-line static and transient stability constrained ATC calculations," *IEEE Power Engineering Society Summer Meeting*, 1, 273–276, 1999
91. D.S. Kirschen, R.N. Allan, G. Strbac, "Contributions of individual generators to loads and flows", *IEEE Transactions on Power Systems*, 12(1), 52–60, 1997
92. X. Wang, X. Wang, "On current trace problem", *Science in China* (E), 30(3), 405–412, 2000
93. J. Bialek, "Topological generation and load distribution factors for supplement cost allocation in transmission open access", *IEEE Transactions on Power Systems*, 12(3), 1185–1193, 1997
94. D.S. Kirschen, G. Strbac, "Tracing active and reactive power between generators and loads using real and imaginary currents," *IEEE Transactions on Power Systems*, 14(4), 1312–1319, 1999
95. Reliability Test System Task Force, "IEEE reliability test system-1996," *IEEE Transactions on Power Systems*, 14(3), 1010–1020, 1999
96. Federal Energy Regulatory Commission, "Open access same-time information system (Formerly Real-time Information Networks) and standards of conduct," Docket no. RM95-9-000, Order 889, 1996
97. North American Electric Reliability Council, "Available transfer capability definition and determination", NERC Planning Standards, June 1996
98. G. Hamoud, "Assessment of available transfer capability of transmission system," *IEEE Transactions on Power Systems*, 15(1), 27–32, 2000
99. G.C. Ejebe, J.G. Waight, M. Sanots-Nieto, et al, "Available transfer capability calculations," *IEEE Transactions on Power Systems*, 13(4), 1521–1527, 1998
100. X. Luo, A.D. Patton, C. Singh, "Real power transfer capability calculations using multi-layer feed-forward neural networks," *IEEE Transactions on Power Systems*, 15(2), 903–908, 2000
101. G.C. Ejebe, J.G. Waight, M.S. Nieto, W.F. Tinney, "Fast calculation of linear available transfer capability," *IEEE Transactions on Power Systems*, 15(3), 1112–1116, 2000
102. M. Shaaban, Y. Ni, F. Wu, "Total transfer capability calculations for competitive power networks using genetic algorithms," *Proceedings of International Conference on DRPT*, City University, London, April 4–7, 2000
103. A.R. Vojdani, "Computing available transmission capability using trace," *EPRI Power System Planning and Operation News*, 1, 1, 1995
104. Y. Xiao, Y.H. Song, "Available transfer capability (ATC) evaluation by stochastic programming," *IEEE Power Engineering Review*, 20(9), 50–52, 2000
105. F. Xia, A.P.S. Meliopoulos, "A methodology for probabilistic simultaneous transfer capability analysis," *IEEE Transactions on Power Systems*, 11(3), 1269–1278, 1996
106. A.B. Rodrigues, M.G. Da Silva, "Solution of simultaneous transfer capability problem by means of Monte Carlo simulation and primal-dual interior-point method," Proceedings of PowerCon International Conference, 2, 1047–1052, 2000
107. X.F. Wang, C.J. Cao, Z.C. Zhou, Experiment on fractional frequency transmission system, *IEEE Transactions on Power Systems*, 21(1), 372–377, 2006
108. N.G. Higorani, "Power electronics in electric utilities: Role of power electronics in future power systems," Proceedings of IEEE, 76(4), 481–482, 1988
109. L. Gyugyi, "Dynamic compensation of AC transmission lines by solid-state synchronous voltage source," *IEEE Transactions on Power Delivery*, 9(2), 904–911, 1994
110. A.A. Edris, R. Aapa, M.H. Baker, et al, "Proposed terms and definitions for flexible AC transmission system (FACTS)," *IEEE Transactions on Power Delivery*, 12(4), 1848–1853, 1997
111. D.A. Braunagel, L.A. Kraft, J.L. Whysong, "Inclusion of DC converter and transmission equation directly in a Newton power flow," *IEEE Transactions on Power Apparatus and Systems*, 95(1), 76–88, 1976

112. J. Arrillaga, P. Bodger, "Integration of HVDC links with fast decoupled load flow solutions," *Proceedings of IEE*, 124(5), 463–468, 1977
113. J. Arrillaga, B. Smith, *AC-DC Power System Analysis*, The Institute of Electrical Engineers, UK, 1998
114. J. Reeve, G. Fahmy, B. Stott, "Versatile load flow method for multiterminal HVDC systems," *IEEE Transactions on Power Apparatus and Systems*, 96(3), 925–932, 1977
115. H. Fudeh, C.M. Ong, "A simple and efficient AC-DC load flow method for miltiterminal DC systems," *IEEE Transactions on Power Apparatus and Systems*, 100(11), 4389–4396, 1981
116. J. Arrillaga, C.P. Arnold, B.J. Harker, *Computer Modeling of Electrical Power Systems*, Wiley, New York, 1983
117. T. Smed, G. Andersson, G.B. Sheblé, L.L. Grigsby, "A new approach to AC/DC power flow," *IEEE Transactions on Power Systems*, 6(3), 1238–1244, 1991
118. G.D. Breuer, J.F. Luini, C.C. Young, "Studies of large AC/DC systems on the digital computer," *IEEE Transactions on Power Apparatus and Systems*, 85(11), 1107–1115, 1966
119. J.F. Clifford, A.H. Schmidt, "Digital representation of a DC transmission system and its control," *IEEE Transactions on Power Apparatus and Systems*, 89(1), 97–105, 1970
120. N. Sato, N.V. David, S.M. Chan, A.L. Burn, J.J. Vithayathil, "Multiterminal HVDC system representation in a transient stability program," *IEEE Transactions on Power Apparatus and Systems*, 99(5), 1927–1936, 1980
121. Working Group 38–01, Task Force no. 2 on SVC, CIGRE Report, *Static Var Compensators*, Ed. by I.A. Erimnez, CIGRE, UK, 1986
122. IEEE Special Stability Controls Working Group, "Static var compensator models for power flow and dynamic performance simulation", *IEEE Transactions on Power Systems*, 9(1), 229–240, 1994
123. L. Gyugyi, N.G. Hinggorani, P.R. Nannery, et al, "Advanced static var compensator using gate turn-off thyristors for utilities applications," *CIGRE Session*, 1990, pp. 23–203
124. Y. Sumi, Y. Harumoto, T. Hasegawa, et al, "New static var control using force-commutated inverters," *IEEE Transactions on Power Apparatus and Systems*, 100(9), 4216–4224, 1981
125. C.W. Edwards, P.R. Nannery, "Advanced static var generator employing GTO thyristors," *IEEE Transactions on Power Delivery*, 3(4), 1622–1627, 1988
126. C. Schauder, M. Gernhardt, E. Stacey, et al, "Development of a ±100Mvar static condensor for voltage control of transmission systems," *IEEE Transactions on Power Delivery*, 10(3), 1486–1496, 1995
127. E.V. Larsen, K. Clark, S.A. Miske Jr., J. Urbanek, "Characteristics and rating consideration of thyristor controlled series compensation," *IEEE Transactions on Power Delivery*, 9(2), 992–1000, 1994
128. G.G. Karady, T.H. Ortmeyer, B.R. Pilvelait, et al, "Continuously regulated series capacitor," *IEEE Transactions on Power Delivery*, 8(3), 1348–1355, 1993
129. L. Gyugyi, C.D. Schauder, K.K. Sen, "Static synchronous series compensator: A solid-state approach to the series compensation of transmission lines," *IEEE Transactions on Power Delivery*, 12(1), 406–417, 1997
130. Y.H. Song, A.T. Johns, *Flexible AC transmission Systems (FACTS)*, IEE Press, London, 1999
131. S. Nyati, M. Eitzmann, J. Kappenmann, et al, "Design issues for a single core transformer thyristor controlled phase-angle regulator," *IEEE Transactions on Power Delivery*, 10(4), 2013–2019, 1995
132. L. Gyugyi, "A unified power flow control concept for flexible AC transmission system," Fifth International Conference on AC and DC Power Transmission, London, Sept'17–20, 1991
133. A. Nabavi-Niaki, M.R. Iravani, "Steady-state and dynamic models of unified power flow controller (UPFC) for power system studies", *IEEE Transactions on Power Systems*, 11(4), 1937–1943, 1996
134. Z.X. Han, "Phase shift and power flow control", *IEEE Transactions on Power Apparatus and Systems*, 101(10), 3790–3795, 1982

135. D.J. Gotham, G.T. Heydt, "Power flow control and studies for systems with FACTS devices," *IEEE Transactions on Power Systems*, 13(1), 60–65, 1998
136. C.R. Fuerte-Esquivel, E. Acha, H. Ambriz-Prez, "A thyristor controlled series compensator model for the power flow solution of practical power networks," *IEEE Transactions on Power Systems*, 15(1), 58–64, 2000
137. C.R. Fuerte-Esquivel, E. Acha, H. Ambriz-Prez, "A comprehensive Newton-Raphson UPFC for the quadratic power flow solution of practical power networks," *IEEE Transactions on Power Systems*, 15(1), 102–109, 2000
138. H. Ambriz-Prez, E. Acha, C.R. Fuerte-Esquivel, "Advanced SVC models for Newton-Raphson load flow and Newton optimal power flow studies," *IEEE Transactions on Power Systems*, 15(1), 129–946, 2000
139. C.R. Fuerte-Esquivel, E. Acha, "A Newton-type algorithm for the controlpower flow in electrical power networks," *IEEE Transactions on Power Systems*, 12(4), 1474–1480, 1997
140. C.R. Fuerte-Esquivel, E. Acha, "Newton-Raphson algorithm for the reliable solution of large power networks with embedded FACTS devices," *IEE Proceedings of Generation, Transmission and Distribution*, 143(5), 447–454, 1996
141. S. Arabi, P. Kundur, "A versatile FACTS device model for power flow and stability simulations," *IEEE Transactions on Power Systems*, 11(4), 1944–1950, 1996
142. C.R. Fuerte-Esquivel, E. Acha, "United power flow controller: A critical comparison of Newton-Raphson UPFC algorithm in power flow studies," *IEE Proceedings of Generation, Transmission and Distribution*, 144(5), 437–444, 1997
143. J.Y. Liu, Y.H. Song, "Strategies for handling UPFC constraints in steady-state power flow and voltage control," *IEEE Transactions on Power Systems*, 15(2), 566–571, 2000
144. W. Fang, H.W. Ngan, "Control setting of unified power flow controllers through a robust load flow calculation", *Proceedings of Generation, Transmission and Distribution*, 146(4), 365–369, 1999
145. H. Sun, D.C. Yu, C. Luo, "A novel method of power flow analysis with Unified Power Flow Controller (UPFC)," IEEE Power Engineering Society Winter Meeting, 4, 2800–2805, 2000
146. A. Blondel, "The two-reaction method for study of oscillatory phenomena in coupled alternators", Revue Générale de Lelectricité, 13, 235–251, February 1923; 515–531, March 1923
147. R.E. Doherty, C.A. Nickle, "Synchronous machines I and II", *AIEE Transactions*, 45, 912–942, 1926
148. R.H. Park, "Two-reaction theory of synchronous machines: Generalized method of analysis - Part I", *AIEE Transactions*, 48, 716–727, 1929; Part II, 52, 352–355, 1933
149. C. Concordia, *Synchronous Machine*, Wiley, New york, 1951
150. G. Shackshaft, P.B. Henser, "Model of generator saturation for use in power system studies", Proceedings of IEE, 126(8), 759–763, 1979
151. G.R. Slemon, *Magnetoelectric Devices*, Wiley, New York, 1966
152. A.E. Fitzgerald, C. Kingsley, *Electric Machinery*, 2nd Edn., McGraw-Hill, New York, 1961
153. P. Kunder, *Power System Stability and Control*, McGraw-Hill, New York, 1994
154. D.W. Olive, "Digital simulation of synchronous machine transients", *IEEE Transactions on Power Apparatus and Systems*, 87(8), 1968
155. M.K. El-Sherbiny, A.M. El-Serafi, "Analysis of dynamic performance of saturated machine and analog simulation", *IEEE Transactions on Power Apparatus and Systems*, 101(7), 1899–1906, 1982
156. D.W. Olive, "New techniques for the calculation of dynamic stability", *IEEE Transactions on Power Apparatus and Systems*, 85(7), 767–777, 1966
157. T.J. Hammons, D.J. Winning, "Comparisons of synchronous machine models in the study of the transient behaviour of electrical power systems", *Proceedings of IEE*, 118, 1442–1458, 1971
158. J. Arrillage, C.P. Arnold, B.J. Harker, *Computer Modeling of Electrical Power Systems*, Wiley, Chichester, 1983

159. IEEE Committee Report, "First benchmark model for computer simulation of subsynchronous resonance", *IEEE Transactions on Power Apparatus and Systems*, 96(5), 1565–1572, 1977
160. IEEE Power Engineering Society, *IEEE Recommended Practice for Excitation System Models for Power System Stability Studies*, IEEE Standards Board, New York, 1992
161. IEEE Working Group Report, "Hydraulic turbine and turbine control models for system dynamic studies", *IEEE Transactions on Power Systems*, PWRS-7(1), 167–179, 1992
162. D.G. Ramey, J.W. Skooglund, "Detailed hydrogovernor representation for system stability studies", *IEEE Transactions on Power Apparatus and Systems*, 89, 106–112, 1970
163. M. Leum, "The development and field experience of a transistor eletric governor for hydro turbines", *IEEE Transactions on Power Apparatus and Systems*, 85, 393–400, 1966
164. IEEE Working Group Report, "Dynamic models for fossil fueled steam units inpower system studies", *IEEE Transactions on Power Systems*, PWRS-6(2), 753–761, 1991
165. IEEE Committee Report, "Dynamic models for steam and hydro turbines in power system studies", *IEEE Transactions on Power Apparatus and Systems*, 92(6), 1904–1915, 1973
166. P. Kundur, D.C. Lee, J.P. Bayne, "Impact of turbine generator overspeed controls on unit performance under system disturbance conditions", *IEEE Transactions on Power Apparatus and Systems*, 104, 1262–1267, 1985
167. M.S. Baldwin, D.P. McFadden, "Power systems performance as affected by turbine-generator controls response during frequency disturbance", *IEEE Transactions on Power Apparatus and Systems*, 100, 2846–2494, 1981
168. IEEE Task Force on Load Representation for Dynamic Performance, "Standard load models for power flow and dynamic performance simulation", *IEEE Transactions on Power Systems*, 10(3), 1302–1313, 1995
169. IEEE Task Force on Load Representation for Dynamic Performance, "Load representation for dynamic performance analysis", *IEEE Transactions on Power Systems*, 8(2), 472–482, 1993
170. IEEE Task Force on Load Representation for Dynamic Performance System Dynamic Performance Subcommittee, Power System Engineering Committee, "Bibliography on load model for power flow and dynamic performance simulation", *IEEE Transactions on Power Systems*, 10(1), 523–538, 1995
171. T. Dovan, T.S. Dillon, C.S. Berger, K.E. Forward,"A microcomputer based on-line identification approach to power system dynamic load modelling", *IEEE Transactions on Power Systems*, 2(3), 529–536, 1987
172. C.W. Talor, "Concepts of under-voltage load shedding for voltage stability", *IEEE Transactions on Power Delivery*, 7(2), 480–488, 1982
173. W.S. Kao, C.J. Lin, C.T. Huang, et al, "Comparison of simulated power system dynamics applying various load models with actual recorded data", *IEEE Power Engineering Society Winter Meeting*, 172–177, 1993
174. W.S. Kao, "The effect of load models on unstable low-frequency oscillation damping in taipower system experience w/wo power system stabilizers", *IEEE Transactions on Power Systems*, 16(3), 463–472, 2001
175. A. Borghetti, R. Caldon, A. Mari, et al, "On dynamic load models for voltage stability studies", *IEEE Transactions on Power Systems*, 12(1), 293–303, 1997
176. F. Nozari, M.D. Kankam, W.W. Price, "Aggregation of induction motors for transient stability load modeling", *IEEE Transactions on Power systems*, 2(4), 1096–1103, 1987
177. P. Kunder, *Power System Stability and Control*, McGraw-Hill, New York, 1994
178. P.M. Anderson, A.A. Fouad, *Power System Control and Stability*, The Iowa State University Press, Iowa, 1977
179. J. Arrillaga, C.P. Arnold, *Computer Analysis of Power Systems*, Wiley, New York, 1990
180. Y.H. Song, A.T. Johns, *Flexible AC Transmission Systems (FACTS)*, The Institution of Electrical Engineers, London, 1999

181. J.A. Momoh, M.E. El-Hawary, *Electric Systems, Dynamics, and Stability with Artificial Intelligence Applications*, Marcel Dekker, New York, 2000
182. W.L. Brogan, *Modern Control Theory*, Prentice Hall, New Jersey, 1991
183. J.J.E. Slotine, W. Li, *Applied Nonlinear Control*, Prentice Hall, New Jersey, 1991
184. C.W. Gear, *Numerical Initial Value Problems in Ordinary Differential Equations*, Prentice Hall, New Jersey, 1971
185. L. Lapidus, J.H. Seinfeld, *Numerical Solution of Ordinary Differential Equations*, Academic Press, New York, 1971
186. J.D. Lambert, *Computational Methods in Ordinary Differential Equations*, Wiley, New York, 1973
187. J.H. Wilkinson, *The Algebraic Eigenvalue Problem*, Clarendon Press, Oxford, 1965
188. G.H. Golub, C.F. Van Loan, *Matrix Computations*, 3rd Edn., The Johns Hopkins University Press, 1996
189. G. Rogers, *Power System Oscillations*, Kluwer, Dordecht, 2000
190. P. Kundur, G.J. Rogers, D.Y. Wong, et al, "A comprehensive computer program package for small signal stability analysis of power systems", *IEEE Transactions on Power Systems*, 5 (4), 1076–1083, 1990
191. S. Aribi, G.J. Rogers, D.Y. Wong, et al, "Small signal stability analysis of SVC and HVDC in AC power systems", *IEEE Transactions on Power Systems*, 6(3), 1147–1153, 1991
192. I.J. Perez-Arriaga, G.C. Verghese, F.C. Schweppe, "Selective modal analysis with applications to electric power systems, Part I: Heuristic introduction, Part II: The dynamic stability problem", *IEEE Transactions on Power Apparatus and Systems*, 101(9), 3117–3134, 1982
193. J.L. Sancha, I.J. Perez-Arriaga, "Selective modal analysis of electric power system oscillatory instability", *IEEE Transactions on Power Systems*, 3(2), 429–438, 1988
194. R.T. Byerly, R.J. Bennon, D.E. Sherman, "Eigenvalue analysis of synchronizing power flow oscillations in large electric power systems", *IEEE Transactions on Power Apparatus and Systems*, 101(1), 235–243, 1982
195. N. Martins, "Efficient eigenvalue and frequency response methods applied to power system small-signal stability studies", *IEEE Transactions on Power Systems*, 1(1), 217–226, 1986
196. D.Y. Wong, G.J. Rogers, B. Porretta, P. Kundur, "Eigenvalue analysis of very large power systems", *IEEE Transactions on Power Systems*, 3(2), 472–480, 1988
197. P.W. Sauer, C. Rajagopalan, M.P. Pai, "An explanation and generalization of the AESOPS and PEALS algorithms", *IEEE Transactions on Power Systems*, 6(1), 293–299, 1991
198. N. Uchida, T. Nagao, "A new Eigen-analysis method of steady-state stability studies for large power systems: S matrix method", *IEEE Transactions on Power Systems*, 3(2), 706–714, 1988
199. W.J. Stewart, A. Jennings, "A simultaneous iteration algorithm for real matrices", *ACM Transactions on Mathematical Software*, 7(2), 184–198, 1981
200. S. Duff, J.A. Scott, "Computing selected eigenvalues of sparse unsymmetric matrices using subspace iteration", *ACM Transactions on Mathematical Software*, 19(2), 137–159, 1993
201. J.A. Scott, "An Arnoldi code for computing selected Eigenvalues of sparse, real, unsymmetric matrices", *ACM Transactions on Mathematical Software*, 21(4), 432–475, 1995
202. A. Semlyen, L. Wang, "Sequential computation of the complete eigensystem for the study zone in small signal stability analysis of large power systems", *IEEE Transactions on Power Systems*, 3(2), 715–725, 1988
203. L. Wang, A. Semlyen, "Application of sparse eigenvalue techniques to the small signal stability analysis of large power systems", *IEEE Transactions on Power Systems*, 5(4), 635–642, 1990
204. D.J. Stadnicki, J.E. Van Ness, "Invariant subspace method for eigenvalue computation", *IEEE Transactions on Power Systems*, 8(2), 572–580, 1993
205. N. Mori, J. Kanno, S. Tsuzuki, "A sparsed–oriented techniques for power system small signal stability analysis with a precondition conjugate residual method", *IEEE Transactions on Power Systems*, 8(3), 1150–1158, 1993

206. G. Angelidis, A. Semlyen, "Efficient calculation of critical eigenvalue clusters in the small signal stability analysis of large power systems", *IEEE Transactions on Power Systems*, 10 (1), 427–432, 1995
207. L.T.G. Lima, L.H. Bezerra, C. Tomei, N. Martins, "New methods for fast small-signal stability assessment of large scale power systems", *IEEE Transactions on Power Systems*, 10(4), 1979–1985, 1995
208. G. Angelidis, A. Semlyen, "Improved methodologies for the calculation of critical eigenvalues in small signal stability analysis", *IEEE Transactions on Power Systems*, 11(3), 1209–1217, 1996
209. J.M. Campagnolo, N. Martins, D.M. Falcao, "Refactored bi-Iteration: A high performance eigensolution method for large power system matrices", *IEEE Transactions on Power Systems*, 11(3), 1228–1235, 1996
210. N. Martins, L.T.G. Lima, H.J.C.P. Pinto, "Computing dominant poles of power system transfer functions", *IEEE Transactions on Power Systems*, 11(1), 162–1170, 1996
211. N. Martins, "The dominant pole spectrum eigenslover", *IEEE Transactions on Power Systems*, 12(1), 245–254, 1997
212. J.M. Campagnolo, N. Martins, J.L.R. Pereira, et al, "Fast small-signal stability assessment using parallel processing", *IEEE Transactions on Power Systems*, 9(2), 949–956, 1994
213. J.M. Campagnolo, N. Martins, D.M. Falcao, "An efficient and robust eigenvalue method for small-signal stability assessment in parallel computers", *IEEE Transactions on Power Systems*, 10(1), 506–511, 1995
214. V. Ajjarapu, "Reducibility and eigenvalue sensitivity for identifying critical generations in multimachine power systems", *IEEE Transactions on Power Systems*, 5(3), 712–719, 1990
215. T. Smed, "Feasible eigenvalue sensitivity for large power systems", *IEEE Transactions on Power Systems*, 8(2), 555–563, 1993
216. H.K. Nam, Y.K. Kim, "A new eigen-sensitivity theory of augmented matrix and its applications to power system stability", *IEEE Transactions on Power Systems*, 15(1), 363–369, 2000
217. K.W. Wang, C.Y. Chung, "Multimachine eigenvalue sensitivities of power system parameters", *IEEE Transactions on Power Systems*, 15(2), 741–747, 2000
218. F.P. Demello, C. Concordia, "Concepts of synchronous machine stability as affected by excitation control", *IEEE Transactions on Power Apparatus and Systems*, 88(4), 316–329, 1969
219. P. Kunder, D.C. Lee, H.M. Zein-el-din, "Power system stabilizers for thermal units: Analytical techniques and on-site validation", *IEEE Transactions on Power Apparatus and Systems*, 100(1), 81–89, 1981
220. M. Klein, G.J. Rogers, P. Kundur, "A fundamental study of inter-area oscillations in power systems", *IEEE Transactions on Power Systems*, 6(3), 914–921, 1991
221. G. Rogers, "Demystifying power system oscillations", *IEEE Computer Application in Power*, 9(3), 30–35, 1996
222. J. Hauer, D. Trudnowski, G. Rogers, et al, "Keeping an eye on power system dynamics", *IEEE Computer Application in Power*, 10(4), 50–54, 1997
223. G. Rogers, "Power system structure and oscillations", *IEEE Computer Application in Power*, 12(2), 14,16,18,20,21, 1999
224. S.K. Starrett, A.A. Fouad, "Nonlinear measures of mode-machine participation", *IEEE Transactions on Power Systems*, 13(2), 389–394, 1998
225. F.L. Pagola, I.J.P. Arriaga, G.C. Verghese, "On sensitivity, residues and participations: applications to oscillatory stability analysis and control", *IEEE Transactions on Power Systems*, 4(1), 278–285, 1989
226. M. Klein, G.J. Rogers, S. Moorty, P. Kundur, "Analytical investigation of factors influencing power system stabilizers performance", *IEEE Transactions on Energy Conversion*, 7(3), 382–390, 1992

227. P. Kundur, M. Klein, G.J. Rogers, M.S. Zywno, "Application of power system stabilizers for enhancement of overall system stability", *IEEE Transactions on Power Systems*, 4(2), 614–626, 1989
228. L. Xu, S. Ahmed-Zaid, "Tuning of power system controllers using symbolic eigensensitivity analysis and linear programming", *IEEE Transactions on Power Systems*, 10(1), 314–322, 1995
229. J.F. Hauer, F. Vakili, "A oscillation detector used in the bpa power system disturbance monitor", *IEEE Transactions on Power Systems*, 5(1), 74–79, 1990

Index